Complex Analytic Geometry

From the Localization Viewpoint

Complex Analytic Geometry

From the Localization Viewpoint

Tatsuo Suwa
Hokkaido University, Japan

World Scientific

NEW JERSEY • LONDON • SINGAPORE • BEIJING • SHANGHAI • HONG KONG • TAIPEI • CHENNAI • TOKYO

Published by

World Scientific Publishing Co. Pte. Ltd.
5 Toh Tuck Link, Singapore 596224
USA office: 27 Warren Street, Suite 401-402, Hackensack, NJ 07601
UK office: 57 Shelton Street, Covent Garden, London WC2H 9HE

Library of Congress Control Number: 2023049056

British Library Cataloguing-in-Publication Data
A catalogue record for this book is available from the British Library.

COMPLEX ANALYTIC GEOMETRY
From the Localization Viewpoint

ISBN 978-981-4374-70-5 (hardcover)
ISBN 978-981-4374-71-2 (ebook for institutions)
ISBN 978-981-4704-29-8 (ebook for individuals)

For any available supplementary material, please visit
https://www.worldscientific.com/worldscibooks/10.1142/8324#t=suppl

Desk Editors: Nimal Koliyat/Lai Fun Kwong

Typeset by Stallion Press
Email: enquiries@stallionpress.com

Preface

Complex analytic geometry is a subject that could be termed, in short, as the study of the sets of common zeros of complex analytic functions. It has a long history — it started with the discovery of complex numbers in the 16th century and then bloomed as the theory of analytic functions of one complex variable, including the Cauchy integral formula, in the 19th century. While the foundation for the case of several variables was laid by H. Cartan, K. Oka et al., the transcendental method was brought in algebraic geometry by G. de Rham, W.V.D. Hodge and K. Kodaira and the modern form of the subject was established in the mid-20th. At about the same time it was enriched by the works of M.F. Atiyah, R. Bott, S.-S. Chern, H. Grauert, P. Griffiths, A. Grothendieck, H. Hironaka, F. Hirzebruch, B. Malgrange, J.W. Milnor, L. Schwartz, J.-P. Serre, I.M. Singer, R. Thom, H. Whitney et al. in various related areas. This trend has been kept growing to the present time and will go on, involving many new ideas and themes. As such, the subject is closely related to many other fields of mathematics and other sciences, where numerous applications have been and will be found.

The subjects taken up in this book are rather classical, however we look at them from the localization viewpoint. This means that we are concerned with, among others, local invariants that arise naturally in complex analytic geometry and their relation with global invariants of the manifold or variety, the Poincaré-Hopf index theorem being a prototype of this. The idea is to look at them as residues associated with the localization of some characteristic classes. Two approaches are taken for this — topological and differential geometric — and the combination of the two brings out further fruitful results. For this, on the one hand we give detailed description of the Alexander duality in combinatorial topology. On the other hand we give thorough presentation of the Čech-de Rham cohomology and integration theory on it. This viewpoint provides us with a way for clearer and more precise presentations of the central concepts as well as fundamental and important results that have been treated only globally so far. It also brings new perspectives into the subject and leads to further results and applications.

Here are some key features of the book:

1. **Detailed description of the dualities and the intersection product in combinatorial topology**
   The Poincaré, Alexander and Lefschetz dualities on C^∞ manifolds are proved using tiangulations and the dual cellular decompositions. The Thom isomorphism and the Thom class are treated in this context. The intersection product is also defined combinatorially (cf. Chapter 4). These are extended to the case of singular varieties as well (cf. Chapters 13 and 14).

2. **Effective use of Čech-de Rham cohomology**
   This cohomology naturally combines the de Rham cohomology and the Čech cohomology of locally constant functions. In the global situation, it gives a canonical correspondence between the two cohomologies, which leads to the canonical de Rham theorem. The relative version is conveniently used to describe the Alexander duality, the Thom class and various localizations in terms of differential forms (cf. Chapter 7). We may also consider Čech-Dolbeault cohomology in a similar manner. In the global situation, this leads to the canonical Dolbeault theorem (cf. Section 11.4). Although it is not treated here, the relative version has many applications including the localization theory for Atiyah classes and explicit representation of Sato hyperfunctions and related operations.

3. **Topological and differential geometric definitions of characteristic classes**
   The Chern classes of a complex vector bundle and the Euler class of an oriented real vector bundle are defined topologically, via obsrtuction theory. For general Chern polynomials the corresponding classes of a complex vector bundle are defined differential geometrically, via the Chern-Weil thoery. Modifying this theory, the classes are naturally defined in the Čech-de Rham cohomology using difference forms for several connections. The relative (localized) version is also discussed in each framework (cf. Chapters 5 and 8). A direct proof is given, including the relative case, of the fact that the differential geometric Chern

class is the image of the topological one by the canonical morphism from the cohomology with integral coefficient to that with complex coefficient (cf. Section 10.5).

4. **Combination of algebra, geometry and analysis**
   The theory of analytic functions of one variable is a model of mathematical theory, where algebra, geometry and analysis are organically combined. We attempted to do similar things in the higher-dimensional case, see in particular Chapters 12 and 13.

The book starts off with basic materials on analytic functions of one or several complex variables in Chapter 1. These functions are also referred to as holomorphic functions and are the ones we are mainly concerned with. We introduce the notion of the germ of a holomorphic function and study the structure of the ring of germs of holomorphic functions. The Weierstrass preparation theorem allows us to treat the ring almost as a polynomial ring. It continues by introducing, in Chapter 2, complex manifolds and analytic varieties, the spaces we work on. We study the relation between germs of varieties and ideals in the ring of germs of holomorphic functions. Particularly noteworthy is the Nullstellensatz (zero locus theorem), which we prove in the case of principal ideals. As for the general case, we quote some structual results and give a proof based on these. We also discuss dimensions of varieties. This chapter includes some remarks on the underlying real structures as well, because of their importance.

In Chapter 3, fiber bundles, in particular vector bundles, are discussed. The homotopical structure of Stiefel manifold is studied for the obstruction theoretical definition of characteristic classes of vector bundles. The cellular structure of Grassmann manifold, which carries a unversal vector bundle, is analyzed. Some notions concerning differentiable structures are also made precise.

We deal with, in Chapter 4, the topics mentioned in the item 1 above, i.e., the Poincaré, Alexander and Lefschetz dualities, the Thom isomorphism and the intersection product in combinatorial topology. In Chapter 5, we define characteristic classes of vector bundles topologically via the obstruction theory. Their localizations by frames and the associated topological residues are also introduced. In Chapter 6, we review differential forms and their integrations. Such topics as integration along fibers and non-singular foliations are also presented.

In Chapter 7, we discuss the Čech-de Rham cohomology in detail (cf. the item 2 above). After the de Rham and Čech cohomologies are reviewed, they are combined to form the Čech-de Rham cohomology. The de Rham cocycles and the Čech cocycles of locally constant functions are naturally thought of as being Čech-de Rham cocycles and this leads to the canonical isomorphism between the de Rham cohomology and the Čech cohomology of locally constant functions. We then present the integration theory on Čech-de Rham cohomology introducing the notion of a system of honeycomb cells. Combining with combinatorial topology, we establish various canonical isomorphisms (Theorem 7.4). The canonical de Rham theorem immediately follows from this. We discuss the relative Čech-de Rham cohomology and prove the relative Čech-de Rham theorem (Corollary 7.8). It is expressed explicitly in terms of integration on topological chains and provides a bridge between topological localizations and differential geometric localizations. It also allows us to describe the dualities in terms of differential forms. The Thom class is expressed explicitly in the relative Čech-de Rham cohomology. As an application, we prove the Lefschetz fixed point formula in this context.

In Chapter 8, characteristic classes of vector bundles are defined differential geometrically via the Chern-Weil theory. The Bott difference forms are then conveniently used to modify the theory to have the classes in Čech-de Rham cohomology. This way of representing characteristic classes is particularly effective in dealing with the localization problem (cf. item 3 above). In Chapter 9, we discuss some topics related to vector bundles with metrics. Among them are harmonic forms (real and complex), Dolbeault cohomology, Kodaira-Serre duality, Hodge structures, Kähler manifolds, Atiyah classes of holomorphic vector bundles and Bott-Chern classes of Hermitian vector bundles.

We explain the fundamental idea of localizing characteristic classes in Chapter 10. We then present the residue theorem and give explicit expressions of the residues. Various types of localizations are exhibited. In particular, the differential geometric localizations of Chern classes of vector bundles by frames is discussed in detail. In Chapter 11, we present various tools and notions that are common in complex analytic geometry and are used in later chapters. In particular, we discuss complex analytic spaces, generalizations of varieties, in some detail.

In Chapter 12, we study the residues associated with localizations of Chern classes by families of sections of holomorphic vector bundles on complex manifolds. The residue is expressed explicitly (Theorem 12.2) in the case the singular set has an expected dimension, called the "proper case".

It shows that, in order to find the residue in the proper case, it suffices to know the residue at an isolated singularity. We give topological, analytic and algebraic expresions of the residue at an isolated singularity. The localized duality (Theorem 12.8) between the Chern class localized by a family of sections of a holomorphic vector bundle and the class of the complex space defined by the family of sections also follows from Theorem 12.2. It is effectively used in the localized intersection theory later in Chapter 14. In Chapter 13, we attempt to do similar things for residues of vector bundles on singular varieties. We discuss Poincaré, Alexander, Lefschetz and Thom morphisms for singular varieties in combinatorial topology. We then present de Rham and Čech-de Rham theories for singular varieties and express the above morphisms in terms of integration of differential forms. We discuss Chern classes and their localizations both from topological and differential geometric viewpoints and show that they are essentially the same. We express the residue explicitly (Theorem 13.9), as in the case of manifolds. It again shows that, in order to find the residue in the proper case, it suffices to know the residue at an isolated singularity. We give topological, analytic and algebraic expresions of the residue at an isolated singularity. We also give examples and discuss related topics. The localized duality (Theorem 13.18) between the Chern class localized by a section of a holomorphic vector bundle and the class of the complex space defined by the section also follows from Theorem 13.9.

We discuss, in Chapter 14, intersection products of the homology classes of complex subspaces in a complex manifold or in a singular variety. They are treated from the residue theoretical viewpoint and the intersection products are localized at the set of intersection. In the case of manifolds, the intersection product in general is defined combinatorially in Chapter 4, as mentioned above. It is an operation dual to the cup product in cohomology. In general, expected formulas do not hold for the intersection product of the classes of subspaces. The intersection products we consider involve, in some form or another, spaces defined by a family of sections of a vector bundle so that localizations of Chern classes and the associated residues are involved. For the residues, expected formulas always hold, thus in the proper case, for subspaces defined by a family of sections, they also hold. As to the intersection product in sigular varieties, we do not have general combinatorial ways of defining it. Thus we consider intersection products of homology classes that come naturally as the images of the Poincaré or Alexander morphism and use the cup product in cohomology. For this, as in the case of manifolds, we consider intersection product involving subspaces defined by a family of sections and develop the theory in parallel with that case. We then specialize the above considerations to the case of intersection products in singular surfaces. Finally we study the case where a variety and a complex space defined by a section intersect in a dimension greater than expected.

In Chapter 15, we discuss the Riemann-Roch theorem in the form A. Grothendieck formulated (cf. Theorem 15.11). It consists of the embedding part and the projection part. We examine the former in detail, as an application of our localization process by the exactness of vector bundle sequences, and prove a prototype of the localized Riemann-Roch theorem for embeddings on the level of Čech-de Rham cocycles (Corollary 15.1). It immediately yields a theorem in the universal situation (Theorem 15.4). We then present three variations of the localized Riemann-Roch theorem for embeddings (Theorems 15.5–15.7). For the projection part, we outline the proof quoting relevant materials. The Grothendieck-Riemann-Roch theorem contains the Hirzebruch-Riemann-Roch theorem as a special case. The latter is also generalized as the Atiyah-Singer index theorem, which we briefly recall in Notes at the end of the chapter.

The book is supplemented with two appendixes, one on commutative algebra and the other on algebraic topology.

Exercises provided in the text are meant to help better understand the materials and should mostly be done rather straightforward.

Here is a word about the use of terminologies "natural" and "canonical". A canonical morphism is the one that is induced from a natural idetification or correspondence of the objects. What is natural is somewhat a subjective matter. For example, we could embed the ring of integers $\mathbb{Z}$ into the complex numbers $\mathbb{C}$ by assigning to 1 in $\mathbb{Z}$ any non-zero complex number. In this case, we think the assignment $1 \mapsto 1$ is natural and the morphism $H_p(X;\mathbb{Z}) \to H_p(X;\mathbb{C})$ induced on the homology of a space X by this identification is canonical. As a typical example appearing in the text, Čech-de Rham cochains naturally combine de Rham cochains (differential forms) and Čech cochains of locally constant functions, as mentioned above, and the isomorphism between the de Rham cohomology and the Čech cohomology given via the Čech-de Rham cohomology is canonical.

Some symbols are used to denote different objects. For example, the letter m is used for the order of a convergent power series, the dimension of a C^∞ manifold and so forth, the letter D for a number of purposes and the notation $(f_1,\dots,f_r)$ for the map with component functions f_i, $i=1,\dots,r$, as well as the ideal generated by the germs of the f_i's. Hopefully it will be clear from the context what it means and there will be no confusion.

Besides calculus and linear algebra, basics of the following are desirable prerequisites:

(1) Analytic functions of one complex variable (reviewed in the text as needed).
(2) Commutative algebra (reviewed in Appendix A).
(3) General topology.
(4) Algebraic topology (reviewed in Appendix B).
(5) Differentiable manifolds (reviewed in the text as needed).

There are numerous literatures on these subjects and the citations made in Notes at the end of each chapter should be supplemented with "for example".

Acknowledgements. This work was supported by JSPS Grants-in-Aid for Scientific Research Grant over the years.

I am indebted to many people through collaboration, discussion and advice in their field of expertise and it is not possible to acknowledge all of them. I owe much to J.-P. Brasselet for the item 1 above, to D. Lehmann for the item 2 and to J. Seade for the item 3. I would also like to express my appreciation to M. Abate, P. Aluffi, D. Angella, C. Bisi, F. Bracci, M. Corrêa, M. Hanamura, N. Honda, T. Izawa, K. Kurano, Y. Matsumoto, J. Noguchi, M. Oka, K. Saito, J. Schürmann, M. Shiota, M. Soares, N. Tardini, A. Tomassini, F. Tovena and S. Yokura. Special thanks are due to T. Ohmoto for his constant interest, encouragement and advice during the preparation of the manuscripts, which took way longer than initially planned. I would like to extend my gratitude to all who were involved in the editorial and production process of the book, especially Kwong Lai Fun of World Scientific and Nimal Koliyat of ACES, for their generous support.

Tatsuo Suwa
Fall 2023

About the Author

Tatsuo Suwa is a Professor Emeritus at Hokkaido University, Japan. He had held academic positions at the University of Tokyo, the University of Michigan, Hokkaido University and Niigata University. He had visiting positions at institutions in a number of countries, including France, Italy, Mexico, Spain and the U.S.A. He has also been invited to give talks at symposiums and schools all over the world. He published a book on residues of singular holomorphic foliations in 1998. He then noticed that the philosophy and technics of localizing characteristic classes used there are applicable to other problems in Complex Analytic Geometry and related areas, some of which have been realized and are presented in this book.

Contents

Chapter 1

Analytic Functions of Several Complex Variables

An analytic function is a function that can locally be expressed as a convergent power series. In the case of one complex variable, it is equivalent to saying that the function is holomorphic, meaning that it is differentiable with respect to the complex variable. As a function of the real and imaginary parts of the variable, it can also be rephrased by saying that the function is continuously differentaible and satisfies the Cauchy-Riemann equations (cf. Theorem 1.1 below).

We start by discussing convergence of multiseries, which is necessary for the case of several variables. We then review basics on analytic functions of one complex variable, notably the Cauchy integral formula, which has a significant influence throughout the book. The case of several variables follows with relation to the one variable case. The terms "analytic" and "holomorphic" will be treated as synonyms hereafter. Regular and singular points of a holomorphic map are defined and the behavior of a holomorphic map near a regular point is clarified (Theorems 1.4, 1.5 and 1.6).

The notion of the germ of a holomorphic function is introduced and the algebraic structure of the ring of germs of holomorphic functions is discussed. The Weierstrass preparation theorem (Theorems 1.10) allows us to treat convergent power series as polynomials to some extent. These are used to analyze the local structure of the set of common zeros of holomorphic functions, i.e., analytic varieties, in the next chapter.

In the following, we denote by $\mathbb{C}$ the field of complex numbers and $\mathbb{C}^n$ the product of n copies of $\mathbb{C}$. An element z in $\mathbb{C}^n$ is expressed as $z = (z_1, \ldots, z_n)$ with each z_i, $i = 1, \ldots, n$, in $\mathbb{C}$ and the set $\mathbb{C}^n$ has a natural complex vector space structure of dimension n. The norm of z is defined by

$$\|z\| = \sqrt{|z_1|^2 + \cdots + |z_n|^2},$$

which makes $\mathbb{C}^n$ a topological vector space.

Recall that, starting from the field $\mathbb{R}$ of real numbers, we may form a real vector space $\mathbb{R}^m$ of dimension m with the Euclidean norm. For a point $z = (z_1, \ldots, z_n)$ in $\mathbb{C}^n$, we write $z_i = x_i + \sqrt{-1}\, y_i$ with x_i and y_i the real and imaginary parts of z_i. Identifying z with $(x_1, y_1, \ldots, x_n, y_n)$, we may identify $\mathbb{C}^n$ with $\mathbb{R}^{2n}$ and by this identification, the above norm in $\mathbb{C}^n$ coincides with the Euclidean norm in $\mathbb{R}^{2n}$.

1.1 Multiseries

Let $\mathbb{N}_0$ denote the set of non-negative integers. Suppose a complex number c_ν is assigned to each $\nu = (\nu_1, \ldots, \nu_n)$ in $\mathbb{N}_0^n$. We call the formal infinite sum

$$\sum_\nu c_\nu = \sum_{\nu_1, \ldots, \nu_n = 0}^{\infty} c_{\nu_1 \ldots \nu_n}$$

a *multiseries*, or a *series* for short.

Definition 1.1. A series $\sum c_\nu$ *converges* to a complex number C if, for every positive number ε, there exists N in $\mathbb{N}_0$ with the following property:

$$\text{for every } (N_1, \ldots, N_n) \text{ in } \mathbb{N}_0^n \text{ with } N_i > N, \left| \sum_{\nu_1=0}^{N_1} \cdots \sum_{\nu_n=0}^{N_n} c_\nu - C \right| < \varepsilon.$$

If this is the case, we write $C = \sum c_\nu$ and call it the sum of the series. We say that $\sum c_\nu$ is *convergent* if it converges to some complex number. Note that if a series $\sum c_\nu$ is convergent, its sum is uniquely determined.

We also say that $\sum c_\nu$ is *absolutely convergent* if the series $\sum |c_\nu|$ is convergent. From the definition we have:

Proposition 1.1. *A series $\sum c_\nu$ is absolutely convergent if and only if the set*

$$\left\{ \sum_{\nu \in \Gamma} |c_\nu| \;\middle|\; \Gamma \subset \mathbb{N}_0^n, \text{ finite} \right\}$$

is bounded. In this case the sum $\sum |c_\nu|$ is the supremum of this set.

In general, for a convergent series $\sum c_\nu$ and a finite set Γ in $\mathbb{N}_0^n$, we set

$$\sum_{\nu \notin \Gamma} c_\nu = \sum_\nu c_\nu - \sum_{\nu \in \Gamma} c_\nu.$$

Thus if a series $\sum c_\nu$ is absolutely convergent, for every positive number ε, there exists a finite set Γ such that $\sum_{\nu \notin \Gamma} |c_\nu| < \varepsilon$.

In the case $n = 1$, i.e., the case of simple series, if a series $\sum c_\nu$ is convergent, the set $\{c_\nu\}$ is bounded. Also in this case, if it is absolutely convergent, it is convergent by the Cauchy criterion. In the case $n > 1$, we may not have the boundedness of $\{c_\nu\}$, even if $\sum c_\nu$ is convergent. However we have:

Proposition 1.2. *If a series $\sum c_\nu$ is absolutely convergent, it is convergent. Moreover the sum $C = \sum c_\nu$ does not depend on the order of summation, i.e., for every bijection $\varphi : \mathbb{N}_0 \to \mathbb{N}_0^n$, the simple series $\sum_{\mu=0}^{\infty} c_{\varphi(\mu)}$ converges to C.*

Proof. Take a bijection $\varphi : \mathbb{N}_0 \to \mathbb{N}_0^n$. By Proposition 1.1, the simple series $\sum_{\mu=0}^{\infty} c_{\varphi(\mu)}$ is absolutely covergent, and thus convergent. We denote its sum by C_φ and show that $\sum c_\nu$ converges to C_φ so that $C_\varphi = C$ and it does not depend on φ. For this, let ε be an arbitrary positive number. Then there exists N_0 in $\mathbb{N}_0$ such that

$$\left|\sum_{\mu=0}^{N_0} c_{\varphi(\mu)} - C_\varphi\right| < \frac{\varepsilon}{2} \quad \text{and} \quad \sum_{\mu=N_0+1}^{\infty} |c_{\varphi(\mu)}| < \frac{\varepsilon}{2}.$$

Set $\Gamma_0 = \varphi(\{0, \ldots, N_0\})$ and choose N in $\mathbb{N}_0$ so that $\Gamma_0 \subset [0, N]^n$. For $(N_1, \ldots, N_n)$ with $N_i > N$, we set $\Gamma_1 = \{\nu \mid 0 \le \nu_i \le N_i\}$. Then we have

$$\left|\sum_{\nu \in \Gamma_1} c_\nu - C_\varphi\right| \le \left|\sum_{\nu \in \Gamma_0} c_\nu - C_\varphi\right| + \sum_{\nu \in \Gamma_1 \setminus \Gamma_0} |c_\nu| < \varepsilon.$$

□

Example 1.1. Let $z = (z_1, \ldots, z_n)$ be a point in $\mathbb{C}^n$ with $0 \ne |z_i| < 1$. For each ν in $\mathbb{N}_0^n$, we set $z^\nu = z_1^{\nu_1} \cdots z_n^{\nu_n}$. By Propositions 1.1 and 1.2, the series $\sum_\nu z^\nu$ is convergent and

$$\sum_\nu z^\nu = \frac{1}{1 - z_1} \cdots \frac{1}{1 - z_n}.$$

By the convention $0^0 = 1$, the above holds for every z with $|z_i| < 1$.

Next we consider series whose terms are functions. Thus let R be a subset in $\mathbb{C}^n$. Suppose we have, for each ν, a complex valued function f_ν defined on R. We denote by $\|f_\nu\|_R$ the supremum of $|f_\nu(z)|$ on R.

Definition 1.2. A series $\sum f_\nu$ is *normally convergent* on R if the series $\sum \|f_\nu\|_R$ is convergent.

Note that if a series $\sum f_\nu$ converges normally on R, the series $\sum f_\nu(z)$ converges absolutely for each z in R and it defines a function F on R:

$$F(z) = \sum_\nu f_\nu(z).$$

Proposition 1.3. *If a series $\sum f_\nu$ converges normally on R, it converges uniformly on R, i.e., we may take N in Definition 1.1 independently of z in R, with c_ν and C replaced by $f_\nu(z)$ and $F(z)$, respectively.*

Proof. Given a positive number ε. Take ε' so that $0 < \varepsilon' < \varepsilon$. By the normal convergence of $\sum f_\nu$, there exists a finite set Γ_0 in $\mathbb{N}_0^n$ such that $\sum_{\nu \notin \Gamma_0} \|f\|_R < \varepsilon'$. Take N in $\mathbb{N}_0$ so that $\Gamma_0 \subset [0, N]^n$. For $(N_1, \ldots, N_n)$ with $N_i > N$, we set $\Gamma_1 = \{\, \nu \mid 0 \le \nu_i \le N_i \,\}$. Then for an arbitrary finite set Γ in N_0^n, we have

$$\Bigl|\sum_{\nu \in \Gamma \setminus \Gamma_1} f_\nu(z)\Bigr| \le \sum_{\nu \in \Gamma \setminus \Gamma_1} \|f\|_R < \varepsilon'.$$

Hence we have

$$\Bigl|\sum_{\nu \in \Gamma_1} f_\nu(z) - F(z)\Bigr| = \Bigl|\sum_{\nu \notin \Gamma_1} f_\nu(z)\Bigr| \le \varepsilon' < \varepsilon.$$

□

From the above we see that, for instance, if f_ν is continuous for each ν, the sum $F = \sum f_\nu$ is also continuous.

Series we are particularly interested in here are "power series". In the following, for $z = (z_1, \ldots, z_n)$ in $\mathbb{C}^n$ and $\nu = (\nu_1, \ldots, \nu_n)$ in $\mathbb{N}_0^n$, we set

$$z^\nu = z_1^{\nu_1} \cdots z_n^{\nu_n}, \quad |\nu| = \nu_1 + \cdots + \nu_n \quad \text{and} \quad \nu! = \nu_1! \cdots \nu_n!.$$

Also, for a point $a = (a_1, \ldots, a_n)$ in $\mathbb{C}^n$ and an n-tuple $\rho = (\rho_1, \ldots, \rho_n)$ of positive real numbers, we set

$$\Delta(a, \rho) = \{\, z \in \mathbb{C}^n \mid |z_i - a_i| < \rho_i \,\}$$

and call it the *polydisk* with center a and polyradius ρ. We may express it as a product of one-dimensional disks:

$$\Delta(a, \rho) = \Delta(a_1, \rho_1) \times \cdots \times \Delta(a_n, \rho_n).$$

For a fixed a, they form a fundamental system of neighborhoods of a.

Definition 1.3. A *power series* with center a is a series $\sum f_\nu$ such that each f_ν is of the form

$$f_\nu(z) = c_\nu (z - a)^\nu, \qquad c_\nu \in \mathbb{C}.$$

In the above, we adopt the convention as in Example 1.1. For power series, we have the following fundamental lemma:

Lemma 1.1. *If a series $\sum c_\nu(z-a)^\nu$ converges absolutely at $z = w$, the series is normally convergent on the closure $\overline{\Delta(a,\rho)}$ of every polydisk $\Delta(a,\rho)$ with $\rho_i < |w_i - a_i|$.*

Proof. Without loss of generality we may assume that $a = (0,\ldots,0)$, the origin in $\mathbb{C}^n$. For ρ as above, we may find an n-tuple $t = (t_1,\ldots,t_n)$ of real numbers with $0 < t_i < 1$ and $\rho_i < t_i|w_i|$. Since the series $\sum|c_\nu w^\nu|$ is convergent, there is a constant K such that $|c_\nu w^\nu| \le K$ for all ν. If z is a point in $\overline{\Delta(0,\rho)}$,

$$|c_\nu z^\nu| \le |c_\nu|\rho^\nu \le Kt^\nu$$

and the lemma follows from Example 1.1. □

Remark 1.1. As is seen in the proof, the same result holds under the assumption that $\{|c_\nu w^\nu|\}$ is bounded. Thus if $n = 1$, only the convergence of the series $\sum c_\nu(z-a)^\nu$ at $z = w$ is sufficient to have the same result.

1.2 Analytic functions of one complex variable

We recall some basics on analytic functions of one complex variable. Let U be an open set in the complex plane $\mathbb{C}$ and f a complex valued function on U.

Definition 1.4. We say that f is *analytic* at a point a in U if there is a power series $\sum_{\nu\ge0} c_\nu(z-a)^\nu$ which converges at each point z in a neighborhood of a and satisfies

$$f(z) = \sum_{\nu\ge0} c_\nu(z-a)^\nu$$

in a neighborhood of a. We say that f is analytic in U if it is analytic at every point in U.

In general, let r be a non-negative integer. Recall that a function of real variables is *of class* C^r, C^r for short, if its partial derivatives exist up to order r and are continuous. By convention, C^0 means continuous. If all the partial derivatives exist, we say it is C^∞.

If we write $z = x + \sqrt{-1}\,y$ with x and y the real and imaginary parts, we may think of f as above as a function of (x,y). If f is analytic in U, it is C^∞ in U.

Definition 1.5. We say that f is *holomorphic* at a point a in U if the limit

$$\lim_{h\to 0}\frac{f(a+h)-f(a)}{h}$$

exists. We say that f is holomorphic in U if it is holomorphic at every point in U.

The above limit, if it exists, is denoted by $\frac{df}{dz}(a)$ and is called the *derivative* of f at a. If f is holomorphic in U, we may think of $\frac{df}{dz}$ as a function on U.

If f is analytic in U, it is holomorphic in U and the derivative is also analytic in U.

Cauchy integral formula: Let f be a holomorphic function on a neighborhood of the closure of a disk Δ in $\mathbb{C}$ and $\gamma = \partial\Delta$ the boundary of Δ oriented counterclockwise. Then for z in Δ,

$$f(z) = \frac{1}{2\pi\sqrt{-1}}\int_\gamma \frac{f(\zeta)\,d\zeta}{\zeta - z}. \tag{1.1}$$

Remark 1.2. The above formula holds more generally if we replace Δ with a relatively compact open set U in $\mathbb{C}$ such that $\overline{U}$ is a real two-dimensional submanifold of $\mathbb{C}$ with piecewise C^∞ boundary ∂U. In this case $\gamma = \partial U$ consists of a finite number of piecewise C^∞ simple closed curves oriented as the boundary of $\overline{U}$ (cf. Section 3.7 below).

This will be generalized later to the case f is C^∞ (Theorem 11.8 below).

For a function f on an open set U in $\mathbb{C}$, we write $f = u + \sqrt{-1}\,v$ with u and v the real and imaginary parts.

Theorem 1.1. *The following are equivalent:*

(1) *f is analytic in U.*
(2) *f is holomorphic in U.*
(3) *f is C^1 in (x,y) and satisfies the "Cauchy-Riemann equations" in U:*

$$\frac{\partial u}{\partial x} = \frac{\partial v}{\partial y}, \qquad \frac{\partial u}{\partial y} = -\frac{\partial v}{\partial x}.$$

Note that, if we introduce the operators

$$\frac{\partial}{\partial z} = \frac{1}{2}\Big(\frac{\partial}{\partial x} - \sqrt{-1}\,\frac{\partial}{\partial y}\Big) \quad\text{and}\quad \frac{\partial}{\partial \bar z} = \frac{1}{2}\Big(\frac{\partial}{\partial x} + \sqrt{-1}\,\frac{\partial}{\partial y}\Big),$$

we may express the Cauchy-Riemann equations as

$$\frac{\partial f}{\partial \bar z} = 0.$$

If this is the case, we have $\frac{\partial f}{\partial z} = \frac{df}{dz}$.

In the following, we use the words "analytic" and "holomorphic" interchangeably. Here are some basic facts that we need. We denote by D a connected open set in $\mathbb{C}$.

Uniqueness of analytic continuation: Let f and g be holomorphic functions on D. If $f = g$ on a neighborhood of a point in D, $f = g$ on D.

Open mapping theorem: Let f be a non-constant holomorphic function on D. Then f is open as a map $D \to \mathbb{C}$.

Maximum principle: Let f be a holomorphic function on D. If there is a point z_0 in D such that $|f(z_0)| \geq |f(z)|$ for all z in a neighborhood of z_0, then f is constant on D. This may be thought of as a special case of the open mapping theorem.

We finish this section by reviewing the following:

Meromorphic functions

Let f and g be holomorphic functions in a neighborhood of a point a in $\mathbb{C}$. Suppose g has no zeros away from a, then f/g is defined and holomorphic away from a. We examin the behavior of f/g near a. If f is not identically equal to zero in a neighborhood of a, we may write

$$f(z) = (z-a)^k f_1(z), \qquad g(z) = (z-a)^{k'} g_1(z)$$

with k and k' non-negative integers and f_1 and g_1 holomorphic functions without zeros in a neighborhood of a. There are two cases:

(1) $k \geq k'$. In this case, we have

$$\frac{f(z)}{g(z)} = (z-a)^{k-k'} \frac{f_1(z)}{g_1(z)}$$

so that f/g extends to a holomorphic function in a neighborhood of a. If $k > k'$, the function has a zero at a. We call $m = k - k'$ the *order of the zero.*

(2) $k < k'$. In this case, we have

$$\frac{f(z)}{g(z)} = \frac{1}{(z-a)^{k'-k}} \frac{f_1(z)}{g_1(z)}.$$

We say that f/g has a *pole* at a and call $m = k' - k$ the *order of the pole.*

Definition 1.6. A *meromorphic function* φ on an open set U in $\mathbb{C}$ is a holomorphic function on the complement of a (possibly empty) set S of isolated points in U such that φ has a pole at each point of S.

Note that in a neighborhood of each point of U, φ may be written $\varphi = f/g$, where f and g are holomorphic and g is not identically equal to zero.

Remark 1.3. 1. In fact, every meromophic function φ on an open set U in $\mathbb{C}$ may be written $\varphi = f/g$, where f and g are holomorphic functions on U and g is not identically equal to zero in any connected component of U. This is a property particular to open sets in $\mathbb{C}$.
2. If we denote by O_U the set of holomorphic functions on an open set U in $\mathbb{C}$, then the usual addition and multiplication of functions give O_U the structure of a ring (in fact a $\mathbb{C}$-algebra). In particular, if $U = D$ is a non-empty connected open set, then O_D is an integral domain (cf. Section A.2), by the uniqueness of analytic continuation. Moreover, by 1 above, the set M_D of meromorphic functions on D may be thought of as the fraction field of O_D.

1.3 Analytic functions of several complex variables

Let f a complex valued function on an open set U in $\mathbb{C}^n$.

Definition 1.7. We say that f is *analytic* at a point a in U if there is a power series

$$\sum c_\nu (z-a)^\nu = \sum_{\nu_1,\dots,\nu_n \geq 0} c_{\nu_1\dots\nu_n}(z_1-a_1)^{\nu_1}\cdots(z_n-a_n)^{\nu_n}$$

which converges absolutely at each point z in a neighborhood of a and satisfies

$$f(z) = \sum c_\nu (z-a)^\nu$$

in a neighborhood of a. We say that f is analytic in U if it is analytic at every point in U.

Note that, in the case $n = 1$, Definitions 1.4 and 1.7 are equivalent (cf. Remark 1.1).

If f is analytic in U, it is continuous in U by Lemma 1.1 and it is analytic in each variable z_i separately, since in a neighborhood each point

a in U we may rearrange the power series as a convergent power series in $z_i - a_i$ by Proposition 1.2. In fact we have:

Theorem 1.2. *For a complex valued function f defined on an open set U in $\mathbb{C}^n$, the following conditions are equivalent:*

(1) *f is analytic in U.*
(2) *f is continuous and is analytic in each variable z_i in U, $i = 1, \ldots, n$.*

Proof. It remains to show that (2) implies (1). Let a be a point in U and choose ρ so that $\overline{\Delta(a, \rho)}$ is contained in U. For each i we denote by γ_i the boundary of $\Delta(a_i, \rho_i)$ oriented counterclockwise. By a repeated use of the Cauchy integral formula (1.1), for z in $\Delta(a, \rho)$,

$$\begin{aligned} f(z) &= \frac{1}{2\pi\sqrt{-1}} \int_{\gamma_1} \frac{f(\zeta_1, z_2, \ldots, z_n)}{\zeta_1 - z_1} d\zeta_1 \\ &= \cdots \\ &= \left(\frac{1}{2\pi\sqrt{-1}}\right)^n \int_{\gamma_1} \frac{d\zeta_1}{\zeta_1 - z_1} \cdots \int_{\gamma_n} \frac{f(\zeta_1, \ldots, \zeta_n)}{\zeta_n - z_n} d\zeta_n \\ &= \left(\frac{1}{2\pi\sqrt{-1}}\right)^n \int \cdots \int_{\gamma} \frac{f(\zeta_1, \ldots, \zeta_n)}{(\zeta_1 - z_1) \cdots (\zeta_n - z_n)} d\zeta_1 \cdots d\zeta_n, \end{aligned}$$

where the last multi-integral is to be performed on $\gamma = \gamma_1 \times \cdots \times \gamma_n$. Note that we used the continuity of f in the last equality.

On the other hand, noting that

$$\frac{1}{\zeta_i - z_i} = \frac{1}{\zeta_i - a_i} \frac{1}{1 - \frac{z_i - a_i}{\zeta_i - a_i}} \quad \text{and} \quad \left|\frac{z_i - a_i}{\zeta_i - a_i}\right| < 1,$$

we have a power series expansion

$$\frac{1}{\zeta_1 - z_1} \cdots \frac{1}{\zeta_n - z_n} = \sum_{\nu_1, \ldots, \nu_n = 0}^{\infty} \frac{(z_1 - a_1)^{\nu_1}}{(\zeta_1 - a_1)^{\nu_1 + 1}} \cdots \frac{(z_n - a_n)^{\nu_n}}{(\zeta_n - a_n)^{\nu_n + 1}},$$

which converges normally with respect to ζ on γ. Hence we may integrate term by term and we have a power series expansion, which converges absolutely at z:

$$f(z) = \sum c_\nu (z - a)^\nu$$

with

$$c_\nu = \left(\frac{1}{2\pi\sqrt{-1}}\right)^n \int \cdots \int_{\gamma} \frac{f(\zeta_1, \ldots, \zeta_n) d\zeta_1 \cdots d\zeta_n}{(\zeta_1 - z_1)^{\nu_1 + 1} \cdots (\zeta_n - z_n)^{\nu_n + 1}}.$$

□

It is known that we may remove the continuity condition in (2) above (Hartogs' theorem).

In general, let f be a complex valued function defined on an open set U in $\mathbb{C}^n$. Writing $z_i = x_i + \sqrt{-1}\,y_i$, we may think of f as a function of the real variables $(x_1, y_1, \ldots, x_n, y_n)$. We recall that (cf. Theorem 1.1) a C^1 function f is analytic in the variable z_i if and only if it satisfies the Cauchy-Riemann equation

$$\frac{\partial f}{\partial \bar{z}_i} = 0. \tag{1.2}$$

If f is analytic in U, by Lemma 1.1, it is C^1 in U, in fact it is C^∞, since we may differentiate it term by term when expressed as a power series and the resulting series also converges normally in the same domain. Thus from Theorem 1.2, we have:

Theorem 1.3. *For a function f defined on U, the following are equivalent:*

(1) *f is analytic in U.*
(2) *f is C^1 and satisfies* (1.2) *in U, for $i = 1, \ldots, n$.*

In the following, we call an analytic function also a *holomorphic* function and use the words "analytic" and "holomorphic" interchangeably as in the case of one variable.

Note that, for a holomorphic function f, the "partial derivative" $\partial f/\partial z_i$ coincides with the derivative of f considered as a holomorphic function of the single variable z_i and, for every ν, the (higher) partial derivative

$$\frac{\partial^\nu f}{\partial z^\nu} = \frac{\partial^{|\nu|} f}{\partial z_1^{\nu_1} \cdots \partial z_n^{\nu_n}}$$

exists and is holomorphic in U. If $f(z) = \sum c_\nu (z-a)^\nu$ is a power series expansion of f, then each coefficient c_ν is given by

$$c_\nu = \frac{1}{\nu!}\frac{\partial^\nu f}{\partial z^\nu}(a).$$

This series is called the *Taylor series* of f at a.

Definition 1.8. 1. Let U be an open set in $\mathbb{C}^n$ and $f : U \to \mathbb{C}^r$ a map. We say that f is *holomorphic* if, writing f componentwise as $f = (f_1, \ldots, f_r)$, each f_i is holomorphic.

2. Let U and U' be two open sets in $\mathbb{C}^n$ and $f : U \to U'$ a map. We say that f is *biholomorphic* if f is bijective and if both f and f^{-1} are holomorphic.

Proposition 1.4. *Let U and U' be open sets in $\mathbb{C}^n$ and $\mathbb{C}^r$, respectively, and let $f : U \to U'$ and $g : U' \to \mathbb{C}$ be maps. If both f and g are holomorphic, the composition $g \circ f$ is also holomorphic.*

Proof. First we note that $g \circ f$ is C^1. If we denote coordinates of $\mathbb{C}^n$ by $(z_1, \ldots, z_n)$ and those of $\mathbb{C}^r$ by $(w_1, \ldots, w_r)$, using the chain rule we have

$$\frac{\partial g \circ f}{\partial \bar{z}_i}(z) = \sum_{j=1}^{r} \frac{\partial g}{\partial w_j}(f(z))\frac{\partial f_j}{\partial \bar{z}_i}(z) + \sum_{j=1}^{r} \frac{\partial g}{\partial \bar{w}_j}(f(z))\frac{\partial \bar{f}_j}{\partial \bar{z}_i}(z).$$

Then we have the proposition as each term in the right-hand side is zero by the Cauchy-Riemann equations. □

For a holomorphic map f from an open set U in $\mathbb{C}^n$ to $\mathbb{C}^r$, we write $f = (f_1, \ldots, f_r)$ and set

$$\frac{\partial(f_1, \ldots, f_r)}{\partial(z_1, \ldots, z_n)} = \begin{pmatrix} \dfrac{\partial f_1}{\partial z_1} & \cdots & \dfrac{\partial f_1}{\partial z_n} \\ \vdots & \ddots & \vdots \\ \dfrac{\partial f_r}{\partial z_1} & \cdots & \dfrac{\partial f_r}{\partial z_n} \end{pmatrix}.$$

We call it the *Jacobian matrix* of f with respect to z. The *rank* of f at a point a in U is the rank of the Jacobian matrix at a.

Definition 1.9. We say that a point a in U is a *regular point* of f if the rank of f at a is maximal possible, i.e., $\min(n, r)$. Otherwise we say that a is a *singular point* , or a *critical point*, of f.

Note that, if a is a regular point of f, there is a neighborhood U_a of a in U such that every point in U_a is a regular point of f.

In the case $n = r$, the determinant of the Jacobian matrix is called the *Jacobian* of f. In this case, a is a regular point of f if and only if

$$\det \frac{\partial(f_1, \ldots, f_n)}{\partial(z_1, \ldots, z_n)}(a) \neq 0.$$

If we denote by u_i and v_i the real and the imaginary parts of f_i, we have:

$$\det \frac{\partial(u_1, v_1, \ldots, u_n, v_n)}{\partial(x_1, y_1, \ldots, x_n, y_n)} = \left| \det \frac{\partial(f_1, \ldots, f_n)}{\partial(z_1, \ldots, z_n)} \right|^2. \tag{1.3}$$

Exercise 1.1. Verify the identity (1.3).

The following three theorems describe the nature of a holomorphic map near its regular point. Thus we consider maps f from a neighborhood of the origin 0 in $\mathbb{C}^n$ to $\mathbb{C}^r$. Without loss of generality, we may assume $f(0) = 0$. We express this situation by writing $f : (\mathbb{C}^n, 0) \to (\mathbb{C}^r, 0)$. Also $f : (\mathbb{C}^n, 0) \to (\mathbb{C}^n, 0)$ being biholomorphic means there exist neighborhoods U and V of 0 such that f is a biholomorphic map of U onto V.

First we prove the following using the corresponding theorem in the real variable case:

Theorem 1.4 (Inverse mapping theorem). *Let $f : (\mathbb{C}^n, 0) \to (\mathbb{C}^n, 0)$ be a holomorphic map. If 0 is a regular point of f, then f is biholomorphic.*

Proof. By (1.3) and the inverse mapping theorem in the real variable case, there exist neighborhoods U and U' of 0 such that f is a bijective map from U onto U' with C^1 inverse f^{-1}. A computation similar to the one in the proof of Proposition 1.4 shows that f^{-1} is also holomorphic. □

From the above, we have the following. Although the proofs are the same as the real case, we give them for the sake of completeness.

Theorem 1.5 (Implicit mapping theorem). *Let $f : (\mathbb{C}^n, 0) \to (\mathbb{C}^k, 0)$ be a holomorphic map. Suppose $n > k$ and 0 is a regular point of f. Renumbering the functions and the variables if necessary, we may assume*

$$\det \frac{\partial(f_1, \ldots, f_k)}{\partial(z_1, \ldots, z_k)}(0) \neq 0.$$

Then there exists a holomorphic map $\varphi : (\mathbb{C}^{n-k}, 0) \to (\mathbb{C}^k, 0)$ such that

$$f(\varphi_1(z_{k+1}, \ldots, z_n), \ldots, \varphi_k(z_{k+1}, \ldots, z_n), z_{k+1}, \ldots, z_n) = 0$$

for $(z_{k+1}, \ldots, z_n)$ in a neighborhood of 0.

Proof. Set $z' = (z_1, \ldots, z_k)$ and $z'' = (z_{k+1}, \ldots, z_n)$ and define a holomorphic map $g : (\mathbb{C}^n, 0) \to (\mathbb{C}^n, 0)$ by $g(z) = (f(z), z'')$. Then 0 is a regular point of g and thus g is biholomorphic by Theorem 1.4. We write $g^{-1} = (a, b)$ with $a : (\mathbb{C}^n, 0) \to (\mathbb{C}^k, 0)$ and $b : (\mathbb{C}^n, 0) \to (\mathbb{C}^{n-k}, 0)$. Then

$$f(a(z), b(z)) = z', \qquad b(z) = z''. \tag{1.4}$$

Thus if we set $\varphi(z'') = a(0, z'')$, φ is the desired holomorphic map. □

Remark 1.4. In the above situation, we may solve the equation

$$f(z_1, \ldots, z_k, \ldots, z_n) = 0$$

for $z_1, \ldots, z_k$ as functions $\varphi_1, \ldots, \varphi_k$ of $(z_{k+1}, \ldots, z_n)$ in a neighborhood of 0 and the set $f^{-1}(0)$ is the graph of the map $\varphi = (\varphi_1, \ldots, \varphi_k)$.

The following gives a local normal form of a map near a regular point:

Theorem 1.6. 1. *Let $f : (\mathbb{C}^d, 0) \to (\mathbb{C}^n, 0)$ be a holomorphic map. Suppose $d \le n$ and 0 is a regular point of f. Setting $k = n - d$ and denoting by $w = (w_1 \ldots, w_d)$ coordinates on $\mathbb{C}^d$, we may assume that*

$$\det \frac{\partial(f_{k+1}, \ldots, f_n)}{\partial(w_1, \ldots, w_d)}(0) \neq 0.$$

Then there exists a biholomorphic map $h : (\mathbb{C}^n, 0) \to (\mathbb{C}^n, 0)$ such that

$$(h \circ f)(w_1, \ldots, w_d) = (0, \ldots, 0, w_1, \ldots, w_d).$$

2. *Let $f : (\mathbb{C}^n, 0) \to (\mathbb{C}^k, 0)$ be a holomorphic map. Suppose $n \ge k$ and 0 is a regular point of f. We may assume that*

$$\det \frac{\partial(f_1, \ldots, f_k)}{\partial(z_1, \ldots, z_k)}(0) \neq 0.$$

Then there exists a biholomorphic map $h : (\mathbb{C}^n, 0) \to (\mathbb{C}^n, 0)$ such that

$$(f \circ h)(z_1, \ldots, z_k, \ldots, z_n) = (z_1, \ldots, z_k).$$

Proof. **1.** We set $z' = (z_1, \ldots, z_k)$ and $z'' = (z_{k+1}, \ldots, z_n)$ and identify $\mathbb{C}^d$ with $\{0\} \times \mathbb{C}^d \subset \mathbb{C}^n$ by identifying w with $(0, z'')$, $z_{k+i} = w_i$. Define a map $g : (\mathbb{C}^n, 0) \to (\mathbb{C}^n, 0)$ by $g(z) = f(z'') + (z', 0)$. Then it is biholomorphic and $h = g^{-1}$ is the desired biholomorphic map.

2. If $n = k$, f is biholomorphic and thus set $h = f^{-1}$. If $n > k$, let $g : (\mathbb{C}^n, 0) \to (\mathbb{C}^n, 0)$ be as in the proof of Theorem 1.5 and set $h = g^{-1}$. Then from (1.4), we see that it is the desired map. □

Remark 1.5. In the case 2 above, we may think of $(f_1, \ldots, f_k, z_{k+1}, \ldots, z_n)$ as a coordinate system on a neighborhood of 0 in $\mathbb{C}^n$.

The following theorems can be proved as in the one variable case or reducing to that case. We denote by D a connected open set in $\mathbb{C}^n$.

Theorem 1.7 (Uniqueness of analytic continuation). *Let f and g be holomorphic functions on D. If $f = g$ on a neighborhood of a point in D then $f = g$ on D.*

Theorem 1.8 (Maximum principle). *Let f be a holomorphic function on D. If there is a point z_0 in D such that $|f(z_0)| \ge |f(z)|$ for all z in a neighborhood of z_0, then f is constant on D.*

1.4 Germs of holomorphic functions

In this section we discuss basic local properties of holomorphic functions.

Let H denote the set of pairs (U, f) of a neighborhood U of 0 and a holomorphic function f on U. We define a relation $\sim$ in H as follows. For two elements (U, f) and (V, g) in H, we say $(U, f) \sim (V, g)$ if there is a neighborhood W of 0 such that $W \subset U \cap V$ and that the restrictions of f and g to W are the same. Then it is easily checked that $\sim$ is an equivalence relation in H. The equivalence class of (U, f) is called the *germ of f at* 0 and is denoted by $\boldsymbol{f}$, or simply by f if there is no fear of confusion.

We let $\mathscr{O}_n$ denote the quotient set $H/\sim$. The set $\mathscr{O}_n$ has the structure of a commutative ring with respect to the operations induced from the addition and multiplication of functions. It has the unity which is the equivalence class of the function constantly equal to 1. We call $\mathscr{O}_n$ the *ring of germs of holomorphic functions at* 0. In fact it contains the field $\mathbb{C}$ as the germs of constant functions and has a natural $\mathbb{C}$-algebra structure.

Remark 1.6. In algebraic terms the ring $\mathscr{O}_n$ is expressed as follows (cf. Section A.1). We may introduce an order relation in the set of neighborhoods of 0 by reverse inclusion, i.e., $U \leq V$ if $V \subset U$. Then it becomes a directed set. For a neighborhood U we denote by O_U the ring of holomorphic functions on U and for $V \subset U$ let $r_{VU} : O_U \to O_V$ be the restriction. Then (O_U, r_{VU}) is a direct system (cf. Definition A.8) and we may write

$$\mathscr{O}_n = \varinjlim_{U \ni 0} O_U,$$

the direct limit on the set of neighborhoods of 0.

If we denote by $\mathbb{C}\{z_1, \ldots, z_n\}$ the set of power series that converge absolutely in some neighborhood of 0, this set also has the structure of a ring. Since $(U, f) \sim (V, g)$ if and only if f and g have the same power series expansion, we may identify $\mathscr{O}_n$ with $\mathbb{C}\{z_1, \ldots, z_n\}$.

We start with some fundamental properties of the ring $\mathscr{O}_n$. For terminologies and basics of commutative algebra, we refer to Appendix A. In the following, we adopt the convention that $\mathbb{C}^0 = \{0\}$ and $\mathscr{O}_0 = \mathbb{C}$.

Proposition 1.5. *The ring $\mathscr{O}_n$ is an integral domain.*

Proof. Suppose $\boldsymbol{fg} = 0$ for germs $\boldsymbol{f}$ and $\boldsymbol{g}$ in $\mathscr{O}_n$. Then there is a connected neighborhood U of 0 such that $f(z)g(z) = 0$ for all z in U. If $\boldsymbol{f} \neq 0$, there is a point z such that $f(z) \neq 0$. By the continuity of f, we may find a neighborhood U_z of z such that $f(z') \neq 0$ for all z' in U_z. Since the restriction of g to U_z must be identically equal to 0, we get $\boldsymbol{g} = 0$ by Theorem 1.7. □

By the above proposition, we may form the fraction field of $\mathscr{O}_n$, which we denote by $\mathscr{M}_n$. Each element in $\mathscr{M}_n$ is expressed as $\boldsymbol{f}/\boldsymbol{g}$ and two expressions $\boldsymbol{f}/\boldsymbol{g}$ and $\boldsymbol{f}'/\boldsymbol{g}'$ stand for the same element if and only if $\boldsymbol{f}\boldsymbol{g}' = \boldsymbol{f}'\boldsymbol{g}$. We call an element of $\mathscr{M}_n$ a germ of a *meromorphic function* at 0 in $\mathbb{C}^n$.

Proposition 1.6. *A germ $\boldsymbol{u}$ in $\mathscr{O}_n$ is a unit if and only if it is the germ of a function u with $u(0) \neq 0$.*

Proof. If u is a holomorphic function in a neighborhood of 0 with $u(0) \neq 0$, then $1/u$ is a holomorphic function in a neighborhood of 0. Thus the germ of u is a unit in $\mathscr{O}_n$. The converse is obvious. □

If we denote by $\mathfrak{m}$ the set of non-units in $\mathscr{O}_n$, it is an ideal in $\mathscr{O}_n$.

Proposition 1.7. *The ideal $\mathfrak{m}$ is the unique maximal ideal in $\mathscr{O}_n$.*

Proof. Let I be an arbitrary ideal in $\mathscr{O}_n$. If I contains a unit, then $I = \mathscr{O}_n$. If I does not contain units, then $I \subset \mathfrak{m}$. □

Note that the quotient $\mathscr{O}_n/\mathfrak{m}$ is canonically isomorphic with $\mathbb{C}$. The isomorphism is given by assigning to the class of $\boldsymbol{f}$ the value $f(0)$. Recall that a ring with a unique maximal ideal is called a *local ring* (cf. Section A.2).

We analyze the structure of the ring $\mathscr{O}_n$ by induction on n. For this, we pay attention to a specific variable, which will be z_n. First, for a germ $\boldsymbol{f}$ in $\mathscr{O}_n$, we write $f = \sum_{|\nu| \geq 0} a_\nu z^\nu$.

Definition 1.10. If $\boldsymbol{f} \neq 0$, the *order* of $\boldsymbol{f}$ is the integer m such that $a_\nu = 0$ for all ν with $|\nu| < m$ and $a_{\nu_0} \neq 0$ for some ν_0 with $|\nu_0| = m$. The order of the germ 0 is defined to be $+\infty$.

Thus the order of $\boldsymbol{f}$ is a positive integer if and only if it is not 0 or a unit. The above notion of order does not depend on the coordinates, while the next one does.

Definition 1.11. The order of $\boldsymbol{f}$ in z_n is the order of $f(0, \ldots, 0, z_n)$ as a power series in z_n. We say that $\boldsymbol{f}$ is *regular in z_n of order m*, if the order m of $\boldsymbol{f}$ in z_n is a positive integer.

Thus $\boldsymbol{f}$ is regular in z_n of order m if and only if $f(0, \ldots, 0, z_n)$ has a zero of multiplicity m at $z_n = 0$.

Lemma 1.2. *If the order of $\boldsymbol{f}$ is m, then we may find a suitable coordinate system $(\zeta_1, \ldots, \zeta_n)$ of $\mathbb{C}^n$ such that the order of $\boldsymbol{f}$ in ζ_n is m.*

Proof. As the case $\boldsymbol{f} = 0$ is obvious, we assume that $\boldsymbol{f} \neq 0$. If we set $f_m = \sum_{|\nu|=m} a_\nu z^\nu$, then $f_m \neq 0$. Thus, there is a point $c = (c_1, \ldots, c_n) \neq 0$ such that $f_m(c) \neq 0$. There exists an $n \times n$ non-singular matrix C whose n-th row is c. If we set $\zeta = zC^{-1}$, $f(\zeta C)$ is of order m in ζ. Since $f_m((0, \ldots, 0, 1)C) = f_m(c) \neq 0$, we see that $f(\zeta C)$ is of order m in ζ_n. □

Remark 1.7. The above arguments work for a finitely many germs in $\mathscr{O}_n$. Thus, a finite number of germs in $\mathscr{O}_n$ whose orders are positive integers can be simultaneously made to be regular in some variable, by a suitable change of coordinates.

We show that the ring $\mathscr{O}_n$ possesses certain properties similar to the ones polynomial rings do. Thus we set $z' = (z_1, \ldots, z_{n-1})$ and let $\mathscr{O}_{n-1}$ denote the ring of germs of holomorphic functions in z'. We consider the ring $\mathscr{O}_{n-1}[z_n]$ of polynomials in z_n with coefficients in $\mathscr{O}_{n-1}$:

$$\mathscr{O}_{n-1}[z_n] = \{\boldsymbol{a}_0 + \boldsymbol{a}_1 z_n + \cdots + \boldsymbol{a}_r z_n^r \mid r \in \mathbb{N}_0, \ \ \boldsymbol{a}_i \in \mathscr{O}_{n-1}\},$$

which is considered as a subring of $\mathscr{O}_n$ and contains $\mathscr{O}_{n-1}$ as a subring.

Definition 1.12. Let m be a positive integer. A *Weierstrass polynomial* in z_n of degree m is a germ $\boldsymbol{h}$ in $\mathscr{O}_{n-1}[z_n]$ of the form

$$\boldsymbol{h} = \boldsymbol{a}_0 + \boldsymbol{a}_1 z_n + \cdots + \boldsymbol{a}_{m-1} z_n^{m-1} + z_n^m,$$

where $\boldsymbol{a}_0, \boldsymbol{a}_1, \ldots, \boldsymbol{a}_{m-1}$ are non-units in $\mathscr{O}_{n-1}$.

Thus a Weierstrass polynomial in z_n is a non-unit in both $\mathscr{O}_{n-1}[z_n]$ and $\mathscr{O}_n$. Note that in the above, $h(0, \ldots, 0, z_n) = z_n^m$. Hence $\boldsymbol{h}$ is regular in z_n of order m. In general, for every germ $\boldsymbol{f}$ in $\mathscr{O}_n$, we may write in a neighborhood of 0

$$f(z) = a_0 + a_1 z_n + \cdots + a_m z_n^m + \cdots$$

with a_i holomorphic functions in z'. The germ $\boldsymbol{f}$ is regular in z_n of order m if and only if $\boldsymbol{a}_0, \boldsymbol{a}_1, \ldots, \boldsymbol{a}_{m-1}$ are non-units in $\mathscr{O}_{n-1}$ and $\boldsymbol{a}_m$ is a unit in $\mathscr{O}_{n-1}$. In this case, $\boldsymbol{a}_m^{-1}(\boldsymbol{a}_0 + \boldsymbol{a}_1 z_n + \cdots + \boldsymbol{a}_m z_n^m)$ is a Weierstrass polynomial in z_n of degree m. The Weierstrass preparation theorem below shows that such a germ $\boldsymbol{f}$ is essentially equal to a Weierstrass polynomial of degree m.

We first prove the following:

Theorem 1.9 (Weierstrass division theorem). *Let $\boldsymbol{h}$ be a Weierstrass polynomial in z_n of degree m. Then, for every germ $\boldsymbol{f}$ in $\mathscr{O}_n$, there exist uniquely determined germs $\boldsymbol{q}$ in $\mathscr{O}_n$ and $\boldsymbol{r}$ in $\mathscr{O}_{n-1}[z_n]$ with* $\deg \boldsymbol{r} < m$ *such that*

$$\boldsymbol{f} = \boldsymbol{q}\boldsymbol{h} + \boldsymbol{r}.$$

Moreover, if $\boldsymbol{f}$ is in $\mathscr{O}_{n-1}[z_n]$, so is $\boldsymbol{q}$.

Proof. We choose $\rho = (\rho_1, \ldots, \rho_n)$, $\rho_i > 0$, small enough so that the germs $\boldsymbol{f}$ and $\boldsymbol{h}$ have representatives on an open set containing $\overline{\Delta(0, \rho)}$. As $\boldsymbol{h}$ is regular in z_n, taking a smaller ρ if necessary, we may assume that $h(z) \neq 0$ for $z = (z_1, \ldots, z_n)$ with $|z_1| < \rho_1, \ldots, |z_{n-1}| < \rho_{n-1}, |z_n| = \rho_n$. We denote by γ the path in the z_n- plane that turns counterclockwise along the circle $|z_n| = \rho_n$. If we set

$$q(z) = \frac{1}{2\pi\sqrt{-1}} \int_\gamma \frac{f(z', \zeta)}{h(z', \zeta)} \frac{d\zeta}{\zeta - z_n},$$

then $q(z)$ is holomorphic in $\Delta(0, \rho)$. If we further set $r(z) = f(z) - q(z)h(z)$, using the Cauchy integral formula

$$f(z) = \frac{1}{2\pi\sqrt{-1}} \int_\gamma \frac{f(z', \zeta)}{\zeta - z_n} d\zeta,$$

we compute

$$r(z) = \frac{1}{2\pi\sqrt{-1}} \int_\gamma \frac{f(z', \zeta)}{h(z', \zeta)} \frac{h(z', \zeta) - h(z', z_n)}{\zeta - z_n} d\zeta.$$

The first function in the integrant does not depend on z_n and the second is a polynomial in z_n of degree less than m, since $\boldsymbol{h}$ is a Weierstrass polynomial in z_n of degree m. Thus we see that $\boldsymbol{r}$ is in $\mathscr{O}_{n-1}[z_n]$ and $\deg \boldsymbol{r} < m$.

To show the uniqueness of $\boldsymbol{q}$ and $\boldsymbol{r}$, suppose $\boldsymbol{f} = \boldsymbol{q}\boldsymbol{h} + \boldsymbol{r} = \boldsymbol{q}'\boldsymbol{h} + \boldsymbol{r}'$. Choose ρ as above. Then we have

$$r'(z', z_n) - r(z', z_n) = (q(z', z_n) - q'(z', z_n))h(z', z_n)$$

in $\Delta(\rho, 0)$. For each fixed a' in $\Delta(\rho', 0)$, $\rho' = (\rho_1, \ldots, \rho_{n-1})$, $h(a', z_n)$ has m zeros in $\Delta(\rho_n, 0)$, counting multiplicities. On the other hand, the left-hand side above is a polynomial in z_n of degree at most $m-1$ and thus $r'(a', z_n) - r(a', z_n) = 0$ in $\Delta(\rho_n, 0)$. Therefore $\boldsymbol{r}' = \boldsymbol{r}$ and also $\boldsymbol{q}' = \boldsymbol{q}$, as $\mathscr{O}_n$ is an integral domain.

The last statement is the division algorithm for polynomials. □

Theorem 1.10 (Weierstrass preparation theorem). *Let $\boldsymbol{f}$ be a germ in $\mathscr{O}_n$ which is regular in z_n of order m. Then there is a unique Weierstrass polynomial $\boldsymbol{h}$ in z_n of degree m such that $\boldsymbol{f} = \boldsymbol{uh}$ with $\boldsymbol{u}$ a unit in $\mathscr{O}_n$.*

Proof. We introduce new variables $w = (w_1, \ldots, w_m)$ and we set $\mathscr{O}_{m+n} = \mathbb{C}\{w, z\}$ and $\mathscr{O}_{m+n-1} = \mathbb{C}\{w, z'\}$. We set

$$p(w, z_n) = w_m + w_{m-1}z_n + \cdots + w_1 z_n^{m-1} + z_n^m$$

and think of it as a Weierstrass polynomial in $\mathscr{O}_{m+n-1}[z_n]$ of degree m. Since $\mathscr{O}_n$ is naturally a subring of $\mathscr{O}_{m+n}$, by the Weierstrass division theorem, there exist unique germs $\boldsymbol{q}$ in $\mathscr{O}_{m+n}$ and $\boldsymbol{r}$ in $\mathscr{O}_{m+n-1}[z_n]$ of degree less than m such that

$$\boldsymbol{f} = \boldsymbol{qp} + \boldsymbol{r}. \tag{1.5}$$

We write

$$r(w, z', z_n) = b_m(w, z') + b_{m-1}(w, z')z_n + \cdots + b_1(w, z')z_n^{m-1}$$

with $\boldsymbol{b}_1, \ldots, \boldsymbol{b}_m$ in $\mathscr{O}_{m+n-1}$. If we set $w = 0$ and $z' = 0$ in (1.5), we get

$$f(0, z_n) = q(0, 0, z_n)z_n^m + b_m(0, 0) + b_{m-1}(0, 0)z_n + \cdots + b_1(0, 0)z_n^{m-1}.$$

Since f is of order m in z_n, $q(0, 0, 0) = c \neq 0$ and

$$b_1(0, 0) = \cdots = b_{m-1}(0, 0) = b_m(0, 0) = 0.$$

If we differentiate the both sides of (1.5) with respect to w_j, noting that $\partial p/\partial w_j = z_n^{m-j}$, we get

$$0 = \frac{\partial q}{\partial w_j}p + qz_n^{m-j} + \frac{\partial b_m}{\partial w_j} + \frac{\partial b_{m-1}}{\partial w_j}z_n + \cdots + \frac{\partial b_1}{\partial w_j}z_n^{m-1}.$$

Setting $w = 0$ and $z' = 0$ again, we see that

$$\frac{\partial b_i}{\partial w_j}(0, 0) = \begin{cases} -c & \text{for } i = j, \\ 0 & \text{for } i > j. \end{cases}$$

Thus

$$\det \frac{\partial(b_1, \ldots, b_m)}{\partial(w_1, \ldots, w_m)}(0, 0) = (-c)^m \neq 0$$

and by Theorem 1.5, there is a holomorphic map $\varphi : (\mathbb{C}^{n-1}, 0) \to (\mathbb{C}^m, 0)$ such that

$$b_1(\varphi(z'), z') = \cdots = b_{m-1}(\varphi(z'), z') = b_m(\varphi(z'), z') = 0.$$

If we substitute $w = \varphi(z')$ in (1.5), we get

$$f(z', z_n) = q(\varphi(z'), z', z_n)p(\varphi(z'), z_n).$$

If we set $u(z', z_n) = q(\varphi(z'), z', z_n)$ and $h(z', z_n) = p(\varphi(z'), z_n)$, then $\boldsymbol{f} = \boldsymbol{uh}$ with $\boldsymbol{u}$ a unit in $\mathscr{O}_n$ and $\boldsymbol{h}$ a Weierstrass polynomial in z_n of degree m. The uniqueness of such $\boldsymbol{u}$ and $\boldsymbol{h}$ is not difficult to see. □

Remark 1.8. By the above theorem and Remark 1.7, for a finite number of germs, each of which is not 0 or a unit in $\mathcal{O}_n$, we may find a coordinate system $(z_1, \ldots, z_n)$ such that each germ is expressed as the product of a unit and a Weierstrass polynomial in z_n.

Next we discuss some important properties of the ring $\mathcal{O}_n$ which follow from the above theorems. First we observe that:

Lemma 1.3. *Let $\boldsymbol{h}$ be a Weierstrass polynomial in $\mathcal{O}_{n-1}[z_n]$. Then $\boldsymbol{h}$ is irreducible in $\mathcal{O}_{n-1}[z_n]$ if and only if it is irreducible in $\mathcal{O}_n$. If it is not irreducible, all of its factors are Weierstrass polynomials in $\mathcal{O}_{n-1}[z_n]$.*

Proof. Suppose $\boldsymbol{h}$ is not irreducible in $\mathcal{O}_n$. Then we may write $\boldsymbol{h} = \boldsymbol{f}_1\boldsymbol{f}_2$ with $\boldsymbol{f}_i$, $i = 1, 2$, non-units in $\mathcal{O}_n$. Since $\boldsymbol{h}$ is regular in z_n, both $\boldsymbol{f}_1$ and $\boldsymbol{f}_2$ are regular in z_n and by the preparation theorem, we may write $\boldsymbol{f}_i = \boldsymbol{u}_i\boldsymbol{h}_i$, $i = 1, 2$, with $\boldsymbol{u}_i$ units in $\mathcal{O}_n$ and $\boldsymbol{h}_i$ Weierstrass polynomials in $\mathcal{O}_{n-1}[z_n]$. By the uniqueness, we have $\boldsymbol{h} = \boldsymbol{h}_1\boldsymbol{h}_2$, and thus $\boldsymbol{h}$ is not irreducible in $\mathcal{O}_{n-1}[z_n]$.

Conversely suppose $\boldsymbol{h}$ is not irreducible in $\mathcal{O}_{n-1}[z_n]$. Then $\boldsymbol{h} = \boldsymbol{g}_1\boldsymbol{g}_2$ with $\boldsymbol{g}_i$, $i = 1, 2$, non-units in $\mathcal{O}_{n-1}[z_n]$. If one of $\boldsymbol{g}_1$ or $\boldsymbol{g}_2$, say $\boldsymbol{g}_1$, were a unit in $\mathcal{O}_n$, we would have $\boldsymbol{u}\boldsymbol{g}_1 = 1$ for some $\boldsymbol{u}$ in $\mathcal{O}_n$ and $\boldsymbol{g}_2 = \boldsymbol{u}\boldsymbol{h}$. From the last part of the division theorem, we would then see that $\boldsymbol{u} \in \mathcal{O}_{n-1}[z_n]$, which contradicts the fact that $\boldsymbol{g}_1$ is a non-unit in $\mathcal{O}_{n-1}[z_n]$. Thus $\boldsymbol{g}_1$ and $\boldsymbol{g}_2$ are non-units in $\mathcal{O}_n$ and $\boldsymbol{h}$ is not irreducible in $\mathcal{O}_n$. □

Theorem 1.11. *The ring $\mathcal{O}_n$ is a unique factorization domain.*

Proof. We proceed by induction on n. First, we have $\mathcal{O}_0 = \mathbb{C}$, which is a unique factorization domain, a UFD for short. We assume that $\mathcal{O}_{n-1}$ is a UFD. Take a germ $\boldsymbol{f}$ in $\mathcal{O}_n$ which is not 0 or a unit. By changing the coordinate system on $\mathbb{C}^n$ if necessary, we may assume that $\boldsymbol{f}$ is regular in z_n. By the Weierstrass preparation theorem, there is a unit $\boldsymbol{u}$ in $\mathcal{O}_n$ such that $\boldsymbol{u}\boldsymbol{f}$ is a Weierstrass polynomial in $\mathcal{O}_{n-1}[z_n]$, which is a UFD by the induction hypothesis and Gauss' lemma (Theorem A.15). Since $\boldsymbol{u}\boldsymbol{f}$ is not 0 or a unit, it can be expressed as a product of irreducible elements in $\mathcal{O}_{n-1}[z_n]$. Then the theorem follows from Lemma 1.3. □

Since a unit in $\mathcal{O}_n$ has m-th roots for every positive integer m, every germ $\boldsymbol{f}$, which is not 0 or a unit, has the irreducible decomposition (cf. (A.11)) of the form

$$\boldsymbol{f} = \boldsymbol{p}_1^{m_1} \cdots \boldsymbol{p}_r^{m_r}.$$

Recall that, in a UFD, an element is irreducible if and only if it is prime.

Theorem 1.12. *The ring $\mathcal{O}_n$ is Noetherian.*

Proof. We proceed by induction on n. We have $\mathcal{O}_0 = \mathbb{C}$, which is Noetherian. We assume that $\mathcal{O}_{n-1}$ is Noetherian.

Take an arbitrary ideal I in $\mathcal{O}_n$. If $I = 0$, we have nothing to prove. If $I \neq 0$, we choose a non-zero element $\boldsymbol{h}$ in I. By the Weierstrass preparation theorem, we may assume that $\boldsymbol{h}$ is a Weierstrass polynomial. Thus $\boldsymbol{h}$ is in $I \cap \mathcal{O}_{n-1}[z_n]$. Since $I \cap \mathcal{O}_{n-1}[z_n]$ is an ideal in $\mathcal{O}_{n-1}[z_n]$, which is Noetherian by the induction hypothesis and the Hilbert basis theorem (Theorem A.17), it has a finite number of generators $\boldsymbol{g}_1, \ldots, \boldsymbol{g}_r$. We show that they also generate the ideal I over $\mathcal{O}_n$. Take an element $\boldsymbol{f}$ in I. Changing the coordinate system of $\mathbb{C}^n$ if necessary, we may assume that $\boldsymbol{f}$ is regular in z_n. Then by the Weierstrass division theorem, we may write $\boldsymbol{f} = \boldsymbol{q}\boldsymbol{h} + \boldsymbol{r}$ for some $\boldsymbol{q}$ in $\mathcal{O}_n$ and $\boldsymbol{r}$ in $\mathcal{O}_{n-1}[z_n]$. Since the germs $\boldsymbol{f}$ and $\boldsymbol{h}$ are in I, so is $\boldsymbol{r}$. Thus $\boldsymbol{r}$ is in $I \cap \mathcal{O}_{n-1}[z_n]$. Since $\boldsymbol{h}$ and $\boldsymbol{r}$ can be written as linear combinations of $\boldsymbol{g}_1, \ldots, \boldsymbol{g}_r$ over $\mathcal{O}_{n-1}[z_n]$, the germ $\boldsymbol{f}$ is a linear combination of $\boldsymbol{g}_1, \ldots, \boldsymbol{g}_r$ over $\mathcal{O}_n$. □

We give another application of Theorem 1.10. For a point z in $\mathbb{C}^n$, let $\mathcal{O}_{n,z}$ be the ring of germs of holomorphic functions at z, which is naturally isomorphic with $\mathcal{O}_n$. For a germ $\boldsymbol{f}$ in $\mathcal{O}_n$ we denote by f_z the germ of f at z near 0.

Proposition 1.8. *If $\boldsymbol{f}$ and $\boldsymbol{g}$ are relatively prime in $\mathcal{O}_n$, then f_z and g_z are relatively prime in $\mathcal{O}_{n,z}$ for all z sufficiently close to* 0.

Proof. If $\boldsymbol{f}$ or $\boldsymbol{g}$ is a unit, the proposition obviously holds. Thus we assume that $\boldsymbol{f}$ and $\boldsymbol{g}$ are non-units. By the preparation theorem, we may assume that they are Weierstrass polynomials in z_n (cf. Remark 1.8) and are relatively prime in $\mathcal{O}_{n-1}[z_n]$ (cf. Lemma 1.3). Thus by Theorem A.16, there exist elements $\boldsymbol{a}$ and $\boldsymbol{b}$ in $\mathcal{O}_{n-1}[z_n]$ and $\boldsymbol{c} \neq 0$ in $\mathcal{O}_{n-1}$ such that

$$\boldsymbol{a}\boldsymbol{f} + \boldsymbol{b}\boldsymbol{g} = \boldsymbol{c}. \tag{1.6}$$

Note that the above equality holds in a neighborhood U of 0 in $\mathbb{C}^n$ where all the germs involved have representatives.

We may assume that U is of the form $U = U' \times U''$ with U' a neighborhood of 0 in $\mathbb{C}^{n-1}$ and U'' a neighborhood of 0 in $\mathbb{C} = \{z_n\}$. As $\boldsymbol{f}$ is regular in z_n, taking a smaller U' if necessary, we may assume that for each z' in U', $f(z', z_n)$ is not identically zero as a function of z_n in U''.

Let a be a point in U. If $f(a) \neq 0$ or $g(a) \neq 0$, f_a and g_a are relatively prime in $\mathscr{O}_{n,a}$. Thus assume that $f(a) = g(a) = 0$. If f_a and g_a have a common factor h_a in $\mathscr{O}_{n,a}$ with $h(a) = 0$, h_a also divides c_a by (1.6) so that the function $h(z', z_n)$ defined near $a = (a', a_n)$ depends only on z'. Thus $h(a', z_n)$ vanishes identically in z_n. This contradicts the fact that $f(a', z_n)$ is not identically zero as a function of z_n. □

We finish this section by discussing the *Riemann extension theorem* in the several variable case.

Theorem 1.13. *Let D be a connected open set in $\mathbb{C}^n$ and g a function holomorphic and not identically 0 in D. We set $V = \{ z \in D \mid g(z) = 0 \}$. If f is a function holomorphic in $D \setminus V$ and is locally bounded, i.e., each point z in D has a neighborhood U such that f is bounded in $(D \setminus V) \cap U$, there is a unique function $\tilde{f}$ holomorphic in D and equal to f on $D \setminus V$.*

Proof. The case $n = 1$ is the classical Riemann extension theorem. Thus we assume that $n \geq 2$ and reduce the problem to this case.

First we note that, by the assumption and Theorem 1.7, V has no interior points. In particular, for every point z in V, the germ g_z is not 0 or a unit. For the existence, it suffices to show that each point z in V has a neighborhood on which there is an extension $\tilde{f}$ of f. Without loss of generality, we may assume $z = 0$. After a change of coordinates if necessary, we may assume that $\boldsymbol{g}$ is regular in z_n, in fact we may assume that $\boldsymbol{g}$ is a Weierstrass polynomial in z_n. We choose $\rho = (\rho_1, \ldots, \rho_n)$, $\rho_i > 0$, small enough so that $\boldsymbol{g}$ has a representative on an open set containing $\overline{\Delta(0, \rho)}$ and that $g(z) \neq 0$ for $z = (z', z_n)$ with $z' \in \Delta(0, \rho')$ and $|z_n| = \rho_n$, where we set $z' = (z_1, \ldots, z_{n-1})$ and $\rho' = (\rho_1, \ldots, \rho_{n-1})$. Denoting by γ the path in the z_n-plane turning counterclockwise along the circle $|z_n| = \rho_n$, we set

$$\tilde{f}(z) = \frac{1}{2\pi\sqrt{-1}} \int_\gamma \frac{f(z', \zeta)}{\zeta - z_n} d\zeta.$$

Then $\tilde{f}$ is holomorphic in $\Delta(0, \rho)$. For each fixed z' in $\Delta(0, \rho')$, $f(z', z_n)$ is holomorphic in z_n in an open set U containing $\overline{\Delta(0, \rho_n)}$ except for a finite number of points (the zeros of $g(z', z_n)$) and is locally bounded. Thus by the Riemann extension theorem in one variable case, $f(z', z_n)$ has a holomorphic extension in U. Therefore by the Cauchy integral formula, $\tilde{f}$ is an extension of f in $\Delta(0, \rho)$. The uniqueness follows from the uniqueness of analytic continuation. □

Corollary 1.1. *If D and V are as in Theorem 1.13, $D \setminus V$ is connected.*

Proof. If $D \setminus V$ is not connected, there are non-empty open sets U_1 and U_2 such that $D \setminus V = U_1 \cup U_2$ and $U_1 \cap U_2 = \emptyset$. The function f, defined by $f(z) = 1$ for $z \in U_1$ and $f(z) = 0$ for $z \in U_2$, has no extensions. □

Notes

For Section 1.2, there are a number of textbooks. Here we list [Cartan (1961)]. See 15.12 Theorem in [Rudin (1987)] for Remark 1.3.1.

We list [Fritzsche and Grauert (2002); Gunning and Rossi (1965)] as general references for the theory of analytic functions of several complex variables. The proof of the Weierstrass preparation theorem (Theorem 1.10) given here is based on the one in [Noguchi and Fukuda (1976)].

Chapter 2

Complex Manifolds and Analytic Varieties

A complex manifold is a space that is locally the set of common zeros of holomorphic functions without singularities. It is also described as a spcae obtained by patching open sets in $\mathbb{C}^n$ together by biholomorphic maps. These are the spaces on which we consider holomorphic functions, maps and related objects in this book. An analytic variety in a complex manifold is a subset of the manifold that is the common zeros of holomorphic functions possibly with singularities. In this chapter we discuss fundamentals on complex manifolds and analytic varieties.

Here is a word about submanifolds and varieties. A submanifold is a variety without singularities. However, when we say that V is a submanifold of a complex manifold M, it is a locally closed set in M, while when we say that V is a variety in M it is a closed set in M.

As in the case of functions, we introduce the notion of the germ of a variety. We study the relation between germs of functions and those of varieties. The important point in this is what is called the "Nullstellensatz" (zero locus theorem). We give a proof in the case of principal ideals (Corollary 2.1) and, for the general case, we quote some structural results and give proofs based on these (Theorem 2.8). We also discuss local and global dimensions of varieties and related topics.

It is important to keep in mind that a complex manifold M carries an underlying real structure. In particular, as a C^∞ manifold, M has the tangent space at each point. We define the holomorphic tangent space and give the relation with the real tangent space (cf. (2.6)).

We also discuss stratification of varieties, which is useful in the study of singular varieties.

2.1 Complex manifolds

Let M be a Hausdorff topological space with a countable basis.

Definition 2.1. A *complex chart*, or simply a *chart*, on M is a pair (U, φ) of an open set U in M and a homeomorphism φ of U onto an open set in $\mathbb{C}^n$ for some $n \geq 0$.

Definition 2.2. A *holomorphic atlas* on M is a collection of charts $\mathcal{A} = \{(U_\alpha, \varphi_\alpha)\}_{\alpha \in I}$ such that $\{U_\alpha\}_{\alpha \in I}$ is a covering of M and that, for each pair (α, β), the map $\varphi_\alpha \circ \varphi_\beta{}^{-1} : \varphi_\beta(U_\alpha \cap U_\beta) \to \varphi_\alpha(U_\alpha \cap U_\beta)$ is holomorphic.

Interchanging α and β, we see that the above $\varphi_\alpha \circ \varphi_\beta{}^{-1}$ is in fact biholomorphic. We also use the convention that a map of the empty set is holomorphic.

For two holomorphic atlases $\mathcal{A}$ and $\mathcal{A}'$, we define a relation $\sim$ saying that $\mathcal{A} \sim \mathcal{A}'$ if their union is a holomorphic atlas. Obviously it is an equivalence relation in the set of holomorphic atlases on M.

Definition 2.3. A *complex structure* on M is an equivalence class of a holomorphic atlas. A *complex manifold* is a space M together with a complex structure.

A complex manifold is often denoted simply by M. The natural number n appearing in Definition 2.1, which is uniquely determined on each connected component of M, is called the (complex) *dimension* of the component. If all the components have dimension n, we say the dimension of M is n.

Remark 2.1. 1. Let X be a topological space and let $\mathcal{U} = \{U_\alpha\}_{\alpha \in I}$ and $\mathcal{V} = \{V_\lambda\}_{\lambda \in J}$ be open coverings of X. Recall that $\mathcal{V}$ is a *refinement* of $\mathcal{U}$ if there exists a map $\iota : J \to I$ such that $V_\lambda \subset U_{\iota(\lambda)}$ for all λ. Every open covering of X is equivalent to a covering whose index set is a subset of the power set of X, in the sense that each one of the two is a refinement of the other. Thus we may talk about the "set" of open coverings of X.

We may define an order relation in the set of open coverings of X by saying that $\mathcal{U} \leq \mathcal{V}$ if $\mathcal{V}$ is a refinement of $\mathcal{U}$. In the following we consider the set of open coverings as an ordered set with this ordering. It is a directed set (cf. Section A.1).

2. Likewise we may talk about the set of atlases on a space M as above.

3. Recall that a topological space X is *paracompact* if it is Hausdorff and if every open covering of X admits a locally finite refinement. Since we assumed that the underlying topological space of a complex manifold M to be Hausdorff with countable basis, we see that M is paracompact and the number of connected components is countable.

4. A complex manifold of dimension 0 is a countable set of points with the discrete topology.

A complex manifold of dimension one is called a *Riemann surface* or a (non-singular) *complex curve.* Also, complex manifold of dimension two is called a (non-singular) *complex surface.* When these terminologies are used, it is usually understood that the manifold is connected.

Let M be a complex manifold. A chart (U, φ) on M is said to be *holomorphic* if it belongs to an atlas representing the complex structure. Let (U, φ) be a holomorphic chart on M and suppose M is of dimension $n \geq 1$. For a point p in U, we call U a *coordinate neighborhood* of p and

$$\varphi(p) = (z_1(p), \ldots, z_n(p))$$

the *local coordinates* of p with respect to φ. Sometimes we identify U with $\varphi(U)$ and p with the point $(z_1(p), \ldots, z_n(p))$ in $\varphi(U) \subset \mathbb{C}^n$. In this case we call $(z_1, \ldots, z_n)$ a *coordinate system* on U.

A complex manifold can be defined by specifying a holomorphic atlas on an appropriate topological space.

Example 2.1. 1. A non-empty open subset in $\mathbb{C}^n$ is an n-dimensional complex manifold with natural global coordinates.

2. *Complex projective space* $\mathbb{P}^n$: We denote by $\zeta = (\zeta_0, \ldots, \zeta_n)$ coordinates on $\mathbb{C}^{n+1} \setminus \{0\}$ and introduce a relation $\sim$ in $\mathbb{C}^{n+1} \setminus \{0\}$ by setting, for $\zeta = (\zeta_0, \ldots, \zeta_n)$ and $\zeta' = (\zeta'_0, \ldots, \zeta'_n)$, $\zeta \sim \zeta'$ if $\zeta' = t\zeta$ for some t in $\mathbb{C}^* = \mathbb{C} \setminus \{0\}$. Obviously this is an equivalence relation and we set $\mathbb{P}^n = (\mathbb{C}^{n+1} \setminus \{0\})/\sim$. Endowed with the quotient topology, it is Hausdorff and connected. The equivalence class of $\zeta = (\zeta_0, \ldots, \zeta_n)$ is denoted by $[\zeta] = [\zeta_0, \ldots, \zeta_n]$. The space $\mathbb{P}^n$ with the following complex structure is called the n-dimensional complex projective space.

First, for each $\alpha = 0, \ldots, n$, the set

$$U_\alpha = \{ [\zeta] \in \mathbb{P}^n \mid \zeta_\alpha \neq 0 \}$$

is a well-defined open set and $\mathbb{P}^n$ is covered by these $n + 1$ open sets. Also the map $\varphi_\alpha : U_\alpha \to \mathbb{C}^n$ given by

$$\varphi_\alpha([\zeta]) = (\zeta_0/\zeta_\alpha, \ldots, \zeta_{\alpha-1}/\zeta_\alpha, \zeta_{\alpha+1}/\zeta_\alpha, \ldots, \zeta_n/\zeta_\alpha) \tag{2.1}$$

is a well-defined homeomorphism. Moreover, it is not difficult to check that for each pair (α, β), the map $\varphi_\alpha \circ \varphi_\beta{}^{-1} : \varphi_\beta(U_\alpha \cap U_\beta) \to \varphi_\alpha(U_\alpha \cap U_\beta)$ is holomorphic. Thus $\mathbb{P}^n$ becomes a complex manifold of dimension n. We call $[\zeta] = [\zeta_0, \ldots, \zeta_n]$ *homogeneous coordinates* on $\mathbb{P}^n$. We denote the right-hand side of (2.1) by $z^\alpha = (z_1^\alpha, \ldots, z_n^\alpha)$:

$$z_i^\alpha = \begin{cases} \zeta_{i-1}/\zeta_\alpha & i = 1, \ldots, \alpha, \\ \zeta_i/\zeta_\alpha & i = \alpha + 1, \ldots, n. \end{cases} \tag{2.2}$$

It is a holomorphic coordinate system on U_α.

We may interpret $\mathbb{P}^n$ as the set of complex lines through 0 (one-dimensional linear subspaces) in $\mathbb{C}^{n+1}$. This viewpoint leads to the construction of a more general class of manifolds, i.e., *Grassmann manifolds*, which will be discussed in Section 3.6 below.

3. If M and M' are complex manifolds of dimensions n and n', respectively, the product $M \times M'$ has naturally the structure of a complex manifold of dimension $n + n'$. If $(z_1, \ldots, z_n)$ is a coordinate system on $U \subset M$ and $(z_1', \ldots, z_{n'}')$ a coordinate system on $U' \subset M'$, then $(z_1, \ldots, z_n; z_1', \ldots, z_{n'}')$ is a coordinate system on $U \times U' \subset M \times M'$.

See Section 3.1 for some more basic examples.

Exercise 2.1. Show that $\mathbb{P}^1$ is homeomorphic to the 2-sphere $\mathbb{S}^2$ and that it may be interpreted as being obtained from the complex plane $\mathbb{C}$ by attaching a "point at ∞", i.e., the *Riemann sphere.*

Exercise 2.2. Let $\mathbb{S}^{2n+1} = \{ (\zeta_0, \ldots, \zeta_n) \in \mathbb{C}^{n+1} \mid |\zeta_0|^2 + \cdots + |\zeta_n|^2 = 1 \}$ be the $(2n+1)$-dimensional unit sphere and π the restriction of the canonical surjection $\mathbb{C}^{n+1} \setminus \{0\} \to \mathbb{P}^n$ to $\mathbb{S}^{2n+1}$. Show that π is surjective (thus $\mathbb{P}^n$ is compact) and find the inverse image $\pi^{-1}(p)$ for each point p in $\mathbb{P}^n$ (cf. Example 3.4 and the expression (3.20) below).

Holomorphic maps

Let M be a complex manifold with an atlas $\{(U_\alpha, \varphi_\alpha)\}$ representing the complex structure. A complex valued function f on an open set U in M is said to be *holomorphic* if, for every α, the function $f \circ \varphi_\alpha^{-1}$ is holomorphic on $\varphi_\alpha(U \cap U_\alpha)$. Also, a map $f : M \to M'$ from M into another complex manifold M' with atlas $\{(V_\lambda, \psi_\lambda)\}$ is said to be *holomorphic*, if for every pair (α, λ), the map $\psi_\lambda \circ f \circ \varphi_\alpha^{-1}$ is holomorphic on $\varphi_\alpha(U_\alpha \cap f^{-1}(V_\lambda))$. A *biholomorphic* map is a bijective holomorphic map f such that f^{-1} is also holomorphic. If there exists a biholomorphic map from M onto M', we say that M is biholomorphic with M' and write $M \simeq M'$.

The following is a consequence of Theorem 1.8:

Proposition 2.1. *Every holomorphic function on a compact connected complex manifold is constant.*

For a holomorphic map $f : M \to M'$, we may define *regular* and *singular points* of f as in Definition 1.9, choosing local coordinate systems, and the definition does not depend on the choice of the coordinate systems. Also the behavior of f near a regular point is given as in Theorem 1.6.

Definition 2.4. Let $f : M \to M'$ be a holomorphic map and let n and n' be dimensions of M and M', respectively.

1. f is an *immersion* if $n \leq n'$ and every point of M is a regular point of f. The map f is an *embedding* if, furthermore, it is a homeomorphism onto $f(M)$ with the topology induced from that of M'.

2. f is a *submersion* if $n \geq n'$ and every point of M is a regular point of f.

Definition 2.5. A holomorphic map $\pi : M \to M'$ is a (holomorphic) *covering map* if it is surjective and if each point of M' has a neighborhood U such that each connected component of $\pi^{-1}(U)$ is mapped biholomorphically onto U by π.

Thus in this case, $\dim M = \dim M'$ and π is immersive and submersive.

Remark 2.2. The set $\{p\}$ consisting of a point p is a connected complex manifold of dimension 0, with a global chart homeomorphic to $\mathbb{C}^0 = \{0\}$. Any map $f : M \to \{p\}$ of a complex manifold M is holomorphic and every point in M is a regular point of f.

Submanifolds

Let M be a complex manifold of dimension n. For non-negative integers d and k with $d + k = n$, we decompose as $\mathbb{C}^n = \mathbb{C}^k \times \mathbb{C}^d$ with $(z_1, \ldots, z_k)$ and $(z_{k+1}, \ldots, z_n)$ as coordinates on $\mathbb{C}^k$ and $\mathbb{C}^d$, respectively.

Definition 2.6. A subset V of M is a (complex) *submanifold* of M if there is an atlas $\{(U_\alpha, \varphi_\alpha)\}$ representing the complex structure of M such that for each α, $\varphi_\alpha(V \cap U_\alpha)$ is an open set in $\mathbb{C}^d = \{0\} \times \mathbb{C}^d \subset \mathbb{C}^n$.

Thus if V is a submanifold of M, with the topology induced from that of M, it is a complex manifold with the complex structure represented by the atlas $\{(V \cap U_\alpha, \varphi_\alpha|_{V\cap U_\alpha})\}$. It is locally closed in M. If $V \neq \emptyset$, the integer d in uniquely determined on each connected component of V and is the dimension of the component. The inclusion $i : V \hookrightarrow M$ is an embedding. Suppose $d \geq 1$ and write $\varphi_\alpha(p) = (z_1^\alpha(p), \cdots, z_n^\alpha(p))$ on U_α. Then for α with $V \cap U_\alpha \neq \emptyset$, $\varphi_\alpha(p) = (0, \ldots, 0, z_{k+1}^\alpha(p), \cdots, z_n^\alpha(p),))$ on $V \cap U_\alpha$ and $(z_{k+1}^\alpha, \cdots, z_n^\alpha)$ is a coordinate system on $V \cap U_\alpha$. Thus, if $k \geq 1$, we may think of V as being given by $z_1^\alpha = \cdots = z_k^\alpha = 0$ in U_α.

2.2 Analytic varieties

Analytic varieties naturally arise as the set of common zeros of holomorphic functions and include submanifolds as a special class.

Let M be a complex manifold.

Definition 2.7. An *analytic variety* in M is a subset V of M such that every point a in M admits a neighborhood U with the property:

(*) there exists a holomorphic map $f : U \to \mathbb{C}^r$ with $V \cap U = f^{-1}(0)$.

We call f as above a *local defining map* of V near a. In the following, an analytic variety will simply be called a variety, if there is no fear of confusion. Note that a variety in M is a closed set in M.

Remark 2.3. 1. Let V be a subset of M. If every point a in V admits a neighborhood U with the property (*), V is a variety in some open set in M. In particular, if V is a closed set in M as above, it is a variety in M.

2. In the above we allow r to be 0 (cf. Remark 2.2). If $r \geq 1$, letting $(f_1, \ldots, f_r)$ be the components of f, we may write

$$V \cap U = \{ z \in U \mid f_1(z) = \cdots = f_r(z) = 0 \}.$$

We call $(f_1, \ldots, f_r)$ as above a *system of local defining functions* of V and $f_1 = \cdots = f_r = 0$ *local equations* for V near a.

3. In particular, M itself and the empty set $\emptyset$ are varieties in M. For M, we may take the map $M \to \mathbb{C}^0$ or the function $M \to \mathbb{C}$ constantly equal to 0 as a defining map. For $\emptyset$, we may take a non-vanishing function, constant function 1 for example, as a defining map.

The first part of the following is obvious from the uniqueness of analytic continuation (Theorem 1.7) and the second part follows from Corollary 1.1.

Theorem 2.1. *Let V be a variety in a connected open set D. If $V \neq D$, it has no interior points as a subset of D. Moreover, $D \setminus V$ is connected.*

Remark 2.4. Recall that a subset A of a topological space X is *nowhere dense* in X, if the closure of A has no interior points in X. Thus, in the above situation, V is nowhere dense in D.

In the following, if we simply say a variety, it means a variety in M, thus it is closed in M. We think of a variety as a topological space with the topology induced from that of M.

Definition 2.8. Let V be a variety and a a point in V. We say that a is a *regular point* of V if there is a local defining map $f : U \to \mathbb{C}^k$ near a such that a is a regular point of f. We say that a is a *singular point* of V if it is not a regular point.

For a variety V, we denote by V_{reg} and $\text{Sing}(V)$, respectively, the sets of regular and singular points of V. Note that V_{reg} is an open set in V. Hence $\text{Sing}(V)$ is a closed set in V (and in M). In fact it is shown that $\text{Sing}(V)$ is again an analytic variety (cf. Theorem 11.5 below). We say that a variety V is *non-singular* if $\text{Sing}(V) = \emptyset$.

Example 2.2. Let V be a submanifold of M. Then it is closed in some open set D. Let $\{(U_\alpha, \varphi_\alpha)\}$ be an atlas as in Definition 2.6. For every point a of V, take U_α containing a. Then $p \circ \varphi_\alpha : U_\alpha \to \mathbb{C}^k$ is a local defining map on U_α, where $p : \mathbb{C}^n \to \mathbb{C}^k$ is the projection, and V is a variety in D. It is non-singular.

Let V be a variety. If a is a regular point of V, by Theorem 1.6, we may assume that $k \leq n$ in Definition 2.8. Moreover, there is a neighborhood U of a such that $V \cap U$ is homeomorphic to an open set in $\mathbb{C}^{n-k}$. In fact, if $k \geq 1$, we may assume that U is a coordinate neighborhood with coordinates $(z_1, \dots, z_n)$ such that $V \cap U = \{z_1 = \cdots = z_k = 0\}$.

Exercise 2.3. Show that V_{reg} has a natural structure of complex submanifold of M.

From the above, we have:

Proposition 2.2. *A subset V of M is a complex submanifold of M if and only if it is a non-singular variety in some open set of M.*

Exercise 2.4. Let $f : M' \to M$ be a holomorphic map of complex manifolds. Show that, if f is an embedding, $f(M')$ may be endowed with a structure of a submanifold of M so that f is a biholomorphic map of M' onto $f(M')$.

Affine algebraic varieties

Consider the complex n-space $\mathbb{C}^n$ with coordinates $(z_1, \ldots, z_n)$ and let, for each $i = 1, \ldots, r$, $P_i(z_1, \ldots, z_n)$ be a polynomial in $(z_1, \ldots, z_n)$. Then the set

$$V = \{ (z_1, \ldots, z_n) \in \mathbb{C}^n \mid P_i(z_1, \ldots, z_n) = 0,\ i = 1, \ldots, r \}$$

is a variety in $\mathbb{C}^n$, called an *affine algebraic variety.*

Example 2.3. 1. Consider $\mathbb{C}^2$ with coordinates (z_1, z_2). We set $P(z_1, z_2) = z_1 z_2$ and $V = \{ (z_1, z_2) \mid P(z_1, z_2) = 0 \}$. Thus V consists of two "complex lines" (z_1 and z_2 "axes") intersecting in $\mathbb{C}^2$ at one point, the origin 0. By definition we see that $V \smallsetminus \{0\} \subset V_{\text{reg}}$, while by looking at the neighborhood structure of 0, we see that 0 is a singular point of V (cf. Exercise 2.5. 1 below). This can be also checked by studying the behavior of the tangent spaces of the regular part. See also Proposition 11.26. Thus $V_{\text{reg}} = V \smallsetminus \{0\}$, which has two connected components each being a one-dimensional complex manifold biholomorphic with $\mathbb{C}^* = \mathbb{C} \smallsetminus \{0\}$.

2. Again we consider $\mathbb{C}^2$. We set $P(z_1, z_2) = z_1^3 - z_2^2$ and let V be the variety defined by P. By definition we see that $V \smallsetminus \{0\} \subset V_{\text{reg}}$, while 0 is a singular point of V (cf. Exercise 2.5. 2). Thus $V_{\text{reg}} = V \smallsetminus \{0\}$, which has one component biholomorphic with $\mathbb{C}^*$. Note that V is homeomorphic with $\mathbb{C}$.

3. Consider $\mathbb{C}^3$ with coordinates (z_1, z_2, z_3). We set $f(z_1, z_2, z_3) = z_1 z_2^2 - z_3^2$ and let V be the variety defined by f. Then V_{reg} is a two-dimensional complex manifold and $\operatorname{Sing}(V)$ is the z_1-axis. The set V is called the Whitney umbrella.

4. Again consider $\mathbb{C}^3$. We set $f(z_1, z_2, z_3) = z_2^2 - z_1^2 z_3^2 - z_1^3$ and let V be the variety defined by f. Then V_{reg} is a two-dimensional complex manifold and $\operatorname{Sing}(V)$ is the z_3-axis.

Exercise 2.5. 1. Let $\mathbb{S}^3 = \{ (z_1, z_2) \mid |z_1|^2 + |z_2|^2 = 1 \}$ be the three-dimensional unit sphere in $\mathbb{C}^2 = \mathbb{R}^4$. Show that, in Example 2.3. 1, the intersection $K = V \cap \mathbb{S}^3$ consists of two circles that are unknotted but link with each other.

2. Show that, in Example 2.3. 2, $K = V \cap \mathbb{S}^3$ is the "(2, 3)-torus knot".

3. In Example 2.3. 2, find an explicit (holomorphic) homeomorphism from $\mathbb{C}$ onto V.

4. Let V be the variety in $\mathbb{C}^4 = \{(z_1, z_2, z_3, z_4)\}$ defined by the three equations:

$$z_1 z_4 - z_2 z_3 = 0, \quad z_2^2 - z_1 z_3 = 0 \quad \text{and} \quad z_3^2 - z_2 z_4 = 0.$$

Find V_{reg} and $\text{Sing}(V)$ (cf. Theorem 11.5, also Example 11.10. 1 below). What is the dimension of V_{reg} ?

Projective algebraic varieties

Consider the n-dimensional complex projective space $\mathbb{P}^n$ with homogeneous coordinates $[\zeta] = [\zeta_0, \ldots, \zeta_n]$ and let, for each $i = 1, \ldots, r$, $P_i(\zeta_0, \ldots, \zeta_n)$ be a homogeneous polynomial in $(\zeta_0, \ldots, \zeta_n)$ of degree p_i. Then the set

$$V = \{\, [\zeta] \in \mathbb{P}^n \mid P_i(\zeta_0, \ldots, \zeta_n) = 0, \ i = 1, \ldots, r \,\}$$

is a well-defined subset of $\mathbb{P}^n$ and is, moreover, a variety in $\mathbb{P}^n$. In fact, in each open set $U_\alpha = \{\, \zeta_\alpha \neq 0 \,\}$, V is defined by the holomorphic functions $f_i = P_i(\zeta_0, \ldots, \zeta_n)/\zeta_\alpha^{p_i}$, $i = 1, \ldots, r$. Such a variety is called a *projective algebraic variety.* Note that, as each U_α may naturally be identified with $\mathbb{C}^n$, $V \cap U_\alpha$ is an affine albegraic variety. A non-singular projective algebraic variety is called a *projective algebraic manifold.*

Example 2.4. 1. Let $\varpi : \mathbb{C}^{n+1} \setminus \{0\} \to \mathbb{P}^n$ be the canonical surjection (cf. Example 2.1. 2). For a $(d+1)$-dimensional subspace V of $\mathbb{C}^{n+1}$, $\varpi(\mathsf{V} \setminus \{0\})$ is a projective algebraic manifold of dimension d, which is biholomorphic with $\mathbb{P}^d$ and is called a d-plane in $\mathbb{P}^n$. In the case $d = n-1$, it is called a *hyperplane.*

2. A projective algebraic variety in $\mathbb{P}^n$ defined by a single (not identically zero) homogeneous polynomial of degree p is called a *hypersurface* of degree p (cf. Section 2.4 below). A hypersurface of degree one is defined by a linear function $\ell(\zeta) = a_0\zeta_0 + \cdots + a_n\zeta_n$ with $(a_0, \ldots, a_n) \neq (0, \ldots, 0)$. Thus it is non-singular and is nothing but a hyperplane. If H is a hyperplane, by a linear change of coordinates, we may assume that it is defined by $\zeta_0 = 0$. Identifying the open set $U_0 = \{\zeta_0 \neq 0\}$ with $\mathbb{C}^n = \{(\zeta_1/\zeta_0, \ldots, \zeta_n/\zeta_0)\}$ and H with $\mathbb{P}^{n-1} = \{[\zeta_1, \ldots, \zeta_n]\}$, we may express $\mathbb{P}^n$ as a disjoint union $\mathbb{P}^n = \mathbb{C}^n \sqcup \mathbb{P}^{n-1}$. This procedure leads to a cellular decomposition of $\mathbb{P}^n$ as $\mathbb{P}^n = \mathbb{C}^n \sqcup \mathbb{C}^{n-1} \sqcup \cdots \sqcup \mathbb{C}^0$ (cf. Section B.2). Using this we may compute the homology of $\mathbb{P}^n$:

$$H_q(\mathbb{P}^n;\mathbb{Z}) = \begin{cases} \mathbb{Z} & \text{for } q = 0, 2, \dots, 2n, \\ 0 & \text{otherwise.} \end{cases} \tag{2.3}$$

For each $i = 0, \dots, n$, the group $H_{2i}(\mathbb{P}^n;\mathbb{Z})$ has a canonical generator u_{2i} which is represented by the cycle $\mathbb{P}^i$. For a hypersurface V of degree p, its homology class is given by $[V] = p \cdot u_{2n-2}$ (cf. Example 12.5 below).

This is a special case of the decomposition of Grassmann manifolds by Schubert cells (cf. Section 3.6).

Exercise 2.6. For complex numbers α, β and γ, let $V_{\alpha,\beta,\gamma}$ be the variety in $\mathbb{P}^2$ defined by

$$V_{\alpha,\beta,\gamma} = \{\, [\zeta_0, \zeta_1, \zeta_2] \in \mathbb{P}^2 \mid \zeta_0\zeta_2^2 - (\zeta_1 - \alpha\zeta_0)(\zeta_1 - \beta\zeta_0)(\zeta_1 - \gamma\zeta_0) = 0 \,\}.$$

1. Show that, if α, β and γ are mutually distinct, then $V_{\alpha,\beta,\gamma}$ has no singular points.

2. Show that, if $\gamma \neq 0$, then $V_{0,0,\gamma}$ has only one singular point at $a = [1,0,0]$, which is equivalent to the one in Example 2.3. 1.

3. Show that $V_{0,0,0}$ has only one singular point at $a = [1,0,0]$, which is equivalent to the one in Example 2.3. 2.

Quasi-projective varieties: A set of the form $V \setminus V'$ in $\mathbb{P}^n$ is called a *quasi-projective variety*, where V and V' are projective algebraic varieties. A *quasi-projective manifold* is a non-singular quasi-projective variety.

Zariski topology: Instead of the usual topology of $\mathbb{C}^n$, we may take as the open sets the complements of the affine algebraic varieties to define a topology on $\mathbb{C}^n$, which is called the *Zariski toplogy.* Likewise we may define the Zariski topology of $\mathbb{P}^n$ by taking as the open sets the complements of the projective algebraic varieties. For a quasi-projective variety V in $\mathbb{P}^n$, it has also the topology induced from the Zariski topology. We denote by V_Z the variety V with Zariski topology. Note that the identity map $V \to V_Z$ is continuous.

Algebraic functions: Let V be a quasi-projective variety. A function $f : V_Z \to \mathbb{C}$ is said to be *algebraically regular* at p in V_Z if there exist a neighborhood U_Z of p and homogeneous polynomials P and Q of the same degree such that Q is nowhere zero on U_Z and $f = P/Q$ on U_Z. It is simply called regular, if there is no fear of confusion. A regular function on an open set in V_Z is a function regular at every point of the open set.

An *algebraic function* on V is the pull-back of a regular function by the identity $V \to V_Z$.

Remark 2.5. It is known that every analytic variety in $\mathbb{P}^n$ is algebraic (Chow's theorem). Also, every compact complex manifold of dimension one (compact Riemann surface) can be embedded in $\mathbb{P}^n$ for some n. Thus it is biholomorphic with a projective algebraic manifold.

Blowing-up

This can be defined in generality, however here we discuss the simplest case to illustrate the idea. See Section 11.6 below for the higher-dimensional case and more discussions on this subject.

Let U be a neighborhood of 0 in $\mathbb{C}^2 = \{(z_1, z_2)\}$ and $\tilde{U}$ the variety in $U \times \mathbb{P}^1$ defined by

$$\tilde{U} = \{ (z, [\zeta]) \in U \times \mathbb{P}^1 \mid z_1\zeta_1 - z_2\zeta_0 = 0 \}.$$

If we set $W_\alpha = \{ [\zeta] \in \mathbb{P}^1 \mid \zeta_\alpha \neq 0 \}$, $\alpha = 0, 1$, the manifold $U \times \mathbb{P}^1$ is covered by two coordinate neighborhoods $U \times W_0$ and $U \times W_1$ with coordinates $(z_1, z_2, \xi) = (z_1, z_2, \zeta_1/\zeta_0)$ and $(z_1, z_2, \eta) = (z_1, z_2, \zeta_0/\zeta_1)$, respectively. In $U \times W_0$, $\tilde{U}$ is given by $z_1\xi - z_2 = 0$ and in $U \times W_1$, by $z_1 - z_2\eta = 0$. Thus $\tilde{U}$ does not have singular points and is a two-dimensional complex manifold, which is covered by two coordinate neighborhoods $\tilde{U}_0 = \tilde{U} \cap (U \times W_0)$ and $\tilde{U}_1 = \tilde{U} \cap (U \times W_1)$ with coordinates (z_1, ξ) and (z_2, η), respectively. If we denote by $\pi : \tilde{U} \to U$ the restriction of the projection $U \times \mathbb{P}^1 \to U$ to $\tilde{U}$, it is a holomorphic map such that $\pi^{-1}(0) \simeq \mathbb{P}^1$ and that its restriction to $\tilde{U} \smallsetminus \pi^{-1}(0)$ is a biholomorphic map onto $U \smallsetminus \{0\}$. This procedure is called the *blowing-up* of U at 0. It replaces 0 with the projective line $\mathbb{P}^1$ to separate the lines through 0 in U.

If V is the variety in Example 2.3. 1, $\pi^{-1}(V)$ is a variety in $\tilde{U}$, which is defined by $z_1^2\xi = 0$ in $\tilde{U}_0$ and by $z_2^2\eta = 0$ in $\tilde{U}_1$. Thus we may write $\pi^{-1}(V) = \tilde{V} \cup \pi^{-1}(0)$, where $\tilde{V}$ consists of two (disjoint) lines, one defined by $\xi = 0$ in $\tilde{U}_0$ and the other by $\eta = 0$ in $\tilde{U}_1$. Notice that $\pi^{-1}(0)$ appears "in double". Also, if V is the variety in Example 2.3. 2, $\pi^{-1}(V)$ is a variety in $\tilde{U}$, which is defined by $z_1^2(z_1 - \xi^2) = 0$ in $\tilde{U}_0$ and by $z_2^2(z_2\eta^3 - 1) = 0$ in $\tilde{U}_1$. Thus we may write $\pi^{-1}(V) = \tilde{V} \cup \pi^{-1}(0)$, where $\tilde{V}$ is the "parabola" given by $z_1 - \xi^2 = 0$ in $\tilde{U}_0$ and by $z_2\eta^3 - 1 = 0$ in $\tilde{U}_1$. Again $\pi^{-1}(0)$ appears in double.

In general, applying this procedure successively, we may "resolve" the singularity of a plane curve (cf. Section 11.6).

2.3 Germs of varieties

In this section, we consider the germs of varieties and discuss the relation between these and the ideals in $\mathcal{O}_n$. In the following, we denote the germ of a function f at 0 also by f.

We first introduce a relation $\sim$ in the set of subsets of $\mathbb{C}^n$. Let A and B be two subsets of $\mathbb{C}^n$. We define $A \sim B$ if there is a neighborhood U of 0 such that $A \cap U = B \cap U$. It is easily checked that this is an equivalence relation. We call the equivalence class of A the *germ of A at* 0 and denote it also by A, unless it is necessary to distinguish the two. Usual operations of sets induce those of germs. Thus for two germs A and B at 0, $A \cap B$, $A \cup B$ and $A \setminus B$ are well-defined. The relation $A \subset B$ is also well-defined.

In Section 1.4, we introduced the ring $\mathcal{O}_n$ of germs of holomorphic functions at 0. Let $f_1, \ldots, f_r$ be germs in $\mathcal{O}_n$. We choose a neighborhood U of 0 such that these germs are represented by holomorphic functions on U, which we also denote by $f_1, \ldots, f_r$. We set

$$V(f_1, \ldots, f_r) = \text{the germ at 0 of } \{ z \in U \mid f_1(z) = \cdots = f_r(z) = 0 \}$$

and call it the *germ of the variety defined by* $f_1, \ldots, f_r$. More generally, let I be an ideal in $\mathcal{O}_n$. By the Noetherian property of $\mathcal{O}_n$ (cf. Theorem 1.12), there exist a finite number of germs $f_1, \ldots, f_r$ such that $I = (f_1, \ldots, f_r)$, the ideal generated by $f_1, \ldots, f_r$. We set $V(I) = V(f_1, \ldots, f_r)$ and call it the germ of the variety defined by I. It is easily checked that it does not depend on the choice of generators of I. Thus each ideal in $\mathcal{O}_n$ defines a germ of a variety at 0.

Conversely suppose we are given a germ V of a variety at 0. We choose a neighborhood U of 0 such that the germ is represented by a variety in U, which we denote also by V. We set

$$I(V) = \{ f \in \mathcal{O}_n \mid f(z) = 0 \text{ for all } z \text{ in } V \text{ and near } 0 \}.$$

It is easily checked that this is an ideal in $\mathcal{O}_n$. Recall that (cf. Section A.2) for an ideal I in $\mathcal{O}_n$, we have the radical $\sqrt{I}$ of I. This is again an ideal in $\mathcal{O}_n$ and contains I. See also Section A.2 for the notions of prime and primary ideals .

Exercise 2.7. 1. For $k = 1, \ldots, n$, we consider the "coordinate functions" $z_1, \ldots, z_k$ as germs in $\mathcal{O}_n$. Show that the ideal $\mathfrak{p} = (z_1, \ldots, z_k)$ is prime and that $I(V(\mathfrak{p})) = \mathfrak{p}$.
2. Show that in the ring $\mathcal{O}_2$, $\mathfrak{q} = (z_1, z_2^2)$ is primary and that $\sqrt{\mathfrak{q}} = (z_1, z_2)$.

We have the following various relations between germs of varieties at 0 and ideals in $\mathcal{O}_n$:

Proposition 2.3. *Let V, V_1 and V_2 be germs of varieties at 0 and I, I_1 and I_2 ideals in $\mathcal{O}_n$.*

1. *If $I_1 \subset I_2$, then $V(I_1) \supset V(I_2)$.*

2. *If $V_1 \subset V_2$, then $I(V_1) \supset I(V_2)$.*

3. $V(I_1) \cap V(I_2) = V(I_1 + I_2)$.

4. $V(I_1) \cup V(I_2) = V(I_1 I_2) = V(I_1 \cap I_2)$.

5. $I(V_1 \cup V_2) = I(V_1) \cap I(V_2)$.

6. $\sqrt{I(V)} = I(V)$.

7. $V(\sqrt{I}) = V(I)$.

8. $V(I(V)) = V$.

9. $I(V(I)) \supset \sqrt{I}$.

Exercise 2.8. Prove Proposition 2.3.

In fact we have the equality in 9 above, which is the content of what is called the "Nullstellensatz" (zero-locus theorem). It will be discussed in the next section. Later in this section we prove this in the case of principal ideals.

Definition 2.9. Let V be a germ of a variety at 0. We say that V is *irreducible* if $V \neq \emptyset$ and if $V = V_1 \cup V_2$ implies $V_1 = V$ or $V_2 = V$, where V_1 and V_2 are germs of varieties at 0.

The following gives an algebraic characterization of the irreducibility:

Theorem 2.2. *A germ V of a variety is irreducible if and only if the ideal $I(V)$ is prime.*

Proof. First we note that $V = \emptyset$ if and only if $I(V) = \mathcal{O}_n$. Thus we exclude this case.

Suppose $I(V)$ is not prime. Then there exist germs f_i in $\mathcal{O}_n$, $i = 1, 2$, such that $f_i \notin I(V)$ and $f_1 f_2 \in I(V)$. We set $V_i = V(f_i) \cap V$. Since $f_i \notin I(V)$, We have $V_i \subsetneqq V$, while $V_1 \cup V_2 = V \cap (V(f_1) \cup V(f_2)) = V \cap V(f_1 f_2) = V$. Thus V is not irreducible.

Conversely, suppose $V = V_1 \cup V_2$ with $V_i \subsetneqq V$, $i = 1, 2$. Then, since $I(V_i) \supsetneqq I(V)$, there exists f_i which is in $I(V_i)$ but not in $I(V)$ for each i. We have $f_1 f_2 \in I(V_1) \cap I(V_2) = I(V)$. Hence $I(V)$ is not prime. □

Theorem 2.3. *Every non-empty germ V of a variety can be expressed as*

$$V = V_1 \cup \cdots \cup V_r,$$

where $V_1, \ldots, V_r$ are germs of varieties such that each V_i is irreducible and that $V_i \not\subset V_j$, if $i \neq j$. Moreover, $V_1, \ldots, V_r$ are uniquely determined by V up to order.

Proof. Let $\mathcal{V}$ denote the set of germs of non-empty varieties that cannot be expressed as a finite union of irreducible germs of varieties. We show that $\mathcal{V} = \emptyset$ by contradiction. Suppose $\mathcal{V} \neq \emptyset$. Then the set of ideals $\{ I(V) \mid V \in \mathcal{V} \}$ would be non-empty and thus would have a maximal element $I(V_0)$ by the Noetherian property of $\mathcal{O}_n$ (cf. Theorem 1.12 and Proposition A.19). Then V_0 would be a minimal element in $\mathcal{V}$. As V_0 is not irreducible, we may write $V_0 = V_1 \cup V_2$ with $V_i \subsetneqq V_0$, $i = 1, 2$. Since V_0 was minimal in $\mathcal{V}$, both V_1 and V_2 can be expressed as finite unions of irreducible germs, which is a contradiction. The uniqueness of the expression is not difficult to see. □

The above expression of the variety V is referred to as the *irreducible decomposition* of V.

Remark 2.6. Let V be a germ of a variety at 0 in $\mathbb{C}^n$. Take a representative, denoted also by V, of V in a neighborhood U of 0. For $z \in U$, we denote by V_z the germ of V at z. Note that, if $z \notin V$, then $V_z = \emptyset$. We remark that even if V is irreducible at 0, V_z may not be irreducible for $z \in V$ arbitrarily close to 0. For example, consider the Whitney umbrella (Example 2.3. 3).

Now we prove the Nullstellensatz for principal ideals and discuss related topics. We say that an ideal I in $\mathcal{O}_n$ is *principal* if it is generated by a single germ, i.e., $I = (f)$ for some f in $\mathcal{O}_n$.

Exercise 2.9. Show that f is irreducible in $\mathcal{O}_n$ if and only if $f \neq 0$ and (f) is prime.

Theorem 2.4. *If p is an irreducible germ in $\mathcal{O}_n$, then $I(V(p)) = (p)$.*

Proof. By the preparation theorem, we may assume that p is a Weierstrass polynomial in z_n of degree m. Since p is irreducible, p and $\frac{\partial p}{\partial z_n}$ are relatively prime and we may write (cf. Theorem A.16)

$$ap + b\frac{\partial p}{\partial z_n} = c,$$

where a and b are in $\mathscr{O}_{n-1}[z_n]$ and c is a non-zero germ in $\mathscr{O}_{n-1}$ (the discriminant of p up to sign). If $p(z', z_n)$ has a multiple root at $z_n = a$ for some z', then $\frac{\partial p}{\partial z_n}(z', z_n) = 0$ so that $c(z') = 0$. Hence $p(z', z_n)$ has m distinct roots for z' with $c(z') \neq 0$. Let h be a germ in $I(V(p))$. Then its representative $h(z', z_n)$ has also m distinct roots in z_n for z' with $c(z') \neq 0$. By the division theorem, we may write

$$h = qp + r, \qquad r \in \mathscr{O}_{n-1}[z_n], \quad \deg r < m.$$

As $\deg r < m$, we have $r(z', z_n) = 0$ away from the zero set of c. Thus $r = 0$ as a germ and h is in the ideal (p). □

Corollary 2.1 (Nullstellensatz for principal ideals). *For a germ f in $\mathscr{O}_n$, we have:*

$$I(V(f)) = \sqrt{(f)}.$$

Proof. The above holds if f is 0 or a unit. Thus we assume that f is not 0 or a unit and let $f = p_1^{m_1} \cdots p_r^{m_r}$ be the irreducible decomposition. On the one hand, $\sqrt{(f)} = (p_1 \cdots p_r)$ and on the other hand, $I(V(f)) = I(V(p_1)) \cap \cdots \cap I(V(p_r))$, which is equal to $(p_1 \cdots p_r)$ by Theorem 2.4. □

Proposition 2.4. *For a germ f which is not 0 or a unit in $\mathscr{O}_n$, the following are equivalent:*

(1) *$V(f)$ is irreducible.*
(2) *There is an irreducible germ p in $\mathscr{O}_n$ such that $f = p^m$ for some positive integer m.*

Proof. Let $f = p_1^{m_1} \cdots p_r^{m_r}$ be the irreducible decomposition of f. If $r \geq 2$, then, since $p_1 \cdots p_r \in \sqrt{(f)}$ but $p_i \notin \sqrt{(f)}$, $\sqrt{(f)}$ is not prime. Thus by Corollary 2.1 and Theorem 2.2, $V(f)$ is not irreducible.

We now suppose that $f = p^m$ with p irreducible. If $f_1 f_2 \in \sqrt{(f)}$, then $(f_1 f_2)^k \in (f)$ for some positve integer k. Thus we may find a germ a in $\mathscr{O}_n$ such that $(f_1 f_2)^k = a p^m$. This implies that either f_1 or f_2 is divisible by p. Hence $f_1 \in \sqrt{(f)}$ or $f_2 \in \sqrt{(f)}$. □

The proof of the following is not difficult.

Theorem 2.5. *Let f be a germ, which is not 0 or a unit, in $\mathscr{O}_n$. If $f = p_1^{m_1} \cdots p_r^{m_r}$ is the irreducible decomposition of f, then*

$$V(f) = V(p_1) \cup \cdots \cup V(p_r)$$

is the irreducible decomposition of $V(f)$.

2.4 Nullstellensatz and dimension

In this section we first quote some fundamental results on algebraic structures concerning prime ideals in $\mathscr{O}_n$ and state their geometric consequences on the variety defined by the ideal. We then discuss the Nullstellensatz and dimensions of varieties based on them.

For $d = 0, \dots, n$, we naturally identify the ring $\mathscr{O}_d = \mathbb{C}\{z_1, \dots, z_d\}$ with a subring of $\mathscr{O}_n = \mathbb{C}\{z_1, \dots, z_n\}$. For a prime ideal $\mathfrak{p}$ in $\mathscr{O}_n$, we consider the canonical morphism

$$\eta : \mathscr{O}_d \longrightarrow \mathscr{O}_n/\mathfrak{p}.$$

We denote by K and L the fraction fields of $\mathscr{O}_d$ and $\mathscr{O}_n/\mathfrak{p}$, respectively. For a germ f in $\mathscr{O}_n$, its class in $\mathscr{O}_n/\mathfrak{p}$ is denoted by $[f]$.

We quote:

Theorem 2.6. *Let $\mathfrak{p}$ be a non-zero prime ideal. Then, after a linear change of coordinates of $\mathbb{C}^n$, there is a uniquely determined integer d, $0 \leq d < n$, such that*

(1) *η is injective and $\mathscr{O}_n/\mathfrak{p}$ is finitely generated over $\mathscr{O}_d$ as an $\mathscr{O}_d$-module,*
(2) *L is generated over K by the single element $[z_{d+1}]$ as a field.*

The statement (1) is proved by a repeated use of the Weierstrass preparation theorem (Theorem 1.10) and (2) follows from the theorem of primitive element in the field theory.

Before we state the following theorem, we recall:

Definition 2.10. A continuous map $f : X \to Y$ of topological spaces is *finite* if it is proper (cf. Section B.1) and if, for every point y in Y, $f^{-1}(y)$ is a (possibly empty) finite set.

From Theorem 2.6, we see that $\mathscr{O}_n/\mathfrak{p}$ is integral over $\mathscr{O}_d$ and that the minimal polynomial $P(X)$ of $[z_{d+1}]$ over K is in $\mathscr{O}_d[X]$ (cf. Proposition A.17). It can be further shown that $P(z_{d+1})$, which is in $\mathfrak{p}$, is a Weierstrass polynomial in z_{d+1}. Let $D(P)$ denote the discriminant of $P(X)$, which is a non-zero element in $\mathscr{O}_d$ (cf. (A.12)). We denote by $\boldsymbol{D}$ the germ at 0 of the variety in $\mathbb{C}^d$ defined by $D(P)$. If we set $V = V(\mathfrak{p})$, the germ of the variety defined by $\mathfrak{p}$, we have the following, which is a consequence of Theorem 2.6:

Theorem 2.7. *For a suitable decomposition $\mathbb{C}^n = \mathbb{C}^d \times \mathbb{C}^{n-d}$, there exist a connected neighborhood U' of 0 in $\mathbb{C}^d$ and a neighborhood U'' of 0 in $\mathbb{C}^{n-d}$ with the following properties:*

(1) *The germ V has a representative in $U = U' \times U''$ and the map*

$$\pi : V \cap U \longrightarrow U',$$

which is the restriction of the projection $\mathbb{C}^n \to \mathbb{C}^d$, is a surjective finite map with $\pi^{-1}(0) = \{0\}$.

(2) *The germ $\boldsymbol{D}$ has a representative in U' and if we set*

$$V^* = V \cap U \setminus \pi^{-1}(\boldsymbol{D}),$$

then V^* is a connected complex submanifold of dimension d of U. Moreover, the closure of V^* in U is V.

(3) *The restriction of π to V^* gives a covering map $V^* \to U' \setminus \boldsymbol{D}$.*

With these, we have the following:

Theorem 2.8 (Nullstellensatz). *For every ideal I in $\mathscr{O}_n$, we have:*

$$I(V(I)) = \sqrt{I}.$$

Proof. The above holds if I is 0 or $\mathscr{O}_n$. Thus we assume that I is not 0 or $\mathscr{O}_n$. By the primary decomposition theorem (Theorem A.19), we see that it suffices to prove the above in the case $I = \mathfrak{p}$ is a prime ideal.

We already know that $I(V(\mathfrak{p})) \supset \sqrt{\mathfrak{p}} = \mathfrak{p}$. Thus let f be a germ in $I(V(\mathfrak{p}))$ and $P(X) = a_0 + a_1 X + \cdots + a_{m-1}X^{m-1} + X^m$ the minimal polynomial of $[f]$ over K. Then $P(X)$ is in fact in $\mathscr{O}_d[X]$ (cf. Proposition A.17) and $P(f)$ is in $\mathfrak{p}$. If $m = 1$, we set $Q(X) = 1$ and if $m > 1$, we set $Q(X) = a_1 + \cdots + a_{m-1}X^{m-2} + X^{m-1}$. Then $P(X) = a_0 + Q(X)X$ and $P(f) = a_0 + Q(f)f$. Let $U = U' \times U''$ be as in Theorem 2.7. Then, by the surjectivity of π, $a_0 = 0$ so that $Q(f)f$ is in $\mathfrak{p}$. If $m = 1$, then f is in $\mathfrak{p}$. If $m > 1$, by the minimality of $P(X)$, $Q(f)$ is not in $\mathfrak{p}$, thus f is in $\mathfrak{p}$. □

Corollary 2.2. 1. *Let I be an ideal in $\mathscr{O}_n$. If I is prime, then $V(I)$ is irreducible. If $V(I)$ is irreducible, then $\sqrt{I}$ is prime.*

2. *Let I be an ideal in $\mathscr{O}_n$ with $I \neq \mathscr{O}_n$. If $I = \mathfrak{q}_1 \cap \cdots \cap \mathfrak{q}_r$ is the primary decomposition of I, then*

$$V(I) = V(\mathfrak{p}_1) \cup \cdots \cup V(\mathfrak{p}_r)$$

is the irreducible decomposition of $V(I)$, where $\mathfrak{p}_i = \sqrt{\mathfrak{q}_i}$.

Proof. The first statement follows from Theorems 2.2 and 2.8 and the second from Proposition 2.3 and Theorem 2.8. □

Remark 2.7. From the above, we see that there is a one-to-one correspondence between the set of irreducible germs of varieties at 0 in $\mathbb{C}^n$ and that of prime ideals in $\mathscr{O}_n$.

Exercise 2.10. Show that, for an ideal I ($\neq \mathscr{O}_n$) in $\mathscr{O}_n$, the complex vector space $\mathscr{O}_n/I$ is finite dimensional if and only if $V(I) = \{0\}$.

Dimension

From the item (2) in Theorem 2.7, we have:

Theorem 2.9. *Let V be an irreducible germ of a variety. Then there is a representative V such that V_{reg} is connected and dense in V.*

Dimension at a point: For the germ V of a non-empty variety at 0 in $\mathbb{C}^n$, we define its *dimension*, denoted by $\dim V$, as follows. If V is irreducible, $\dim V$ is the integer d appearing in Theorems 2.6 and 2.7. It is the dimension of the complex manifold V_{reg}. In general, if $V = V_1 \cup \cdots \cup V_r$ is the irreducible decomposition of V, we set $\dim V = \max_{1 \leq i \leq r} \dim V_i$. We say that V is *pure dimensional* if all the components V_i have the same dimension.

We also define the codimension by $\operatorname{codim} V = n - \dim V$.

Note that, if $V = V_1 \cup \cdots \cup V_r$ is the irreducible decomposition of V, then there is the corresponding decomposition $V_{\mathrm{reg}} = C_1 \cup \cdots \cup C_r$ of V_{reg} into its connected components C_i. Each C_i is a complex manifold whose closure coincides with V_i. If V is irreducible, i.e., $r = 1$, then $C_1 = V_{1,\mathrm{reg}} = V_{\mathrm{reg}}$. In general, C_i is contained in $V_{i,\mathrm{reg}}$, however may not coincide with $V_{i,\mathrm{reg}}$. For example, consider the germ at 0 of the variety in Example 2.3. 1.

Remark 2.8. The above definition of $\dim V$ coincides with the Krull dimension of $\mathscr{O}_n/I(V)$ (cf. Section A.2).

If V is a germ of a variety at 0, taking a representative V of V, we denote by V_z the germ of V at z near 0, as before.

From definition we have the following two propositions:

Proposition 2.5. *Let V be the germ of a non-empty variety at 0. There is a neighborhood U of 0 and a representative V of V in U such that*

$$\dim V_z \leq \dim V \qquad \text{for } z \in V \cap U.$$

The equality holds if V is pure dimensional.

Proposition 2.6. *For two germs V and V' of non-empty varieties at 0 with $V \subset V'$, we have $\dim V \leq \dim V'$.*

Theorem 2.10. *Let f be a germ which is not 0 or a unit in $\mathscr{O}_n$. Then $V(f)$ is of pure dimension $n-1$.*

Proof. In view of Theorem 2.5, it suffices to show that $\dim V(p) = n-1$ for every irreducible germ p. From Theorem 2.9, we see that p and $V(p)$ have representatives in a neighborhood U of 0 in $\mathbb{C}^n$ such that $V(p)_{\mathrm{reg}}$ is a connected complex submanifold of U and is dense in $V(p)$. We show that it is of dimension $n-1$. First note that $\dim V(p)_{\mathrm{reg}} \leq n-1$, since otherwise p would be 0. We may assume that p is a Weierstrass polynomial in z_n so that p and $\frac{\partial p}{\partial z_n}$ are relatively prime, i.e., $\frac{\partial p}{\partial z_n} \notin (p)$. By Theorem 2.4, $\{z \in V(p) \mid \frac{\partial p}{\partial z_n}(z) \neq 0\}$ is non-empty and is an $(n-1)$-dimensional complex manifold in $V(p)_{\mathrm{reg}}$. Thus $\dim V(p)_{\mathrm{reg}} = n-1$. □

More generally, we quote the following:

Theorem 2.11. *Let V be an irreducible germ of a variety of dimension d and f a non-unit in $\mathscr{O}_n$. If $f \notin I(V)$, then $V \cap V(f)$ is of pure dimension $d-1$.*

Using this we have:

Theorem 2.12. *Let V be the germ of a variety of pure dimension $n-1$. Then there is a germ f which is not 0 or a unit in $\mathscr{O}_n$ such that $I(V) = (f)$.*

Proof. First we suppose that V is irreducible so that $I(V)$ is prime (cf. Theorem 2.2). As $\dim V = n-1$, $I(V) \neq 0$. Thus there is a germ f in $I(V)$ which is not 0 or a unit. Since $I(V)$ is prime, it contains an irreducible component p of f. We claim that $I(V) = (p)$. Take an arbitrary germ g in $I(V)$. Suppose $g \notin (p)$ which is equal to $I(V(p))$ by Theorem 2.4. Then $\dim V(p) \cap V(g) = n-2$ by Theorem 2.11, which is a contradiction as it contains V.

In general, let $V = V_1 \cup \cdots \cup V_r$ be the irreducible decomposition of V. Then $I(V_i) = (p_i)$ for some irreducible germ p_i, $i = 1, \ldots, r$, so that $I(V) = (p_1 \cdots p_r)$. □

Remark 2.9. From the proofs of Theorems 2.10 and 2.12, we see that there is a one-to-one correspondence between the set of irreducible germs of varieties of codimension one at 0 in $\mathbb{C}^n$ and that of prime principal ideals in $\mathscr{O}_n$.

Let $f : M \to M'$ be a holomorphic map of complex manifolds and V a variety in M. For a point z' in M', $V \cap f^{-1}(z')$ is a variety in M.

In this situation, we quote the following two theorems:

Theorem 2.13 (Upper semi-continuity). *For any point a in V, there is a neighborhood U of a in M such that*

$$\dim(V \cap f^{-1}f(z))_z \leq \dim(V \cap f^{-1}f(a))_a \qquad \textit{for } z \in V \cap U.$$

Theorem 2.14 (Proper mapping theorem). *If $f|_V$ is proper, $f(V)$ is a variety in M' and, for $z' \in f(V)$,*

$$\dim f(V)_{z'} = \max\{\dim V_z - \dim(V \cap f^{-1}f(z))_z \mid z \in V,\ f(z) \textit{ near } z'\}.$$

Global dimension: A variety V in a complex manifold M is said to be *globally irreducible*, or simply *irreducible* if there is no fear of confusin, if it is non-empty and cannot be expressed as a union of two varieties V_1 and V_2 in M with $V_1, V_2 \neq V$. This notion should be distinguished from the irreducibility at a point (Definition 2.9). For example the variety $V_{0,0,\gamma}$ of Exercise 2.6. 2 is globally irreducible, but not irreducible at a. Note that every non-empty variety is expressed as a locally finite union of irreducible varieties. Note also that V is irreducible if and only if the regular part V_{reg} is connected. Hence, for an irreducible variety V and a point z in V, the dimension of V at z remains constant. We call it the *dimension* of V. In general, we say that V is pure dimensional, if all the irreducible components of V have the same dimension.

A variety V of pure dimension $n - 1$ in a complex manifold M of dimension n is called a *hypersurface* in M. By Theorems 2.10 and 2.12, it is equivalent to saying that V is locally defined by a single (not identically zero) holomorphic function with zeros.

We finish this section by quoting a generalization of Theorem 1.8 and discussing some of its consequences and related topics.

Theorem 2.15. *Let V be a connected variety in a complex manifold M and f a holomorphic function on a neighborhood of V. If there is a point z_0 in V such that $|f(z_0)| \geq |f(z)|$ for all z in a neighborhood of z_0 in V, then f is constant on V.*

Corollary 2.3. *A compact variety V in a coordinate neighborhood of a complex manifold is a finite set of points.*

Proof. The absolute value of a coordinate function attains its maximum on each connected component of V. Thus the function is constant on each component. □

From Corollary 2.3, we have:

Proposition 2.7. *Let V be a variety in a coordinate neighborhood U of a complex manifold M and $f : U \to M'$ a holomorphic map into another complex manifold M'. If $f|_V$ is proper, then $f|_V$ is finite.*

As a related topic, from Theorem 2.14, we have:

Proposition 2.8. *Let $f : M \to M'$ be a holomorphic map of complex manifolds and V a variety in M. If $f|_V$ is finite, then $f(V)$ is a variety in M' and, for $z' \in f(V)$,*

$$\dim f(V)_{z'} = \max\{\dim V_z \mid z \in V,\ f(z) \text{ near } z'\}.$$

In particular, if V is pure d-dimensional, so is $f(V)$.

2.5 Underlying real structures

C^r manifolds

Let r denote a non-negative integer or ∞. Recall that a C^r *structure* and a C^r *manifold* of dimension m are defined by replacing, in Definitions 2.1, 2.2 and 2.3, $\mathbb{C}^n$ with $\mathbb{R}^m$ and "holomorphic maps" with "C^r maps". Also, a C^r function on a C^r manifold or a C^r map between C^r manifolds is defined by replacing "holomorphic" with "C^r" in the above. A C^0 manifold is also called a *topological manifold* and a C^r manifold, $1 \le r \le \infty$, a *differentiable manifold* (of class C^r).

If $r \ge 1$, we may consider the Jacobian matrix of a C^r map f and define regular and singular points of f as in Definition 1.9. We also have the inverse mapping theorem (cf. Theorem 1.4), in fact the proof given there goes back to the real case via (1.3), and the implicit mapping theorem (cf. Theorems 1.5 and 1.6). The notions of C^r immersion, embedding and submersion are defined as in Definition 2.4. A C^r *submanifold* is also defined as in Definition 2.6. Thus it is what is called a "regular submanifold".

By restricting to real variables in the definition of analytic functions above, we may consider real analytic functions, which is also referred to as C^ω *functions*. Accordingly, we may define *real analytic manifolds*, or C^ω *manifolds* for short.

Identifying $\mathbb{C}^n$ with $\mathbb{R}^{2n}$, a complex manifold of dimension n may be naturally thought of as a C^r manifold of dimension $2n$ for every r with $0 \leq r \leq \infty$ or $r = \omega$.

For real manifolds we mostly consider the C^∞ case, since on the one hand it is known that every C^r structure, $r \geq 1$, contains a unique C^∞, in fact C^ω, structure and on the other hand some tools such as partition of unity are available in the C^∞ case, but not in the C^ω case.

Real tangent space

Let M be a C^∞ manifold of dimension m and $\{(U_\alpha, \varphi_\alpha)\}$ a C^∞ atlas representing the C^∞ structure. For a ponit x in U_α, we write $\varphi_\alpha(x) = (x_1^\alpha, \ldots, x_m^\alpha)$ and call $(x_1^\alpha, \ldots, x_m^\alpha)$ local coordinates of x relative to φ_α, as in the case of complex manifolds. We recall the following:

Definition 2.11. Let x be a point in M. A *tangent vector* on M at x is an $\mathbb{R}$-linear functional v on the set of C^∞ functions near x satisfying the derivation law:

$$v(fg) = v(f)g(x) + f(x)v(g)$$

for two C^∞ functions f and g near x.

We denote by $T_{\mathbb{R},x}M$ the set of tangent vectors at x and call it the (real) *tangent space* of M at x. It has a natural $\mathbb{R}$-vector space structure. Suppose that $(x_1, \ldots, x_m)$ is a coordinate system near x. Then for each $i = 1, \ldots, m$, we have the tangent vector $\left(\frac{\partial}{\partial x_i}\right)_x$ at x, which is defined by $\left(\frac{\partial}{\partial x_i}\right)_x(f) = \frac{\partial f}{\partial x_i}(x)$. It will be simply written $\frac{\partial}{\partial x_i}$, if there is no fear of confusion. Then we may take $(\frac{\partial}{\partial x_1}, \ldots, \frac{\partial}{\partial x_m})$ as a basis of $T_{\mathbb{R},x}M$.

Differential: Let $f : M \to M'$ be a C^∞ map of C^∞ manifolds. For each point x in M, we have an $\mathbb{R}$-linear map, called the *differential* of f at x:

$$f_* : T_{\mathbb{R},x}M \longrightarrow T_{\mathbb{R},f(x)}M'.$$

It is defined by $f_*(v)(h) = v(h \circ f)$, for v in $T_{\mathbb{R},x}M$ and a C^∞ function h in a neighborhood of $f(x)$. It will be denoted by $f_{*,x}$ if it is necessary to make the point explicit.

Let m and m' be dimensions of M and M', respectively. Choose local coordinate systems $(x_1 \ldots, x_m)$ around x and $(y_1, \ldots, y_{m'})$ around $f(x)$ and set $f_j = y_j \circ f$. Then in terms of the bases $(\frac{\partial}{\partial x_1}, \ldots, \frac{\partial}{\partial x_m})$ of $T_{\mathbb{R},x}M$

and $(\frac{\partial}{\partial y_1}, \dots, \frac{\partial}{\partial y_{m'}})$ of $T_{\mathbb{R},f(x)}M'$, the differential f_* is represented by the Jacobian matrix evaluated at x:

$$\frac{\partial(f_1, \dots, f_{m'})}{\partial(x_1, \dots, x_m)}(x).$$

Thus f is an immersion if and only if $f_{*,x}$ is injective for every x in M and f is a submersion if and only if $f_{*,x}$ is surjective for every x in M (cf. Definition 2.4 with "holomorphic" replaced by "C^∞").

If $M \xrightarrow{f} M' \xrightarrow{f'} M''$ is a sequence of C^∞ maps, we have

$$(f' \circ f)_{*,x} = f'_{*,f(x)} \circ f_{*,x}. \tag{2.4}$$

Remark 2.10. As one of the fundamental facts about C^∞ maps, there is Sard's theorem: if we set $S = \{x \in M \mid f_{*,x} \text{ is not surjective}\}$ for a C^∞ map $f : M \to M'$, then $f(S)$ has measure zero in M'.

Normal space: Let V be a C^∞ submanifold of M with $i : V \hookrightarrow M$ the inclusion. Then for each point x in V, $i_* : T_{\mathbb{R},x}, V \to T_{\mathbb{R},x}M$ is injective so that we may identify $T_{\mathbb{R},x}V$ with a subspace of $T_{\mathbb{R},x}M$. The quotient space $N_{\mathbb{R},V,x} = T_{\mathbb{R},x}M/T_{\mathbb{R},x}V$ is called the (real) *normal space* of V in M at x.

Exercise 2.11. Let V be as above and x a point of V. Suppose V is of codimension k' and is defined by a system of C^∞ functions $(f_1, \dots, f_{k'})$ near x. Let $(x_1, \dots, x_m)$ be a coordinate system in a neighborhood of x in M. We identify $T_{\mathbb{R},x}M$ with $\mathbb{R}^m = \{(\xi_1, \dots, \xi_m)\}$ by taking $(\frac{\partial}{\partial x_1}, \dots, \frac{\partial}{\partial x_m})$ as its basis. Show that, in $\mathbb{R}^m$, $T_{\mathbb{R},x}V$ is given by

$$\sum_{j=1}^{m} \frac{\partial f_i}{\partial x_j}(x) \cdot \xi_j = 0, \qquad i = 1, \dots, k'.$$

Product manifold: Let M and M' be C^∞ manifolds of dimensions m and m', respectively. If $(x_1, \dots, x_m)$ is a coordinate system near x in M and $(x'_1, \dots, x'_{m'})$ a coordinate system near x' in M', then $(x_1, \dots, x_m; x'_1, \dots, x'_{m'})$ is a coordinate system near (x, x') in $M \times M'$. From this we see that there is a natural isomorphism

$$T_{\mathbb{R},(x,x')}(M \times M') \simeq T_{\mathbb{R},x}M \oplus T_{\mathbb{R},x'}M'. \tag{2.5}$$

Holomorphic tangent space

Let M be a complex manifold of dimension n. If $(z_1, \dots, z_n)$ is a holomorphic coordinate system on a neighborhood U of a point z in M, then writing $z_i = x_i + \sqrt{-1}\,y_i$ with x_i and y_i the real and the imaginary parts of z_i, $(x_1, y_1, \ldots, x_n, y_n)$ is a C^∞ coordinate system on U. We may think of the vectors

$$\frac{\partial}{\partial z_i} = \frac{1}{2}\Big(\frac{\partial}{\partial x_i} - \sqrt{-1}\,\frac{\partial}{\partial y_i}\Big) \quad\text{and}\quad \frac{\partial}{\partial \bar{z}_i} = \frac{1}{2}\Big(\frac{\partial}{\partial x_i} + \sqrt{-1}\,\frac{\partial}{\partial y_i}\Big)$$

as being in the complexification $T^c_{\mathbb{R},z}M = \mathbb{C} \otimes_{\mathbb{R}} T_{\mathbb{R},z}M$ of the real tangent space $T_{\mathbb{R},z}M$ of M at z.

Definition 2.12. The *holomorphic tangent space* T_zM and the *antiholomorphic tangent space* $\overline{T}_zM$ of M at z are the subspaces of the complex vector space $T^c_{\mathbb{R},z}M$ spanned by $(\frac{\partial}{\partial z_1}, \ldots, \frac{\partial}{\partial z_n})$ and $(\frac{\partial}{\partial \bar{z}_1}, \ldots, \frac{\partial}{\partial \bar{z}_n})$, respectively.

Obviously we have a decomposition of complex vector spaces:

$$T^c_{\mathbb{R},z}M = T_zM \oplus \overline{T}_zM. \tag{2.6}$$

Exercise 2.12. 1. Show that T_zM and $\overline{T}_zM$ do not depend on the choice of the holomorphic coordinates $(z_1, \ldots, z_n)$.

2. Show that if $f : M \to M'$ is a holomorphic map of complex manifolds, its differential f_* (extended to the complexifications) leaves the holomorphic and antiholomorphic components invariant, i.e., $f_*(T_zM) \subset T_{f(z)}M'$ and $f_*(\overline{T}_zM) \subset \overline{T}_{f(z)}M'$.

Let V be a variety in M and $i : V \hookrightarrow M$ the inclusion. If z is a regular point of V, we may define T_zV and the differential $i_* : T_zV \to T_zM$ is injective (cf. Exercise 2.3). Thus we may identify T_zV with a subspace of T_zM. We call the quotient space T_zM/T_zV the *holomorphic normal space* of V in M at z.

Remark 2.11. Let V be a complex vector space. Its complex conjugate $\overline{\mathsf{V}}$ is a complex vector space such that the underlying set and the additive structure are the same as those of V but that the scalar multiplication, denoted by $*$, is defined by $c * v = \bar{c}v$ for $c \in \mathbb{C}$ and $v \in \mathsf{V}$.

Let R be a subring of $\mathbb{R}$ and M an R-module. Also let V be a complex subspace of the complexification $\mathsf{M}^c = \mathbb{C} \otimes_{\mathsf{R}} \mathsf{M}$ of M. For an element $v = \sum c_i \otimes x_i$, $c_i \in \mathbb{C}$, $x_i \in \mathsf{M}$, of V, we set $\bar{v} = \sum \bar{c}_i \otimes x_i$. Then the set $\{\bar{v} \mid v \in V\}$ is a complex subspace of M^c which is canonically isomorphic with $\overline{V}$ by the correspondence $\bar{v} \leftrightarrow v$. In particular, in the case $\mathsf{M} = T_{\mathbb{R},z}M$ with $\mathsf{R} = \mathbb{R}$, we may identify the space $\overline{T}_zM$ with the complex conjugate of T_zM via the correspondence

$$\sum_{i=1}^{n} c_i \frac{\partial}{\partial \bar{z}_i} \longleftrightarrow \sum_{i=1}^{n} c_i * \frac{\partial}{\partial z_i} = \sum_{i=1}^{n} \bar{c}_i \frac{\partial}{\partial z_i}.$$

Stratifications

Stratification is a way to study singular varieties by decomposing them into a union of submanifolds.

Definition 2.13. Let S be a closed set in a C^∞ manifold M. A *stratification* of S is a family $\{X_\alpha\}_{\alpha\in I}$ of C^∞ submanifolds X_α of M, called *strata*, satisfying the following conditions:

(1) $S = \bigsqcup_{\alpha\in I} X_\alpha$ (disjoint union),
(2) each point of S has a neighborhood which intersects with only a finite number of X_α,
(3) if $X_\alpha \cap \overline{X_\beta} \neq \emptyset$, then $X_\alpha \subset \overline{X_\beta}$.

Let $\{X_\alpha\}_{\alpha\in I}$ be a stratification of S.

Definition 2.14. 1. We say that a pair (X_α, X_β) of strata satisfies the *Whitney condition* at $x \in X_\alpha \cap \overline{X_\beta}$ if the following holds:

(*) let $\{x_n\}$ and $\{y_n\}$ be sequences of points in X_α and X_β, respectively, such that (1) both converge to x, (2) the line joining x_n and y_n converges to a line L and (3) the tangent space $T_{\mathbb{R},y_n} X_\beta$ converges to a space T, then $L \subset T$.

2. We say that $\{X_\alpha\}_{\alpha\in I}$ is a *Whitney stratification* if every pair (X_α, X_β) satisfies the Whitney condition at each point in $X_\alpha \cap \overline{X_\beta}$.

In (*) above, we consider everything in a local coordinate system around x and think of lines and spaces as being in real Grassmann manifolds (cf. Section 3.6 below).

The following is known:

Theorem 2.16. *Every analytic variety in a complex manifold M admits a Whitney stratification whose strata are complex submanifolds of M.*

Fundamental properties of Whitney stratifications include the existence of controlled tube systems (cf. Theorem 13.1 below) and the Thom isotopy lemma, which states that, if $\{X_\alpha\}$ is a Whitney stratification of S, then S is "locally topologically trivial" along each stratum X_α.

Exercise 2.13. Find a Whitney stratification for each of the varieties V in Example 2.3.

Topological dimension

Let X be a topological space. An open covering $\mathcal{U}$ of X is said to have *order* $m+1$ if there exist $m+1$ sets in $\mathcal{U}$ whose intersection is non-empty and the intersection of more than $m+1$ sets in $\mathcal{U}$ is always empty.

Definition 2.15. A space X is *finite-dimensional* if there is an integer m such that every open covering of X admits a refinement having order at most $m+1$. The *topological dimension* of X is the smallest value of such an m.

It is known that a topological manifold of dimension m is of topological dimension m.

Notes

We list [Griffiths and Harris (1978); Kodaira (1986)] as general references on complex manifolds. For analytic varieties, particularly for the details of Section 2.4, we refer to [Gunning and Rossi (1965); Narasimhan (1966); Noguchi (2016)]. As to knots and links that appear in Exercise 2.5, we refer to [Rolfsen (2003)], see also [Milnor (1968)].

See [Hartshorne (1977)] for the corresponding theory in algebraic geometry of the materials in Section 2.3. The Nullstellensatz in algebraic category is referred to as Hilbert Nullstellensatz and that in analytic category as Rückert Nullstellensatz. As to Remark 2.8 and related subjects, we refer to [de Jong and Pfister (2000)].

For fundamental materials on C^∞ manifolds in Section 2.5 and the subsequent chapters, we refer to [Lee (2013); Matsushima (1972)]. In particular, for Remark 2.1.3 on the topology of manifolds, see Ch. II, §15, Theorem 1 in [Matsushima (1972)]. A C^∞ manifold is often called a smooth manifold. However we avoid the use of the terminology "smooth" here, as it sometimes means "non-singular" in other areas such as algebraic geometry and complex analytic geometry.

The fact that every C^r structure, $r \geq 1$, contains a unique C^∞ structure is proved in [Whitney (1936)]. There it is also proved that every C^r structure, $r \geq 1$, contains a C^ω structure. The uniqueness follows from a result in [Grauert (1958)]. See [Shiga (1976)] for details on this matter.

For stratifications, we refer to [Gibson, Wirthmüller, du Plessis and Looijenga (1976); Mather (2012); Shiota (1997)]. Theorem 2.16 in the text is due to [Whitney (1965)].

As to the topological dimension, we refer to [Munkres (1975)] and references therein. It is also called the (Lebesgue) covering dimension.

Chapter 3

Vector Bundles

The tangent space of a manifold "linearly approximates" the manifold at each point. The notion of the tangent bundle naturally comes up as the collection of these spaces. In fact, such objects arise here and there. In this chapter, we review general theory of fiber bundles.

After recalling group actions and related topics, we introduce fiber bundles. Among them is a special class of bundles whose fibers are vector spaces, as in the case of tangent bunles. These are called vector bundles and are studied in some detail. We introduce and discuss the structure of the Stiefel manifold $W(N,r)$, the set of r linearly independent vectors in $\mathbb{C}^N$. It is used for the obstruction theoretical definition of chracteristic classes of vector bundles in Chapter 5. We then discuss the Grassmann manifold $G(N,r)$, the set of linear subspaces of dimension r in $\mathbb{C}^N$. It is a generalization of the projective space and is the orbit space of the natural action of the general linear group $GL(r,\mathbb{C})$ on $W(N,r)$. It admits the tautological vector bundle, of which $W(N,r)$ is the associated principal $GL(r,\mathbb{C})$-bundle. We study the cellular structure of $G(r,N)$. The real case is also discussed.

We also make precise such notions as orientability, C^∞ maps of subsets, manifolds with boundary, tubular neighborhoods and transversality.

3.1 Group actions

Recall that a *topological group* is a group G with a topological structure such that the group operations $G \to G$ given by $g \mapsto g^{-1}$ and $G \times G \to G$ given by $(g,h) \mapsto gh$ are continuous. Let X be a topological space and G a topological group. We denote by e the unity in G.

Definition 3.1. A continuous *left action* of G on X is a continuous map

$$\alpha : G \times X \longrightarrow X$$

such that, denoting $\alpha(g, x)$ by gx,

(1) $ex = x$ for all $x \in X$,
(2) $g(hx) = (gh)x$ for all $g, h \in G$ and $x \in X$.

In this case, we say that G acts continuously on X from the left. A left action is also denoted by (G, X). We may also define a *right action* (X, G) symmetrically to the left one.

Let (G, X) be a left action. For a point x in X, the subset

$$Gx = \{ gx \mid g \in G \}$$

of X is called the G-*orbit* of x. If we introduce a relation $\sim$ in X by saying that $x' \sim x$ if $x' = gx$ for some g in G, this is an equivalence relation and the equivalence class of x coincides with Gx. The quotient set with the quotient topology is called the *orbit space* of the action (G, X) and is denoted by $G \backslash X$. For a right action (X, G), the orbit and the orbit space will be denoted by xG and X/G.

A *fixed point* of an element g of G is a point x such that $gx = x$. For a point x in X, the set of elements in G that leaves x fixed:

$$G_x = \{ g \in G \mid gx = x \}$$

is a subgroup of G, called the *stabilizer* of x in G. Note that, if X is Hausdorff, the subgroup G_x is closed for every x. We say that the action is *free* if $G_x = \{e\}$ for all x. We also say that the action is *effective* if $\bigcap_{x \in X} G_x = \{e\}$.

Let (G, X) and (G, X') be two actions. We say that a map $f : X \to X'$ is G-*equivariant* if $f(gx) = gf(x)$ for all (g, x) in $G \times X$.

Homogeneous spaces

We say that an action (G, X) is *transitive* if it has only a single orbit. In this case, we say that X is a *homogeneous space* of G. For example, for a subgroup G' of G, the right coset space

$$G/G' = \{ gG' \mid g \in G \},$$

which is the orbit space of the natural right action of G' on G, is a homogeneous space of G with the natural left action of G. We recall:

Theorem 3.1. *Let X be a locally compact Hausdorff space and G a locally compact topological group with countable basis. If X is a homogeneous space of G, then for every x in X, the map*

$$G/G_x \longrightarrow X \qquad \text{given by } gG_x \mapsto gx$$

is a G-equivariant homeomorphism.

For r with $1 \leq r \leq \infty$ or $r = \omega$, we may also define C^r actions, taking a real Lie group, a C^r manifold and a C^r map as G, X and α above. Note that a real Lie group is always C^ω. Likewise we may consider holomorphic actions, taking a complex Lie group, a complex manifold and a holomorphic map. Here a complex Lie group is a group with a complex manifold structure such that the group operations are holomorphic.

As standard real Lie groups, we have the real general linear group $GL(p, \mathbb{R})$, the subgroup $GL^+(p, \mathbb{R})$ of $GL(p, \mathbb{R})$ consisting of matrices with positive determinant, the orthogonal group $O(p)$, the special orthogonal group $SO(p)$ and the unitary group $U(q)$ (cf. Section 9.1 below). The complex general linear group $GL(q, \mathbb{C})$ is a typical complex Lie group. In particular, $GL(1, \mathbb{C})$ is the multiplicative group $\mathbb{C}^*$ of non-zero complex numbers.

The following are proved similarly as the corresponding statements for the real case:

Theorem 3.2. *Let G be a complex Lie group and G' a closed complex Lie subgroup of G. Then G/G' admits a complex structure such that the canonical projection $G \to G/G'$ and the natural action $G \times G/G' \to G/G'$ are holomorphic.*

Theorem 3.3. *Let M be a complex manifold and G a complex Lie group acting holomorphically on M. If M is a homogeneous space of G, then for every x in M, G_x is a closed complex Lie subgroup of G and the map*

$$G/G_x \longrightarrow M \quad \text{given by} \quad gG_x \mapsto gx$$

is a G-equivariant biholomorphic map.

Example 3.1. The usual multiplication of matrices on the column vectors defines a left action of $GL(p, \mathbb{R})$ on $\mathbb{R}^p$. This action induces actions of $O(p)$ on the unit ball $\mathbb{B}^p = \{\, {}^t(x_1, \ldots, x_p) \in \mathbb{R}^p \mid |x_1|^2 + \cdots + |x_p|^2 \leq 1 \,\}$ and on its boundary $\partial\mathbb{B}^p$, which is the unit shere $\mathbb{S}^{p-1}$ (cf. Example 3.9 below). The action $(O(p), \mathbb{S}^{p-1})$ is transitive and the stabilizer of ${}^t(1, 0, \ldots, 0)$ may be identified with $O(p-1)$ so that we have a C^ω diffeomorphism

$\mathbb{S}^{p-1} \simeq O(p)/O(p-1)$. If $p > 1$, $SO(p)$ also acts transitively on $\mathbb{S}^{p-1}$ and we have $\mathbb{S}^{p-1} \simeq SO(p)/SO(p-1)$.

Likewise $GL(q, \mathbb{C})$ acts on $\mathbb{C}^q$ from the left. This action induces actions of $U(q)$ on $\mathbb{B}^{2q} = \{\, {}^t(z_1, \ldots, z_q) \in \mathbb{C}^q \mid |z_1|^2 + \cdots + |z_q|^2 \leq 1 \,\}$ and on its boundary $\partial\mathbb{B}^{2q} = \mathbb{S}^{2q-1}$. The action $(U(q), \mathbb{S}^{2q-1})$ is transitive and the stabilizer of ${}^t(1, 0, \ldots, 0)$ may be identified with $U(q-1)$ so that we have a C^ω diffeomorphism $\mathbb{S}^{2q-1} \simeq U(q)/U(q-1)$. See Section 3.5 below for a more general case.

Properly discontinuous actions

We say that an action (G, X) is *proper* if the map

$$G \times X \longrightarrow X \times X \qquad \text{given by} \quad (g, x) \mapsto (gx, x)$$

is proper, i.e., the inverse image of an arbitrary compact set is compact. A *properly discontinuous action* is a proper action of a discrete group. If an action (G, X) is properly discontinuous, then the orbit space $G\backslash X$ is Hausdorff. We also have:

Lemma 3.1. *Let X be a locally compact Hausdorff space and G a discrete group. Then an action (G, X) is properly discontinuous if and only if for every compact set K of X, the set $\{\, g \in G \mid gK \cap K \neq \emptyset \,\}$ is finite.*

Using the above, the following is proved as in the real case:

Theorem 3.4. *Suppose G acts on a complex manifold W holomorphically. If the action is properly discontinuously and free, the orbit space $M = G\backslash W$ has the structure of a complex manifold such that the canonical surjection $\pi : W \to M$ is a holomorphic covering map* (cf. Definition 2.5).

Example 3.2. *Complex tori*: Let $(\omega_1, \ldots, \omega_{2n})$ be a $2n$-tuple of elements in $\mathbb{C}^n$ linearly independent over $\mathbb{R}$ and $G = \langle \omega_1, \ldots, \omega_{2n} \rangle$ the free Abelian group of rank $2n$ generated by the ω_i's. Then G acts on $\mathbb{C}^n$ by translations and the action is properly discontinuous and free. The quotient $T = G\backslash\mathbb{C}^n$ is called a complex torus. It is diffeomorphic to the product $\mathbb{S}^1 \times \cdots \times \mathbb{S}^1$ of $2n$ copies of a circle $\mathbb{S}^1$.

Exercise 3.1. Show that a one-dimensional complex torus is biholomorphic with the one given by $G = \langle 1, \omega \rangle$ with $\Im\omega > 0$ and that it is also biholomorphic with the quotient of $\mathbb{C}^* = \mathbb{C} \setminus \{0\}$ by the infinite cyclic group $\langle \eta \rangle$ generated by $\zeta \mapsto \eta\zeta$ for some complex number η with $0 < |\eta| < 1$ (in fact, $\eta = e^{2\pi\sqrt{-1}\,\omega}$).

The last viewpoint can be generalized to the higher dimensional case:

Example 3.3. *Hopf manifolds*: Suppose $n \geq 2$. Let η_i be complex numbers with $0 < |\eta_i| < 1$, $i = 1, \ldots, n$, and G the infinite cyclic group generated by an element g which acts on $W = \mathbb{C}^n \smallsetminus \{0\}$ by

$$g(z_1, \ldots, z_n) = (\eta_1 z_1, \ldots, \eta_n z_n).$$

Then the action is properly discontinuous and free so that $M = G\backslash W$ is a complex manifold of dimension n, called a Hopf manifold.

Exercise 3.2. Show that a Hopf manifold is diffeomorphic with $\mathbb{S}^1 \times \mathbb{S}^{2n-1}$.

Thus a Hopf manifold M admits a "C^∞ fiber bundle structure" over $\mathbb{P}^{n-1}$ with fiber the real torus $\mathbb{S}^1 \times \mathbb{S}^1$ (cf. Exercise 2.2 and the subsequent sections). In particular, if $\eta_1 = \cdots = \eta_n = \eta$, there is a holomorphic map $M \to \mathbb{P}^{n-1}$, which gives a "holomorphic fiber bundle structure" with fiber the complex torus $\langle\eta\rangle\backslash\mathbb{C}^*$ (cf. Exercise 3.5 below).

3.2 Fiber bundles

Let G be a topological group and F a topological space with a continuous effective left action of G. Also let $\pi : E \to X$ be a continuous map of topological spaces.

Definition 3.2. An *F-trivialization* of π is a pair (U, ψ) of an open set U in X and a homeomorphism

$$\psi : \pi^{-1}(U) \xrightarrow{\sim} F \times U$$

with $p \circ \psi = \pi$, where $p : F \times U \to U$ is the projection. In this case we also say that ψ is a trivialization of E on U.

Definition 3.3. A *system of F-trivializations with group G* of π is a collection $\mathcal{T} = \{(U_\alpha, \psi_\alpha)\}_{\alpha \in I}$ of F-trivializations with the following properties:

(1) $\{U_\alpha\}_{\alpha \in I}$ is a covering of X,
(2) for each pair (α, β), there exists a continuous map $g^{\alpha\beta} : U_\alpha \cap U_\beta \to G$ such that $\psi_\alpha \circ {\psi_\beta}^{-1} : F \times (U_\alpha \cap U_\beta) \to F \times (U_\alpha \cap U_\beta)$ is of the form

$$\psi_\alpha \circ {\psi_\beta}^{-1}(\xi^\beta, x) = (g^{\alpha\beta}(x)\xi^\beta, x), \qquad (\xi^\beta, x) \in F \times (U_\alpha \cap U_\beta),$$

where we denote elements of $F \times U_\alpha$ by (ξ^α, x).

Note that the maps $g^{\alpha\beta}$ are uniquely determined by $\mathcal{T}$ as we assumed the action of G on F to be effective.

For two systems $\mathcal{T}$ and $\mathcal{T}'$ as above, we write $\mathcal{T} \sim \mathcal{T}'$ if their union is a system of trivializations. Clearly it is an equivalence relation in the set of systems of F-trivializations.

Definition 3.4. A *fiber bundle structure with fiber F and group G* on the map $\pi : E \to X$ is an equivalence class of systems of F-trivializations of π with group G. A *fiber bundle* is a map $\pi : E \to X$ together with a fiber bundle structure.

A fiber bundle will simply be denoted by $\pi : E \to X$. We also say that E is a fiber bundle on X. We call E the *total space*, X the *base space* and, for $x \in X$, $E_x = \pi^{-1}(x)$ the *fiber* at x, which is homeomorphic with F. Sometimes G is referred to as the *structure group* and F the *typical fiber*. If the bundle is defined by a system $\mathcal{T} = \{(U_\alpha, \psi_\alpha)\}$ as in Definition 3.3, we call the collection $\{g^{\alpha\beta}\}$ the *system of transition maps* associated with $\mathcal{T}$. In this case we simply say that $\{g^{\alpha\beta}\}$ is a system of transition maps of E. From the effectiveness of the action of G, we have

$$g^{\alpha\beta}(x)g^{\beta\gamma}(x) = g^{\alpha\gamma}(x), \qquad x \in U_\alpha \cap U_\beta \cap U_\gamma. \tag{3.1}$$

In particular, $g^{\alpha\alpha}(x) = e$, the unity in G, and $g^{\beta\alpha}(x) = (g^{\alpha\beta}(x))^{-1}$.

Conversely, if we are given an open covering $\{U_\alpha\}$ of X and a collection $\{g^{\alpha\beta}\}$ of continuous maps $g^{\alpha\beta} : U_\alpha \cap U_\beta \to G$ satisfying (3.1), we may construct a fiber bundle as follows. We consider the disjoint union $\mathcal{X} = \bigsqcup_\alpha (F \times U_\alpha)$ and define a relation $\sim$ as follows: for (ξ^α, x^α) in $F \times U_\alpha$ and (ξ^β, x^β) in $F \times U_\beta$, $(\xi^\alpha, x^\alpha) \sim (\xi^\beta, x^\beta)$ if

$$x^\alpha = x^\beta (= x) \quad \text{and} \quad \xi^\alpha = g^{\alpha\beta}(x)\xi^\beta.$$

Then it is easy to see that this is an equivalence relation in $\mathcal{X}$. Let $E = \mathcal{X}/\sim$ be the quotient space with the quotient topology and denote the class of (ξ^α, x^α) by $[\xi^\alpha, x^\alpha]$. The map $\pi : E \to X$ given by $[\xi^\alpha, x^\alpha] \mapsto x^\alpha$ is well-defined and continuous. Since

$$(F \times U_\alpha)/\sim = F \times U_\alpha,$$

the map $\psi_\alpha : \pi^{-1}(U_\alpha) \to F \times U_\alpha$ given by $[\xi^\alpha, x^\alpha] \mapsto (\xi^\alpha, x^\alpha)$ is an F-trivialization on U_α and $\{(U_\alpha, \psi_\alpha)\}$ defines a fiber bundle structure on E with $\{g^{\alpha\beta}\}$ as a system of transition maps.

We say that a fiber bundle $\pi : E \to X$ is *trivial*, if the fiber bundle structure defining it contains the "global F-trivialization" (X, ψ), i.e., a homeomorphism $\psi : E \xrightarrow{\sim} F \times X$ commuting with the projections. We refer to $F \times X$ as a *product bundle.*

Let $\pi : E \to X$ and $\pi' : E' \to X$ be two fiber bundles with fiber F and group G and let $\mathcal{T} = \{(U_\alpha, \psi_\alpha)\}_{\alpha \in I}$ and $\mathcal{T}' = \{(V_\lambda, \chi_\lambda)\}_{\lambda \in J}$ be systems of trivializations representing the bundle structures of E and E', respectively.

Definition 3.5. An *isomorphism* $\varphi : E \to E'$ is a continuous map with:

(1) $\pi = \pi' \circ \varphi$,
(2) for each pair (α, λ), there exists a continuous map $g^{\lambda\alpha} : U_\alpha \cap V_\lambda \to G$ such that $\chi_\lambda \circ \varphi \circ {\psi_\alpha}^{-1} : F \times (U_\alpha \cap V_\lambda) \to F \times (U_\alpha \cap V_\lambda)$ is of the form

$$\chi_\lambda \circ \varphi \circ {\psi_\alpha}^{-1}(\xi^\alpha, x) = (g^{\lambda\alpha}(x)\xi^\alpha, x), \qquad (\xi^\alpha, x) \in F \times (U_\alpha \cap V_\lambda).$$

Note that the above definition does not depend on the choices of trivializations $\mathcal{T}$ and $\mathcal{T}'$. Also, an isomorphism is bijective and φ^{-1} is an isomorphism. The composition of two isomorphisms is an isomorphism. If the total spaces E and E' are the same, the identity map 1_E is an isomorphism if and only if the bundle structures are the same. We say that two bundles E and E' are *isomorphic* and write $E \simeq E'$ if there exists an isomorphism between them.

Exercise 3.3. Let E and E' be fiber bundles on X as above.

1. Suppose E and E' are defined by systems of trivializations $\mathcal{T}$ and $\mathcal{T}'$ on the same coveing $\mathcal{U} = \{U_\alpha\}$. Let $\{g^{\alpha\beta}\}$ and $\{{g'}^{\alpha\beta}\}$ be the systems of transition maps associated with $\mathcal{T}$ and $\mathcal{T}'$, respectively. Show that $E \simeq E'$ if and only if, for each α, there exists a continuous map $g^\alpha : U_\alpha \to G$ such that

$$g^{\alpha\beta}(x) = g^\alpha(x)^{-1}{g'}^{\alpha\beta}(x)g^\beta(x).$$

In particular, the bundle E defined by a system $\{g^{\alpha\beta}\}$ is trivial if and only if, for each α, there exists a continuous map $g^\alpha : U_\alpha \to G$ such that

$$g^{\alpha\beta}(x) = g^\alpha(x)^{-1}g^\beta(x).$$

2. Let $\mathcal{V} = \{V_\lambda\}_{\lambda \in J}$ be a refinement of $\mathcal{U}$, i.e., there is a map $\iota : J \to I$ such that $V_\lambda \subset U_{\iota(\lambda)}$ for all λ (cf. Remark 2.1.1). Show that $\{g^{\iota(\lambda)\iota(\mu)}\}$ and $\{g^{\alpha\beta}\}$ define isomorphic bundle structures.

3. Let $\{g^{\alpha\beta}\}$ and $\{{g'}^{\lambda\mu}\}$ be two systems on $\mathcal{U}$ and $\mathcal{V}$, respectively. Show that they define isomorphic structures if and only if there exists a common refinement $\mathcal{W}$ of $\mathcal{U}$ and $\mathcal{V}$ such that the restrictions are as in *1.*

Definition 3.6. A *section* of a fiber bundle $\pi : E \to X$ on a subset A of X is a continuous map $s : A \to E$ such that $\pi \circ s = 1_A$.

A section s on A can be described as follows. Let $\{(U_\alpha, \psi_\alpha)\}$ be a system of trivializations defining the bundle. Then we may write

$$\psi_\alpha(s(x)) = (s^\alpha(x), x) \qquad \text{for } x \in A \cap U_\alpha$$

with $s^\alpha : A \cap U_\alpha \to F$ a continuous map. For each point x in $A \cap U_\alpha \cap U_\beta$, we have

$$s^\alpha(x) = g^{\alpha\beta}(x) s^\beta(x). \tag{3.2}$$

Conversely suppose we have a collection $\{s^\alpha\}$ of continuous maps satisfying (3.2). Then setting $s(x) = \psi_\alpha^{-1}(s^\alpha(x), x)$ for x in $A \cap U_\alpha$, we have a section s on A.

Note that a "global section" $s : X \to E$ defines a homeomorphism of X onto its image $s(X)$, the inverse being $\pi|_{s(X)}$.

Definition 3.7. A *principal G-bundle* is a fiber bundle with group G and fiber $F = G$ with the natural left action of G.

Exercise 3.4. Show that a principal bundle is trivial if and only if it admits a global section.

Let G' be a subgroup of a topological group G. Then it is a topological group with the topology as a subspace of G. Let $\pi : E \to X$ be a fiber bundle with group G.

Definition 3.8. We say that the structure group of the bundle may be *reduced to* G' if the bundle structure contains a trivialization such that, in the associated system of transition maps $\{g^{\alpha\beta}\}$, each $g^{\alpha\beta}$ maps $U_\alpha \cap U_\beta$ into G'.

If $E \to X$ is a fiber bundle with fiber F and group G and if F' is a space with a left G action, using a system $\{g^{\alpha\beta}\}$ for E, we may construct a new bundle $E' \to X$ with fiber F', as in the way described above. We call such a bundle a *bundle associated with E*.

Remark 3.1. For r with $1 \le r \le \infty$ or $r = \omega$, we have the notion of a C^r or holomorphic fiber bundle by assuming that G be a real or complex Lie group, F and X C^r or complex manifolds, the action of G on F C^r or holomorphic and the transition maps C^r or holomorphic. In this case E admits naturally the structure of a C^r or complex manifold so that π is a C^r or holomorphic submersion. We may also talk about C^r or holomorphic sections.

Example 3.4. Let $\varpi : \mathbb{C}^{n+1} \setminus \{0\} \to \mathbb{P}^n$ be the canonical surjection (cf. Example 2.1.2). We show that it has the structure of a holomorphic principal $\mathbb{C}^*$-bundle. We denote by $\zeta = (\zeta_0, \ldots, \zeta_n)$ coordinates on $\mathbb{C}^{n+1} \setminus \{0\}$ so that $\varpi(\zeta) = [\zeta] = [\zeta_0, \ldots, \zeta_n]$ are homogeneous coordinates on $\mathbb{P}^n$. For each $\alpha = 0, \ldots, n$, let U_α be the open set in $\mathbb{P}^n$ given by $\zeta_\alpha \neq 0$. Then we have a trivialization

$$\psi_\alpha : \varpi^{-1}(U_\alpha) \xrightarrow{\sim} \mathbb{C}^* \times U_\alpha, \qquad \zeta \mapsto (\zeta_\alpha, [\zeta]).$$

We have $\psi_\alpha \circ \psi_\beta^{-1}(\zeta_\beta, [\zeta]) = (\zeta_\alpha/\zeta_\beta \cdot \zeta_\beta, [\zeta])$. Thus $\mathbb{C}^{n+1} \setminus \{0\}$ is a principal $\mathbb{C}^*$-bundle with a system of transition functions $\{g^{\alpha\beta}\}$ given by $g^{\alpha\beta} = \zeta_\alpha/\zeta_\beta$. The associated line bundle is the tautological bundle on $\mathbb{P}^n$ (cf. Section 9.3 below).

Exercise 3.5. Show that a Hopf manifold $M = \langle \eta \rangle \backslash (\mathbb{C}^n \setminus \{0\})$ (cf. Example 3.3) admits the structure of a holomorphic fiber bundle on $\mathbb{P}^{n-1}$ with fiber a one-dimensional complex torus $\langle \eta \rangle \backslash \mathbb{C}^*$.

The following is proved similarly as the corresponding statement in the real case:

Theorem 3.5. *Let G be a complex Lie group and G' a closed complex Lie subgroup of G. Then $\pi : G \to G/G'$ admits a local holomorphic section so that it becomes a holomorphic G'-principal bundle.*

Homotopy groups

Here we briefly review homotopy groups of topological spaces and state some fundamental theorems.

Let X be a topological space and x_0 a point in X. We denote by $\pi_0(X, x_0)$ the set of path components of X with the component of x_0 as distinguished element. For a positive integer i, we consider the i-cube I^i, $I = [0, 1]$, and denote by $\pi_i(X, x_0)$ the set of homotopy classes of maps $(I^i, \partial I^i) \to (X, x_0)$ relative to ∂I^i, or equivalently, the set of homotopy classes of maps $(\mathbb{S}^i, s_0) \to (X, x_0)$ relative to a point s_0 in $\mathbb{S}^i$, a sphere of dimension i. For two maps $f, g : (I^i, \partial I^i) \to (X, x_0)$ we define a map $f * g : (I^i, \partial I^i) \to (X, x_0)$ by

$$f * g(t_1, \ldots, t_i) = \begin{cases} f(2t_1, t_2, \ldots, t_i) & 0 \leq t_1 \leq \frac{1}{2}, \\ g(2t_1 - 1, t_2, \ldots, t_i) & \frac{1}{2} \leq t_1 \leq 1. \end{cases}$$

Then it induces a group structure on $\pi_i(X, x_0)$ with the class of the constant map as the identity. We call it the i-th *homotopy group* of X with base point x_0. It is Abelian if $i \geq 2$. Thus we use the additive notation for $\pi_i(X, x_0)$, $i \geq 2$, while we use the multiplicative notation for $\pi_1(X, x_0)$, unless it is Abelian. By convention, $\pi_0(X, x_0) = 0$ means that X is path connected, in which case $\pi_i(X, x_0)$ is determined up to isomorphisms, independent of x_0, and $\pi_i(X, x_0)$ will simply be denoted by $\pi_i(X)$.

If $\varphi : (X, x_0) \to (Y, y_0)$ is a map of topological spaces, it induces a morphism $\varphi_* : \pi_i(X, x_0) \to \pi_i(Y, y_0)$, which assigns to the class of a map $f : (I^i, \partial I^i) \to (X, x_0)$ the class of $\varphi \circ f$.

Example 3.5. For a p-sphere $\mathbb{S}^p$, we have

$$\pi_i(\mathbb{S}^p) = \begin{cases} 0 & i = 0, \dots, p-1, \\ \mathbb{Z} & i = p. \end{cases}$$

Note that $\pi_p(\mathbb{S}^p)$ has a canonical generator which can be thought of either as the homotopy class of the map $(I^p, \partial I^p) \to (\mathbb{S}^p, s_0)$ collapsing ∂I^p to a point s_0 or as the homotopy class of the identity map of $\mathbb{S}^p$.

Also, as $\mathbb{C}^q \setminus \{0\}$ deformation retracts to $\mathbb{S}^{2q-1}$,

$$\pi_i(\mathbb{C}^q \setminus \{0\}) = \begin{cases} 0 & i = 0, \dots, 2q-2, \\ \mathbb{Z} & i = 2q-1. \end{cases}$$

The group $\pi_{2q-1}(\mathbb{C}^q \setminus \{0\})$ has a canonical generator, i.e., the homotopy class of the inclusion $\mathbb{S}^{2q-1} \hookrightarrow \mathbb{C}^q \setminus \{0\}$.

Homotopy exact sequence: Let $\pi : E \to X$ be a fiber bundle with fiber F. Choose a point x_0 in X and identify F with $\pi^{-1}(x_0)$ so that we have an inclusion $\iota : F \hookrightarrow E$. Then choose a point e_0 in F and consider the homotopy groups of F, E and X with base points e_0, e_0 and x_0, respectively. Here we quote the fact that a fiber bundle has the homotopy lifting property for cubes, i.e., every homotopy of maps $f : (I^i, \partial I^i) \to (X, x_0)$ can be "lifted" to a homotopy of maps $\tilde{f} : (I^i, \partial I^i, 0) \to (E, F, e_0)$, $0 = (0, \dots, 0)$. From this we see that there is a morphism $\partial_* : \pi_i(X, x_0) \to \pi_{i-1}(F, e_0)$. If $i \geq 2$, it assigns to the class of f the class of $\tilde{f}|_{\partial I^i} : (\partial I^i, 0) \to (F, e_0)$, noting that ∂I^i is homeomorphic to $\mathbb{S}^{i-1}$. In the case $i = 1$, $\partial I = \{0, 1\}$ and ∂_* assigns to the class of f the path component of $\tilde{f}(1)$. With these we have the following, where base points are suppressed:

Theorem 3.6. *For a fiber bundle $\pi : E \to X$ with fiber F, we have an exact sequence of homotopy groups:*

$$\cdots \longrightarrow \pi_i(F) \xrightarrow{\iota_*} \pi_i(E) \xrightarrow{\pi_*} \pi_i(X) \xrightarrow{\partial_*} \pi_{i-1}(F) \longrightarrow \cdots \longrightarrow \pi_0(X).$$

In the above, the exactness of a sequence of maps involving π_0 means that the image equals the inverse image of the distinguished element.

Hurewicz theorem: Since $(I^i, \partial I^i)$ is homeomorphic with $(\mathbb{B}^i, \mathbb{S}^{i-1})$, we have $H_i(I^i, \partial I^i; \mathbb{Z}) \simeq \mathbb{Z}$ for $i \geq 1$, and it has a canonical generator μ_i (cf. Example B.5.2 and Remark B.13.2). We have a natural morphism

$$h_i : \pi_i(X) \longrightarrow H_i(X; \mathbb{Z}),$$

which assigns to the homotopy class of $f : (I^i, \partial I^i) \to (X, x_0)$, the homology class $f_*\mu_i$ in $H_i(X, x_0; \mathbb{Z}) = H_i(X; \mathbb{Z})$.

Now let $k \geq 0$. We say that a space X is *k-connected* if $\pi_i(X) = 0$ for $i = 0, \ldots, k$.

Theorem 3.7. *Suppose X is k-connected.*

1. *If $k = 0$, then h_1 is an epimorphism with the commutator $[\pi_1(X), \pi_1(X)]$ as its kernel.*

2. *If $k \geq 1$, then h_{k+1} is an isomorphism.*

3.3 Vector bundles

In the following, we denote by $\mathbb{K}$ either $\mathbb{R}$ or $\mathbb{C}$ and express the elements of $\mathbb{K}^l$ by column vectors, unless otherwise stated. Thus we may represent the group of linear transformations of $\mathbb{K}^l$ by $GL(l, \mathbb{K})$ acting from the left. Recall that it has the structure of a real or complex Lie group according as $\mathbb{K}$ is $\mathbb{R}$ or $\mathbb{C}$.

Definition 3.9. A *vector bundle* is a fiber bundle $\pi : E \to X$ with fiber $\mathbb{K}^l$ and group $GL(l, \mathbb{K})$ (cf. Definition 3.4). We say that it is real or complex according as $\mathbb{K}$ is $\mathbb{R}$ or $\mathbb{C}$.

Thus a vector bundle structure is represented by a system of trivializations $\mathcal{T} = \{(U_\alpha, \psi_\alpha)\}$ as follows:

(1) for each α, ψ_α is a homeomorphism

$$\psi_\alpha : \pi^{-1}(U_\alpha) \xrightarrow{\sim} \mathbb{K}^l \times U_\alpha \qquad \text{with} \quad p \circ \psi_\alpha = \pi,$$

where $p : \mathbb{K}^l \times U_\alpha \to U_\alpha$ is the projection,

(2) for each pair (α, β), there is a continuous map

$$g^{\alpha\beta} : U_\alpha \cap U_\beta \longrightarrow GL(l, \mathbb{K}) \qquad \text{with}$$

$$\psi_\alpha \circ \psi_\beta^{-1}(\xi^\beta, x) = (g^{\alpha\beta}(x)\xi^\beta, x) \qquad \text{for} \quad (\xi^\beta, x) \in \mathbb{K}^l \times (U_\alpha \cap U_\beta).$$

Note that each fiber $E_x = \pi^{-1}(x)$, $x \in X$, has the structure of a vector space of dimension l over $\mathbb{K}$ so that every trivialization restricted to E_x is a $\mathbb{K}$-linear isomorphism. We call l the *rank* of E. A vector bundle of rank one is called a *line bundle.*

We call $\{g^{\alpha\beta}\}$ the *system of transition matrices* associated with $\mathcal{T}$. For $\xi \in E_x$, we may write $\psi_\alpha(\xi) = (\xi^\alpha, x)$ with $\xi^\alpha \in \mathbb{K}^l$. We call ξ^α the *fiber coordinate* of ξ relative to ψ_α. Thus if $x \in U_\alpha \cap U_\beta$,

$$\xi^\alpha = g^{\alpha\beta}(x)\xi^\beta.$$

Morphisms: Let E and F be vector bundles on X.

Definition 3.10. A *morphism* $\varphi : E \to F$ is a continuous map commuting with the projections such that the induced map $\varphi_x : E_x \to F_x$ on each fiber is $\mathbb{K}$-linear.

Let l and k be the ranks of E and F. Also let $\{g^{\alpha\beta}\}$ and $\{h^{\alpha\beta}\}$ be systems of transition matrices for E and F, respectively. Then a morphism $\varphi : E \to F$ is represented by a collection $\{\varphi_\alpha\}$ of maps

$$\varphi_\alpha : \mathbb{K}^l \times U_\alpha \longrightarrow \mathbb{K}^k \times U_\alpha$$

of the form $\varphi_\alpha(\xi^\alpha, x) = (h^\alpha(x)\xi^\alpha, x)$, where h^α is a continuous map from U_α into $\mathrm{Hom}(\mathbb{K}^l, \mathbb{K}^k)$, the space of linear maps of $\mathbb{K}^l$ into $\mathbb{K}^k$, satisfying

$$h^\alpha(x)g^{\alpha\beta}(x) = h^{\alpha\beta}(x)h^\beta(x).$$

In the above each h^α is represented by a $k \times l$ matrix valued function.

We say that the morphism φ is an *isomorphism* if it is furthermore a homeomorphism. In this case φ induces a $\mathbb{K}$-isomorphism on each fiber. We also say that E and F are isomorphic (or E is isomorphic with F), and write $E \simeq F$, if there is an isomorphism of E onto F.

Remark 3.2. This definition of isomorphism is equivalent to the one given in Definition 3.5 in the case of vector bundles.

From the above local description of φ, we have the following:

Proposition 3.1. *Let E and F be two vector bundles on X of rank l and systems of transition matrices $\{g^{\alpha\beta}\}$ and $\{h^{\alpha\beta}\}$, respectively, on an open covering $\{U_\alpha\}$. Then E and F are isomorphic if and only if there exists a continuous map $g^\alpha : U_\alpha \to GL(l, \mathbb{K})$ for each α, such that*

$$g^{\alpha\beta}(x) = g^\alpha(x)^{-1}h^{\alpha\beta}(x)g^\beta(x) \qquad \textit{for } x \in U_\alpha \cap U_\beta.$$

In the case the total spaces of E and F are the same, the vector bundle structures defined by two systems $\{g^{\alpha\beta}\}$ and $\{h^{\alpha\beta}\}$ are isomorphic if and only if the corresponding systems of trivializations define the same bundle structure (see Exercise 3.3). Thus a vector bundle is trivial if and only if it is isomorphic to the product $\mathbb{K}^l \times X$.

We say that a sequence of morphisms of vector bundles

$$E \xrightarrow{\varphi} F \xrightarrow{\psi} G$$

is *exact* if, for each x in X, the induced sequence $E_x \overset{\varphi_x}{\to} F_x \overset{\psi_x}{\to} G_x$ is exact, i.e., $\operatorname{Ker}\psi_x = \operatorname{Im}\varphi_x$.

Sub and quotient bundles: Let $\pi : E \to X$ be a vector bundle of rank l. For $l' \leq l$, we think of $\mathbb{K}^{l'}$ as a subspace of $\mathbb{K}^l$ by identifying ${}^t(a_1, \ldots, a_{l'})$ with ${}^t(0, \ldots, 0, a_1, \ldots, a_{l'})$.

Definition 3.11. A *subbundle* of E is a subset E' of E with the following property: there exists a system $\{(U_\alpha, \psi_\alpha)\}$ of trivialization for E such that each ψ_α maps $\pi'^{-1}(U_\alpha)$ onto $\mathbb{K}^{l'} \times U_\alpha$, $\pi' = \pi|_{E'}$.

In this case, each transition matrix $g^{\alpha\beta}$ is of the form

$$g^{\alpha\beta} = \begin{pmatrix} g''^{\alpha\beta} & 0 \\ * & g'^{\alpha\beta} \end{pmatrix}, \tag{3.3}$$

where $g'^{\alpha\beta} : U_\alpha \cap U_\beta \to GL(l', \mathbb{K})$ and $g''^{\alpha\beta} : U_\alpha \cap U_\beta \to GL(l - l', \mathbb{K})$. The map $\pi' : E' \to X$ has a vector bundle structure of rank l' with $\{g'^{\alpha\beta}\}$ a system of transition matrices.

The bundle with fiber $\mathbb{K}^{l-l'}$ and system $\{g''^{\alpha\beta}\}$ is called the *quotient bundle* of E by E' and is denoted by E/E'. Note that there is a surjective morphism $\varphi : E \to E/E'$ so that the sequence

$$0 \longrightarrow E' \xrightarrow{\iota} E \xrightarrow{\varphi} E/E' \longrightarrow 0$$

is exact, where ι denotes the inclusion.

In general, if we may choose a system $\{g^{\alpha\beta}\}$ of transition matrices of a vector bundle E with fiber $\mathbb{K}^l$ so that each $g^{\alpha\beta}$ is of the form (3.3), then E admits a subbundle with $\{g'^{\alpha\beta}\}$ as a system of transition matrices.

Exercise 3.6. Let $\varphi : E \to F$ be a morphism of vector bundles. Show that, if the rank of the restriction φ_x of φ to each fiber E_x, $x \in X$, is constant, then the kernel $\operatorname{Ker}\varphi = \bigsqcup_{x\in M} \operatorname{Ker}\varphi_x$ and the image $\operatorname{Im}\varphi = \bigsqcup_{x\in M} \operatorname{Im}\varphi_x$ of φ are subbundles of E and F, respectively. Show also that the quotient bundle $E/\operatorname{Ker}\varphi$ is isomorphic with $\operatorname{Im}\varphi$. The quotient $F/\operatorname{Im}\varphi$ is called the cokernel of φ and is denoted by $\operatorname{Coker}\varphi$.

Pull-back: Let $f : Y \to X$ be a continuous map of topological spaces and $\pi : E \to X$ a vector bundle of rank l. We define the *pull-back* f^*E of E by f by

$$f^*E = \{ (\xi, y) \in E \times Y \mid \pi(\xi) = f(y) \}. \tag{3.4}$$

Denoting by $\tilde{f}$ and ϖ the restrictions of the projections onto the first and second factors, we have the commutative diagram:

$$\begin{array}{ccc} f^*E & \xrightarrow{\tilde{f}} & E \\ {\scriptstyle\varpi}\downarrow & & \downarrow{\scriptstyle\pi} \\ Y & \xrightarrow{f} & X. \end{array}$$

We see that $\varpi : f^*E \to X$ has a vector bundle structure of rank l. Indeed, if $\{(U_\alpha, \psi_\alpha)\}$ is a system of trivializations of E, we write $\psi_\alpha(\xi) = (\xi^\alpha(\xi), x)$ and define $\chi_\alpha : \varpi^{-1}(f^{-1}(U_\alpha)) \to \mathbb{K}^l \times f^{-1}(U_\alpha)$ by $\chi_\alpha(\xi, y) = (\xi^\alpha(\xi), y)$. Then $\{(f^{-1}(U_\alpha), \chi_\alpha)\}$ is a system of trivializations of ϖ. Note that $(f^*E)_y = E_{f(y)}$.

From the definition we have the following:

Proposition 3.2. 1. *If 1_X is the identity map of X and if E is a vector bundle on X, there is a canonical isomorphism*

$$1_X^*E \simeq E \qquad \text{given by } (\xi, x) \leftrightarrow \xi.$$

2. *If $Z \xrightarrow{g} Y \xrightarrow{f} X$ is a sequence of continuous maps and if E is a vector bundle on X, there is a canonical isomorphism*

$$(f \circ g)^*E \simeq g^*(f^*E) \qquad \text{given by } (\xi, z) \leftrightarrow (\xi, y, z),\ y = g(z).$$

If A is a subspace of X with the inclusion $i : A \hookrightarrow X$ and if $\pi : E \to X$ is a vector bundle on X, i^*E is called the *restriction* of E to A and is denoted by $E|_A$. Note that its total space coincides with $\pi^{-1}A$.

This notion of pull-back can be defined for other fiber bundles as well.

Sections and frames: We recall Definition 3.6 for the notion of a section. A vector bundle $\pi : E \to X$ always admits the *zero section*, i.e., the map $s_0 : X \to E$ that assigns to each point x in X the zero of the vector space E_x. The zero section s_0 is a homeomorphism of X onto its image Σ, with the restriction of π to Σ its inverse.

Let l be the rank of E.

Definition 3.12. A *frame* of E on an open set U in X is an ordered family $e^{(l)} = (e_1, \dots, e_l)$ of l sections e_i of E on U linearly independent at each point of U.

Let $(v_1, \ldots, v_l)$ be a basis of $\mathbb{K}^l$. If $\psi : \pi^{-1}(U) \to \mathbb{K}^l \times U$ is a trivialization of E, it defines a frame $e^{(l)} = (e_1, \ldots, e_l)$ on U by

$$e_i(x) = \psi^{-1}(v_i, x).$$

Conversely a frame $e^{(l)} = (e_1, \ldots, e_l)$ defines a trivialization ψ by

$$\psi(\xi) = \Big(\sum_{i=1}^{l} a_i v_i, x\Big) \qquad \text{for } \xi = \sum_{i=1}^{l} a_i e_i(x).$$

Suppose $(v_1, \ldots, v_l)$ is the standard basis of K^l. If $\{(U_\alpha, \psi_\alpha)\}$ is a system of trivializations and if $\{g^{\alpha\beta}\}$ is the associated system of transition matrices, for the corresponding system of frames $\{e_\alpha^{(l)}\}$, we may write:

$$e_\beta^{(l)} = e_\alpha^{(l)} g^{\alpha\beta}. \tag{3.5}$$

Let $f : Y \to X$ be a continuous map and $\pi : E \to X$ a vector bundle. If s is a section of E on an open set U in X, we have the pull-back f^*s, which is a section of f^*E (cf. (3.4)) on $f^{-1}(U)$ defined by

$$(f^*s)(y) = (s(f(y)), y) \qquad \text{for } y \in f^{-1}(U). \tag{3.6}$$

Vector bundles with other structures: A vector bundle E is C^r or holomorphic if it is so as a fiber bundle (see Remark 3.1), in the latter $\mathbb{K}$ being assumed to be $\mathbb{C}$. We may also talk about C^r or holomorphic sections. The set of C^r sections of E on U is denoted by $C^r(U; E)$. This has a natural structure of vector space by the operations defined by $(s_1+s_2)(x) = s_1(x) + s_2(x)$ and $(cs)(x) = cs(x)$ for s_1, s_2 and s in $C^r(U; E)$, c in $\mathbb{K}$ and x in U. The set of holomorphic sections of E on U is denoted by $\Gamma(U; E)$. This has the structure of a complex vector space.

Furthermore, if M is a quasi-projective manifold (cf. Section 2.2), we may consider a regular vector bundle on M_Z. An *algebraic vector bundle* on M is the pull-back of a regular vector bundle by the identity $M \to M_Z$. In particular, it is a holomorphic vector bundle.

Remark 3.3. The following is known:

(V_1) Let E be a vector bundle on X and f, $g : Y \to X$ two continuous maps with Y paracompact. If f and g are homotopic, then $f^*E \simeq g^*E$.

(V_2) The "topological classification" and the "C^∞ classification" of the vector bundles on a C^∞ manifold M are the same. Namely, if E is a continuous vector bundle on M, there exists a C^∞ vector bundle that is isomorphic with E as a continuous bundle and two C^∞ vector bundles are isomorphic as continuous bundles if and only if they are as C^∞ bundles.

Algebraic operations

If we are given some vector bundles, we may construct new ones by algebraic operations. Thus let E and F be vector bundles on X with fibers $\mathbb{K}^l$ and $\mathbb{K}^k$, respectively. We choose a suitable covering $\{U_\alpha\}$ of X and let $\{g^{\alpha\beta}\}$ and $\{h^{\alpha\beta}\}$ be systems of transition matrices for E and F, respectively.

Direct sum $E \oplus F$: We set $E \oplus F = \bigsqcup_{x\in X} E_x \oplus F_x$ with the obviously defined map onto X. It is not difficult to see that it has the structure of a vector bundle with $\mathbb{K}^l \oplus \mathbb{K}^k \simeq \mathbb{K}^{l+k}$ as fiber and $\{g^{\alpha\beta} \oplus h^{\alpha\beta}\}$ as a system of transition maps. By the natural isomorphism $\mathbb{K}^l \oplus \mathbb{K}^k \simeq \mathbb{K}^{l+k}$, $g^{\alpha\beta} \oplus h^{\alpha\beta}$ corrresponds to the matrix

$$\begin{pmatrix} g^{\alpha\beta} & 0 \\ 0 & h^{\alpha\beta} \end{pmatrix}.$$

The direct sum is sometimes called the *Whitney sum.*

In general, let

$$0 \longrightarrow E' \xrightarrow{\iota} E \xrightarrow{\varphi} E'' \longrightarrow 0 \tag{3.7}$$

be an exact sequence of vector bundle morphisms. We say that the sequence *splits* if there exists a morphism $\eta : E'' \to E$ such that $\varphi \circ \eta = 1_{E''}$, or equivalently, $\zeta : E \to E'$ such that $\zeta \circ \iota = 1_{E'}$. A morphism η or ζ as above is called a *splitting* of (3.7). If this is the case, we have an isomorphism $E \simeq E' \oplus E''$, the correspondence being given by assigning to each ξ in E the element $(\zeta(\xi), \varphi(\xi))$ in $E' \oplus E''$.

Proposition 3.3. *If* (3.7) *is an exact sequence of vector bundles on a paracompact space X, there exists a splitting.*

Proof. Let $\mathcal{U} = \{U_\alpha\}$ be a locally finite covering of X such that the bundles are trivial on each U_α. Choosing suitable frames, we see that there exists a splitting η_α on each U_α. Let $\{\rho_\alpha\}$ be a partition of unity subordinate to $\mathcal{U}$. Then $\eta = \sum \rho_\alpha \eta_\alpha$ gives a splitting of (3.7). □

Remark 3.4. If (3.7) is an exact sequence of C^∞ vector bundles on a C^∞ manifold, there exists a C^∞ splitting.

Product $E \times F$: The product $E \times F$ of the total spaces has a natural vector bundle structure on $X \times X$. Indeed, we may write $E \times F = \bigsqcup_{(x_1,x_2)\in X\times X}(E_{x_1} \times F_{x_2})$. As a vector space, $E_{x_1} \times F_{x_2} = E_{x_1} \oplus F_{x_2}$ and the local trivial structures on E and F define that on $E \times F$.

If we let $d : X \hookrightarrow X \times X$ denote the diagonal embedding $x \mapsto (x, x)$, then there is a canonical isomorphism

$$d^*(E \times F) \simeq E \oplus F. \tag{3.8}$$

We may as well define the product of vector bundles on distinct spaces. Namely, if E and F are vector bundles on X and Y, respectively, $E \times F$ is the vector bundle on $X \times Y$ given by $E \times F = \bigsqcup_{(x,y)\in X\times Y}(E_x \times F_y)$.

Morphism $\mathrm{Hom}(E, F)$: Set $\mathrm{Hom}(E, F) = \bigsqcup_{x\in X} \mathrm{Hom}(E_x, F_x)$ with the obviously defined map onto X. Then it has the structure of a vector bundle with $\mathrm{Hom}(\mathbb{K}^l, \mathbb{K}^k) \simeq \mathbb{K}^{kl}$ as fiber and $\{\mathrm{Hom}(g^{\alpha\beta}, h^{\alpha\beta})\}$ as a system of transition maps.

Exercise 3.7. Show that, if we represent elements in $\mathrm{Hom}(\mathbb{K}^l, \mathbb{K}^k)$ by $k \times l$ matrices, the bundle $\mathrm{Hom}(E, F)$ is defined by the trivialization $\{(U_\alpha, \psi_\alpha)\}$ such that, if $(A^\alpha, x) = \psi_\alpha \circ \psi_\beta^{-1}(A^\beta, x)$, A^α is given by

$$A^\alpha = \mathrm{Hom}(g^{\alpha\beta}, h^{\alpha\beta})(x)A^\beta = h^{\alpha\beta}(x) \cdot A^\beta \cdot (g^{\alpha\beta}(x))^{-1}.$$

In particular if $F = \mathbb{K} \times X$, we denote $\mathrm{Hom}(E, F)$ by E^* and call it the *dual bundle* of E. This is the vector bundle with fiber $(\mathbb{K}^l)^* \simeq \mathbb{K}^l$. If we express the elements in $(\mathbb{K}^l)^*$ by row vectors, then a system of transition matrices is given by $\{(g^{\alpha\beta})^{-1}\}$ with multiplication from the right and if we express them by column vectors, then it is given by $\{{}^t(g^{\alpha\beta})^{-1}\}$ with multiplication from the left. There is a canonical isomorphism $(E^*)^* \simeq E$.

Tensor product $E \otimes F$: We set $E \otimes F = \bigsqcup_{x\in X} E_x \otimes F_x$. Then it has the structure of a vector bundle with $\mathbb{K}^l \otimes \mathbb{K}^k \simeq \mathbb{K}^{lk}$ as fiber and $\{g^{\alpha\beta} \otimes h^{\alpha\beta}\}$ as a system of transition maps. The natural isomorphisms $\mathrm{Hom}(E_x, F_x) \simeq E_x^* \otimes F_x$ on the fibers induces a natural isomorphism

$$\mathrm{Hom}(E, F) \simeq E^* \otimes F.$$

Exterior power $\bigwedge^r E$: We set $\bigwedge^r E = \bigsqcup_{x\in X} \bigwedge^r E_x$. Then it has the structure of a vector bundle with $\bigwedge^r \mathbb{K}^l \simeq \mathbb{K}^{\binom{l}{r}}$ as fiber and $\{\bigwedge^r g^{\alpha\beta}\}$ as a system of transition maps. In particular, we denote $\bigwedge^l E$ by $\det E$ and call it the *determinant* (*bundle*) of E. It is a vector bundle of rank one defined by the system $\{\det g^{\alpha\beta}\}$. In general, we have a natural isomorphism

$$\bigwedge^r(E \oplus F) \simeq \bigoplus_{p+q=r} \bigwedge^p E \otimes \bigwedge^q F.$$

Complex conjugate $\overline{E}$: We assume that $\mathbb{K} = \mathbb{C}$ and set $\overline{E} = \bigsqcup_{x \in X} \overline{E}_x$, where $\overline{E}_x$ denotes the complex conjugate of E_x (cf. Remark 2.11). Then it has the structure of a complex vector bundle with $\mathbb{C}^l$ as fiber and $\{\overline{g^{\alpha\beta}}\}$ as a system of transition matrices.

Complexification E^c: We assume that $\mathbb{K} = \mathbb{R}$ and set $E^c = \bigsqcup_{x \in X} E^c_x$, $E^c_x = \mathbb{C} \otimes_{\mathbb{R}} E_x$. Then it has the structure of a complex vector bundle with $\mathbb{C}^l$ as fiber and $\{g^{\alpha\beta}\}$ as a system of transition matrices.

Complex bundle as real bundle: The complex vector space $\mathbb{C}^l$ is naturally considered as a real vector space of dimension $2l$ and this defines a natural morphism

$$\rho : GL(l, \mathbb{C}) \longrightarrow GL(2l, \mathbb{R}). \tag{3.9}$$

Thus if E is a complex vector bundle of rank l, it has the structure of a real vector bundle of rank $2l$. If $\{g^{\alpha\beta}\}$ is a system of transition matrices for E, then $\{\rho \circ g^{\alpha\beta}\}$ is a system of transition matrices of E as a real bundle.

3.4 Tangent bundle and vector fields

Real tangent bundle

Let M be a C^∞ manifold of dimension m. We may give a natural vector bundle structure on the (disjoint) union $T_{\mathbb{R}}M = \bigsqcup_{x \in M} T_{\mathbb{R},x}M$ of the tangent spaces of M as follows. First, define $\pi : T_{\mathbb{R}}M \to M$ by assigning to each tangent vector its base point. Then let $\{U_\alpha\}$ be a covering of M by coordinate neighborhoods U_α with coordinates $(x^\alpha_1, \dots, x^\alpha_m)$. Taking $(\frac{\partial}{\partial x^\alpha_1}, \dots, \frac{\partial}{\partial x^\alpha_m})$ as a basis of $T_{\mathbb{R},x}M$ for each x in U_α, we have a bijection $\psi_\alpha : \pi^{-1}(U_\alpha) \to \mathbb{R}^m \times U_\alpha$. Since we have the relation

$$\frac{\partial}{\partial x^\beta_j} = \sum_{i=1}^m \frac{\partial x^\alpha_i}{\partial x^\beta_j}(x) \frac{\partial}{\partial x^\alpha_i}, \qquad j = 1, \dots, m, \tag{3.10}$$

for x in $U_\alpha \cap U_\beta$, we see that $\psi_\alpha \circ \psi_\beta^{-1}(\xi, x) = (t^{\alpha\beta}(x)\xi, x)$ for (ξ, x) in $\mathbb{R}^m \times (U_\alpha \cap U_\beta)$, where

$$t^{\alpha\beta} = \frac{\partial(x^\alpha_1, \dots, x^\alpha_m)}{\partial(x^\beta_1, \dots, x^\beta_m)} = \begin{pmatrix} \dfrac{\partial x^\alpha_1}{\partial x^\beta_1} & \cdots & \dfrac{\partial x^\alpha_1}{\partial x^\beta_m} \\ \vdots & \ddots & \vdots \\ \dfrac{\partial x^\alpha_m}{\partial x^\beta_1} & \cdots & \dfrac{\partial x^\alpha_m}{\partial x^\beta_m} \end{pmatrix}.$$

Therefore $T_{\mathbb{R}}M$ admits the structure of a real vector bundle of rank m with $\{t^{\alpha\beta}\}$ as a system of transition matrices. We call it the (real) *tangent bundle* of M.

The bundle dual to the tangent bundle $T_{\mathbb{R}}M$ is called the (real) *cotangent bundle* and is denoted by $T^*_{\mathbb{R}}M$.

Exercise 3.8. Let M and M' be C^∞ manifolds. Consider the product manifold $M \times M'$ and let p and p' denote the projections onto the first and the second factors. Show that there is a natural isomorphism

$$T_{\mathbb{R}}(M \times M') \simeq p^* T_{\mathbb{R}}M \oplus (p')^* T_{\mathbb{R}}M',$$

which gives the isomorphism (2.5) on each fiber.

Vector fields: A *vector field* v on an open set U in M is a section of $T_{\mathbb{R}}M$ on U. Thus it is expressed as, on each $U \cap U_\alpha$,

$$v = \sum_{i=1}^{m} f_i^\alpha(x) \frac{\partial}{\partial x_i^\alpha},$$

where the f_i^α's are functions on $U_\alpha \cap U$. In $U \cap U_\alpha \cap U_\beta$, we have $f^\alpha = t^{\alpha\beta} f^\beta$, $f^\alpha = {}^t(f_1^\alpha, \ldots, f_m^\alpha)$. The vector field v is C^∞ if each f_i^α is.

For a C^∞ vector field v and a C^∞ function f on U, $v(f)$ is a C^∞ function on U. The bracket operation assigns to a pair (u, v) of C^∞ vector fields on U a C^∞ vector field $[u, v]$ on U satisfying

$$[u, v](f) = u(v(f)) - v(u(f)).$$

Note that $[u, v]$ is $\mathbb{C}$-bilinear and alternating in (u, v) and that

$$[fu, gv] = fg[u, v] + fu(g)v - gv(f)u$$

for C^∞ functions f and g. We also have the *Jacobi identity*

$$[u, [v, w]] + [v, [w, u]] + [w, [u, v]] = 0$$

for every triple (u, v, w) of vector fields.

Real normal bundle: If V is a submanifold of codimension k' of M, we may cover V with coordinate neighborhoods U_α on M with coordinates $(x_1^\alpha, \ldots, x_m^\alpha)$ such that

$$V \cap U_\alpha = \{x \in U_\alpha \mid x_1^\alpha = \cdots = x_{k'}^\alpha = 0\}.$$

Then the restriction $t^{\alpha\beta}|_V$ of $t^{\alpha\beta}$ to $V \cap U_\alpha \cap U_\beta$ is of the form

$$t^{\alpha\beta}|_V = \begin{pmatrix} t''^{\alpha\beta} & 0 \\ * & t'^{\alpha\beta} \end{pmatrix}, \tag{3.11}$$

where $t'^{\alpha\beta}$ and $t''^{\alpha\beta}$ denote the Jacobian matrices

$$\frac{\partial(x^\alpha_{k'+1},\dots,x^\alpha_m)}{\partial(x^\beta_{k'+1},\dots,x^\beta_m)} \quad \text{and} \quad \frac{\partial(x^\alpha_1,\dots,x^\alpha_{k'})}{\partial(x^\beta_1,\dots,x^\beta_{k'})},$$

respectively, both restricted to V. Since the restriction of $(x^\alpha_{k'+1},\dots,x^\alpha_m)$ to V form a coordinate system on $V \cap U_\alpha$, we see that $T_{\mathbb{R}}V$ is a subbundle of $T_{\mathbb{R}}M|_V$. We call the quotient bundle the (real) *normal bundle* of V in M and denote it by $N_{\mathbb{R},V}$. It is defined by the system of transition matrices $\{t''^{\alpha\beta}\}$ and we have an exact sequence

$$0 \longrightarrow T_{\mathbb{R}}V \longrightarrow T_{\mathbb{R}}M|_V \xrightarrow{\varpi} N_{\mathbb{R},V} \longrightarrow 0. \tag{3.12}$$

Exercise 3.9. Let $f : M \to M'$ be a submersion of C^∞ manifolds (cf. Definition 2.4 with "holomorphic" replaced by "C^∞"). Show that $T_{\mathbb{R}}M$ admits as a subbundle the bundle $T_{\mathbb{R}}f$ of vectors that are "tangent to the fibers of f" so that there is an exact sequence of vector bundles on M:

$$0 \longrightarrow T_{\mathbb{R}}f \longrightarrow T_{\mathbb{R}}M \longrightarrow f^*T_{\mathbb{R}}M' \longrightarrow 0.$$

Exercise 3.10. Let $\pi : E \to M$ be a C^∞ vector bundle and Σ the image of the zero section $s_0 : M \to E$.

1. Show that Σ is a submanifold of E, s_0 is a diffeomorphism of M onto Σ and that there is a canonical isomorphism $s_0^*N_{\mathbb{R},\Sigma} \simeq E$. Thus pulling back the sequence (3.12) for $(M,V) = (E,\Sigma)$ by s_0, we have the exact sequence of vector bundles on M:

$$0 \longrightarrow T_{\mathbb{R}}M \longrightarrow s_0^*T_{\mathbb{R}}E \longrightarrow E \longrightarrow 0.$$

2. Applying the sequence in Exercise 3.9, we have the exact sequence

$$0 \longrightarrow T_{\mathbb{R}}\pi \longrightarrow T_{\mathbb{R}}E \longrightarrow \pi^*T_{\mathbb{R}}M \longrightarrow 0.$$

Show that there is a canonical isomorphism $s_0^*T_{\mathbb{R}}\pi \simeq E$. Thus pulling back the above sequence by s_0, we have the exact sequence

$$0 \longrightarrow E \longrightarrow s_0^*T_{\mathbb{R}}E \longrightarrow T_{\mathbb{R}}M \longrightarrow 0.$$

3. Show that there is in fact a canonical isomorphism

$$s_0^*T_{\mathbb{R}}E \simeq E \oplus T_{\mathbb{R}}M.$$

Exercise 3.11. Let M be a C^∞ manifold. Show that the diagonal

$$\Delta = \{\, (x,x) \in M \times M \mid x \in M \,\}$$

is a closed submanifold of $M \times M$ and the map $\iota : M \to M \times M$ given by $\iota(x) = (x,x)$ is a diffeomorphism onto Δ. Show moreover that there is an isomorphism $\iota^*N_{\mathbb{R},\Delta} \simeq T_{\mathbb{R}}M$.

Holomorphic tangent bundle

Let M be a complex manifold of dimension n and $\{U_\alpha\}$ a covering of M by coordinate neighborhoods U_α with holomorphic coordinates $(z_1^\alpha, \dots, z_n^\alpha)$. Then, as in the real case, the union $TM = \bigsqcup_{z\in M} T_z M$ of the holomorphic parts of the complexified tangent spaces of M admits the structure of a complex vector bundle of rank n with $\{\tau^{\alpha\beta}\}$ given by

$$\tau^{\alpha\beta} = \frac{\partial(z_1^\alpha, \dots, z_n^\alpha)}{\partial(z_1^\beta, \dots, z_n^\beta)},$$

as a system of transition matrices. Since, for each pair (α, β), $\tau^{\alpha\beta}$ is a holomorphic map from $U_\alpha \cap U_\beta$ into $GL(n, \mathbb{C})$, TM is a holomorphic bundle. We call it the *holomorphic tangent bundle* of M. Note that $\overline{T}M = \bigsqcup_{z\in M} \overline{T}_z M$ may be identified with the complex conjugate of TM (cf. Remark 2.11 and Section 3.3).

The bundle complex dual to TM is called the *holomorphic cotangent bundle* of M and is denoted by T^*M.

Definition 3.13. The line bundle $\det T^*M$ is called the *canonical bundle* of M and is denoted by K_M.

We give some fundamental facts on the tangent bundles of a complex manifold. Let $T_{\mathbb{R}}^c M = \mathbb{C} \otimes_{\mathbb{R}} T_{\mathbb{R}} M$ be the complexification of $T_{\mathbb{R}} M$.

Proposition 3.4. *There is a natural isomorphism*

$$T_{\mathbb{R}}^c M \simeq TM \oplus \overline{T}M,$$

which induces the decomposition (2.6) *on each fiber.*

Proof. We cover M by coordinate neighborhoods U_α with coordinates $(z_1^\alpha, \dots, z_n^\alpha)$. If we write $z_i^\alpha = x_i^\alpha + \sqrt{-1}\, y_i^\alpha$, then the transition matrix $t^{\alpha\beta}$ of the vector bundle $T_{\mathbb{R}}^c M$ on $U_\alpha \cap U_\beta$ is the Jacobian matrix of $(x_1^\alpha, y_1^\alpha, \dots, x_n^\alpha, y_n^\alpha)$ with respect to $(x_1^\beta, y_1^\beta, \dots, x_n^\beta, y_n^\beta)$. It is not difficult to show that this bundle is isomorphic with the one defined by the transition matrix which is equal to the Jacobian matrix of $(z_1^\alpha, \dots, z_n^\alpha, \bar{z}_1^\alpha, \dots, \bar{z}_n^\alpha)$ with respect to $(z_1^\beta, \dots, z_n^\beta, \bar{z}_1^\beta, \dots, \bar{z}_n^\beta)$. Because of the Cauchy-Riemann equations, this matrix is equal to $\begin{pmatrix} \tau^{\alpha\beta} & 0 \\ 0 & \overline{\tau^{\alpha\beta}} \end{pmatrix}$. □

Proposition 3.5. *We have $TM \simeq T_{\mathbb{R}} M$ as real bundles.*

Proof. We cover M by coordinate neighborhoods U_α as in Proposition 3.4. Then again by the Cauchy-Riemann equations, we see that $\rho \circ \tau^{\alpha\beta}$ (cf. (3.9)) is identical with the Jacobian matrix of $(x_1^\alpha, y_1^\alpha, \ldots, x_n^\alpha, y_n^\alpha)$ with respect to $(x_1^\beta, y_1^\beta, \ldots, x_n^\beta, y_n^\beta)$. □

Let v be a section of TM. On a coordinate neighborhood U with coordinates $(z_1, \ldots, z_n)$, v is expressed as

$$v = \sum_{i=1}^{n} f_i(z) \frac{\partial}{\partial z_i}$$

with f_i complex valued functions on U.

The following is not difficult to see:

Proposition 3.6. *Let v be as above. If we write $f_i = u_i + \sqrt{-1}\, v_i$ with u_i and v_i real valued functions, then under the isomorphism of Proposition 3.5, v corresponds to the real vector field given by*

$$\sum_{i=1}^{n} u_i(x, y) \frac{\partial}{\partial x_i} + \sum_{i=1}^{n} v_i(x, y) \frac{\partial}{\partial y_i}.$$

Definition 3.14. A *holomorphic vector field* is a holomorphic section of TM.

Thus v as above being holomorphic means that each f_i is holomorphic. We denote by $\Gamma(U; TM)$ the space of holomorphic vector fields on U.

Exercise 3.12. 1. Give detailed proofs of Propositions 3.4–3.6.

2. We may naturally extend the bracket operation [,] to the space $C^\infty(U; T_{\mathbb{R}}^{\mathbb{C}} M)$. Show that, if we identify TM with a subbundle of $T_{\mathbb{R}}^{\mathbb{C}} M$ by Propositions 3.4, the subspaces $C^\infty(U; TM)$ and $\Gamma(U; TM)$ of $C^\infty(U; T_{\mathbb{R}}^{\mathbb{C}} M)$ are both stable by [,].

3. Show that the isomorphism of Proposition 3.5 is compatible with the bracket operations on TM and on $T_{\mathbb{R}} M$.

Holomorphic normal bundle If V is a complex submanifold of M, TV may naturally be thought of as a subbundle of $TM|_V$, as in the real case. We call the quotient the *holomorphic normal bundle* of V in M and denote it by N_V so that we have the exact sequence

$$0 \longrightarrow TV \longrightarrow TM|_V \xrightarrow{\varpi} N_V \longrightarrow 0. \tag{3.13}$$

Note that, by Proposition 3.5, N_V is isomorphic with $N_{\mathbb{R},V}$ as a real bundle.

Note also that the statements in Exercises 3.8–3.11 are valid if we replace "C^∞" and various real bundles with "holomorphic" and the corresponding holomorphic bundles.

Line bundle associated with a hypersurface

Let M be a complex manifold and V a hypersurface (possibly with singularities) in M. We may take an open covering $\{U_\alpha\}$ of M so that in each U_α, V is defined by a "reduced equation" $f^\alpha = 0$, i.e., the germ of f^α at each point in $V \cap U_\alpha$ is reduced (cf. Section 2.3). Note that if $V \cap U_\alpha = \emptyset$, then we may take a non-zero constant as f^α. Then, for each pair (α, β), $g^{\alpha\beta} = f^\alpha/f^\beta$ is a non-vanishing holomorphic function, i.e., a holomorphic function without zeros, on $U_\alpha \cap U_\beta$ and the system $\{g^{\alpha\beta}\}$ satisfies the cocycle condition.

Definition 3.15. The *line bundle associated with* V, denoted by L_V, is the line bundle defined by the cocycle $\{g^{\alpha\beta}\}$.

Note that L_V is a holomorphic bundle and admits a natural holomorphic section whose zero set is exactly V, i.e., the section represented by the collection $\{f^\alpha\}$ (cf. the paragraph after Definition 3.6).

Exercise 3.13. Show that, if V is a non-singular hypersurface in M, there is a natural isomorphism $L_V|_V \simeq N_V$.

Example 3.6. *Hyperplane bundle*: Let $\mathbb{P}^n$ be the n-dimensional projective space with homogeneous coordinates $[\zeta_0, \ldots, \zeta_n]$ and $\mathcal{U} = \{U_\alpha\}$ the covering of $\mathbb{P}^n$ given by $U_\alpha = \{\zeta_\alpha \neq 0\}$. If H is a hyperplane (cf. Example 2.4) in $\mathbb{P}^n$ defined by a linear function $\ell(\zeta)$, it is defined by the holomorphic function $\ell(\zeta)/\zeta_\alpha$ on U_α. The associated line bundle L_H is then defined by the system of transition functions $\{g^{\alpha\beta}\}$ with $g^{\alpha\beta} = \zeta_\beta/\zeta_\alpha$ on the covering $\mathcal{U}$. Thus it is uniquely determined modulo isomorphisms, independently of H. It is called the *hyperplane bundle* and is denoted by H_n.

If V is a hypersurface of degree p, we see that L_V is defined by the system of transition functions $\{(\zeta_\beta/\zeta_\alpha)^p\}$ so that $L_V = H_n^{\otimes p}$, p times tensor product of H_n.

Exercise 3.14. Let H be the hyperplane in $\mathbb{P}^n$ defined by $\zeta_0 = 0$ and p_0 the point $[1, 0, \ldots, 0]$. We identify H with $\mathbb{P}^{n-1} = \{[\zeta_1, \ldots, \zeta_n]\}$ and define a map $\pi : \mathbb{P}^n \smallsetminus \{p_0\} \to \mathbb{P}^{n-1}$ by $[\zeta] \mapsto [\zeta_1, \ldots, \zeta_n]$. Show that it has the structure of a holomorphic line bundle isomorphic with H_{n-1}.

We discuss in Section 11.6 below more generally the line bundle associated with a "divisor".

3.5 Stiefel manifold

In this and the subsequent sections, the elements of $\mathbb{C}^N$ are denoted by column vectors. Also we let $e_i = {}^t(0,\dots,0,\overset{i}{1},0\dots,0)$, $i = 1,\dots,N$. An *r-frame* in $\mathbb{C}^N$ is an ordered family of linearly independent r vectors in $\mathbb{C}^N$, $1 \le r \le N$.

Definition 3.16. The *Stiefel manifold* of r-frames in $\mathbb{C}^N$, denoted by $W(N,r)$, is the set of r-frames in $\mathbb{C}^N$.

Let $M(N,r)$ denote the set of $N \times r$ complex matrices, which may be identified with $\mathbb{C}^{Nr}$. Since an r-frame is expressed by an $N \times r$ matrix of rank r, $W(N,r)$ may be thought of as the set of $N \times r$ matrices of rank r, which is an open set in $M(N,r)$. Thus $W(N,r)$ is naturally a complex manifold of dimension Nr. The complex Lie group $GL(N,\mathbb{C})$ acts transitively on $W(N,r)$ from the left with the stabilizer at $(e_1,\dots,e_r)$ the subgroup $G_{N,N-r}$ of $GL(N,\mathbb{C})$ consisting of matrices of the form

$$\begin{pmatrix} I_r & * \\ 0 & P' \end{pmatrix}, \qquad P' \in GL(N-r,\mathbb{C})$$

so that $W(N,r)$ is a homogeneous space (cf. Theorem 3.3):

$$W(N,r) \simeq GL(N,\mathbb{C})/G_{N,N-r}. \tag{3.14}$$

Let $W_0(N,r)$ denote the subset of $W(N,r)$ consisting of orthonormal r-frames in $\mathbb{C}^N$ with respect to the standard Hermitian metric. The unitary group $U(N)$ acts on $W_0(N,r)$ transitively from the left with the stabilizer at $(e_1,\dots,e_r)$ the set of unitary matrices of the form

$$\begin{pmatrix} I_r & 0 \\ 0 & U' \end{pmatrix}, \qquad U' \in U(N-r)$$

so that $W_0(N,r)$ is a homogeneous space:

$$W_0(N,r) \simeq U(N)/U(N-r).$$

We may think of the above as describing the C^∞ manifold structure of $W_0(N,r)$. Thus it is compact, connected and of dimension $(2N-r)r$. Note that the Gram-Schmidt process gives a deformation retraction

$$W(N,r) \longrightarrow W_0(N,r).$$

In particular, $W(N,1) = \mathbb{C}^N \setminus \{0\}$ and $W_0(N,1) = \mathbb{S}^{2N-1}$ so that

$$\mathbb{S}^{2N-1} \simeq U(N)/U(N-1),$$

which is the situation considered in Example 3.1.

Fiber bundle structure: Suppose $r > 1$. Taking the first $(r-1)$ vectors of each r-frame we have a map

$$\pi : W(N,r) \longrightarrow W(N,r-1). \tag{3.15}$$

Proposition 3.7. *The map* (3.15) *has a fiber bundle structure with fiber* $\mathbb{C}^{r-1} \times (\mathbb{C}^q \setminus \{0\})$, $q = N - r + 1$.

Proof. We think of $W(N,r-1)$ as the set of $N \times (r-1)$ matrices of rank $r-1$. Set

$$\mathcal{I} = \{\, (i_1, \dots, i_{r-1}) \mid 1 \le i_1 < \cdots < i_{r-1} \le N \,\}$$

and for $A' \in W(N,r-1)$ and $I \in \mathcal{I}$, denote by A'_I the $(r-1) \times (r-1)$ matrix consisting of $r-1$ rows of A' corresponding to I. Then $W(N,r-1)$ is covered by the open sets U_I given by

$$U_I = \{\, A' \in W(N,r-1) \mid \det A'_I \neq 0 \,\}.$$

If we represent an element of $W(N,r)$ by an $N \times r$ matrix $A = (a_{ij})$ of rank r, the map π assigns to A the $N \times (r-1)$ matrix A' consisting of the first $r-1$ columns of A. For $I \in \mathcal{I}$, we denote by I^* the complement of I in $(1, \dots, N)$. For each $I = (i_1, \dots, i_{r-1}) \in \mathcal{I}$ and $i \in I^*$, let $A_{I,i}$ denote the $r \times r$ matrix consisting of the $i_1, \dots, i_{r-1}$ and i-th rows of A. Note that, for a fixed I, at least one of the $\det A_{I,i}$'s is non-zero, as the rank of A is r. With these, we have a trivialization

$$\psi_I : \pi^{-1}(U_I) \xrightarrow{\sim} \mathbb{C}^{r-1} \times (\mathbb{C}^q \setminus \{0\}) \times U_I$$

given by $\psi_I(A) = ({}^t(a_{i_1 r}, \dots, a_{i_{r-1} r}), (\det A_{I,i})_{i \in I^*}, A')$ and the system $\{(U_I, \psi_I)\}_{I \in \mathcal{I}}$ defines a fiber bundle structure on π. □

Remark 3.5. 1. The inverse image of $(e_1, \dots, e_{r-1})$ by π consists of matrices of the form $(e_1, \dots, e_{r-1}, a)$, where $a = {}^t(a_1, \dots, a_{r-1}, a_r, \dots, a_N)$ with ${}^t(a_1, \dots, a_{r-1}) \in \mathbb{C}^{r-1}$ and $(a_r, \dots, a_N) \in \mathbb{C}^q \setminus \{0\}$.

2. If we use (3.14), we see that the fiber of π is given by $G_{N,N-r+1}/G_{N,N-r}$, which is biholomorphic with $\mathbb{C}^{r-1} \times (GL(q,\mathbb{C})/G_{q,q-1}) \simeq \mathbb{C}^{r-1} \times W(q,1) = \mathbb{C}^{r-1} \times (\mathbb{C}^q \setminus \{0\})$.

Consider the composition of inclusions

$$\iota : \mathbb{S}^{2q-1} \hookrightarrow \mathbb{C}^q \setminus \{0\} \hookrightarrow W(N,r),$$

where the last one is the inclusion of $\mathbb{C}^q \setminus \{0\}$ as a fiber of the bundle $\pi : W(N,r) \to W(N,r-1)$.

Proposition 3.8. *The inclusion* ι *induces an isomorphism*

$$\iota_* : \pi_i(\mathbb{S}^{2q-1}) \xrightarrow{\sim} \pi_i(W(N,r)) \qquad \textit{for } \; i = 0, \dots, 2q-1.$$

Proof. This can be seen by considering the sequence of maps

$$W(N,r) \longrightarrow W(N,r-1) \longrightarrow \cdots \longrightarrow W(N,1) = \mathbb{C}^N \setminus \{0\}$$

and applying Proposition 3.7 and Theorem 3.6. □

Exercise 3.15. Give a detailed proof of Proposition 3.8.

From Proposition 3.8 we have:

$$\pi_i(W(N,r)) \simeq \begin{cases} 0 & i = 0, \ldots, 2q-2, \\ \mathbb{Z} & i = 2q-1. \end{cases} \tag{3.16}$$

Note that $\pi_{2q-1}(W(N,r))$ has a canonical generator, i.e., the homotopy class of ι (cf. Example 3.5).

These can be seen also by retracting everything from the beginning to $W_0(N,r)$, as $W_0(N,r) \to W_0(N,r-1)$ is a fiber bundle with fiber $\mathbb{S}^{2q-1}$.

By (3.16) and Theorem 3.7 we have:

$$H_i(W(N,r);\mathbb{Z}) \simeq \begin{cases} 0 & i = 1, \ldots, 2q-2, \\ \mathbb{Z} & i = 0,\ 2q-1. \end{cases} \tag{3.17}$$

Note that $H_{2q-1}(W(N,r);\mathbb{Z})$ has a canonical generator, i.e., $\iota_* \nu_{2q-1}$, where ν_{2q-1} is the canonical generator of $H_{2q-1}(\mathbb{S}^{2q-1};\mathbb{Z})$ (cf. Remark B.13. 2).

Real Stiefel manifold

We may also consider the real Stielfel manifold $V(N',r)$ of r-frames in $\mathbb{R}^{N'}$. As in the complex case, $V(N',r)$ has naturally the structure of a C^∞ manifold of dimension $N'r$. The real Lie group $GL(N',\mathbb{R})$ acts transitively on $V(N',r)$ and we have:

$$V(N',r) \simeq GL(N',\mathbb{R})/G_{N',N'-r},$$

where $G_{N',N'-r}$ is the subgroup of $GL(N',\mathbb{R})$ consisting of matrices of the form

$$\begin{pmatrix} I_r & * \\ 0 & P' \end{pmatrix}, \qquad P' \in GL(N'-r,\mathbb{R}).$$

Let $V_0(N',r)$ denote the subset of $V(N',r)$ consisting of orthonormal r-frames in $\mathbb{R}^{N'}$ with respect to the standard Euclidian metric. The orthogonal group $O(N')$ acts on $V_0(N',r)$ transitively from left so that $V_0(N',r)$ is a homogeneous space:

$$V_0(N',r) \simeq O(N')/O(N'-r).$$

If $r < N'$, $SO(N')$ also acts on $V_0(N', r)$ transitively so that

$$V_0(N', r) \simeq SO(N')/SO(N' - r).$$

We may think of the two representation of $V_0(N', r)$ as describing its C^∞ manifold structure.

Note that the Gram-Schmidt process gives a deformation retraction $V(N', r) \to V_0(N', r)$.

In particular, $V(N', 1) = \mathbb{R}^{N'} \setminus \{0\}$ and $V_0(N', 1) = \mathbb{S}^{N'-1}$ so that

$$\mathbb{S}^{N'-1} \simeq O(N')/O(N' - 1).$$

Moreover if $N' > 1$, $\mathbb{S}^{N'-1} \simeq SO(N')/SO(N' - 1)$.

Suppose $r < N'$ and set $q' = N' - r + 1$. By arguments somewhat more complicated than the complex case, we have $\pi_i(V(N', r)) = 0$ for $i \leq q' - 2$ and

$$\pi_{q'-1}(V(N', r)) = \begin{cases} \mathbb{Z} & \text{if } q' - 1 \text{ is even or } r = 1, \\ \mathbb{Z}_2 & \text{if } q' - 1 \text{ is odd and } r > 1. \end{cases} \tag{3.18}$$

Note that it has a canonical generator.

3.6 Grassmann manifold

An *r-plane* in $\mathbb{C}^N$ is an r-dimensional linear subspace of $\mathbb{C}^N$.

Definition 3.17. The *Grassmann manifold* of r-planes in $\mathbb{C}^N$, denoted by $G(N, r)$, is the set of r-planes in $\mathbb{C}^N$.

We will see that $G(N, r)$ has a natural complex structure of dimension $(N - r)r$.

An r-plane in $\mathbb{C}^N$ is determined by an $N \times r$ matrix of rank r, whose column vectors forming its basis, i.e., by an element of the Stiefel manifold $W(N, r)$. Two matrices A and A' in $W(N, r)$ define the same space if and only if there exists P in $GL(r, \mathbb{C})$ such that $A' = AP$. Thus $G(N, r)$ is the orbit space of the right action of $GL(r, \mathbb{C})$ on $W(N, r)$:

$$G(N, r) = W(N, r)/GL(r, \mathbb{C}).$$

The orbit of A is denoted by $[A]$. Let $\mathcal{I}$ be the set of r-tuples of integers given by

$$\mathcal{I} = \{ (i_1, \ldots, i_r) \mid 1 \leq i_1 < \cdots < i_r \leq N \}.$$

For an element $I = (i_1, \ldots, i_r)$ of $\mathcal{I}$, we denote by A_I the $r \times r$ matrix consisting of the $i_1, \ldots, i_r$-th rows of A. Note that the property that $\det A_I \neq 0$ does not depend on the choice of the representative of $[A]$. The space $G(N, r)$ is covered by $\binom{N}{r}$ open sets U_I given by

$$U_I = \{ [A] \mid \det A_I \neq 0 \}.$$

Each class $[A]$ has a unique representative A° with $A^\circ_I = I_r$, the identity matrix. Denoting by I^* the complement of I in $(1, \ldots, N)$, the entries of $A^\circ_{I^*}$ gives a chart

$$\varphi_I : U_I \xrightarrow{\sim} M(N - r, r) = \mathbb{C}^{(N-r)r}.$$

More precisely, for the class of A, $A^\circ = A(A_I)^{-1}$ so that

$$\varphi_I([A]) = (A(A_I)^{-1})_{I^*} = A_{I^*}(A_I)^{-1}.$$

To see that $\varphi_I \circ \varphi_{I'}^{-1}$ is holomorphic wherever it is defined, take an arbitrary matrix $\tilde{A}$ in $M(N - r, r)$ and let A be the $N \times r$ matrix with $A_{I'} = I_r$ and $A_{(I')^*} = \tilde{A}$. If $[A] \in U_I$, then $\varphi_I \circ \varphi_{I'}^{-1}(\tilde{A}) = A_{I^*}(A_I)^{-1}$, which is holomorphic in $\tilde{A}$. Thus these charts define a complex structure on $G(N, r)$ of dimension $(N - r)r$.

In particular if $r = 1$, we have $A = {}^t(a_1, \ldots, a_N)$ and the correspondence $[A] \mapsto [\zeta_0, \ldots, \zeta_{N-1}] = [a_1, \ldots, a_N]$ gives a biholomorphic map of $G(N, 1)$ onto $\mathbb{P}^{N-1}$.

The Grassmann manifolds can be viewed also in the following manner. Let $e_i = {}^t(0, \ldots, 0, \overset{i}{1}, 0 \ldots, 0)$, $i = 1, \ldots, N$, as in Section 3.5. The complex Lie group $GL(N, \mathbb{C})$ acts on $G(N, r)$ holomorphically and transitively from the left and the stabilizer at the space $[e_1, \ldots, e_r]$ spanned by $(e_1, \ldots, e_r)$ is $GL(r, N - r; \mathbb{C})$, which is the set of matrices of the form

$$\begin{pmatrix} P & * \\ 0 & Q \end{pmatrix}, \qquad P \in GL(r, \mathbb{C}), \quad Q \in GL(N - r, \mathbb{C})$$

and is a closed complex Lie subgroup of $GL(N, \mathbb{C})$. Thus we have a biholomorphic map

$$GL(N, \mathbb{C})/GL(r, N - r; \mathbb{C}) \xrightarrow{\sim} G(N, r)$$

(cf. Theorem 3.3). We could alternatively define the complex structure of $G(N, r)$ from this viewpoint.

By the Gram-Schmidt process, the unitary group $U(N)$ also acts transitively on $G(N, r)$ with stabilizer at $[e_1, \ldots, e_r]$ being $U(r) \times U(N - r)$. Thus we have a C^∞ diffeomorphism

$$G(N, r) \xrightarrow{\sim} U(N)/U(r) \times U(N - r),$$

which shows that $G(N, r)$ is connected and compact.

Schubert cells

Recall that every $N \times r$ matrix A of rank r can be made to the following "reduced form" $\hat{A}$ by elementary transformations on the columns, i.e., by multiplying a suitable matrix in $GL(r, \mathbb{C})$ from the right:

(*) For each j, $1 \le j \le r$, we look the j-th column of $\hat{A}$ from the bottom and let $\hat{a}_{i_j j}$ be the first non-zero entry. Then $i_1 < \cdots < i_r$ and $\hat{A}_I = I_r$, thus $\hat{a}_{i_j j} = 1$.

Note that $\hat{A}$ is uniquely determined by A. Also A and A' determine the same element in $G(N, r)$ if and only if $\hat{A} = \hat{A}'$. We say that $\hat{A}$ is of type I if $\hat{A}_I = I_r$. If we denote by $\boldsymbol{e}_I$ the set of $N \times r$ reduced matrices of type I, it is naturally identified with $\mathbb{C}^{d_I}$, $d_I = \sum_{j=1}^r (i_j - j)$. We have a bijection

$$\psi : G(N, r) \xrightarrow{\sim} \bigsqcup_{I \in \mathcal{I}} \boldsymbol{e}_I$$

by assigning $\hat{A}$ to $[A]$, which may be thought of as giving a cellular decomposition of $G(N, r)$ (cf. Section B.2). The $\boldsymbol{e}_I$'s are called *Schubert cells.* The largest cell is for $I = (N - r + 1, \ldots, N)$ and is of complex dimension $(N - r)r$ and the smallest cell is for $I = (1, \ldots, r)$ and is of complex dimension 0.

We may also describe the Schubert cells as follows. For $i = 1, \ldots, N$, let V_i denote the subspace of $\mathbb{C}^N$ spanned by $e_1, \ldots, e_i$. We set $V_0 = \{0\}$. For $I = (i_1, \ldots, i_r)$ in $\mathcal{I}$, consider the set

$$E_I = \{\, L \in G(N, r) \mid \dim L \cap V_i = j,\ i_j \le i < i_{j+1},\ 0 \le j \le r \,\},$$

where we use the convention that $i_0 = 0$ and $i_{r+1} = N + 1$. Then we have $E_I = \psi^{-1}(\boldsymbol{e}_I)$. For an $N \times r$ matrix A of rank r and $i = 1, \ldots, N$, we denote by $A^{(i)}$ the $(N - i) \times r$ matrix obtained from A by removing the first i rows. We set $A^{(0)} = A$. Then from the relation

$$\dim(L + V_i) = r + i - \dim L \cap V_i,$$

we see that each element in E_I is represented by a matrix A such that

$$\operatorname{rank} A^{(i)} = r - j, \quad i_j \le i < i_{j+1}, \quad 0 \le j \le r,$$

where we set $\operatorname{rank} A^{(N)} = 0$. Thus the closure of E_I is given by

$$\overline{E_I} = \{\, L \in G(N, r) \mid \dim L \cap V_{i_j} \ge j,\ 1 \le j \le r \,\}$$

and each element is represented by a matrix A such that

$$\operatorname{rank} A^{(i_j)} \le r - j, \quad 1 \le j \le r.$$

This shows that $\overline{E_I}$ is a subvariety of $G(N, r)$. These are called *Schubert varieties*, or *Schubert cycles*. If we define an relation in $\mathcal{I}$ by saying that $I \geq I'$ if $i_j \geq i'_j$ for $j = 1, \ldots, r$, we may write

$$\overline{E_I} = \bigcup_{I' \leq I} E_{I'}.$$

Since there are no cells in real odd dimensions, we have the following:

Theorem 3.8. *The homology $H_*(G(N, r); \mathbb{Z})$ is the free $\mathbb{Z}$-module with basis the classes of the Schubert cycles.*

Dual Grassmann manifold: We set $\mathsf{W} = \mathbb{C}^N$ and $\mathsf{W}^* = \mathrm{Hom}(\mathbb{C}^N, \mathbb{C})$. Then $\mathsf{W}^* \simeq \mathbb{C}^N$. We denote vectors in W^* by row vectors, which acts on column vectors by multiplication from the left. We may consider the Grassmann manifold $G(\mathsf{W}^*, N - r)$ of $(N - r)$-planes in W^*. An element in W^* is represented by an $(N - r) \times N$ matrix B of rank $N - r$, whose rows forming its basis. Such matrices B and B' represents the same element if and only if there is $Q \in GL(N - r, \mathbb{C})$ such that $B' = QB$. We have a natural biholomorphic map

$$\boldsymbol{D} : G(\mathsf{W}, r) \xrightarrow{\sim} G(\mathsf{W}^*, N - r), \tag{3.19}$$

which assigns to an r-plane L, the $(N - r)$-plane $(\mathsf{W}/L)^*$. In terms of matrices, $[A]$ is assigned to $[B]$ with $BA = 0$.

For B, we may define its reduced form $\hat{B}$ by interchanging rows and columns in (*): Every $(N - r) \times N$ matrix B of rank $N - r$ can be made to the following "reduced form" $\hat{B}$ by elementary transformations on the rows, i.e., by multiplying a suitable matrix in $GL(N - r, \mathbb{C})$ from the left:

(**) For each i, $1 \leq i \leq N - r$, we look the i-th row of $\hat{B}$ from the left and let $\hat{b}_{j_i i}$ be the first non-zero entry. Then $j_1 < \cdots < j_{N-r}$ and $\hat{B}_J = I_{N-r}$, thus $\hat{b}_{j_i i} = 1$.

For $J = (j_1, \ldots, j_{N-r})$ we denote by $\boldsymbol{f}_J$ the cell in $G(\mathsf{W}^*, N - r)$ corresponding to J. By linear algebra we see the following:

Proposition 3.9. *The cell $\boldsymbol{f}_J$ corresponds to $\boldsymbol{e}_I$ by $\boldsymbol{D}$ in* (3.19) *if and only if $J = I^*$.*

Example 3.7. For $G(N, 1) = \mathbb{P}^{N-1}$ we have the cellular decomposition

$$G(N, 1) = \boldsymbol{e}_{(N)} \sqcup \cdots \sqcup \boldsymbol{e}_{(1)},$$

where $\boldsymbol{e}_{(i)} \simeq \mathbb{C}^{i-1}$.

Example 3.8. For $G(4,2)$ we have the cellular decomposition

$$G(4,2) = e_{(3,4)} \sqcup e_{(2,4)} \sqcup e_{(1,4)} \sqcup e_{(2,3)} \sqcup e_{(1,3)} \sqcup e_{(1,2)},$$

where $e_{(3,4)} \simeq \mathbb{C}^4$, $e_{(2,4)} \simeq \mathbb{C}^3$, $e_{(1,4)} \simeq e_{(2,3)} \simeq \mathbb{C}^2$, $e_{(1,3)} \simeq \mathbb{C}$ and $e_{(1,2)} \simeq \{0\}$. We also have

$$\begin{aligned} \overline{e_{(2,4)}} &= G(4,2) \setminus e_{(3,4)} \\ \overline{e_{(1,4)}} &= e_{(1,4)} \sqcup e_{(1,3)} \sqcup e_{(1,2)} \simeq G(3,1) \simeq \mathbb{P}^2, \\ \overline{e_{(2,3)}} &= e_{(2,3)} \sqcup e_{(1,3)} \sqcup e_{(1,2)} \simeq G(3,2) \simeq \mathbb{P}^2, \end{aligned}$$

and

$$\overline{e_{(1,4)}} \cap \overline{e_{(2,3)}} = \overline{e_{(1,3)}} = e_{(1,3)} \sqcup e_{(1,2)} \simeq G(2,1) \simeq \mathbb{P}^1.$$

The cycle $\overline{e_{(2,4)}}$ is a subvariety of dimension 3 given as follows:

$$\begin{aligned} &\overline{e_{(2,4)}} \cap U_{(3,4)} = \emptyset, \\ &\overline{e_{(2,4)}} \cap U_{(i_1,i_2)} \simeq \mathbb{C}^3 \quad \text{for} \quad (i_1,i_2) = (1,3),(1,4),(2,3) \text{ and } (2,4). \end{aligned}$$

$\overline{e_{(2,4)}} \cap U_{(1,2)}$ is given by, in the canonical coordinate system

$${}^t\begin{pmatrix} 1 & 0 & a_{31} & a_{41} \\ 0 & 1 & a_{32} & a_{42} \end{pmatrix}, \quad \text{by} \quad \begin{vmatrix} a_{31} & a_{32} \\ a_{41} & a_{42} \end{vmatrix} = 0,$$

which has an isolated singularity at

$${}^t\begin{pmatrix} 1 & 0 & 0 & 0 \\ 0 & 1 & 0 & 0 \end{pmatrix}.$$

Exercise 3.16. Verify the statements in Examples 3.7 and 3.8.

Universal bundle

There are some natural complex vector bundles on the Grassmann manifold. First, there is the *tautological bundle* $\pi : S \to G(N,r)$, i.e., the bundle whose fiber over a point L in $G(N,r)$ is L itself as a vector space. It is of rank r and can be described as

$$S = \{\, (z,L) \in \mathbb{C}^N \times G(N,r) \mid z \in L \,\}$$

with π restriction of the projection to the first factor (see Section 9.3 below for more explicit description as a vector bundle in the case $r = 1$). Thus S is a subbundle of the trivial bundle $\mathbb{I}^N = \mathbb{C}^N \times G(N,r)$. The quotient $Q = \mathbb{I}^N / S$ is called the *universal bundle*.

As noted above, the Lie group $GL(r,\mathbb{C})$ acts on the Stiefel manifold $W(N,r)$ from right and the orbit space is the Grassmann manifold:

$$G(N,r) = W(N,r)/GL(r,\mathbb{C}) \simeq GL(N,\mathbb{C})/GL(r, N-r;\mathbb{C}).$$

In fact $W(N,r)$ is a principal $GL(r,\mathbb{C})$-bundle associated with the tautological bundle on $G(N,r)$.

The group $U(r)$ acts on $W_0(N,r)$ from the right and there is a natural injection $W_0(N,r)/U(r) \to G(N,r)$, which is in fact surjection by the Gram-Schmidt process. Thus we may express

$$G(N,r) = W_0(N,r)/U(r) = U(N)/U(r) \times U(N-r).$$

In particular, as $G(N,1) = \mathbb{P}^{N-1}$, we have

$$\mathbb{P}^{N-1} = \mathbb{S}^{2N-1}/\mathbb{S}^1. \tag{3.20}$$

Thus the canonical projection $\mathbb{S}^{2N-1} \to \mathbb{P}^{N-1}$ is a principal $\mathbb{S}^1$-bundle.

By (3.19), the tautological bundle S_r on $G(\mathsf{W},r)$ corresponds to the universal quotient bundle Q_r on $G(\mathsf{W}^*, N-r)$.

Proposition 3.10. *Let $E \to X$ be a complex vector bundle of rank l on a topological space X. If there exists N global sections $(s_1,\dots,s_N)$ of E that span the fiber over every point of X, then there is a map $f: X \to G(N,r)$, $r = N-l$, such that $E = f^*Q$.*

Proof. Let $\mathbb{I}^N = \mathbb{C}^N \times X$ be the product bundle and $\varphi : \mathbb{I}^N \to E$ the bundle morphism defined by $\varphi({}^t(a_1,\dots,a_N),x) = \sum_{i=1}^N a_i s_i(x)$. Then by assumption, it is surjective and $\operatorname{Ker}\varphi$ is a subbundle of $\mathbb{I}^N$ of rank $r = N-l$. Let $f : X \to G(N,r)$ be defined by $f(x) = \operatorname{Ker}\varphi_x$. Then we see that $E = f^*Q$. □

Note that if X admits a finite covering $\{U_\alpha\}$ such that E is trivial on each U_α, then E has global sections with the property as above.

We discuss more about the universal bundle in Section 5.6 below (cf. Theorem 5.4 and the subsequent paragraph).

Real Grassmann manifold

We may also consider the real Grassmann manifold $G(N',r)$ of r-planes in $\mathbb{R}^{N'}$. As in the complex case, $G(N',r)$ is the orbit space of the right action of $GL(r,\mathbb{R})$ on the real Stiefel manifold $V(N',r)$:

$$G(N',r) = V(N',r)/GL(r,\mathbb{R})$$

and we may define a C^ω structure on $G(N',r)$ of dimension $(N'-r)r$.

The Grassmann manifolds can be viewed also in the following manner. The real Lie group $GL(N',\mathbb{R})$ acts on $G(N',r)$ real analytically and transitively from the left and the stabilizer at the space $[e_1,\dots,e_r]$ spanned by $(e_1,\dots,e_r)$ is $GL(r,N-r;\mathbb{R})$. Thus we have a C^ω diffeomorphism

$$GL(N',\mathbb{R})/GL(r,N'-r;\mathbb{R}) \xrightarrow{\sim} G(N',r).$$

By the Gram-Schmidt process, the orthogonal group $O(N')$ also acts transitively on $G(N',r)$ with stabilizer at $[e_1,\dots,e_r]$ being $O(r)\times O(N'-r)$. Thus we have a C^ω diffeomorphism

$$G(N',r) \xrightarrow{\sim} O(N')/O(r)\times O(N'-r),$$

which shows that $G(N',r)$ is compact. Also $SO(N')$ acts transitively on $G(N',r)$ and we have a C^ω diffeomorphism

$$G(N',r) \xrightarrow{\sim} SO(N')/H,$$

where H is the subgroup of $SO(N')$ of matrices of the form

$$\begin{pmatrix} P & 0 \\ 0 & Q \end{pmatrix}, \quad P\in O(N'),\ Q\in O(N'-r),\ \det P\cdot\det Q = 1.$$

Thus $G(N',r)$ is connected.

3.7 Some topics on differentiable manifolds

Orientability

Orientations of a real vector space: Let V be a real vector space of dimension l'. Assume that $l'\geq 1$ for the moment. We introduce a relation $\sim$ in the set of ordered bases of V by saying that $(e_1,\dots,e_{l'})\sim(e'_1,\dots,e'_{l'})$, if there exists a matrix P in $GL^+(l',\mathbb{R})$ such that $(e'_1,\dots,e'_{l'}) = (e_1,\dots,e_{l'})P$. Then it is an equivalence relation and there are exactly two equivalence classes, each of which is called an *orientation* of V. An *oriented vector space* is a vector space together with a prescribed orientation. If V is an oriented vector space, we say a basis $(e_1,\dots,e_{l'})$ of V positive or negative according as it belongs or not to the orientation. We adopt the following:

Convention 3.1. We orient $\mathbb{R}^{l'}$ so that the standard basis $(e_1,\dots,e_{l'})$, $e_i = {}^t(0,\dots,0,\overset{i}{1},0\dots,0)$, is positive.

If $\mathsf{V}=0$, by convention, we think of it as having two orientations.

Orientability of C^∞ manifolds: Let M be a C^∞ manifold of dimension m. We say that M is orientable if it is possible to specify an orientation of the tangent space at each point of M so that the specified orientations are compatible with the C^∞ structure in the following sense. For the moment we assume $m > 0$ and let $\{(U_\alpha, \varphi_\alpha)\}$ be a C^∞ atlas representing the C^∞ structure of M. For a point x in U_α, we write $\varphi_\alpha(x) = (x_1^\alpha, \dots, x_m^\alpha)$. Then $(\frac{\partial}{\partial x_1^\alpha}, \dots, \frac{\partial}{\partial x_m^\alpha})$ form a basis of $T_{\mathbb{R},x}M$. If $x \in U_\alpha \cap U_\beta$, $(\frac{\partial}{\partial x_1^\beta}, \dots, \frac{\partial}{\partial x_m^\beta})$ also form a basis of $T_{\mathbb{R},x}M$ and the matrix of base change is given by $t^{\alpha\beta}(x) = \partial(x_1^\alpha, \dots, x_m^\alpha)/\partial(x_1^\beta, \dots, x_m^\beta)(x)$ (cf. (3.10)).

Definition 3.18. A C^∞ manifold M is *orientable* if M admits a C^∞ atlas $\mathcal{A} = \{(U_\alpha, \varphi_\alpha)\}$ with the following property:

(*) for every pair (α, β) and x in $U_\alpha \cap U_\beta$, $t^{\alpha\beta}(x)$ is in $GL^+(m, \mathbb{R})$.

For two atlases $\mathcal{A}$ and $\mathcal{A}'$ with the property (*) above, we write $\mathcal{A} \sim \mathcal{A}'$, if their union has the property (*). Then it is an equivalence relation in the set of such atlases. We call an equivalence class an *orientation* of M. Note that if M is connected and is orientable, there are exactly two orientations of M. We say that M is *oriented*, if it is orientable and an orientation is specified.

Suppose M is oriented. A coordinate system on a connected open set is said to be positive or negative according as it belongs or not to an atlas representing the orientation. Also, an m-form ω is said to be positive or negative according as, when we write $\omega = f(x)\, dx_1 \wedge \cdots \wedge dx_m$ with positive coordinate system $(x_1, \dots, x_m)$, $f(x)$ is a (real) positive or negative valued function. For each point x in M, let $(x_1, \dots, x_m)$ be a positive coordinate system near x. We orient $T_{\mathbb{R},x}M$ so that $(\frac{\partial}{\partial x_1}, \dots, \frac{\partial}{\partial x_m})$ is a positive basis. Then the orientation does not depend on the choice of the positive coordinate system.

In the case $m = 0$, we think of M as being orientable with each connected component, which is a point, having two orientations.

A diffeomorphism $f : M \to M$ of an oriented manifold M is said to be *orientation preserving* if, for each x in M, $f_* : T_{\mathbb{R},x}M \to T_{\mathbb{R},f(x)}M$ transforms a positive basis to a positive basis.

Note that $\mathbb{R}^m$ is orientable as a C^∞ manifold. In the following, we always orient $\mathbb{R}^m$ so that the canonical coordinate system $(x_1, \dots, x_m)$ is positive. This is consistent with the orientation of $\mathbb{R}^m$ as a vector space (cf. Convention 3.1).

If M is a complex manifold, it is always orientable (cf. (1.3)). We orient M so that, if $(z_1, \ldots, z_n)$ is a complex coordinate system on M, $(x_1, y_1, \ldots, x_n, y_n)$ is a positive coordinate system, where $z_i = x_i + \sqrt{-1}\, y_i$, $i = 1, \ldots, n$.

C^∞ *maps of subsets*

We give the following:

Definition 3.19. Let M and M' be C^∞ manifolds and A a subset of M. A map $f : A \to M'$ is C^∞ if, for every point a of A, there is a neighborhood U of a in M such that $f|_{A \cap U}$ may be extended to a C^∞ map $U \to M'$.

The following is proved by the partition of unity argument:

Proposition 3.11. *Let A be a subset of a C^∞ manifold M. If $f : A \to \mathbb{R}^k$ is C^∞, f may be extended to a C^∞ map in a neighborhood of A.*

From the continuity of partial derivatives we have:

Proposition 3.12. *Let A be a subset of $\mathbb{R}^m$ and $f : A \to \mathbb{R}^k$ a C^∞ map. Suppose there exists an open set D in $\mathbb{R}^m$ such that $D \subset A \subset \overline{D}$. Then the values at each point of A of the partial derivatives of the components of an extension of f do not depend on the chosen extension.*

Manifolds with boundary

We set

$$\mathbb{H}^m = \{\, (x_1, \ldots, x_m) \in \mathbb{R}^m \mid x_1 \leq 0 \,\}.$$

Note that C^∞ maps of open sets in $\mathbb{H}^m$ are defined as in Definition 3.19.

Let R be a Hausdorff topological space with a countable basis.

Definition 3.20. A C^∞ *manifold with boundary* is a space R as above together with a C^∞ structure represented by an C^∞ atlas $\{(U_\alpha, \varphi_\alpha)\}$ on R, as in the case of C^∞ manifolds, except each φ_α is a homeomorphism onto an open set in $\mathbb{R}^m$ or in $\mathbb{H}^m$.

The integer m as above is called the dimension of R. The union of the inverse images of $\{x_1 = 0\}$ in $\mathbb{H}^m$ by the φ_α's is called the boundary of R and is denoted by ∂R. The interior $\operatorname{Int} R$ of R is the set $R \smallsetminus \partial R$, which is a usual C^∞ manifold of dimension m. We do not exclude the case $\partial R = \emptyset$.

Sometimes we use the terminology "a manifold possibly with boundary" to emphasize this.

For R as above, we readily see that there is a C^∞ manifold M of dimension m containing R with the following property: for every point p of ∂R there is a coordinate neighborhood U in M with coordinates $(x_1, \dots, x_m)$ such that $R \cap U = \{ q \in U \mid x_1(q) \leq x_1(p) \}$. If $\partial R \neq \emptyset$, it is an $(m-1)$-dimensional C^∞ submanifold of M. In fact if $(x_1, \dots, x_m)$ is a coordinate system as above, then $(x_2, \dots, x_m)$ is a coordinate system on $\partial R \cap U$. In this case we say that R is a *manifold with boundary in M*. Note that such a manifold M is uniquely determined in a neighborhood of R. We say that R is *orientable* if we may choose M as above so that it is orientable. In this case ∂R is also orientable. We adopt the following:

Convention 3.2. If M is oriented so that a coordinate system $(x_1, \dots, x_m)$ as above is positive, we orient ∂R so that $(x_2, \dots, x_m)$ is positive.

Example 3.9. The closed unit m-ball

$$\mathbb{B}^m = \{ x \in \mathbb{R}^m \mid \|x\|^2 = |x_1|^2 + \cdots + |x_m|^2 \leq 1 \}$$

is an m-dimensioanl manifold with boundary. It inherits the orientation of $\mathbb{R}^m$. The unit $(m-1)$-sphere $\mathbb{S}^{m-1} = \{ x \in \mathbb{R}^m \mid \|x\|^2 = 1 \}$ is its boundary $\partial \mathbb{B}^m$. It is an $(m-1)$-dimensional manifold and is oriented according to Convention 3.2. This is consistent with the orientation of $\mathbb{B}^m$ as a closed cell (cf. Remark B.13. 2).

Remark 3.6. For each $i = 1, \dots, m$, we may choose the coordinate system $(x_1, \dots, x_m)$ on M so that $R \cap U = \{ q \in U \mid x_i(q) \leq x_i(p) \}$. In this case ∂R is given by $x_i = 0$ and $(x_1, \dots, x_{i-1}, x_{i+1}, \dots, x_m)$ is a coordinate system on ∂R. Moreover, if M is oriented and if $(x_1, \dots, x_m)$ is positive, ∂R is oriented so that $(x_1, \dots, x_{i-1}, x_{i+1}, \dots, x_m)$ is positive or negative according as i is odd or even.

Let R be a C^∞ manifold with boundary in M and let M' be another C^∞ manifold. A C^∞ map $f : R \to M'$ is defined as in Definition 3.19. Thus it induces a C^∞ map $\partial R \to M'$, which will be denoted by ∂f.

We quote the following:

Theorem 3.9 (Collar neighborhood theorem). *Let R be a C^∞ manifold with boundary. Then ∂R has a neighborhood which is diffeomorphic with $\partial R \times [0, 1)$.*

One of the consequences of this is that the inclusion of $\operatorname{Int} R$ into R is a homotopy equivalence.

Let M be a C^∞ manifold of dimension m. For non-negative integers d' and k' with $d'+k'=m$, we decompose as $\mathbb{R}^m = \mathbb{R}^{k'} \times \mathbb{R}^{d'}$ with $(x_1, \ldots, x_{k'})$ and $(x_{k'+1}, \ldots, x_m)$ as coordinates on $\mathbb{R}^{k'}$ and $\mathbb{R}^{d'}$, respectively. We identify $\mathbb{R}^{d'}$ with $\{0\} \times \mathbb{R}^{d'} \subset \mathbb{R}^m$. We also set

$$\mathbb{H}^{d'} = \{ (x_{k'+1}, \ldots, x_m) \in \mathbb{R}^{d'} \mid x_{k'+1} \leq 0 \}.$$

Definition 3.21. A subset R of M is a C^∞ *submanifold of M with boundary*, if there is an atlas $\{(U_\alpha, \varphi_\alpha)\}$ representing the C^∞ structure of M such that for each α, $\varphi_\alpha(R \cap U_\alpha)$ is an open set in $\mathbb{R}^{d'}$ or in $\mathbb{H}^{d'}$.

If R is as above, it is a d'-dimensional C^∞ manifold with boundary ∂R, which is the union of the inverse images of $\{x_{k'+1} = 0\}$ in $\mathbb{H}^{d'}$ by the φ_α's. For such an R, there exists a d'-dimensional C^∞ submanifold (without boundary) V of M containing R with the following property: for every point p of ∂R there is a coordinate neighborhood U in M with coordinates $(x_1, \ldots, x_m)$ such that $V \cap U = \{q \in U \mid x_1(q) = \cdots = x_{k'}(q) = 0\}$ and that $R \cap U = \{ q \in V \cap U \mid x_{k'+1}(q) \leq x_{k'+1}(p) \}$. If $\partial R \neq \emptyset$, it is a $(d'-1)$-dimensional C^∞ submanifold of V. In fact if $(x_1, \ldots, x_m)$ is a coordinate system as above, then $(x_{k'+2}, \ldots, x_m)$ is a coordinate system on $\partial R \cap U$. In this case we say that R is a C^∞ manifold with boundary in V, or in M. The manifold R is orientable if we may choose V as above so that it is orientable. In this case ∂R is also orientable. If R is oriented, ∂R is oriented according to Convention 3.2.

Note that ∂R is not equal to the boundary of R in M, in generel.

Remark 3.7. In some literature, a "closed manifold" means a compact manifold without boundary. Here if we say that R is a closed submanifold (with boundary) in M, it means that it is a closed subset of M.

Orientability of fiber bundles

Let (G, F) be an effective left action of a Lie group G on a C^∞ manifold F possibly with boundary (cf. Definition 3.20). If F has a boundary, we assume that each element of G preserves the boundary. Let $\pi : T \to M$ be a C^∞ fiber bundle on a C^∞ manifold M with fiber F and group G. Note that, if the dimensions of M and F are m and l', respectively, then T is a C^∞ manifold of dimension $m + l'$. If F has a boundary ∂F, but not M, T is a manifold with boundary ∂T which has the structure of a fiber bundle on M with fiber ∂F and group G. We set $\partial\pi = \pi|_{\partial T}$. We now suppose that F is oriented (cf. Definition 3.18).

Definition 3.22. The bundle $\pi : T \to M$ is *orientable*, if the structure group may be reduced to a sub-Lie group G' of G such that each element in G' preserves the orientation of F.

We call a system of trivialization $\mathcal{T}$ of π *admissible*, if it gives a reduction as above. For two admissible systems $\mathcal{T}$ and $\mathcal{T}'$, we write $\mathcal{T} \sim \mathcal{T}'$ if their union is also admissible. Then it is an equivalence relation in the set of admissible system of trivializations. An *orientation* of the bundle $\pi : T \to M$ is defined to be an equivalence class of admissible systems. We say that T is *oriented*, if it is orientable and an orientation is specified.

Suppose the bundle T is oriented. We say a trivialization (U, ψ) of T, with U a connected open set in M, positive or negative according as it belongs or not to an admissible system representing the orientation. In this case, each fiber is oriented so that a positive trivialization induces an orientation preserving diffeomorphism of the fiber onto F.

Note that a trivialization (U, ψ) of T induces an isomorphism

$$T_{\mathbb{R},t}T \underset{\psi_*}{\xrightarrow{\sim}} T_{\mathbb{R},(y,x)}(F \times U) \simeq T_{\mathbb{R},y}F \oplus T_{\mathbb{R},x}M, \qquad \psi(t) = (y, x). \quad (3.21)$$

If $\pi : T \to M$ is an oriented bundle and if M is oriented, then the total space T is orientable.

Convention 3.3. We orient the total space T so that the orientation of the fiber followed by that of M gives the orientation of T, i.e., so that the isomorphism (3.21) is orientation preserving for a positive trivialization ψ.

Exercise 3.17. Show that, by the above convention, the orientation of ∂T as the boundary of T coincides with the one as the total space of the fiber bundle $\partial\pi : \partial T \to M$.

Let $\pi : E \to M$ be a real C^∞ vector bundle of rank l'. We may think of the fiber V as being oriented once we fix an isomorphism $\mathsf{V} \simeq \mathbb{R}^{l'}$. Thus we may talk about the orientability of the vector bundle E (cf. Definition 3.22). In particular, if the fiber of E is $\mathbb{R}^{l'}$, E is orientable if and only if the structure group may be reduced to $GL^+(l', \mathbb{R})$ (cf. Remark 9.1.2 below). Since the morphism ρ in (3.9) below maps $GL(l, \mathbb{C})$ into $GL^+(2l, \mathbb{R})$, a complex vector bundle is always orientable.

Note that the manifold M is orientable if and only if its tangent bundle $T_{\mathbb{R}}M$ is orientable as a bundle (cf. Definitions 3.18 and 3.22). In this case, we take the orientation of the bundle $T_{\mathbb{R}}M$ so that, if $(x_1, \ldots, x_m)$ is a positive coordinate system on M, the frame $(\frac{\partial}{\partial x_1}, \ldots, \frac{\partial}{\partial x_m})$ determines a positive trivialization of $T_{\mathbb{R}}M$.

Let V be a submanifold of codimension k' of M. Letting $N_{\mathbb{R},V}$ be the normal bundle of V in M, we have the exact sequence (3.12).

Proposition 3.13. *If M and V are orientable, the bundle $N_{\mathbb{R},V}$ is orientable.*

Proof. There is a C^∞ atlas $\{(U_\alpha, (x_1^\alpha, \ldots, x_m^\alpha))\}$ on M such that $V \cap U_\alpha$ is given by $x_1^\alpha = \cdots = x_{k'}^\alpha = 0$. We specify orientations of M and V. Changing the signs of some variables, if necessary, we may assume that $(x_1^\alpha, \ldots, x_m^\alpha)$ is a positive coordinate system on M for every α and that the restriction of $(x_{k'+1}^\alpha, \ldots, x_m^\alpha)$ to V is a positive coordinate system on V for every α with $V \cap U_\alpha \neq \emptyset$. Thus $N_{\mathbb{R},V}$ is orientable (cf. (3.11)). □

Convention 3.4. If M and V are oriented and if $(x_1, \ldots, x_m)$ and $(x_{k'+1}, \ldots, x_m)$ are positive coordinate systems on M and V, we orient the bundle $N_{\mathbb{R},V}$ so that the frame $(\varpi(\frac{\partial}{\partial x_1}), \ldots, \varpi(\frac{\partial}{\partial x_{k'}}))$ determines a positive trivialization.

The total space $N_{\mathbb{R},V}$ is then oriented according to Convention 3.3.

Tubular neighborhoods

Let M be a C^∞ manifold and V a submanifold of M with the normal bundle $p : N_{\mathbb{R},V} \to V$.

We quote the following:

Theorem 3.10 (Tubular neighborhood theorem). *There exist a neighborhood U of V in M, a neighborhood W of the image Z of the zero section in $N_{\mathbb{R},V}$ and a diffeomorphism τ of U onto W such that $\tau(V) = Z$ and that $(p \circ \tau)|_V = 1_V$.*

If we take an open ball bundle (cf. Remark 9.1.1 below) as W, then $r = p \circ \tau : U \to V$ is a C^∞ deformation retraction. Such a neighborhood U is referred to as a *tubular neighborhood* of V in M.

Remark 3.8. 1. If M and V are oriented, we orient $N_{\mathbb{R},V}$ as described after Proposition 3.13. Then may we take τ so that it is orientation preserving.

2. In the case V is a complex submanifold of a complex manifold M, we can say more about τ (cf. Proposition 15.8 below).

Transversality

Let $f : M \to M'$ be a C^∞ map of C^∞ manifolds and V' a submanifold of codimension k' of M'. We look for a condition for $f^{-1}V'$ to be a submanifold of M. Noting that the problem is local, let x be a point in $f^{-1}V'$. There is a neighborhood U' of $f(x)$ in M' and a C^∞ submersion $g : U' \to \mathbb{R}^{k'}$ such that $V' \cap U' = g^{-1}(0)$. We then have $f^{-1}V' \cap f^{-1}U' = (g \circ f)^{-1}(0)$. Thus $f^{-1}V'$ is a submanifold in a neighborhood of x if and only if $(g \circ f)_{*,x}$ is surjective. Since $(g \circ f)_{*,x} = g_{*,f(x)} \circ f_{*,x}$ (cf. (2.4)) and $g_{*,f(x)}$ is surjective and $\operatorname{Ker} g_{*,f(x)} = T_{\mathbb{R},f(x)}V'$, $(g \circ f)_{*,x}$ is surjective if and only if $\operatorname{Im} f_{*,x}$ and $T_{\mathbb{R},f(x)}V'$ span $T_{\mathbb{R},f(x)}M'$. Thus we introduce the following:

Definition 3.23. We say that f is *transverse* to V' if

$$\operatorname{Im} f_{*,x} + T_{\mathbb{R},f(x)}V' = T_{\mathbb{R},f(x)}M' \qquad \text{for all } \ x \in f^{-1}V'.$$

In this case, we write $f \pitchfork V'$.

As we observed, if $f \pitchfork V'$, then $f^{-1}V'$ is a submanifold of M and, if $f^{-1}V' \neq \emptyset$,

$$\operatorname{codim} f^{-1}V' = \operatorname{codim} V'.$$

In particular, if M is a submanifold of M' and if $i : M \hookrightarrow M'$ is the inclusion, $i^{-1}V' = M \cap V'$ and $i \pitchfork V'$ if and only if

$$T_{\mathbb{R},y}M + T_{\mathbb{R},y}V' = T_{\mathbb{R},y}M' \qquad \text{for all } \ y \in M \cap V'.$$

In this case we also say M is transverse to V' and write $M \pitchfork V'$. If this is the case, then $M \cap V'$ is a submanifold of M' and, if $M \cap V' \neq \emptyset$,

$$\operatorname{codim}(M \cap V') = \operatorname{codim} M + \operatorname{codim} V'.$$

Exercise 3.18. Let M be a C^∞ manifold and V a submanifold of codimension k' of M. Show that, for every point x in V, there is a k'-dimensional submanifold of M which intersects V transversally at x.

Let R be a C^∞ manifold with boundary and $f : R \to M'$ a C^∞ map. Also let V' be a submanifold of M'. If both f and $\partial f : \partial R \to M'$ are transverse to V', then $f^{-1}V'$ is a manifold with boundary, $(\partial f)^{-1}(V')$ is a submanifold of ∂R and $\partial(f^{-1}V') = f^{-1}V' \cap \partial R = (\partial f)^{-1}(V')$ as sets (cf. Proposition 3.15 below). Here we quote the following:

Theorem 3.11 (Transversality homotopy theorem). *For every C^∞ map $f : R \to M'$ and a submanifold V' of M', there exists a C^∞ map $f' : R \to M'$ homotopic to f such that $f' \pitchfork V'$ and $\partial f' \pitchfork V'$.*

Orientation: Let $f : M \to M'$ and V' be as above. Let m and m' be dimensions of M and M' and k' the codimension of V'. Suppose $f \pitchfork V'$ and $f^{-1}V' \neq \emptyset$ so that $f^{-1}V'$ is a submanifold of codimension k' of M.

Proposition 3.14. *In the above situation, if M, M' and V' are orientable, $f^{-1}V'$ is also orientable.*

Proof. We specify orientations of M, M' and V' and let $\{(U'_\lambda, \varphi_\lambda)\}$ be a C^∞ atlas representing the orientation of M' such that, writing $\varphi_\lambda(y) = (y_1^\lambda, \ldots, y_{m'}^\lambda)$, $V' \cap V_\lambda$ is given by $y_1^\lambda = \cdots = y_{k'}^\lambda = 0$ and that the restriction of $(y_{k'+1}^\lambda, \ldots, y_{m'}^\lambda)$ to V' is a positive coordinate system on V'. Set $x_i^\lambda = y_i^\lambda \circ f$, $i = 1, \ldots, k'$. We may choose a coordinate system $(x_1^\lambda, \ldots, x_m^\lambda)$ so that it is positive. Thus $f^{-1}V'$ is orientable. □

Convention 3.5. We orient the manifold $f^{-1}V'$ so that the restriction of $(x_{k'+1}^\lambda, \ldots, x_m^\lambda)$ is a positive coordinate system.

Note that this convention is consistent with Convention 3.4. With this convention, we have

Proposition 3.15. *Let $f : R \to M'$ be a C^∞ map. If $f \pitchfork V'$ and $\partial f \pitchfork V'$,*

$$\partial(f^{-1}V') = (-1)^{k'}(\partial f)^{-1}V',$$

as oriented manifolds.

Proof. We take $(x_1, \ldots, x_m)$ as in the proof of Proposition 3.14 so that R is given by $x_{k'+1} \leq 0$. If we take the orientation of $f^{-1}V'$ as above, the restriction of $(x_{k'+2}, \ldots, x_m)$ is a positive coordinate system on $\partial(f^{-1}V')$ (cf. Convention 3.2).

On the other hand, $(x_1, \ldots, x_{k'}, x_{k'+2}, \ldots, x_m)$ is a coordinate system on ∂R which is positive or negative according as k' is even or odd (cf. Remark 3.6). Thus $(x_{k'+2}, \ldots, x_m)$ is a coordinate system on $(\partial f)^{-1}V'$ positive or negative according as k' is even or odd. □

In the above situation, let R' be an $(m' - k')$-dimensional C^∞ manifold with boundary in V'.

Proposition 3.16. *If $f \pitchfork R'$ and $f \pitchfork \partial R'$, then $f^{-1}R'$ is a manifold with boundary in M and*

$$\partial(f^{-1}R') = f^{-1}(\partial R'),$$

as oriented manifolds.

Proof. If V' is locally defined by $y_1 = \cdots = y_{k'} = 0$, $f^{-1}V'$ is defined by $y_1 \circ f = \cdots = y_{k'} \circ f = 0$. Let $(x_1, \ldots, x_m)$ be as before so that the restriction of $(x_{k'+1}, \ldots, x_m)$ is a positive coordinate system on $f^{-1}V'$. Let R' be defined by $y_{k'+1} \leq 0$. From the condition $f \pitchfork \partial R'$, we may set $x_{k'+1} = y_{k'+1} \circ f$ and $\partial(f^{-1}R')$ is given by $x_{k'+1} \leq 0$. Then $(x_{k'+2}, \ldots, x_m)$ is a positive coordinate system on $\partial(f^{-1}R')$ (cf. Convention 3.2). On the other hand, $f^{-1}(\partial R')$ is oriented so that it is a positive coordinate system (cf. Convention 3.5). □

Slices: Let M be a C^∞ manifold of dimension m and V a submanifold of codimension k'.

Definition 3.24. A *slice* of V in M at $x \in V$ is a k'-dimensional submanifold $\mathbb{D}$ of M containing x, transverse to V at x and diffemorphic with an open k'-ball.

Taking a coordinate system $(x_1, \ldots, x_m)$ on M around x so that V is given by $x_1 = \cdots = x_{k'} = 0$, we see that a slice always exists.

If M and V are oriented, we always orient a slice $\mathbb{D}$ so that its orientation followed by that of V gives the orientation of M.

Likewise, if M is a complex manifold of dimension n and V a complex submanifold of codimension k, we may define a *complex slice* to be a k-dimensional submanifold $\mathbb{D}$ of M, containing x, transverse to V at x and diffemorphic with an open $2k$-ball.

Ehresmann fibration theorem

Definition 3.25. Let $f : M \to M'$ be a C^∞ map.

1. We say that f is a C^∞ *trivial fibration* if it admits a trivial C^∞ fiber bundle structure (cf. Section 3.2).

2. The map f is a C^∞ *locally trivial fibration*, if each point of M' has a neighborhood U such that $f|_{f^{-1}(U)} : f^{-1}(U) \to U$ is a C^∞ trivial fibration.

Thus f being trivial fibration means that there exist a C^∞ manifold F and a diffeomorphism $\psi : M \xrightarrow{\sim} F \times M'$ such that $f = p \circ \psi$, where

$p : F \times M' \to M'$ is the projection. In this case, each fiber $f^{-1}(x)$, $x \in M'$, is diffeomorphic to F. We quote the following:

Theorem 3.12. *Let $f : M \to M'$ be a C^∞ map. If it is a proper surjective submersion, it is a C^∞ locally trivial fibration.*

Remark 3.9. 1. Suppose $f : M \to M'$ is a locally trivial fibration. If each fiber $f^{-1}(x)$, $x \in M'$, is diffeomorphic to a fixed manifold F, for example this is always the case if M' is connected, we may think of $f : M \to M'$ as a fiber bundle with fiber F and group the group of diffeomorphisms of F.

2. The above theorem holds if we replace M with a C^∞ manifold R with boundary, provided that f restricted to ∂R is also a submersion. In this case each fiber is a C^∞ manifold with boundary.

Notes

For the proofs of the theorems stated in Section 3.1, see [Kobayashi and Nomizu (1963); Matsushima (1972)].

We list [Steenrod (1951)] as a fundamental reference for fiber bundles. For detailed discussions of Theorem 3.5, see §7 in there. As to the homotopy exact sequence and the Hurewicz theorem, we refer to §17 and §15 in there, see also Ch. 7 of [Spanier (1966)]. For Remark 3.3, we refer to §11 in there, see also §4 of [Hirzebruch (1966)].

For Stiefel and Grassmann manifolds, we also refer to [Steenrod (1951)]. See [Fulton (1984); Ikeda (2018)] and the references therein for the so-called Schubert calculus in Grassmann manifolds.

We refer to [Guillemin and Pollack (1974); Lee (2013)] as to Theorems 3.9–3.11. Theorem 3.12 is due to [Ehresmann (1950)].

Chapter 4

Dualities and Thom Class

In the case of manifolds, the theory of combinatorial topology becomes particularly rich, as there is a deep relation between the (co)homology defined by a triangulation and that defined by its dual cellular decomposition. We investigate such relations and prove the Poincaré, Alexander and Lefschetz dualities. The Alexander dualty is a localized version of the Poincaré duality and particularly important to describe localizations in various settings. The Lefschetz duality is for manifolds with boundary and may be thought of as a special case of the Alexander duality.

We then prove the Thom isomorphism for a submanifold whose normal bundle is oriented and define the Thom class for such a submanifold. If the manifolds are oriented, the Thom isomorphism is directly related to the Poincaré and Alexander isomorphisms. As a special case, we define the Thom class of an oriented real vector bundle. This turns out to be the localized version of the Euler class we discuss in the next chapter.

We explicitly define the intersection product in homology in terms of combinatorial topology. This is an operation dual to the cup product in cohomology and will particularly be exploited in Chapter 14 below.

We refer to Appendix B for basic materials on algebraic topology. Throughout this chapter, we take $\mathbb{Z}$ as the coefficient ring of homology and cohomology, unless otherwise stated. Also, we let M denote a C^∞ manifold of dimension m.

4.1 Algebraic topology on manifolds

C^∞ triangulations

Let X be a topological space. A *triangulation* of X is a pair (K, h) of a simplicial complex K and a homeomorphism $h : |K| \to X$. We abbreviate this by saying that $h : |K| \to X$ is a triangulation, or K is a triangulation of X. Let Y be a subspace of X. We say that the triangulation is compatible with Y if there is a subcomplex L of K such that the restriction of h to $|L|$ is a triangulation of Y.

Definition 4.1. A triangulation $h : |K| \to M$ of a C^∞ manifold M is C^∞ if, for every simplex $\boldsymbol{s}$ of K, $h|_{\boldsymbol{s}}$ is C^∞ and its rank at each point of $\boldsymbol{s}$ is equal to $\dim \boldsymbol{s}$.

In the above, we think of $\boldsymbol{s}$ as being in the affine space spanned by $\boldsymbol{s}$ and $h|_{\boldsymbol{s}}$ being C^∞ means C^∞ in the sense of Definition 3.19. Its rank is well-defined by Proposition 3.12.

Note that the above definitions also make sense if we replace M with a C^∞ manifold with boundary.

The following is known:

(T_1) Every C^∞ manifold M admits a C^∞ triangulation. In fact, if R is a closed C^∞ submanifold of M possibly with boundary, there is a C^∞ triangulation of M compatible with R and ∂R.

(T_2) If K_1 and K_2 are C^∞ triangulations of M, there exist subdivisions of K_1 and K_2 that are simplicially isomorphic.

In this chapter we do not explicitly use the fact that the triangulations we consider are C^∞, although it is essential to have a differentiable structure on M for the above facts.

Homology via triangulation and dual cellular decomposition

We take a triangulation (K_0, h) of M and let K denote the barycentric subdivision of K_0. We further let K' be the barycentric subdivision of K, i.e., the second barycentric subdivision of K_0. We take the second barycentric subdivision so that the star of a K_0-subcomplex L of K_0 relative to K' has the same homotopy type as the polyhedron $|L|$ of L (cf. Proposition B.26). In the following, sometimes a simplex $\boldsymbol{s}$ of K is identified with $h(\boldsymbol{s})$ and $|K|$ is identified with M.

Dual cellular decomposition: For a p-simplex $\boldsymbol{s}$ of K, we denote by $\boldsymbol{s}^*$ the union of $(m-p)$-simplices of K' intersecting with $\boldsymbol{s}$ at its barycenter $b_{\boldsymbol{s}}$. It is a regular closed $(m-p)$-cell in $|K|$, called the cell dual to $\boldsymbol{s}$. The intersection of $\boldsymbol{s}$ and $\boldsymbol{s}^*$ consists of the one point $b_{\boldsymbol{s}}$. The cells dual to simplices in K form a cellular decomposition of $|K| = M$, which will be denoted by K^*.

Orientations of simplices and cells: In order to describe the homology and cohomology of M via triangulation or dual cellular decomposition, we fix orientations of simplices of K and cells of K^*. As to the orientations of simplices of K', we impose the following conditions. Thus let $\boldsymbol{t}$ be a p-simplex of K'.

(1) If $\boldsymbol{t} \subset \boldsymbol{s}$, a p-simplex of K, the orientation of $\boldsymbol{t}$ is the same as that of $\boldsymbol{s}$.
(2) If $\boldsymbol{t} \subset \boldsymbol{s}'^*$, a p-cell of K^*, the orientation of $\boldsymbol{t}$ is the same as that of $\boldsymbol{s}'^*$.

Note that for $\boldsymbol{t}$ not satisfying either of the above assumptions, there is still freedom of choice of the orientation.

Homology and cohomology of $\boldsymbol{M}$: We denote by $H_p(M)$ the p-th singular homology of M (cf. Section B.1). An important feature in the case of a manifold is that it can be computed using either the triangulation K or the cellular decomposition K^* in the following sense. Thus let $(C^K_\bullet(M), \partial)$ be the chain complex with $C^K_p(M)$ the free Abelian group generated by the oriented p-simplices in K and $\partial : C^K_p(M) \to C^K_{p-1}(M)$ the boundary operator defined by

$$\partial(v_0, \dots, v_p) = \sum_{i=0}^{p} (-1)^i (v_0, \dots, \widehat{v_i}, \dots, v_p),$$

for an oriented simplex $\boldsymbol{s} = (v_0, \dots, v_p)$ with vertices $v_0, \dots, v_p$, and extended linearly (cf. (B.29)). We denote by $H^K_p(M)$ the p-th homology of $C^K_\bullet(M)$. Denoting by $S_p(M)$ the group of singular p-chains of M, there is a natural chain morphism

$$\eta^K_\bullet : C^K_\bullet(M) \longrightarrow S_\bullet(M), \tag{4.1}$$

which is defined as follows. For an oriented p-simplex $\boldsymbol{s} = (v_0, \dots, v_p)$, let $\varphi_{\boldsymbol{s}} : \Delta^p \to \boldsymbol{s}$ be the affine map with $\varphi_{\boldsymbol{s}}(P_i) = v_i$, $i = 0, \dots, p$. Then η^K_p assigns to $\boldsymbol{s}$ the singular simplex $\sigma = h \circ \varphi_{\boldsymbol{s}} : \Delta^p \to \boldsymbol{s} \to M$. Note that $\eta^K_\bullet$ above is injective. It induces an isomorphism on the homology level (cf. (B.30)):

$$\eta^K_* : H^K_p(M) \xrightarrow{\sim} H_p(M).$$

Also if we denote by $(C_\bullet^{K^*}(M), \partial)$ the chain complex with $C_p^{K^*}(M)$ the free Abelian group generated by the oriented p-cells in K^*, we have a natural isomorphism:

$$\eta_*^{K^*} : H_p^{K^*}(M) \xrightarrow{\sim} H_p(M), \tag{4.2}$$

where $H_p^{K^*}(M)$ is the p-th homology of $C_\bullet^{K^*}(M)$ (cf. Theorem B.22). In our case we have injective chain morphisms

$$C_\bullet^{K^*}(M) \xrightarrow{\iota_\bullet} C_\bullet^{K'}(M) \xrightarrow{\eta_\bullet^{K'}} S_\bullet(M), \tag{4.3}$$

where $\iota_\bullet$ is the morphism that regards a K^*-chain as a K'-chain, and the composition induces the isomorphism (4.2).

Also the singular cohomology $H^p(M)$ of M can be computed either from the cochain complex $(C_K^\bullet(M), \delta)$ with $C_K^p(M) = \mathrm{Hom}(C_p^K(M), \mathbb{Z})$ or from the cochain complex $(C_{K^*}^\bullet(M), \delta)$ with $C_{K^*}^p(M) = \mathrm{Hom}(C_p^{K^*}(M), \mathbb{Z})$. That is to say that the transposes of the chain morphisms $\eta_\bullet^K$ and $\eta_\bullet^{K^*}$ induce isomorphisms

$$\eta_K^* : H^p(M) \xrightarrow{\sim} H_K^p(M) \quad \text{and} \quad \eta_{K^*}^* : H^p(M) \xrightarrow{\sim} H_{K^*}^p(M). \tag{4.4}$$

Denoting by $\check{C}_\bullet^K(M)$ and $\check{S}_\bullet(M)$ the chain complexes of locally finite chains of K and of locally finite singular chains of M, respectively, we have a chain morphism $\check{\eta}_\bullet^K : \check{C}_\bullet^K(M) \to \check{S}_\bullet(M)$ as in (4.1). It induces an isomorphism on the homology level (cf. (B.38)):

$$\check{\eta}_*^K : \check{H}_p^K(M) \xrightarrow{\sim} \check{H}_p(M). \tag{4.5}$$

Likewise, considering the complex $\check{C}_\bullet^{K^*}(M)$ of locally finite chains of K^*, we have a canonical isomorphism $\check{\eta}_*^{K^*} : \check{H}_p^{K^*}(M) \xrightarrow{\sim} \check{H}_p(M)$ (cf. (B.37)).

In the following, $\langle \ , \ \rangle$ denotes the paring of chains and cochains, i.e., the Kronecker product (cf. Section A.1).

4.2 Poincaré, Alexander and Lefschetz dualities

We prove the dualities in the case of manifolds, which will be generalized to the case of singular varieties in Section 13.2 below.

Let M, K_0, K, K' and K^* be as in Section 4.1. In this section we assume that M is oriented and take orientations of the simplices and cells so that they satisfy the conditions (1) and (2) in Section 4.1 and that they are furthermore compatible with that of M in the following sense:

(3) The orientation of each m-simplex is the same as that of M.
(4) For every p-simplex $\boldsymbol{s}$ of K, $0 < p < m$, the orientation of $\boldsymbol{s}^*$ followed by the orientation of $\boldsymbol{s}$ gives the orientation of M.

As a consequence we have:

(*) Let $\boldsymbol{s}$ be a p-simplex in K and let $\boldsymbol{t}$ and $\boldsymbol{t}'$ be an $(m-p)$-simplex and a p-simplex of K'. If $\boldsymbol{t} \subset \boldsymbol{s}^*$ and $\boldsymbol{t}' \subset \boldsymbol{s}$ and if $\boldsymbol{t}$ and $\boldsymbol{t}'$ span an m-simplex $\boldsymbol{t}_0$ of K', the orientation of $\boldsymbol{t}$ followed by that of $\boldsymbol{t}'$ gives the orientation of $\boldsymbol{t}_0$, which is the same as that of M.

Poincaré duality

We define a morphism

$$P : C^p_{K^*}(M) \longrightarrow \check{C}^K_{m-p}(M) \qquad \text{by} \;\; P(u) = \sum_{\boldsymbol{s}} \langle \boldsymbol{s}^*, u \rangle\, \boldsymbol{s} \tag{4.6}$$

for a p-cochain u of K^*, where the sum is taken over all $(m-p)$-simplices $\boldsymbol{s}$ of M. For a p-cell $\boldsymbol{s}^*$ of K^* we denote by $\vartheta(\boldsymbol{s}^*)$ the cochain in $C^p_{K^*}(M)$ dual to $\boldsymbol{s}^*$, i.e., for every p-cell $\boldsymbol{s}'^*$ of K^*,

$$\langle \boldsymbol{s}'^*, \vartheta(\boldsymbol{s}^*) \rangle = \begin{cases} 1 & \text{if } \;\boldsymbol{s}'^* = \boldsymbol{s}^*, \\ 0 & \text{otherwise.} \end{cases}$$

Then $C^p_{K^*}(M)$ is the free Abelian group generated by the $\vartheta(\boldsymbol{s}^*)$'s and the morphism P sends $\vartheta(\boldsymbol{s}^*)$ to $\boldsymbol{s}$. Thus P is in fact an isomorphism. We now prove that it is compatible with boundary and coboundary operators so that it induces an isomorphism between the corresponding cohomology and homology.

For an oriented p-simplex $\boldsymbol{t}$ of K', we denote by $\vartheta(\boldsymbol{t})$ the cochain in $C^p_{K'}(M)$ dual to $\boldsymbol{t}$, defined similarly as above. Let $M_{K'}$ denote the sum of all m-simplices of K', which is an m-cycle.

Ordering of vertices of K': Let $\mathcal{V}$ denote the set of vertices of K'. Recall that every element v of $\mathcal{V}$ is the barycenter of a simplex of K, which is uniquely determined by v. We denote it by $\boldsymbol{s}_v$ and introduce an order relation in $\mathcal{V}$ by saying that $v \leq v'$ if $\boldsymbol{s}_v \succ \boldsymbol{s}_{v'}$. Then it is a simplicial ordering (cf. Definition B.14). We may express each p-simplex $\boldsymbol{t}$ of K' as

$$\boldsymbol{t} = \varepsilon(v_0, \dots, v_p), \qquad v_0 < \cdots < v_p,$$

where $\varepsilon = \pm 1$ and the sign is to be determined according to the prescribed orientation of $\boldsymbol{t}$. The cup and cap products are then defined by (B.31) and (B.32).

Remark 4.1. Let $\boldsymbol{t} = \varepsilon(v_0, \ldots, v_p)$, $v_0 < \cdots < v_p$, be a p-simplex of K'. For a p-simplex $\boldsymbol{s}_1$ of K, $\boldsymbol{t} \subset \boldsymbol{s}_1$ if and only if $v_0 = b_{\boldsymbol{s}_1}$ and in this case v_i is the barycenter of a $(p-i)$-simlpex of K which is a face of $\boldsymbol{s}_1$. Also for an $(m-p)$-simplex $\boldsymbol{s}_2$ of K, $\boldsymbol{t} \subset \boldsymbol{s}_2^*$ if and only if $v_p = b_{\boldsymbol{s}_2}$ and in this case there is an m-simplex $\boldsymbol{s}$ of K such that $\boldsymbol{s}_2 \prec \boldsymbol{s}$ and that v_i is the barycenter of an $(m-i)$-simplex of K which is a face of $\boldsymbol{s}$.

Lemma 4.1 (Key lemma). *For every p-simplex $\boldsymbol{t}$ of K', we have:*

$$M_{K'} \frown \vartheta(\boldsymbol{t}) = \begin{cases} \boldsymbol{s} & \text{if } \boldsymbol{t} \subset \boldsymbol{s}^* \text{ for some } (m-p)\text{-simplex } \boldsymbol{s} \text{ of } K, \\ 0 & \text{otherwise,} \end{cases}$$

as K'-chains.

Proof. Let $\boldsymbol{t}_0$ be an arbitrary m-simplex of K'. Note that, for each p, there exists a unique $(m-p)$-simplex $\boldsymbol{s}$ of K such that $\boldsymbol{t}_0$ is spanned by a p-simplex $\boldsymbol{t}_0^{(1)}$ of K' in $\boldsymbol{s}^*$ and an $(m-p)$-simplex $\boldsymbol{t}_0^{(2)}$ of K' in $\boldsymbol{s}$. By our orientation convention (*), we arrange the vertices as $\boldsymbol{t}_0 = \varepsilon(v_0, \ldots, v_m)$, $v_0 < \cdots < v_m$, so that $\boldsymbol{t}_0^{(1)} = \varepsilon_1(v_0, \ldots, v_p)$, $\boldsymbol{t}_0^{(2)} = \varepsilon_2(v_p, \ldots, v_m)$ and $\varepsilon = \varepsilon_1\varepsilon_2$. Note that $v_p = b_{\boldsymbol{s}}$ (cf. Remark 4.1).

Let $\boldsymbol{t}$ be a p-simplex of K'. If it is not in the dual cell of any $(m-p)$-simplex of K, then $\boldsymbol{t}_0 \frown \vartheta(\boldsymbol{t}) = 0$ for every m-simplex $\boldsymbol{t}_0$ of K' so that $M_{K'} \frown \vartheta(\boldsymbol{t}) = 0$. Thus suppose $\boldsymbol{t} \subset \boldsymbol{s}^*$ for some $(m-p)$-simplex $\boldsymbol{s}$ of K. Then we have

$$M_{K'} \frown \vartheta(\boldsymbol{t}) = \sum_{\boldsymbol{t}_0^{(1)} = \boldsymbol{t}} \boldsymbol{t}_0 \frown \vartheta(\boldsymbol{t}) = \sum_{\boldsymbol{t}_0^{(1)} = \boldsymbol{t}} \boldsymbol{t}_0^{(2)} = \boldsymbol{s}.$$

□

Recall that, as each cell in K^* may be thought of as a K'-chain, there is a natural monomorphism $\iota : C_p^{K^*}(M) \to C_p^{K'}(M)$, whose transpose is denoted by ι^*. Likewise there is a natural monomorphism $\check{C}_{m-p}^{K}(M) \to \check{C}_{m-p}^{K'}(M)$, which is denoted by κ.

Corollary 4.1. *The following diagram is commutative:*

$$\begin{array}{ccc} C^p_{K^*}(M) & \xrightarrow{\ P\ } & \check{C}^K_{m-p}(M) \\ \uparrow{\scriptstyle \iota^*} & & \downarrow{\scriptstyle \kappa} \\ C^p_{K'}(M) & \xrightarrow{M_{K'}\frown} & \check{C}^{K'}_{m-p}(M). \end{array} \tag{4.7}$$

Lemma 4.2. *We have*

$$P\delta u = (-1)^{p+1}\partial P(u) \qquad \textit{for} \quad u \in C^p_{K^*}(M).$$

Proof. Noting that, in (4.7), ι^* is surjective and compatible with coboundary operators and that κ is injective and compatible with boundary operators, the lemma follows from the property of cap product (B.17). □

Combining with the isomorphisms $\eta^*_{K^*}$ in (4.4) and $\breve{\eta}^K_*$ in (4.5), we have proved:

Theorem 4.1 (Poincaré duality). *For an oriented C^∞ manifold M of dimension m, the isomorphism P of* (4.6) *induces an isomorphism*

$$P_M : H^p(M) \xrightarrow{\sim} \breve{H}_{m-p}(M).$$

We denote by $[M]$ the class in $\breve{H}_m(M)$ corresponding to $[M_{K'}]$ by the isomorphism $\breve{H}^{K'}_m(M) \simeq \breve{H}_m(M)$. Note that it does not depend on the choice of the triangulation by the property (T_2) of C^∞ triangulations (cf. Section 4.1). By Corollary 4.1, we may write

$$P_M(\alpha) = [M] \frown \alpha \qquad \text{for } \alpha \in H^p(M).$$

Remark 4.2. 1. If M is connected, the cycle $M_{K'}$ is called the *fundamental cycle* of M in K' and the class $[M]$ the *fundamental class* of M. In this case, there is a canonical isomorphism $\breve{H}_m(M) \simeq \mathbb{Z}$, the class $[M]$ corresponding to 1.

2. Let $\boldsymbol{s}$ be an $(m-p)$-simplex of K. We may write $\partial \boldsymbol{s} = \sum_{i=0}^{m-p} \varepsilon_i \boldsymbol{s}^{(i)}$, where the $\boldsymbol{s}^{(i)}$'s are the $(m-p-1)$-faces of $\boldsymbol{s}$ and $\varepsilon_i = \pm 1$, the sign being determined according to the prescribed orientations. Then, if we set $\vartheta((\partial \boldsymbol{s})^*) = \sum_{i=0}^{m-p} \varepsilon_i \vartheta({\boldsymbol{s}^{(i)}}^*)$, by Lemma 4.2, we have

$$\delta\vartheta(\boldsymbol{s}^*) = (-1)^{p+1}\vartheta((\partial \boldsymbol{s})^*).$$

Dual description: We define a morphism

$$P' : C^p_K(M) \longrightarrow \breve{C}^{K^*}_{m-p}(M) \qquad \text{by } P'(u) = \sum_{\boldsymbol{s}} \langle \boldsymbol{s}, u \rangle\, \boldsymbol{s}^* \tag{4.8}$$

for a p-cochain u, where the sum is taken over all p-simplices $\boldsymbol{s}$ of M. For a p-simplex $\boldsymbol{s}$ of K, we denote by $\vartheta(\boldsymbol{s})$ the cochain in $C^p_K(M)$ dual to $\boldsymbol{s}$. Then P' sends $\vartheta(\boldsymbol{s})$ to $\boldsymbol{s}^*$ and is an isomorphism. We now prove that it is compatible with boundary and coboundary operators so that it induces an isomorphism between the corresponding cohomology and homology.

We keep the orientation conventions and ordering of vertices of K'. The following is proved as Lemma 4.1, except the cap product becomes right. We give a proof for the sake of completeness.

Lemma 4.3. *For every p-simplex* $\boldsymbol{t}$ *of* K', *we have:*

$$\vartheta(\boldsymbol{t}) \frown M_{K'} = \begin{cases} \boldsymbol{s}^* & \text{if } \boldsymbol{t} \subset \boldsymbol{s} \text{ for some } p\text{-simplex } \boldsymbol{s} \text{ of } K, \\ 0 & \text{otherwise} \end{cases}$$

as K'*-chains.*

Proof. Let $\boldsymbol{t}_0$ be an arbitrary m-simplex of K'. Note that, for each p, there exists a unique p-simplex $\boldsymbol{s}$ of K such that $\boldsymbol{t}_0$ is spanned by an $(m-p)$-simplex $\boldsymbol{t}_0^{(1)}$ of K' in $\boldsymbol{s}^*$ and a p-simplex $\boldsymbol{t}_0^{(2)}$ of K' in $\boldsymbol{s}$. By our orientation convention (*), we arrange the vertices as $\boldsymbol{t}_0 = \varepsilon(v_0, \dots, v_m)$ so that $\boldsymbol{t}_0^{(1)} = \varepsilon_1(v_0, \dots, v_{m-p})$, $\boldsymbol{t}_0^{(2)} = \varepsilon_2(v_{m-p}, \dots, v_m)$ and $\varepsilon = \varepsilon_1\varepsilon_2$.

Let $\boldsymbol{t}$ be a p-simplex of K'. If it is not in any p-simplex of K, then $\vartheta(\boldsymbol{t}) \frown \boldsymbol{t}_0 = 0$ for every m-simplex $\boldsymbol{t}_0$ of K' so that $\vartheta(\boldsymbol{t}) \frown M_{K'} = 0$. Thus suppose $\boldsymbol{t} \subset \boldsymbol{s}$ for some p-simplex $\boldsymbol{s}$ of K. Then we have

$$\vartheta(\boldsymbol{t}) \frown M_{K'} = \sum_{\boldsymbol{t}_0^{(2)} = \boldsymbol{t}} \vartheta(\boldsymbol{t}) \frown \boldsymbol{t}_0 = \sum_{\boldsymbol{t}_0^{(2)} = \boldsymbol{t}} \boldsymbol{t}_0^{(1)} = \boldsymbol{s}^*.$$

□

Recall that, as each simplex in K may be thought of as a K'-chain, there is a natural monomorphism $\iota' : C_p^K(M) \to C_p^{K'}(M)$, whose transpose is denoted by ι'^*. Likewise there is a natural monomorphism $\check{C}_{m-p}^{K^*}(M) \to \check{C}_{m-p}^{K'}(M)$, which is denoted by κ'. From the above lemma, we have the following commutative diagram:

$$\begin{array}{ccc} C_K^p(M) & \xrightarrow{P'} & \check{C}_{m-p}^{K^*}(M) \\ {\scriptstyle \iota'^*}\uparrow & & \downarrow{\scriptstyle \kappa'} \\ C_{K'}^p(M) & \xrightarrow{\frown M_{K'}} & \check{C}_{m-p}^{K'}(M). \end{array} \tag{4.9}$$

Thus we have (cf. Lemma 4.2, except we use (B.24) instead of (B.17))

$$P'\delta u = (-1)^{m-p}\partial P'(u) \qquad \text{for} \quad u \in C_K^p(M).$$

and an isomorphism

$$P' : H^p(M) \xrightarrow{\sim} \check{H}_{m-p}(M).$$

Remark 4.3. The above isomorphism is given by the right cap product with $[M]$ and differs from P by a sign of $(-1)^{p(m-p)}$.

The case M is compact: In the case M is compact, using the homology and cohomology with $\mathbb{C}$-coefficient (in fact we may use $\mathbb{Q}$ as the coefficient), the Poincaré duality is described as follows. If M is compact,

$$\breve{H}_{m-p}(M;\mathbb{C}) = H_{m-p}(M;\mathbb{C}) \simeq H^{m-p}(M;\mathbb{C})^*$$

and from (B.20) we have:

Proposition 4.1. *If M is compact, P_M sends a class α in $H^p(M;\mathbb{C})$ to the class a in $H_{m-p}(M;\mathbb{C})$ such that*

$$\langle a, \beta\rangle = \langle [M], \alpha \smile \beta\rangle \qquad \text{for every } \beta \text{ in } H^{m-p}(M;\mathbb{C}).$$

Letting $\varepsilon_* : H_0(M;\mathbb{C}) \to \mathbb{C}$ be the augmentation, we have (cf. (B.19)):

Corollary 4.2. *If M is compact, the pairing*

$$H^p(M;\mathbb{C}) \times H^{m-p}(M;\mathbb{C}) \xrightarrow{\smile} H^m(M;\mathbb{C}) \xrightarrow{[M]\frown} H_0(M;\mathbb{C}) \xrightarrow{\varepsilon_*} \mathbb{C}$$

is non-degenerate and induces the Poincaré duality with $\mathbb{C}$-coefficient.

Alexander duality

Let S be a closed set in M. Suppose that there is a triangulation K_0 of M such that S is a K_0-subcomplex of M, i.e., K_0 is compatible with S. Recall that the *star* $S_{K'}(S)$ of S in K' is the union of simplices of K' intersecting with S, i.e., the union of cells of K^* intersecting with S. Let $O_{K'}(S) = S_{K'}(S) \smallsetminus \partial S_{K'}(S)$ denote the *open star*. Note that there is a proper deformation retraction $S_{K'}(S) \to S$ and a deformation retraction $O_{K'}(S) \to S$ (cf. Proposition B.26).

We have:

Proposition 4.2. *For a simplex $\boldsymbol{s}$ of K, the following three conditions are equivalent:*

$$(1)\ \boldsymbol{s} \subset S, \qquad (2)\ \boldsymbol{s}^* \cap S \neq \emptyset, \qquad (3)\ \boldsymbol{s}^* \cap O_{K'}(S) \neq \emptyset.$$

Exercise 4.1. Verify the above.

Noting that $M \smallsetminus O_{K'}(S)$ is a K^*-subcomplex of M, we may write

$$C^p_{K^*}(M, M \smallsetminus O_{K'}(S)) = \{\, u \in C^p_{K^*}(M) \mid \langle \boldsymbol{s}^*, u\rangle = 0 \ \text{ for } \boldsymbol{s} \not\subset S \,\}.$$

They form a subcomplex of $C^\bullet_{K^*}(M)$. Denoting by $H^p_{K^*}(M, M \smallsetminus O_{K'}(S))$ its cohomology, we have a natural isomorphism:

$$\eta^*_{K^*} : H^p(M, M \smallsetminus O_{K'}(S)) \xrightarrow{\sim} H^p_{K^*}(M, M \smallsetminus O_{K'}(S)).$$

Note that there is a natural isomorphism

$$H^p(M, M \setminus O_{K'}(S)) \simeq H^p(M, M \setminus S).$$

Now in the sum in (4.6), if u is in $C^p_{K^*}(M, M \setminus O_{K'}(S))$, only $(m-p)$-simplices in S appear. Thus P in (4.6) induces an isomorphism

$$A : C^p_{K^*}(M, M \setminus O_{K'}(S)) \xrightarrow{\sim} \check{C}^K_{m-p}(S). \tag{4.10}$$

Since $(C^\bullet_{K^*}(M, M \setminus O_{K'}(S)), \delta)$ is a subcomplex of $(C^\bullet_{K^*}(M), \delta)$ and $(\check{C}^K_\bullet(S), \partial)$ is a subcomplex of $(\check{C}^K_\bullet(M), \partial)$, from Lemma 4.2, we see that A is also compatible with boundary and coboundary operators. Thus we have:

Theorem 4.2 (Alexander duality). *For a closed set S which is a subcomplex of M with respect to some triangulation, the isomorphism* (4.10) *induces an isomorphism*

$$A_{M,S} : H^p(M, M \setminus S) \xrightarrow{\sim} \check{H}_{m-p}(S).$$

Remark 4.4. Noting that $S_{K'}(S) \setminus O_{K'}(S) = \partial S_{K'}(S)$ and that $S_{K'}(S)$ and $\partial S_{K'}(S)$ are K^*-subcomplexes of M, we may identify the cochain group $C^p_{K^*}(M, M \setminus O_{K'}(S))$ with $C^p_{K^*}(S_{K'}(S), \partial S_{K'}(S))$. Then the Alexander isomorphism is induced from the isomorphism

$$C^p_{K^*}(S_{K'}(S), \partial S_{K'}(S)) \xrightarrow{\sim} \check{C}^K_{m-p}(S),$$

which is the restriction of P in (4.6).

We now try to express A in terms of cap product. We set

$$C^p_{K'}(M, M \setminus O_{K'}(S)) = \{\, u \in C^p_{K'}(M) \mid \langle \boldsymbol{t}, u\rangle = 0 \ \text{ for } \boldsymbol{t} \text{ with } \boldsymbol{t} \cap S = \emptyset \,\}.$$

Then from Lemma 4.1, we have the commutative diagram:

$$\begin{array}{ccc} C^p_{K^*}(M, M \setminus O_{K'}(S)) & \xrightarrow{A} & \check{C}^K_{m-p}(S) \\ \uparrow{\scriptstyle \iota^*} & & \downarrow{\scriptstyle \kappa} \\ C^p_{K'}(M, M \setminus O_{K'}(S)) & \xrightarrow{M_{K'} \frown} & \check{C}^{K'}_{m-p}(S), \end{array} \tag{4.11}$$

where ι^* and κ are defined as in (4.7). Thus A is again represented by the cap product with the cycle M on the level of chains and cochains. To pass on to the level of homology and cohomology, we proceed as follows.

First, we may identify $C^p_{K'}(M, M \smallsetminus O_{K'}(S))$ with $C^p_{K'}(S_{K'}(S), \partial S_{K'}(S))$ as in the case of K^*-cochain groups in Remark 4.4. We have the cap product (cf. (B.39))

$$\check{C}^{K'}_r(S_{K'}(S), \partial S_{K'}(S)) \times C^p_{K'}(S_{K'}(S), \partial S_{K'}(S)) \xrightarrow{\frown} \check{C}^{K'}_{r-p}(S_{K'}(S)). \quad (4.12)$$

With this, the second horizontal morphism in (4.11) is the cap product with the relative cycle $S_{K'}(S)$, the sum of m-simplices of K' in $S_{K'}(S)$, in $\check{C}^{K'}_m(S_{K'}(S), \partial S_{K'}(S))$. Here we note that there is a commutative diagram

$$\begin{array}{ccc} \check{C}^{K'}_m(S_{K'}(S)) & \xrightarrow{\partial} & \check{C}^{K'}_{m-1}(S_{K'}(S)) \\ \downarrow{=} & & \downarrow \\ \check{C}^{K'}_m(S_{K'}(S), \partial S_{K'}(S)) & \xrightarrow{\underline{\partial}} & \check{C}^{K'}_{m-1}(S_{K'}(S), \partial S_{K'}(S)). \end{array}$$

In the above, $\underline{\partial}$ denotes the boundary morphism for relative cochain groups and the vertical arrows are the canonical surjections, of which the first one is the identity, as $\check{C}^{K'}_m(\partial S_{K'}(S)) = 0$. For the chain $S_{K'}(S)$, $\underline{\partial} S_{K'}(S) = 0$, while $\partial S_{K'}(S) \neq 0$ so that it is a cycle only as a relative chain. For this reason we denote $S_{K'}(S)$ by $\underline{S_{K'}(S)}$ when considered as a relative chain.

Noting that

$$H^p_{K'}(S_{K'}(S), \partial S_{K'}(S)) \simeq H^p(M, M \smallsetminus S) \quad \text{and} \quad \check{H}_{r-p}(S_{K'}(S)) \simeq \check{H}_{r-p}(S),$$

(4.12) represents the cap product

$$\check{H}_r(S_{K'}(S), \partial S_{K'}(S)) \times H^p(M, M \smallsetminus S) \xrightarrow{\frown} \check{H}_{r-p}(S).$$

Thus we may write

$$A(\alpha) = [\underline{S_{K'}(S)}] \frown \alpha \qquad \text{for } \alpha \in H^p(M, M \smallsetminus S). \quad (4.13)$$

From this we see that the isomorphism A is independent of the triangulation for which S is a subcomplex (cf. (T_2) in Section 4.1).

From the properties of cup and cap products (cf. (B.18), (B.22)), we have:

Proposition 4.3. *Let M and S be as above and $i : S \hookrightarrow M$ the inclusion. For any class $\gamma \in H^q(M)$, the following diagram is commutative:*

$$\begin{array}{ccc} H^p(M, M \smallsetminus S) & \xrightarrow[A]{\sim} & \check{H}_{m-p}(S) \\ {\scriptstyle (\ )\smile\gamma}\downarrow & & \downarrow{\scriptstyle (\ )\frown i^*\gamma} \\ H^r(M, M \smallsetminus S) & \xrightarrow[A]{\sim} & \check{H}_{m-r}(S), \end{array}$$

where $r = p + q$.

Remark 4.5. 1. On the level of chains and cochains, the product (4.12) depends on the ordering of the vertices. For the cycle $\underline{S_{K'}(S)}$ and our specific choice of ordering as in Lemma 4.1, the receiving chain group can be made to be $\check{C}^{K'}_{m-p}(S)$.

2. The reason we use the relative homology class of $S_{K'}(S)$ instead of the homology class of M is as follows. Observe that we have the cap product

$$\check{C}^{K'}_r(M) \times C^p_{K'}(M, M \setminus O_{K'}(S)) \xrightarrow{\frown} \check{C}^{K'}_{r-p}(S_{K'}(S)).$$

To see this, let $\boldsymbol{t}$ be an r-simplex of K'. Then for every ordering $(v_0, \dots, v_r)$ of the vertices of $\boldsymbol{t}$, if $(v_p, \dots, v_r) \not\subset S_{K'}(S)$, then $(v_0, \dots, v_p) \subset M \setminus O_{K'}(S)$ so that $\boldsymbol{t} \frown u = 0$ for u in $C^p_{K'}(M, M \setminus O_{K'}(S))$. It induces

$$\check{H}_r(M) \times H^p_{K'}(M, M \setminus O_{K'}(S)) \xrightarrow{\frown} \check{H}_{r-p}(S).$$

The problem is that there is no corresponding cap product in the singular theory.

3. In the case S is compact, $S_{K'}(S)$ is also compact. We have

$$\check{H}_r(S_{K'}(S), \partial S_{K'}(S)) = H_r(S_{K'}(S), \partial S_{K'}(S)) \simeq H_r(M, M \setminus S). \qquad (4.14)$$

For the isomorphism above, let $X_1 = S_{K'}(S)$ and $X_2 = M \setminus S$ in (B.9) and notice that $S_{K'}(S) \setminus S$ deformation retracts to $\partial S_{K'}(S)$. Thus we have the cap product

$$H_r(M, M \setminus S) \times H^p(M, M \setminus S) \xrightarrow{\frown} H_{r-p}(S)$$

and we may write

$$A(\alpha) = [\underline{M}] \frown \alpha \qquad \text{for } \alpha \in H^p(M, M \setminus S),$$

where $[\underline{M}]$ denotes the class in $H_m(M, M \setminus S)$ that corresponds to the class $[\underline{S_{K'}(S)}]$ in $H_m(S_{K'}(S), \partial S_{K'}(S))$ by the isomorphism (4.14).

4. Note that, while $\check{H}_r(S_{K'}(S), \partial S_{K'}(S))$ makes sense as $\partial S_{K'}(S)$ is a closed set on $S_{K'}(S)$, $\check{H}_r(M, M \setminus S)$ is not defined in general, as $M \setminus S$ is an open set.

Dual description: Let K_0, K, K' and K^* be as before, except here we assume K_0 is a barycentric subdivision of a triangulation of M so that $O_K(S)$ deformation retracts to S. Noting that $M \setminus O_K(S)$ is a sub-K complex of M, the restriction of P' in (4.8) gives a morphism

$$A' : C^p_K(M, M \setminus O_K(S)) \longrightarrow \check{C}^{K^*}_{m-p}(S_{K'}(S)). \qquad (4.15)$$

From Lemma 4.3, we have the commutative diagram:

$$\begin{array}{ccc} C^p_K(M, M \smallsetminus O_K(S)) & \xrightarrow{A'} & \check{C}^{K^*}_{m-p}(S_{K'}(S)) \\ \uparrow{\scriptstyle \iota'^*} & & \downarrow{\scriptstyle \kappa'} \\ C^p_{K'}(M, M \smallsetminus O_K(S)) & \xrightarrow{\frown M_{K'}} & \check{C}^{K'}_{m-p}(S_{K'}(S)). \end{array} \tag{4.16}$$

Thus we have an isomorphism

$$A' : H^p(M, M \smallsetminus S) \xrightarrow{\sim} \breve{H}_{m-p}(S).$$

Remark 4.6. The above isomorphism differs from A by a sign of $(-1)^{p(m-p)}$ (cf. Remark 4.3).

Lefschetz duality

Let R be a C^∞ manifold of dimension m with boundary ∂R in M. We may assume that R and ∂R are K_0-subcomplexes of M. We apply the above by setting $S = R$. Since we have canonical isomorphisms

$$H^p(M, M \smallsetminus O_{K'}(R)) \simeq H^p(S_{K'}(R), \partial S_{K'}(R)) \simeq H^p(R, \partial R), \tag{4.17}$$

we have:

Theorem 4.3 (Lefschetz duality). *For a manifold R of dimension m with boundary ∂R in M, we have an isomorphism*

$$L : H^p(R, \partial R) \xrightarrow{\sim} \breve{H}_{m-p}(R).$$

Note that the above isomorphism is given by the cap product with the relative class $[\underline{R}]$ of R in $\breve{H}_m(R, \partial R)$. It is the class corresponding to $[\underline{S_{K'}(R)}]$ in the isomorphism $\breve{H}_m(R, \partial R) \simeq \breve{H}_m(S_{K'}(R), \partial S_{K'}(R))$.

Remark 4.7. The above theorem holds if R is only piecewise C^∞ (cf. Definition 5.10 below).

From the construction we have:

Proposition 4.4. *For a subcomplex S in the interior of R, we have the commutative diagram:*

$$\begin{array}{ccccc} H^p(M, M \smallsetminus S) & \xrightarrow{j^*} & H^p(R, \partial R) & \xrightarrow{j'^*} & H^p(M) \\ \wr\downarrow{\scriptstyle A} & & \wr\downarrow{\scriptstyle L} & & \wr\downarrow{\scriptstyle P} \\ \breve{H}_{m-p}(S) & \xrightarrow{\imath_*} & \breve{H}_{m-p}(R) & \xrightarrow{\imath'_*} & \breve{H}_{m-p}(M), \end{array}$$

where $j : (R, \partial R) \hookrightarrow (M, M \smallsetminus S)$, $\imath : S \hookrightarrow R$ *and* $\imath' : R \hookrightarrow M$ *denote the inclusions and* j'^* *denotes the isomorphism in* (4.17) *followed by the canonical morphism* $H^p(M, M \smallsetminus O_{K'}(R)) \to H^p(M)$.

Suppose S has a finite number of connected components $(S_\lambda)_\lambda$. Then $\check{H}_{m-p}(S) = \bigoplus_\lambda \check{H}_{m-p}(S_\lambda)$ and each class α in $H^p(M, M \smallsetminus S)$ determines a class a_λ in $\check{H}_{m-p}(S_\lambda)$ via the Alexander isomorphism.

Corollary 4.3. *In the above situation,*

$$\sum_\lambda (\imath_\lambda)_* a_\lambda = [\underline{R}] \frown j^*\alpha \qquad \text{in } \check{H}_{m-p}(R),$$

where $\imath_\lambda : S_\lambda \hookrightarrow R$ *denotes the inclusion.*

In many cases we do not consider the auxiliary manifold R and take M itself as R. In this situation we have the commutative diagram

$$\begin{array}{ccc} H^p(M, M \smallsetminus S) & \xrightarrow{j^*} & H^p(M) \\ \wr\big\downarrow A & & \wr\big\downarrow P \\ \check{H}_{m-p}(S) & \xrightarrow{\imath_*} & \check{H}_{m-p}(M) \end{array} \qquad (4.18)$$

and the right-hand side of the identity in Corollary 4.3 is written $[M] \frown j^*\alpha$ in $\check{H}_{m-p}(M)$.

The above is the basis of the "residue theorem" we discuss in later sections (cf. Theorems 5.2, 10.1 and 10.6).

Remark 4.8. Let M and S be as above. The isomorphism (4.6) induces an isomorphism

$$C^p_{K^*}(M \smallsetminus O_{K'}(S)) \xrightarrow{\sim} \check{C}^K_{m-p}(M, S),$$

which in turn induces another Alexander duality

$$H^p(M \smallsetminus S) \xrightarrow{\sim} \check{H}_{m-p}(M, S).$$

The exact sequence

$$\cdots \to H^{p-1}(M \smallsetminus S) \xrightarrow{\delta^*} H^p(M, M \smallsetminus S) \xrightarrow{j^*} H^p(M) \to H^p(M \smallsetminus S) \to \cdots$$

is dual to

$$\cdots \to \check{H}_{m-p+1}(M, S) \xrightarrow{\partial_*} \check{H}_{m-p}(S) \xrightarrow{\imath_*} \check{H}_{m-p}(M) \to \check{H}_{m-p}(M, S) \to \cdots .$$

Push-forward in cohomology: Let $f : M \to M'$ be a map of oriented manifolds M and M' of dimensions m and m', respectively. We define a morphism $f_* : H^{m-p}(M) \to H^{m'-p}(M')$ so that the following diagram is commutative:

$$\begin{array}{ccc} H^{m-p}(M) & \xrightarrow[P]{\sim} & \breve{H}_p(M) \\ {\scriptstyle f_*}\downarrow & & \downarrow{\scriptstyle f_*} \\ H^{m'-p}(M') & \xrightarrow[P]{\sim} & \breve{H}_p(M'). \end{array} \tag{4.19}$$

4.3 Thom isomorphism and Thom class

Thom class of a submanifold

Let M be a C^∞ manifold of dimension m and V a closed submanifold of dimension d'. We set $k' = m - d'$. We take a triangulation K_0 of M compatible with V and let K and K' be as in the previous section. We denote by K_V the triangulation of V induced from K and by K'_V its barycentric subdivision. Then K'_V is the set of simplices of K' that are in V. We denote by K^* and K^*_V the cellular decompositions of M and V dual to K and K_V, respectively (cf. Section 4.1). Note that, for a p-simplex $\boldsymbol{s}$ of K in V, its dual $\boldsymbol{s}^*$ in K^* is an $(m-p)$-cell and that its dual $\boldsymbol{s}^*_V$ in K^*_V is a $(d'-p)$-cell. They are related by $\boldsymbol{s}^*_V = \boldsymbol{s}^* \cap V$ as sets. In the case M and V are orientable, there is a more precise formula (cf. Lemma 4.5 below).

The simplices and cells of K, K' and K^* are oriented so that the conditions (1) and (2) in Section 4.1 are satisfied. The simplices of K_V and K'_V are oriented as simplices of K and K', respectively. In order to describe the homology and cohomology of V, we impose similar conditions for simplices and cells of K_V, K'_V and K^*_V. The condition corresponding to (1) is automatically satisfied. Thus we impose:

$(2)_V$ Let $\boldsymbol{t}$ be a p-simplex of K'_V. If $\boldsymbol{t} \subset \boldsymbol{s'}^*_V$, a p-cell of K^*_V, the orientation of $\boldsymbol{t}$ is the same as that of $\boldsymbol{s'}^*_V$.

For a simplex $\boldsymbol{s}$ in K_V we would like to relate the orientation of $\boldsymbol{s}^*$ and that of $\boldsymbol{s}^*_V$. For this we further impose the following condition. To state it, let $\boldsymbol{s}$ be an $(m-p)$-simplex of K in V. We take a d'-simplex $\boldsymbol{s}_1$ of K_V so that $\boldsymbol{s} \prec \boldsymbol{s}_1$. Then there exist a p-simplex $\boldsymbol{t}$ of K' in $\boldsymbol{s}^*$, a k'-simplex $\boldsymbol{t}_1$ of K' in $\boldsymbol{s}^*_1$ and a $(p-k')$-simplex $\boldsymbol{t}_V$ of K'_V in $\boldsymbol{s}^*_V$ such that $\boldsymbol{t}_1$ and $\boldsymbol{t}_V$ span

$\boldsymbol{t}$. Note that the simplices $\boldsymbol{t}$, $\boldsymbol{t}_1$ and $\boldsymbol{t}_V$ have the same orientations as $\boldsymbol{s}^*$, $\boldsymbol{s}_1^*$ and $\boldsymbol{s}_V^*$, respectively (cf. (2) and $(2)_V$).

(5) The simplices and cells are oriented so that the orientation of $\boldsymbol{s}_1^*$ followed by the orientation of $\boldsymbol{s}_V^*$ gives the orientation of $\boldsymbol{s}^*$.

Note that, by the tubular neighborhood theorem (Theorem 3.10), the above can be done consistently with other conditions, in particular independently of the choice of $\boldsymbol{s}_1$, if the normal bundle $N_{\mathbb{R},V}$ of V in M is orientable (cf. Sections 3.4 and 3.7). In this case we may think of $\boldsymbol{s}_1^*$ as being in the fiber direction and we orient the normal bundle $N_{\mathbb{R},V}$ accordingly. Note that the above condition (5) is consistent with Convention 3.3.

Now we consider the morphism

$$T_V : C^p_{K_V^*}(V) \longrightarrow C^{k'+p}_{K^*}(M, M \smallsetminus O_{K'}(V)), \qquad u_V \mapsto u,$$

where u is given by, for each $(d'-p)$-simplex $\boldsymbol{s}$ of K (cf. Proposition 4.2),

$$\langle \boldsymbol{s}^*, u\rangle = \begin{cases} \langle \boldsymbol{s}_V^*, u_V\rangle & \text{if } \boldsymbol{s} \subset V, \\ 0 & \text{otherwise}. \end{cases} \tag{4.20}$$

It is clearly an isomorphism. Moreover, with the assumption that the bundle $N_{\mathbb{R},V}$ is oriented as the condition (5), we have

$$\delta \circ T_V = (-1)^{k'} T_V \circ \delta \tag{4.21}$$

and thus the following:

Theorem 4.4. *If the normal bundle of V in M is oriented, the above T_V induces an isomorphism*

$$T_V : H^p(V) \xrightarrow{\sim} H^{k'+p}(M, M \smallsetminus V).$$

We call T_V the *Thom isomorphism* for V.

In the following, we assume that the normal bundle $N_{\mathbb{R},V}$ is oriented as above.

Definition 4.2. The *Thom class* of V in M, denoted by $\Psi_{M,V}$ or simply by Ψ_V, is the image of [1] in $H^0(V)$ by T_V:

$$\Psi_V = T_V([1]) \in H^{k'}(M, M \smallsetminus V),$$

where [1] denotes the class of the cocycle that assigns 1 to each 0-cell (cf. Remark B.9).

From the definition we have:

Proposition 4.5. *The Thom class Ψ_V is represented by a cocycle that assigns* 1 *or* 0 *to each oriented k'-cell $\boldsymbol{s}^*$ forming a basis of $C_{k'}^{K^*}(M)$ according as $\boldsymbol{s}^*$ intersects with V or not.*

Remark 4.9. 1. The classes in $H^{k'}(M, M \smallsetminus V)$ may also be represented by cocycles in $C_{K'}^{k'}(M, M \smallsetminus O_{K'}(V))$. If ψ is a cocycle representing the Thom class, we have

$$\langle \iota \boldsymbol{s}^*, \psi \rangle = \langle \boldsymbol{s}^*, \iota^* \psi \rangle = \begin{cases} 1 & \text{if } \ \boldsymbol{s} \subset V, \\ 0 & \text{otherwise}, \end{cases}$$

where $\iota : C_{k'}^{K^*}(M) \to C_{k'}^{K'}(M)$ denotes a natural monomorphism and ι^* : $C_{K'}^{k'}(M, M \smallsetminus O_{K'}(V)) \to C_{K^*}^{k'}(M, M \smallsetminus O_{K'}(V))$ a natural epimorphism.

2. In the above, we do not assume that M or V to be orientable. In the case they are, we have the dulalities on M and V and we have the commutative diagrams as in Propositions 4.9 below.

Thom class of an oriented vector bundle

Let $\pi : E \to M$ be a C^∞ oriented real vector bundle of rank l'. We denote by Σ the image of the zero section $s_0 : M \to E$. Note that s_0 is a diffeomorphism of M onto Σ and that $s_0^* N_{\mathbb{R},\Sigma} = E$ (cf. Exercise 3.10). We apply the above considerations by letting M, V and k' be E, Σ and l', respectively. We also replace K with K_E. Thus K_E is a triangulation of E compatible with Σ. The *Thom isomorphism*

$$T_E : H^p(M) \xrightarrow{\sim} H^{l'+p}(E, E \smallsetminus \Sigma)$$

for E is defined by $T_E = T_\Sigma \circ (s_0^*)^{-1}$ so that the diagram

$$\begin{array}{ccc} H^p(M) & \xrightarrow[T_E]{\sim} & H^{l'+p}(E, E \smallsetminus \Sigma) \\ {\scriptstyle \wr}\uparrow{\scriptstyle s_0^*} & \nearrow_{T_\Sigma}^{\sim} & \\ H^p(\Sigma). & & \end{array}$$

is commutative. The *Thom class* Ψ_E of E is defined by $\Psi_E = T_E([1])$, which is in fact the Thom class Ψ_Σ of Σ.

The Thom isomorphism may also be written (cf. (4.25), Remark 4.11 and (7.69) below)

$$T_E(\alpha) = \Psi_E \smile \pi^*(\alpha) \qquad \text{for } \ \alpha \in H^p(M). \tag{4.22}$$

Rephrasing Proposition 4.5, we have:

Proposition 4.6. *The Thom class $\Psi_E \in H^{l'}(E, E \setminus \Sigma)$ is represented by a cocycle that assigns* 1 *or* 0 *to the dual l'-cell $\boldsymbol{s}^*$ in E of each m-simplex $\boldsymbol{s}$ in K_E according as $\boldsymbol{s}$ is in Σ or not.*

Example 4.1. If M is a point, then $E = \mathbb{R}^{l'}$ and we have

$$T_{\mathbb{R}^{l'}} : H^0(\{0\}) \xrightarrow{\sim} H^{l'}(\mathbb{R}^{l'}, \mathbb{R}^{l'} \setminus 0),$$

the both sides being canonically isomorphic with $\mathbb{Z}$. The Thom class $\Psi_{\mathbb{R}^{l'}}$ is the canonical generator of $H^{l'}(\mathbb{R}^{l'}, \mathbb{R}^{l'} \setminus 0)$ (cf. Example B.6. 2 and Remark B.13. 2) and is represented by the cocycle assigning the value 1 to the dual cell in $\mathbb{R}^{l'}$ of the vertex 0 and the value 0 to the other l'-cells.

The relation between the Thom class of E and that of each fiber is given as follows. For each point x in M, let $i_x : (E_x, E_x \setminus 0) \hookrightarrow (E, E \setminus \Sigma)$ be the inclusion. We have the commutative diagram:

$$\begin{array}{ccccccc}
 & H^{l'-1}(E_x \setminus 0) & \xrightarrow{\delta^*} & H^{l'}(E_x, E_x \setminus 0) & & & \\
 & \uparrow i_x^* & & \uparrow i_x^* & & & \\
\longrightarrow & H^{l'-1}(E \setminus \Sigma) & \xrightarrow{\delta^*} & H^{l'}(E, E \setminus \Sigma) & \xrightarrow{j^*} & H^{l'}(E) & \longrightarrow \\
 & & & & & \wr\uparrow \pi^* & \\
 & & & & & H^{l'}(M). &
\end{array} \tag{4.23}$$

Note that δ^* in the first row is an epimorphism for $l' = 1$ and an isomorphism for $l' > 1$. From the above description, we have:

Proposition 4.7. *A class Ψ in $H^{l'}(E, E \setminus \Sigma)$ coincides with Ψ_E if and only if $i_x^* \Psi = \Psi_{E_x}$ for all x in M.*

We see below (cf. (5.21)) that $(\pi^*)^{-1} j^* \Psi_E = e(E)$, the Euler class of E.

Pull-back: Let $f^* : M' \to M$ be a C^∞ map of C^∞ manifolds. Then the pull-back bundle f^*E (cf. (3.4)) is orientable and has a natural orientation. The image of its zero section is given by $(\tilde{f})^{-1}\Sigma$. We have the Thom class Ψ_{f^*E} in $H^{l'}(f^*E, f^*E \setminus (\tilde{f})^{-1}\Sigma)$. There is a canonical morphism

$$(\tilde{f})^* : H^{l'}(E, E \setminus \Sigma) \longrightarrow H^{l'}(f^*E, f^*E \setminus (\tilde{f})^{-1}\Sigma).$$

The following is a consequence of Proposition 4.7:

Proposition 4.8. *In the above situation, we have:*

$$\Psi_{f^*E} = (\tilde{f})^* \Psi_E.$$

Remark 4.10. 1. Let M be a C^∞ manifold and V a closed submanifold of M with the normal bundle $p : N_{\mathbb{R},V} \to V$. By the tubular neighborhood theorem (Theorem 3.10), there exist a neighborhood U of V in M, a neighborhood W of the image Z of the zero section in $N_{\mathbb{R},V}$ and a diffeomorphism $\tau : (U, V) \to (W, Z)$ with $(p \circ \tau)|_V = 1_V$. We have a composition of isomorphisms

$$H^p(N_{\mathbb{R},V}, N_{\mathbb{R},V} \smallsetminus Z) \simeq H^p(W, W \smallsetminus Z) \xrightarrow[\tau^*]{\sim} H^p(U, U \smallsetminus V) \simeq H^p(M, M \smallsetminus V).$$

If $N_{\mathbb{R},V}$ is oriented, the Thom class $\Psi_{N_{\mathbb{R},V}}$ corresponds to Ψ_V.

2. We discuss the Thom isomorphism and the Thom class in terms of differential forms in Section 7.9 below, where they are treated in cohomology with $\mathbb{C}$-coefficients.

Product formula

Suppose we have two pairs of topological spaces (X, A) and (Y, B) with A and B open sets in X and Y, respectively. We denote by $(X, A) \times (Y, B)$ the pair $(X \times Y, (A \times Y) \cup (X \times B))$. For example,

$$(\mathbb{R}^{l'_1}, \mathbb{R}^{l'_1} \smallsetminus 0) \times (\mathbb{R}^{l'_2}, \mathbb{R}^{l'_2} \smallsetminus 0) = (\mathbb{R}^{l'}, \mathbb{R}^{l'} \smallsetminus 0), \qquad l' = l'_1 + l'_2.$$

We have the cup product (cf. Appendix B)

$$H^p(X \times Y, A \times Y) \times H^q(X \times Y, X \times B) \xrightarrow{\smile} H^{p+q}((X, A) \times (Y, B)).$$

Letting

$$p_1 : (X \times Y, A \times Y) \longrightarrow (X, A) \quad \text{and} \quad p_2 : (X \times Y, X \times B) \longrightarrow (Y, B)$$

be the projections, we define the *cross product*

$$H^p(X, A) \times H^q(Y, B) \xrightarrow{\times} H^{p+q}((X, A) \times (Y, B)) \text{ by } \alpha \times \beta = p_1^*(\alpha) \smile p_2^*(\beta).$$

Recall that $H^1(\mathbb{R}, \mathbb{R} \smallsetminus 0)$ has a canonical generator e. We denote by $e^{l'} \in H^{l'}(\mathbb{R}^{l'}, \mathbb{R}^{l'} \smallsetminus 0)$ the l'-fold cross product $e \times \cdots \times e$.

We quote:

Theorem 4.5. *For any pair (X, A) with A open in X, the assignment $\alpha \mapsto \alpha \times e^{l'}$ defines an isomorphism*

$$H^p(X, A) \xrightarrow{\sim} H^{p+l'}((X, A) \times (\mathbb{R}^{l'}, \mathbb{R}^{l'} \smallsetminus 0)).$$

From this, we have (cf. Example 4.1):

Corollary 4.4. 1. $e^{l'} = \Psi_{\mathbb{R}^{l'}}$*, the Thom class of* $\mathbb{R}^{l'}$.

2. $\Psi_{\mathbb{R}^{l'_1}} \times \Psi_{\mathbb{R}^{l'_2}} = \Psi_{\mathbb{R}^{l'}}$, $l' = l'_1 + l'_2$.

For $i = 1, 2$, let E_i be a C^∞ real oriented vector bundle of rank l'_i on M. Letting Σ_i be (the image of) the zero section of E_i, we have the Thom class Ψ_{E_i} in $H^{l'_i}(E_i, E_i \smallsetminus \Sigma_i)$. Noting that the zero section of the product bundle $E_1 \times E_2$ (cf. Section 3.3) is $((E_1 \smallsetminus \Sigma_1) \times E_2) \cup (E_1 \times (E_2 \smallsetminus \Sigma_2))$, we have the Thom class $\Psi_{E_1 \times E_2}$ in $H^{l'}((E_1, E_1 \smallsetminus \Sigma_1) \times (E_2, E_2 \smallsetminus \Sigma_2))$, $l' = l'_1 + l'_2$.

Theorem 4.6. *In the above situation,*

$$\Psi_{E_1 \times E_2} = \Psi_{E_1} \times \Psi_{E_2}.$$

Proof. This follows from Proposition 4.7 and Corollary 4.4. 2. □

Let E_i and Σ_i be as above, $i = 1, 2$. We set $E = E_1 \oplus E_2$ and let Σ be the zero section of E. Letting $d : M \hookrightarrow M \times M$ be the diagonal embedding, we identify E with $d^*(E_1 \times E_2)$ by (3.8). Then we have the commutative diagram

$$\begin{array}{ccc} E & \xrightarrow{\tilde{d}} & E_1 \times E_2 \\ \downarrow & & \downarrow \\ M & \xrightarrow{d} & M \times M. \end{array}$$

We have a canonical morphism

$$(\tilde{d})^* : H^{l'}((E_1, E_1 \smallsetminus \Sigma_1) \times (E_2, E_2 \smallsetminus \Sigma_2)) \longrightarrow H^{l'}(E, E \smallsetminus \Sigma).$$

From Theorem 4.6 and Proposition 4.8, we have

Corollary 4.5. *We have*

$$\Psi_E = (\tilde{d})^*(\Psi_{E_1} \times \Psi_{E_2}).$$

We discuss more on this subject in Sections 5.6, 8.3, 10.3, 10.4 and 14.1 below.

Poincaré, Alexander and Thom isomorphisms

Let M be a C^∞ manifold of dimension m and V a closed submanifold of dimension d', as above. We assume that M and V are oriented. In order to describe the dualities for M we impose the conditions (1), (2) in Section 4.1 and (3), (4) in Section 4.2. We also impose the corresponding conditions for the simplices and cells of K_V, K'_V and K^*_V. Note that the one corresponding to (1) is already satisfied. Besides $(2)_V$ we further impose:

$(3)_V$ The orientation of each d'-simplex of K_V is the same as that of V.

$(4)_V$ For every p-simplex $\boldsymbol{s}$, $0 < p < d'$, of K_V, the orientation of $\boldsymbol{s}_V^*$ followed by the orientation of $\boldsymbol{s}$ gives the orientation of V.

Since we assumed that M and V to be oriented, the normal bundle $N_{\mathbb{R},V}$ is orientable (cf. Proposition 3.13) and is oriented according to Convention 3.4. The total space of $N_{\mathbb{R},V}$ is then oriented according to Convention 3.3. By identifying neighborhoods of V in M and $N_{\mathbb{R},V}$ by the tubular neighborhood theorem, we may rephrase the above conventions as (cf. Remark 3.8. 1):

Convention 4.1. We orient the bundle $N_{\mathbb{R},V}$ so that the orientation of the fiber of $N_{\mathbb{R},V}$ followed by that of V gives the orientation of M.

We also impose the condition (5), which is consistent with the above convention. Then the following diagram is commutative:

$$\begin{array}{ccc} C^p_{K_V^*}(V) & \xrightarrow[T]{\sim} & C^{k'+p}_{K^*}(M, M \smallsetminus O_{K'}(V)) \\ \wr\big\downarrow P & \underset{A}{\overset{\sim}{\swarrow}} & \\ \check{C}^K_{d'-p}(V). & & \end{array}$$

Thus we have:

Proposition 4.9. *In the case M and V are oriented, we have the commutative diagram:*

$$\begin{array}{ccc} H^p(V) & \xrightarrow[T]{\sim} & H^{k'+p}(M, M \smallsetminus V) \\ \wr\big\downarrow P & \underset{A}{\overset{\sim}{\swarrow}} & \\ \check{H}_{d'-p}(V). & & \end{array}$$

From the above, we see that

$$A(\Psi_V) = [V], \tag{4.24}$$

the fundamental class of V.

Note that there is the cup product

$$H^{k'}(M, M \smallsetminus V) \times H^p(M) \xrightarrow{\smile} H^{k'+p}(M, M \smallsetminus V).$$

Denoting by $i : V \hookrightarrow M$ the inclusion, we have:

Proposition 4.10. *The following diagram is commutative:*

$$\begin{array}{ccc} H^p(M) & \xrightarrow{\Psi_V \smile (\ )} & H^{k'+p}(M, M \smallsetminus V) \\ i^* \big\downarrow & & \wr \big\downarrow A_{M,V} \\ H^p(V) & \xrightarrow[P_V]{\sim} & \check{H}_{d'-p}(V), \end{array}$$

Proof. For a class $\alpha \in H^p(M)$, by (4.13), (B.18) and (B.22),

$$\begin{aligned} A_{M,V}(\Psi_V \smile \alpha) &= [\underline{S_{K'}(V)}] \frown (\Psi_V \smile \alpha) = ([\underline{S_{K'}(V)}] \frown \Psi_V) \frown i^*\alpha \\ &= [V] \frown i^*\alpha = P_V(i^*\alpha). \end{aligned}$$

□

Combining with Proposition 4.9, we have:

Corollary 4.6. *For a class α in $H^p(M)$, we have*

$$T_V(i^*\alpha) = \Psi_V \smile \alpha.$$

Suppose U is a neighborhood of V in M with a deformation retraction $r : U \to V$, for example, the open star $O_{K'}(V)$ of V in K' or a tubular neighborhood of V (cf. Theorem 3.10). Then we have isomorphisms $H^{k'+p}(M, M \smallsetminus V) \simeq H^{k'+p}(U, U \smallsetminus V)$ (excision) and $r^* : H^p(V) \xrightarrow{\sim} H^p(U)$. From Corollary 4.6, we have:

$$T_V(\alpha_V) = \Psi_V \smile r^*(\alpha_V) \qquad \text{for } \alpha_V \in H^p(V). \tag{4.25}$$

Remark 4.11. In fact, the identity in (4.25) holds without the assumption that M and V are orientable.

If S is a K_0-subcomplex of M, setting $Z = S \cap V$, we have the cup product

$$H^{k'}(M, M \smallsetminus V) \times H^p(M, M \smallsetminus S) \xrightarrow{\smile} H^{k'+p}(M, M \smallsetminus Z).$$

Proposition 4.11. *The following diagram is commutative:*

$$\begin{array}{ccc} H^p(M, M \smallsetminus S) & \xrightarrow{\Psi_V \smile (\ )} & H^{k'+p}(M, M \smallsetminus Z) \\ {\scriptstyle i^*}\downarrow & & \wr\downarrow {\scriptstyle A_{M,Z}} \\ H^p(V, V \smallsetminus Z) & \xrightarrow[A_{V,Z}]{\sim} & \breve{H}_{d'-p}(Z). \end{array}$$

Proof. For a class $\alpha \in H^p(M, M \smallsetminus S)$, by (4.13), (B.18) and (B.22),

$$A_{M,Z}(\Psi_V \smile \alpha) = [\underline{S_{K'}(Z)}] \frown (\Psi_V \smile \alpha) = ([\underline{S_{K'}(Z)}] \frown \Psi_V) \frown i^*\alpha.$$

Thus the proposition follows if we show $[\underline{S_{K'}(Z)}] \frown \Psi_V = [\underline{S_{K'_V}(Z)}]$. For this, if we let ψ be a cocycle in $C^{k'}_{K'}(M, M \smallsetminus O_{K'}(V))$ representing the Thom class, the class $[\underline{S_{K'}(Z)}] \frown \Psi_V$ is represented by the cycle

$$\sum_{\boldsymbol{t} \subset S_{K'}(Z)} \boldsymbol{t} \frown \psi, \tag{4.26}$$

where the sum is taken over m-simplices $\boldsymbol{t}$ of K' in $S_{K'}(Z)$. For such a simplex $\boldsymbol{t}$, we write $\boldsymbol{t} = \varepsilon(v_0, \dots, v_{k'}, \dots, v_m)$ and set $\boldsymbol{t}_1 = \varepsilon_1(v_0, \dots, v_{k'})$ and $\boldsymbol{t}_2 = \varepsilon_2(v_{k'}, \dots, v_m)$, where ε, ε_1 and ε_2 are ± 1 and the sign is to be determined accoding to the prescribed orientation. In the above, $\boldsymbol{t}_2$ is in a d'-simplex $\boldsymbol{s}$ of K, $\boldsymbol{t}_1$ is in its dual k'-cell $\boldsymbol{s}^*$ and $v_m \in Z$.

Then the cycle in (4.26) is written

$$\sum_{\boldsymbol{s}} \langle \iota \boldsymbol{s}^*, \psi \rangle \, \boldsymbol{t}_2,$$

where the sum is taken over d'-simplices $\boldsymbol{s}$ of K that intersect with Z and d'-simplices $\boldsymbol{t}_2$ of K' in $\boldsymbol{s}$ that intersect with Z. By Remark 4.9.1, it is equal to $\sum_{\boldsymbol{s} \subset V} \boldsymbol{t}_2$ and represents the class $[S_{K'_V}(Z)]$. □

Gysin morphism: In the above situation, let $i : V \hookrightarrow M$ denote the inclusion. The composition

$$H^p(V) \xrightarrow{T_V} H^{k'+p}(M, M \smallsetminus V) \xrightarrow{j^*} H^{k'+p}(M)$$

is referred to as the *Gysin morphism* and denoted by i_* so that we have the commutative diagram (cf. (4.18))

$$\begin{array}{ccc} H^p(V) & \xrightarrow{i_*} & H^{k'+p}(M) \\ \wr \downarrow P_V & & \wr \downarrow P_M \\ \breve{H}_{d'-p}(V) & \xrightarrow{i_*} & \breve{H}_{d'-p}(M). \end{array} \tag{4.27}$$

4.4 Intersection product

Let M be an oriented and connected C^∞ manifold of dimension m. We take a triangulation K_0 of M and let K, K' and K^* be as in Section 4.1.

Intersection product of a dual cell and a simplex

Let $\boldsymbol{s}_1$ be an oriented $(m-r)$-simplex and $\boldsymbol{s}_2$ an oriented s-simplex of K. We denote by $\frown_r$ and $\frown_l$ the right and left cap products of chains and cochains of K' (cf. Appendix B).

Definition 4.3. The *intersection product* $\boldsymbol{s}_1^* \cdot \boldsymbol{s}_2$ is an $(r+s-m)$-chain of K' defined by

$$\boldsymbol{s}_1^* \cdot \boldsymbol{s}_2 = (\vartheta(\boldsymbol{t}_1) \frown_r M_{K'}) \frown_l \vartheta(\boldsymbol{t}_2),$$

where $\boldsymbol{t}_1$ and $\boldsymbol{t}_2$ are an $(m-r)$-simplex and an $(m-s)$-simplex of K', respectively, such that $\boldsymbol{t}_1 \subset \boldsymbol{s}_1$ and $\boldsymbol{t}_2 \subset \boldsymbol{s}_2^*$.

Note that we may also write (cf. (B.26))

$$\boldsymbol{s}_1^* \bullet \boldsymbol{s}_2 = \vartheta(\boldsymbol{t}_1) \frown_r (M_{K'} \frown_l \vartheta(\boldsymbol{t}_2)).$$

By Lemmas 4.3 and 4.1,

$$\vartheta(\boldsymbol{t}_1) \frown_r M_{K'} = \boldsymbol{s}_1^* \quad \text{and} \quad M_{K'} \frown_l \vartheta(\boldsymbol{t}_2) = \boldsymbol{s}_2 \tag{4.28}$$

as K'-chains. Thus we see that the definition does not depend on the choice of $\boldsymbol{t}_1$ or $\boldsymbol{t}_2$.

We examine the above definition more closely. First note that we may write $\boldsymbol{t}_2 = \varepsilon_2(v_0, \ldots, v_{m-s})$, $v_{m-s} = b_{\boldsymbol{s}_2}$ (cf. Remark 4.1). We have

$$\boldsymbol{s}_1^* \bullet \boldsymbol{s}_2 = \boldsymbol{s}_1^* \frown_l \vartheta(\boldsymbol{t}_2) = \sum_{\boldsymbol{t}_2 \prec \boldsymbol{t} \subset \boldsymbol{s}_1^*} \boldsymbol{t} \frown_l \vartheta(\boldsymbol{t}_2), \tag{4.29}$$

where the sum is taken over the r-simplices $\boldsymbol{t}$ of K' in $\boldsymbol{s}_1^*$ containing $\boldsymbol{t}_2$ as a face. An r-simplex $\boldsymbol{t}$ of K' in $\boldsymbol{s}_1^*$ may be written $\boldsymbol{t} = \varepsilon(v'_0, \ldots, v'_r)$, $v'_r = b_{\boldsymbol{s}_1}$. The condition $\boldsymbol{t}_2 \prec \boldsymbol{t}$ is given by $(v'_0, \ldots, v'_{m-s}) = (v_0, \ldots, v_{m-s})$ and in this case, $\boldsymbol{t} \frown_l \vartheta(\boldsymbol{t}_2) = \varepsilon\varepsilon_2(v_{m-s}, \ldots, v'_r) = \varepsilon\varepsilon_2(b_{\boldsymbol{s}_2}, \ldots, b_{\boldsymbol{s}_1})$. Thus

$$\boldsymbol{s}_1^* \bullet \boldsymbol{s}_2 = \sum_{\boldsymbol{t}_2 \prec \boldsymbol{t} \subset \boldsymbol{s}_1^*} \varepsilon\varepsilon_2(v'_{m-s}, v'_{m-s+1}, \ldots, v'_{r-1}, v'_r), \quad v'_{m-s} = b_{\boldsymbol{s}_2},\ v'_r = b_{\boldsymbol{s}_1}.$$

This shows that $\boldsymbol{s}_1^* \bullet \boldsymbol{s}_2$ is non-zero if and only if $\boldsymbol{s}_1 \prec \boldsymbol{s}_2$.

Remark 4.12. 1. The support of $\boldsymbol{s}_1^* \bullet \boldsymbol{s}_2$ is the set theoretical intersection of $\boldsymbol{s}_1^*$ and $\boldsymbol{s}_2$.

2. We may write $P'\alpha'^*\vartheta(\boldsymbol{t}_1) = \boldsymbol{s}_1^*$ and $P\alpha^*\vartheta(\boldsymbol{t}_2) = \boldsymbol{s}_2$ (cf. (4.9), (4.7) and (4.28)).

Exercise 4.2. Show that in the case $\boldsymbol{s}_1 = \boldsymbol{s}_2 = \boldsymbol{s}$, we have

$$\boldsymbol{s}^* \bullet \boldsymbol{s} = b_{\boldsymbol{s}}.$$

Note that $\partial \boldsymbol{s}_1^*$ is a linear combination of cells of K^* and that $\partial \boldsymbol{s}_2$ is a linear combination of simplices of K.

Lemma 4.4. *We have*

$$\partial(\boldsymbol{s}_1^* \bullet \boldsymbol{s}_2) = (-1)^{m-s}(\partial \boldsymbol{s}_1^*) \bullet \boldsymbol{s}_2 + \boldsymbol{s}_1^* \bullet \partial \boldsymbol{s}_2.$$

Proof. By the first identity in (4.29) and (B.17),

$$\begin{aligned}\partial(\boldsymbol{s}_1^* \bullet \boldsymbol{s}_2) &= (-1)^{m-s}((\partial \boldsymbol{s}_1^*) \frown_l \vartheta(\boldsymbol{t}_2) - \boldsymbol{s}_1^* \frown_l \delta\vartheta(\boldsymbol{t}_2)) \\ &= (-1)^{m-s}((\partial \boldsymbol{s}_1^*) \bullet \boldsymbol{s}_2 - \boldsymbol{s}_1^* \frown_l \delta\vartheta(\boldsymbol{t}_2)).\end{aligned}$$

By (B.33), we have

$$s_1^* \frown_l \delta\vartheta(t_2) = \sum_{t_2 \prec t \subset s_1^*} t \frown_l \delta\vartheta(t_2)$$
$$= (-1)^{m-s+1} \sum_{t_2 \prec t \subset s_1^*} \varepsilon\varepsilon_2(v'_{m-s+1}, \dots, v'_{r-1}, v'_r), \quad v'_r = b_{s_1}, \tag{4.30}$$

where we use the expressions $t_2 = \varepsilon_2(v_0, \dots, v_{m-s})$, $v_{m-s} = b_{s_2}$, and $t = \varepsilon(v'_0, \dots, v'_r)$, $v'_r = b_{s_1}$, as above.

We may write

$$\partial s_2 = \sum_{i=0}^{s} \varepsilon^{(i)} s_2^{(i)},$$

where, for each i, $s_2^{(i)}$ is an $(s-1)$-simplex of K and $\varepsilon^{(i)}$ is to be determined accoding to the prescribed orientation. We may take $t_2^{(i)} \subset s_2^{(i)*}$ so that $t_2 \prec t_2^{(i)}$. Thus we may write $t_2^{(i)} = \varepsilon_2^{(i)}(v_0, \dots, v_{m-s}, v_{m-s+1}^{(i)})$, $v_{m-s+1}^{(i)} = b_{s_2^{(i)}}$.

Note that, since $v'_{m-s} = v_{m-s} = b_{s_2}$, in (4.30), $v'_{m-s+1} = b_{s_2^{(i)}}$ for some i with $t_2^{(i)} \prec t$.

On the other hand we have

$$s_1^* \bullet \partial s_2 = \sum_{i=0}^{s} \varepsilon^{(i)} (s_1^* \bullet s_2^{(i)}) = \sum_{i=0}^{s} \varepsilon^{(i)} \sum_{t_2^{(i)} \prec t \subset s_1^*} t \frown_l \vartheta(t_2^{(i)})$$
$$= \sum_{i=0}^{s} \varepsilon^{(i)} \sum_{t_2^{(i)} \prec t \subset s_1^*} \varepsilon\varepsilon_2^{(i)}(v'_{m-s+1}, \dots, v'_r), \quad v'_{m-s+1} = b_{s_2^{(i)}}, v'_r = b_{s_1}.$$

Note that the first sum is taken for i such that $s_1 \prec s_2^{(i)}$.

Now we claim that $\varepsilon_i \varepsilon_2^{(i)} = \varepsilon_2$ for each i, which will prove the lemma. We take an s-simplex t' of K' in s of the form $t' = \varepsilon'(v_{m-s}, v_{m-s+1}, w_{m-s+2}, \dots, w_m)$. Then $\varepsilon'\varepsilon^{(i)}(v_{m-s+1}, w_{m-s+2}, \dots, w_m)$ is an $(s-1)$-simplex of K' in $s_2^{(i)}$. By our orientation convention (*), $\varepsilon_2\varepsilon' = \varepsilon_2^{(i)}\varepsilon'\varepsilon^{(i)}$. □

Corollary 4.7. *For a locally finite r-chain c^* of K^* and a locally finite s-chain c of K,*

$$\partial(c^* \bullet c) = (-1)^{m-s}(\partial c^*) \bullet c + c^* \bullet \partial c.$$

Noting that the intersection of an r-chain of K^* and an s-chain of K is an $(r+s-m)$-chain of K', we have a bilinear map

$$\check{C}_r^{K^*}(M) \times \check{C}_s^{K}(M) \xrightarrow{\bullet} \check{C}_{r+s-m}^{K'}(M).$$

By Corollary 4.7, it induces the intersection product in homology:

$$\check{H}_r(M) \times \check{H}_s(M) \xrightarrow{\bullet} \check{H}_{r+s-m}(M).$$

Recall that we have the Poincaré duality (Theorem 4.1):

$$P_M \colon H^p(M) \xrightarrow{\sim} \check{H}_{m-p}(M).$$

In the following, we denote P_M by P if there is no fear of confusion. By Definition 4.3 and Remark 4.12. 2, we have the following:

Theorem 4.7. *The following diagram is commutative:*

$$\begin{array}{ccc} H^p(M) \times H^q(M) & \xrightarrow{\smile} & H^{p+q}(M) \\ {\scriptstyle P\times P}\downarrow\wr & & \wr\downarrow{\scriptstyle P} \\ \check{H}_r(M) \times \check{H}_s(M) & \xrightarrow{\bullet} & \check{H}_{r+s-m}(M), \end{array}$$

where $p+r=m$ and $q+s=m$.

Thus the intersection product may also be written

$$a \cdot b = P(P^{-1}a \smile P^{-1}b), \qquad a \in \check{H}_r(M),\ b \in \check{H}_s(M).$$

From the properties of the cup product, we see that

$$a \cdot b = (-1)^{(m-r)(m-s)} b \cdot a. \tag{4.31}$$

Since P is given by the cap product with the fundamental calss $[M]$, we may also write

$$a \cdot b = [M] \frown (P^{-1}a \smile P^{-1}b) = a \frown P^{-1}b. \tag{4.32}$$

In the case M is compact and connected, $\check{H}_0(M) = H_0(M) \simeq \mathbb{Z}$, the isomorphism given by the augmentation. Thus if $r+s=m$, we may think of $a \cdot b$ as an integer.

We also have, for three homology classes a, b and c,

$$(a \cdot b) \cdot c = a \cdot (b \cdot c) \tag{4.33}$$

Localized intersection product

Let S_1 and S_2 be subcomplexes of M and set $S = S_1 \cap S_2$. Then we have the intersection product

$$\check{C}_r^{K^*}(S_{K'}(S_1)) \times \check{C}_s^{K}(S_2) \xrightarrow{\bullet} \check{C}_{r+s-m}^{K'}(S_{K'}(S)).$$

By Corollary 4.7, it induces the *localized intersection product* in homology:

$$\check{H}_r(S_1)\times\check{H}_s(S_2)\xrightarrow{(\ \bullet\ )_S}\check{H}_{r+s-m}(S). \tag{4.34}$$

In general, for a K_0-subcomplex S of M, we have the Alexander duality (Theorem 4.2):

$$A_{M,S}:H^p(M,M\smallsetminus S)\xrightarrow{\sim}\check{H}_{m-p}(S).$$

Let S_1 and S_2 be subcomplexes of M and set $S=S_1\cap S_2$ as before. Then we have the cup product

$$H^p(M,M\smallsetminus S_1)\times H^q(M,M\smallsetminus S_2)\xrightarrow{\smile}H^{p+q}(M,M\smallsetminus S).$$

As Theorem 4.7, we have:

Theorem 4.8. *The following diagram is commutative:*

$$\begin{array}{ccc}
H^p(M,M\smallsetminus S_1)\times H^q(M,M\smallsetminus S_2) & \xrightarrow{\smile} & H^{p+q}(M,M\smallsetminus S)\\
{\scriptstyle A_1\times A_2}\downarrow\wr & & \wr\downarrow{\scriptstyle A}\\
\check{H}_r(S_1)\times\check{H}_s(S_2) & \xrightarrow{(\ \bullet\ )_S} & \check{H}_{r+s-m}(S),
\end{array}$$

where $p+r=m$, $q+s=m$ and A, A_1 and A_2 denote the Alexander isomorphisms for S, S_1 and S_2.

Thus the localized intersection product may also be written

$$(a\bullet b)_S=A(A_1^{-1}a\smile A_2^{-1}b),\qquad a\in\check{H}_r(S_1),\ b\in\check{H}_s(S_2).$$

In the case S is compact and connected, $\check{H}_0(S)=H_0(S)\simeq\mathbb{Z}$. Thus if $r+s=m$ we may think of $(a\bullet b)_S$ as an integer.

We also have formulas corresponding to (4.31) and (4.33).

In general from definition we have:

Proposition 4.12. *Denoting by $\imath_1:S_1\hookrightarrow M$, $\imath_2:S_2\hookrightarrow M$ and $\imath:S\hookrightarrow M$ the inclusions,*

$$(\imath_1)_*a\bullet(\imath_2)_*b=\imath_*((a\bullet b)_S)\qquad \textit{in}\ \check{H}_{r+s-m}(M).$$

Remark 4.13. The diagrams in Theorems 4.7 and 4.8 are compatible via the diagram (4.18).

Suppose S has a finite number of connected components $(S_\lambda)_\lambda$. Then $\check{H}_{r+s-m}(S)=\bigoplus_\lambda\check{H}_{r+s-m}(S_\lambda)$ and the class $(a\bullet b)_S$ determines a class $(a\bullet b)_{S_\lambda}$ in $\check{H}_{r+s-m}(S_\lambda)$ for each λ. We have

$$(a\bullet b)_S=\sum_\lambda(i_\lambda)_*(a\bullet b)_{S_\lambda}\qquad\text{in}\ \check{H}_{r+s-m}(S), \tag{4.35}$$

where $i_\lambda:S_\lambda\hookrightarrow S$ denotes the inclusion.

Intersection product with a submanifold

Let V be a closed submanifold of dimension d' of M with $i : V \hookrightarrow M$ the inclusion. We assume that V is oriented and that the normal bundle of V is orientable and is oriented as in Convention 4.1. We set $k' = m - d'$. We may assume that V is a K_0-subcomplex of M.

Convention 4.2. Let c be an r-cycle in M. The class of c in $\breve{H}_r(M)$ is denote by $[c]$, while the class of c in $\breve{H}_r(S)$ is simply denoted by c, where $S = |c|$ is the support of c.

Thus the class of V in $\breve{H}_{d'}(M)$ is denoted by $[V]$, while that in $\breve{H}_{d'}(V)$ by V.

We consider the intersection of V and a homology class in M or in a subcomplex S of M. From the definition of the intersection product and various orientation conventions, we have (cf. the first paragraph of Section 4.3):

Lemma 4.5. *If we think of V as a d'-chain of K, for every simplex $\boldsymbol{s}$ of K_V, we have*

$$\boldsymbol{s}_V^* = \boldsymbol{s}^* \cdot V$$

as K_V'-chains.

Exercise 4.3. Verify the above.

First localization: Letting S_1 and S_2 be V and M, respectively, in (4.34), we have the intersection product localized at V:

$$\breve{H}_r(V) \times \breve{H}_s(M) \xrightarrow{(\ \cdot\ )_V} \breve{H}_{r+s-m}(V).$$

Thus, for a class a in $\breve{H}_{m-p}(M)$, we have the product $(V \cdot a)_V$ in $\breve{H}_{d'-p}(V)$. It is sent to $[V] \cdot a$ by $i_* : \breve{H}_{d'-p}(V) \to \breve{H}_{d'-p}(M)$.

Second localization: Let S be a subcomplex of M with respect to K_0 and set $Z = V \cap S$. Letting S_1 and S_2 be V and S, respectively, in (4.34), we have the intersection product localized at Z:

$$\breve{H}_r(V) \times \breve{H}_s(S) \xrightarrow{(\ \cdot\ )_Z} \breve{H}_{r+s-m}(Z).$$

Thus, for a class a in $\breve{H}_{m-p}(S)$, we have the product $(V \cdot a)_Z$ in $\breve{H}_{d'-p}(Z)$. If Z has a finite number of connected components, we have the product componentwise and a formula as (4.35).

The following proposition follows from Propositions 4.10, 4.11, Theorems 4.7 and 4.8:

Proposition 4.13. *The following diagrams are commutative:*

$$\begin{array}{ccc} H^p(M) & \xrightarrow[P_M]{\sim} & \check{H}_{m-p}(M) \\ {\scriptstyle i^*}\big\downarrow & & \big\downarrow{\scriptstyle (V\bullet\ )_V} \\ H^p(V) & \xrightarrow[P_V]{\sim} & \check{H}_{d'-p}(V), \end{array} \qquad \begin{array}{ccc} H^p(M, M\setminus S) & \xrightarrow[A_{M,S}]{\sim} & \check{H}_{m-p}(S) \\ {\scriptstyle i^*}\big\downarrow & & \big\downarrow{\scriptstyle (V\bullet\ )_Z} \\ H^p(V, V\setminus Z) & \xrightarrow[A_{V,Z}]{\sim} & \check{H}_{d'-p}(Z). \end{array}$$

Remark 4.14. The two diagrams above are compatible via the diagram (4.18) applied for the pairs (M, S) and (V, Z).

Intersection product with a map

In view of the above, we define intersection products in a more general situation where V is not necessarily a submanifold of M.

Thus let V be an oriented C^∞ manifold of dimension d' and $f : V \to M$ a C^∞ map.

Definition 4.4. We define the intersection product $V\bullet_f$ so that the first diagram below is commutative. Also, for a subcomplex S in M, we set $Z = f^{-1}(S)$. Suppose Z is a subcomplex with respect to some triangulation of V. We then define the localized intersection product $(V\bullet_f\ )_Z$ so that the second diagram is commutative:

$$\begin{array}{ccc} H^p(M) & \xrightarrow[P_M]{\sim} & \check{H}_{m-p}(M) \\ {\scriptstyle f^*}\big\downarrow & & \big\downarrow{\scriptstyle (V\bullet_f\ )_V} \\ H^p(V) & \xrightarrow[P_V]{\sim} & \check{H}_{d'-p}(V), \end{array} \qquad \begin{array}{ccc} H^p(M, M\setminus S) & \xrightarrow[A_{M,S}]{\sim} & \check{H}_{m-p}(S) \\ {\scriptstyle f^*}\big\downarrow & & \big\downarrow{\scriptstyle (V\bullet_f\ )_Z} \\ H^p(V, V\setminus Z) & \xrightarrow[A_{V,Z}]{\sim} & \check{H}_{d'-p}(Z). \end{array}$$

Thus, for a class a in $\check{H}_{m-p}(S)$, we have the product $(V\bullet_f a)_Z$ in $\check{H}_{d'-p}(Z)$. If Z has a finite number of connected components, we have the product componentwise and a formula as (4.35).

Remark 4.15. 1. The two diagrams above are compatible via the diagram (4.18) applied for the pairs (M, S) and (V, Z).

2. If V is a submanifold of M and if $f = i : V \hookrightarrow M$ is the inclusion, $(V\bullet_i\ )_V$ and $(V\bullet_i\ )_Z$ coincide with $(V\bullet\ )_V$ and $(V\bullet\ )_Z$, respectively, defined before.

3. The above products, in the case V or Z is compact, are expressed using differential forms in Section 7.8 below (cf. Propositions 7.13 and 7.20).

4. See Section 7.11 below for an application.

Notes

For C^∞ triangulations, we refer to [Munkres (1966)]. The statements (T_1) and (T_2) in Section 4.1 are due to [Whitehead (1940)].

As to the dualities, we followed the descriptions in [Brasselet (1981)], except for the orientation convention, see also [Suwa (2008)]. In fact there are four ways to present the duality combinatorially. Thus let M be an oriented C^∞ manifold and let S, K and K^* be as above. We also denote by C_K^p and $C_{K^*}^p$ the groups of p-cochains of K and K^*, respectively, and by $\check{C}_{m-p}^K$ and $\check{C}_{m-p}^{K^*}$ the groups of locally finite $(m-p)$-chains of K and K^*, respectively.

For the duality morphisms (in fact isomorphisms) there are two possibilities:

(i) $P : C_K^p(M) \longrightarrow \check{C}_{m-p}^{K^*}(M), \quad A : C_K^p(M, M \smallsetminus S) \longrightarrow \check{C}_{m-p}^{K^*}(S_{K'}(S)),$
(ii) $P : C_{K^*}^p(M) \longrightarrow \check{C}_{m-p}^{K}(M), \quad A : C_{K^*}^p(M, M \smallsetminus S) \longrightarrow \check{C}_{m-p}^{K}(S).$

The morphisms in (ii) are the ones in Section 4.2 above and the ones in (i) are their "duals".

For orientation conventions, denoting by $\boldsymbol{s}$ a simplex in K, there are two possibilities:

(a) orientation of $\boldsymbol{s}$ followed by that of $\boldsymbol{s}^*$ gives the orientation of M,
(b) orientation of $\boldsymbol{s}^*$ followed by that of $\boldsymbol{s}$ gives the orientation of M.

The convention (b) is the one we adopted (cf. (4) in Section 4.2).

Thus there are four possible combinations (ia), (ib), (iia) and (iib). The convention (ia) is the one adopted in most of the literature, where usually only P is defined. For the definition of A in (i) see Section 7.2 of [Suwa (2008)]. The one in [Brasselet (1981)] is (iia) and ours is (iib). The convention (ii) for the morphism is particularly relevant for A, since the chains involved are in S and this is essential in the localization problem we later deal with. While in (i), the chains are in a neighborhood of S, which deformation retracts to S, and are a little obscured. The reason we use (b) instead of (a) is that if $\boldsymbol{s}$ is in S, $\boldsymbol{s}^*$ usualy appears in a direction normal to S and we wished to make the convention compatible with Convention 4.1.

On the level of (co)homology, the duality is given by the left cap product with $[M]$ in the conventions (ia) and (iib), while it is given by the right cap product with $[M]$ in (ib) and (iia) (cf. Remark B.11). The difference between them is a sign of $(-1)^{p(m-p)}$.

For the paragraph on the product formula of Thom classes, in particular, Theorem 4.5, we refer to [Milnor and Stasheff (1974)]. We also refer to this for Remark 4.11.

Chapter 5

Chern Classes and Localization via Obstruction Theory

Characteristic classes of a fiber bundle are invariants of the bundle that measure to what extent the bundle differs from the trivial bundle. There are several ways to define them. In this chapter, we review the Chern classes of complex vector bundles and the Euler class of oriented real vector bundles on cell complexes from the obstruction theoretical viewpoint. In this framework, characteristic classes are obstructions to constructing certain frames of the bundle, the obstruction being given as the mapping degree of the frame on the boundary of each cell.

We start with discussing the index of a family of sections at its singularity. We then explain how the Chern classes of a complex vector bundle are defined. A similar construction applies to the Euler class of an oriented real vector bundle. We also discuss the case of relative classes, where a frame is already given on a certain subset. This leads to the localization of characteristic classes by frames.

After introducing piecewise C^∞ manifolds and pseudo-manifolds, we discuss the topological localization of Chern classes by a frame and the associated residues that arise via the Alexander duality. The reidue theorem in this setting is stated (Theorem 5.2) and the residues are expressed explicitly (Theorem 5.3).

We make similar considerations for the Euler class. The Poincaré-Hopf theorem (Corollary 5.4) naturally comes up in this setting. In fact, it is a prototype of the residue theorem. We prove that the Thom class is a localized Euler class (Theorem 5.7).

For basic materials on algebraic topology, we refer to Appendix B and the references therein. Throughout this chapter we take $\mathbb{Z}$ as the coefficient ring of homology and cohomology. Also maps, vector bundles and sections are all continuous, unless otherwise stated.

5.1 Index of a family of sections

Let X be a topological space and E a complex vector bundle of rank l on X.

Definition 5.1. An *r-section* on a subset A of X is an ordered family $s^{(r)} = (s_1, \dots, s_r)$ of r sections of E on A. A *singular point* of $s^{(r)}$ is a point where the s_i's fail to be linearly independent over $\mathbb{C}$. An *r-frame* is an r-section without singularities.

A 1-section is nothing but a section. In this case the singular points are the zeros of the section. Recall that an l-frame is simply called a frame (cf. Definition 3.12).

In the next section, we define the q-th Chern class $c^q(E)$ of E to be the primary obstruction to constructing an r-frame of E, $r = l - q + 1$. For this purpose, we introduce the notion of the *index* of an r-section at a point where it is singular or is not defined.

Mapping degree

Let $W(l,r)$ be the Stiefel manifold of r-frames in $\mathbb{C}^l$ (cf. Section 3.5). Recall that $H_{2q-1}(W(l,r)) \simeq \mathbb{Z}$, $q = l - r + 1$ (cf. (3.17)). It has a canonical generator w_{2q-1}, i.e., there is a natural inclusion $\iota : \mathbb{S}^{2q-1} \hookrightarrow W(l,r)$ and $w_{2q-1} = \iota_* \nu_{2q-1}$ with ν_{2q-1} the canonical generator of $H_{2q-1}(\mathbb{S}^{2q-1}) \simeq \mathbb{Z}$. In particular, if $r = 1$, then $q = l$ and $W(l,1) = \mathbb{C}^l \smallsetminus \{0\}$, which deformation retracts to $\mathbb{S}^{2l-1}$.

Suppose we have a map $\varphi : \mathbb{S}^{2q-1} \to W(l,r)$. Then it induces a morphism

$$\varphi_* : H_{2q-1}(\mathbb{S}^{2q-1}) \longrightarrow H_{2q-1}(W(l,r))$$

and we may write

$$\varphi_*(\nu_{2q-1}) = d \cdot w_{2q-1}$$

with d an integer uniquely determined by the homotopy class of φ.

Definition 5.2. The above integer d is called the *mapping degree* of φ and is denoted by $\deg \varphi$.

Remark 5.1. 1. Note that $\pi_{2q-1}(W(l,r)) \simeq \mathbb{Z}$, which has the canonical generator $[\iota]$, $\iota : \mathbb{S}^{2q-1} \hookrightarrow W(l,r)$ being as above. A map φ as above defines an element in $\pi_{2q-1}(W(l,r))$, which is of the form $d \cdot [\iota]$ with $d = \deg \varphi$.

Let $\mathbb{B}^{2q}$ denote a closed $2q$-ball whose boundary is $\mathbb{S}^{2q-1}$. Then $\deg\varphi = 0$ if and only if φ can be extended to a map $\mathbb{B}^{2q} \to W(l,r)$ and we may think of $\deg\varphi$ as the *obstruction* to extending φ to $\mathbb{B}^{2q}$.

2. The mapping degree is also expressed in terms of cohomology. For this, note that both $H^{2q-1}(\mathbb{S}^{2q-1})$ and $H^{2q-1}(W(l,r))$ are isomorphic with $\mathbb{Z}$. Let ν_{2q-1}^* and w_{2q-1}^* denote their generators dual to ν_{2q-1} and w_{2q-1}, respectively, i.e., the ones with $\langle \nu_{2q-1}, \nu_{2q-1}^* \rangle = 1$ and $\langle w_{2q-1}, w_{2q-1}^* \rangle = 1$. For a map $\varphi : \mathbb{S}^{2q-1} \to W(l,r)$, we have the induced morphism

$$\varphi^* : H^{2q-1}(W(l,r)) \longrightarrow H^{2q-1}(\mathbb{S}^{2q-1})$$

and we may write (cf. B.8):

$$\varphi^*(w_{2q-1}^*) = \deg\varphi \cdot \nu_{2q-1}^*.$$

3. Here is a differential geometric interpretation of the mapping degree (cf. Section 10.5 below for details).

We have $H^{2q-1}((W(l,r);\mathbb{C}) \simeq \mathbb{C}$ with a canonical generator (cf. Theorem B.20). By the de Rham theorem (Theorem 7.5 below), the generator is represented by a differential form ω_{2q-1} on $W(l,r)$ and, for a C^∞ map $\varphi : \mathbb{S}^{2q-1} \to W(l,r)$, we may express its degree as (cf. Proposition 10.13):

$$\deg\varphi = \int_{\mathbb{S}^{2q-1}} \varphi^*\omega_{2q-1}.$$

An explicit expression of ω_{2q-1} is given in Proposition 7.28 below. In particular, if $r = 1$, then $q = l$ and $W(l,1) = \mathbb{C}^l \smallsetminus \{0\}$. In this case we may take as ω_{2q-1} the Bochner-Martinelli form β_l (cf. Section 7.10). Note that

$$\beta_1 = \frac{1}{2\pi\sqrt{-1}}\frac{dz}{z},$$

the Cauchy form.

Index of an r-frame

We now consider the bundle $W(E,r)$ of r-frames of E on X. This is a bundle associated with E whose fiber $W(E,r)_x$ at each point x of X is diffeomorphic with $W(l,r)$. An r-frame of E is nothing but a section of $W(E,r)$.

Suppose we have a regular oriented $2q$-cell e in X (cf. Section B.2). Letting $\mathbb{B}^{2q}$ be the unit closed ball with canonical orientation (cf. Example 3.9 and Remark B.13. 2), we take a characteristic map $\chi : \mathbb{B}^{2q} \to \overline{e}$

so that it is a homeomorphism determining the orientation of $\boldsymbol{e}$. We may assume that E is trivial on $\overline{\boldsymbol{e}}$. Suppose we are given an r-frame $s^{(r)}$ of E on $\overline{\boldsymbol{e}}\setminus\boldsymbol{e}$. Although it is not necessary, in order to fix the idea, we extend $s^{(r)}$ to an r-frame on $\overline{\boldsymbol{e}}\setminus\{a\}$, where a is a point in $\boldsymbol{e}$. It is always possible as $\overline{\boldsymbol{e}}\setminus\{a\}$ deformation retracts to $\overline{\boldsymbol{e}}\setminus\boldsymbol{e}$. Denoting by $\mathbb{S}^{2q-1}$ the boundary of $\mathbb{B}^{2q}$, which is a canonically oriented $(2q-1)$-sphere, we have a composition of maps:

$$\varphi : \mathbb{S}^{2q-1} \xrightarrow{\partial\chi} \overline{\boldsymbol{e}}\setminus\boldsymbol{e} \xrightarrow{s^{(r)}} W(E,r)|_{\overline{\boldsymbol{e}}} \simeq (W(l,r)\times\overline{\boldsymbol{e}}) \xrightarrow{p} W(l,r), \qquad (5.1)$$

where $\partial\chi = \chi|_{\mathbb{S}^{2q-1}}$ and p is the projection onto the first factor.

Definition 5.3. The *index* $I(s^{(r)},a)$ of $s^{(r)}$ at a is defined by

$$I(s^{(r)},a) = \deg\varphi.$$

Note that the definition does not depend on the order of the members of $s^{(r)}$, the choice of a in $\boldsymbol{e}$ or the trivialization of E.

Example 5.1. Let U be a neighborhood of 0 in $\mathbb{C}=\{z\}$ and $E=\mathbb{C}\times U$ the product bundle on U. Also let $\mathbb{B}^2$ be a closed 2-ball in U with center 0. For an integer m, let s be the frame of E on $\mathbb{S}^1=\partial\mathbb{B}^2$ given by $s(z)=(z^m,z)$. Then, noting that $W(1,1)=\mathbb{C}^*$ and the map in (5.1) is given by $z\mapsto z^m$, we see that $I(s,0)=m$.

More generally, if the frame s on $\mathbb{S}^1$ is given by $s(z)=(f(z),z)$ with f a non-vanishing C^∞ function, we have (cf. Remark 5.1.3):

$$I(s,0) = \frac{1}{2\pi\sqrt{-1}}\int_{\mathbb{S}^1}\frac{df}{f}.$$

The index in Definition 5.3 may also be described as follows. Let $s^{(r)}$ be an r-frame of E on $\overline{\boldsymbol{e}}\setminus\boldsymbol{e}$, with $r>1$. We write as $s^{(r)}=(s^{(r-1)},s_r)$, individualizing the last section. The $(r-1)$-frame $s^{(r-1)}$ determines a map

$$\varphi_1 : \mathbb{S}^{2q-1} \longrightarrow W(l,r-1)$$

similarly as (5.1). Since $\pi_{2q-1}(W(l,r-1))=0$, the map φ_1 extends to a map $\mathbb{B}^{2q}\to W(l,r-1)$. Thus $s^{(r-1)}$ can be extended to an $(r-1)$-frame on $\overline{\boldsymbol{e}}$ and generates a trivial complex subbundle $\mathbb{I}^{r-1}$ of rank $r-1$ of $E|_{\overline{\boldsymbol{e}}}$. This way we have an exact sequence

$$0 \longrightarrow \mathbb{I}^{r-1} \longrightarrow E|_{\overline{\boldsymbol{e}}} \xrightarrow{\rho} E' \longrightarrow 0,$$

where E' is a complex vector bundle of rank q on $\overline{e}$. We denote by s'_r the section of E' given by $\rho \circ s_r$, which is non-vanishing on $\overline{e} \setminus e$. Consider the composition (cf. (5.1)):

$$\varphi' : \mathbb{S}^{2q-1} \xrightarrow{\partial \chi} \overline{e} \setminus e \xrightarrow{s'_r} W(E', 1) \simeq W(q, 1) \times \overline{e} \xrightarrow{p} W(q, 1).$$

Then by definition, $I(s'_r, a) = \deg \varphi'$.

Proposition 5.1. *In the above situation,*

$$I(s^{(r)}, a) = I(s'_r, a).$$

Proof. As we noted, the map φ_1 extends to a map $\mathbb{B}^{2q} \to W(l, r-1)$, which is also denoted by φ_1. We may assume that for every point x in $\mathbb{B}^{2q}$, $\varphi_1(x)$ is given by the matrix $(e_1, \dots, e_{r-1})$, where $e_i = {}^t(0, \dots, \overset{i}{1}, \dots, 0)$. This means that we take the trivialization $W(E, r)|_{\overline{e}} \simeq W(l, r) \times \overline{e}$ in (5.1) so that for each point x in $\mathbb{S}^{2q-1}$, $\varphi(x)$ is given by a matrix of the form $(e_1, \dots, e_{r-1}, a)$, where $a = {}^t(a_1, \dots, a_{r-1}, a_r \dots, a_l)$ with ${}^t(a_1, \dots, a_{r-1})$ in $\mathbb{C}^{r-1}$ and $(a_r \dots, a_l)$ in $\mathbb{C}^q \setminus \{0\} = W(q, 1)$. This also means that we take the trivialization of $W(E', 1)$ so that $\varphi'(x) = (a_r \dots, a_l)$.

Recall that the map $\pi : W(l, r) \to W(l, r-1)$ given by taking the first $(r-1)$ vectors has a fiber bundle structure with fiber $\mathbb{C}^{r-1} \times W(q, 1)$ and that the inclusion of $W(q, 1)$ into a fiber of π induces an isomorphism of the $(2q-1)$-st homotopy groups (cf. Propositions 3.7 and 3.8). The proposition then follows from Remark 3.5. 1. □

5.2 Chern classes of a complex vector bundle

We review the obstruction theoretic construction of the Chern classes of complex vector bundles on cell complexes.

Let $(X, \{e_\lambda\})$ be a regular cell complex (cf. Section B.2). Recall that the singular homology $H_p(X)$ of X can be computed from the chain complex $(C_\bullet(X), \partial)$ with $C_p(X)$ the free Abelian group generated by the oriented p-cells in X. Also the singular cohomology $H^p(X)$ can be computed from the cochain complex $(C^\bullet(X), \delta)$ with $C^p(X) = \mathrm{Hom}(C_p(X), \mathbb{Z})$ and δ the transpose of ∂. This means that, if we denote by $H_p(C_\bullet(X))$ the p-th homology of $(C_\bullet(X), \partial)$ and by $H^p(C^\bullet(X))$ the p-th cohomology of $(C^\bullet(X), \delta)$, there are canonical isomorphisms (cf. Theorem B.22):

$$H_p(C_\bullet(X)) \simeq H_p(X), \qquad H^p(C^\bullet(X)) \simeq H^p(X).$$

Let E be a (continuous) complex vector bundle of rank l on X and r an integer with $1 \leq r \leq l$. We set $q = l - r + 1$ as before. We try to construct an r-frame, i.e., a section of $W(E, r)$, on each skeleton X^p of X starting from the 0-skeleton and then extending it inductively to larger skeletons. On the way we get a $2q$-cochain γ as the obstruction to the construction. It is shown that γ is in fact a cocycle and defines a class in $H^{2q}(X)$, which will be the q-th Chern class $c^q(E)$ of E.

First, it is always possible to construct a section $s^{(r)}$ of $W(E, r)$ on X^0. Let $\boldsymbol{e}$ be a p-cell. If a section $s^{(r)}$ of $W(E, r)$ is given on $\overline{\boldsymbol{e}} \setminus \boldsymbol{e}$, it defines a map as (5.1), replacing $\mathbb{S}^{2q-1}$ with $\mathbb{S}^{p-1} \simeq \overline{\boldsymbol{e}} \setminus \boldsymbol{e}$. Thus $s^{(r)}$ defines an element in $\pi_{p-1}(W(l, r))$. If $p \leq 2l - 2r + 1 = 2q - 1$, the section $s^{(r)}$ can be extended to $\overline{\boldsymbol{e}}$, since the homotopy group vanishes. This way we may construct a section $s^{(r)}$ of $W(E, r)$ on X^{2q-1}. Then we reach to an "obstruction" when $p = 2q$. Namely, for each $2q$-cell $\boldsymbol{e}$, the r-frame $s^{(r)}$ on $\overline{\boldsymbol{e}} \setminus \boldsymbol{e}$ is extended to an r-frame on $\overline{\boldsymbol{e}}$ possibly except for a point a in $\boldsymbol{e}$ and the obstruction is given by the index $I(s^{(r)}, a)$. We define a cochain γ by

$$\langle \boldsymbol{e}, \gamma \rangle = I(s^{(r)}, a) \tag{5.2}$$

for each $2q$-cell $\boldsymbol{e}$ and then extending it linearly.

Proposition 5.2. *The cochain γ is a cocycle.*

Proof. It suffices to show that, for every $(2q+1)$-cell $\boldsymbol{e}'$,

$$\langle \boldsymbol{e}', \delta\gamma \rangle = 0. \tag{5.3}$$

Let Y denote the subcomplex of X consisting of the cells in $\overline{\boldsymbol{e}'}$. Since Y^{2q-1} is $(2q-2)$-connected, by Theorem 3.7, we have an isomorphism $h : \pi_{2q-1}(Y^{2q-1}) \xrightarrow{\sim} H_{2q-1}(Y^{2q-1})$. We also have $H_{2q-1}(Y^{2q-1}) = Z_{2q-1}(Y^{2q-1}) = Z_{2q-1}(Y)$. On the other hand, using the fact that the bundle $E|_{\overline{\boldsymbol{e}'}}$ is trivial, the given r-section $s^{(r)}$ on X^{2q-1} defines a map $s_*^{(r)} : \pi_{2q-1}(Y^{2q-1}) \to \pi_{2q-1}(W(l, r)) \simeq \mathbb{Z}$. We denote by α the composition of the maps

$$C_{2q}(Y) \xrightarrow{\partial} Z_{2q-1}(Y) \xrightarrow{h^{-1}} \pi_{2q-1}(Y^{2q-1}) \xrightarrow{s_*^{(r)}} \pi_{2q-1}(W(l, r)) \simeq \mathbb{Z}.$$

By definition, for a $2q$-chain c in $C_{2q}(Y)$, $\langle c, \gamma \rangle = \alpha(c)$. Since $C_{2q-2}(Y)$ is a free Abelian group, $Z_{2q-1}(Y)$ is a direct summand of $C_{2q-1}(Y)$. Thus $s_*^{(r)} \circ h^{-1}$ extends to a map $u : C_{2q-1}(Y) \to \pi_{2q-1}(W(l, r)) \simeq \mathbb{Z}$, which may be thought of as a cochain in $C^{2q-1}(Y)$. Then we have, for every chain c in $C_{2q}(Y)$, $\langle c, \gamma \rangle = \alpha(c) = \langle \partial c, u \rangle = \langle c, \delta u \rangle$ so that $\gamma = \delta u$ in $C^{2q}(Y)$ and (5.3) follows. $\square$

Remark 5.2. 1. It is also shown that different choices of extensions in each step leads to cocycles that are cohomologous.

2. Precisely speaking, γ is a cochain with coefficients in the "local system formed by $\pi_{2q-1}(W(E_x, r))$". Since the group $GL(l, \mathbb{C})$ is connected, this system is in fact trivial and we may naturally think of γ as being in $C^{2q}(X)$.

Definition 5.4. The *q-th topological Chern class* $c^q_{\text{top}}(E)$ of E is the class of γ in $H^{2q}(X)$.

Throughout this chapter, we denote $c^q_{\text{top}}(E)$ simply by $c^q(E)$ and call it the q-th Chern class of E, unless otherwise is stated. The *total Chern class of* E is defined by

$$c^*(E) = 1 + c^1(E) + \cdots + c^l(E),$$

which is an element in the cohomology ring $H^*(X)$ and is invertible.

Remark 5.3. 1. If E admits an r-frame on X, then clearly $c^q(E) = 0$. Conversely if $c^q(E) = 0$, then it is possible to construct an r-frame of E on X^{2q}, but not on X in general. Thus $c^q(E)$ is referred to as the *primary obstruction* to constructing an r-frame of E.

2. The classes constructed above are in fact "functorial", i.e., for a continuous map $f : X' \to X$ of cell complexes and a complex vector bundle E on X,

$$c^q(f^*E) = f^*c^q(E) \qquad \text{in } H^{2q}(X').$$

This is proved by taking a "cellular approximation" of f.

In particular, the above construction of $c^q(E)$ does not depend on the choice of cellular decomposition of X.

3. In Chapter 8 below, we define differential geometric Chern classes for a C^∞ complex vector bundle on a C^∞ manifold via the Chern-Weil theory and discuss, in Chapter 10, the relation between the classes defined from two different approaches. We prove the above funtoriality in the framework of Chern-Weil theory.

5.3 Euler class of an oriented real vector bundle

Let E be an oriented real vector bundle of rank l' on a regular cell complex $(X, \{e_\lambda\})$. We assume again that each cell is oriented. If $l' = 1$, the orientability of E implies that E is a trivial bundle. Thus we assume that $l' > 1$ in the following. In view of (3.18), we may perform a similar construction to obtain the Euler class of an oriented real vector bundle.

We denote by $V(l',1)$ the real Stiefel manifold of 1-frames in $\mathbb{R}^{l'}$, which is in fact $\mathbb{R}^{l'} \setminus \{0\}$ so that it has the homotopy type of $\mathbb{S}^{l'-1}$. Thus $V(l',1)$ is $(l'-2)$-connected and $\pi_{l'-1}(V(l',1)) \simeq \mathbb{Z}$, which has a canonical generator. Hence we have $H_{l'-1}(V(l',1)) \simeq \mathbb{Z}$ with a canonical generator and, for a map $\varphi : \mathbb{S}^{l'-1} \to V(l',1)$, we my define its degree, denoted by $\deg\varphi$, as in Definition 5.2.

We denote by $V(E,1)$ the bundle of 1-frames of E on X. Then the Euler class $e(E)$ of E is defined to be the obstruction to constructing a section of $V(E,1)$, i.e., a non-vanishing section of E, on $X^{l'}$. It will be defined *a priori* in the cohomology with coefficients in the local system formed by $\pi_{l'-1}(V(E_x,1))$. Since $GL^+(l',\mathbb{R})$ is connected, this system is in fact trivial and we may naturally think of the obstruction as being in the integral cohomology $H^{l'}(X;\mathbb{Z})$, as in the case of Chern classes.

We may define the index of a section at its singularity as before. Thus let $\boldsymbol{e}$ be an l'-cell in X, on which E is trivial. Suppose we have a non-vanishing section s of E on $\overline{\boldsymbol{e}} \setminus \{a\}$, a being a point in $\boldsymbol{e}$. We have a composition (cf. (5.1)):

$$\varphi : \mathbb{S}^{l'-1} \xrightarrow{\partial\chi} \overline{\boldsymbol{e}} \setminus \boldsymbol{e} \xrightarrow{s} V(E,1)|_{\overline{\boldsymbol{e}}} \simeq (V(l',1) \times \overline{\boldsymbol{e}}) \xrightarrow{p} V(l',1). \tag{5.4}$$

The *index* of s at a, denoted by $I(s,a)$, is the mapping degree of φ:

$$I(s,a) = \deg\varphi. \tag{5.5}$$

We try to construct a section s of $V(E,1)$ on each skeleton X^p of X inductively from the 0-skeleton, as in the case of Chern classes. Continuing this process, we reach to an "obstruction" when $p = l'$. Namely, for each l'-cell $\boldsymbol{e}$, the 1-frame s on $\overline{\boldsymbol{e}} \setminus \boldsymbol{e}$ is extended to a 1-frame on $\overline{\boldsymbol{e}}$ possibly except for a point a in $\boldsymbol{e}$ and the obstruction is given by the index $I(s,a)$. We may define a cochain γ in $C^{l'}(X)$ by setting $\gamma(\boldsymbol{e}) = I(s,a)$ for each l'-cell $\boldsymbol{e}$ and then extending it linearly. This cochain is in fact a cocycle as in the case of Chern classes.

Definition 5.5. The *Euler class* $e(E)$ of E is the class of γ in $H^{l'}(X)$.

This class is independent of the choices of various objects used in the definition.

Remark 5.4. 1. As in the case of Chern classes, $e(E) = 0$ if and only if E admits a non-vanishing section on $X^{l'}$. The Euler class is also functorial (cf. Remark 5.3).

2. If E is a complex vector bundle of rank l, we may think of it as a naturally oriented real bundle of rank $2l$ and $W(l,1)$ may be identified with $V(2l,1)$. In this case, we have

$$e(E) = c^l(E),$$

the top Chern class of E.

3. For an oriented real vector bundle E of even rank $l' = 2k'$, it is known that

$$(-1)^{k'} c^{l'}(E^c) = e(E) \smile e(E). \tag{5.6}$$

4. The use of (3.18) in general case of frames leads to the obstruction theoretic definition of Stiefel-Whitney classes of real vector bundles.

5.4 Relative classes

Let $(X, \{e_\lambda\})$ be a regular cell complex and A a subcomplex of X (cf. Section B.2).

Relative Chern classes

Let E be a complex vector bundle of rank l on X and suppose we are already given an r-frame $s^{(r)}$ of E on the $2q$-skeleton A^{2q} of A, $q = l - r + 1$. We follow the procedure described in Section 5.2 starting with this frame to obtain an r-frame $\tilde{s}^{(r)}$ on $X^{2q} \smallsetminus \{\text{isolated points}\}$. To be more precise, the above frame $\tilde{s}^{(r)}$ has the following property for a $2q$-cell $\boldsymbol{e}$ in X:

$$\begin{cases} \text{if } \boldsymbol{e} \subset A, \text{ then } \tilde{s}^{(r)} \text{ is defined and equals } s^{(r)} \text{ on } \overline{\boldsymbol{e}}, \\ \text{if } \boldsymbol{e} \not\subset A, \text{ then } \tilde{s}^{(r)} \text{ is defined on } \overline{\boldsymbol{e}} \smallsetminus \{a\}, \text{ where } a \text{ is a point in } \boldsymbol{e}. \end{cases}$$

For every $2q$-cell $\boldsymbol{e}$ in A, we have $I(\tilde{s}^{(r)}, a) = I(s^{(r)}, a) = 0$, a being a point in $\boldsymbol{e}$. Thus the cocycle γ defined by (5.2) for $\tilde{s}^{(r)}$ is in $C^{2q}(X, A)$. It represents a class in $H^{2q}(X, A)$, which is denoted by $c^q(E, s^{(r)})$ and is called the *relative Chen class* defined by $s^{(r)}$. The class $c^q(E, s^{(r)})$ depends on $s^{(r)}$, but not on the choice of the extension $\tilde{s}^{(r)}$ of $s^{(r)}$ (cf. Remark 5.2.1). Its image by the canonical morphism $H^{2q}(X, A) \to H^{2q}(X)$ is the usual Chern class $c^q(E)$.

Relative Euler class

Let E be an oriented real vector bundle of rank l' on X and suppose we are already given a non-vanishing section s of E on $A^{l'}$. We follow the procedure described in Section 5.2 for the Euler class starting with this section to obtain a non-vanishing section $\tilde{s}$ on $X^{l'} \setminus \{\text{isolated points}\}$, as in the case of relative Chern classes. Then, for every l'-cell e in A, we have $I(\tilde{s}, a) = I(s, a) = 0$. Thus the cocycle γ for $\tilde{s}$ defining the Euler class $e(E)$ is in fact in $C^{l'}(X, A)$. It represents a class in $H^{l'}(X, A)$, which we denote by $e(E, s)$ and call the relative Euler class defined by s. Its image by the canonical morphism $H^{l'}(X, A) \to H^{l'}(X)$ is the Euler class $e(E)$.

5.5 Piecewise-linear manifolds and pseudo-manifolds

PL manifolds and smoothings

Polyhedra: Let v be a point and B a subset of $\mathbb{R}^N$. The *cone* with vertex v and base B is the set

$$v * B = \{\, x \in \mathbb{R}^N \mid x = (1-t)v + tb,\ b \in B,\ \ 0 \le t \le 1 \,\}.$$

Definition 5.6. A *polyhedron* is a subset P of $\mathbb{R}^N$ such that each point a in P has a neighborhood S in P of the form $S = a * L$ with L compact.

A neighborhood S as above is called a *star* of a in P and L a *link*. For example, $\mathbb{R}^m$ and $\mathbb{H}^m = \{\, (x_1, \dots, x_m) \in \mathbb{R}^m \mid x_1 \le 0 \,\}$ are polyhedra. An open set of a polyhedron is a polyhedron. The polyhedron $|K|$ of a simplicial complex K is a polyhedron. For a point a in $|K|$, $S_K(a)$ is a star and $L_K(a)$ a link of a. In fact we have:

Proposition 5.3. *A subset P of $\mathbb{R}^N$ is a polyhedron if and only if it is the polyhedron of some simplicial complex K in $\mathbb{R}^N$, i.e., $P = |K|$.*

We call K as above a *simplicial decomposition* of P. In this case the identity map $|K| \to P$ is a triangulation of P.

Definition 5.7. A map $f : P \to P'$ between polyhedra in $\mathbb{R}^N$ and $\mathbb{R}^{N'}$, respectively, is *piecewise-linear*, *PL* for short, if P admits a simplicial decomposition K such that for each simplex s of K, $f|_s$ is affine.

It is equivalent to saying that the graph $\Gamma_f = \{\, (x, f(x)) \mid x \in P \,\}$ of f is a polyhedron in $\mathbb{R}^{N+N'}$. Thus if a PL map $f : P \to P'$ is a homeomorphism,

f^{-1} is also PL. A simplicial map $f : |K| \to |K'|$ of simplicial complexes is PL. If $f : P \to P'$ is PL and proper, there exist simplicial decompositions K and K' of P and P', respectively, such that $f : |K| \to |K'|$ is simplicial.

PL manifolds: A *PL manifold with boundary* is defined by replacing "C^∞" with "PL" in Definition 3.20. Thus let Q be a Hausdorff space with countable basis.

Definition 5.8. A *PL structure* on Q is the equivalence class of an atlas $\{(U_\alpha, \varphi_\alpha)\}$ such that

(1) for each α, φ_α is a homeomorphism onto an open set in $\mathbb{R}^m$ or in $\mathbb{H}^m$,
(2) for each pair (α, β), $\varphi_\alpha \circ \varphi_\beta^{-1} : \varphi_\beta(U_\alpha \cap U_\beta) \to \varphi_\alpha(U_\alpha \cap U_\beta)$ is PL.

The integer m as above is called the dimension of Q. The boundary ∂Q of Q is the set of points in Q that correspond to $x_1 = 0$ in $\mathbb{H}^m$. If $\partial Q \neq \emptyset$, it is an $(m-1)$-dimensional PL manifold without boundary.

A PL manifold Q has a triangulation $h : |K| \to Q$ such that each map $h^{-1} \circ \varphi_\alpha^{-1} : \varphi_\alpha(U_\alpha) \to |K|$ is PL.

Let M be a PL manifold without boundary. A *PL submanifold with boundary* in M is defined as in Definition 3.21, replacing "C^∞" with "PL". Thus if Q is a PL submanifold with boundary, it is a PL manifold with boundary.

Remark 5.5. A polyhedron P is a PL manifold of dimension m if and only if, for each point $a \in P$, there is a triangulation K of P such that a is a vertex of K and that the link $L_K(a)$ of a is a PL $(m-1)$-sphere $\partial \Delta^m$.

Piecewise C^∞ homeomorphism: Let P be a polyhedron and R a C^∞ manifold with boundary.

Definition 5.9. A homeomorphism $h : P \to R$ is piecewise C^∞, PD for short, if P has a simplicial decomposition K such that h is a C^∞ triangulation (cf. Definition 4.1).

In this case, the atlas on $P = |K|$ given by the open stars of vertices in K makes P into a PL manifold.

Let Q be a PL manifold with boundary and R a C^∞ manifold with boundary. A homeomorphism $h : Q \to R$ is PD if Q has a triangulation $h' : |K| \to Q$ such that $h \circ h' : |K| \to R$ is PD.

Remark 5.6. We may rephrase the statements (T_1) and (T_2) in Section 4.1 by saying that every C^∞ manifold M admits a PD homeomorphism $Q \to M$ and the PL manifold Q is unique up to PL isomorphism.

Smoothings of PL manifolds: Let Q be a PL manifold with boundary. If σ is a C^∞ structure on (the underling topological space of) Q, we denote by Q_σ the C^∞ manifold defined by σ. A *smoothing* of Q is a C^∞ structure σ on Q that makes the identity $1_Q : Q \to Q_\sigma$ PD. Thus if $h : Q \to R$ is a PD homeomorphism onto a C^∞ manifold R, the C^∞ structure induced from that of R by h is a smoothing of Q. We also call $h : Q \to R$ a smoothing.

Let Q be a PL manifold and $Q_1 \subset Q \setminus \partial Q$ a PL submanifold of codimension k. We say that Q_1 is *flat* if it has a neighborhood U in Q such that (U, Q_1) is PL homeomorphic to $(Q_1 \times \mathbb{R}^k, Q_1 \times \{0\})$.

We quote:

Theorem 5.1. *Let Q be a PL manifold and $Q_1 \subset Q \setminus \partial Q$ a flat PL submanifold. Also let R be a C^∞ manifold and $h : Q \to R$ a smoothing. Then there is a PD isotopy $h_t : Q \to R$ such that $h_0 = h$ and that $h_1(Q_1)$ is a C^∞ submanifold of R. Moreover, if Q_1 is compact, for every neighborhood N of Q_1 and every ε, h_t can be chosen to be an ε-isotopy relative to $Q \setminus N$.*

Here we recall some terminologies in the above (cf. Section B.1). In general, if X and Y are topological spaces, an *isotopy* $h_t : X \to Y$ is a homotopy $\{h_t\}_{0 \le t \le 1}$ such that $H : X \times I \to Y \times I$, $H(x,t) = (h_t(x), t)$, is a homeomorphism, $I = [0,1]$.

If Q is a PL manifold and R a C^∞ manifold, a *PD isotopy* is an isotopy $h_t : Q \to R$ such that the extension $H' : Q \times \mathbb{R} \to R \times \mathbb{R}$ of H given by $H'(x,t) = (h_0(x), t)$, for $t \le 0$, and $H'(x,t) = (h_1(x), t)$, for $t \ge 1$, is a PD homeomorphism. Let $\varepsilon : Q \to \mathbb{R}_+ = \{\, a \in \mathbb{R} \mid a > 0 \,\}$ be a continuous function. An isotopy h_t is an *ε-isotopy* if $d(h_t(x), h_0(x)) < \varepsilon(x)$, for all $x \in Q$ and $t \in I$, where $d(\ ,\ )$ denotes a metric on R.

Orientability: Let Q be a PL manifold of dimension m and $h : |K| \to Q$ a triangulation. Then, for each $(m-1)$-simplex $\boldsymbol{s}$ of K, there exist either one or two m-simlices which contain $\boldsymbol{s}$ as a face. The PL manifold Q is said to be *orientable* if it admits a triangulation K with the following property: the m-simplices in K can be oriented so that, if $\boldsymbol{s}$ is an $(m-1)$-simplex in K and if $\boldsymbol{s}_1$ and $\boldsymbol{s}_2$ are two simplices containing $\boldsymbol{s}$ as a face, then the prescribed orientations of $\boldsymbol{s}_1$ and $\boldsymbol{s}_2$ induce opposite orientations of $\boldsymbol{s}$.

If Q is orientable, it has two orientations, i.e., there are two ways to orient the m-simplices in the above manner. Once we fix an orientation, we say that Q is oriented. In this case ∂Q is naturally oriented.

Piecewise C^∞ manifolds: Let M be a C^∞ manifold.

Definition 5.10. A subset R of M is a *piecewise C^∞ submanifold of M with boundary*, if there exist a C^∞ triangulation (K, h) of M and a subcomplex L of K such that $h(|L|) = R$ and that $|L|$ is a PL submanifold of $|K|$ with boundary.

In the above, we also say that R is a piecewise C^∞ manifold with boundary in M.

Note that a closed C^∞ submanifold with boundary is a piecewise C^∞ submanifold with boundary.

Let R be a piecewise C^∞ submanifold of M with boundary. The dimension of R is defined to be that of $|L|$, L being as in Definition 5.10. Let $d' = \dim R$. If $\partial R \neq \emptyset$, it is a $(d'-1)$-dimensional piecewise C^∞ manifold without boundary.

Definition 5.11. A point a in R is said to be *general*, if there exist (K, h) and L as above such that a is the image by h of an interior point of a d'-simplex of L.

The set of general points in R is a d'-dimensional C^∞ submanifold of M without boundary.

Let R be as above. We say that R is orientable if $|L|$ is orientable as a PL-manifold. This is equivalent to saying that the set of general points in R is orientable as a C^∞ manifold. If this is the case, ∂R is also orientable. We may think of the set of general points of ∂R as the boundary of a C^∞ manifold with boundary whose interior is the set of general points of R. If R is oriented, ∂R is oriented according to Convention 3.2.

Pseudo-manifolds

Definition 5.12. A *pseudo-manifold* of dimension d' in M is a subcomplex X of M, with respect to some triangulation of M, satisfying the following conditions:

(1) Every simplex in X is a face of some d'-simplex in X.
(2) Every $(d'-1)$-simplex is the face of exactly two d'-simplices.

The pseudo-manifold X is *orientable* if

(3) The d'-simplices in X can be oriented so that, if s is a $(d'-1)$-simplex in X and if s_1 and s_2 are the two simplices that contain s in their

boundary, then the prescribed orientations of $\boldsymbol{s}_1$ and $\boldsymbol{s}_2$ induce opposite orientations of $\boldsymbol{s}$.

An orientable pseudo-manifold X is said to be *oriented*, once orientations of d'-simplices in X satisfying (3) above are fixed. We say that X is *irreducible* if $X \smallsetminus X^{d'-2}$ is connected, where $X^{d'-2}$ denotes the $(d'-2)$-skeleton of X. A *general point* of X is a point in $X \smallsetminus X^{d'-2}$. If X is an oriented pseudo-manifold, the union of all the d'-simplices in X is a cycle. In particular, if X is irreducible, it is called the *fundamental cycle* of X and its class in $\breve{H}_{d'}(X)$ the *fundamental class* of X. In this case, there is a canonical isomorphism $\breve{H}_{d'}(X) \simeq \mathbb{Z}$, the fundamental class corresponding to 1.

In general, we have a decomposition into irreducible components:

$$X = \bigcup_i X_i.$$

If X is oriented, then $\breve{H}_{d'}(X)$ is generated by the fundamental classes of the X_i's.

Example 5.2. 1. A closed submanifold of a C^∞ manifold is a non-singular pseudo-manifold.

2. An analytic variety V in a complex manifold is an oriented pseudo-manifold. The irreducible decomposition of V as a variety gives that as a pseudo-manifold. A general point of V is nothing but a non-singular point of V.

5.6 Localization and topological residues

In this section, we let M be a C^∞ manifold of dimension m and let K_0, K, K' and K^* be as in Section 4.1.

Duals of characteristic classes

Suppose M is oriented and connected so that we have the Poincaré duality (Theorem 4.1).

Chern classes: Let E be a complex vector bundle of rank l on M. We use the dual cellular decomposition K^* to define the Chern classes of E (cf. Section 5.2). Denoting by $s^{(r)}$ an r-frame already constructed on the

$(2q-1)$-skeleton $(K^*)^{2q-1}$ of K^*, the Poincaré dual of $c^q(E)$ is represented in $\check{H}_{m-2q}(M)$ by the cycle

$$C = \sum_{\boldsymbol{s}} I(s^{(r)}, b_{\boldsymbol{s}})\, \boldsymbol{s}, \tag{5.7}$$

where the sum is taken over all the $(m-2q)$-simplices $\boldsymbol{s}$ of K and $I(s^{(r)}, b_{\boldsymbol{s}})$ denotes the index of $s^{(r)}$ on $\boldsymbol{s}^* \cap (K^*)^{2q-1}$ at the barycenter $b_{\boldsymbol{s}}$ (cf. Definition 5.3, we may take $b_{\boldsymbol{s}}$ as a there).

Euler class: Let E be an oriented real vector bundle of rank l' on M. Denoting by s a non-singular section already constructed on the $(l'-1)$-skeleton of K^*, the Poincaré dual of $e(E)$ is represented in $\check{H}_{m-l'}(M)$ by the cycle

$$C = \sum_{\boldsymbol{s}} I(s, b_{\boldsymbol{s}})\, \boldsymbol{s}, \tag{5.8}$$

where the sum is taken over all the $(m-l')$-simplices $\boldsymbol{s}$ of K.

Localized Chern classes

Let E be a complex vector bundle of rank l on M and S a K_0-subcomplex of M. We apply the considerations of Section 5.4 by letting A be the sub-K^*-complex $M \setminus O_{K'}(S)$. Thus suppose we are already given an r-frame $s^{(r)}$ of E on the $2q$-skeleton of $M \setminus O_{K'}(S)$, $q = l - r + 1$. Then we have the relative class in $H^{2q}(M, M \setminus O_{K'}(S)) \simeq H^{2q}(M, M \setminus S)$, which is denoted by $c^q_{S,\mathrm{top}}(E, s^{(r)})$ and is called the *topological localization* of $c^q(E)$ by $s^{(r)}$. It will also be denoted by $c^q_S(E, s^{(r)})$, $c^q_{\mathrm{top}}(E, s^{(r)})$ or $c^q(E, s^{(r)})$, if there is no fear of confusion. The class $c^q_{\mathrm{top}}(E, s^{(r)})$ depends on $s^{(r)}$, but not on the choice of the extension $\tilde{s}^{(r)}$ of $s^{(r)}$. Its image by the canonical morphism $H^{2q}(M, M \setminus S) \to H^{2q}(M)$ is the usual Chern class $c^q(E)$.

Recall that, in the above construction, for an $(m-2q)$-simplex $\boldsymbol{s}$ of K, the following are equivalent (cf. Proposition 4.2):

$$(1)\ \boldsymbol{s} \not\subset S, \qquad (2)\ \boldsymbol{s}^* \cap S = \emptyset, \qquad (3)\ \boldsymbol{s}^* \subset M \setminus O_{K'}(S).$$

Note that the point a may be assumed to be the barycenter $b_{\boldsymbol{s}}$ of $\boldsymbol{s}$.

Suppose M is oriented so that we have the Alexander isomorphism (Theorem 4.2):

$$A : H^{2q}(M, M \setminus S) \xrightarrow{\sim} \check{H}_{m-2q}(S).$$

Definition 5.13. The *topological residue* $\mathrm{TRes}_{c^q}(s^{(r)}, E; S)$ *of* $s^{(r)}$ *for* $c^q(E)$ *at* S is the image of $c^q_{\mathrm{top}}(E, s^{(r)})$ by A.

Remark 5.7. In order to have the above localization and residue, it suffices to have $s^{(r)}$ on $\partial S_{K'}(S) \cap (K^*)^{2q}$ (cf. Remark 4.4).

Suppose that S has only a finite number of connected components $(S_\lambda)_\lambda$. Then we have a decomposition

$$\breve{H}_{m-2q}(S) = \bigoplus_\lambda \breve{H}_{m-2q}(S_\lambda)$$

and accordingly we have the residue $\mathrm{TRes}_{c^q}(s^{(r)}, E; S_\lambda)$ in $\breve{H}_{m-2q}(S_\lambda)$ for each λ.

We have (cf. (5.7)):

Proposition 5.4. *In the above situation, the residue* $\mathrm{TRes}_{c^q}(s^{(r)}, E; S_\lambda)$ *is represented by the cycle*

$$C_\lambda = \sum_{\boldsymbol{s}} I(\tilde{s}^{(r)}, b_{\boldsymbol{s}})\, \boldsymbol{s}, \tag{5.9}$$

where the sum is taken over the $(m-2q)$*-simplices of* K *in* S_λ*.*

Note that the class of the cycle does not depend on the choice of $\tilde{s}^{(r)}$.

In particular, if $2q = m$ and if S_λ is compact, $\breve{H}_0(S_\lambda) = H_0(S_\lambda) \simeq \mathbb{Z}$, the isomorphism given by the augmentation (cf. Appendix B). Thus we may regard $\mathrm{TRes}_{c^q}(s^{(r)}, E; S_\lambda)$ as an integer and we have

$$\mathrm{TRes}_{c^q}(s^{(r)}, E; S_\lambda) = \sum_{\boldsymbol{s}} I(\tilde{s}^{(r)}, b_{\boldsymbol{s}}), \tag{5.10}$$

where the sum is taken over all the 0-simplices $\boldsymbol{s}$ of K in S_λ, in fact $b_{\boldsymbol{s}} = \boldsymbol{s}$.

Let R be an m-dimensional manifold possibly with boundary in M. We may assume that R and ∂R are K_0-subcomplexes of M. In the above considerations, we let S be R. Thus suppose we are given an r-frame $s^{(r)}$ of E on the $2q$-skeleton of $M \smallsetminus O_{K'}(R)$. Then we have the relative class in $H^{2q}(M, M \smallsetminus O_{K'}(R)) \simeq H^{2q}(R, \partial R)$, which is denoted by $c^q_{R,\mathrm{top}}(E, s^{(r)})$, or simply by $c^q_R(E, s^{(r)})$. Its image by the canonical morphism $H^{2q}(R, \partial R) \to H^{2q}(M)$ is the usual Chern class $c^q(E)$.

Remark 5.8. To have the above relative class $c^q_R(E, s^{(r)})$, it suffices to have $s^{(r)}$ on $\partial S_{K'}(R) \cap (K^*)^{2q}$ (cf. Remark 5.7).

Suppose M is oriented so that we have the Lefschetz isomorphism (Theorem 4.3):

$$L : H^{2q}(R, \partial R) \xrightarrow{\sim} \breve{H}_{m-2q}(R).$$

Recall that it is given by the left cap product with the class $[\underline{R}]$. The class $[\underline{R}] \frown c_R^q(E, s^{(r)})$ is represented by a cycle as in the right-hand side of (5.9), where the sum is taken over the $(m-2q)$-simplices of K in R.

In particular, if $2q = m$ and if R is compact, $\check{H}_0(R) = H_0(R)$ and there is the augmentation $\varepsilon_* : H_0(R) \to \mathbb{Z}$. In this case, $\varepsilon_*([\underline{R}] \frown c_R^q(E, s^{(r)}))$ is equal to the Kronecker product $\langle [\underline{R}], c_R^q(E, s^{(r)}) \rangle$ (cf. (B.19)). It is an integer given by

$$\sum_{\boldsymbol{s}} I(\tilde{s}^{(r)}, b_{\boldsymbol{s}}),$$

where the sum is taken over all the 0-simplices $\boldsymbol{s}$ of K in R, in fact $b_{\boldsymbol{s}} = \boldsymbol{s}$.

Coming back to the previous situation, let S be a K_0-subcomplex of M and $s^{(r)}$ an r-frame of E on the $2q$-skeleton of $M \setminus O_{K'}(S)$. Let R be as above and suppose $S \subset \operatorname{Int} R$. The r-frame $s^{(r)}$ restricts to an r-frame, denoted also by $s^{(r)}$, on the $2q$-skeleton of $M \setminus O_{K'}(R)$. Thus we have the relative class $c_R^q(E, s^{(r)})$ in $H^{2q}(R, \partial R)$ as above.

From Corollary 4.3, we have the following "residue theorem":

Theorem 5.2. *Let M be an oriented C^∞ manifold of dimension m and E a complex vector bundle of rank l on M. Also let S be a subcomplex of M with a finite number of connected components $(S_\lambda)_\lambda$ and $s^{(r)}$ an r-frame of E on $2q$-skeleton of $M \setminus O_{K'}(S)$, $q = l - r + 1$. In this situation, it holds:*

1. *For each λ, we have the residue $\operatorname{TRes}_{c^q}(s^{(r)}, E; S_\lambda)$ in $\check{H}_{m-2q}(S_\lambda)$, which is represented by the cycle (5.9).*

2. *If S is in the interior of a manifold R of dimension m possibly with boundary in M,*

$$\sum_\lambda (\imath_\lambda)_* \operatorname{TRes}_{c^q}(s^{(r)}, E; S_\lambda) = [\underline{R}] \frown c_R^q(E, s^{(r)}) \qquad \text{in } \check{H}_{m-2q}(R),$$

where $\imath_\lambda : S_\lambda \hookrightarrow R$ denotes the inclusion.

Remark 5.9. 1. If we take M as R in 2 above, the right-hand side is expressed as $[M] \frown c^q(E)$ in $\check{H}_{m-2q}(M)$.

2. In the case $2q = m$, if R is compact, $\check{H}_0(R) = H_0(R)$. In this case each S_λ is compact and, appling the augmentation $\varepsilon_* : H_0(R) \to \mathbb{Z}$ to the both sides of the equality in 2 in Theorem 5.2, we have (cf. (B.2))

$$\sum_\lambda \operatorname{TRes}_{c^q}(s^{(r)}, E; S_\lambda) = \langle [\underline{R}], c_R^q(E, s^{(r)}) \rangle$$

as integers. Suppose M is compact. If we take M as R, the right-hand side is expressed as $\langle [M], c^q(E) \rangle$.

Transverse residues: Let M, K_0, K, K^* and S be as above. Suppose the maximum dimension of the simplices of K in S is $m-2q$ and let S' be an oriented submanifold of M of dimension $m-2q$ which is contained in S. We may assume that the orientations of simplices in K are compatible with that of S'. Let x be a point in S' and $\mathbb{D}$ a slice of S' in M at x (cf. Definition 3.24). We may assume that x is the barycenter $b_{\boldsymbol{s}}$ of some $(m-2q)$-simplex $\boldsymbol{s}$ of K in S' and that $\boldsymbol{s}^*$ is in $\mathbb{D}$ (cf. Theorem 5.1). We may also extend the triangulation K' on $\boldsymbol{s}^*$ throughout $\mathbb{D}$. Let $s^{(r)}$ be an r-frame of E on the $2q$-skeleton of $M \smallsetminus O_{K'}(S)$, as before. Restricting E and $s^{(r)}$ to $\mathbb{D}$, we have the localization $c^q_{\text{top}}(E|_{\mathbb{D}}, s^{(r)}|_{\mathbb{D}})$ and the residue $\text{TRes}_{c^q}(s^{(r)}|_{\mathbb{D}}, E|_{\mathbb{D}}; x)$, which correspond to each other by the Alexander isomorphism

$$H^{2q}(\mathbb{D}, \mathbb{D} \smallsetminus \{x\}) \simeq H_0(\{x\}).$$

As $H_0(\{x\}) \simeq \mathbb{Z}$, we may think of $\text{TRes}_{c^q}(s^{(r)}|_{\mathbb{D}}, E|_{\mathbb{D}}; x)$ as an integer, which is referred to as the *transverse residue* at x. In fact it is given by

$$\text{TRes}_{c^q}(s^{(r)}|_{\mathbb{D}}, E|_{\mathbb{D}}; x) = I(s^{(r)}, b_{\boldsymbol{s}}).$$

Remark 5.10. 1. If we denote by $\iota : \mathbb{D} \hookrightarrow M$ the inclusion, we have the commutative diagram (cf. Proposition 4.13):

$$\begin{array}{ccc} H^{2q}(M, M \smallsetminus S) & \xrightarrow[A_{M,S}]{\sim} & \check{H}_{m-2q}(S) \\ \downarrow{\scriptstyle \iota^*} & & \downarrow{\scriptstyle (\mathbb{D}\,\bullet\ )_x} \\ H^{2q}(\mathbb{D}, \mathbb{D} \smallsetminus \{x\}) & \xrightarrow[A_{\mathbb{D},x}]{\sim} & \check{H}_0(\{x\}) \end{array}$$

so that we may write

$$\text{TRes}_{c^q}(s^{(r)}|_{\mathbb{D}}, E|_{\mathbb{D}}; x) = \big(\mathbb{D} \bullet \text{TRes}_{c^q}(s^{(r)}, E; S)\big)_x.$$

2. By Remarks 4.4 and 5.7, in the above notation, $\text{TRes}_{c^q}(s^{(r)}|_{\boldsymbol{s}^*}, E|_{\boldsymbol{s}^*}; b_{\boldsymbol{s}})$ makes sense and is equal to $\text{TRes}_{c^q}(s^{(r)}|_{\mathbb{D}}, E|_{\mathbb{D}}; x)$.

As a function of x, $\text{TRes}_{c^q}(s^{(r)}|_{\mathbb{D}}, E|_{\mathbb{D}}; x)$ is locally constant. Thus, if S' is connected, it is constant. From Proposition 5.4, we have (cf. Convention 4.2):

Proposition 5.5. *If S_λ is an $(m-2q)$-dimensional submanifold of M,*

$$\text{TRes}_{c^q}(s^{(r)}, E; S_\lambda) = \text{TRes}_{c^q}(s^{(r)}|_{\mathbb{D}}, E|_{\mathbb{D}}; x) \cdot S_\lambda \qquad \text{in } \check{H}_{m-2q}(S_\lambda),$$

where x is a point in S_λ and $\mathbb{D}$ a slice of S_λ at x.

The above expression of the residue is generalized to the case S_λ is a pseudo-manifold. Thus let S_λ be a connected component of S as above and suppose it is an oriented pseudo-manifold of dimension $m-2q$ (cf. Definition 5.12). If $\boldsymbol{s}$ is an $(m-2q)$-simplex of K in S_λ, $\boldsymbol{s}^*$ is a $2q$-cell such that $\boldsymbol{s}^* \cap S_\lambda = \{b_{\boldsymbol{s}}\}$. Thus we have the transverse residue $\mathrm{TRes}_{c^q}(s^{(r)}|_{\boldsymbol{s}^*}, E|_{\boldsymbol{s}^*}; b_{\boldsymbol{s}})$, which is equal to $I(s^{(r)}, b_{\boldsymbol{s}})$ (cf. Remark 5.10. 2). From Proposition 5.4, we have:

Theorem 5.3. *Suppose S_λ is an oriented pseudo-manifold of dimension $m-2q$. Then the residue $\mathrm{TRes}_{c^q}(s^{(r)}, E; S_\lambda)$ in $\check{H}_{m-2q}(S_\lambda)$ is represented by the cycle*

$$\sum_{\boldsymbol{s}} \mathrm{TRes}_{c^q}(s^{(r)}|_{\boldsymbol{s}^*}, E|_{\boldsymbol{s}^*}; b_{\boldsymbol{s}}) \cdot \boldsymbol{s},$$

where $\boldsymbol{s}$ runs through all the $(m-2q)$-simplices of K in S_λ.

Corollary 5.1. *In the above situation, suppose that the set of general points of S_λ has the structure of a C^∞ submanifold of dimension $m-2q$ of M and that, in the irreducible decomposition $S_\lambda = \bigcup_i S_{\lambda,i}$, the set $\{S_{\lambda,i}\}_i$ is locally finite. Then we have:*

$$\mathrm{TRes}_{c^q}(s^{(r)}, E; S_\lambda) = \sum_i \mathrm{TRes}_{c^q}(s^{(r)}|_{\mathbb{D}}, E|_{\mathbb{D}}; x_{\lambda,i}) \cdot [S_{\lambda,i}] \qquad \text{in } \check{H}_{m-2q}(S_\lambda),$$

where $x_{\lambda,i}$ is a general point of $S_{\lambda,i}$ and $\mathbb{D}$ a slice of $S_{\lambda,i}$ at $x_{\lambda,i}$.

In the above, $[S_{\lambda,i}]$ denotes the class of $S_{\lambda,i}$ in $\check{H}_{m-2q}(S_\lambda)$. In fact $\check{H}_{m-2q}(S_\lambda)$ is a free Abelian group generated by these classes.

In the case S_λ is a submanifold, the above reduces to Proposition 5.5.

Remark 5.11. In Chapter 10 below we discuss localization problems in various settings mainly from the differential geometric viewpoint and give a general residue theorem (Theorem 10.1). The localization of Chern classes by frames in this context is treated in Section 10.3, where the differential geometric counterpart of Theorem 5.2 is given (cf. Theorem 10.6). We then prove that these two are essentially the same (cf. Theorem 10.13). In Chapters 12 and 13, we discuss these residues in detail.

We now give some fundamental examples.

Hyperplane bundle: Let $\mathbb{P}^n$ be the n-dimensional complex projective space with homogeneous coordinates $[\zeta_0, \dots, \zeta_n]$. We denote by U_α the open set in $\mathbb{P}^n$ defined by $\zeta_\alpha \neq 0$, $\alpha = 0, \dots, n$. Let H be the hyperplane defined by $\zeta_0 = 0$. The hyperplane bundle L_H is the line bundle defined by the transition functions $\{\zeta_\beta/\zeta_\alpha\}$ (cf. Example 3.6, where it is denoted by H_n). It admits a section s represented by the collection (s_α) given by $s_\alpha = \zeta_0/\zeta_\alpha$. The zero set of s is exactly H so that we have the localization $c^1(L_H, s)$ in $H^2(\mathbb{P}^n, \mathbb{P}^n \setminus H)$ of $c^1(L_H)$ and the associated residue $\mathrm{TRes}_{c^1}(s, L_H; H)$, which is the image of $c^1(L_H, s)$ by the Alexander isomorphism

$$H^2(\mathbb{P}^n, \mathbb{P}^n \setminus H) \xrightarrow{\sim} H_{2n-2}(H).$$

Proposition 5.6. *We have:*

$$\mathrm{TRes}_{c^1}(s, L_H; H) = [H] \qquad \text{in } H_{2n-2}(H).$$

Proof. By Proposition 5.5, it suffices to show that the transverse residue is equal to 1. Thus let x be a point in H and assume that it is in U_α, $\alpha \neq 0$. We have a trivialization $L_H|_{U_\alpha} \simeq \mathbb{C} \times U_\alpha$ and s is expressed as $[\zeta] \mapsto (z, [\zeta])$, $z = \zeta_0/\zeta_\alpha$. Let $\mathbb{D}$ be a complex 1-dimensional disk transverse to H at x. We may think of z as a coordinate on $\mathbb{D}$ and we see that the transverse residue is 1 (cf. Example 5.1). □

The above is a special case of Theorem 12.2 below.

Corollary 5.2. *The first Chern class $c^1(L_H)$ of the hyperplane bundle is the Poincaré dual of $[H]$.*

Universal bundle: More generally we consider the universal bundle Q on the Grassmann manifold $G(N, r)$ of r-planes in $\mathbb{C}^N$, which is of rank $l = N - r$ (cf. Section 3.6). For $q = 1, \ldots, l$, we have the q-th Chern class $c^q(Q)$ of Q. Let I^q denote the r-tuple of integers $(l+1-q, l+2, l+3, \ldots, l+r = N)$. Note that the Schubert variety $\overline{E_{I^q}}$ has complex dimension $lr - q$ and each element is represented by a matrix A such that the rank of $A^{(l+1-q)}$, which is an $(r + q - 1) \times r$ matrix, is less than or equal to $r - 1$. We have the exact sequence of vector bundles on $G(N, r)$:

$$0 \longrightarrow S \longrightarrow \mathbb{I}^N \xrightarrow{\pi} Q \longrightarrow 0. \tag{5.11}$$

Let s_i be the section of Q defined by $s_i = \pi(e_i)$, $i = 1, \ldots, N$, and set $s^{(k)} = (s_1, \ldots, s_k)$, $k = l - q + 1$.

Theorem 5.4. *For $q = 1, \ldots, l$, the singular set of $s^{(k)}$ is $\overline{E_{I^q}}$ and*

$$\mathrm{TRes}_{c^q}(s^{(k)}, Q; \overline{E_{I^q}}) = [\overline{E_{I^q}}] \qquad \text{in } H_{2(lr-q)}(\overline{E_{I^q}}).$$

Proof. If we think of $G(N, r)$ as the union of the e_I's, then $(s_1, \ldots, s_k)$ fails to be linearly independent exactly on $\overline{e_{I^q}}$. Moreover, in the coordinate neighborhood U_{I^q}, $\overline{E_{I^q}}$ is given by $a_{l+2-q,1} = \cdots = a_{l+1,1} = 0$. □

Corollary 5.3. *The Schubert class* $[\overline{E_{I^q}}]$ *is the Poincaré dual of* $c^q(Q)$.

Remark 5.12. More generally, let $I = (i_1, \dots, i_r)$ be an arbitrary element in $\mathcal{I}$. For each $j = 1, \dots, r$, let λ_j be defined by $\lambda_j = N - r + j - i_j$ and set

$$\Delta_I(Q) = \det(c^{\lambda_j + i - j}(Q))_{1 \leq i,j \leq r}.$$

In the above, the product is the cup product. Since the Chern classes are of even degree, we may define the determinant as in the case of usual numbers. Then we have (Giambelli's formula):

$$[G(N,r)] \frown \Delta_I(Q) = [\overline{E_I}],$$

i.e., the Schubert class $[\overline{E_I}]$ is the Poincaré dual of $\Delta_I(Q)$.

In view of Theorem 3.8, we see that the cohomology $H^*(G(N,r);\mathbb{Z})$ is the graded $\mathbb{Z}$-algebra $\mathbb{Z}[c^1(Q), \dots, c^l(Q)]$, the polynomial ring over $\mathbb{Z}$ freely generated by the Chern classes of the universal bundle.

Localized Euler class

Let E be an oriented real vector bundle of rank l' on M and S a K_0-subcomplex of M. Suppose we are already given a non-vanishing section s of E on the l'-skeleton of $M \smallsetminus O_{K'}(S)$. Then we have the relative class in $H^{l'}(M, M \smallsetminus O_{K'}(S)) \simeq H^{l'}(M, M \smallsetminus S)$, which we denote by $e(E,s)$ and call the *localization of* $e(E)$ *by* s. Its image by the canonical morphism $H^{l'}(M, M \smallsetminus S) \to H^{l'}(M)$ is the Euler class $e(E)$.

Suppose M is oriented so that we have the Alexander isomorphism

$$A : H^{l'}(M, M \smallsetminus S) \xrightarrow{\sim} \check{H}_{m-l'}(S).$$

Definition 5.14. The *topological residue* $\mathrm{TRes}_e(s, E; S)$ *of* s *for* $e(E)$ *at* S is the image of $e(E,s)$ by A.

Remark 5.13. As in the case of Chern classes, it suffices to have s on $\partial S_{K'}(S) \cap (K^*)^{l'}$ to have the localization and residue (cf. Remark 5.7).

If S has a finite number of connected components $(S_\lambda)_\lambda$ we have the residue $\mathrm{TRes}_e(s, E; S_\lambda)$ in $\check{H}_{m-l'}(S_\lambda)$ for each λ. It is represented by the cycle (cf. (5.8))

$$C_\lambda = \sum_{\boldsymbol{s}} I(\tilde{s}, b_{\boldsymbol{s}})\, \boldsymbol{s}, \tag{5.12}$$

where the sum is taken over the $(m - l')$-simplices of K in S_λ.

In particular, if $l' = m$ and if S_λ is compact, $\check{H}_0(S_\lambda) = H_0(S_\lambda) \simeq \mathbb{Z}$ and we may regard $\mathrm{TRes}_e(s, E; S_\lambda)$ as an integer and we have

$$\mathrm{TRes}_e(s, E; S_\lambda) = \sum_{\boldsymbol{s}} I(\tilde{s}, b_{\boldsymbol{s}}), \tag{5.13}$$

where the sum is taken over all the 0-simplices $\boldsymbol{s}$ of K in S_λ, in fact $b_{\boldsymbol{s}} = \boldsymbol{s}$.

Let R be an m-dimensional manifold possibly with boundary in M. In the above considerations, we let S be R. Thus suppose we are given a non-vanishing section s of E on the l'-skeleton of $M \smallsetminus O_{K'}(R)$. Then we have the relative class $e_R(E, s)$ in $H^{l'}(R, \partial R)$.

Remark 5.14. Similar remark as in Remark 5.8 apply to the case of Euler class.

If M is oriented, we have the Lefschetz isomorphism:

$$L : H^{l'}(R, \partial R) \xrightarrow{\sim} \check{H}_{m-l'}(R).$$

The class $[\underline{R}] \frown e_R(E, s)$ is represented by a cycle as in the right-hand side of (5.12), where the sum is taken over the $(m - l')$-simplices of K in R.

In particular, if $l' = m$ and if R is compact, $\check{H}_0(R) = H_0(R)$ and there is the augmentation $\varepsilon_* : H_0(R) \to \mathbb{Z}$. We have $\varepsilon_*([\underline{R}] \frown e_R(E, s)) = \langle [\underline{R}], e_R(E, s) \rangle$. It is an integer given by

$$\sum_{\boldsymbol{s}} I(\tilde{s}, b_{\boldsymbol{s}}),$$

where the sum is taken over all the 0-simplices $\boldsymbol{s}$ of K in R, in fact $b_{\boldsymbol{s}} = \boldsymbol{s}$.

Coming back to the previous situation, let S be a K_0-subcomplex of M and s a non-vanishing section of E on the l'-skeleton of $M \smallsetminus O_{K'}(S)$. Let R be as above and suppose $S \subset \mathrm{Int}\, R$. The section s restricts to a non-vanishing section on the l'-skeleton of $M \smallsetminus O_{K'}(R)$. Thus we have the relative class $e_R(E, s)$ in $H^{l'}(R, \partial R)$ as above.

We have the residue theorem as Theorem 5.2 for the Euler class:

Theorem 5.5. *In the above situation, it holds:*

1. *For each λ, we have the residue* $\mathrm{TRes}_e(s, E; S_\lambda)$ *in* $\check{H}_{m-l'}(S_\lambda)$, *which is represented by the cycle* (5.12).

2. *We have*

$$\sum_\lambda (\imath_\lambda)_* \mathrm{TRes}_e(s, E; S_\lambda) = [\underline{R}] \frown e_R(E, s) \qquad \text{in } \check{H}_{m-l'}(R).$$

Remark 5.15. Similar remarks as in Remark 5.9 apply to this case, replacing c^q, $2q$ and $s^{(r)}$ with e, l' and s, respectively.

Transverse residues: As in the case of Chern classes, we may consider the transverse residue. For simplicity we consider the case S_λ is a submanifold of dimension $m - l'$. Let x be a point in S_λ and $\mathbb{D}$ a slice of S_λ in M at x. Then restricting E and s to $\mathbb{D}$, we have the transverse residue $\mathrm{TRes}_e(s|_{\mathbb{D}}, E|_{\mathbb{D}}, x)$, which is an integer independent of x. We have (cf. Proposition 5.5):

Proposition 5.7. *If S_λ is an $(m - l')$-dimensional submanifold of M,*

$$\mathrm{TRes}_e(s, E; S_\lambda) = \mathrm{TRes}_e(s|_{\mathbb{D}}, E|_{\mathbb{D}}; x) \cdot S_\lambda \qquad \textit{in } \check{H}_{m-l'}(S_\lambda).$$

In the case S_λ is a pseudo-manifold, we have expressions of the residues similar to the ones in Theorem 5.3 and Corollary 5.1.

Poincaré-Hopf theorem

Euler-Poincaré characteristic: Let X be a topological space with finite homology type and b_p the p-th Betti number (cf. (B.13). Here we set $A = \emptyset$). Suppose $b_p = 0$ for $p > p_0$. Then the alternating sum

$$\chi(X) = \sum_{p=0}^{p_0} (-1)^p \, b_p$$

is called the *Euler-Poincaré characteristic* of X.

If $X = |K|$ is the polyhedron of a finite simplicial complex K, then X satisfies the above conditions and it is shown that its Euler-Poincaré characteristic is also given as

$$\chi(X) = \sum_{p \geq 0} (-1)^p \, k_p,$$

where k_p is the number of p-simplices in K.

If M is a compact C^∞ manifold, it admits a triangulation by a finite simplicial complex and we have

$$\chi(M) = \sum_{p=0}^{m} (-1)^p \, b_p = \sum_{p=0}^{m} (-1)^p \, k_p, \tag{5.14}$$

where m is the dimension of M. Note that, if m is odd, then $\chi(M) = 0$ by the Poincaré duality.

Likewise we may consider the Euler-Poincaré characteristic $\chi(R)$ of a compact manifold R with boundary.

Euler class of the tangent bundle: Let M be a C^∞ manifold of dimension m. If $E = T_{\mathbb{R}}M$ is the tangent bundle of M, then $l' = m$ so that $(K^*)^{l'} = M$. The sections of $T_{\mathbb{R}}M$ are vector fields.

Definition 5.15. Let a be a point in M and U a neighborhood of a. For a non-vanishing vector field v on $U \smallsetminus \{a\}$, its *Poincaré-Hopf index* $\mathrm{PH}(v,a)$ at a is the index $I(v,a)$ as defined in (5.5).

Now suppose that M is oriented so that $T_{\mathbb{R}}M$ is also oriented. In this case, we define the *Euler class* $e(M)$ of M to be the Euler class of $T_{\mathbb{R}}M$. In the above, we may think of a as a vertex of K and may write (cf. (5.13))

$$\mathrm{PH}(v,a) = \mathrm{TRes}_e(v, T_{\mathbb{R}}M; a).$$

More generally, let S be a K_0-subcomplex of M. For a compact connected component S_λ of S and a non-vanishing vector field v on $U \smallsetminus S_\lambda$, where U is a neighborhood of S_λ, we define (cf. (5.13))

$$\mathrm{PH}(v, S_\lambda) = \mathrm{TRes}_e(v, T_{\mathbb{R}}M; S_\lambda). \tag{5.15}$$

Let R be a compact manifold of dimension m with boundary in M. For a non-vanishing vector field v on a neighborhood of ∂R, we have the relative class $e_R(T_{\mathbb{R}}M, v)$ in $H^m(R, \partial R)$ (cf. Remark 5.14). We then define

$$\mathrm{PH}(v, R) = \langle [\underline{R}], e_R(T_{\mathbb{R}}M, v)\rangle. \tag{5.16}$$

Then for a compact subcomplex S of M in the interior of R and a non-vanishing vector field v on $M \smallsetminus S$, by Theorem 5.5 (see also Remark 5.15), we have

$$\sum_\lambda \mathrm{PH}(v, S_\lambda) = \mathrm{PH}(v, R).$$

In the above situation, if M is compact, we have

$$\sum_\lambda \mathrm{PH}(v, S_\lambda) = \langle [M], e(M)\rangle. \tag{5.17}$$

On the other hand, there exists a vector field v_0 having a singularity of index 1 at the barycenter of each even-dimensional simplex and a singularity of index -1 at the barycenter of each odd-dimensional simplex, and for v_0 we have (cf. (5.14))

$$\sum_{\boldsymbol{s}} \mathrm{PH}(v_0, b_{\boldsymbol{s}}) = \chi(M),$$

where $\boldsymbol{s}$ runs through all the simplices in K. From (5.17), we have

$$\chi(M) = \langle [M], e(M)\rangle. \tag{5.18}$$

Thus we have:

Theorem 5.6. *Let M be a compact oriented C^∞ manifold. For a vector field v defined and non-vanishing away from a compact subsomplex S with a finite number of connected components S_λ,*

$$\sum_\lambda \mathrm{PH}(v, S_\lambda) = \chi(M).$$

Corollary 5.4 (Poincaré-Hopf theorem). *Let M be as above. For a vector field v defined and non-vanishing on M, except for a finite number of points $a_1, \dots, a_r$,*

$$\sum_{i=1}^{r} \mathrm{PH}(v, a_i) = \chi(M).$$

Remark 5.16. Let R be a C^∞ manifold of dimension m with boundary in M. For a point p in ∂R, we take a coordinate system $(U, (x_1, \dots, x_m))$ around p in M so that $R \cap U$ is given by $x_1 \leq 0$. We say that a tangent vector $v = \sum_{i=1}^m a_i \frac{\partial}{\partial x_i}$ at p is *pointing outward*, or *inward*, if $a_1 > 0$ or $a_1 < 0$, respectively. Suppose R as above is compact and oriented. For a vector field v on a neighborhood of ∂R in M pointing outward at every point of ∂R, we have

$$\mathrm{PH}(v, R) = \chi(R).$$

On the other hand, if v is pointing inward at every point of ∂R, we have

$$\mathrm{PH}(v, R) = \chi(R) - \chi(\partial R).$$

Thus the above results are generalized for such vector fields.

Case of complex vector bundles: From the construction we have the following (cf. Remark 5.4. 2):

Proposition 5.8. *If E is a complex vector bundle of rank l, we may think of it as an oriented real vector bundle of rank $2l$ and we have*

$$e(E, s) = c^l(E, s) \qquad \text{in } H^{2l}(M, M \smallsetminus S).$$

If M is oriented, we also have

$$\mathrm{TRes}_e(s, E; S_\lambda) = \mathrm{TRes}_{c^l}(s, E; S_\lambda) \qquad \text{in } \breve{H}_{m-2l}(S_\lambda).$$

Suppose M is a complex manifold of dimension n. Then the q-th *Chern class* $c^q(M)$ *of* M is defined to be the q-th Chern class $c^q(TM)$ of the holomorphic tangent bundle TM. Recall that (cf. Proposition 3.5) TM can be naturally identified with the real tangent bundle $T_{\mathbb{R}}M$. Thus a section of TM can be considered as either a complex vector field or a real vector field.

As a special case of Proposition 5.8, we have:

Proposition 5.9. *Let v be a section of TM defined and non-vanishing on a neighborhood of a, possibly except for at a. Then its Poincaré-Hopf index* $\mathrm{PH}(v, a)$ *as defined in Definition 5.15 coinsides with the index as defined in Definition 5.3 (with $r = 1$).*

We also see that, since the top Chern class $c^n(M)$ is the primary obstruction to constructing a (non-vanishing) vector field, it coincides with the Euler class $e(M)$ of $T_{\mathbb{R}}M$:

$$c^n(M) = e(M).$$

In particular, if M is compact, (5.18) reads

$$\chi(M) = \langle [M], c^n(M) \rangle. \tag{5.19}$$

Example 5.3. Let $M = \mathbb{P}^1$ be the Riemann sphere with homogeneous coordinates $[\zeta_0, \zeta_1]$ (cf. Exercise 2.1). For $i = 0, 1$, let U_i be the open set given by $\zeta_i \neq 0$. On U_0, $z = \zeta_1/\zeta_0$ is a coordinate and on U_1, $z' = \zeta_0/\zeta_1$ is a coordinate. Let $a_0 = [1, 0]$ and $a_1 = [0, 1]$. For every integer d, we consider the holomorphic vector field v on $\mathbb{P}^1 \smallsetminus \{a_0, a_1\}$ given by $v = z^d \frac{\partial}{\partial z} = -(z')^{2-d} \frac{\partial}{\partial z'}$. We have (cf. Example 5.1)

$$\mathrm{PH}(v, a_0) + \mathrm{PH}(v, a_1) = d + 2 - d = 2,$$

confirming that $\chi(\mathbb{P}^1) = 2$.

Later in Section 8.2, we represent the Chern classes of a complex vector bundle by differential forms using connections for the bundle. If M is compact, (5.19) may be written (cf. Proposition 7.11 below)

$$\chi(M) = \int_M c^n(M), \tag{5.20}$$

which is referred to as the "Gauss-Bonnet formula".

Thom class as a localized Euler class

Let M be a C^∞ manifold of dimension m and $\pi : E \to M$ an oriented real vector bundle of rank l'. We apply the above consideration to the "diagonal section" of the pull-back π^*E. Recall that (cf. (3.4)) the pull-back of E to itself is a vector bundle on E given by

$$\pi^*E = \{\, (\xi_1, \xi_2) \in E \times E \mid \pi(\xi_1) = \pi(\xi_2)\,\}.$$

We think of it as a vector bundle $\varpi : \pi^*E \to E$ on the second factor with ϖ the restriction of the projection. We denote by Σ the image of the zero section of $\pi : E \to M$, which is narurally diffeomorphic with M. The bundle π^*E admits the diagonal section s_Δ defined by $s_\Delta(\xi) = (\xi, \xi)$ for ξ in E, whose zero set is Σ. Thus we have the localization $e(\pi^*E, s_\Delta)$ in $H^{l'}(E, E \smallsetminus \Sigma)$ of $e(\pi^*E)$ by s_Δ.

Suppose M is oriented so that Σ is also oriented. We orient the total space E so that the orientation of the fiber followed by that of Σ gives the orientation of E (cf. Convention 3.3). Then we have the corresponding residue $\mathrm{TRes}_e(s_\Delta, \pi^*E; \Sigma)$ in $\check{H}_m(\Sigma)$ (note that E is $(m+l')$-dimensional).

Recall that we have the Thom class Ψ_E of E in $H^{l'}(E, E \smallsetminus \Sigma)$.

Theorem 5.7. *In the above situation,*

1. $e(\pi^*E, s_\Delta) = \Psi_E$ *in* $H^{l'}(E, E \smallsetminus \Sigma)$.

2. *If M is oriented,*

$$\mathrm{TRes}_e(s_\Delta, \pi^*E; \Sigma) = \Sigma \qquad \textit{in} \ \ \check{H}_m(\Sigma).$$

Proof. It suffices to prove the statement 1 as the statement 2 follows directly from 1.

This time we take a triangulation K_0 of E compatible with Σ. Let γ be a cocycle representing $e(\pi^*E, s_\Delta)$. For an m-simplex $\boldsymbol{s}$ not in Σ, we have $\langle \boldsymbol{s}^*, \gamma \rangle = 0$, as the dual l'-cell $\boldsymbol{s}^*$ does not intersect Σ. Suppose $\boldsymbol{s}$ is in Σ and let U be a neighborhood of $b_{\boldsymbol{s}}$ such that we have a trivialization $E|_U \simeq \mathbb{R}^{l'} \times U$. Then we have a trivialization

$$\pi^*E|_{\pi^{-1}(U)} \simeq \mathbb{R}^{l'} \times \pi^{-1}(U) \simeq \mathbb{R}^{l'} \times \mathbb{R}^{l'} \times U$$

and the section s_Δ is expressed as $(\xi, x) \mapsto (\xi, \xi, x)$. As the l'-cell $\boldsymbol{s}^*$ may be thought of as being in the fiber of E at $b_{\boldsymbol{s}}$, we have $\langle \boldsymbol{s}^*, \gamma \rangle = 1$ (cf. (5.5)). By Proposition 4.6 we have the theorem. □

Remark 5.17. 1. By the "functoriality" of the obstruction cocycles (cf. Remark 5.4. 1) and Theorem 5.7, we have $\pi^*e(E) = e(\pi^*E) = j^*\Psi_E$. As the map $\pi : E \to M$ is a deformation retraction, it induces an isomorphism $\pi^* : H^{l'}(M) \xrightarrow{\sim} H^{l'}(E)$ and we have

$$e(E) = (\pi^*)^{-1} j^* \Psi_E. \tag{5.21}$$

2. The Thom class is a *universal lozalization* of the Euler class in the following sense. Given a section $s : M \to E$ of E with the zero set S, which is a subcomplex of M. We have the induced morphism

$$s^* : H^{l'}(E, E \smallsetminus \Sigma) \longrightarrow H^{l'}(M, M \smallsetminus S).$$

By the functoriality of relative obstruction cocycles, we have

$$e(E, s) = s^* e(\pi^* E, s_\Delta) = s^* \Psi_E.$$

If M is oriented, we have the commutative diagram:

$$\begin{array}{ccc} H^{l'}(E, E \smallsetminus \Sigma) & \xrightarrow[A]{\sim} & \check{H}_m(\Sigma) \\ \downarrow{\scriptstyle s^*} & & \downarrow{\scriptstyle (M \,\bullet_s\,)_S} \\ H^{l'}(M, M \smallsetminus S) & \xrightarrow[A]{\sim} & \check{H}_{m-l'}(S), \end{array} \tag{5.22}$$

where $(M \bullet_s \ )_S$ denotes the intersection product with s localized at S (cf. Definition 4.4). Thus we may express the residue as

$$\mathrm{TRes}_e(s, E; S) = (M \bullet_s \Sigma)_S.$$

If E is a complex vector bundle of rank l, we have the topological localization $c^l_{\mathrm{top}}(\pi^* E, s_\Delta)$ in $H^{2l}(E, E \smallsetminus M)$, which coincides with $e(\pi^* E, s_\Delta)$ (cf. Proposition 5.8) so that we have:

Corollary 5.5. *For a complex vector bundle E of rank l,*

1. $c^l_{\mathrm{top}}(\pi^* E, s_\Delta) = \Psi_E$ *in* $H^{2l}(E, E \smallsetminus \Sigma)$.

2. *If M is oriented,*

$$\mathrm{TRes}_{c^l}(s_\Delta, \pi^* E; \Sigma) = \Sigma \qquad \textit{in } \ \check{H}_m(\Sigma).$$

Remark 5.18. 1. Remark 5.17. 2 applies with e and l' replaced by c^l and $2l$, i.e., the Thom class of a complex vector bundle is a universal localization of the top Chern class. Namely, If $E \to M$ is a complex vector bundle of rank l and s a section of E on M with the zero set S, which is a subcomplex of M, then

$$c^l_{\mathrm{top}}(E, s) = s^* \Psi_E. \tag{5.23}$$

We come back to this point and review this from differential geometric viewpoint in Section 10.4 below (cf. also Section 7.9).

2. The previous example of hyperplane bundle (Proposition 5.6) may be thought of as a special case of Corollary 5.5. Let $\mathbb{P}^n$ be the projective space with homogeneous coordinates $[\zeta_0, \dots, \zeta_n]$. Also let H be the hyperplane defined by $\zeta_0 = 0$ and p the point $[1, 0, \dots, 0]$. Recall that (cf. Exercise 3.14) the map $\pi : \mathbb{P}^n \setminus \{p\} \to H$, $[\zeta] \mapsto [0, \zeta_1, \dots, \zeta_n]$, may be identified with the hyperplane bundle $\pi : H_{n-1} \to \mathbb{P}^{n-1}$. With this identification, $\pi^* H_{n-1}$ is the restriction of $H_n \to \mathbb{P}^n$ to H_{n-1} and the diagonal section s_Δ is the restriction of the canonical section to H_{n-1}:

$$\begin{array}{ccccc} H_n & \hookleftarrow & \pi^* H_{n-1} & \longrightarrow & H_{n-1} \\ \downarrow & & \downarrow & & \downarrow \pi \\ \mathbb{P}^n & \hookleftarrow & H_{n-1} & \xrightarrow{\pi} & \mathbb{P}^{n-1}. \end{array}$$

Moreover, by excision we have $H^2(\mathbb{P}^n, \mathbb{P}^n \setminus H) = H^2(H_{n-1}, H_{n-1} \setminus H)$.

Whitney sum formula for localized classes

Let E be an oriented real vector bundle of rank l' on M. Also let s be a section of E on M whose zero set S is a K_0-subcomplex of M. In this situation, we have the localization $e(E, s)$ in $H^{l'}(M, M \setminus S)$. If M is oriented, we have the residue $\mathrm{TRes}_e(s, E; S)$ in $\check{H}_{m-l'}(S)$ as the image of $e(E, s)$ by the Alexander duality

$$H^{l'}(M, M \setminus S) \xrightarrow{\sim} \check{H}_{m-l'}(S).$$

Note that, if s is the zero section, then $S = M$ and $e(E, s)$ is the Euler class $e(E)$.

For $i = 1, 2$, let E_i be an oriented real vector bundle of rank l'_i on M. Also let s_i be a section of E_i on M whose zero set S_i is a K_0-subcomplex of M. We set $E = E_1 \oplus E_2$, $l' = l'_1 + l'_2$, and $s = s_1 \oplus s_2$. Then the zero set S of s is given by $S = S_1 \cap S_2$. In this situation, we have the localizations $e(E_i, s_i)$ in $H^{l'_i}(M, M \setminus S_i)$, $i = 1, 2$, and $e(E, s)$ in $H^{l'}(M, M \setminus S)$. We also have the cup product

$$H^{l'_1}(M, M \setminus S_1) \times H^{l'_2}(M, M \setminus S_2) \xrightarrow{\smile} H^{l'}(M, M \setminus S).$$

Theorem 5.8. *In the above situation,*

1. $e(E, s) = e(E_1, s_1) \smile e(E_2, s_2)$.

2. *If M is oriented,*

$$\mathrm{TRes}_e(s, E; S) = (\mathrm{TRes}_e(s_1, E_1; S_1) \cdot \mathrm{TRes}_e(s_2, E_2; S_2))_S \quad \textit{in } \check{H}_{m-l'}(S).$$

Proof. The statement 1 follows from Corillary 4.5 and Remark 5.17. 2 and the statement 2 from Theorem 4.8. □

In the case of complex vector bundles, we have:

Corollary 5.6. *In the above situation, suppose E_i is a complex vector bundle of rank l_i, $i = 1, 2$, and set $l = l_1 + l_2$.*

1. $c^l_{\rm top}(E, s) = c^{l_1}_{\rm top}(E_1, s_1) \smile c^{l_2}_{\rm top}(E_2, s_2)$.

2. *If M is oriented,*

$$\mathrm{TRes}_{c^l}(s, E; S) = (\mathrm{TRes}_{c^{l_1}}(s_1, E_1; S_1) \cdot \mathrm{TRes}_{c^{l_2}}(s_2, E_2; S_2))_S \quad \text{in } \check{H}_{m-2l}(S).$$

Following is an application of the above, which will be stated for complex vector bundles and Chern classes, however, it holds for real oriented vector bundles and Euler classes as well.

Suppose we have an exact sequence of complex vector bundles on M:

$$0 \longrightarrow E' \longrightarrow E \longrightarrow E'' \longrightarrow 0.$$

We denote by l', l and l'' the (complex) ranks of E', E and E'', respectively. Let s be a section of E' whose zero set S is a K_0-subcomplex of M. We may think of s also as a section of E so that we have the localizations $c^{l'}(E', s)$ and $c^l(E, s)$. Moreover, if M is oriented, we have the residues $\mathrm{TRes}_{c^{l'}}(s, E'; S)$ in $\check{H}_{m-2l'}(S)$ and $\mathrm{TRes}_{c^l}(s, E; S)$ in $\check{H}_{m-2l}(S)$.

Proposition 5.10. *In the above situation, we have:*

$$\mathrm{TRes}_{c^l}(s, E; S) = \mathrm{TRes}_{c^{l'}}(s, E'; S) \frown i^* c^{l''}(E'') \qquad \text{in } \check{H}_{m-2l}(S),$$

where $i : S \hookrightarrow M$ denotes the inclusion.

Proof. By Proposition 3.3, E is isomorphic with $E' \oplus E''$ and s corresponds to $s \oplus 0$. Thus by Corollary 5.6, $c^l(E, s) = c^{l'}(E', s) \smile c^{l''}(E)$. The proposition then follows from Proposition 4.3, letting $p = 2l'$, $r = 2l$ and $\gamma = c^{l''}(E'')$. □

Example 5.4. Let $\mathbb{P}^2$ be the projective plane with homogeneous coordinates $[\zeta_0, \zeta_1, \zeta_2]$ and set $M = \mathbb{P}^2 \smallsetminus \{[0, 0, 1]\}$, i.e., the total space of the hyperplane bundle on $\mathbb{P}^1 = \{[\zeta_0, \zeta_1]\}$. The manifold M is covered by two open sets $U^{(i)}$ given by $\zeta_i \neq 0$, $i = 0, 1$. On $U^{(0)}$, we set $z_1 = \zeta_1/\zeta_0$ and $z_2 = \zeta_2/\zeta_0$, and on $U^{(1)}$, we set $z'_1 = \zeta_0/\zeta_1$ and $z'_2 = \zeta_2/\zeta_1$. Consider the vector field v (section of TM) given by

$$v = z_2 \frac{\partial}{\partial z_2} = z'_2 \frac{\partial}{\partial z'_2}.$$

The singular set S of v is the projective line $\zeta_2 = 0$. We wish to find $\mathrm{Res}_{c^2}(v, TM; S)$, which is in $H_0(S)$ and is equal to the Poincaré-Hopf index of v at S. Let $\pi : M \to S$ be the bundle map so that we have the exact sequence

$$0 \longrightarrow T\pi \longrightarrow TM \longrightarrow \pi^*TS \longrightarrow 0.$$

We may think of v as a section of $T\pi$ and its transverse residue is 1. By Proposition 5.5, we have $\mathrm{TRes}_{c^1}(v, T\pi; S) = S$. Using Proposition 5.10, we have

$$\mathrm{TRes}_{c^2}(v, TM; S) = S \frown i^*c^1(\pi^*TS) = S \frown c^1(TS) = \chi(S) = 2.$$

Note that v can be extended to a vector field on $\mathbb{P}^2$, which is given by

$$v = -z_1'' \frac{\partial}{\partial z_1''} - z_2'' \frac{\partial}{\partial z_2''}$$

in the coordinates $z_1'' = \zeta_0/\zeta_2$ and $z_2'' = \zeta_1/\zeta_2$. Thus v has a singularity of index 1 at $[0, 0, 1]$ and the above result is consistent with the Poincaré-Hopf theorem, as $\chi(\mathbb{P}^2) = 3$.

Remark 5.19. Later we show the Whitney sum formula for the total Chern classes (cf. (8.15)) as well as the formula corresponding to the one in Proposition 5.10 (cf. Theorem 10.7 and Corollary 10.2) in the framework of Chern-Weil theory. See also Corollary 10.5 for the differential geometric counterpart of Corollary 5.6.

The Whitney sum formula is further refined in Section 14.1 below.

Intermediate Thom classes

The identity in Corollary 5.5. 1 suggests the possibility of defining the "intermediate Thom classes" for a complex vector bundle by localizing appropriate Chern classes.

Let M be a C^∞ manifold of dimension m and $\pi : E \to M$ a complex vector bundle of rank l. Let $E^r = E \oplus \cdots \oplus E$ be the direct sum of r copies of E with the projection $\rho : E^r \to M$. Consider the "fiber product":

$$\begin{array}{ccc} \rho^*E & \longrightarrow & E \\ \big\downarrow{\scriptstyle \varpi} & & \big\downarrow{\scriptstyle \pi} \\ E^r & \xrightarrow{\ \rho\ } & M. \end{array} \qquad (5.24)$$

Thus ρ^*E is a vector bundle of rank l on E^r given by

$$\rho^*E = \{\, (\xi, (\xi_1, \ldots, \xi_r)) \in E \times E^r \mid \pi(\xi) = \rho(\xi_1, \ldots, \xi_r) \,\}.$$

Note that the condition $\pi(\xi) = \rho(\xi_1, \dots, \xi_r)$ means that $\xi_1, \dots, \xi_r$ and ξ are all in the same fiber of π. The bundle ρ^*E has a natural r-section, i.e., the "diagonal r-section" $s_\Delta^{(r)} = (s_1, \dots, s_r)$ given by

$$s_i(\xi_1, \dots, \xi_r) = (\xi_i, (\xi_1, \dots, \xi_r)).$$

Let $W(l,r)$ denote the Stiefel manifold of ordered r frames in $\mathbb{C}^l$ (cf. Section 3.5). It is an open subset in $(\mathbb{C}^l)^r = \mathbb{C}^{lr}$ and is given by $W(l,r) = (\mathbb{C}^l)^r \smallsetminus \Sigma_0$, where

$$\Sigma_0 = \{ (v_1, \dots, v_r) \in (\mathbb{C}^l)^r \mid v_1 \wedge \cdots \wedge v_r = 0 \}. \tag{5.25}$$

If we represent elements in $\mathbb{C}^{lr}$ by $l \times r$ matrices, Σ_0 is the common zero set of all the $r \times r$ minors and has codimension $l - r + 1$.

Let $W(E,r)$ denote the Stiefel bundle of r frames in E, which is a fiber bundle with fiber $W(l,r)$ associated with E and is an open subset in E^r. We have $W(E,r) = E^r \smallsetminus \Sigma$, where

$$\Sigma = \{ (\xi_1, \dots, \xi_r) \in E^r \mid \xi_1 \wedge \cdots \wedge \xi_r = 0 \}. \tag{5.26}$$

It coincides with the singular set of $s_\Delta^{(r)}$: $\Sigma = S(s_\Delta^{(r)})$. Thus we have the localization $c^q(\rho^*E, s_\Delta^{(r)}) \in H^{2q}(E^r, E^r \smallsetminus \Sigma)$ by $s_\Delta^{(r)}$ of $c^q(\rho^*E) \in H^{2q}(E^r)$, $q = l - r + 1$.

Definition 5.16. The *topological q-th Thom class* Ψ_E^q of E is defined by

$$\Psi_E^q = c_{\text{top}}^q(\rho^*E, s_\Delta^{(r)}) \qquad \text{in } H^{2q}(E^r, E^r \smallsetminus \Sigma).$$

Note that $\Psi_E^l = \Psi_E$ is the Thom class defined previously. We have a diagram as (4.23) with l', E_x, $E_x \smallsetminus 0$, E, $E \smallsetminus M$ and π replaced by $2q$, E_x^r, $E_x^r \smallsetminus \Sigma_x$, E^r, $E^r \smallsetminus \Sigma$ and ρ, respectively. Noting that δ^* is an isomorphism and $E_x^r \smallsetminus \Sigma_x \simeq W(l,r)$, we have $H^{2q}(E_x^r, E_x^r \smallsetminus \Sigma_x) \simeq H^{2q-1}(W(l,r)) \simeq \mathbb{Z}$. We also have:

$$c^q(E) = (\rho^*)^{-1} j^* \Psi_E^q.$$

As in the case of Thom class, the intermediate Thom class Ψ_E^q may be thought of as a universal localization of $c^q(E)$. In general, if we have a collection $s^{(r)} = (s_1, \dots, s_r)$ of r sections of E, we may think of it as a section of $\rho : E^r \to M$, which induces a morphism

$$s^{(r)*} : H^{2q}(E^r, E^r \smallsetminus \Sigma) \longrightarrow H^{2q}(M, M \smallsetminus S),$$

where $S = S(s^{(r)})$. We have

$$c^q(E, s^{(r)}) = s^{(r)*} \Psi_E^q.$$

If M is oriented, we have a commutative diagram:

$$\begin{array}{ccc} H^{2q}(E^r, E^r \setminus \Sigma) & \xrightarrow[A]{\sim} & \breve{H}_{m+2lr-2q}(\Sigma) \\ {\scriptstyle s^{(r)*}} \downarrow & & \downarrow {\scriptstyle (M \bullet_{s^{(r)}} \ )_S} \\ H^{2q}(M, M \setminus S) & \xrightarrow[A]{\sim} & \breve{H}_{m-2q}(S). \end{array}$$

We have $\mathrm{TRes}_{c^q}(\rho^* E, s_\Delta^{(r)}; \Sigma) = A(\Psi_E^q) = \Sigma$ and

$$\mathrm{TRes}_{c^q}(E, s^{(r)}; S) = (M \bullet_{s^{(r)}} \Sigma)_S.$$

Remark 5.20. In Section 10.4 below we define the classes corresponding to the above via differential geometric localization. Thus we refer the above classes as topological intermediate Thom classes.

Notes

Obstruction theory for characteristic classes of fiber bundles is thoroughly explained in [Steenrod (1951)], to which we refer for details of the materials in this chapter. The descriptions given here are modifications of those in [Brasselet, Lehmann, Seade and Suwa (2001)].

As a general reference for characteristic classes, we list [Milnor and Stasheff (1974)], to which we refer for the equality (5.6), see also [Hirzebruch (1966)].

For piecewise-linear manifolds, we refer to [Hirsch and Mazur (1974); Rourke and Sanderson (1972)]. In particular, Theorem 5.1 is due to [Hirsch and Mazur (1974)].

See [Brasselet, Seade and Suwa (2009)] and references therein for Remark 5.16.

Let E be a complex vector bundle of rank l on an oriented C^∞ manifold M. Also let s be a section of E with zero set S. Then $\mathrm{TRes}_{c^l}(s, E; S)$ corresponds to what is called the "localized top Chern class" defined in the algebraic category in [Fulton (1984)]. The formula in Proposition 5.10 above corresponds to the one in Example 14.1.3 in there and the formulas in Corollary 5.6 to the one in Example 17.4.8.

Chapter 6

Differential Forms

Differential forms are objects that represent cochains explicitly in the cohomology theory of C^∞ manifolds, the exterior derivative corresponding to the coboundary operator and the dual operation being given by integration. In fact, if we pass on to the cohomology, the cohomology defined by differential forms is canonically isomorphic with the (singular or simplicial) cohomology of the manifold. That is the content of the de Rham theorem, which we prove in the next chapter. In this setting, the exterior product corresponds to the cup product.

In this chapter, we discuss some topics related to differential forms on C^∞ and complex manifolds. Integration and integration along fibers are discussed. We also recall the Frobenius theorem and non-singular holomorphic foliations.

6.1 Vector fields and differential forms

Let M be a C^∞ manifold and U an open set in M. Recall that a vector field on U is a section on U of the tangent bundle $T_{\mathbb{R}}M$ of M (cf. Section 3.4). Differential forms are alternating forms on the space of vector fields:

Definition 6.1. A complex valued *differential p-form* of class C^∞, a C^∞ p-form for short, on U is a C^∞ section of the bundle $\bigwedge^p(T^c_{\mathbb{R}}M)^*$ on U.

We denote by $A^p(U)$ the set of C^∞ p-forms on U, which is by definition $C^\infty(U;\bigwedge^p(T^c_{\mathbb{R}}M)^*)$. The set $A^0(U)$ is thought of as the set of C^∞ functions on U and it has the structure of a $\mathbb{C}$-algebra. For each p, $A^p(U)$ has the structure of an $A^0(U)$-module.

Let $\{U_\alpha\}$ be a covering of M by coordinate neighborhoods U_α with coordinates $(x_1^\alpha, \dots, x_m^\alpha)$. For a point x in U_α, we denote by $(dx_1^\alpha, \dots, dx_m^\alpha)$ the basis of $T^*_{\mathbb{R},x} M$ dual to the basis $(\frac{\partial}{\partial x_1^\alpha}, \dots, \frac{\partial}{\partial x_m^\alpha})$ of $T_{\mathbb{R},x} M$. A p-form ω on U is expressed as, on $U \cap U_\alpha$,

$$\omega = \sum_{1 \le i_1 < \cdots < i_p \le m} f^\alpha_{i_1,\dots,i_p}(x)\, dx^\alpha_{i_1} \wedge \cdots \wedge dx^\alpha_{i_p}, \tag{6.1}$$

where $f^\alpha_{i_1,\dots,i_p}$ are C^∞ functions on $U \cap U_\alpha$.

Exterior product: We have the *exterior product*

$$A^p(U) \times A^q(U) \longrightarrow A^{p+q}(U),$$

which assigns to $\omega \in A^p(U)$ and $\theta \in A^q(U)$, the form $\omega \wedge \theta$ given by $(\omega \wedge \theta)(x) = \omega(x) \wedge \theta(x)$ for $x \in U$. It is bilinear in ω and θ and satisfies

$$\omega \wedge \theta = (-1)^{pq} \theta \wedge \omega.$$

As said above, we may think of $A^p(U)$ as the space of alternating p-forms on $C^\infty(U; T^c_{\mathbb{R}} M)$. In particular, for p 1-forms $\omega_1, \dots, \omega_p$ and p vector fields $v_1, \dots, v_p$, we have

$$\omega_1 \wedge \cdots \wedge \omega_p(v_1, \dots, v_p) = \det(\omega_i(v_j)).$$

Exterior derivative: The *exterior derivative*

$$d = d^p : A^p(U) \longrightarrow A^{p+1}(U)$$

is defined as follows: for $p = 0$,

$$df(v) = v(f)$$

and for $p > 0$,

$$\begin{aligned} d\omega(v_1, \dots, v_{p+1}) = & \sum_{i=1}^{p+1} (-1)^{i+1} v_i(\omega(v_1, \dots, \widehat{v_i}, \dots, v_{p+1})) \\ & + \sum_{i<j} (-1)^{i+j} \omega([v_i, v_j], v_1, \dots, \widehat{v_i}, \dots, \widehat{v_j}, \dots, v_{p+1}). \end{aligned}$$

In particular, if $p = 1$,

$$d\omega(u, v) = u(\omega(v)) - v(\omega(u)) - \omega([u, v]).$$

The operator d is $\mathbb{C}$-linear and satisfies:

$$d^p \circ d^{p-1} = 0 \tag{6.2}$$

and

$$d(\omega \wedge \theta) = d\omega \wedge \theta + (-1)^p \omega \wedge d\theta \tag{6.3}$$

for $\omega \in A^p(U)$ and $\theta \in A^q(U)$.

In terms of local coordinates as before it is given by, on $U \cap U_\alpha$,

$$df = \sum_{i=1}^{m} \frac{\partial f}{\partial x_i^\alpha}(x) dx_i^\alpha$$

for $f \in A^0(U)$ and

$$d\omega = \sum_{1 \le i_1 < \cdots < i_p \le m} df_{i_1,\dots,i_p}^\alpha \wedge dx_{i_1}^\alpha \wedge \cdots \wedge dx_{i_p}^\alpha$$

for $\omega \in A^p(U)$ expressed as (6.1).

Pull-back: Let $f : M' \to M$ be a C^∞ map of C^∞ manifolds. For an open set U in M, we have the "pull-back"

$$f^* : A^p(U) \longrightarrow A^p(f^{-1}(U)), \tag{6.4}$$

which is a $\mathbb{C}$-linear map with the following properties:

(1) $f^* h = h \circ f, \qquad h \in A^0(U)$,
(2) $f^*(\omega \wedge \theta) = f^*\omega \wedge f^*\theta, \qquad \omega \in A^p(U), \quad \theta \in A^q(U)$,
(3) $f^* d\omega = d\, f^*\omega, \qquad \omega \in A^p(U)$.

If we denote by $\langle\ ,\ \rangle$ the canonical pairing between the elements of the tangent and cotangent spaces, we have

$$f^*\langle f_* v, \omega\rangle = \langle v, f^*\omega\rangle$$

for $v \in T^c_{\mathbb{R},x} M'$ and $\omega \in (T^c_{\mathbb{R},f(x)} M)^*$.

Remark 6.1. The above notion of pull-back should be distinguished from that of the pull-back of a section (cf. (3.6)). Although a from ω on M' may be though of as a section of the bundle $\bigwedge^p(T^c_{\mathbb{R}}M')^*$, the pull-back $f^*\omega$ as a form is not a section of the bundle $f^*(\bigwedge^p(T^c_{\mathbb{R}}M')^*)$.

Forms with coefficients in a vector bundle: For a complex vector bundle E on M and an open set U in M, we set $A^p(U;E) = C^\infty(U; \bigwedge^p(T^c_{\mathbb{R}}M)^* \otimes E)$. An element in $A^p(U;E)$, called a *differential p-form with coefficients in E*, is expressed locally as a finite sum $\sum \omega_i \otimes s_i$ with ω_i p-forms and s_i sections of E. The exterior product induces a bilinear map

$$A^p(U) \times A^q(U;E) \longrightarrow A^{p+q}(U;E).$$

Suppose $f : M \to M'$ is a C^∞ map and $\pi : E \to M'$ a C^∞ complex vector bundle. For an open set U of M', we may define the pull-back

$$f^* : A^p(U;E) \longrightarrow A^p(f^{-1}(U); f^*E) \tag{6.5}$$

by $\sum \omega_i \otimes s_i \mapsto \sum f^*\omega_i \otimes f^*s_i$, where $f^*\omega_i$ denotes the pull-back as a from (cf. Remark 6.1). If $p = 0$, it coincides with the pull-back of sections (cf. (3.6)) and if $E = \mathbb{C} \times M'$ is the product bundle, it coincides with the pull-back of forms.

Differential forms on complex manifolds

Let M be a complex manifold of dimension n. Recall that the holomorphic cotangent bundle is the vector bundle T^*M dual to the holomorphic tangent bundle TM. By (3.4), we have a natural isomorphism

$$(T^c_{\mathbb{R}}M)^* \simeq T^*M \oplus \overline{T}^*M. \tag{6.6}$$

Hence we have an isomorphism

$$\bigwedge^r (T^c_{\mathbb{R}}M)^* \simeq \bigoplus_{p+q=r} \bigwedge^p T^*M \otimes \bigwedge^q \overline{T}^*M.$$

Definition 6.2. A *differential form of type* (p, q), a (p, q)*-form* for short, is a section of $\bigwedge^p T^*M \otimes \bigwedge^q \overline{T}^*M$.

A differential r-form is expressed as a sum of (p, q)-forms with $p+q = r$. Suppose that a point z in M is in a coordinate neighborhood U_α with coordinates $(z^\alpha_1, \dots, z^\alpha_n)$. We write $z^\alpha_i = x^\alpha_i + \sqrt{-1}\, y^\alpha_i$ and identify $T^*_z M$ and $\overline{T}^*_z M$ with subspaces of $(T_{\mathbb{R},z}^{c} M)^*$ by the isomorphism (6.6). Then, if we set

$$dz^\alpha_i = dx^\alpha_i + \sqrt{-1}\, dy^\alpha_i \quad \text{and} \quad d\bar{z}^\alpha_i = dx^\alpha_i - \sqrt{-1}\, dy^\alpha_i,$$

a straightforward computation shows that $dz^\alpha_1, \dots, dz^\alpha_n$ are in $T^*_z M$ and form a basis, which is dual to the basis $(\frac{\partial}{\partial z^\alpha_1}, \dots, \frac{\partial}{\partial z^\alpha_n})$ of $T_z M$ and that $d\bar{z}^\alpha_1, \dots, d\bar{z}^\alpha_n$ are in $\overline{T}^*_z M$ and form a basis, which is dual to the basis $(\frac{\partial}{\partial \bar{z}^\alpha_1}, \dots, \frac{\partial}{\partial \bar{z}^\alpha_n})$ of $\overline{T}_z M$. As bases of $\bigwedge^p T^*_z M$ and $\bigwedge^q \overline{T}^*_z M$, we may take the collections

$$(dz^\alpha_{i_1} \wedge \cdots \wedge dz^\alpha_{i_p})_{1 \le i_1 < \cdots < i_p \le n} \quad \text{and} \quad (d\bar{z}^\alpha_{j_1} \wedge \cdots \wedge d\bar{z}^\alpha_{j_q})_{1 \le j_1 < \cdots < j_q \le n},$$

respectively. Thus a (p, q)-form ω is written as, on U_α,

$$\omega = \sum_{\substack{1 \le i_1 < \cdots < i_p \le n \\ 1 \le j_1 < \cdots < j_q \le n}} f^\alpha_{i_1,\dots,i_p,j_1,\dots,j_q}(z)\, dz^\alpha_{i_1} \wedge \cdots \wedge dz^\alpha_{i_p} \wedge d\bar{z}^\alpha_{j_1} \wedge \cdots \wedge d\bar{z}^\alpha_{j_q}, \tag{6.7}$$

where $f^\alpha_{i_1,\dots,i_p,j_1,\dots,j_q}$ are C^∞ functions on U_α. By setting $I = (i_1, \dots, i_p)$ and $J = (j_1, \dots, j_q)$ we may write (6.7) in short as

$$\omega = \sum f^\alpha_{IJ}(z)\, dz^\alpha_I \wedge d\bar{z}^\alpha_J. \tag{6.8}$$

In particular, a $(p,0)$-form ω can be written as, on U_α,

$$\omega = \sum_{1 \le i_1 < \cdots < i_p \le n} f^\alpha_{i_1,\ldots,i_p}(z)\, dz^\alpha_{i_1} \wedge \cdots \wedge dz^\alpha_{i_p}.$$

If each $f^\alpha_{i_1,\ldots,i_p}$ is holomorphic, we say that ω is a *holomorphic p-form.* It is nothing but a holomorphic section of $\bigwedge^p T^*M$.

We denote by $A^{p,q}(U)$ the set of (p,q)-forms on an open set U in M. For each (p,q), it is an $A^0(U)$-module and we have the decomposition

$$A^r(U) = \bigoplus_{p+q=r} A^{p,q}(U).$$

Thus we may express the exterior derivative d as a sum

$$d = \partial + \bar\partial$$

with

$$\partial : A^{p,q}(U) \longrightarrow A^{p+1,q}(U) \quad \text{and} \quad \bar\partial : A^{p,q}(U) \longrightarrow A^{p,q+1}(U).$$

From (6.2), we have

$$\partial \circ \partial = 0, \quad \bar\partial \circ \bar\partial = 0 \quad \text{and} \quad \partial \circ \bar\partial + \bar\partial \circ \partial = 0. \tag{6.9}$$

The operators ∂ and $\bar\partial$ are given, on a coordinate neighborhood U_α, as follows. First, for a C^∞ function f,

$$\partial f = \sum_{i=1}^n \frac{\partial f}{\partial z^\alpha_i} dz^\alpha_i \quad \text{and} \quad \bar\partial f = \sum_{i=1}^n \frac{\partial f}{\partial \bar z^\alpha_i} d\bar z^\alpha_i.$$

If ω is a (p,q)-form expressed as (6.8),

$$\partial\omega = \sum \partial f^\alpha_{IJ} \wedge dz^\alpha_I \wedge d\bar z^\alpha_J \qquad \text{and} \qquad \bar\partial\omega = \sum \bar\partial f^\alpha_{IJ} \wedge dz^\alpha_I \wedge d\bar z^\alpha_J.$$

If we denote by $\Gamma(U; \bigwedge^p T^*M)$ the set of holomorphic p-forms on U, it is a sub-$\mathbb{C}$-vector space of $A^{p,0}(U)$. A form ω in $A^{p,0}(U)$ is in $\Gamma(U; \bigwedge^p T^*M)$ if and only if $\bar\partial\omega = 0$ and we have:

$$d = \partial : \Gamma(U; \textstyle\bigwedge^p T^*M) \longrightarrow \Gamma(U; \textstyle\bigwedge^{p+1} T^*M).$$

For a C^∞ complex vector bundle E on M and an open set U in M, we set $A^{p,q}(U; E) = C^\infty(U; \bigwedge^p T^*M \otimes \bigwedge^q \overline{T}^* M \otimes E)$. An element in $A^{p,q}(U;E)$, called a differential (p,q)-form with coefficients in E, is expressed locally as a finite sum $\sum \omega_i \otimes s_i$ with ω_i (p,q)-forms and s_i sections of E.

Suppose E is holomorphic. Then, we may define

$$\bar{\partial} : A^0(U;E) \longrightarrow A^{0,1}(U;E)$$

as follows. Let $e^l = (e_1, \dots, e_l)$ be a local holomorphic frame of E. For $s \in A^0(U;E)$, we may write $s = \sum_{i=1}^l f_i e_i$. If we set $\bar{\partial} s = \sum_{i=1}^l \bar{\partial} f_i \otimes e_i$, then we see that it does not depend on the choice of holomorphic frame and defined an element in $A^{0,1}(U;E)$. We have $\bar{\partial}(fs) = \bar{\partial} f \otimes s + f\bar{\partial} s$, for $f \in A^0(U)$. Furthermore we may define

$$\bar{\partial} : A^{p,q}(U;E) \longrightarrow A^{p,q+1}(U;E) \tag{6.10}$$

so that it satisfies

$$\bar{\partial}(\omega \otimes s) = \bar{\partial}\omega \otimes s + (-1)^{p+q}\omega \wedge \bar{\partial} s$$

for a (p,q)-form ω and a section s of E. Note that

$$\bar{\partial} \circ \bar{\partial} = 0.$$

If $f : M \to M'$ is a holomorphic map of complex manifolds, then f^* preserves the types so that it induces

$$f^* : A^{p,q}(U) \longrightarrow A^{p,q}(f^{-1}(U)).$$

Moreover, it commutes with ∂ and $\bar{\partial}$.

Adjunction formula: Let V be a non-singular hypersurface in M. From (3.13), we have the exact sequence

$$0 \longrightarrow N_V^* \longrightarrow T^*M|_V \longrightarrow T^*V \longrightarrow 0,$$

which together with the isomorphism $N_V \simeq L_V|_V$ (cf. Exercise 3.13) implies

$$K_V \simeq (K_M \otimes L_V)|_V, \tag{6.11}$$

where K_V and K_M denote the canonical bundles of V and M (cf. Definition 3.13). The above is referred to as the *adjunction formula* (cf. Theorem 14.3 below for a related result).

Let $\{U_\alpha\}$ be a covering of M such that V is given by $f_\alpha = 0$ on U_α. Then a holomorphic section ω of $K_M \otimes L_V$ is represented by a collection $\{\omega_\alpha\}$ such that ω_α is a holomorphic n-form on U_α and $\omega_\alpha / f_\alpha = \omega_\beta / f_\beta$ on $U_\alpha \cap U_\beta$, i.e., we may think of ω as a meromorphic n-form with a pole along V. We may assume that U_α is a coordinate neighborhood and that f_α is one of the coordinates on U_α. Thus we may write $\omega_\alpha = df_\alpha \wedge \omega'_\alpha$. Then we see that $i^*\omega'_\alpha = i^*\omega'_\beta$ in $U_\alpha \cap U_\beta \cap V$ so that it defines an $(n-1)$-form on V, which corresponds to ω under the above isomorphism. The correspondence $\omega \mapsto \omega'$ is called the *Poincaré residue map*.

6.2 Integration of differential forms

In this section we let M denote a C^∞ manifold of dimension m and R a commutative ring with unity.

Integration via singular homology

C^∞ singular homology and cohomology: Let Δ^p be the standard p-simplex in $\mathbb{R}^p$ (cf. Section B.1). A C^∞ *singular p-simplex* of M is a map $\sigma : \Delta^p \to M$ which is C^∞ in the sense of Definition 3.19. We denote by $S_p^{\mathrm{ci}}(M;\mathsf{R})$ the free R-module generated by all the C^∞ singular p-simplices. Then $S_\bullet^{\mathrm{ci}}(M;\mathsf{R})$ is a subcomplex of $S_\bullet(M;\mathsf{R})$ and, if we denote by $H_p^{\mathrm{ci}}(M;\mathsf{R})$ the p-th homology of $S_\bullet^{\mathrm{ci}}(M;\mathsf{R})$, it is known that the inclusion $S_\bullet^{\mathrm{ci}}(M;\mathsf{R}) \hookrightarrow S_\bullet(M;\mathsf{R})$ induces an isomorphism

$$H_p^{\mathrm{ci}}(M;\mathsf{R}) \xrightarrow{\sim} H_p(M;\mathsf{R}). \tag{6.12}$$

Also, if we set $S_{\mathrm{ci}}^p(M;\mathsf{R}) = \mathrm{Hom}(S_p^{\mathrm{ci}}(M;\mathsf{R}),\mathsf{R})$ and denote by $H_{\mathrm{ci}}^p(M;\mathsf{R})$ the p-th cohomology of the complex $S_{\mathrm{ci}}^\bullet(M;\mathsf{R})$, the canonical cochain morphism $S^\bullet(M;\mathsf{R}) \to S_{\mathrm{ci}}^\bullet(M;\mathsf{R})$ induces an isomorphism

$$H^p(M;\mathsf{R}) \xrightarrow{\sim} H_{\mathrm{ci}}^p(M;\mathsf{R}). \tag{6.13}$$

Remark 6.2. The proofs for the isomorphisms (6.12) and (6.13) are based on the *Whitney approximation theorem*, which asserts that every continuous map $f : M \to M'$ of C^∞ manifolds is homotopic to a C^∞ map and that, if f is C^∞ on a closed set A in M, the homotopy can be taken to be relative to A.

Integration on singular chains: Let $\sigma : \Delta^p \to M$ be a C^∞ singular p-simplex of M. Then, from definition, $\sigma|_{\mathrm{Int}\,\Delta^p}$ is C^∞, where $\mathrm{Int}\,\Delta^p$ denotes the interior of Δ^p in $\mathbb{R}^p$, and for each point a of $\partial\Delta^p$, there is a neighborhood U of a in $\mathbb{R}^p$ such that $\sigma|_{\Delta^p\cap U}$ has a C^∞ extension on U. We may take U so that the image of the extension is in a coordinate neighborhood in M. Then, in view of Proposition 3.12, for a p-form ω on a neighborhood of $\sigma(\Delta^p)$ in M, we may define the integration of ω on σ by

$$\int_\sigma \omega = \int_{\Delta^p} \sigma^*\omega,$$

where Δ^p is oriented as in Remark B.13. 1.

Also for a finite C^∞ singular p-chain $c = \sum a_\sigma \sigma$, $a_\sigma \in \mathbb{C}$, of M and a p-form ω on a neighborhood of $\bigcup \sigma(\Delta^p)$ in M, we define

$$\int_c \omega = \sum a_\sigma \int_\sigma \omega. \tag{6.14}$$

Then we have:

Theorem 6.1 (Stokes theorem). *For a C^∞ singular p-chain c and a $(p-1)$-form ω,*

$$\int_c d\omega = \int_{\partial c} \omega.$$

From definitions, we have:

Proposition 6.1. *Let $f : M \to M'$ be a C^∞ map of C^∞ manifolds. For a p-form ω on M' and a C^∞ singular p-chain c of M,*

$$\int_c f^*\omega = \int_{f_*c} \omega.$$

Integration via triangulation

Let (K, h) be a C^∞ triangulation of M (cf. Section 4.1). We regard the orientation conventions in Sections 4.1 and 4.2. Let $\boldsymbol{s}$ be an oriented p-simplex of K. Then we may think of $h|_{\boldsymbol{s}} : \boldsymbol{s} \to M$ as a C^∞ singular p-simplex. For a C^∞ p-form ω on a neighborhood of $h(\boldsymbol{s})$ in M, we set

$$\int_{\boldsymbol{s}} \omega = \int_{\boldsymbol{s}} (h|_{\boldsymbol{s}})^*\omega, \tag{6.15}$$

which is to say that we identify $\boldsymbol{s}$ with $h(\boldsymbol{s})$ in the C^∞ sense. For a finite p-chain $c = \sum a_{\boldsymbol{s}}\boldsymbol{s}$, $a_{\boldsymbol{s}} \in \mathbb{C}$, and a C^∞ p-form ω on a neighborhood of $|c|$ in M, we define the integral $\int_c \omega$ of ω on c as (6.14) with σ replaced by $\boldsymbol{s}$. Then we have the Stokes theorem as before. The integrations are compatible with the natural morphism $C_p^K(M;\mathbb{C}) \to S_p^{\mathrm{ci}}(M;\mathbb{C})$.

Let R be a closed p-dimensional piecewise C^∞ submanifold of M possibly with boundary. We may assume that (K, h) is as in Definition 5.10. Suppose that R is oriented. Thus ∂R is orientable and is oriented according to Convention 3.2, modified for piecewise C^∞ manifolds. If R is compact, we may think of it as a finite p-chain and, for a p-form ω on a neighborhood of R, we have the integral

$$\int_R \omega = \sum \int_{\boldsymbol{s}} \omega,$$

where the sum is taken over all p-simplices $\boldsymbol{s}$ in R with the orientation same as that of R. The definition does not depend on the choice of the C^∞ triangulation because of the uniqueness after subdivision. In this case we may think of ∂R as a finite $(p-1)$-chain and thus we may integrate a $(p-1)$-form on ∂R. From Theorem 6.1, we have the following, which is also referred to as the Stokes theorem:

Theorem 6.2. *Let R be an oriented compact p-dimensional piecewise C^∞ submanifold of M possibly with boundary. For an $(p-1)$-form ω in a neighborhood of R,*

$$\int_R d\omega = \int_{\partial R} \omega.$$

In the case M itself is oriented and compact, we may define $\int_M \omega$ for an m-form ω on M.

If we denote by $-R$ the manifold R with opposite orientation, we have

$$\int_{-R} \omega = -\int_R \omega.$$

Exercise 6.1. Let $f : M \to M'$ be a C^∞ map.

1. Show that the graph $\Gamma_f = \{ (x, y) \in M \times M' \mid y = f(x) \}$ of f has the structure of a submanifold of $M \times M'$ such that the map $\tilde{f} : M \to \Gamma_f$ given by $x \mapsto (x, f(x))$ is a diffeomorphism.

2. Suppose M is a compact oriented manifold of dimension m. We orient Γ_f so that $\tilde{f}$ is an orientation preserving diffeomorphism. Show that for forms ω and θ on M and M' of degrees p and $m-p$, we have

$$\int_{\Gamma_f} \pi_1^*\omega \wedge \pi_2^*\theta = \int_M \omega \wedge f^*\theta,$$

where π_1 and π_2 denote the projections of $M \times M'$ onto the first and second factors.

6.3 Integration along fibers

Let M be a C^∞ manifold of dimension m without boundary and F a compact oriented C^∞ manifold of dimension l' possibly with boundary ∂F. Also, let $\pi : T \to M$ be an oriented C^∞ fiber bundle with fiber F. In this situation, we have the *integration along fibers* of π, *fiber integration* for short, which is a $\mathbb{C}$-linear map

$$\pi_* : A^{l'+p}(T) \longrightarrow A^p(M)$$

defined as follows. If F has a boundary, the space $A^{l'+p}(T)$ is to be understood as described below. We take a covering $\{U_\alpha\}$ of M so that there is a positive trivialization $\psi_\alpha : \pi^{-1}(U_\alpha) \xrightarrow{\sim} F \times U_\alpha$ for each α. Taking a smaller U_α, if necessary, we may assume that each U_α is a coordinate neighborhood with coordinates $(x_1^\alpha, \dots, x_m^\alpha)$. We think of F as being inside of some oriented manifold $\tilde{F}$ of the same dimension. Then we may think of T as being in a fiber bundle $\tilde{\pi} : \tilde{T} \to M$ with fiber $\tilde{F}$ and each ψ_α as being the restriction of a trivialization $\tilde{\psi}_\alpha : (\tilde{\pi})^{-1}(U_\alpha) \xrightarrow{\sim} \tilde{F} \times U_\alpha$ of $\tilde{T}$ on U_α. In fact, we define the integration along fibers as a map

$$\pi_* : A^{l'+p}(\tilde{T}) \longrightarrow A^p(M).$$

We cover F by coordinate neighborhoods $\{V_\lambda\}$. Let $(t_1^\lambda, \dots, t_{l'}^\lambda)$ be a positive coordinate system on V_λ for each λ and suppose that, if $\partial F \cap V_\lambda \neq \emptyset$, then $F \cap V_\lambda = \{t_1^\lambda \leq 0\}$. Then recall that $(t_2^\lambda, \dots, t_{l'}^\lambda)$ is a coordinate system on $\partial F \cap V_\lambda$ and ∂F is oriented so that the form $dt_2^\lambda \wedge \cdots \wedge dt_{l'}^\lambda$ is positive. If we set $W_{\alpha,\lambda} = (\tilde{\psi}_\alpha)^{-1}(V_\lambda \times U_\alpha)$, it is a coordinate neighborhood on $\tilde{T}$ with coordinates $(t_1^\lambda, \dots, t_{l'}^\lambda, x_1^\alpha, \dots, x_m^\alpha)$ and T is covered by $\{W_{\alpha,\lambda}\}$. Let ω be a p-form on (a neighborhood of) T. If we write ω explicitly on $W_{\alpha,\lambda}$, the sum of the terms involving $dt^\lambda = dt_1^\lambda \wedge \cdots \wedge dt_{l'}^\lambda$ is of the form

$$\sum_I f_I^{\alpha,\lambda}(t^\lambda, x^\alpha)\, dt^\lambda \wedge dx_I^\alpha,$$

where the sum runs through the p-tuples of integers $I = (i_1, \dots, i_p)$ with $1 \leq i_1 < \cdots < i_p \leq m$, $f_I^{\alpha,\lambda}(t^\lambda, x^\alpha)$ are C^∞ functions on $W_{\alpha,\lambda}$ and $dx_I^\alpha = dx_{i_1}^\alpha \wedge \cdots \wedge dx_{i_p}^\alpha$. It is not difficult to show that, for each I,

$$f_I^{\alpha,\lambda}(t^\lambda, x^\alpha)\, dt^\lambda = f_I^{\alpha,\mu}(t^\mu, x^\alpha)\, dt^\mu \quad \text{in} \quad W_{\alpha,\lambda} \cap W_{\alpha,\mu}. \tag{6.16}$$

Hence the collection $\{f_I^{\alpha,\lambda}(t^\lambda, x^\alpha)\, dt^\lambda\}$ defines an l'-form $\omega_I^\alpha(t, x^\alpha)$ on $(\tilde{\psi})^{-1}(U_\alpha)$. By integrating $\omega_I^\alpha(t, x^\alpha)$ on each fiber of π, we obtain a function $F_I^\alpha(x^\alpha)$ on U_α. It is further shown that

$$\sum_I F_I^\alpha(x^\alpha)\, dx_I^\alpha = \sum_I F_I^\beta(x^\beta)\, dx_I^\beta \quad \text{in} \quad U_\alpha \cap U_\beta. \tag{6.17}$$

Thus we obtain a p-from on M, which is denoted by $\pi_*\omega$ and is called the integral of ω along fibers of π. The definition does not depend on the choice of coordinates and trivializations involved.

If we further assume, in the above situation, that M is orientable, then the total space T is also orientable. If M is oriented so that $(x_1^a, \dots, x_m^\alpha)$ as above is positive, then we orient T so that the form $dt^\lambda \wedge dx^\alpha$ is positive, where we set $dx^\alpha = dx_1^\alpha \wedge \cdots \wedge dx_m^\alpha$. Thus ∂T is oriented so that the form $dt_2^\lambda \wedge \cdots \wedge dt_{l'}^\lambda \wedge dx^\alpha$ is positive. This coincides with the orientation of ∂T as a fiber bundle over M with fiber ∂F.

With these, we have the following fundamental formulas.

Proposition 6.2 (Projection formula I). *Let M be a C^∞ manifold of dimension m and $\pi : T \to M$ an oriented fiber bundle with fiber a compact oriented manifold F of dimension l' possibly with boundary.*

1. *For ω in $A^{l'+p}(\tilde{T})$ and θ in $A^q(M)$,*

$$\pi_*(\omega \wedge \pi^*\theta) = \pi_*\omega \wedge \theta.$$

2. *If M is compact and oriented, for ω in $A^{l'+p}(\tilde{T})$ and θ in $A^{m-p}(M)$,*

$$\int_T \omega \wedge \pi^*\theta = \int_M \pi_*\omega \wedge \theta.$$

Proposition 6.3. *In the situation of the previous proposition,*

$$\pi_* \circ d + (-1)^{l'+1} d \circ \pi_* = (\partial\pi)_* \circ i^*,$$

where i denotes the inclusion $\partial T \hookrightarrow T$.

Exercise 6.2. Show (6.16), (6.17) and Propositions 6.2 and 6.3.

6.4 Frobenius theorem and non-singular foliations

We start by considering everything in a neighborhood of the origin 0 in $\mathbb{C}^n$. Let $\mathscr{O}_n$ be the ring of germs of holomorphic functions at 0 (cf. Section 1.4). As in the case of functions, we may define germs of vector fields and differential forms. We denote by Θ_n and Ω_n^p, respectively, the sets of germs of holomorphic vector fields and holomorphic p-forms at 0. Each of them has a natural structure of $\mathscr{O}_n$-module. We denote Ω_n^1 simply by Ω_n. A germ v in Θ_n is written as $v = \sum_{i=1}^n f_i \cdot \partial/\partial z_i$ with f_i germs in $\mathscr{O}_n$ and we see that Θ_n is a free $\mathscr{O}_n$-module of rank n having $(\partial/\partial z_1, \dots, \partial/\partial z_n)$ as a basis. Also, a germ ω in Ω_n^p is written as $\omega = \sum_{1 \le i_1 < \cdots < i_p \le n} f_{i_1,\dots,i_p} dz_{i_1} \wedge \cdots \wedge dz_{i_p}$ with $f_{i_1,\dots,i_p}$ germs in $\mathscr{O}_n$ and we see that Ω_n^p is a free $\mathscr{O}_n$-module of rank $\binom{n}{p}$ having $(dz_{i_1} \wedge \cdots \wedge dz_{i_p})_{1 \le i_1 < \cdots < i_p \le n}$ as a basis. The exterior product

$$\Omega_n^p \times \Omega_n^q \longrightarrow \Omega_n^{p+q}$$

sending (ω, θ) to $\omega \wedge \theta$ and the exterior derivative

$$d : \Omega_n^p \longrightarrow \Omega_n^{p+1}$$

are well-defined. Since Ω_n is the dual $\mathscr{O}_n$-module of Θ_n, there is a natural pairing

$$\Theta_n \times \Omega_n \longrightarrow \mathscr{O}_n, \qquad (v, \omega) \mapsto \langle v, \omega \rangle.$$

Now we take a system $(\omega_1, \ldots, \omega_q)$ of germs in Ω_n. Each germ ω_i is represented by a holomorphic 1-form in a neighborhood of 0. In the following, we assume that

$$\omega_1 \wedge \cdots \wedge \omega_q(0) \neq 0. \tag{6.18}$$

Thus $\omega_1(z), \ldots, \omega_q(z)$ are linearly independent over $\mathbb{C}$ at each point z in a neighborhood of 0.

Definition 6.3. A system $(\omega_1, \ldots, \omega_q)$ of germs in Ω_n with (6.18) is said to be *integrable* if there exist germs θ_{ij} in Ω_n, for $i, j = 1, \ldots, q$, such that

$$d\omega_i = \sum_{j=1}^{q} \theta_{ij} \wedge \omega_j, \qquad 1 \leq i \leq q.$$

The following is proved as in the real case:

Theorem 6.3 (Frobenius theorem). *A system* $(\omega_1, \ldots, \omega_q)$ *of germs in* Ω_n *satisfying* (6.18) *is integrable if and only if there exist germs* f_i *and* g_{ij} *in* $\mathcal{O}_n$, $i, j = 1, \ldots, q$, *such that* $\det(g_{ij})(0) \neq 0$ *and that*

$$\omega_i = \sum_{j=1}^{q} g_{ij}\, df_j, \qquad 1 \leq i \leq q.$$

Note that the condition $\det(g_{ij})(0) \neq 0$ implies that $\det(g_{ij})$ is a unit in $\mathcal{O}_n$ and thus (g_{ij}) is an invertible matrix. Hence the Frobenius theorem shows that if a system $(\omega_1, \ldots, \omega_q)$ is integrable, the sub-$\mathcal{O}_n$-module of Ω_n generated by $\omega_1, \ldots, \omega_q$ is the same as the one generated by $df_1, \ldots, df_q$. In other words, the system of differential equations

$$\omega_1 = \cdots = \omega_q = 0$$

is equivalent to

$$df_1 = \cdots = df_q = 0,$$

which can be easily solved to get equations

$$f_1 = c_1, \ldots, \ f_q = c_q, \qquad c_i \in \mathbb{C}.$$

From (6.18), we see that

$$df_1 \wedge \cdots \wedge df_q(0) \neq 0.$$

Hence 0 is a regular point of $(f_1, \ldots, f_q)$. By the implicit function theorem, we may choose $(f_1, \ldots, f_q)$ as a part of a local coordinate system about 0 and for each $(c_1, \ldots, c_q)$, the set $\{ z \mid f_i(z) = c_i \}$ is a submanifold of dimension $n-q$. This is an integral submanifold of the system $(\omega_1, \dots, \omega_q)$ in the sense that each ω_i vanishes identically on this submanifold.

We may as well state the Frobenius theorem in terms of vector fields. In general, if G is a sub-$\mathscr{O}_n$-module of Ω_n, we set

$$G^a = \{ v \in \Theta_n \mid \langle v, \omega \rangle = 0, \text{ for all } \omega \in G \}$$

and call it the annihilator of G. This is a sub-$\mathscr{O}_n$-module of Θ_n. Also, for a sub-$\mathscr{O}_n$-module F of Θ_n, we set

$$F^a = \{ \omega \in \Omega_n \mid \langle v, \omega \rangle = 0, \text{ for all } v \in F \}$$

and call it the annihilator of F. This is a sub-$\mathscr{O}_n$-module of Ω_n.

Note that the sub-$\mathscr{O}_n$-module of Ω_n generated by $\omega_1, \dots, \omega_q$ satisfying (6.18) is a free submodule of rank q. The proof of the following is also the same as in the real case.

Lemma 6.1. **1.** *Let $(\omega_1, \dots, \omega_q)$ be a system of germs in Ω_n satisfying* (6.18) *and G the sub-$\mathscr{O}_n$-module of Ω_n generated by $\omega_1, \dots, \omega_q$. If we set $F = G^a$, then F is a free sub-$\mathscr{O}_n$-module of Θ_n of rank $p = n-q$ and admits a basis $v_1, \dots, v_p$ satisfying*

$$v_1 \wedge \cdots \wedge v_p(0) \neq 0. \tag{6.19}$$

2. *Let $(v_1, \dots, v_p)$ be a system of germs in Θ_n satisfying* (6.19) *and F the sub-$\mathscr{O}_n$-module of Θ_n generated by $v_1, \dots, v_p$. If we set $G = F^a$, then G is a free sub-$\mathscr{O}_n$-module of Ω_n of rank $q = n-p$ and admits a basis $\omega_1, \dots, \omega_q$ satisfying* (6.18).

3. *In the above correspondence, the system $(\omega_1, \dots, \omega_q)$ is integrable if and only if*

$$[v_i, v_j] = \sum_{k=1}^{p} a_{ij}^k v_k \qquad \textit{for some } \; a_{ij}^k \in \mathscr{O}_n, \quad i, j, k = 1, \dots, p.$$

Remark 6.3. In particular, if G is generated by $(dz_{p+1}, \dots, dz_n)$, $F = G^a$ is generated by $(\partial/\partial z_1, \dots, \partial/\partial z_p)$ and vice versa.

Next we introduce global (non-singular) foliations on a complex manifold. Let M be a complex manifold of dimension n.

Definition 6.4. Let F be a holomorphic subbundle of TM of rank p. We say that F is *involutive* (or *integrable*) if, for each point in M, there exist a neighborhood U and a holomorphic frame $(v_1, \dots, v_p)$ of F over U satisfying

$$[v_i, v_j] = \sum_{k=1}^{p} a_{ij}^k v_k, \qquad i, j, k = 1, \dots, p,$$

for some holomorphic functions a_{ij}^k on U.

Let F be as above and $G = F^a$ its annihilator, i.e., G is the holomorphic subbundle of T^*M such that

$$\Gamma(U;G) = \{\,\omega \in \Gamma(U;T^*M) \mid \langle v,\omega\rangle = 0, \text{ for all } v \in \Gamma(U,F)\,\}.$$

Then G is a subbundle of T^*M of rank $q = n - p$ and is integrable in the sense that, for each point in M, there exist a neighborhood U and a holomorphic frame $(\omega_1,\ldots,\omega_q)$ of G over U satisfying

$$d\omega_i = \sum_{j=1}^{q} \theta_{ij} \wedge \omega_j, \qquad i = 1,\ldots,q,$$

for some holomorphic 1-forms θ_{ij} on U. Conversely let G be an integrable holomorphic subbundle G of T^*M of rank q. Then its annihilator $F = G^a$ is an involutive subbundle of TM of rank $p = n - q$.

By the Frobenius theorem, an involutive subbundle F of TM corresponds to a structure on M defined by a particular coordinate covering $\{U_\alpha\}$, i.e., we may find a coordinate system $(z_1^\alpha,\ldots,z_n^\alpha)$ on each U_α such that $\partial z_i^\alpha/\partial z_j^\beta = 0$ in $U_\alpha \cap U_\beta$, for $i = p+1,\ldots,n$ and $j = 1,\ldots,p$. Such a structure is called a (non-singular) *holomorphic foliation* on M. We call F and G, respectively, the tangent and conormal bundles the foliation and p and q, respectively, the dimension and the codimension of the foliation. Sometimes we refer to F or G simply as a foliation.

A (complex) submanifold L of M is called an integral submanifold of the foliation if

$$\iota_*(T_pL) \subset F_p$$

for all p in L, where $\iota : L \hookrightarrow M$ denotes the inclusion. This is equivalent to saying that

$$\iota^*\omega_i = 0, \qquad i = 1,\ldots,q$$

for the system $(\omega_1,\ldots,\omega_q)$ as above. A p-dimensional integral submanifold L is maximal if L' is a p-dimensional integral submanifold containing L, then $L' = L$. Such a submanifold L is called a *leaf* of the foliation. From the Frobenius theorem, we get the following theorem.

Theorem 6.4. *If M admits a foliation, then M is the disjoint union of leaves of the foliation.*

Notes

We list [Lee (2013); Matsushima (1972)] as general references for this chapter. For the topics involving complex structures, we refer to [Griffiths and Harris (1978)].

See [Lee (2013)] for a proof of the isomorphism in (6.12). The Whitney approximation theorem mentioned in Remark 6.2 is due to [Whitney (1936)].

Chapter 7

Čech-de Rham Cohomology

We introduce the de Rham cohomology, Čech cohomologies of locally constant functions and of differential forms. We then combine them to form the Čech-de Rham cohomology, through which we have a canonical correspondence between the de Rham cohomology and the Čech cohomology of locally constant functions. We define integration on Čech-de Rham cochains and establish various canonical isomorphisms (Theorem 7.4). The canonical de Rham theorem (Theorem 7.5) then immediately follows.

We then discuss the relative Čech-de Rham cohomology, which is particularly suited to describe various localizations. We prove the relative version of the de Rham theorem (Corollary 7.8).

We give description of the Poincaré, Alexander and Lefschetz dualities in terms of differential forms. We define the Thom class in the relative Čech-de Rham cohomology. As an application we prove the Lefschetz fixed point formula.

For algebraic background on homology and cohomology, we refer to Section A.1.

7.1 de Rham cohomology

Let M be a C^∞ manifold of dimension m and $A^p(M)$ the $\mathbb{C}$-vector space of complex valued C^∞ p-forms on M.

By (6.2), the exterior derivative defines a complex of $\mathbb{C}$-vector spaces:

$$0 \longrightarrow A^0(M) \xrightarrow{d^0} A^1(M) \xrightarrow{d^1} \cdots \xrightarrow{d^{m-1}} A^m(M) \longrightarrow 0,$$

which is called the *de Rham complex* of M and is denoted by $(A^\bullet(M), d)$. We set $d^{-1} = 0$.

Definition 7.1. The p-th *de Rham cohomology* of M is the p-th cohomology of $(A^\bullet(M), d)$:

$$H^p_d(M) = \operatorname{Ker} d^p / \operatorname{Im} d^{p-1}.$$

For a closed p-form ω, its class in $H^p_d(M)$ is denoted by $[\omega]$. Since $\operatorname{Ker} d^0 = \mathbb{C}(M)$, the complex valued locally constant functions on M, we have:

$$H^0_d(M) = \mathbb{C}(M).$$

In the case M is connected, $H^0_d(M) = \mathbb{C}$.

One of the main aims of this chapter is to prove the de Rham theorem which states that $H^p_d(M)$ is canonically isomorphic with the singular cohomology $H^p(M; \mathbb{C})$ of M (cf. Theorem 7.5 below). The following is one of the essential ingredients for the proof of this:

Lemma 7.1 (Poincaré lemma). *Let ω be a p-form on $\mathbb{R}^m$, $p > 0$. If $d\omega = 0$, there exists a $(p-1)$-form θ on $\mathbb{R}^m$ such that $\omega = d\theta$.*

Proof. For a p-tuple of integers $(i_1, \ldots, i_p)$ with $1 \le i_1 < \cdots < i_p \le m$, we set

$$\psi_{i_1,\ldots,i_p} = \sum_{\nu=1}^{p} (-1)^{\nu-1} x_{i_\nu} dx_{i_1} \wedge \cdots \wedge \widehat{dx_{i_\nu}} \wedge \cdots \wedge dx_{i_p}$$

so that $d\psi_{i_1,\ldots,i_p} = p\,dx_{i_1} \wedge \cdots \wedge dx_{i_p}$. We define

$$h = h^p : A^p(\mathbb{R}^m) \longrightarrow A^{p-1}(\mathbb{R}^m)$$

by setting, for $\omega = f(x) dx_{i_1} \wedge \cdots \wedge dx_{i_p}$,

$$h^p(\omega) = \int_0^1 t^{p-1} f(tx) dt \cdot \psi_{i_1,\ldots,i_p}$$

and extending them $\mathbb{C}$-linearly. Then we compute $d \circ h + h \circ d = 1$, which implies that, if $d\omega = 0$, then $\omega = d\,h\omega$. □

Corollary 7.1. *The de Rham complex of $\mathbb{R}^m$ is acyclic, i.e.,*

$$H^p_d(\mathbb{R}^m) = 0 \qquad \text{for } p > 0.$$

Remark 7.1. The proof of Lemma 7.1 is valid for every open set U in $\mathbb{R}^m$ that is "star shaped" relative to some point in $\mathbb{R}^m$.

Functoriality: If $f : M' \to M$ is a C^∞ map of C^∞ manifolds, we have the pull-back $f^* : A^p(M) \to A^p(M')$ for each p (cf. (6.4)). They commute with the exterior derivative. Thus it induces a morphism

$$f_d^* : H_d^p(M) \longrightarrow H_d^p(M').$$

It is functorial in the sense that:

(1) for the identity map 1_M of M, $(1_M)_d^*$ is the identity map of $H_d^p(M)$,
(2) for a sequence of C^∞ maps $M'' \xrightarrow{g} M' \xrightarrow{f} M$, $(f \circ g)_d^* = g_d^* \circ f_d^*$.

In particular, if f is a C^∞ diffeomorphism, f_d^* is an isomorphism. For example we have:

Proposition 7.1. *If M is diffeomorphic with $\mathbb{R}^m$, the Poincaré lemma holds on M and the de Rham complex of M is acyclic.*

We will see that this pull-back corresponds to the pull-back in the singular cohomology under the isomorphism of Theorem 7.5 (cf. Proposition 7.5 below).

7.2 Čech cohomology

Let M be a C^∞ manifold and $\mathcal{U} = \{U_\alpha\}_{\alpha \in I}$ an open covering of M.

Čech cohomology of locally constant functions

Let R be a commutative ring with unity. For an open set U in M, we denote by $\mathsf{R}(U)$ the ring of locally constant functions on U with values in R. We set $\mathsf{R}(\emptyset) = 0$.

For $\alpha_0, \dots, \alpha_p$ in I, we set $U_{\alpha_0 \dots \alpha_p} = \bigcap_{\nu=0}^p U_{\alpha_\nu}$. We define an R-module $C^p(\mathcal{U}; \mathsf{R})$ as the direct product:

$$C^p(\mathcal{U}; \mathsf{R}) = \prod_{(\alpha_0, \dots, \alpha_p) \in I^{p+1}} \mathsf{R}(U_{\alpha_0 \dots \alpha_p}).$$

An element c in $C^p(\mathcal{U}; \mathsf{R})$ is called a *p-cochain* of locally constant functions on $\mathcal{U}$. It assigns to each $(\alpha_0, \dots, \alpha_p)$ in I^{p+1} an element $c_{\alpha_0 \dots \alpha_p}$ in $\mathsf{R}(U_{\alpha_0 \dots \alpha_p})$. The *coboundary operator*

$$\delta = \delta^p : C^p(\mathcal{U}; \mathsf{R}) \longrightarrow C^{p+1}(\mathcal{U}; \mathsf{R})$$

is defined by

$$(\delta c)_{\alpha_0 \dots \alpha_{p+1}} = \sum_{\nu=0}^{p+1} (-1)^\nu c_{\alpha_0 \dots \widehat{\alpha_\nu} \dots \alpha_{p+1}}, \tag{7.1}$$

where "$\widehat{\ }$" means the symbol under it is to be omitted. Note also that $c_{\alpha_0 \dots \widehat{\alpha_\nu} \dots \alpha_{p+1}}$ in the right-hand side means its restriction to $U_{\alpha_0 \dots \alpha_{p+1}}$. We do not write this explicitly to avoid complications, however this is one thing that should not be forgotten. The operators have the property $\delta^p \circ \delta^{p-1} = 0$ and define a complex of R-modules:

$$0 \longrightarrow C^0(\mathcal{U}; \mathsf{R}) \xrightarrow{\delta^0} C^1(\mathcal{U}; \mathsf{R}) \xrightarrow{\delta^1} \cdots \xrightarrow{\delta^{p-1}} C^p(\mathcal{U}; \mathsf{R}) \xrightarrow{\delta^p} \cdots,$$

which is called the *Čech complex* of locally constant functions with values in R on $\mathcal{U}$ and is denoted by $(C^\bullet(\mathcal{U}; \mathsf{R}), \delta)$. We set $\delta^{-1} = 0$.

Definition 7.2. The p-th *Čech cohomology* of locally constant functions on $\mathcal{U}$ is the p-th cohomology of $(C^\bullet(\mathcal{U}; \mathsf{R}), \delta)$:

$$H^p(\mathcal{U}; \mathsf{R}) = \operatorname{Ker} \delta^p / \operatorname{Im} \delta^{p-1}.$$

For a p-cocycle c, its class in $H^p(\mathcal{U}; \mathsf{R})$ is denoted by $[c]$. Since $\operatorname{Ker} \delta^0 = \mathsf{R}(M)$, we have:

$$H^0(\mathcal{U}; \mathsf{R}) = \mathsf{R}(M).$$

In particular, if M is connected, $H^0(\mathcal{U}; \mathsf{R}) = \mathsf{R}$.

Exercise 7.1. Verify the identity $\delta^p \circ \delta^{p-1} = 0$.

Simplicial coverings: Let K be a triangulation of M and I the set of vertices in K. For each α in I, let $U_\alpha = O_K(\alpha)$, the open star of α in K. Then $\mathcal{U} = \{U_\alpha\}$ is an open covering of M, called the *simplicial covering* associated with K. Note that the set of simplicial coverings is cofinal in the set of open coverings of M (cf. Remark 2.1. 1).

Remark 7.2. The coboundary operator δ is define as in (7.1) so that it is compatible with the boundary operator of simplicial chains in the case $\mathcal{U}$ is a simplicial covering. In fact in this case, $H^p(\mathcal{U}; \mathsf{R})$ is isomorphic with the simplicial cohomology of M and in turn with the singular cohomology $H^p(M; \mathsf{R})$ of M (cf. Theorem 7.3 below).

Čech cohomology of forms

For an open set U in M, let $A^q(U)$ denote the $\mathbb{C}$-vector space of C^∞ q-forms on U. We set $A^q(\emptyset) = 0$. Replacing $\mathsf{R}(U)$ with $A^q(U)$ in the above, we have $H^p(\mathcal{U}; A^q)$, the Čech cohomology of q-forms on $\mathcal{U}$. Thus we define $C^p(\mathcal{U}; A^q)$ to be the direct product:

$$C^p(\mathcal{U}; A^q) = \prod_{(\alpha_0,\dots,\alpha_p)\in I^{p+1}} A^q(U_{\alpha_0\dots\alpha_p}).$$

An element ξ in $C^p(\mathcal{U}; A^q)$, called a p-cochain of q-forms on $\mathcal{U}$, assigns to each $(\alpha_0, \dots, \alpha_p)$ in I^{p+1} an element $\xi_{\alpha_0\dots\alpha_p}$ in $A^q(U_{\alpha_0\dots\alpha_p})$. The coboundary operator

$$\delta = \delta^p : C^p(\mathcal{U}; A^q) \longrightarrow C^{p+1}(\mathcal{U}; A^q)$$

is defined as (7.1). It holds again that $\delta^p \circ \delta^{p-1} = 0$.

Definition 7.3. The p-th *Čech cohomology* of q-forms on $\mathcal{U}$ is the p-th cohomology of $(C^\bullet(\mathcal{U}; A^q), \delta)$:

$$H^p(\mathcal{U}; A^q) = \operatorname{Ker} \delta^p / \operatorname{Im} \delta^{p-1}.$$

Since $\operatorname{Ker} \delta^0 = A^q(M)$, we have

$$H^0(\mathcal{U}; A^q) = A^q(M).$$

Lemma 7.2. *Suppose $\mathcal{U}$ is locally finite. For a Čech p-cocycle ξ, $p > 0$, there exists η in $C^{p-1}(\mathcal{U}; A^q)$ such that $\xi = \delta\eta$.*

Proof. Let $\{\rho_\alpha\}$ be a C^∞ partition of unity subordinate to $\mathcal{U}$. For $p > 0$, we define

$$h = h^p : C^p(\mathcal{U}; A^q) \longrightarrow C^{p-1}(\mathcal{U}; A^q)$$

by setting, for ξ in $C^p(\mathcal{U}; A^q)$, $(h\xi)_{\alpha_0\dots\alpha_{p-1}} = \sum_\alpha \rho_\alpha \xi_{\alpha\alpha_0\dots\alpha_{p-1}}$. Then we see that $\delta h + h\delta = 1$, which implies that, if $\delta\xi = 0$, then $\xi = \delta h \xi$. □

Corollary 7.2. *If $\mathcal{U}$ is locally finite, the Čech complex of q-forms on $\mathcal{U}$ is acyclic, i.e.,*

$$H^p(\mathcal{U}; A^q) = 0 \qquad \textit{for} \;\; p > 0.$$

Remark 7.3. Since a C^∞ manifold M is paracompact, the set of locally finite coverings is cofinal in the set of open coverings of M, i.e., for every covering $\mathcal{U}$, there exists a locally finite refinement of $\mathcal{U}$ (cf. Remark 2.1.3).

Alternating cochains

Let S denote either R or A^q. A cochain ξ in $C^p(\mathcal{U}; S)$ is said to be *alternating* if:

(1) $\xi_{\alpha_0 \dots \alpha_p} = 0$, whenever two of the indices $\alpha_0, \dots, \alpha_p$ are equal,
(2) $\xi_{\alpha_{\rho(0)} \dots \alpha_{\rho(p)}} = \operatorname{sgn} \rho \cdot \xi_{\alpha_0 \dots \alpha_p}$, for every permutation ρ of $\{0, \dots, p\}$.

The alternating p-cochains form a subcomplex of $(C^\bullet(\mathcal{U}; S), \delta)$, which is denoted by $(C^\bullet_{\mathrm{alt}}(\mathcal{U}; S), \delta)$. We denote by $H^p_{\mathrm{alt}}(\mathcal{U}; S)$ its p-th cohomology.

Proposition 7.2. *The inclusion $C^\bullet_{\mathrm{alt}}(\mathcal{U}; S) \hookrightarrow C^\bullet(\mathcal{U}; S)$ induces an isomorphism*

$$H^p_{\mathrm{alt}}(\mathcal{U}; S) \xrightarrow{\sim} H^p(\mathcal{U}; S).$$

Proof. Choose a total ordering on I and let

$$\pi : C^p(\mathcal{U}; S) \longrightarrow C^p_{\mathrm{alt}}(\mathcal{U}; S)$$

be defined as follows. Take an element ξ in $C^p(\mathcal{U}; S)$. For $(\alpha_0, \dots, \alpha_p)$ in I^{p+1}, we set

$$\pi(\xi)_{\alpha_0 \dots \alpha_p} = \begin{cases} \operatorname{sgn} \rho \cdot \xi_{\alpha_{\rho(0)} \dots \alpha_{\rho(p)}} & \text{if the } \alpha_i\text{'s are distinct,} \\ 0 & \text{otherwise,} \end{cases}$$

where ρ is the permutation such that $\alpha_{\rho(0)} < \dots < \alpha_{\rho(p)}$. Then it can be shown that π is a cochain equivalence. □

Reduced expression

In practice, it is more convenient to use the following expression of cochains to avoid the presence of redundant terms.

Definition 7.4. The index set I of a covering $\mathcal{U} = \{U_\alpha\}_{\alpha \in I}$ of M is *properly ordered* if it is ordered so that if $U_{\alpha_0 \dots \alpha_p} \neq \emptyset$, the induced order on the subset $\{\alpha_0, \dots, \alpha_p\}$ is total.

For instance we may endow I with a total ordering to fulfill this condition. Also, if $\mathcal{U}$ is a simplicial covering with I the set of vertices, every simplicial order in I satisfies the condition.

For a covering $\mathcal{U} = \{U_\alpha\}_{\alpha \in I}$ with I as above, we set

$$I^{(p)} = \{ (\alpha_0, \dots, \alpha_p) \mid \alpha_0 < \dots < \alpha_p, \ \alpha_\nu \in I \}.$$

Let S denote either R or A^q and denote by $\hat{C}^p(\mathcal{U}; S)$ the direct product:

$$\hat{C}^p(\mathcal{U}; S) = \prod_{(\alpha_0, \dots, \alpha_p) \in I^{(p)}} S(U_{\alpha_0 \dots \alpha_p}).$$

We may define the coboundary operator $\delta : \hat{C}^p(\mathcal{U}; S) \to \hat{C}^{p+1}(\mathcal{U}; S)$ as (7.1) and we have $\delta^{p+1} \circ \delta^p = 0$. Thus we have a complex $(\hat{C}^\bullet(\mathcal{U}; S), \delta)$. We denote by $\hat{H}^p(\mathcal{U}; S)$ its p-th cohomology.

Convention 7.1. Let ξ be an element in $\hat{C}^p(\mathcal{U}; S)$. For every $(\alpha_0, \dots, \alpha_p)$ in I^{p+1}, we set

$$\xi_{\alpha_0 \dots \alpha_p} = \begin{cases} \operatorname{sgn} \rho \cdot \xi_{\alpha_{\rho(0)} \dots \alpha_{\rho(p)}} & \text{if } U_{\alpha_0 \dots \alpha_p} \neq \emptyset \text{ and the } \alpha_i\text{'s are distinct,} \\ 0 & \text{otherwise,} \end{cases}$$

where ρ is the permutation such that $\alpha_{\rho(0)} < \cdots < \alpha_{\rho(p)}$.

By the above convention, $\hat{C}^p(\mathcal{U}; S)$ may be identified with $C^p_{\text{alt}}(\mathcal{U}; S)$ and the coboundary operators are the same. Thus from Proposition 7.2, we have an isomorphism

$$\hat{H}^p(\mathcal{U}; S) = H^p_{\text{alt}}(\mathcal{U}; S) \xrightarrow{\sim} H^p(\mathcal{U}; S). \tag{7.2}$$

An immediate consequence of the above is that, if the index set I of the covering $\mathcal{U}$ consists of k elements, then $H^p(\mathcal{U}; S) = 0$ for $p \geq k$.

7.3 Čech-de Rham cohomology

Let M be a C^∞ manifold of dimension m. For an open covering $\mathcal{U}$ of M, we have the group $C^p(\mathcal{U}; A^q)$ and the operator $\delta : C^p(\mathcal{U}; A^q) \to C^{p+1}(\mathcal{U}; A^q)$ as defined in the previous section. This together with the exterior derivative $d : C^p(\mathcal{U}; A^q) \to C^p(\mathcal{U}; A^{q+1})$ makes $C^\bullet(\mathcal{U}; A^\bullet)$ a double complex (cf. Section A.1). The simple complex associated with it is denoted by $(A^\bullet(\mathcal{U}), D)$ and is called the *Čech-de Rham complex*, *ČdR complex* for short, of $\mathcal{U}$.

Definition 7.5. The r-th *Čech-de Rham cohomology* $H^r_D(\mathcal{U})$ of $\mathcal{U}$ is the r-th cohomology of $(A^\bullet(\mathcal{U}), D)$.

We examine this cohomology more closely. First, we have

$$A^r(\mathcal{U}) = \bigoplus_{p+q=r} C^p(\mathcal{U}; A^q)$$

and an element ξ in $A^r(\mathcal{U})$, i.e., a ČdR r-cochain, may be expressed as $\xi = (\xi^{(p)})_{0 \leq p \leq r}$ with $\xi^{(p)} \in C^p(\mathcal{U}; A^{r-p})$. We write $\xi^{(p)}_{\alpha_0 \dots \alpha_p}$ also $\xi_{\alpha_0 \dots \alpha_p}$. Then the total differential $D : A^r(\mathcal{U}) \to A^{r+1}(\mathcal{U})$ is given by

$$(D\xi)^{(p)} = \delta \xi^{(p-1)} + (-1)^p d\xi^{(p)}, \qquad 0 \leq p \leq r+1, \tag{7.3}$$

where we set $\xi^{(-1)} = 0$ and $\xi^{(r+1)} = 0$. In particular, for small values of p,

$$(D\xi)_{\alpha_0} = d\xi_{\alpha_0}, \quad (D\xi)_{\alpha_0\alpha_1} = \xi_{\alpha_1} - \xi_{\alpha_0} - d\xi_{\alpha_0\alpha_1} \quad \text{and}$$
$$(D\xi)_{\alpha_0\alpha_1\alpha_2} = \xi_{\alpha_1\alpha_2} - \xi_{\alpha_0\alpha_2} + \xi_{\alpha_0\alpha_1} + d\xi_{\alpha_0\alpha_1\alpha_2}.$$

In general the condition for ξ being a cocycle is given by

$$\begin{cases} d\xi^{(0)} = 0, \\ \delta\xi^{(p-1)} + (-1)^p d\xi^{(p)} = 0, \qquad 1 \leq p \leq r, \\ \delta\xi^{(r)} = 0. \end{cases} \tag{7.4}$$

Obviously we have

$$H^0_d(M) = H^0_D(\mathcal{U}) = H^0(\mathcal{U}; \mathbb{C}),$$

all of them being equal to $\mathbb{C}(M)$.

Remark 7.4. 1. The Čech-de Rham cohomology $H^r_D(\mathcal{U})$ is in fact the hypercohomology of the complex of sheaves of differential forms on $\mathcal{U}$ (cf. Definition 11.8 below).

2. As in the case of Čech cohomology, we may consider the group of alternating cochains:

$$A^r_{\text{alt}}(\mathcal{U}) = \bigoplus_{p+q=r} C^p_{\text{alt}}(\mathcal{U}; A^q).$$

The operator D leaves $A^\bullet_{\text{alt}}(\mathcal{U})$ invariant and, as in Proposition 7.2, we see that the cohomology of $(A^\bullet_{\text{alt}}(\mathcal{U}), D)$ is naturally isomorphic with $H^r_D(\mathcal{U})$.

We may also consider the reduced expression:

$$\hat{A}^r(\mathcal{U}) = \bigoplus_{p+q=r} \hat{C}^p(\mathcal{U}; A^q).$$

The operator D is also defined on $\hat{A}^\bullet(\mathcal{U})$ and, by Convention 7.1, $\hat{A}^\bullet(\mathcal{U})$ may be identified with $A^\bullet_{\text{alt}}(\mathcal{U})$. Thus denoting by $\hat{H}^r_D(\mathcal{U})$ the cohomology of $\hat{A}^\bullet_D(\mathcal{U})$, we have a natural isomorphism:

$$\hat{H}^r_D(\mathcal{U}) \xrightarrow{\sim} H^r_D(\mathcal{U}). \tag{7.5}$$

Example 7.1. Let $\mathcal{U} = \{U_0, U_1\}$ be an open covering of M consisting of two open sets with $0 < 1$. We have

$$\hat{A}^r(\mathcal{U}) = A^r(U_0) \oplus A^r(U_1) \oplus A^{r-1}(U_{01}).$$

Thus an element ξ in $\hat{A}^r(\mathcal{U})$ is given by a triple $\xi = (\xi_0, \xi_1, \xi_{01})$ with ξ_i an r-form on U_i, $i = 0, 1$, and ξ_{01} an $(r-1)$-form on U_{01}. The differential $D : \hat{A}^r(\mathcal{U}) \to \hat{A}^{r+1}(\mathcal{U})$ is given by

$$D\xi = (d\xi_0, d\xi_1, \xi_1 - \xi_0 - d\xi_{01}).$$

de Rham-ČdR correspondence

There is a natural morphism

$$\varphi_d : A^r(M) \longrightarrow C^0(\mathcal{U}; A^r) \subset A^r(\mathcal{U}), \tag{7.6}$$

which assigns to an r-form ω the cochain ξ given by $\xi_\alpha = \omega|_{U_\alpha}$. It is injective and the image of ω in $A^r(\mathcal{U})$ will also be denoted by ω.

Theorem 7.1. *If $\mathcal{U}$ is locally finite, the morphism* (7.6) *induces an isomorphism*

$$\varphi_d : H_d^r(M) \xrightarrow{\sim} H_D^r(\mathcal{U}).$$

Proof. As an element in $A^r(\mathcal{U})$, $\delta\omega = 0$ so that we have $d\omega = D\omega$ and we see that the inclusion induces a morphism $\varphi_d : H_d^r(M) \to H_D^r(\mathcal{U})$.

First we show that φ_d is surjective. For this, we claim that, given a ČdR r-cocycle ξ, $r \geq 1$, there exist a closed r-form ω on M and a ČdR $(r-1)$-cochain η such that

$$\xi = \omega + D\eta. \tag{7.7}$$

To prove this, by the last equality in (7.4) and Lemma 7.2, there exists $\eta^{(r-1)}$ in $C^{r-1}(\mathcal{U}; A^0)$ such that $\xi^{(r)} = \delta\eta^{(r-1)}$. Then from the middle in (7.4) and Lemma 7.2, we can successively find $\eta^{(p)}$, $0 \leq p \leq r-1$, satisfying

$$\xi^{(p)} = \delta\eta^{(p-1)} + (-1)^p d\eta^{(p)}, \qquad 1 \leq p \leq r-1.$$

Finally, from the identities $\delta\xi^{(0)} - d\xi^{(1)} = 0$ and $\xi^{(1)} = \delta\eta^{(0)} - d\eta^{(1)}$, we have $\delta(\xi^{(0)} - d\eta^{(0)}) = 0$. Noting that the kernel of $\delta : C^0(\mathcal{U}; A^r) \to C^1(\mathcal{U}; A^r)$ is $A^r(M)$, we have $\xi^{(0)} - d\eta^{(0)} = \omega$, $\omega \in A^r(M)$. From the first in (7.4), ω is closed. Thus, for $\eta = (\eta^{(p)})$, we have (7.7).

Now we show that φ_d is injective. Thus suppose

$$\omega = D\eta$$

for some $(r-1)$-chain η. The above is equivalent to

$$\omega = d\eta^{(0)}, \quad \delta\eta^{(p-1)} + (-1)^p d\eta^{(p)} = 0, \ 1 \leq p \leq r-1, \ \text{and} \ \delta\eta^{(r-1)} = 0.$$

As in the above algorithm, we may find an $(r-2)$-chain σ such that $\eta^{(r-1)} = \delta\sigma^{(r-2)}$ and

$$\eta^{(p)} = \delta\sigma^{(p-1)} + (-1)^p d\sigma^{(p)}, \qquad 1 \leq p \leq r-2.$$

From $\delta\eta^{(0)} - d\eta^{(1)} = 0$ and $\eta^{(1)} = \delta\sigma^{(0)} - d\sigma^{(1)}$, we have $\delta(\eta^{(0)} - d\sigma^{(0)}) = 0$. Thus $\theta = \eta^{(0)} - d\sigma^{(0)}$ is an $(r-1)$-form on M and from $\omega = d\eta^{(0)}$, we have $\omega = d\theta$. □

In view of Remark 7.3, we can make the following assumption without loss of generality:

Assumption 7.1. Hereafter the coverings we consider are locally finite.

Corollary 7.3. *The Čech-de Rham cohomology $H^r_D(\mathcal{U})$ is determined uniquely modulo canonical isomorphisms, independently of $\mathcal{U}$.*

Exercise 7.2. Considering the reduced expression (cf. Remark 7.4. 2), we have the isomorphisms:

$$H^r_d(M) \xrightarrow[\varphi_d]{\sim} H^r_D(\mathcal{U}) \xleftarrow{\sim} \hat{H}^r_D(\mathcal{U}),$$

where the second is the one in (7.5). Let $\mathcal{U} = \{U_0, U_1\}$ be an open covering of M and $\{\rho_0, \rho_1\}$ a C^∞ partition of unity subordinate to $\mathcal{U}$. Show that, for a given ČdR cocycle $\xi = (\xi_0, \xi_1, \xi_{01})$ in $\hat{A}^r(\mathcal{U})$ (cf. Example 7.1),

$$\omega = \rho_0\xi_0 + \rho_1\xi_1 - d\rho_0 \wedge \xi_{01}$$

is a closed form on M representing the class in $H^r_d(M)$ that corresponds to the class $[\xi]$ in $\hat{H}^r_D(\mathcal{U})$.

Čech-ČdR correspondence

There is a natural morphism

$$\varphi_\delta : C^r(\mathcal{U}; \mathbb{C}) \longrightarrow C^r(\mathcal{U}; A^0) \subset A^r(\mathcal{U}). \tag{7.8}$$

It is injective and we denote the image of an r-chain c in $A^r(\mathcal{U})$ also by c.

Definition 7.6. We say that an open covering $\mathcal{U}$ of M is *good*, if every non-empty finite intersection $U_{\alpha_0,\dots,\alpha_p}$ is diffeomorphic with $\mathbb{R}^m$.

Note that any simplicial covering is good so that every C^∞ manifold M admits a good covering and that the set of good coverings is cofinal in the set of coverings of M.

Theorem 7.2. *If $\mathcal{U}$ is good, the morphism* (7.8) *induces an isomorphism*

$$\varphi_\delta : H^r(\mathcal{U}; \mathbb{C}) \xrightarrow{\sim} H^r_D(\mathcal{U}).$$

Proof. As a chain in $A^r(\mathcal{U})$, $dc = 0$ so that $\delta c = Dc$. Thus the inclusion induces a morphism $\varphi_\delta : H^r(\mathcal{U}; \mathbb{C}) \to H^r_D(\mathcal{U})$.

We show that it is surjective. For this, given a ČdR r-cocycle ξ, $r \geq 1$, there exist an r-cocycle c in $C^r(\mathcal{U}; \mathbb{C})$ and a ČdR $(r-1)$-cochain ζ such that

$$\xi = c + D\zeta. \tag{7.9}$$

This is done as in the proof of Theorem 7.1. This time, from the first equality in (7.4) and Proposition 7.1, there exists $\zeta^{(0)}$ in $C^0(\mathcal{U}; A^{r-1})$ such that $\xi^{(0)} = d\zeta^{(0)}$. Then from the middle in (7.4) and Proposition 7.1, we can successively find $\zeta^{(p)}$, $0 \leq p \leq r-1$, satisfying

$$\xi^{(p)} = \delta\zeta^{(p-1)} + (-1)^p d\zeta^{(p)}, \qquad 1 \leq p \leq r-1.$$

From $\delta\xi^{(r-1)} + (-1)^r d\xi^{(r)} = 0$ and $\xi^{(r-1)} = \delta\zeta^{(r-2)} + (-1)^{r-1} d\zeta^{(r-1)}$, we have $d(\xi^{(r)} - \delta\zeta^{(r-1)}) = 0$. Since the kernel of $d : C^r(\mathcal{U}; A^0) \to C^r(\mathcal{U}; A^1)$ is $C^r(\mathcal{U}; \mathbb{C})$, we have $\xi^{(r)} - \delta\zeta^{(r-1)} = c$, $c \in C^r(\mathcal{U}; \mathbb{C})$. From the last in (7.4), c is a cocycle. Thus for $\zeta = (\zeta^{(p)})$, we have (7.9). The proof for injectivity is similar. □

de Rham-Čech correspondence

From Theorems 7.1 and 7.2, we have:

Corollary 7.4. *For a good covering $\mathcal{U}$, there exist canonical isomorphisms:*

$$H^r_d(M) \underset{\varphi_d}{\xrightarrow{\sim}} H^r_D(\mathcal{U}) \underset{\varphi_\delta}{\xleftarrow{\sim}} H^r(\mathcal{U}; \mathbb{C}).$$

In the above, we think of a closed r-form ω and a Čech r-cocycle c as ČdR cocycles. Thus the classes $[\omega]$ and $[c]$ correspond in the above isomorphism, if and only if there exists an $(r-1)$-ČdR cochain χ such that

$$\omega - c = D\chi \qquad \text{as ČdR cocycles.} \tag{7.10}$$

Such a χ is given by $\chi = \zeta - \eta$ with η and ζ as in the proofs of Theorem 7.1 and Theorem 7.2, respectively. The relation (7.10) is rephrased as, for $\chi^{(p)}$ in $C^p(\mathcal{U}; A^{r-p-1})$, $0 \leq p \leq r-1$,

$$\begin{aligned} &\omega = d\chi^{(0)}, \qquad c = -\delta\chi^{(r-1)}, \\ &\delta\chi^{(p-1)} + (-1)^p d\chi^{(p)} = 0, \qquad 1 \leq p \leq r-1. \end{aligned}$$

Here we write down the relation explicitly for small values of r:

(1) if $r = 1$, for χ in $C^0(\mathcal{U}; A^0)$,

$$\omega = d\chi_\alpha \ \text{ on } U_\alpha \quad \text{and} \quad c_{\alpha\beta} = -\chi_\beta + \chi_\alpha \ \text{ on } U_{\alpha\beta},$$

(2) if $r = 2$, for $\chi^{(0)}$ in $C^0(\mathcal{U}; A^1)$ and $\chi^{(1)}$ in $C^1(\mathcal{U}; A^0)$,

$$\omega = d\chi^{(0)}_\alpha \ \text{ on } U_\alpha, \quad d\chi^{(1)}_{\alpha\beta} = \chi^{(0)}_\beta - \chi^{(0)}_\alpha \ \text{ on } U_{\alpha\beta} \ \text{ and}$$
$$c_{\alpha\beta\gamma} = -\chi^{(1)}_{\beta\gamma} + \chi^{(1)}_{\alpha\gamma} - \chi^{(1)}_{\alpha\beta} \ \text{ on } U_{\alpha\beta\gamma}.$$

The above correspondence may be illustrated in the following diagram:

$$
\begin{array}{ccccccccc}
 & & & & & & & & \omega \\
 & & & & & & & & \cap \\
 & & A^0(X) & \xrightarrow{d^0} & A^1(X) & \xrightarrow{d^1} & \cdots & \xrightarrow{d^{r-1}} & A^r(X) \xrightarrow{d^r} \cdots \\
 & & \downarrow & & \downarrow & & & & \downarrow \\
 & C^0(\mathcal{U};\mathbb{C}) \hookrightarrow & C^0(\mathcal{U};A^0) & \xrightarrow{d^0} & C^0(\mathcal{U};A^1) & \xrightarrow{d^1} & \cdots & \xrightarrow{d^{r-1}} & C^0(\mathcal{U};A^r) \xrightarrow{d^r} \cdots \\
 & \downarrow \delta^0 & \downarrow \delta^0 & & \downarrow \delta^0 & & & & \downarrow \delta^0 \\
 & C^1(\mathcal{U};\mathbb{C}) \hookrightarrow & C^1(\mathcal{U};A^0) & \xrightarrow{-d^0} & C^1(\mathcal{U};A^1) & \xrightarrow{-d^1} & \cdots & \xrightarrow{-d^{r-1}} & C^1(\mathcal{U};A^r) \xrightarrow{-d^r} \cdots \\
 & \downarrow \delta^1 & \downarrow \delta^1 & & \downarrow \delta^1 & & & & \downarrow \delta^1 \\
 & \vdots & \vdots & & \vdots & & & & \vdots \\
 & \downarrow \delta^{r-1} & \downarrow \delta^{r-1} & & \downarrow \delta^{r-1} & & & & \\
c \in & C^r(\mathcal{U};\mathbb{C}) \hookrightarrow & C^r(\mathcal{U};A^0) & \xrightarrow{(-1)^r d^0} & C^r(\mathcal{U};A^1) & \xrightarrow{(-1)^r d^1} & \cdots & & \\
 & \downarrow \delta^r & \downarrow \delta^r & & \downarrow \delta^r & & & & \\
 & \vdots & \vdots & & \vdots & & & &
\end{array}
\tag{7.11}
$$

7.4 Integration of Čech-de Rham cochains

In this section, we define integration on the Čech-de Rham cohomology and for this purpose we introduce the notion of a system of honeycomb cells.

In the following, we let M denote a C^∞ manifold of dimension m.

Special cases

We discuss some special cases before we proceed to the general case. Here we assume M to be compact and oriented.

Integration on de Rham cohomology: Recall that we have the integration $\int_M : A^m(M) \to \mathbb{C}$. By the Stokes theorem (Theorem 6.2), it induces the integration

$$\int_M : H_d^m(M) \longrightarrow \mathbb{C}. \tag{7.12}$$

If M is connected, the above is an isomorphism, which is the content of the Poincaré duality with $\mathbb{C}$-coefficient in this case (cf. Theorem 7.12 below).

Covering with two open sets: Let $\mathcal{U} = \{U_0, U_1\}$ be an open covering of M. For $i = 0, 1$, let R_i be an m-dimensional manifold with C^∞ boundary in U_i such that $R_0 \cup R_1 = M$ and that $\operatorname{Int} R_0 \cap \operatorname{Int} R_1 = \emptyset$, where $\operatorname{Int} R_i$ denotes the interior of R_i. We set $R_{01} = \partial R_0$, which is equal to $-\partial R_1$, the boundary of R_1 with opposite orientation. Let $\hat{A}^r(\mathcal{U})$ be as in Example 7.1 and define

$$\int_M : \hat{A}^m(\mathcal{U}) \longrightarrow \mathbb{C} \qquad \text{by} \quad \int_M \xi = \int_{R_0} \xi_0 + \int_{R_1} \xi_1 + \int_{R_{01}} \xi_{01}. \tag{7.13}$$

Then we see that

(1) if ξ is a coboundary, then $\int_M \xi = 0$,
(2) if ξ is a cocycle, then the right-hand side does not depend on the choice of $\{R_0, R_1\}$ as above.

Thus the above integration induces the integration

$$\int_M : H^m_D(\mathcal{U}) \longrightarrow \mathbb{C}.$$

Again, if M is connected, the above is an isomorphism, which is another expression of the Poincaré duality in this case.

Exercise 7.3. Verify the above statements (1) and (2) (cf. Proposition 7.3 below for general case).

System of honeycomb cells

Let $\mathcal{U} = \{U_\alpha\}_{\alpha \in I}$ be an open covering of M.

Definition 7.7. A *system of honeycomb cells*, or a *honeycomb system* for short, adapted to $\mathcal{U}$ is a collection $\{R_\alpha\}_{\alpha \in I}$ of m-dimensional piecewise C^∞ manifolds R_α with boundary in M (cf. Definition 5.10) satisfying:

(1) for each α, $R_\alpha \subset U_\alpha$ and $M = \bigcup_\alpha R_\alpha$,
(2) if $\alpha \neq \beta$, then $\operatorname{Int} R_\alpha \cap \operatorname{Int} R_\beta = \emptyset$,
(3) if $U_{\alpha_0 \dots \alpha_p} \neq \emptyset$ and if the α_ν's are distinct, then $R_{\alpha_0 \dots \alpha_p}$ is an $(m-p)$-dimensional piecewise C^∞ manifold with boundary,
(4) if the set $\{\alpha_0, \dots, \alpha_p\}$ is maximal, then $R_{\alpha_0 \dots \alpha_p}$ has no boundary.

In the above, we set $R_{\alpha_0 \dots \alpha_p} = \bigcap_{\nu=0}^p R_{\alpha_\nu}$, which is equal to $\bigcap_{\nu=0}^p \partial R_{\alpha_\nu}$ by (2). Also, $\{\alpha_0, \dots, \alpha_p\}$ being maximal means that if $U_{\alpha, \alpha_0, \dots, \alpha_p} \neq \emptyset$, then α is in $\{\alpha_0, \dots, \alpha_p\}$.

Remark 7.5. If $\{R_\alpha\}_{\alpha\in I}$ satisfies, instead of (1),

(1)′ for each α, $R_\alpha \subset U_\alpha$,

then we say that it is *subordinate to* $\mathcal{U}$.

Orientation: Suppose M is oriented and I is properly ordered (cf. Definition 7.4). In this case, we orient $R_{\alpha_0\dots\alpha_p}$ by the following rules:

(1) each R_α has the same orientation as M and its boundary is oriented as in Convention 3.2.
(2) for $(\alpha_0,\dots,\alpha_p)$ in $I^{(p)}$, $p \geq 1$, $R_{\alpha_0\dots\alpha_p}$ is oriented as a part of $\partial R_{\alpha_0\dots\alpha_{p-1}}$,
(3) for $(\alpha_0,\dots,\alpha_p)$ in I^{p+1}, we set $R_{\alpha_0\dots\alpha_p} = \operatorname{sgn}\rho \cdot R_{\alpha_{\rho(0)}\dots\alpha_{\rho(p)}}$, if $U_{\alpha_0\dots\alpha_p} \neq \emptyset$ and the α_i's are distinct, where ρ is the permutation such that $\alpha_{\rho(0)} < \cdots < \alpha_{\rho(p)}$. Otherwise we set $R_{\alpha_0\dots\alpha_p} = \emptyset$.

With the above convention, we may write:

$$\partial R_{\alpha_0\dots\alpha_p} = \sum_{\alpha\in I} R_{\alpha_0\dots\alpha_p\alpha}. \tag{7.14}$$

Example 7.2. 1. Let K be a C^∞ triangulation of M and $\mathcal{U} = \{U_\alpha\}$ the associated simplicial covering of M (cf. Section 7.2). Thus, for each vertex α in K, $U_\alpha = O_K(\alpha)$, the open star of α in K. Let R_α be the dual cell of the vertex α. Then $\{R_\alpha\}$ is a honeycomb system adapted to $\mathcal{U}$.

2. If $\mathcal{U} = \{U_0, U_1\}$ and if $\{R_0, R_1\}$ is a honeycomb system adapted to $\mathcal{U}$, then R_{01} is an $(m-1)$-dimensional piecewise C^∞ submanifold of M without boundary, separating the two open sets $\operatorname{Int} R_0$ and $\operatorname{Int} R_1$. If M is oriented, we have $\partial R_0 = R_{01}$ and $\partial R_1 = R_{10} = -R_{01}$.

3. Let S be a closed set in M. Suppose we are given a stratification $S = \bigsqcup_{\alpha\in I'} X_\alpha$ of S by submanifolds X_α of M (cf. Definition 2.13). We introduce an order relation on I' by saying that $\alpha > \beta$ if $X_\alpha \subset \overline{X_\beta}\smallsetminus X_\beta$. Let $I = I' \cup \{0\}$ and assume $0 < \alpha$ for every α in I'. Let $U_0 = M\smallsetminus S$ and, for each $\alpha \in I'$, let U_α be a tubular neighborhood of X_α in M (cf. Theorem 3.10). We denote by $\mathcal{U}$ the covering $\{U_\alpha\}_{\alpha\in I}$ of M. If we choose each U_α sufficiently small, then I is properly ordered (cf. Definition 7.4). We set $T_0 = U_0$ and, for each α in I', we let T_α be a closed ball bundle in U_α. Moreover, for each α in I, we set

$$R_\alpha = T_\alpha \cap \bigcap_{\alpha'>\alpha} (M\smallsetminus \operatorname{Int} T_{\alpha'}).$$

Then, if $\{X_\alpha\}$ is a Whitney stratification, $\{R_\alpha\}$ form a honeycomb system adapted to $\mathcal{U}$ (cf. Section 13.1 below).

Remark 7.6. Let $\mathcal{U}$ be an open covering of a C^∞ manifold M of dimension m. It is conjectured that, if $\mathcal{U}$ has order at most $m+1$, then there exists a honeycomb system adapted to $\mathcal{U}$. Thus, if this is true, every open covering of M admits a refinement having an adapted honeycomb system (cf. Section 2.5).

Integration on ČdR cohomology

In this paragraph we assume M to be compact and oriented. Let $\mathcal{U} = \{U_\alpha\}_{\alpha\in I}$ be an open covering of M with I a finite properly ordered set. Here we assume that there exists a honeycomb system $\{R_\alpha\}$ adapted to $\mathcal{U}$. Noting that each R_α is compact, as M is compact, we define the integration

$$\int_M : A^m(\mathcal{U}) \longrightarrow \mathbb{C} \qquad \text{by} \quad \int_M \xi = \sum_{p=0}^{m} \sum_{(\alpha_0,\dots,\alpha_p)\in I^{(p)}} \int_{R_{\alpha_0\dots\alpha_p}} \xi_{\alpha_0\dots\alpha_p}, \tag{7.15}$$

which depends *a priori* on the choice of the honeycomb system $\{R_\alpha\}$.

Proposition 7.3. 1. *If $\xi = D\eta$, then $\int_M \xi = 0$.*
2. *If $D\xi = 0$, then $\int_M \xi$ does not depend on the choice of $\{R_\alpha\}$.*

Proof. **1.** Setting $\eta^{(-1)} = 0$ and $\eta^{(m)} = 0$, we have (cf. (7.3))

$$(D\eta)^{(p)} = \delta\eta^{(p-1)} + (-1)^p d\eta^{(p)}, \qquad 0 \le p \le m.$$

Thus

$$\begin{aligned}\int_M D\eta = &\sum_{p=1}^{m} \sum_{(\alpha_0,\dots,\alpha_p)\in I^{(p)}} \int_{R_{\alpha_0\dots\alpha_p}} (\delta\eta^{(p-1)})_{\alpha_0\dots\alpha_p} \\ &+ \sum_{p=0}^{m-1} (-1)^p \sum_{(\alpha_0,\dots,\alpha_p)\in I^{(p)}} \int_{R_{\alpha_0\dots\alpha_p}} d\eta^{(p)}_{\alpha_0\dots\alpha_p}.\end{aligned} \tag{7.16}$$

Using the Stokes theorem and (7.14), we compute

$$\sum_{(\alpha_0,\dots,\alpha_p)\in I^{(p)}} \int_{R_{\alpha_0\dots\alpha_p}} d\eta^{(p)}_{\alpha_0\dots\alpha_p} = \sum_{(\alpha_0,\dots,\alpha_p)\in I^{(p)}} \sum_{\alpha\in I} \int_{R_{\alpha_0\dots\alpha_p\alpha}} \eta^{(p)}_{\alpha_0\dots\alpha_p}. \tag{7.17}$$

For $(\alpha_0,\dots,\alpha_p) \in I^{(p)}$ and $\alpha \in I$ with $U_{\alpha_0\dots\alpha_p\alpha} \neq \emptyset$, there are three possibilities: (1) $\alpha < \alpha_0$, (2) $\alpha_i < \alpha < \alpha_{i+1}$, $0 \le i \le p-1$, or (3) $\alpha_p < \alpha$. We have

$$R_{\alpha_0\dots\alpha_p\alpha} = \begin{cases} (-1)^{p+1} R_{\alpha\alpha_0\dots\alpha_p} & \text{case (1)}, \\ (-1)^{p-i} R_{\alpha_0\dots\alpha_i\alpha\alpha_{i+1}\dots\alpha_p} & \text{case (2)}, \\ R_{\alpha_0\dots\alpha_p\alpha} & \text{case (3)}. \end{cases}$$

Thus we may express the right-hand side of (7.17) as

$$\begin{aligned}(-1)^{p+1} & \sum_{(\alpha_0\ldots\alpha_{p+1})\in I^{(p+1)}} \int_{R_{\alpha_0\cdots\alpha_{p+1}}} \sum_{i=0}^{p+1}(-1)^i \eta^{(p)}_{\alpha_0,\ldots,\hat{\alpha}_i,\ldots,\alpha_{p+1}} \\ &= (-1)^{p+1} \sum_{(\alpha_0\ldots\alpha_{p+1})\in I^{(p+1)}} \int_{R_{\alpha_0\cdots\alpha_{p+1}}} (\delta\eta^{(p)})_{\alpha_0,\ldots,\alpha_{p+1}}.\end{aligned}$$

Hence from (7.16), we have $\int_M D\eta = 0$.

2. If $D\xi = 0$, there is an m-form ω on M such that $\xi = \omega + D\eta$ for some $(m-1)$-ČdR cochain η (cf. (7.7)). The statement then follows from 1. □

Remark 7.7. 1. The above may also be proved using the Stokes theorem for ČdR cochains (cf. Lemma 7.4 and (7.5) below).

2. It suffices to consider reduced cochains to define the above integration (cf. Remark 7.4. 2).

From the above we see that the integration (7.15) induces the integration on the cohomology

$$\int_M : H^m_D(\mathcal{U}) \longrightarrow \mathbb{C}.$$

Let φ_d and φ_δ be the inclusions (7.6) and (7.8). By restricting (7.15), we have the commutative diagram:

$$\begin{array}{ccccc} A^m(M) & \xrightarrow[\varphi_d]{} & A^m(\mathcal{U}) & \xleftarrow[\varphi_\delta]{} & C^m(\mathcal{U};\mathbb{C}) \\ & {\scriptstyle \int_M}\searrow & \downarrow {\scriptstyle \int_M} & \swarrow{\scriptstyle \int_M} & \\ & & \mathbb{C}. & & \end{array}$$

The integral

$$\int_M : A^m(M) \longrightarrow \mathbb{C}.$$

coincides with the usual one and

$$\int_M : C^m(\mathcal{U};\mathbb{C}) \longrightarrow \mathbb{C} \qquad \text{is given by } \ c \mapsto \sum_{(\alpha_0,\ldots,\alpha_m)\in I^{(m)}} c_{\alpha_0\cdots\alpha_m}.$$

From the above we have:

Proposition 7.4. *We have the commutative diagram:*

$$\begin{array}{ccccc} H^m_d(M) & \xrightarrow[\varphi_d]{\sim} & H^m_D(\mathcal{U}) & \xleftarrow[\varphi_\delta]{} & H^m(\mathcal{U};\mathbb{C}) \\ & {\scriptstyle \int_M}\searrow & \downarrow {\scriptstyle \int_M} & \swarrow{\scriptstyle \int_M} & \\ & & \mathbb{C}, & & \end{array}$$

where φ_d is the isomorphism of Theorem 7.1 and, if $\mathcal{U}$ is good, φ_δ is an isomorphism, which is the one in Theorem 7.2.

This will be refined in Section 7.5 below to yield the de Rham theorem.

Exercise 7.4. Let $\mathcal{U} = \{U_0, U_1\}$ be an open covering of M. For a ČdR cocycle ξ in $\hat{A}^m(\mathcal{U})$, let ω be the form as given in Exercise 7.2. Show by direct computations that

$$\int_M \xi = \int_M \omega.$$

Remark 7.8. If $\mathcal{U}$ is the simplicial covering associated with some triangulation of M (cf. Section 7.2), there is always a honeycomb system adapted to $\mathcal{U}$ (cf. Example 7.2. 1). In this case φ_δ in Proposition 7.4 is an isomorphism.

7.5 Combination with combinatorial topology

Throughout this section, we let M denote a C^∞ manifold of dimension m.

Integration of ČdR cochains

C^∞ singular chains: Let R be a commutative ring with unity. We denote by $S^{\mathrm{ci}}_\bullet(M;\mathsf{R})$ and $S^\bullet_{\mathrm{ci}}(M;\mathsf{R})$ the chain and cochain complexes of C^∞ singular simplices of M with coefficients in R (cf. Section 6.2). For their homology $H^{\mathrm{ci}}_p(M;\mathsf{R})$ and cohomology $H^p_{\mathrm{ci}}(M;\mathsf{R})$, we have canonical isomorphisms (cf. (6.12) and (6.13)):

$$H^{\mathrm{ci}}_p(M;\mathsf{R}) \xrightarrow{\sim} H_p(M;\mathsf{R}) \quad \text{and} \quad H^p(M;\mathsf{R}) \xrightarrow{\sim} H^p_{\mathrm{ci}}(M;\mathsf{R}).$$

In the following, we do not distinguish the two homologies or the two cohomologies to avoid complications of presentation.

Transverse singular simplices: Let $\mathcal{U} = \{U_\alpha\}_{\alpha\in I}$ be an open covering of M and $\{R_\alpha\}$ a honeycomb system adapted to $\mathcal{U}$.

Definition 7.8. A C^∞ singular simplex $\sigma : \Delta \to M$ is *transverse* to $\{R_\alpha\}$, if σ and all its faces are transverse to $R_{\alpha_0\cdots\alpha_p}$ for all $(\alpha_0\cdots\alpha_p) \in I^{(p)}$.

In the above, "transverse to $R_{\alpha_0\dots\alpha_p}$" means "transverse to the set of general points in $R_{\alpha_0\dots\alpha_p}$" (cf. Definition 5.11).

We denote by $S^{\mathrm{tr}}_\bullet(M;\mathsf{R})$ and $S^\bullet_{\mathrm{tr}}(M;\mathsf{R})$ the chain and cochain complexes of C^∞ singular simplices transverse to $\{R_\alpha\}$ of M with coefficients in R. Their homology and cohomology are denoted by $H^{\mathrm{tr}}_p(M;\mathsf{R})$ and $H^p_{\mathrm{tr}}(M;\mathsf{R})$. We quote the following fact, which is proved using Theorem 3.11:

Lemma 7.3. *The inclusion* $S_\bullet^{\rm tr}(M;\mathsf{R}) \hookrightarrow S_\bullet^{\rm ci}(M;\mathsf{R})$ *induces isomorphisms*

$$H_p^{\rm tr}(M;\mathsf{R}) \xrightarrow{\sim} H_p^{\rm ci}(M;\mathsf{R}) \quad \textit{and} \quad H^p_{\rm ci}(M;\mathsf{R}) \xrightarrow{\sim} H^p_{\rm tr}(M;\mathsf{R}).$$

Integration of ČdR cochains: Let $\mathcal{U}$ be an open covering of M and suppose that there exists a honeycomb system $\{R_\alpha\}$ adapted to $\mathcal{U}$. Let $\sigma : \Delta^r \to M$ be a C^∞ singular r-simplex transverse to $\{R_\alpha\}$. Recall that the standard simplex $\Delta^r \subset \mathbb{R}^r$ is oriented as in Remark B.13.1. For every $(\alpha_0 \dots \alpha_p)$, $\sigma^{-1}R_{\alpha_0\dots\alpha_p}$ is an $(r-p)$-dimensional piecewise C^∞ manifold with boundary in (a neighborhood of) Δ^r. We orient these manifolds by the following rules (cf. Propositions 3.15 and 3.16):

(1) for each α, $\sigma^{-1}R_\alpha$ has the same orientation as Δ^r, and its boundary is oriented according to Convention 3.2, modified for piecewise C^∞ manifolds,

(2) for $(\alpha_0,\dots,\alpha_p)$ in I^{p+1}, $\sigma^{-1}R_{\alpha_0\dots\alpha_p}$ is oriented inductively by the following identity:

$$\partial(\sigma^{-1}R_{\alpha_0\dots\alpha_p}) = (-1)^p(\partial\sigma)^{-1}R_{\alpha_0\dots\alpha_p} + \sigma^{-1}(\partial R_{\alpha_0\dots\alpha_p}), \tag{7.18}$$

$$\sigma^{-1}(\partial R_{\alpha_0\dots\alpha_p}) = \sum_{\alpha\in I} \sigma^{-1}R_{\alpha_0\dots\alpha_p\alpha}.$$

With these we define the integration of an r-ČdR cochain $\xi = (\xi^{(p)})$ on a singular r-simplex σ transverse to $\{R_\alpha\}$ by (cf. (7.15))

$$\int_\sigma \xi = \int_{\Delta^r} \sigma^*\xi = \sum_{p=0}^{r} \sum_{(\alpha_0,\dots,\alpha_p)\in I^{(p)}} \int_{\sigma^{-1}R_{\alpha_0\dots\alpha_p}} \sigma^*\xi^{(p)}_{\alpha_0\dots\alpha_p}.$$

Then we have:

Lemma 7.4 (ČdR Stokes theorem). *For an* $(r-1)$*-ČdR cochain* η *and a singular* r*-simplex* σ *transverse to* $\{R_\alpha\}$*,*

$$\int_\sigma D\eta = \int_{\partial\sigma} \eta.$$

Proof. Setting $\eta^{(-1)} = 0$ and $\eta^{(r)} = 0$, we have (cf. (7.3))

$$(D\eta)^{(p)} = \delta\eta^{(p-1)} + (-1)^p d\eta^{(p)}, \qquad 0 \le p \le r.$$

Thus

$$\int_\sigma D\eta = \sum_{p=1}^{r} \sum_{(\alpha_0,\dots,\alpha_p)\in I^{(p)}} \int_{\sigma^{-1}R_{\alpha_0\dots\alpha_p}} \sigma^*(\delta\eta^{(p-1)})_{\alpha_0\dots\alpha_p}$$

$$+ \sum_{p=0}^{r-1}(-1)^p \sum_{(\alpha_0,\dots,\alpha_p)\in I^{(p)}} \int_{\sigma^{-1}R_{\alpha_0\dots\alpha_p}} \sigma^* d\eta^{(p)}_{\alpha_0\dots\alpha_p}. \tag{7.19}$$

Using the usual Stokes theorem and (7.18), we compute

$$
\begin{aligned}
(-1)^p & \sum_{(\alpha_0,\dots,\alpha_p)\in I^{(p)}} \int_{\sigma^{-1}R_{\alpha_0\dots\alpha_p}} \sigma^* d\eta^{(p)}_{\alpha_0\dots\alpha_p} \\
&= \sum_{(\alpha_0,\dots,\alpha_p)\in I^{(p)}} \int_{(\partial\sigma)^{-1}R_{\alpha_0\dots\alpha_p}} \sigma^* \eta^{(p)}_{\alpha_0\dots\alpha_p} \\
&\quad - \sum_{(\alpha_0,\dots,\alpha_{p+1})\in I^{(p+1)}} \int_{\sigma^{-1}R_{\alpha_0\dots\alpha_{p+1}}} \sigma^*(\delta\eta^{(p)})_{\alpha_0\dots\alpha_{p+1}}.
\end{aligned}
$$

Hence the right-hand side of (7.19) is equal to

$$\sum_{p=0}^{r-1} \sum_{(\alpha_0,\dots,\alpha_p)\in I^{(p)}} \int_{(\partial\sigma)^{-1}R_{\alpha_0\dots\alpha_p}} \sigma^* \eta^{(p)}_{\alpha_0\dots\alpha_p} = \int_{\partial\sigma} \eta.$$

□

We have a morphism

$$\chi_D : A^r(\mathcal{U}) \longrightarrow S^r_{\mathrm{tr}}(M;\mathbb{C}) \qquad \text{given by } \langle \sigma, \chi_D(\xi)\rangle = \int_\sigma \xi. \tag{7.20}$$

By the above lemma, it induces a morphism

$$\chi_D : H^r_D(\mathcal{U}) \longrightarrow H^r(M;\mathbb{C}).$$

Remark 7.9. In this and the next paragraphs, we do not assume that M is orientable.

Integration via triangulation: Let K be a C^∞ triangulation of M and denote by $C^K_\bullet(M;\mathsf{R})$ and $C^\bullet_K(M;\mathsf{R})$ the chain and cochain complexes of oriented simplices in K with coefficients in R as in Section 4.1, where R is assumed to be $\mathbb{Z}$. If we denote by $H^K_p(M;\mathsf{R})$ and $H^p_K(M;\mathsf{R})$ their p-th homology and cohomology, the natural injective chain morphism $C^K_\bullet(M;\mathsf{R}) \xrightarrow{\eta^K_\bullet} S^{\mathrm{ci}}_\bullet(M;\mathsf{R}) \hookrightarrow S_\bullet(M;\mathsf{R})$ induces isomorphisms

$$\eta^K_* : H^K_p(M;\mathsf{R}) \xrightarrow{\sim} H_p(M;\mathsf{R}) \quad \text{and} \quad \eta^*_K : H^p(M;\mathsf{R}) \xrightarrow{\sim} H^p_K(M;\mathsf{R}).$$

Let $\mathcal{U} = \{U_\alpha\}_{\alpha\in I}$ be an open covering of M and $\{R_\alpha\}$ a honeycomb system adapted to $\mathcal{U}$.

Definition 7.9. We say that $\{R_\alpha\}$ is adapted to K, if for each r-simplex s of K and $(\alpha_0,\dots,\alpha_p)$ in $I^{(p)}$, $s \cap R_{\alpha_0\dots\alpha_p}$ is an $(r-p)$-chain with respect to some subdivision of K.

For example, if $\mathcal{U}$ is the simplicial covering associated with K, then the system $\{R_\alpha\}$ as in Example 7.2.1 is adapted to $\mathcal{U}$ and K.

Let $\{R_\alpha\}$ be a honeycomb system adapted to $\mathcal{U}$ and K. We orient $\boldsymbol{s} \cap R_{\alpha_0 \dots \alpha_p}$ by the following rules:

(1) each $\boldsymbol{s} \cap R_\alpha$ has the same orientation as $\boldsymbol{s}$ and its boundary is oriented according to Convention 3.2,
(2) for $(\alpha_0, \dots, \alpha_p)$ in I^{p+1}, $\boldsymbol{s} \cap R_{\alpha_0 \dots \alpha_p}$ is oriented inductively by the following identity:

$$\partial(\boldsymbol{s} \cap R_{\alpha_0 \dots \alpha_p}) = (-1)^p \partial \boldsymbol{s} \cap R_{\alpha_0 \dots \alpha_p} + \boldsymbol{s} \cap \partial R_{\alpha_0 \dots \alpha_p}, \tag{7.21}$$
$$\boldsymbol{s} \cap \partial R_{\alpha_0 \dots \alpha_p} = \sum_{\alpha \in I} \boldsymbol{s} \cap R_{\alpha_0 \dots \alpha_p \alpha}.$$

With these we define the integration of an r-ČdR cochain $\xi = (\xi^{(p)})$ on an r-simplex $\boldsymbol{s}$ by

$$\int_{\boldsymbol{s}} \xi = \sum_{p=0}^{r} \sum_{(\alpha_0, \dots, \alpha_p) \in I^{(p)}} \int_{\boldsymbol{s} \cap R_{\alpha_0 \dots \alpha_p}} \xi^{(p)}_{\alpha_0 \dots \alpha_p}. \tag{7.22}$$

Remark 7.10. In the case $\mathcal{U}$ consists of two open sets, we have

$$\int_{\boldsymbol{s}} \xi = \int_{\boldsymbol{s} \cap R_0} \xi_0 + \int_{\boldsymbol{s} \cap R_1} \xi_1 + \int_{\boldsymbol{s} \cap R_{01}} \xi_{01}$$

for an r-simplex $\boldsymbol{s}$ and an r-ČdR cochain $\xi = (\xi_0, \xi_1, \xi_{01})$.

The following is proved as Lemma 7.4.

Lemma 7.5. *For an $(r-1)$-ČdR cochain η and an r-simplex $\boldsymbol{s}$ of K,*

$$\int_{\boldsymbol{s}} D\eta = \int_{\partial \boldsymbol{s}} \eta.$$

It is also referred to as the ČdR Stokes theorem.

We have a morphism

$$\psi_D : A^r(\mathcal{U}) \longrightarrow C^r_K(M; \mathbb{C}) \qquad \text{given by } \langle \boldsymbol{s}, \psi_D(\xi) \rangle = \int_{\boldsymbol{s}} \xi. \tag{7.23}$$

By the above lemma, it induces a morphism

$$\psi_D : H^r_D(\mathcal{U}) \longrightarrow H^r(M; \mathbb{C}).$$

Note that we have the commutative diagrams:

$$\begin{array}{ccc} A^r(\mathcal{U}) & & \\ {\scriptstyle \chi_D} \downarrow & \searrow^{\psi_D} & \\ S^r_{\mathrm{ci}}(M; \mathbb{C}) & \xrightarrow[\eta^r_K]{} & C^r_K(M; \mathbb{C}), \end{array} \qquad \begin{array}{ccc} H^r_D(\mathcal{U}) & & \\ {\scriptstyle \chi_D} \downarrow & \searrow^{\psi_D} & \\ H^r(M; \mathbb{C}) & \xrightarrow[\eta^{*}_K]{\sim} & H^r_K(M; \mathbb{C}). \end{array} \tag{7.24}$$

de Rham case

Recall that there is a natural inclusion $\varphi_d : A^r(M) \hookrightarrow C^0(\mathcal{U}; A^r) \subset A^r(\mathcal{U})$ (cf. (7.6)). Restricting χ_D and ψ_D to $A^r(M)$, the diagrams (7.24) become:

$$\begin{array}{ccc} A^r(M) & & \\ {\scriptstyle \chi_d}\downarrow & \overset{\psi_d}{\searrow} & \\ S^r_{\mathrm{ci}}(M;\mathbb{C}) & \underset{\eta^r_K}{\longrightarrow} & C^r_K(M;\mathbb{C}), \end{array} \qquad \begin{array}{ccc} H^r_d(M) & & \\ {\scriptstyle \chi_d}\downarrow & \overset{\psi_d}{\searrow} & \\ H^r(M;\mathbb{C}) & \underset{\eta^*_K}{\xrightarrow{\sim}} & H^r_K(M;\mathbb{C}). \end{array} \tag{7.25}$$

In the first diagram, χ_d and ψ_d are given by

$$\langle \sigma, \chi_d(\omega)\rangle = \int_\sigma \omega, \qquad \langle \boldsymbol{s}, \psi_d(\omega)\rangle = \int_{\boldsymbol{s}} \omega$$

and coincides with the ones defined in Section 6.2. Note that the ČdR Stokes theorem reduces to the usual one. We will see that, in the second diagram, ψ_d and thus χ_d are in fact isomorphisms (de Rham theorem, cf. Theorem 7.5 below).

Čech case

Recall that there is a natural inclusion $\varphi_\delta : C^r(\mathcal{U};\mathbb{C}) \hookrightarrow C^r(\mathcal{U}; A^0) \subset A^r(\mathcal{U})$ (cf. (7.8)). Restricting χ_D and ψ_D to $C^r(\mathcal{U};\mathbb{C})$, the diagrams (7.24) become:

$$\begin{array}{ccc} C^r(\mathcal{U};\mathbb{C}) & & \\ {\scriptstyle \chi_\delta}\downarrow & \overset{\psi_\delta}{\searrow} & \\ S^r_{\mathrm{ci}}(M;\mathbb{C}) & \underset{\eta^r_K}{\longrightarrow} & C^r_K(M;\mathbb{C}), \end{array} \qquad \begin{array}{ccc} H^r(\mathcal{U};\mathbb{C}) & & \\ {\scriptstyle \chi_\delta}\downarrow & \overset{\psi_\delta}{\searrow} & \\ H^r(M;\mathbb{C}) & \underset{\eta^*_K}{\xrightarrow{\sim}} & H^r_K(M;\mathbb{C}). \end{array} \tag{7.26}$$

In the first diagram, χ_δ and ψ_δ are given by

$$\langle \sigma, \chi_\delta(c)\rangle = \sum c_{\alpha_0\ldots\alpha_r}, \qquad \langle \boldsymbol{s}, \psi_\delta(c)\rangle = \sum c_{\alpha_0\ldots\alpha_r},$$

where the first sum is taken over $(\alpha_0, \ldots, \alpha_r) \in I^{(r)}$ with the condition $\sigma(\Delta^r) \cap R_{\alpha_0\ldots\alpha_r} \neq \emptyset$ and the second sum with the condition $\boldsymbol{s} \cap R_{\alpha_0\ldots\alpha_r} \neq \emptyset$. We will show that, in the second diagram, ψ_δ is an isomorphism, if $\mathcal{U}$ is the simplicial covering associated with K (cf. Section 7.2).

Although the following is a direct consequence of the definitions, we state it, as it is fundamental:

Lemma 7.6. *If $\mathcal{U}$ is the simplicial covering associated with K, the morphism $\psi_\delta : C^r(\mathcal{U};\mathbb{C}) \to C^r_K(M;\mathbb{C})$ in (7.26) is given by*

$$\langle (\alpha_0, \ldots, \alpha_r), \psi_\delta(c)\rangle = c_{\alpha_0\ldots\alpha_r}.$$

For a commutative ring R with unity, we may also define a morphism

$$\psi_\delta : C^r(\mathcal{U}; \mathsf{R}) \to C^r_K(M; \mathsf{R}) \tag{7.27}$$

the same way.

Theorem 7.3. *For the simplicial covering $\mathcal{U}$ associated with K, the morphism ψ_δ induces an isomorphism*

$$\psi_\delta : H^r(\mathcal{U}; \mathsf{R}) \xrightarrow{\sim} H^r_K(M; \mathsf{R}).$$

Proof. Consider the diagram

$$\begin{array}{ccc} C^r(\mathcal{U}; \mathsf{R}) & \xrightarrow{\ \pi\ } & C^r_{\mathrm{alt}}(\mathcal{U}; \mathsf{R}) \\ & \underset{\psi_\delta}{\searrow} & \wr \downarrow \psi^{\mathrm{a}}_\delta \\ & & C^r_K(M; \mathsf{R}), \end{array}$$

where π is the morphism as in the proof of Proposition 7.2 and ψ^{a}_δ is the restriction of ψ_δ. The diagram is commutative and ψ^{a}_δ is an isomorphism. Moreover the morphisms are all compatible with the coboundary operators. Thus we have the above isomorphism, in view of Proposition 7.2. □

Setting $\chi_\delta = (\eta^*_K)^{-1} \circ \psi_\delta$, we have:

Corollary 7.5. *For a simplicial covering $\mathcal{U}$, there is a canonical isomorphism:*

$$\chi_\delta : H^r(\mathcal{U}; \mathsf{R}) \xrightarrow{\sim} H^r(M; \mathsf{R}).$$

Summary

Summarizing the above we have the commutative diagram:

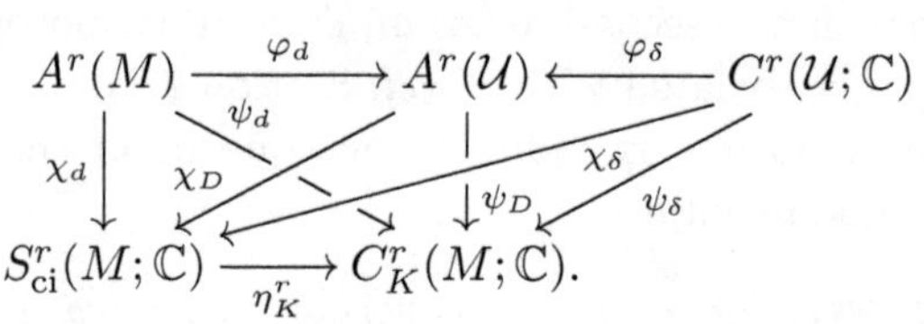

(7.28)

The morphisms in (7.28) are all compatible with the coboundary operators and we immediately have:

Theorem 7.4 (Fundamental theorem). *For a C^∞ manifold M, an open covering $\mathcal{U}$ of M and a C^∞ triangulation K of M, we have a commutative diagram:*

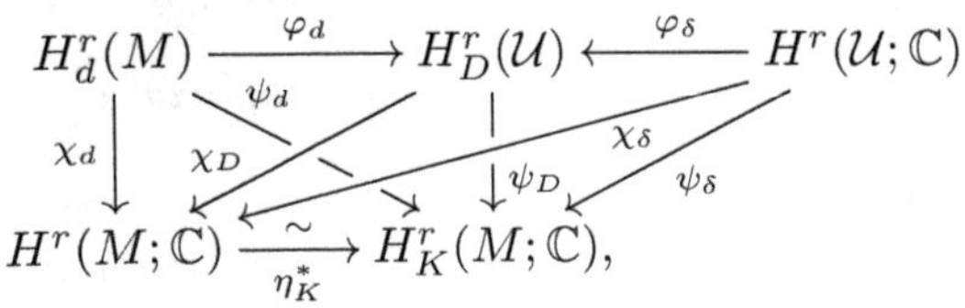

(7.29)

where

(1) φ_d, ψ_d *and* χ_d *always exist and* φ_d *is an isomorphism* (Theorem 7.1),
(2) φ_δ *always exists and is an isomorphism, if* $\mathcal{U}$ *is good* (Theorem 7.2),
(3) χ_D *and* χ_δ *exist, if there is a honeycomb system adapted to* $\mathcal{U}$,
(4) ψ_D *and* ψ_δ *exist, if there is a honeycomb system adapted to* $\mathcal{U}$ *and* K,
(5) ψ_δ *and* χ_δ *exist and are isomorphisms, if* $\mathcal{U}$ *is the simplicial covering associated with* K (Theorem 7.3 and Corollary 7.5 with $\mathsf{R} = \mathbb{C}$). *In this case,* ψ_D *and* χ_D *also exist and are isomorphisms.*

7.6 de Rham theorem and related topics

Let M be a C^∞ manifold and $\mathcal{U}$ an open covering of M. As corollaries of Theorem 7.4, we have the following theorems.

First, considering the case where $\mathcal{U}$ is the simplicial covering associated with K and noting that $\psi_d = \psi_\delta \circ \varphi_\delta^{-1} \circ \varphi_d$, we see that

$$\psi_d : H^r_d(M) \xrightarrow{\sim} H^r_K(M;\mathbb{C}) \tag{7.30}$$

is an isomorphism. Combined with the isomorphism η_K^*, we have:

Theorem 7.5 (de Rham theorem). *For a C^∞ manifold M, there is an isomorphism*

$$\chi_d : H^r_d(M) \xrightarrow{\sim} H^r(M;\mathbb{C}),$$

which is given as in (7.25).

Theorem 7.6 (Čech-de Rham theorem). *For a C^∞ manifold M and any covering $\mathcal{U}$ of M, there is an isomorphism*

$$\chi_d \circ \varphi_d^{-1} : H^r_D(\mathcal{U}) \xrightarrow{\sim} H^r(M;\mathbb{C}),$$

which coincides with χ_D in (7.24), *if there is a honeycomb system adapted to $\mathcal{U}$.*

Theorem 7.7 (Čech theorem). *For a C^∞ manifold M and a good covering $\mathcal{U}$ of M, there is an isomorphism*

$$\chi_d \circ \varphi_d^{-1} \circ \varphi_\delta : H^r(\mathcal{U};\mathbb{C}) \xrightarrow{\sim} H^r(M;\mathbb{C}),$$

which coincides with χ_δ in (7.26), *if $\mathcal{U}$ is a simplicial covering.*

Remark 7.11. 1. The point in the above is that we use the isomorphism $H_d^r(M) \simeq H_D^r(\mathcal{U})$ to say that $H_D^r(\mathcal{U})$ is independent of $\mathcal{U}$ modulo isomorphism and, in the case $\mathcal{U}$ is simplicial, we have isomorphisms $H_D^r(\mathcal{U}) \simeq H^r(\mathcal{U};\mathbb{C}) \simeq H^r(M;\mathbb{C})$.

2. In the isomorphism of Theorem 7.5 for $r = 0$, the class [1] in $H^0(M;\mathbb{C})$ (cf. Remark B.9) corresponds to the class in $H_d^0(M)$ of the function constantly equal to 1.

As we see in the following, we have in fact isomorphisms of graded rings.

Product structures

Čech-de Rham: For a covering $\mathcal{U}$ of M, we define a product, called the *cup product*

$$A^r(\mathcal{U}) \times A^s(\mathcal{U}) \xrightarrow{\smile} A^{r+s}(\mathcal{U}) \tag{7.31}$$

by assigning to (ξ,η) in $A^r(\mathcal{U}) \times A^s(\mathcal{U})$ the ČdR cochain $\xi \smile \eta$ in $A^{r+s}(\mathcal{U})$ given by

$$(\xi \smile \eta)_{\alpha_0\dots\alpha_p} = \sum_{\nu=0}^{p} (-1)^{(r-\nu)(p-\nu)} \xi_{\alpha_0\dots\alpha_\nu} \wedge \eta_{\alpha_\nu\dots\alpha_p}.$$

Then $\xi \smile \eta$ is linear in ξ and η and we have

$$D(\xi \smile \eta) = D\xi \smile \eta + (-1)^r \xi \smile D\eta. \tag{7.32}$$

Thus it induces a product

$$H_D^r(\mathcal{U}) \times H_D^s(\mathcal{U}) \xrightarrow{\smile} H_D^{r+s}(\mathcal{U}). \tag{7.33}$$

Here we write down the above product explicitly for small values of p:

$$(\xi \smile \eta)_{\alpha_0} = \xi_{\alpha_0} \wedge \eta_{\alpha_0},$$
$$(\xi \smile \eta)_{\alpha_0\alpha_1} = (-1)^r \xi_{\alpha_0} \wedge \eta_{\alpha_0\alpha_1} + \xi_{\alpha_0\alpha_1} \wedge \eta_{\alpha_1},$$
$$(\xi \smile \eta)_{\alpha_0\alpha_1\alpha_2} = \xi_{\alpha_0} \wedge \eta_{\alpha_0\alpha_1\alpha_2} + (-1)^{r-1} \xi_{\alpha_0\alpha_1} \wedge \eta_{\alpha_1\alpha_2} + \xi_{\alpha_0\alpha_1\alpha_2} \wedge \eta_{\alpha_2}.$$

de Rham: There is a natural inclusion $\varphi_d : A^r(M) \hookrightarrow C^0(\mathcal{U}; A^r) \subset A^r(\mathcal{U})$ (cf. (7.6)) and the restriction of (7.31) coincides with the exterior product

$$A^r(M) \times A^s(M) \xrightarrow{\wedge} A^{r+s}(M).$$

It induces the product

$$H_d^r(M) \times H_d^s(M) \xrightarrow{\wedge} H_d^{r+s}(M)$$

given by $([\omega], [\theta]) \mapsto [\omega \wedge \theta]$. We have

$$[\omega] \wedge [\theta] = (-1)^{rs} [\theta] \wedge [\omega] \tag{7.34}$$

and this product makes the direct sum $H_d^*(M) = \bigoplus_r H_d^r(M)$ a graded ring. The product corresponds to the cup product in $H^*(M; \mathbb{C})$ under the isomorphism of Theorem 7.5 (cf. Theorem 7.9 below).

Čech: There is a natural inclusion $\varphi_\delta : C^r(\mathcal{U}; \mathbb{C}) \hookrightarrow C^r(\mathcal{U}; A^0) \subset A^r(\mathcal{U})$ (cf. (7.8)) and the product (7.31) restricts to

$$C^r(\mathcal{U}; \mathbb{C}) \times C^s(\mathcal{U}; \mathbb{C}) \xrightarrow{\smile} C^{r+s}(\mathcal{U}; \mathbb{C}) \tag{7.35}$$

given by

$$(c \smile c')_{\alpha_0 \dots \alpha_{r+s}} = c_{\alpha_0 \dots \alpha_r} \cdot c'_{\alpha_r \dots \alpha_{r+s}}.$$

Let R be a commutative ring with unity. We may also define a product

$$C^r(\mathcal{U}; \mathsf{R}) \times C^s(\mathcal{U}; \mathsf{R}) \xrightarrow{\smile} C^{r+s}(\mathcal{U}; \mathsf{R})$$

the same way. Then we have

$$\delta(c \smile c') = \delta c \smile c' + (-1)^r c \smile \delta c'$$

so that it induces a product in cohomology:

$$H^r(\mathcal{U}; \mathsf{R}) \times H^s(\mathcal{U}; \mathsf{R}) \xrightarrow{\smile} H^{r+s}(\mathcal{U}; \mathsf{R}). \tag{7.36}$$

This product makes the direct sum $H^*(\mathcal{U}; \mathsf{R}) = \bigoplus_r H^r(\mathcal{U}; \mathsf{R})$ a graded ring.

Theorem 7.8. *For the simplicial covering $\mathcal{U}$ associated with a triangulation K of M, the morphism ψ_δ in (7.27) induces an isomorphism*

$$\psi_\delta : H^*(\mathcal{U}; \mathsf{R}) \xrightarrow{\sim} H_K^*(M; \mathsf{R}),$$

where the product in $H_K^*(M; \mathsf{R})$ is the cup product in topology.

Proof. By Theorem 7.3, it suffices to show that ψ_δ is compatible with the products. Let I be the set of vertices of K and endow it with a simplicial ordering (cf. Definition B.14). Then for an ordered $(r+s)$-simplex $\boldsymbol{s} = (\alpha_0, \dots, \alpha_{r+s})$, $\alpha_0 < \dots < \alpha_{r+s}$,

$$\begin{aligned}\langle \boldsymbol{s}, \psi_\delta(c \smile c')\rangle &= (c \smile c')_{\alpha_0 \dots \alpha_{r+s}} = c_{\alpha_0 \dots \alpha_r} \cdot c'_{\alpha_r \dots \alpha_{r+s}} \\ &= \langle \boldsymbol{s}, \psi_\delta(c) \smile \psi_\delta(c')\rangle,\end{aligned}$$

which proves the theorem. □

Corollary 7.6. *For a simplicial covering $\mathcal{U}$, there is a canonical isomorphism:*

$$\chi_\delta : H^*(\mathcal{U}; \mathsf{R}) \xrightarrow{\sim} H^*(M; \mathsf{R}),$$

where the product in $H^*(M; \mathsf{R})$ is the cup product.

Theorem 7.9. *For a C^∞ manifold M, an open covering $\mathcal{U}$ of M and a triangulation K of M, we have a commutative diagram of graded rings:*

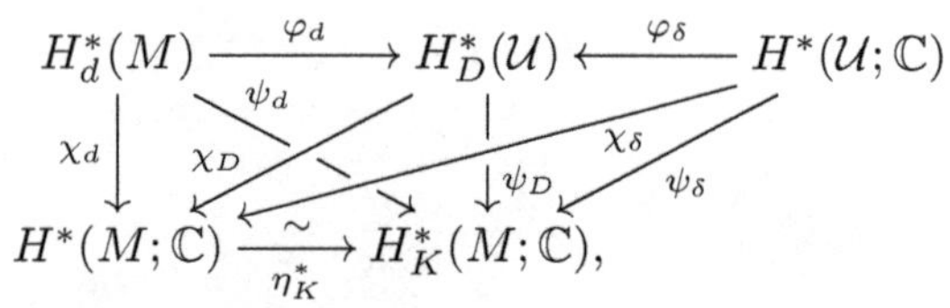

where

(1) φ_d, ψ_d *and* χ_d *always exist and are isomorphisms,*
(2) φ_δ *always exists and is an isomorphism, if* $\mathcal{U}$ *is good,*
(3) χ_D *and* χ_δ *exist and are isomorphisms, if there is a honeycomb system adapted to* $\mathcal{U}$*,*
(4) ψ_D *and* ψ_δ *exist and are isomorphisms, if there is a honeycomb system adapted to* $\mathcal{U}$ *and* K*.*

Proof. By Theorems 7.4–7.7, it suffices to show that the morphisms are compatible with the product structures. First, φ_d and φ_δ are compatible with the products. If $\mathcal{U}$ is the simplicial covering associated with K, ψ_δ is compatible with the products (Theorem 7.8 with $\mathsf{R} = \mathbb{C}$). Then the others are also compatible with the products. □

Remark 7.12. Recall that, if $\mathcal{U}$ is the simplicial covering associated with K, there is a honeycomb system adapted to $\mathcal{U}$ and K.

Pull-backs

Let $f : M' \to M$ be a C^∞ map of C^∞ manifolds.

de Rham: As is seen in Section 7.1, the pull-back $f^* : A^r(M) \to A^r(M')$ on forms induces a morphism $f_d^* : H_d^r(M) \to H_d^r(M')$. Since it preserves the exterior product, it is in fact a morphism of graded rings:

$$f_d^* : H_d^*(M) \longrightarrow H_d^*(M'). \tag{7.37}$$

On the other hand, we have the pull-back $f^* : H^*(M;\mathbb{C}) \to H^*(M';\mathbb{C})$ in the singular cohomology (cf. Section B.1).

Proposition 7.5. *The pull-back* (7.37) *corresponds to the pull-back in the singular cohomology under the isomorphism of Theorem 7.5.*

Proof. It suffices to show the commutativity of the diagram:

$$\begin{array}{ccc} A^r(M) & \xrightarrow{\chi_d} & S^r_{\mathrm{ci}}(M;\mathbb{C}) \\ \downarrow{\scriptstyle f^*} & & \downarrow{\scriptstyle f^*} \\ A^r(M') & \xrightarrow{\chi'_d} & S^r_{\mathrm{ci}}(M';\mathbb{C}). \end{array}$$

Let ω be a form in $A^r(M)$ and σ a C^∞ singular r-simplex in M'. We compute

$$\langle \sigma, \chi'_d f^*\omega\rangle = \int_{\Delta^r} \sigma^* f^*\omega = \int_{\Delta^r} (f\circ\sigma)^*\omega = \langle f_*\sigma, \chi_d\omega\rangle = \langle \sigma, f^*\chi_d\omega\rangle.$$

□

Čech: Let $\mathcal{U} = \{U_\alpha\}_{\alpha\in I}$ and $\mathcal{V} = \{V_\lambda\}_{\lambda\in J}$ be open coverings of M and M', respectively. We assume that there is a map $\psi : J \to I$ such that $f(V_\lambda) \subset U_{\psi(\lambda)}$ for all λ in J. Then we define a $\mathbb{C}$-linear map

$$f_\psi^* : C^r(\mathcal{U};\mathsf{R}) \longrightarrow C^r(\mathcal{V};\mathsf{R})$$

by assigning to c in $C^r(\mathcal{U};\mathsf{R})$ the element $f_\psi^*(c)$ in $C^r(\mathcal{V};\mathsf{R})$ given by

$$f_\psi^*(c)_{\lambda_0\dots\lambda_r} = c_{\psi(\lambda_0)\dots\psi(\lambda_r)}.$$

Then the map f_ψ^* induces a $\mathbb{C}$-linear map

$$f^* : H^r(\mathcal{U};\mathsf{R}) \longrightarrow H^r(\mathcal{V};\mathsf{R}),$$

which is independent of ψ.

Čech-de Rham: Let $\mathcal{U} = \{U_\alpha\}_{\alpha\in I}$ and $\mathcal{V} = \{V_\lambda\}_{\lambda\in J}$ be open coverings of M and M', respectively. We assume that there is a map $\psi : J \to I$ such that $f(V_\lambda) \subset U_{\psi(\lambda)}$ for all λ in J. Then we define a $\mathbb{C}$-linear map

$$f_\psi^* : A^r(\mathcal{U}) \longrightarrow A^r(\mathcal{V})$$

by assigning to ξ in $A^r(\mathcal{U})$ the element $f_\psi^*(\xi)$ in $A^r(\mathcal{V})$ given by

$$f_\psi^*(\xi)_{\lambda_0\ldots\lambda_p} = f^*\xi_{\psi(\lambda_0)\ldots\psi(\lambda_p)}.$$

Then the map f_ψ^* induces a $\mathbb{C}$-linear map

$$f^* : H_D^r(\mathcal{U}) \longrightarrow H_D^r(\mathcal{V}), \tag{7.38}$$

which is compatible, via the isomorphism of Theorem 7.1, with the pull-back on the de Rham cohomology (7.37). In particular it is independent of ψ. Moreover by Proposition 7.5, it corresponds to the pull-back in the singular cohomology under the isomorphism of Theorem 7.6.

Isomorphisms in dual cells

We take a C^∞ triangulation K of M and let K^* denote the cellular decomposition dual to K, as in Section 4.1. We could express the above isomorphisms using dual cells, as they are K'-chains.

We have morphisms

$$A^r(M) \xrightarrow{\psi'_d} C^r_{K'}(M;\mathbb{C}) \xrightarrow{\iota^*} C^r_{K^*}(M;\mathbb{C}),$$

where ψ'_d is the one in (7.25) with K replaced by K' and ι^* is the transpose of ι in (4.3). If we denote the composition $\iota^* \circ \psi_d$ by ψ_d^*, it is given by

$$\langle \boldsymbol{s}^*, \psi_d^*(\omega)\rangle = \int_{\boldsymbol{s}^*} \omega$$

and it induces an isomorphism

$$\psi_d^* : H_d^r(M) \xrightarrow{\sim} H^r_{K^*}(M;\mathbb{C}).$$

Combined with $\eta^*_{K^*}$ in (4.4) with $\mathbb{C}$-coefficient, we have the de Rham isomorphism in Theorem 7.5.

For any covering $\mathcal{U}$ of M, we then have an isomorphism

$$\psi_d^* \circ \varphi_d^{-1} : H_D^r(\mathcal{U}) \xrightarrow{\sim} H^r_{K^*}(M;\mathbb{C}),$$

where φ_d is the isomorphism of Theorem 7.1. Combined with $\eta^*_{K^*}$, we have the Čech-de Rham isomorphism in Theorem 7.6. If there is a honeycomb system adapted to $\mathcal{U}$ and K' (cf. Definition 7.9), this is expressed in terms of integration of ČdR cocycles on dual cells. This can be done for an arbitrary covering, however we do it for coverings consisting of two open sets for simplicity. In fact this is the case we later deal with. Thus let $\mathcal{U} = \{U_0, U_1\}$ be an open covering of M. We assume that there exists a honeycomb system $\{R_0, R_1\}$ adapted to $\mathcal{U}$ and K'. Then, for each r-cell $\boldsymbol{s}^*$, $\boldsymbol{s}^* \cap R_i$, $i = 0, 1$, are r-chains and $\boldsymbol{s}^* \cap R_{01}$ is an $(r-1)$-chain with respect to some subdivision of K'. We consider the composition

$$A^r(\mathcal{U}) \xrightarrow{\psi'_D} C^r_{K'}(M;\mathbb{C}) \xrightarrow{\iota^*} C^r_{K^*}(M;\mathbb{C}), \tag{7.39}$$

where ψ'_D is the morphism in (7.23) with K replaced by K', and set $\psi^*_D = \iota^* \circ \psi'_D$. Then it is given by, for $\xi = (\xi_0, \xi_1, \xi_{01})$ in $A^r(\mathcal{U})$,

$$\langle \boldsymbol{s}^*, \psi^*_D(\xi)\rangle = \int_{\boldsymbol{s}^*} \xi = \int_{\boldsymbol{s}^* \cap R_0} \xi_0 + \int_{\boldsymbol{s}^* \cap R_1} \xi_1 + \int_{\boldsymbol{s}^* \cap R_{01}} \xi_{01}, \tag{7.40}$$

where the orientations of $\boldsymbol{s}^* \cap R_i$ are naturally determined from that of $\boldsymbol{s}^*$ and we endow with $\boldsymbol{s}^* \cap R_{01}$ the orientation as a part of the boundary of $\boldsymbol{s}^* \cap R_0$. Then ψ^*_D induces an isomorphism $\psi^*_D : H^r_D(M) \xrightarrow{\sim} H^r_{K^*}(M;\mathbb{C})$, which is equal to $\psi^*_d \circ \varphi_d^{-1}$.

7.7 Relative Čech-de Rham cohomology

We introduce the relative Čech-de Rham cohomology in the simplest situation, although it can be done in more general settings, for example see the proof of Theorem 7.11 below.

Let M be a C^∞ manifold of dimension m and S a closed set in M. Letting $U_0 = M \smallsetminus S$ and U_1 an open neighborhood of S, we consider the covering $\mathcal{U} = \{U_0, U_1\}$ of M. In the following, we consider the reduced expression of ČdR cochains (cf. Remark 7.4.2, also Example 7.1) and we denote $\hat{A}^r(\mathcal{U})$ simply by $A^r(\mathcal{U})$.

We set

$$A^r(\mathcal{U}, U_0) = \{\, \xi = (\xi_0, \xi_1, \xi_{01}) \in A^r(\mathcal{U}) \mid \xi_0 = 0 \,\}.$$

Then we see that $(A^\bullet(\mathcal{U}, U_0), D)$ is a subcomplex of $(A^\bullet(\mathcal{U}), D)$. Note that we may write

$$A^r(\mathcal{U}, U_0) = A^r(U_1) \oplus A^{r-1}(U_{01})$$

and $D : A^r(\mathcal{U}, U_0) \to A^{r+1}(\mathcal{U}, U_0)$ is given by $D(\xi_1, \xi_{01}) = (d\xi_1, \xi_1 - d\xi_{01})$.

Definition 7.10. The *r-th relative Čech-de Rham cohomology* $H^r_D(\mathcal{U}, U_0)$ of the pair $(\mathcal{U}, U_0)$ is the r-th cohomology of $(A^\bullet(\mathcal{U}, U_0), D)$.

Later we show that, if S is a subcomplex of M with respect to some triangulation of M, then $H^r_D(\mathcal{U}, U_0)$ is canonically isomorphic with the relative singular cohomology $H^r(M, M \smallsetminus S; \mathbb{C})$ (cf. Corollary 7.8 below).

First we note that there is an exact sequence of complexes

$$0 \longrightarrow A^\bullet(\mathcal{U}, U_0) \xrightarrow{j^\bullet} A^\bullet(\mathcal{U}) \xrightarrow{i^\bullet} A^\bullet(U_0) \longrightarrow 0,$$

where $j^r(\xi_1, \xi_{01}) = (0, \xi_1, \xi_{01})$ and $i^r(\xi_0, \xi_1, \xi_{01}) = \xi_0$. This gives rise to a long exact sequence

$$\cdots \longrightarrow H^{r-1}_d(U_0) \xrightarrow{\delta^*} H^r_D(\mathcal{U}, U_0) \xrightarrow{j^*} H^r_D(\mathcal{U}) \xrightarrow{i^*} H^r_d(U_0) \longrightarrow \cdots. \tag{7.41}$$

Recall that there is a canonical isomorphism $\varphi_d : H^r_d(M) \xrightarrow{\sim} H^r_D(\mathcal{U})$, which is induced by $\omega \mapsto (\omega|_{U_0}, \omega|_{U_1}, 0)$ (cf. Theorem 7.1). Its inverse assigns to the class of (ξ_0, ξ_1, ξ_{01}) the class of $\rho_0\xi_0 + \rho_1\xi_1 - d\rho_0 \wedge \xi_{01}$ with (ρ_0, ρ_1) a partition of unity subordinate to $\mathcal{U}$ (cf. Exercise 7.2).

Exercise 7.5. Show that δ^* assigns to the class of a closed $(r-1)$-form θ on U_0 the class of $(0, -\theta)$ (cf. Proposition A.11).

The cohomology $H^r_D(\mathcal{U}, U_0)$ is determined by the local structure of M near S in the following sense.

We consider a special covering $\mathcal{U}^\star = \{U_0, U_1^\star\}$ of M given by $U_0 = M \smallsetminus S$ and $U_1^\star = M$.

Definition 7.11. The cohomology $H^r_D(\mathcal{U}^\star, U_0)$ is called the r-th *relative de Rham cohomology* of the pair $(M, M \smallsetminus S)$ and is denoted by $H^r_D(M, M \smallsetminus S)$.

Remark 7.13. The relative de Rham cohomology is a special type of the relative Čech-de Rham cohomology where $U_1 = M$.

Proposition 7.6. *For every covering $\mathcal{U} = \{U_0, U_1\}$ as above, the morphism $A^r(\mathcal{U}^\star, U_0) \to A^r(\mathcal{U}, U_0)$ given by restriction of forms induces an isomorphism*

$$H^r_D(M, M \smallsetminus S) \xrightarrow{\sim} H^r_D(\mathcal{U}, U_0).$$

Proof. The morphism $A^r(\mathcal{U}^\star) \to A^r(\mathcal{U})$ given by restriction of forms is compatible with the coboundary operators D and induces a morphism $H^r_D(\mathcal{U}^\star) \to H^r_D(\mathcal{U})$. This is in fact an isomorphism, since it decomposes as $\varphi_d \circ (\varphi_d^\star)^{-1} : H^r_D(\mathcal{U}^\star) \xrightarrow{\sim} H^r_d(M) \xrightarrow{\sim} H^r_D(\mathcal{U})$, where φ_d and $\varphi_d^\star$ are the isomorphisms in Theorem 7.1 for the coverings $\mathcal{U}$ and $\mathcal{U}^\star$. The morphism $A^r(\mathcal{U}^\star, U_0) \to A^r(\mathcal{U}, U_0)$ also induces a morphism $H^r_D(M, M \smallsetminus S) \to H^r_D(\mathcal{U}, U_0)$. Comparing the sequences (7.41) for $\mathcal{U}^\star$ and $\mathcal{U}$, we see that this is an isomorphism by the five lemma (Lemma A.6). □

Corollary 7.7. *The cohomology $H^r_D(\mathcal{U}, U_0)$ is determined uniquely modulo canonical isomorphisms, independently of the choice of U_1.*

Proposition 7.7 (Excision). *Let S be a closed set in M. For every open set U in M containing S, there is a canonical isomorphism*

$$H^r_D(M, M \smallsetminus S) \xrightarrow{\sim} H^r_D(U, U \smallsetminus S).$$

Proof. We denote by $\mathcal{U}$ the covering of M consisting of $U_0 = M \smallsetminus S$ and $U_1 = U$ and by $\mathcal{W}$ the covering of U consisting of $W_0 = U \smallsetminus S$ and $W_1 = U$. Then we may identify $A^r(\mathcal{U}, U_0)$ and $A^r(\mathcal{W}, W_0)$ and we have, by Proposition 7.6, a canonical isomorphism

$$H^r_D(M, M \smallsetminus S) \xrightarrow{\sim} H^r_D(\mathcal{U}, U_0) = H^r_D(\mathcal{W}, W_0) = H^r_D(U, U \smallsetminus S). \qquad \square$$

Integration: Now we assume that M is oriented and S is a compact set in M. Letting $U_0 = M \smallsetminus S$ and U_1 a neighborhood of S in M and consider the covering $\mathcal{U} = \{U_0, U_1\}$. Let R_1 be a compact m-dimensional manifold with boundary in U_1 containing S in its interior and set $R_0 = M \smallsetminus \operatorname{Int} R_1$. Then $\{R_0, R_1\}$ is a honeycomb system adapted to $\mathcal{U}$. The integration (7.15), see also (7.13), which is not defined in general for $A^m(\mathcal{U})$ unless M is compact, makes sense on $A^m(\mathcal{U}, U_0)$:

$$\int_M : A^m(\mathcal{U}, U_0) \longrightarrow \mathbb{C} \qquad \text{given by } \xi \mapsto \int_{R_1} \xi_1 + \int_{R_{01}} \xi_{01}. \tag{7.42}$$

Then the conditions (1) and (2) after (7.13) also hold and it induces the integration

$$\int : H^m_D(\mathcal{U}, U_0) \longrightarrow \mathbb{C}. \tag{7.43}$$

We will see that, if S is a subcomplex of M with respect to some triangulation, then $H^m_D(\mathcal{U}, U_0)$ is canonically isomorphic with the relative cohomology $H^m(M, M \smallsetminus S; \mathbb{C})$ (cf. Corollary 7.8 below). In this case the integration (7.43) is written $\int_{\underline{M}}$ (cf. Remark 4.5. 3). If, moreover, S is connected, the morphism (7.43) is an isomorphism, which is interpreted as a special case of the Alexander duality with $\mathbb{C}$-coefficient (cf. Proposition 7.16 below).

Let R be an m-dimensional compact manifold with boundary in M. Letting O be a neighborhood of ∂R in M, we apply the above by setting $S = R \smallsetminus O$. As to the covering $\mathcal{U} = \{U_0, U_1\}$, we let $U_0 = M \smallsetminus S$ and U_1 a neighborhood of R. Let $R_1 = R$ and $R_0 = M \smallsetminus \operatorname{Int} R$. Then $\{R_0, R_1\}$ is a honeycomb system adapted to $\mathcal{U}$. Then, Noting that $R_{01} = -\partial R$, we have the integration

$$\int : H^m_D(\mathcal{U}, U_0) \longrightarrow \mathbb{C} \qquad \text{given by } [\xi] \mapsto \int_R \xi_1 - \int_{\partial R} \xi_{01}. \tag{7.44}$$

If $S = R \smallsetminus O$ is a subcomplex of M, there is a canonical isomorphism $H^m_D(\mathcal{U}, U_0) \simeq H^m(M, M \smallsetminus S; \mathbb{C})$, which is isomorphic with $H^m(R, \partial R; \mathbb{C})$, if S has the same homotopy type as R. If, moreover, R is connected, the integration morphism (7.44) is an isomorphism, which is a special case of the Lefschetz duality with $\mathbb{C}$-coefficent (cf. Remark 7.18 and Proposition 7.19 below).

Remark 7.14. In most of the situations we consider, the set S as above is given beforehand. However, if $\{U_0, U_1\}$ is a covering of M in general, we may consider the cohomology $H^r_D(\mathcal{U}, U_0)$ by regarding $M \smallsetminus U_0$ as S.

Relative Čech-de Rham theorem

Let S be a closed set in M. Letting $U_0 = M \smallsetminus S$ and U_1 a neighborhood of S in M, we consider the covering $\mathcal{U} = \{U_0, U_1\}$, as before. If $\{R_0, R_1\}$ is a honeycomb system adapted to $\mathcal{U}$, the morphism (7.20)

$$\chi_D : A^r(\mathcal{U}) \longrightarrow S^r_{\mathrm{tr}}(M; \mathbb{C})$$

is given by

$$\langle \sigma, \chi_D(\xi) \rangle = \int_\sigma \xi = \int_{\sigma^{-1}R_0} \sigma^*\xi_0 + \int_{\sigma^{-1}R_1} \sigma^*\xi_1 + \int_{\sigma^{-1}R_{01}} \sigma^*\xi_{01}.$$

Suppose now that S is a subcomplex of M with respect to a C^∞ triangulation K_0 of M and let K, K' and K^* be as in Section 4.1. We may assume without loss of generality that U_1 contains the star $S_{K'}(S)$ of S in K'.

Definition 7.12. A honeycomb system $\{R_0, R_1\}$ is adapted to $\mathcal{U}$ and S if it is adapted to $\mathcal{U}$ and if $R_1 \subset O_{K'}(S)$.

For example, if we set $R_1 = S_{K''}(S)$, the star of S in the barycentric subdivision K'' of K', and $R_0 = M \smallsetminus \mathrm{Int}\, R_1$, then $\{R_0, R_1\}$ is adapted to $\mathcal{U}$ and S.

Theorem 7.10. *If $\{R_0, R_1\}$ is adapted to $\mathcal{U}$ and S, then the morphism χ_D induces a morphism*

$$\chi_D : A^r(\mathcal{U}, U_0) \longrightarrow S^r_{\mathrm{tr}}(M, M \smallsetminus O_{K'}(S); \mathbb{C}).$$

Moreover, it is compatible with the coboundary operators.

Proof. First, let $\{R_0, R_1\}$ be an arbitrary honeycomb system adapted to $\mathcal{U}$. Then for an element $\xi = (\xi_1, \xi_{01})$ in $A^r(\mathcal{U}, U_0)$ and a singular r-simplex σ,

$$\langle \sigma, \chi_D(\xi) \rangle = \int_{\sigma^{-1}R_1} \sigma^* \xi_1 + \int_{\sigma^{-1}R_{01}} \sigma^* \xi_{01}.$$

Now suppose $\{R_0, R_1\}$ is adapted to S. Then, for a simplex σ whose image is in $M \smallsetminus O_{K'}(S)$, $\sigma^{-1}R_1 = \emptyset$ so that the two terms above vanish. The last part follows from Lemma 7.4. □

Corollary 7.8 (Relative Čech-de Rham theorem). *Let S be a closed set in M. If S is a subcomplex of M with respect to some triangulation of M, there is a canonical isomorphism:*

$$\chi_D : H_D^r(\mathcal{U}, U_0) \xrightarrow{\sim} H^r(M, M \smallsetminus S; \mathbb{C}).$$

Proof. The morphism χ_D in Theorem 7.10 induces a morphism

$$H_D^r(\mathcal{U}, U_0) \to H_{\mathrm{tr}}^r(M, M \smallsetminus O_{K'}(S); \mathbb{C}) \simeq H^r(M, M \smallsetminus S; \mathbb{C}).$$

Comparing (7.41) with the exact sequence

$$\cdots \to H^{r-1}(M \smallsetminus S) \to H^r(M, M \smallsetminus S) \to H^r(M) \to H^r(M \smallsetminus S) \to \cdots,$$

and using the five lemma, we see that it is in fact an isomorphism. □

Remark 7.15. 1. The isomorphism χ_D in Corollary 7.8 is independent of the choice of the honeycomb system adapted to $\mathcal{U}$ and S.

2. In Corollary 7.8, considering the case of the covering $\mathcal{U}^*$, we see that there is a canonical isomorphism $H_D^r(M, M \smallsetminus S) \xrightarrow{\sim} H^r(M, M \smallsetminus S; \mathbb{C})$. This is compatible with the excisions in Proposition 7.7 and (B.10).

3. In fact, for every closed set S in M, there is a canonical isomorphism $H_D^r(\mathcal{U}, U_0) \xrightarrow{\sim} H^r(M, M \smallsetminus S; \mathbb{C})$. The point in the above description is that the isomorphism is expressed as the integration of ČdR cocycles on singular chains.

Pull-back: Let M and M' be C^∞ manifolds and let S and S' be subcomplexes of M and M', respectively, with respect to some triangulations of M and M'. Let $f : (M', M' \smallsetminus S') \to (M, M \smallsetminus S)$ be a C^∞ map. We set $U_0 = M \smallsetminus S$ and $U'_0 = M' \smallsetminus S'$. Letting U_1 and U'_1 be neighborhoods of S and S' such that $f(U'_1) \subset U_1$, we consider coverings $\mathcal{U} = \{U_0, U_1\}$ and $\mathcal{U}' = \{U'_0, U'_1\}$ of M and M'. On the one hand we have the pull-back $f^* : H_D^r(\mathcal{U}, U_0) \to H_D^r(\mathcal{U}', U'_0)$ as defined in (7.38) and on the other hand we have the pull-back $f^* : H^r(M, M \smallsetminus S; \mathbb{C}) \to H^r(M', M' \smallsetminus S'; \mathbb{C})$ on the singular cohomology. We readily see the following:

Proposition 7.8. *The following diagram is commutative:*

$$\begin{array}{ccc} H^r_D(\mathcal{U},U_0) & \xrightarrow[\chi_D]{\sim} & H^r(M,M\smallsetminus S;\mathbb{C}) \\ {\scriptstyle f^*}\downarrow & & \downarrow{\scriptstyle f^*} \\ H^r_D(\mathcal{U}',U_0') & \xrightarrow[\chi_D']{\sim} & H^r(M',M'\smallsetminus S';\mathbb{C}). \end{array}$$

Isomorphism via triangulation: We could also use a triangulation to define the isomorphism in Corollary 7.8. We choose a honeycomb system $\{R_0,R_1\}$ so that it is adapted to $\mathcal{U}$, K' and S (cf. Definitions 7.9 and 7.12). For example, we may set $R_1=S_{K''}(S)$ and $R_0=M\smallsetminus\operatorname{Int}R_1$. Then we have the morphism (7.23) for K':

$$\psi'_D : A^r(\mathcal{U}) \longrightarrow C^r_{K'}(M;\mathbb{C}).$$

It induces a morphism

$$\psi'_D : A^r(\mathcal{U},U_0) \longrightarrow C^r_{K'}(M,M\smallsetminus O_{K'}(S);\mathbb{C}), \tag{7.45}$$

given by

$$\langle \boldsymbol{t},\psi'_D(\xi)\rangle = \int_{\boldsymbol{t}}\xi = \int_{\boldsymbol{t}\cap R_1}\xi_1 + \int_{\boldsymbol{t}\cap R_{01}}\xi_{01},$$

for a relative r-ČdR cochain $\xi=(\xi_1,\xi_{01})$ and an r-simplex $\boldsymbol{t}$ of K'. Then ψ'_D induces an isomorphism

$$\psi'_D : H^r_D(\mathcal{U},U_0) \xrightarrow{\sim} H^r_{K'}(M,M\smallsetminus O_{K'}(S);\mathbb{C}).$$

Combined with $H^r_{K'}(M,M\smallsetminus O_{K'}(S);\mathbb{C})\simeq H^r(M,M\smallsetminus S;\mathbb{C})$, we have the isomorphism χ_D of Corollary 7.8.

Isomorphism in dual cells: We consider the morphism (7.39)

$$\psi^*_D : A^r(\mathcal{U}) \longrightarrow C^r_{K^*}(M;\mathbb{C}),$$

which is given by (7.40). If we choose a honeycomb system $\{R_0,R_1\}$ so that it is adapted to $\mathcal{U}$, K' and S, we see that it induces a morphism

$$\psi^*_D : A^r(\mathcal{U},U_0) \longrightarrow C^r_{K^*}(M,M\smallsetminus O_{K'}(S);\mathbb{C}),$$

given by

$$\langle \boldsymbol{s}^*,\psi^*_D(\xi)\rangle = \int_{\boldsymbol{s}^*}\xi = \int_{\boldsymbol{s}^*\cap R_1}\xi_1 + \int_{\boldsymbol{s}^*\cap R_{01}}\xi_{01}, \tag{7.46}$$

for an r-cochain $\xi=(\xi_1,\xi_{01})$ and an r-cell $\boldsymbol{s}^*$. Then ψ^*_D induces an isomorphism

$$\psi^*_D : H^r_D(\mathcal{U},U_0) \xrightarrow{\sim} H^r_{K^*}(M,M\smallsetminus O_{K'}(S);\mathbb{C}).$$

Combined with $H^r_{K^*}(M,M\smallsetminus O_{K'}(S);\mathbb{C})\simeq H^r(M,M\smallsetminus S;\mathbb{C})$, we have the isomorphism of Corollary 7.8.

Cup product: Let M, S and $\mathcal{U} = \{U_0, U_1\}$ be as above. In the product $A^r(\mathcal{U}) \times A^s(\mathcal{U}) \xrightarrow{\smile} A^{r+s}(\mathcal{U})$ of (7.31), we have, for $\xi = (\xi_0, \xi_1, \xi_{01})$ and $\eta = (\eta_0, \eta_1, \eta_{01})$,

$$\xi \smile \eta = (\xi_0 \wedge \eta_0, \xi_1 \wedge \eta_1, (-1)^r \xi_0 \wedge \eta_{01} + \xi_{01} \wedge \eta_1).$$

Thus, if ξ is a relative ČdR cochain, i.e., if $\xi_0 = 0$, then

$$\xi \smile \eta = (0, \xi_1 \wedge \eta_1, \xi_{01} \wedge \eta_1) \tag{7.47}$$

and we have a product $A^r(\mathcal{U}, U_0) \times A^s(\mathcal{U}) \xrightarrow{\smile} A^{r+s}(\mathcal{U}, U_0)$, which induces a product

$$H^r_D(\mathcal{U}, U_0) \times H^s_D(\mathcal{U}) \xrightarrow{\smile} H^{r+s}_D(\mathcal{U}, U_0).$$

Suppose now that S is a subcomplex of M with respect to a triangulation K_0 of M. Then there is a honeycomb system adapted to $\mathcal{U}$ (cf. the paragraph after Definition 7.12) and thus there is a canonical isomorphism $\chi_D : H^s_D(\mathcal{U}) \xrightarrow{\sim} H^s(M; \mathbb{C})$ (Theorem 7.6). There is also a canonical isomorphism $\chi_D : H^r_D(\mathcal{U}, U_0) \xrightarrow{\sim} H^r(M, M \smallsetminus S; \mathbb{C})$ (Corollary 7.8).

Recall that we have the cup product (cf. Section B.1)

$$H^r(M, M \smallsetminus S; \mathbb{C}) \times H^s(M; \mathbb{C}) \xrightarrow{\smile} H^{r+s}(M, M \smallsetminus S; \mathbb{C}).$$

Theorem 7.11. *In the above situation, the following diagram is commutative:*

$$\begin{array}{ccc} H^r_D(\mathcal{U}, U_0) \times H^s_D(\mathcal{U}) & \xrightarrow{\quad\smile\quad} & H^{r+s}_D(\mathcal{U}, U_0) \\ {\scriptstyle \chi_D \times \chi_D}\downarrow\wr & & \wr\downarrow{\scriptstyle \chi_D} \\ H^r(M, M \smallsetminus S; \mathbb{C}) \times H^s(M; \mathbb{C}) & \xrightarrow{\smile} & H^{r+s}(M, M \smallsetminus S; \mathbb{C}). \end{array}$$

Proof. Recall that in the diagram of Theorem 7.9, if $\mathcal{U}$ is the simplicial covering associated with K, all the morphisms exist and are isomorphisms.

Let K_0 be as above and K' the second barycentric subdivision of K_0, as before. Let I be the set of vertices of K' and we consider the simplicial covering $\mathcal{W} = \{W_\alpha\}_{\alpha \in I}$ associated with K'. Let $(A^\bullet(\mathcal{W}), \mathcal{D})$ denote the Čech-de Rham complex on $\mathcal{W}$. Also let I_0 denote the set of vertices not in S and I_1 the set of vertices in S. We set $\mathcal{W}_0 = \{W_\alpha\}_{\alpha \in I_0}$,

$$C^p(\mathcal{W}, \mathcal{W}_0; A^q) = \{\, \zeta \in C^p(\mathcal{W}; A^q) \mid \zeta_{\alpha_0 \dots \alpha_p} = 0, \ \text{if } \alpha_i \in I_0, \ 0 \le i \le p \}$$

and $A^r(\mathcal{W}, \mathcal{W}_0) = \bigoplus_{p+q=r} C^p(\mathcal{W}, \mathcal{W}_0; A^q)$. Then $A^\bullet(\mathcal{W}, \mathcal{W}_0)$ is a subcomplex of $A^\bullet(\mathcal{W})$, whose cohomology is denoted by $H^r_D(\mathcal{W}, \mathcal{W}_0)$.

There is a natural inclusion

$$\varphi_D : A^r(\mathcal{U}, U_0) \hookrightarrow C^0(\mathcal{W}, \mathcal{W}_0; A^r) \oplus C^1(\mathcal{W}, \mathcal{W}_0; A^{r-1}) \subset A^r(\mathcal{W}, \mathcal{W}_0),$$

which assigns to a cochain (ξ_1, ξ_{01}) the cochain ζ given by

$$\zeta_\alpha = \begin{cases} \xi_1 & \text{if } \alpha \in I_1 \\ 0 & \text{otherwise,} \end{cases} \qquad \zeta_{\alpha_0\alpha_1} = \begin{cases} \xi_{01} & \text{if } \alpha_0 \in I_0,\ \alpha_1 \in I_1 \\ 0 & \text{otherwise.} \end{cases}$$

It is compatible with the differentials D and $\mathcal{D}$ and induces a morphism

$$\varphi_D : H^r_D(\mathcal{U}, U_0) \longrightarrow H^r_{\mathcal{D}}(\mathcal{W}, \mathcal{W}_0),$$

which is shown to be an isomorphism, as Theorem 7.1.

Next we set

$$C^r(\mathcal{W}, \mathcal{W}_0; \mathbb{C}) = \{\, c \in C^r(\mathcal{W}; \mathbb{C}) \mid c_{\alpha_0 \dots \alpha_r} = 0, \text{ if } \alpha_i \in I_0,\ 0 \le i \le r \,\}.$$

Then $C^\bullet(\mathcal{W}, \mathcal{W}_0; \mathbb{C})$ is a subcomplex of $(C^\bullet(\mathcal{W}; \mathbb{C}), \delta)$, whose cohomology is denoted by $H^r(\mathcal{W}, \mathcal{W}_0; \mathbb{C})$. There is a natural inclusion

$$\varphi_\delta : C^r(\mathcal{W}, \mathcal{W}_0; \mathbb{C}) \hookrightarrow C^r(\mathcal{W}, \mathcal{W}_0; A^0) \subset A^r(\mathcal{W}, \mathcal{W}_0).$$

It is compatible with δ and $\mathcal{D}$ and induces a morphism

$$\varphi_\delta : H^r(\mathcal{W}, \mathcal{W}_0; \mathbb{C}) \longrightarrow H^r_{\mathcal{D}}(\mathcal{W}, \mathcal{W}_0),$$

which is shown to be an isomorphism, as Theorem 7.2.

Since there is a honeycomb system adapted to $\mathcal{W}$, K' and S, there is a morphism $\psi'_{\mathcal{D}} : A^r(\mathcal{W}) \to C^r_{K'}(M; \mathbb{C})$ (cf. (7.23)), which induces $\psi'_{\mathcal{D}} : A^r(\mathcal{W}, W_0) \to C^r_{K'}(M, M \smallsetminus O_{K'}(S); \mathbb{C})$. The composition $\psi'_{\mathcal{D}} \circ \varphi_D$ is the morphism ψ'_D of (7.45), which induces an isomorphism on the cohomology level. We have the morphism $\psi'_\delta : C^r(\mathcal{W}, \mathcal{W}_0; \mathbb{C}) \to C^r_{K'}(M, M \smallsetminus O_{K'}(S); \mathbb{C})$ as the composition $\psi'_{\mathcal{D}} \circ \varphi_\delta$. It induces an isomorphism on the cohomology level as Theorem 7.3. Thus we have the commutative diagram (cf. (7.28))

$$\begin{array}{ccccc} A^r(\mathcal{U}, U_0) & \xrightarrow{\varphi_D} & A^r(\mathcal{W}, \mathcal{W}_0) & \xleftarrow{\varphi_\delta} & C^r(\mathcal{W}, \mathcal{W}_0; \mathbb{C}) \\ & {\scriptstyle \psi'_D} \searrow & \downarrow {\scriptstyle \psi'_{\mathcal{D}}} & \swarrow {\scriptstyle \psi'_\delta} & \\ & & C^r_{K'}(M, M \smallsetminus O_{K'}(S); \mathbb{C}), & & \end{array}$$

which induces the commutative diagram

$$\begin{array}{ccccc} H^r_D(\mathcal{U}, U_0) & \xrightarrow{\varphi_D} & H^r_{\mathcal{D}}(\mathcal{W}, \mathcal{W}_0) & \xleftarrow{\varphi_\delta} & H^r(\mathcal{W}, \mathcal{W}_0; \mathbb{C}) \\ & {\scriptstyle \psi'_D} \searrow & \downarrow {\scriptstyle \psi'_{\mathcal{D}}} & \swarrow {\scriptstyle \psi'_\delta} & \\ & & H^r_{K'}(M, M \smallsetminus O_{K'}(S); \mathbb{C}). & & \end{array}$$

In the above, φ_D, ψ'_D, φ_δ and ψ'_δ are isomorphisms. Thus $\psi'_{\mathcal{D}}$ is also an isomorphism. The product $H^r_D(\mathcal{U}, U_0) \times H^s_D(\mathcal{U})$ corresponds to the product $H^r_{\mathcal{D}}(\mathcal{W}, W_0) \times H^s_{\mathcal{D}}(\mathcal{W})$, which in turn corresponds to the cup product $H^r(M, M \smallsetminus S; \mathbb{C}) \times H^s(M; \mathbb{C})$ via the isomorphisms of Theorem 7.9. $\square$

Remark 7.16. The formula (7.47) shows that there is a product

$$A^r(\mathcal{U}, U_0) \times A^s(U_1) \xrightarrow{\smile} A^{r+s}(\mathcal{U}, U_0),$$

which induces a product

$$H^r_D(\mathcal{U}, U_0) \times H^s_d(U_1) \xrightarrow{\smile} H^{r+s}_D(\mathcal{U}, U_0).$$

Denoting by $\mathcal{U}_1$ the covering $\{U_{01}, U_1\}$ of U_1, we have canonical isomorphisms $H^r_D(\mathcal{U}, U_0) \xleftarrow{\sim} H^r_D(M, M \smallsetminus S) \xrightarrow{\sim} H^r_D(U_1, U_1 \smallsetminus S) = H^r_D(\mathcal{U}_1, U_{01})$ (cf. Propositions 7.6 and 7.7). There is also the de Rham isomorphism $\chi_d : H^s_d(U_1) \xrightarrow{\sim} H^s(U_1; \mathbb{C})$ (Theorem 7.5). Thus we have the commutative diagram:

$$\begin{array}{ccc}
H^r_D(\mathcal{U}, U_0) \times H^s_d(U_1) & \xrightarrow{\quad\smile\quad} & H^{r+s}_D(\mathcal{U}, U_0) \\
{\scriptstyle \chi_D \times \chi_d}\big\downarrow\wr & & \wr\big\downarrow{\scriptstyle \chi_D} \\
H^r(U_1, U_1 \smallsetminus S; \mathbb{C}) \times H^s(U_1; \mathbb{C}) & \xrightarrow{\smile} & H^{r+s}(U_1, U_1 \smallsetminus S; \mathbb{C}).
\end{array}$$

More generally, let S_i be a closed set in M, $i = 1, 2$. Let $U_0^{(i)} = M \smallsetminus S_i$ and $U_1^{(i)}$ a neighborhood of S_i and consider the covering $\mathcal{U}^{(i)} = \{U_0^{(i)}, U_1^{(i)}\}$ of M. We set $S = S_1 \cap S_2$, $U_0 = M \smallsetminus S$ and U_1 an open neighborhood of S contained in $U_1^{(1)} \cap U_1^{(2)}$ and consider the covering $\mathcal{U} = \{U_0, U_1\}$ of M. The set U_0 is covered by two open sets $U_0^{(1)}$ and $U_0^{(2)}$. Let $\{\rho^{(1)}, \rho^{(2)}\}$ be a partition of unity subordinate to the covering. We define a product

$$A^r(\mathcal{U}^{(1)}, U_0^{(1)}) \times A^s(\mathcal{U}^{(2)}, U_0^{(2)}) \xrightarrow{\smile} A^{r+s}(\mathcal{U}, U_0) \qquad \text{by}$$

$$\xi \smile \eta = (\xi_1 \wedge \eta_1, \rho^{(1)} \xi_{01} \wedge \eta_1 + (-1)^r (\rho^{(2)} \xi_1 \wedge \eta_{01} - d\rho^{(1)} \wedge \xi_{01} \wedge \eta_{01}). \quad (7.48)$$

Then we see that the equality (7.32) also holds and we have the product

$$H^r_D(\mathcal{U}^{(1)}, U_0^{(1)}) \times H^s_D(\mathcal{U}^{(2)}, U_0^{(2)}) \xrightarrow{\smile} H^{r+s}_D(\mathcal{U}, U_0). \qquad (7.49)$$

It is not difficule to see that (7.49) does not depend on the choice of the partition of unity $\{\rho^{(1)}, \rho^{(2)}\}$.

In particular, if $S_2 = M$, we may set $\rho^{(1)} \equiv 1$ and $\rho^{(2)} \equiv 0$ and (7.48) reduces to (7.47).

Proposition 7.9. *If S_1 and S_2 are K_0-subcomplexes of M, the following diagram is commutative:*

$$\begin{array}{ccc}
H^r_D(\mathcal{U}^{(1)}, U_0^{(1)}) \times H^s_D(\mathcal{U}^{(2)}, U_0^{(2)}) & \xrightarrow{\quad\smile\quad} & H^{r+s}_D(\mathcal{U}, U_0) \\
{\scriptstyle \chi_D \times \chi_D}\big\downarrow\wr & & \wr\big\downarrow{\scriptstyle \chi_D} \\
H^r(M, M \smallsetminus S_1; \mathbb{C}) \times H^s(M, M \smallsetminus S_2; \mathbb{C}) & \xrightarrow{\smile} & H^{r+s}(M, M \smallsetminus S; \mathbb{C}).
\end{array}$$

7.8 Description of dualities in differential forms

We describe the Poincaré, Alexander and Lefschetz dualities with $\mathbb{C}$-coefficient using differential forms. We also discuss some related topics along the way.

Let M be a C^∞ manifold of dimension m. Also let K_0 be a C^∞ triangulation of M and let K, K' and K^* be as in Section 4.1.

Absolute case

Cap product: Let $\psi_d : A^p(M) \to C^p_K(M;\mathbb{C})$ be as in (7.25). We define the cap product

$$\check{C}^K_r(M;\mathbb{C}) \times A^p(M) \xrightarrow{\frown} \check{C}^K_{r-p}(M;\mathbb{C}) \qquad \text{by } \ c \frown \omega = c \frown \psi_d(\omega) \quad (7.50)$$

for a locally finite r-chain c of K and a p-form ω with $r \geq p$. More explicitly, we endow the set of vertices of K with a simplicial ordering and for an ordered r-simplex $\boldsymbol{s} = (\alpha_0, \dots, \alpha_r)$, $\alpha_0 < \cdots < a_r$, we set $\boldsymbol{s}_1 = (\alpha_0, \dots, \alpha_p)$ and $\boldsymbol{s}_2 = (\alpha_p, \dots, \alpha_r)$. Then

$$\boldsymbol{s} \frown \omega = \Big(\int_{\boldsymbol{s}_1} \omega\Big)\, \boldsymbol{s}_2. \qquad (7.51)$$

By (B.17), we have the cap product

$$\check{H}_r(M;\mathbb{C}) \times H^p_d(M) \xrightarrow{\frown} \check{H}_{r-p}(M;\mathbb{C}).$$

For a closed p-form ω and a closed q-form θ, we may write, by Theorem 7.9,

$$\psi_d(\omega \wedge \theta) = \psi_d(\omega) \smile \psi_d(\theta) + \delta u, \qquad u \in C^{p+q-1}_K(M;\mathbb{C}).$$

Then for an r-cycle c we compute, using (B.17) and (B.18),

$$c \frown (\omega \wedge \theta) = (c \frown \omega) \frown \theta + (-1)^{p+q}\, \partial(c \frown u). \qquad (7.52)$$

Thus on the level of homology and cohomology,

$$[c] \frown [\omega \wedge \theta] = ([c] \frown [\omega]) \frown [\theta].$$

Note that, if c is a finite chain in (7.50), so is $c \frown \omega$. Thus we have the cap product

$$C^K_r(M;\mathbb{C}) \times A^p(M) \xrightarrow{\frown} C^K_{r-p}(M;\mathbb{C}),$$

which induces

$$H_r(M;\mathbb{C}) \times H^p_d(M) \xrightarrow{\frown} H_{r-p}(M;\mathbb{C}).$$

Proposition 7.10. *Let c be a finite r-chain.*
1. *For an r-form ω,*

$$\varepsilon(c \frown \omega) = \int_c \omega,$$

where $\varepsilon : C_0^K(M;\mathbb{C}) \to \mathbb{C}$ denotes the augmentation.

2. *Let ω and θ be a closed p-form and a closed q-form, respectively. If c is a cycle and if $p+q=r$,*

$$\int_c \omega \wedge \theta = \int_{c \frown \omega} \theta.$$

Proof. The statement 1 follows from (7.51) and 2 follows from (7.52) and 1, noting that the composition $C_1^K(M;\mathbb{C}) \xrightarrow{\partial} C_0^K(M;\mathbb{C}) \xrightarrow{\varepsilon} \mathbb{C}$ is the zero map. □

We can do exactly the same things for ČdR cochains, using the morphism $\psi_D : A^r(\mathcal{U}) \to C_K^r(M;\mathbb{C})$ in (7.23) instead of ψ_d.

Recall that, if M is compact and oriented, we have the integration $\int_M : H_d^m(M) \to \mathbb{C}$ (cf. (7.12)).

Proposition 7.11. *Suppose M is compact and oriented. Then, for a class α in $H^m(M;\mathbb{C})$ and the class $[\omega]$ in $H_d^m(M)$ corresponding to α under the isomorphism of Theorem 7.5, we have*

$$\langle [M], \alpha \rangle = \int_M [\omega].$$

Proof. This follows from Proposition 7.10. 1 and (B.19). □

The right-hand side above is in fact $\int_M \omega$.

Poincaré duality: We assume that M is oriented and use the orientation conventions in Sections 4.1 and 4.2. From the Poincaré duality with $\mathbb{C}$-coefficient (cf. Theorem 4.1) and the de Rham theorem (Theorem 7.5), we have:

Theorem 7.12. *There is a canonical isomorphism:*

$$P : H_d^p(M) \xrightarrow{\sim} \check{H}_{m-p}(M;\mathbb{C}). \tag{7.53}$$

It is also referred to as the Poincaré duality. The isomorphism is induced by the composition

$$A^p(M) \xrightarrow{\psi_d^*} C_{K^*}^p(M;\mathbb{C}) \longrightarrow \check{C}_{m-p}^K(M;\mathbb{C}),$$

where the second one is (4.6) with $\mathbb{C}$-coefficient. It assigns to a p-form ω the $(m-p)$-chain

$$\sum_{\boldsymbol{s}} c_{\boldsymbol{s}} \boldsymbol{s}, \qquad c_{\boldsymbol{s}} = \int_{\boldsymbol{s}^*} \omega,$$

where $\boldsymbol{s}$ runs through all the oriented $(m-p)$-simplices of K.

In particular consider the case $p = m$. If M is compact, $\breve{H}_0(M;\mathbb{C}) = H_0(M;\mathbb{C})$ and there is the augmentation $\varepsilon_* : H_0(M;\mathbb{C}) \to \mathbb{C}$. In this case, we have

$$\varepsilon_* P([\omega]) = \int_M \omega, \tag{7.54}$$

for any closed m-form ω, which restates the equality in Proposition 7.11. If, moreover, M is connected, ε_* is an isomorphism so that (7.53) may be thought of as given by the integration

$$\int_M : H_d^m(M) \longrightarrow \mathbb{C}.$$

The isomorphism (7.53) is also induced from the composition (cf. Corollary 4.1)

$$A^p(M) \xrightarrow{\psi'_d} C^p_{K'}(M;\mathbb{C}) \xrightarrow{M\frown} \breve{C}^{K'}_{m-p}(M;\mathbb{C}), \qquad \omega \mapsto M \frown \omega.$$

Thus, from Proposition 7.10.2, we have:

Proposition 7.12. *Suppose M is compact. Then in (7.53), a class $[\omega]$ in $H_d^p(M)$ corresponds to the class of an $(m-p)$-cycle c such that*

$$\int_c \theta = \int_M \omega \wedge \theta \qquad \text{for every closed } (m-p)\text{-form } \theta \text{ on } M.$$

In the case $p = m$, we recover (7.54).

Noting that, if M is compact,

$$\breve{H}_{m-p}(M;\mathbb{C}) = H_{m-p}(M;\mathbb{C}) \simeq H^{m-p}(M;\mathbb{C})^* \simeq H_d^{m-p}(M)^*,$$

we have

Corollary 7.9. *If M is compact, the pairing*

$$A^p(M) \times A^{m-p}(M) \xrightarrow{\wedge} A^m(M) \xrightarrow{\int_M} \mathbb{C}$$

induces a non-degenerate pairing

$$H_d^p(M) \times H_d^{m-p}(M) \xrightarrow{\wedge} H_d^m(M) \xrightarrow{\int_M} \mathbb{C}$$

that defines the Poincaré duality (7.53).

These describe Proposition 4.1 and Corollary 4.2 in terms of differential forms.

Intersection product: Intersection products (cf. Section 4.4) may also be described in terms of differential forms, using Theorems 4.7 and 7.12.

We consider the case M is compact. For an $(m-p)$-cycle c and a p-cycle c' on M, we have the intersection product $[c]\cdot[c']$ in $\check{H}_0(M;\mathbb{C}) = H_0(M;\mathbb{C})$. We take a closed p-form ω and a closed $(m-p)$-form θ such that $P([\omega]) = [c]$ and $P([\omega']) = [c']$. Then we have (cf. (4.32) and Proposition 7.10)

$$\varepsilon_*([c]\cdot[c']) = \int_M \omega\wedge\omega' = \int_c \omega',$$

where $\varepsilon_* : H_0(M;\mathbb{C}) \to \mathbb{C}$ denotes the augmentation. If M is connected, we may drop ε_*.

The intersection product with a map is described as follows. Let V be a compact oriented C^∞ manifold of dimension d' and $f : V \to M$ a C^∞ map (M may not be compact). From Propositions 7.5 and 7.12 we have:

Proposition 7.13. *Let a be a class in $\check{H}_{m-p}(M;\mathbb{C})$ and ω a closed p-form on M with $P([\omega]) = a$. Then the intersection product $(V\bullet_f a)_V$ in $H_{d'-p}(V;\mathbb{C})$ is represented by a cycle c' such that*

$$\int_{c'}\varphi = \int_V f^*\omega\wedge\varphi \qquad \textit{for every closed } (d'-p)\textit{-form } \varphi \textit{ on } V.$$

In particular if $p = d'$, we have

$$\varepsilon_*(V\bullet_f a)_V = \int_V f^*\omega,$$

where $\varepsilon_* : H_0(V;\mathbb{C}) \to \mathbb{C}$ denotes the augmentation. If V is connected, we may drop ε_*.

Note that, in the case V is a submanifold of M with the inclusion $i : V \hookrightarrow M$, $(V\bullet_i a)_V = (V\cdot a)_V$ is the intersection product with V in $H_{d'-p}(V;\mathbb{C})$.

We discuss the following topic before we go further to the relative case.

Push-forward: Let $\pi : T \to M$ be an oriented fiber bundle with fiber F (cf. Section 3.7). If F is compact we have the integration along the fibers $\pi_* : A^{l'+p}(T) \to A^p(M)$, where l' is the dimension of F (cf. Section 6.3). By Proposition 6.3, if $\partial F = \emptyset$, π_* induces a $\mathbb{C}$-linear map

$$\pi_* : H_d^{l'+p}(T) \longrightarrow H_d^p(M).$$

On the other hand, if M is oriented, the total space T is also oriented and we have the push-forward $\pi_* : H^{l'+p}(T) \to H^p(M)$ as defined in (4.19).

Proposition 7.14. *In the above situation, the following diagram is commutative:*

$$\begin{array}{ccc} H_d^{l'+p}(T) & \xrightarrow{\sim} & H^{l'+p}(T;\mathbb{C}) \\ {\scriptstyle \pi_*}\downarrow & & \downarrow{\scriptstyle \pi_*} \\ H_d^p(M) & \xrightarrow{\sim} & H^p(M;\mathbb{C}), \end{array}$$

where the horizontal isomorphisms are the de Rham isomorphisms.

Exercise 7.6. Prove Proposition 7.14.

See Section 7.9 for the case $\partial F \neq \emptyset$.

Relative case

Let S be a closed set in M and $\mathcal{U} = \{U_0, U_1\}$ a covering of M with $U_0 = M \smallsetminus S$ and U_1 a neighborhood of S as before.

Cap product: Suppose that S is a subcomplex of M with respect to K_0. Then there is a honeycomb system $\{R_0, R_1\}$ adapted to $\mathcal{U}$, K' and S. Recall that there is the cap product (cf. (B.39), (4.12))

$$\check{C}_r^{K'}(S_{K'}(S), \partial S_{K'}(S);\mathbb{C}) \times C_{K'}^p(S_{K'}(S), \partial S_{K'}(S);\mathbb{C}) \xrightarrow{\frown} \check{C}_{r-p}^{K'}(S_{K'}(S);\mathbb{C}).$$

Note that $\check{C}_r^{K'}(S_{K'}(S), \partial S_{K'}(S);\mathbb{C}) = \check{C}_r^{K'}(S_{K'}(S);\mathbb{C})/\check{C}_r^{K'}(\partial S_{K'}(S);\mathbb{C})$. We denote by $\underline{c}$ the class of a locally finite chain c and refer to it as a locally finite relative chain.

We have the morphism $\psi'_D : A^p(\mathcal{U}, U_0) \to C_{K'}^p(M, M \smallsetminus O_{K'}(S);\mathbb{C})$ in (7.45). Noting that $C_{K'}^p(M, M \smallsetminus O_{K'}(S);\mathbb{C})$ may be identified with $C_{K'}^p(S_{K'}(S), \partial S_{K'}(S);\mathbb{C})$, we define the cap product

$$\check{C}_r^{K'}(S_{K'}(S), \partial S_{K'}(S);\mathbb{C}) \times A^p(\mathcal{U}, U_0) \xrightarrow{\frown} \check{C}_{r-p}^{K'}(S_{K'}(S);\mathbb{C}) \qquad (7.55)$$

by $\underline{c} \frown \xi = \underline{c} \frown \psi'_D(\xi)$ for a locally finite relative r-chain $\underline{c}$ and a relative p-ČdR cochain ξ.

To express it more explicitly, endow the set of vertices I of K' a simplicial ordering and for an ordered r-simplex $\boldsymbol{t} = (\alpha_0, \dots, \alpha_r)$, $\alpha_0 < \cdots < a_r$, we set $\boldsymbol{t}_1 = (\alpha_0, \dots, \alpha_p)$ and $\boldsymbol{t}_2 = (\alpha_p, \dots, \alpha_r)$. For $\underline{c}$ as above, we write $c = \sum a_{\boldsymbol{t}}\, \boldsymbol{t}$, locally finite sum with $\boldsymbol{t}$ ordered r-simplices and $a_{\boldsymbol{t}}$ complex numbers. Then, for a relative p-ČdR cochain $\xi = (\xi_1, \xi_{01})$,

$$\underline{c} \frown \xi = \sum a_{\boldsymbol{t}} \Big(\int_{\boldsymbol{t}_1 \cap R_1} \xi_1 + \int_{\boldsymbol{t}_1 \cap R_{01}} \xi_{01} \Big) \cdot \boldsymbol{t}_2. \qquad (7.56)$$

Note that the right-hand side does not depend on the representative $c = \sum a_t \boldsymbol{t}$, since, if $\boldsymbol{t}$ is in $\partial S_{K'}(S)$, then $\boldsymbol{t}_1 \cap R_1 = \emptyset$.

By (B.17), the product (7.55) represents the cap product

$$\check{H}_r(S_{K'}(S), \partial S_{K'}(S); \mathbb{C}) \times H^p_D(\mathcal{U}, U_0) \xrightarrow{\frown} \check{H}_{r-p}(S; \mathbb{C}).$$

We apply Theorem 7.11 with $M = U_1$ (cf. Remark 7.16). For a relative p-ČdR coycle ξ and a closed q-form η_1 on U_1, we may write

$$\psi'_D(\xi \smile \eta_1) = \psi'_D(\xi) \smile \psi'_d(\eta_1) + \delta u, \qquad u \in C^{p+q-1}_{K'}(U_1, U_1 \smallsetminus O_{K'}(S); \mathbb{C}).$$

Then for a relative r-cycle $\underline{c}$ in $\check{C}^{K'}_r(S_{K'}(S), \partial S_{K'}(S); \mathbb{C})$, we compute, using (B.17) and (B.18),

$$\underline{c} \frown (\xi \smile \eta_1) = (\underline{c} \frown \xi) \frown \eta_1 + (-1)^{p+q}\, \partial(\underline{c} \frown u). \tag{7.57}$$

Thus on the level of homology and cohomology,

$$[\underline{c}] \frown ([\xi] \smile [\eta_1]) = ([\underline{c}] \frown [\xi]) \frown [\eta_1].$$

Note that, if c is a finite chain in $C^{K'}_r(S_{K'}(S); \mathbb{C})$ and ξ is a relative p-ČdR cochain, then $\underline{c} \frown \xi$ is a finite chain in $C^{K'}_{r-p}(S_{K'}(S); \mathbb{C})$. In particular, if $p = r$, writing $c = \sum a_t \boldsymbol{t}$, we have the integration (cf. (7.22), Remark 7.10)

$$\int_c \xi = \sum a_t \Big(\int_{\boldsymbol{t} \cap R_1} \xi_1 + \int_{\boldsymbol{t} \cap R_{01}} \xi_{01} \Big).$$

The following is a relative version of Proposition 7.10:

Proposition 7.15. *Let c be a finite r-chain in $C^{K'}_r(S_{K'}(S); \mathbb{C})$.*

1. *For a relative r-ČdR cochain ξ,*

$$\varepsilon(\underline{c} \frown \xi) = \int_c \xi,$$

where $\varepsilon : C^{K'}_0(S_{K'}(S); \mathbb{C}) \to \mathbb{C}$ denotes the augmentation.

2. *Let ξ be a relative p-ČdR cocycle and η_1 a closed q-form on U_1. If $\underline{c}$ is a relative cycle and if $p + q = r$,*

$$\int_c \xi \smile \eta_1 = \int_{\underline{c} \frown \xi} \eta_1.$$

Proof. The statement 1 follows from (7.56) and 2 follows from (7.57) and 1, noting that the composition $C^{K'}_1(S_{K'}(S); \mathbb{C}) \xrightarrow{\partial} C^{K'}_0(S_{K'}(S); \mathbb{C}) \xrightarrow{\varepsilon} \mathbb{C}$ is the zero map. □

Alexander duality: We now assume that M is oriented and use the orientation conventions in Sections 4.1 and 4.2. From the Alexander duality with $\mathbb{C}$-coefficient (cf. Theorem 4.2) and the relative Čech-de Rham theorem (Corollary 7.8), we have:

Theorem 7.13. *If S is a subcomplex with respect to a triangulation K_0 of M, there is a canonical isomorphism:*

$$A : H^p_D(\mathcal{U}, U_0) \xrightarrow{\sim} \check{H}_{m-p}(S; \mathbb{C}). \tag{7.58}$$

It is also referred to as the Alexander duality. Let K, K' and K^* be as before and $\{R_0, R_1\}$ a honeycomb system adapted to $\mathcal{U}$, K' and S (cf. Definitions 7.9 and 7.12). Then the isomorphism is induced from the composition

$$A^p(\mathcal{U}, U_0) \xrightarrow{\psi^*_D} C^p_{K^*}(M, M \smallsetminus O_{K'}(S); \mathbb{C}) \longrightarrow \check{C}^K_{m-p}(S; \mathbb{C}),$$

where the first one is defined as in (7.46) and the second one is (4.10) with $\mathbb{C}$-coefficient. It assigns to a relative p-ČdR cochain $\xi = (\xi_1, \xi_{01})$ the chain

$$\sum_{\boldsymbol{s}} c_{\boldsymbol{s}}\, \boldsymbol{s}, \qquad c_{\boldsymbol{s}} = \int_{\boldsymbol{s}^* \cap R_1} \xi_1 + \int_{\boldsymbol{s}^* \cap R_{01}} \xi_{01}, \tag{7.59}$$

where $\boldsymbol{s}$ runs through all the oriented $(m-p)$-simplices of K in S.

In particular consider the case $p = m$. If S is compact, $\check{H}_0(S; \mathbb{C}) = H_0(S; \mathbb{C})$ and there is the augmentation $\varepsilon_* : H_0(S; \mathbb{C}) \to \mathbb{C}$. In this case, since the m-cells $\boldsymbol{s}^*$ appearing in (7.59) cover $S_{K'}(S)$, we have

$$\varepsilon_* A([\xi]) = \int_{S_{K'}(S)} \xi = \int_{R_1} \xi_1 + \int_{R_{01}} \xi_{01}, \tag{7.60}$$

which coincides with the integral $\int_M \xi$ in (7.42). If, moreover, S is connected, ε_* is an isomorphism and we have:

Proposition 7.16. *If S is compact and connected, in the Alexander duality*

$$H^m_D(\mathcal{U}, U_0) \xrightarrow{\sim} H_0(S; \mathbb{C}) \simeq \mathbb{C},$$

the class of $\xi = (\xi_1, \xi_{01})$ is sent to the complex number $\int_M \xi$.

The isomorphism (7.58) is also induced from the composition (cf. (4.11))

$$A^p(\mathcal{U}, U_0) \xrightarrow{\psi'_D} C^p_{K'}(S_{K'}(S), \partial S_{K'}(S); \mathbb{C}) \longrightarrow \check{C}^{K'}_{m-p}(S_{K'}(S); \mathbb{C}),$$

where the second map is the cap product with the relative cycle $S_{K'}(S)$. Note that for a particular choice of ordering of the vertices of K', the receiving group can be $\check{C}^{K'}_{m-p}(S; \mathbb{C})$.

Suppose that S is compact. In this case $\underline{S_{K'}(S)}$ is a finite relative cycle and letting c be $S_{K'}(S)$ in Proposition 7.15. 2, we have:

Proposition 7.17. *Suppose S is compact. Then in* (7.58), *a class* $[\xi]$ *in* $H^p_D(\mathcal{U}, U_0)$ *corresponds to the class of an* $(m-p)$*-cycle* c *in* S *such that*

$$\int_c \eta_1 = \int_{R_1} \xi_1 \wedge \eta_1 + \int_{R_{01}} \xi_{01} \wedge \eta_1$$

for every closed $(m-p)$*-form* η_1 *on* U_1.

In the case $p = m$, we recover (7.60).

If S is compact and if U_1 has the same homotopy type as S (for example, the open star $O_K(S)$ of S in K), we have

$$\check{H}_{m-p}(S;\mathbb{C}) = H_{m-p}(S;\mathbb{C}) \simeq H_{m-p}(U_1;\mathbb{C}) \simeq H^{m-p}_d(U_1)^*.$$

Corollary 7.10. *If S is compact and if U_1 has the same homotopy type as S, the pairing* (*cf. Remark 7.16*)

$$A^p(\mathcal{U}, U_0) \times A^{m-p}(U_1) \overset{\smile}{\longrightarrow} A^m(\mathcal{U}, U_0) \xrightarrow{\int_M} \mathbb{C}$$

induces a non-degenerate pairing

$$H^p_D(\mathcal{U}, U_0) \times H^{m-p}_d(U_1) \overset{\smile}{\longrightarrow} H^m_D(\mathcal{U}, U_0) \xrightarrow{\int_M} \mathbb{C}$$

that defines the Alexander duality (7.58).

Remark 7.17. Let M and S be as above. If a stratification of S is given, then, instead of the covering as above, we could take a covering as described in Example 7.2. 3 to obtain more "precise" results.

Before we proceed to the Lefschetz duality, we consider the following situation. Let R be an m-dimensional compact manifold possibly with boundary in M. We may assume that R and ∂R are K_0-subcomplex of M. By setting $S = R$ in the above, we have the integration, which we denote by $\int_{\underline{R}}$:

$$\int_{\underline{R}} : H^m_D(\mathcal{U}, U_0) \longrightarrow \mathbb{C}. \tag{7.61}$$

As noted above, it coincides with the integration (7.43) for $S = R$. We also have a canonical isomorphism (cf. Corollary 7.8):

$$H^p_D(\mathcal{U}, U_0) \simeq H^p(R, \partial R; \mathbb{C}).$$

If R is connected, the integration morphism (7.61) is an isomorphism, which is a special case of the Lefschetz duality with $\mathbb{C}$-coefficent (cf. Proposition 7.19 below).

In this situation, we have a generalization of Proposition 7.11:

Proposition 7.18. *Suppose M is oriented and R is compact. Then, for a class α in $H^m(R, \partial R; \mathbb{C})$ and the class $[\xi]$ in $H^m_D(\mathcal{U}, U_0)$ corresponding to α under the above isomorphism, we have*

$$\langle [\underline{R}], \alpha \rangle = \int_{\underline{R}} [\xi].$$

Proof. This follows from Proposition 7.15.1 and (B.19). □

Remark 7.18. In the equality above, the right-hand side is given by $\int_R \xi = \int_{R_1} \xi_1 + \int_{R_{01}} \xi_{01}$ with $\{R_0, R_1\}$ a honeycomb system adapted to $\mathcal{U}$. If ξ_{01} is defined on a neighborhood of ∂R, by the Stokes theorem, we may write (cf. (7.44))

$$\int_{\underline{R}} [\xi] = \int_R \xi_1 - \int_{\partial R} \xi_{01}.$$

Lefschetz duality: Suppose M is oriented. If R is an m-dimensional manifold with boundary in M, the Lefschetz isomorphism with $\mathbb{C}$-coefficient (cf. Theorem 4.3)

$$L : H^p_D(\mathcal{U}, U_0) \xrightarrow{\sim} \breve{H}_{m-p}(R; \mathbb{C}) \tag{7.62}$$

is obtained by setting $S = R$ in Theorem 7.13. Thus it is induced from the morphism assigning to $\xi = (\xi_1, \xi_{01})$, the cycle as given in (7.59).

In particular, in the case $p = m$ and R is compact, we have an expression as (7.60):

$$\varepsilon_* L([\xi]) = \int_R \xi \tag{7.63}$$

for any m-ČdR cocycle ξ, which restates the equality in Proposition 7.18. If, moreover, R is connected, we have a statement as in Proposition 7.16 for the Lefschetz duality:

Proposition 7.19. *If R is compact and connected, in the Lefschetz duality*

$$H^m_D(\mathcal{U}, U_0) \xrightarrow{\sim} H_0(R; \mathbb{C}) \simeq \mathbb{C},$$

the class of $\xi = (\xi_1, \xi_{01})$ is sent to $\int_R \xi$.

If R is compact, we have statements corresponding to Proposition 7.17 and Corollary 7.10, setting $S = R$.

Remark 7.19. With the expressions as above of the duality theorems, we have a diagram similar to the one in Proposition 4.4 with $\mathbb{C}$-coefficient.

Example 7.3. We describe the Alexander duality for the pair $(\mathbb{R}^m, \{0\})$ in terms of relative Čech-de Rham cohomology. By the Alexander duality,

$$H^p(\mathbb{R}^m, \mathbb{R}^m \setminus 0; \mathbb{C}) \simeq H_{m-p}(\{0\}; \mathbb{C}) = \begin{cases} \mathbb{C} & p = m, \\ 0 & p \neq m, \end{cases}$$

where we denote $\mathbb{R}^m \setminus \{0\}$ simply by $\mathbb{R}^m \setminus 0$. Let $\mathcal{U} = \{U_0, U_1\}$ be the covering of $\mathbb{R}^m$ given by $U_0 = \mathbb{R}^m \setminus 0$ and $U_1 = \mathbb{R}^m$. Then we have $H^p(\mathbb{R}^m, \mathbb{R}^m \setminus 0; \mathbb{C}) \simeq H^p_D(\mathcal{U}, U_0)$. Also, let R_1 be the closed unit ball with center 0 and $R_0 = \mathbb{R}^m \setminus \operatorname{Int} R_1$ so that $R_{01} = -\partial R_1 = -\mathbb{S}^{m-1}$. Any class of $H^m(\mathbb{R}^m, \mathbb{R}^m \setminus 0; \mathbb{C})$ is represented by a cocycle $\xi = (\xi_1, \xi_{01})$ in $A^m(\mathcal{U}, U_0)$ and the isomorphism $H^m(\mathbb{R}^m, \mathbb{R}^m \setminus 0; \mathbb{C}) \xrightarrow{\sim} \mathbb{C}$ is given by $[\xi] \mapsto \int_{R_1} \xi_1 + \int_{R_{01}} \xi_{01}$.

We may make the expression more explicit. First, as $H^p(\mathbb{R}^m; \mathbb{C}) = 0$ for $p \geq 1$, we have the exact sequence:

$$0 \longrightarrow H^{m-1}(\mathbb{R}^m; \mathbb{C}) \xrightarrow{i^*} H^{m-1}(\mathbb{R}^m \setminus 0; \mathbb{C}) \xrightarrow{\delta^*} H^m(\mathbb{R}^m, \mathbb{R}^m \setminus 0; \mathbb{C}) \longrightarrow 0,$$

where $i : \mathbb{R}^m \setminus 0 \hookrightarrow \mathbb{R}^m$ is the inclusion.

(1) The case $m = 1$. We have $H^0(\mathbb{R}; \mathbb{C}) \simeq \mathbb{C}$, $H^1(\mathbb{R}, \mathbb{R} \setminus 0; \mathbb{C}) \simeq \mathbb{C}$ and $H^0(\mathbb{R} \setminus 0; \mathbb{C}) \simeq \mathbb{C}^2$ and each element in $H^0(\mathbb{R} \setminus 0; \mathbb{C}) \simeq H^0_d(U_0)$ is represented by a locally constant function f on $\mathbb{R} \setminus 0$, which is written $f = (c_+, c_-)$ with c_+ and c_-, respectively, the values of f on the positive and negative parts. Thus every element in $H^1(\mathbb{R}, \mathbb{R} \setminus 0; \mathbb{C})$ is represented by an element of the form $(0, -f)$ with f as above. It corresponds to $-\int_{R_{01}} f = c_+ - c_-$ by the isomorphism $H^1(\mathbb{R}, \mathbb{R} \setminus 0; \mathbb{C}) \simeq \mathbb{C}$.

(2) The case $m > 1$. We have $H^{m-1}(\mathbb{R}^m; \mathbb{C}) = 0$ and δ^* is an isomorphism. It assigns to the class $[\theta]$ in $H^{m-1}(\mathbb{R}^m \setminus 0; \mathbb{C}) \simeq H^{m-1}_d(\mathbb{R}^m \setminus 0)$ of a closed $(m-1)$-form θ on $\mathbb{R}^m \setminus 0$ the class of $(0, -\theta)$ in $H^m(\mathbb{R}^m, \mathbb{R}^m \setminus 0; \mathbb{C})$. Thus every class $[\xi]$ in $H^m(\mathbb{R}^m, \mathbb{R}^m \setminus 0; \mathbb{C})$ is represented by a cocycle of this form and the isomorphism $H^m(\mathbb{R}^m, \mathbb{R}^m \setminus 0; \mathbb{C}) \xrightarrow{\sim} \mathbb{C}$ is given by

$$[\xi] \mapsto \int_{\mathbb{S}^{m-1}} \theta.$$

Localized intersection product: Localized intersection products (cf. Section 4.4) may be described in terms of differential forms, using Theorems 4.8 and 7.13. Here we discuss the case of intersection product with a map (cf. Definition 4.4, where the homology and cohomology are with $\mathbb{Z}$-coefficient).

Let M, S and $\mathcal{U}$ be as before and suppose S is a subcomplex of M. Let V be an oriented C^∞ manifold of dimension d' and $f : V \to M$ a C^∞ map. We set $Z = f^{-1}(S)$ and let $\mathcal{W} = \{W_0, W_1\}$ be a covering of V with $W_0 = V \smallsetminus Z$ and W_1 a neighborhood of Z in V such that $f(W_1) \subset U_1$. Suppose that Z is a subcomplex of V with respect to some triangulation L_0 of V and let L' be the second barycentric subdivision of L_0. Then there is a honeycomb system $\{Q_0, Q_1\}$ adapted to $\mathcal{W}$, L' and Z (cf. Definitions 7.9 and 7.12). For example, we may set $Q_1 = S_{L''}(Z)$ and $Q_0 = V \smallsetminus \operatorname{Int} Q_1$.

For a class c in $\check{H}_{m-p}(S;\mathbb{C})$, we have the localized intersection product $(V \bullet_f c)_Z$ in $\check{H}_{d'-p}(Z;\mathbb{C})$.

From Propositions 7.8 and 7.17 we have:

Proposition 7.20. *In the above situation, suppose that Z is compact. Let a be a class in $\check{H}_{m-p}(S;\mathbb{C})$ and $\xi = (\xi_1, \xi_{01})$ a cocycle in $A^p(\mathcal{U}, U_0)$ with $A([\xi]) = a$. Then the localized intersection product $(V \bullet_f a)_Z$ in $H_{d'-p}(Z;\mathbb{C})$ is represented by a cycle c' in Z such that*

$$\int_{c'} \varphi = \int_{Q_1} f^*\xi_1 \wedge \varphi + \int_{Q_{01}} f^*\xi_{01} \wedge \varphi$$

for every closed $(d'-p)$-form φ on W_1.

In particular if $p = d'$, we have

$$\varepsilon_*(V \bullet_f a)_Z = \int_{Q_1} f^*\xi_1 + \int_{Q_{01}} f^*\xi_{01},$$

where $\varepsilon_* : H_0(Z;\mathbb{C}) \to \mathbb{C}$ denotes the augmentation.

Suppose Z has a finite number of connected components $(Z_\lambda)_\lambda$. Then $H_{d'-p}(Z;\mathbb{C}) = \bigoplus_\lambda H_{d'-p}(Z_\lambda;\mathbb{C})$ and, for a class a in $\check{H}_{m-p}(S;\mathbb{C})$, the class $(V \bullet_f a)_Z$ determines a class $(V \bullet_f a)_{Z_\lambda}$ in $H_{d'-p}(Z_\lambda;\mathbb{C})$ for each λ. For each λ, we take an open neighborhood W_λ of Z_λ in W_1 so that $W_\lambda \cap W_\mu = \emptyset$ if $\lambda \neq \mu$. We may assume that $Q_1 \subset \bigcup_\lambda W_\lambda$ and set $Q_\lambda = Q_1 \cap W_\lambda$ and $Q_{\lambda 0} = Q_{10} \cap W_\lambda (= -\partial Q_\lambda)$. Then, letting ξ be as in Proposition 7.20, the localized intersection product $(V \bullet_f a)_{Z_\lambda}$ in $H_{d'-p}(Z_\lambda;\mathbb{C})$ is represented by a cycle c'_λ such that

$$\int_{c'_\lambda} \varphi = \int_{Q_\lambda} f^*\xi_1 \wedge \varphi + \int_{Q_{0\lambda}} f^*\xi_{01} \wedge \varphi$$

for every closed $(d'-p)$-form φ on W_λ.

In particular if $p = d'$, $H_0(Z_\lambda;\mathbb{C}) \simeq \mathbb{C}$ and $(V \bullet_f a)_{Z_\lambda}$ may be thought of as a number given by

$$(V \bullet_f a)_{Z_\lambda} = \int_{Q_\lambda} f^*\xi_1 + \int_{Q_{0\lambda}} f^*\xi_{01}. \tag{7.64}$$

7.9 Thom class in Čech-de Rham cohomology

The Thom isomorphism and the Thom class are introduced in Section 4.3 from the topological viewpoint in cohomology with $\mathbb{Z}$-coefficient. In this section we express them in terms of differential forms via the relative Čech-de Rham theorem (Corollary 7.8). Thus the cohomologies involved here are with $\mathbb{C}$-coefficient.

Throughout this section, M denotes a C^∞ manifold of dimension m.

Case of submanifolds

Let V be a closed submanifold of dimension d' of M. We set $k' = m - d'$. Assuming that the normal bundle of V in M is oriented, we have the Thom isomorphism (Theorem 4.4). It also holds with $\mathbb{C}$-coefficient and there is an isomorphism:

$$T_V^c : H^p(V;\mathbb{C}) \xrightarrow{\sim} H^{k'+p}(M, M \smallsetminus V;\mathbb{C}).$$

We set $\Psi_V^c = T_V^c([1])$ and call it the *complexified Thom class* of V. The class Ψ_V^c is the image of the Thom class Ψ_V in Definition 4.2 by the canonical morphism $H^{k'}(M, M \smallsetminus V;\mathbb{Z}) \to H^{k'}(M, M \smallsetminus V;\mathbb{C})$.

Letting $U_0 = M \smallsetminus V$ and U_1 a neighborhood of V, we consider the covering $\mathcal{U} = \{U_0, U_1\}$ of M. Then we have a canonical isomorphism (cf. Corollary 7.8):

$$H^{k'}(M, M \smallsetminus V;\mathbb{C}) \simeq H_D^{k'}(\mathcal{U}, U_0).$$

The class in $H_D^{k'}(\mathcal{U}, U_0)$ corresponding to Ψ_V^c is also denoted by Ψ_V^c. Let K_0, K, K', K^* and K_V^* be as in Section 4.3. If we choose a honeycomb system $\{R_0, R_1\}$ so that it is adapted to $\mathcal{U}$, K' and V (cf. Definitions 7.9 and 7.12), then by Proposition 4.5, the class Ψ_V^c is represented by a cocycle (ψ_1, ψ_{01}) in $A^{k'}(\mathcal{U}, U_0)$ such that, for each oriented k'-cell $\boldsymbol{s}^*$ forming a basis of $C_{k'}^{K^*}(M;\mathbb{C})$,

$$\int_{\boldsymbol{s}^*\cap R_1} \psi_1 + \int_{\boldsymbol{s}^*\cap R_{01}} \psi_{01} = \begin{cases} 1 & \text{if } \boldsymbol{s}^* \cap V \neq \emptyset, \\ 0 & \text{otherwise} \end{cases} \tag{7.65}$$

(cf. (7.46)). In fact the above characterizes the Thom class.

Remark 7.20. In the case M and V are oriented, we have a commutative diagram as in Proposition 4.9 with $\mathbb{C}$-coefficient and we have $A(\Psi_V^c) = [V]$ in $\check{H}_{d'}(V;\mathbb{C})$ (cf. (4.24)). Since $\check{H}_{d'}(V;\mathbb{Z}) \to \check{H}_{d'}(V;\mathbb{C})$ is injective, so is $H^{k'}(M, M \smallsetminus V;\mathbb{Z}) \to H^{k'}(M, M \smallsetminus V;\mathbb{C})$. Thus we may identify Ψ_V^c with Ψ_V and denote Ψ_V^c by Ψ_V in this case.

If we choose U_1 to be a tubular neighborhood of V with a C^∞ deformation retraction $r : U_1 \to V$, then we have the composition of isomorphisms

$$H^p(V;\mathbb{C}) \simeq H_d^p(V) \xrightarrow[r^*]{\sim} H_d^p(U_1),$$

where the first one is the de Rham isomorphism (Theorem 7.5). We may express the Thom isomorphism as (cf (4.25) and Remark 4.11)

$$T_V^c : H_d^p(V) \xrightarrow{\sim} H_D^{k'+p}(\mathcal{U}, U_0), \qquad T_V^c([\theta]) = \Psi_V^c \smile r^*([\theta]), \tag{7.66}$$

where $\Psi_V^c \smile r^*([\theta])$ is represented by $(\psi_1 \wedge r^*\theta, \psi_{01} \wedge r^*\theta)$ (cf. Remark 7.16).

Case of oriented vector bundles

Let $\pi : E \to M$ be a C^∞ oriented real vector bundle of rank l'. We denote by Σ the image of the zero section $s_0 : M \to E$. Let K_E be a triangulation of the total space E compatible with Σ. We identify M with Σ by s_0 and denote by K the triangulation of M induced by K_E. If we denote by K_E^* and K^* the cellular decompositions of E and M dual to K_E and K, the Thom isomorphism with $\mathbb{C}$-coefficient

$$T_E^c : H^p(M;\mathbb{C}) \xrightarrow{\sim} H^{l'+p}(E, E \smallsetminus \Sigma;\mathbb{C})$$

is induced from the isomorphism

$$T_E^c : C_{K^*}^p(M;\mathbb{C}) \xrightarrow{\sim} C_{K_E^*}^{l'+p}(E, E \smallsetminus O_{K_E'}(\Sigma);\mathbb{C}), \qquad u \mapsto u_E,$$

where u_E is given by, for each $(m-p)$-simplex $\boldsymbol{s}$ in K_E,

$$\langle \boldsymbol{s}_E^*, u_E \rangle = \begin{cases} \langle \boldsymbol{s}^*, u \rangle & \text{if } \ \boldsymbol{s} \subset M, \\ 0 & \text{otherwise.} \end{cases}$$

In the above, $\boldsymbol{s}_E^*$ and $\boldsymbol{s}^*$ are the dual cells of $\boldsymbol{s}$ in K_E^* and K^*, respectively, so that $\boldsymbol{s}^* = \boldsymbol{s}_E^* \cap M$ (cf. (4.20)). The *complexified Thom class* Ψ_E^c of E is defined by $\Psi_E^c = T_E^c([1])$. It is the image of Ψ_E by the canonical morphism $H^{l'}(E, E \smallsetminus \Sigma;\mathbb{Z}) \to H^{l'}(E, E \smallsetminus \Sigma;\mathbb{C})$.

We show that, if we use the relative Čech-de Rham cohomology, the inverse of the Thom isomorphism in given by integration along fibers. Thus we set $W_0 = E \smallsetminus \Sigma$. Letting $W_1 = E$, in fact we may take arbitrary neighborhood of Σ in E as W_1 by Corollary 7.7, we consider the covering $\mathcal{W} = \{W_0, W_1\}$ of (the total sapce of) E. Let T_1 be a closed ball bundle in W_1, with respect to some Riemannian metric on E (cf. Remark 9.1 below), and let $T_0 = E \smallsetminus \operatorname{Int} T_1$. Then $\{T_0, T_1\}$ is a honeycomb system adapted to $\mathcal{W}$. We denote by π_1 and π_{01}, respectively, the restrictions of π to T_1 and T_{01}. Thus $\pi_1 : T_1 \to M$ is a closed l'-ball bundle and $\pi_{01} : T_{01} \to M$ an $(l'-1)$-sphere bundle. Note that the orientation of T_{01} is opposite to that of the boundary of T_1.

Definition 7.13. The *fiber integration* on ČdR cochains

$$\pi_* : A^{l'+p}(\mathcal{W}, W_0) \longrightarrow A^p(M)$$

is defined by

$$\pi_* \xi = (\pi_1)_* \xi_1 + (\pi_{01})_* \xi_{01} \qquad \text{for } \xi = (\xi_1, \xi_{01}).$$

From Proposition 6.2, we have the following:

Proposition 7.21 (Projection formula II). *In the above situation,*

1. *For ξ in $A^{l'+p}(\mathcal{W}, W_0)$ and θ in $A^q(M)$,*

$$\pi_*(\xi \smile \pi^*\theta) = \pi_*\xi \wedge \theta,$$

where $\pi^*\theta$ is considered as an element in $A^q(W_1)$.

2. *If M is compact and oriented, for ξ in $A^{l'+p}(\mathcal{W}, W_0)$ and θ in $A^{m-p}(M)$,*

$$\int_E \xi \smile \pi^*\theta = \int_M \pi_*\xi \wedge \theta.$$

Also, using Proposition 6.3 and noting that $(\partial\pi_1)_* = -(\pi_{01})_*$, we have:

$$\pi_* \circ D + (-1)^{l'+1} d \circ \pi_* = 0. \tag{7.67}$$

From this we see that the fiber integration induces a morphism

$$\pi_* : H^{l'+p}(E, E \smallsetminus \Sigma; \mathbb{C}) \simeq H_D^{l'+p}(\mathcal{W}, W_0) \longrightarrow H_d^p(M) \simeq H^p(M; \mathbb{C}). \tag{7.68}$$

Theorem 7.14. *The morphism (7.68) is an isomorphism. In fact, it is the inverse of the Thom isomorphism T_E^c with $\mathbb{C}$-coefficient.*

Proof. By Proposition 7.21. 2, we see that the following diagram is commutative:

$$\begin{array}{ccc}
A^{l'+p}(\mathcal{W}, W_0) & \xrightarrow{\quad \pi_* \quad} & A^p(M) \\
{\scriptstyle \psi_D}\downarrow & & \downarrow{\scriptstyle \psi_d} \\
C^{l'+p}_{K_E^*}(E, E \smallsetminus O_{K'_E}(\Sigma); \mathbb{C}) & \xleftarrow{\;T_E^c\;} & C^p_{K^*}(M; \mathbb{C}).
\end{array}$$

□

From Proposition 7.21.1, we recover the expression (4.22) in terms of differential forms:

$$T_E^c(\alpha) = \Psi_E^c \smile \pi^*(\alpha) \qquad \text{for } \alpha \in H^p(M;\mathbb{C}). \tag{7.69}$$

Thus, if Ψ_E^c is represented by a cocycle $(\psi_1, \psi_{01}) \in A^{l'}(\mathcal{W}, W_0)$, T_E^c is induced in cohomology by the map $\theta \mapsto (\psi_1 \wedge \pi^*\theta, \psi_{01} \wedge \pi^*\theta)$ from $A^p(M)$ into $A^{l'+p}(\mathcal{W}, W_0)$ (cf. (7.66)).

We now describe the Thom class in terms of Čech-de Rham cohomology. First note that, from Theorem 7.14, we have:

Corollary 7.11. *A class Ψ in $H^{l'}(\mathcal{W}, W_0)$ coincides with the Thom class Ψ_E^c if and only if*

$$\pi_*\Psi = 1.$$

Proposition 7.22. *The Thom class Ψ_E^c in $H^{l'}(\mathcal{W}, W_0)$ is represented by a cocycle in $A^{l'}(\mathcal{W}, W_0)$ of the form*

$$(\pi^*\epsilon, -\psi),$$

where ϵ is a closed l'-form on M and ψ is an $(l'-1)$-form on W_{01} such that $d\psi = -\pi^*\epsilon$ in W_{01} and $-(\pi_{01})_*\psi = 1$.

Proof. Suppose Ψ_E^c is represented by a cocycle (ψ_1, ψ_{01}) in $A^{l'}(\mathcal{W}, W_0)$. Then, from $0 = D(\psi_1, \psi_{01}) = (d\psi_1, \psi_1 - d\psi_{01})$, we see that ψ_1 is a closed l'-form on W_1 and ψ_{01} is an $(l'-1)$-form on W_{01} such that $d\psi_{01} = \psi_1$. Since π induces an isomorphism between the de Rham cohomologies of W_1 and M, we have $\psi_1 = \pi^*\epsilon + d\varphi_1$ for some closed l'-form ϵ on M and an $(l'-1)$-form φ_1 on W_1. If we set $\psi = -\psi_{01} + \varphi_1$, then it is an $(l'-1)$-form on W_{01} and from $(\psi_1, \psi_{01}) = (\pi^*\epsilon, -\psi) + D(\varphi_1, 0)$, we see that Ψ_E^c is represented by $(\pi^*\epsilon, -\psi)$. We have $d\psi = -\pi^*\epsilon$ in W_{01} and, moreover, from $(\pi_1)_*\psi_1 + (\pi_{01})_*\psi_{01} = 1$, we have $-(\pi_{01})_*\psi = 1$. □

The form ψ above is called a *global angular form.* In particular, if M is a point, then $E = \mathbb{R}^{l'}$ and we have

$$H_D^{l'}(\mathcal{W}, W_0) = H^{l'}(\mathbb{R}^{l'}, \mathbb{R}^{l'} \smallsetminus \{0\}; \mathbb{C}) \simeq \mathbb{C}.$$

The Thom class Ψ_E^c is then represented by a cocycle $(0, -\psi)$ with

$$\int_{\mathbb{S}^{l'-1}} \psi = 1,$$

i.e., ψ is an *angular form* (cf. Section 7.10 below).

In the diagram (4.23) with $\mathbb{C}$-coefficients, we see that the global angular form ψ restricts to an angular form on each fiber E_x and we recover Proposition 4.7 using differential forms.

We readily have:

Proposition 7.23. *Let $E = \mathbb{R}^{l'} \times M$ be the product bundle and $\rho : E \to \mathbb{R}^{l'}$ the projection onto the fiber direction. Then the Thom class $\Psi_E^{\mathbb{C}}$ is represented by a cocycle in $A^{l'}(\mathcal{W}, W_0)$ of the form*

$$(0, -\rho^*\psi),$$

where ψ is an angular form on $\mathbb{R}^{l'} \smallsetminus \{0\}$.

If $E = \mathbb{C}^l \times M$, $\Psi_E^{\mathbb{C}}$ is represented by a cocycle $(0, -\rho^*\beta_l)$ with β_l the Bochner-Martinelli form on $\mathbb{C}^l$ (cf. Sections 7.10 and 10.4 below).

The Euler class in terms of differential forms naturally arises in this context:

Proposition 7.24. *The form ϵ in Proposition 7.22 represents the Euler class $e(E)$ in $H_d^{l'}(M) \simeq H^{l'}(M; \mathbb{C})$.*

Proof. The class of ϵ in $H_d^{l'}(M)$ corresponds, by π^*, to the class of $H^{l'}(\mathcal{W})$ represented by $(\pi^*\epsilon, \pi^*\epsilon, 0)$. In view of (5.21), the proposition follows from $(0, \pi^*\epsilon, -\psi) - (\pi^*\epsilon, \pi^*\epsilon, 0) = D(\psi, 0, 0)$ in $A^{l'}(\mathcal{W})$. □

Proposition 7.25. *The Euler class $e(E)$ vanishes if and only if there is a closed $(l-1)$-form ψ' on W_{01} such that $\Psi_E^{\mathbb{C}}$ is represented by $(0, -\psi')$.*

Proof. If $e(E) = 0$, there is an $(l'-1)$-form η on M such that $\epsilon = d\eta$. Then $(\pi^*\epsilon, -\psi) = (0, -\psi') + D(\pi^*\eta, 0)$, where $\psi' = \psi + \pi^*\eta$. The converse is obvious. □

In particular, if E is trivial, we have $e(E) = 0$. More generally, we recover the following:

Proposition 7.26. *If E admits a non-vanishing section, then $e(E) = 0$.*

Proof. A non-vanishing section s can be thought of as a section of $W_{01} = E \smallsetminus \Sigma$. Since $\pi^*\epsilon = -d\psi$ in W_{01}, we have $\epsilon = -d\, s^*\psi$ on M. □

7.10 Angular form and Bochner-Martinelli form

We give an explicit closed $(m-1)$-form generating $H_d^{m-1}(\mathbb{R}^m \smallsetminus \{0\}) \simeq H^{m-1}(\mathbb{S}^{m-1}; \mathbb{C}) \simeq \mathbb{C}$. As we have seen, it is in the core of the Thom class of a real oriented vector bundle. In the complex case, we have the Bochner-Martinelli form on $\mathbb{C}^n \smallsetminus \{0\}$, which is a closed $(2n-1)$-form generating $H_d^{2n-1}(\mathbb{C}^n \smallsetminus \{0\}) \simeq H^{2n-1}(\mathbb{S}^{2n-1};\mathbb{C}) \simeq \mathbb{C}$. In fact it is a $\bar\partial$-closed $(n,n-1)$-form. As we see later, it is in the core of the Thom class of a complex vector bundle (cf. Section 10.4).

Consider the forms on $\mathbb{R}^m = \{(x_1,\dots,x_m)\}$ given by

$$\Phi(x) = dx_1 \wedge \cdots \wedge dx_m, \quad \Phi_i(x) = (-1)^{i-1} x_i\, dx_1 \wedge \cdots \wedge \widehat{dx_i} \wedge \cdots \wedge dx_m$$

so that $d\Phi_i(x) = \Phi(x)$. Also let C'_m be the constant given by

$$C'_m = \begin{cases} \frac{(k-1)!}{2\pi^k} & m = 2k, \\ \frac{(2k)!}{2^{2k+1}\pi^k k!} & m = 2k+1. \end{cases}$$

Then the form

$$\psi_m = C'_m \frac{\sum_{i=1}^m \Phi_i(x)}{\|x\|^m}$$

is a closed $(m-1)$-form on $\mathbb{R}^m \smallsetminus \{0\}$ such that

$$\int_{\mathbb{S}^{m-1}} \psi_m = 1, \tag{7.70}$$

where $\mathbb{S}^{m-1}$ is the unit sphere, in fact it may be a sphere of arbitrary radius. It is called the *angular form* on $\mathbb{R}^m$.

If we identify $\mathbb{C}^n$ with $\mathbb{R}^{2n}$, then $\psi_{2n} = (\beta_n + \overline{\beta_n})/2$, where

$$\beta_n = C_n \frac{\sum_{i=1}^n \overline{\Phi_i(z)} \wedge \Phi(z)}{\|z\|^{2n}}, \qquad C_n = (-1)^{\frac{n(n-1)}{2}} \frac{(n-1)!}{(2\pi\sqrt{-1})^n}.$$

We call β_n the *Bochner-Martinelli form* on $\mathbb{C}^n$. Note that

$$\beta_1 = \frac{1}{2\pi\sqrt{-1}} \frac{dz}{z},$$

the *Cauchy form* on $\mathbb{C}$. The form β_n is a closed form of type $(n,n-1)$ on $\mathbb{C}^n \smallsetminus \{0\}$, real on $\mathbb{S}^{2n-1}$ and

$$\int_{\mathbb{S}^{2n-1}} \beta_n = 1. \tag{7.71}$$

This follows from (7.70), however will be reproved in Proposition 9.18 below using the following expression:

Proposition 7.27. *The Bochner-Martinelli form on $\mathbb{C}^n$ is given by*

$$\beta_n = \left(\frac{1}{2\pi\sqrt{-1}}\right)^n (\partial \log \|z\|^2) \wedge (\bar\partial\partial \log \|z\|^2)^{n-1}.$$

Proof. Using the notation $(dz, z) = \sum \bar{z}_i dz_i$ and $(dz, dz) = \sum dz_i \wedge d\bar{z}_i$, we compute

$$\partial \log \|z\|^2 = \frac{(dz, z)}{(z, z)} \quad \text{and} \quad \bar{\partial}\partial \log \|z\|^2 = -\frac{(dz, dz)}{(z, z)} + \frac{(dz, z) \wedge (z, dz)}{(z, z)^2}.$$

Since $(dz, z) \wedge (dz, z) = 0$, we have

$$(\partial \log \|z\|^2) \wedge (\bar{\partial}\partial \log \|z\|^2)^{n-1} = (-1)^{n-1} \frac{(dz, z) \wedge (dz, dz)^{n-1}}{(z, z)^n}. \tag{7.72}$$

For each i, the term in $(dz, dz)^{n-1}$ not containing dz_i is given by

$$(-1)^{\frac{n(n-1)}{2}} (n-1)! d\bar{z}_1 \wedge \cdots \wedge \widehat{d\bar{z}_i} \wedge \cdots \wedge d\bar{z}_n \wedge dz_1 \wedge \cdots \wedge \widehat{dz_i} \wedge \cdots \wedge dz_n.$$

Hence the right-hand side of (7.72) is given by

$$(-1)^{\frac{n(n-1)}{2}} (n-1)! \frac{\sum \overline{\Phi_i(z)} \wedge \Phi(z)}{\|z\|^{2n}}.$$

Comparing with the definition of β_n, we have the proposition. □

Remark 7.21. We may write

$$\beta_n = \partial \gamma_n, \qquad \gamma_n = \left(\frac{1}{2\pi\sqrt{-1}}\right)^n \log \|z\|^2 (\bar{\partial}\partial \log \|z\|^2)^{n-1}.$$

Cohomology of Stiefel manifolds

Let $W(N, r)$ be the Stiefel manifold of r-frames in $\mathbb{C}^N$ (cf. Section 3.5). Recall that the map $\pi : W(N, r) \to W(N, r-1)$, which is given by taking the first $(r-1)$ vectors, has a fiber bundle structure with fiber $\mathbb{C}^{r-1} \times (\mathbb{C}^q \smallsetminus \{0\})$, $q = N - r + 1$. Let $\iota : \mathbb{C}^q \smallsetminus \{0\} \hookrightarrow W(N, r)$ be the inclusion of $\mathbb{C}^q \smallsetminus \{0\}$ into a fiber of π. Then it induces an isomorphism $\iota_* : H_{2q-1}(\mathbb{C}^q \smallsetminus \{0\}; \mathbb{Z}) \xrightarrow{\sim} H_{2q-1}(W(N, r); \mathbb{Z})$. It also induces an isomorphism $\iota^* : H^{2q-1}(W(N, r); \mathbb{C}) \to H^{2q-1}(\mathbb{C}^q \smallsetminus \{0\}; \mathbb{C}) \simeq \mathbb{C}$. By the de Rham theorem (Theorem 7.5) and Proposition 7.5, we have an isomorphism

$$\iota^* : H_d^{2q-1}(W(N, r)) \xrightarrow{\sim} H_d^{2q-1}(\mathbb{C}^q \smallsetminus \{0\}) \simeq \mathbb{C}.$$

We have seen that the class of the Bochner-Martinelli form β_q generates $H_d^{2q-1}(\mathbb{C}^q \smallsetminus \{0\})$. Thus we have:

Proposition 7.28. *The de Rham cohomology $H_d^{2q-1}(W(N, r))$ is generated by $(\iota^*)^{-1}[\beta_q]$.*

7.11 Lefschetz fixed point formula

In order to illustrate the use of intersection products introduced in Definition 4.4 and the effectiveness of the representation of the Thom class as in Propositions 7.22 and 7.23, we give a proof of this classical formula, see in particular Theorem 7.16 below. In this section, the homology and cohomology are with $\mathbb{Z}$-coefficients, unless otherwise stated.

Let M be a compact connected oriented C^∞ manifold of dimension m. We set $W = M \times M$ and endow it with the orientation so that it is the orientation of the first factor followed by that of the second. Denoting by $\Delta = \{ (x,x) \mid x \in M \}$ the diagonal in W with inclusion $i : \Delta \hookrightarrow W$, we have the commutative diagram:

$$\begin{array}{ccc} H^m(W, W \smallsetminus \Delta) & \xrightarrow{j^*} & H^m(W) \\ \wr \downarrow A & & \wr \downarrow P \\ H_m(\Delta) & \xrightarrow{i_*} & H_m(W). \end{array}$$

Note that Δ is an m-dimensional closed submanifold of W and that the map $\iota : M \to \Delta$ given by $\iota(x) = (x,x)$, $x \in M$, is a diffeomorphism. We orient Δ so that ι is orientation preserving. Let Δ and $[\Delta]$ denote the classes of Δ in $H_m(\Delta)$ and $H_m(W)$, respectively (cf. Convention 4.2). The Thom class Ψ_Δ of Δ in W is given by $A^{-1}(\Delta)$ (cf. (4.24)). We set $\eta_\Delta = P^{-1}([\Delta])$ so that $\eta_\Delta = j^* \Psi_\Delta$.

Let $f : M \to M$ be a C^∞ map and $\Gamma_f = \{ (x,y) \in W \mid y = f(x) \}$ its graph in W. We orient Γ_f so that the map

$$\tilde{f} : M \longrightarrow \Gamma_f \subset W = M \times M$$

given by $\tilde{f}(x) = (x, f(x))$ is an orientation preserving diffeomorphism.

The Lefschetz fixed point formula for f consists basicaly in computing the intersection number $\Gamma_f \cdot \Delta$ in two ways, i.e., globally and locally.

First we consider the global case. For this, we consider the diagram (cf. Definition 4.4):

$$\begin{array}{ccc} H^m(W) & \xrightarrow[P_W]{\sim} & H_m(W) \\ \downarrow \tilde{f}^* & & \downarrow (M \cdot_{\tilde{f}}\ )_M \\ H^m(M) & \xrightarrow[P_M]{\sim} & H_0(M). \end{array}$$

Definition 7.14. The *global fixed point index* $\Lambda(f)$ of f is defined as

$$\Lambda(f) = (M \cdot_{\tilde{f}} [\Delta])_M \qquad \text{in } H_0(M) \simeq \mathbb{Z}.$$

We may write

$$\Lambda(f) = P_M(\tilde{f}^*\eta_\Delta) = \langle M, \tilde{f}^*\eta_\Delta \rangle. \tag{7.73}$$

Note that $\tilde{f}$ induces an isomorphism $\tilde{f}_* : H_0(M) \xrightarrow{\sim} H_0(\Gamma_f)$ and $\Lambda(f)$ corresponds to $(\Gamma_f \cdot [\Delta])_{\Gamma_f}$ in $H_0(\Gamma_f)$, which is sent to $[\Gamma_f] \cdot [\Delta]$ in $H_0(W) \simeq \mathbb{Z}$ by the canonical morphism $H_0(\Gamma_f) \to H_0(W)$.

Now we consider the local contributions from the fixed point set of f:

$$\mathrm{Fix}(f) = \{\, x \in M \mid f(x) = x \,\}.$$

Setting $S = \mathrm{Fix}(f)$, we have the commutative diagram (cf. Definition 4.4):

$$\begin{array}{ccc} H^m(W, W \smallsetminus \Delta) & \xrightarrow[A_{W,\Delta}]{\sim} & H_m(\Delta) \\ \big\downarrow \tilde{f}^* & & \big\downarrow (M \cdot_{\tilde{f}}\ )_S \\ H^m(M, M \smallsetminus S) & \xrightarrow[A_{M,S}]{\sim} & H_0(S). \end{array}$$

Definition 7.15. The *local fixed point index* $\Lambda(f, S)$ of f at S is defined as

$$\Lambda(f, S) = (M \cdot_{\tilde{f}} \Delta)_S \qquad \text{in} \quad H_0(S).$$

We may write

$$\Lambda(f, S) = A(\tilde{f}^*\Psi_\Delta).$$

Note that $\tilde{f}$ induces a morphism $\tilde{f}_* : H_0(S) \to H_0(\Gamma_f)$ and it sends $\Lambda(f, S)$ to $(\Gamma_f \cdot [\Delta])_{\Gamma_f}$.

Suppose $S = \mathrm{Fix}(f)$ has a finite number of connected components $(S_\lambda)_\lambda$. Then we have $H_0(S) = \bigoplus_\lambda H_0(S_\lambda)$ and accordingly we have the local fixed point index $\Lambda(f, S_\lambda)$ in $H_0(S_\lambda) \simeq \mathbb{Z}$. We have (cf. Remark 4.15. 1):

$$\Lambda(f) = \sum_\lambda \Lambda(f, S_\lambda). \tag{7.74}$$

We try to look the both sides more closely. For the left-hand side, let

$$H^p(f) : H^p(M; \mathbb{C}) \longrightarrow H^p(M; \mathbb{C})$$

be the morphism induced by f on the p-th cohomology with $\mathbb{C}$-coefficients.

Definition 7.16. The *Lefschetz number* $L(f)$ of f is defined as

$$L(f) = \sum_{p=0}^{m} (-1)^p \operatorname{tr} H^p(f).$$

Note that this number turns out to be an integer, as the following theorem shows:

Theorem 7.15. *In the above situation, we have*

$$\Lambda(f) = L(f).$$

Proof. Let $\{u_{p,i}\}_i$ be a basis of $H^p(M;\mathbb{C})$ and $\{u^*_{m-p,j}\}_j$ the basis of $H^{m-p}(M;\mathbb{C})$ dual to $\{u_{p,i}\}$ via the Poincaré duality (cf. Corollary 4.2):

$$\langle M, u_{p,i} \smile u^*_{m-p,j}\rangle = \delta_{ij}.$$

If we set $v_{p,i,j} = \pi_1^* u_{p,i} \smile \pi_2^* u^*_{m-p,j}$, then $\{v_{p,i,j}\}_{p,i,j}$ form a basis of $H^m(W;\mathbb{C})$ by the Künneth formula (cf. (B.27)). We try to find η_Δ in terms of this basis. Thus we write $\eta_\Delta = \sum a_{p,i,j}\, v_{p,i,j}$ with $a_{p,i,j} \in \mathbb{C}$. Set $v^*_{m-p,i,j} = \pi_1^* u^*_{m-p,i} \smile \pi_2^* u_{p,j}$ and consider the identity (cf. (B.20))

$$\langle W, \eta_\Delta \smile v^*_{m-p,i,j}\rangle = \langle [\Delta], v^*_{m-p,i,j}\rangle.$$

The left-hand side is equal to $(-1)^{(p+1)(m-p)} a_{p,i,j}$ and the right-hand side is

$$\langle \iota_* M, v^*_{m-p,i,j}\rangle = \langle M, \iota^* v^*_{m-p,i,j}\rangle = \langle M, u^*_{m-p,i} \smile u_{p,j}\rangle = (-1)^{p(m-p)}\delta_{ij}.$$

Thus $a_{p,i,j} = (-1)^{m-p}\,\delta_{ij}$ and

$$\eta_\Delta = \sum_{p=0}^{m} (-1)^{m-p} \sum_i v_{p,i,i}.$$

If we write $f^* u^*_{m-p,i} = \sum h_{j,i}\, u^*_{m-p,j}$ with $h_{i,j} \in \mathbb{C}$, by (7.73), we have

$$\Lambda(f) = \sum_{p=0}^{m} (-1)^{m-p} \sum_i h_{i,i} = \sum_{p=0}^{m} (-1)^p \operatorname{tr} H^p(f).$$

□

Now we look at the right-hand side of (7.74). Let $W_0 = W \smallsetminus \Delta$ and W_1 a neighborhood of Δ in W. We consider the Čech-de Rham cohomology of the covering $\mathcal{W} = \{W_0, W_1\}$ of W. The class Ψ_Δ is represented by a cocycle (ψ_1, ψ_{01}) in $A^m(\mathcal{W}, W_0)$. Suppose $S = \mathrm{Fix}(f)$ has only a finite number of connected components $(S_\lambda)_\lambda$ as before. If we set $U_0 = M \smallsetminus S$, we have $\tilde{f}(U_0) \subset W_0$. For each λ, we take an open neighborhood U_λ of S_λ such that $\tilde{f}(U_\lambda) \subset W_1$ and that $U_\lambda \cap U_\mu = \emptyset$ if $\lambda \neq \mu$. For each λ, we take a compact submanifold R_λ of dimension m with C^∞ boundary in U_λ, containing S_λ in its interior. We set $R_{0\lambda} = -\partial R_\lambda$. Then we have (cf. (7.64))

$$\Lambda(f, S_\lambda) = \int_{R_\lambda} \tilde{f}^* \psi_1 + \int_{R_{0\lambda}} \tilde{f}^* \psi_{01}. \tag{7.75}$$

From (7.74) and Theorem 7.15, we have:

Proposition 7.29. *In the above situation,*

$$L(f) = \sum_{\lambda} \Lambda(f, S_\lambda),$$

where $L(f)$ is as in Definition 7.16 and $\Lambda(f, S_\lambda)$ is given by (7.75).

To compute (7.75), we proceed as follows. We may take as W_1 a tubular neighborhood of Δ in W with a C^∞ deformation retraction $\pi : W_1 \to \Delta$. If x and y are positive coordinate systems on the first and the second factors M, $(x - y, x)$ is a positive coordinate system on W_1 and we may think of $x - y$ as being a fiber coordinate system of π and x a base coordinate system. Thus $\pi : W_1 \to \Delta$ is isomorphic to (a neighborhood of the zero section of) the normal bundle $N_{\mathbb{R},\Delta}$ of Δ in W, with all the orientations taken into account. If we identify W_1 with $N_{\mathbb{R},\Delta}$, the Thom class Ψ_Δ may be thought of as the Thom class of the vector bundle $N_{\mathbb{R},\Delta}$. Thus the class Ψ_Δ is represented by a cocycle of the form $(\pi^*\epsilon, -\psi)$ in $A^m(\mathcal{W}, W_0)$ (cf. Proposition 7.22).

Now we find an explicit expression for the local index $\Lambda(f, S_\lambda)$ in the case S_λ consists of a point a.

Let U be a coordinate neighborhood around a with coordinates $x = (x_1, \dots, x_m)$ in M. Also let $\mathbb{B}^m$ be a small closed ball of dimension m around a in U such that $f(\mathbb{B}^m) \subset U$, which may be naturally thought of as a subset of $\mathbb{R}^m$. Denoting by 1 the identity map of $\mathbb{B}^m$, we may consider the map $1 - f : \mathbb{B}^m \to \mathbb{R}^m$ whose image is the origin 0 in $\mathbb{R}^m$ only at a. The boundary $\partial\mathbb{B}^m$ is homeomorphic to the unit sphere $\mathbb{S}^{m-1}$ and we have the map

$$\gamma : \partial\mathbb{B}^m \longrightarrow \mathbb{S}^{m-1} \qquad \text{defined by } \gamma(x) = \frac{x - f(x)}{\|x - f(x)\|}.$$

We denote the degree of this map by $\deg(1 - f, a)$.

Theorem 7.16. *In the above situation*

$$\Lambda(f, a) = \deg(1 - f, a).$$

Proof. We may take $\mathbb{B}^m$ as R_λ in (7.75). Since $R_{0\lambda} = -\partial\mathbb{B}^m$, we have

$$\Lambda(f, a) = \int_{\mathbb{B}^m} \tilde{f}^*\psi_1 - \int_{\partial\mathbb{B}^m} \tilde{f}^*\psi_{01}. \tag{7.76}$$

Recall that Ψ_Δ may be naturally identified with the Thom class of the normal bundle $N_{\mathbb{R},\Delta}$, which is trivial over $\iota(U)$; $N_{\mathbb{R},\Delta}|_{\iota(U)} \simeq \mathbb{R}^m \times \iota(U)$.

Let $\rho : N_{\mathbb{R},\Delta}|_{\iota(U)} \to \mathbb{R}^m$ denote the projection onto the fiber direction. Also let ψ_m be an angular form on $\mathbb{R}^m \smallsetminus \{0\}$. Then on $\iota(U)$ the Thom class of N_Δ is represented by the cocycle (cf. Proposition 7.23)

$$(0, -\rho^* \psi_m).$$

Let π_1 and π_2 denote the projections of $W = M \times M$ onto the first and second factors, respectively. We set $x = \pi_1^* x$ and $y = \pi_2^* y$ on $U \times U$. By our orientation convention, in a neighborhood of $\iota(a)$ in N_Δ, we may take $x - y$ as fiber coordinates and x as base coordinates of the bundle N_Δ so that we may write $\rho(x, y) = x - y$. Then (7.76) becomes

$$\Lambda(f, a) = \int_{\partial \mathbb{B}^m} (\rho \circ \tilde{f})^* \psi_m = \int_{\partial \mathbb{B}^m} (1 - f)^* \psi_m,$$

which is nothing but $\deg(1 - f, a)$. □

In the above situation, let $J_f(a)$ denote the Jacobian matrix of f at a. A fixed point a of the pair f is said to be *non-degenerate* if

$$\det(I - J_f(a)) \neq 0,$$

where I denotes the identity matrix of rank m.

Corollary 7.12. *If a is a non-degenerate fixed point,*

$$\Lambda(f, a) = \operatorname{sgn} \det(I - J_f(a)).$$

Combining the above, we have:

Theorem 7.17 (Lefschetz fixed point formula). *Let M be a compact oriented C^∞ manifold and $f : M \to M$ a C^∞ map with only isolated fixed points. Then we have*

$$L(f) = \sum_{a \in \mathrm{Fix}(f)} \deg(1 - f, a).$$

Remark 7.22. 1. We may interpret the local fixed point index $\Lambda(f, S_\lambda)$ as the (Poincaré-Hopf type) index of a section of a certain vector bundle. Let U_λ be as above and consider the map $g_\lambda = \pi \circ \tilde{f} : U_\lambda \to \Delta$. We have the pull-back bundle

$$g_\lambda^* N_{\mathbb{R},\Delta} = \{ (x, \nu) \in U_\lambda \times N_{\mathbb{R},\Delta} \mid g_\lambda(x) = \pi(\nu) \}.$$

If we define $s_\lambda : U_\lambda \to g_\lambda^* N_{\mathbb{R},\Delta}$ by $s_\lambda(x) = (x, \tilde{f}(x))$, it is a section of $g_\lambda^* N_{\mathbb{R},\Delta}$ on U_λ which is non-vanishing away from S_λ. Then it is not difficult to see that $\Lambda(f, S_\lambda)$ is equal to the topological residue $\mathrm{TRes}_e(s_\lambda, g_\lambda^* N_{\mathbb{R},\Delta}; S_\lambda)$ of the section s_λ at S_λ (cf. Definition 5.14).

2. Let v be a C^∞ vector field on a C^∞ manifold M as above. Associated with v, there exists a 1-parameter family $\{\varphi_t \mid t \in \mathbb{R}\}$ of C^∞ diffeomorphisms of M, i.e., φ_t is a C^∞ diffeomorphism of M for each $t \in \mathbb{R}$ with the properties that $\varphi_0 = 1_M$, the identity map of M, and that, for each $x \in M$, the vector $v(x)$ is tangent to the curve $\{\varphi_t(x) \mid t \in \mathbb{R}\}$ at $t = 0$.

If we let $f = \varphi_t$ in the above, $L(f) = L(1_M) = \chi(M)$. Moreover, the zeros of v correspond to the fixed points of f and, if S_λ is a compact connected component of the zero set of v, the Poincaré-Hopf index $\mathrm{PH}(v, S_\lambda)$ (cf. (5.15)) coincides with the local fixed point index $\Lambda(f, S_\lambda)$ above. Thus Theorem 5.6 and Corollary 5.4 are special cases of Proposition 7.29 and Theorem 7.17, respectively.

Notes

We list [Bott and Tu (1982)] as a basic reference for Sections 7.1, 7.2 and 7.3. For Čech cohomology, we also refer to [Hirzebruch (1966); Serre (1955b)]. The idea of combining de Rham and Čech cochains is due to [Weil (1952)]. It was then made to be a cohomology theory in [Bott and Tu (1982)].

The integration theory on Čech-de Rham cohomology is developed in [Lehmann (1991)]. Our orientation convention is different from the one there, for example compare the identity (7.18) in the text and Lemma 2 in [Lehmann (1991)]. See also this reference for the proof of Lemma 7.3 and the conjecture mentioned in Remark 7.6. We refer to [Suwa (1998)] for these materials and for the treatment of the Thom isomorphism and the Thom class in the context of Čech-de Rham cohomology. As to the relation with combinatorial topology, we refer to [Suwa (2008)]. See [Suwa (2023)] concerning Remark 7.15. 3. For the Bochner-Martinelli form, we refer to [Griffiths and Harris (1978)].

The Lefschetz fixed point formula was initiated in [Lefschetz (1937)] and there have been many literatures on this subject; here we list [Guillemin and Pollack (1974); Vick (1994)]. It has been generalized to the fixed point formula for elliptic complexes in [Atiyah and Bott (1967-8)]. It is studied in [Goresky and MacPherson (1985)] in the context of intersection homology. The significance of the approach as above is the use of the expression of the Thom calss in the relative Čech-de Rham cohomology. See [Bisi, Bracci, Izawa and Suwa (2016); Brasselet and Suwa (2021)] for generalizations in the context of coincidence theory.

Chapter 8

Chern-Weil Theory Adapted to Čech-de Rham Cohomology

The Chern-Weil theory gives a way to represent characteristic classes of vector bundles by differential forms through the use of connections. In this chapter we review this theory for complex vector bundles and modify it to have the classes in Čech-de Rham cohomology.

We first introduce the notion of a connection for a vector bundle. Intuitively, it measures the rate of change of a section in each tangential direction. We then define the curvature of a connection and, for an invariant polynomial, the characteristic form, which is shown to be closed and defines a class in the de Rham cohomology and in turn a class in the (singular or simplicial) cohomology with $\mathbb{C}$-coefficient via the de Rham theorem. We discuss difference forms, which are constructed if we have several connections for a bundle. In particular, this is used to show that the class defined as above does not depend on the choice of the connection. It is proved, in Section 10.5 below, that the class defined this way is the image of the corresponding class defined via obstruction theory by the canonical morphism from the cohomology with $\mathbb{Z}$-coefficient to that with $\mathbb{C}$-coefficient.

We then explain how to define characteristic classes in Čech-de Rham cohomology, where difference forms play a key rolle. This way of representing characteristic classes is particularly effective in dealing with the localization problem, which we discuss in Chapter 10. One of the advantages is that we may define localizations for general invariant polynomial on the cocycle level and it leads to local invariants that do not come from classes in homology with $\mathbb{Z}$-coefficients, for example, some residues of singular holomorphic foliations.

Throughout this chapter, we let M denote a C^∞ manifold of dimension m. Also, the vector bundles we consider are C^∞. For an open set U in M, we denote by $A^0(U)$ the $\mathbb{C}$-algebra of C^∞ functions on U as before. For a complex vector bundle E on M, we set $A^p(U;E) = C^\infty(U; \bigwedge^p(T^c_{\mathbb{R}}M)^* \otimes E)$, the $A^0(U)$-module of C^∞ p-forms with coefficients in E. Thus $A^0(U;E)$ is the $A^0(U)$-module of C^∞ sections of E and for the product bundle $E = \mathbb{C} \times M$, $A^p(U;E) = A^p(U)$, the $A^0(U)$-module of p-forms on U.

8.1 Connections

Let E be a C^∞ complex vector bundle of rank l on M.

Definition 8.1. A *connection* for E is a $\mathbb{C}$-linear map

$$\nabla : A^0(M;E) \longrightarrow A^1(M;E)$$

satisfying the Leibniz rule:

$$\nabla(fs) = df \otimes s + f\nabla(s) \qquad \text{for } f \in A^0(M) \text{ and } s \in A^0(M;E).$$

Example 8.1. The exterior derivative

$$d : A^0(M) \longrightarrow A^1(M)$$

is a connection for the product bundle $\mathbb{C} \times M$. More generally, if E is trivial, choosing a frame and using d we may define a connection for E.

Lemma 8.1. *A connection ∇ is a local operator, i.e., if a section s is identically 0 on an open set U, so is $\nabla(s)$.*

Proof. Let x be a point in U and U_x a neighborhood of x in U. Take a C^∞ function f such that $f \equiv 1$ on $\overline{U_x}$ and $f \equiv 0$ on $M \setminus U$. Since the section fs is identically 0 on M, so is $\nabla(fs) = df \otimes s + f\nabla(s)$. Evaluating the both sides at x, we have $\nabla(s)(x) = 0$. □

Thus the restriction of ∇ to an open set U makes sense and it is a connection for $E|_U$. If $e^{(l)} = (e_1, \dots, e_l)$ is a frame of E on U, for every section s, writing $s = \sum_{i=1}^l f_i e_i$, we have

$$\nabla(s) = \sum_{i=1}^{l} (df_i \otimes e_i + f_i \nabla(e_i)), \quad \text{on } U. \tag{8.1}$$

Thus $\nabla(s)$ is determined uniquely by the values $\nabla(e_i)$.

Definition 8.2. Let $e^{(r)} = (e_1, \dots, e_r)$ be a C^∞ r-frame of E on an open set U. We say that a connection ∇ is *trivial with respect to* $e^{(r)}$, or $e^{(r)}$-*trivial* for short, on U if $\nabla(e_i) = 0$ for $i = 1, \dots, r$.

Thus in Example 10.1, d is trivial with respect to an arbitrary (non-zero) constant section. If E is trivial and if ∇ is trivial with respect to some frame, it is the one in Example 10.1, as can be seen from (8.1).

From the definition we have the following:

Lemma 8.2. *Let $\nabla_1, \dots, \nabla_k$ be connections for E and $f_1, \dots, f_k$ C^∞ functions on M with $\sum_{i=1}^k f_i \equiv 1$. Then $\sum_{i=1}^k f_i \nabla_i$ is a connection for E.*

Using the above, we have:

Proposition 8.1. *Every vector bundle E admits a connection.*

Proof. Let $\mathcal{U} = \{U_\alpha\}$ be a locally finite open covering of M such that E is trivial on each U_α. On each U_α, there exists a connection ∇_α trivial with respect to some frame. Let $\{\rho_\alpha\}$ be a C^∞ partition of unity subordinate to $\mathcal{U}$. Then $\nabla = \sum \rho_\alpha \nabla_\alpha$ is a connection for E. □

Lemma 8.3. *Let ∇ be a connection for E on an open set U of M. For every open set U' with $\overline{U'} \subset U$, there exists a connection $\tilde{\nabla}$ for E on M such that $\tilde{\nabla} = \nabla$ on U'. Moreover, if ∇ is $e^{(r)}$-trivial on U for an r-frame $e^{(r)}$ on M, we may find $\tilde{\nabla}$ which is $e^{(r)}$-trivial on M.*

Proof. Take an open set U'' so that $\overline{U'} \subset U'' \subset \overline{U''} \subset U$ and let f be a C^∞ function on M such that $f \equiv 0$ on $M \smallsetminus U''$ and $f \equiv 1$ on $\overline{U'}$. Let ∇' be an arbitrary connection for E on M and set $\tilde{\nabla} = f\nabla + (1-f)\nabla'$. Then it has the desired property. For the rest, in the proof of Proposition 8.1, on each U_α choose a frame so that $e^{(r)}$ is a part of it and let ∇_α be the connection trivial with respect to this frame. Use the connection constructed this way as ∇'. □

If ∇ is a connection for E, it induces a $\mathbb{C}$-linear map

$$\nabla : A^1(M; E) \longrightarrow A^2(M; E)$$

satisfying

$$\nabla(\omega \otimes s) = d\omega \otimes s - \omega \wedge \nabla(s) \qquad \text{for } \omega \in A^1(M) \text{ and } s \in A^0(M; E).$$

The composition

$$K = \nabla \circ \nabla : A^0(M; E) \longrightarrow A^2(M; E)$$

is called the *curvature* of ∇. It is easily checked that

$$K(fs) = fK(s) \qquad \text{for } f \in A^0(M) \text{ and } s \in A^0(M; E).$$

From this we see that the curvature K may be thought of as an element in $A^2(M; \mathrm{Hom}(E, E))$ so that it is locally represented by an $l \times l$ matrix whose entries are differential 2-forms. On the other hand the connection ∇ itself may not be thought of as an element in $A^1(M; \mathrm{Hom}(E, E))$. However, the fact that it is a local operator allows us to represent it locally by a matrix whose entries are 1-forms.

Thus suppose that ∇ is a connection for E and that E is trivial on U. If $e^{(l)} = (e_1, \ldots, e_l)$ is a frame of E on U, we may write, for $i = 1, \ldots, l$,

$$\nabla(e_i) = \sum_{j=1}^{l} \theta_{ji} \otimes e_j, \qquad \theta_{ij} \in A^1(U).$$

We call $\theta = (\theta_{ij})$ the *connection matrix* of ∇ with respect to $e^{(l)}$. Note that ∇ is $e^{(l)}$-trivial, if and only if $\theta = 0$.

For the curvature K of ∇, we may write

$$K(e_i) = \sum_{j=1}^{l} \kappa_{ji} \otimes e_j, \qquad \kappa_{ij} \in A^2(U).$$

We call $\kappa = (\kappa_{ij})$ the *curvature matrix* of ∇ with respect to $e^{(l)}$. From the definition we compute $\kappa_{ij} = d\theta_{ij} + \sum_{k=1}^{l} \theta_{ik} \wedge \theta_{kj}$, which we write as

$$\kappa = d\theta + \theta \wedge \theta. \tag{8.2}$$

From this, we get

$$d\kappa = \kappa \wedge \theta - \theta \wedge \kappa, \tag{8.3}$$

which is referred to as the *Bianchi identity.*

If $e'^{(l)} = (e'_1 \ldots, e'_l)$ is another frame of E on U', we have $e'_i = \sum_{j=1}^{l} p_{ji} e_j$ for some C^∞ functions p_{ij} on $U \cap U'$. The matrix $P = (p_{ij})$ is non-singular at each point of $U \cap U'$. If we denote by θ' and κ' the connection and curvature matrices of ∇ with respect to $e'^{(l)}$,

$$\theta' = P^{-1} \cdot dP + P^{-1}\theta P \quad \text{and} \quad \kappa' = P^{-1}\kappa P \quad \text{in} \quad U \cap U'. \tag{8.4}$$

Note that the second relation signifies the fact that K is an element in $A^2(M, \mathrm{Hom}(E, E))$ (cf. Exercise 3.7).

Remark 8.1. If ∇ and ∇' are connections for E, the difference $\nabla' - \nabla$ may be thought of as an element in $A^1(M, \mathrm{Hom}(E, E))$.

8.2 Characteristic classes of complex vector bundles

Invariant polynomials

Let $M(l,\mathbb{C})$ denote the vector space of $l\times l$ complex matrices. A *polynomial* φ on $M(l,\mathbb{C})$ is a function $\varphi(A)$ of $A\in M(l,\mathbb{C})$ which is a polynomial in the entries of A. It is said to be *invariant*, or *symmetric*, if

$$\varphi(P^{-1}AP)=\varphi(A)\qquad \text{for all } P\in GL(l,\mathbb{C}).$$

If we define a function σ_q, for $q=1,2,\dots,l$, by

$$\sigma_q(A)=\operatorname{tr}\big(\textstyle\bigwedge^q A\big),$$

it is an invariant polynomial, homogeneous of degree q. Here we recall that the exterior power $\bigwedge^q A$ is defined as follows. We set

$$\mathcal{I}=\{\,(i_1,\dots,i_q)\mid 1\le i_1<\dots<i_q\le l\,\},\tag{8.5}$$

which has $N=\binom{l}{q}$ elements, and endow $\mathcal{I}$ with the lexicographic order. Then $\bigwedge^q A$ is the $N\times N$ matrix whose (I,J) entry is $\det A_{I,J}$, where $A_{I,J}$ denotes the $q\times q$ minor of A consisting the rows and columns corresponding to I and J. In particular,

$$\sigma_1(A)=\operatorname{tr}(A)\quad\text{and}\quad \sigma_l(A)=\det(A).$$

We may also write

$$\det(I_l+A)=1+\sigma_1(A)+\dots+\sigma_l(A),$$

where I_l denotes the identity matrix of rank l. We call σ_q the q-th *elementary invariant polynomial.*

It is known that:

(P$_1$) every invariant polynomial is a polynomial in the elementary invariant polynomials,

(P$_2$) for a polynomial φ homogeneous of degree k, there exists a symmetric k-linear form

$$\tilde{\varphi}:M(l,\mathbb{C})\times\dots\times M(l,\mathbb{C})\longrightarrow\mathbb{C},$$

called the *polarization* of φ, such that $\varphi(A)=\tilde{\varphi}(A,\dots,A)$. If φ is invariant, $\tilde{\varphi}$ is invariant in the sense that

$$\tilde{\varphi}(P^{-1}A_1P,\dots,P^{-1}A_kP)=\tilde{\varphi}(A_1,\dots,A_k)\qquad\text{for all } P\in GL(l,\mathbb{C}).$$

The invariance of $\tilde{\varphi}$ may be expressed as follows. If we set $P = e^{tB}$, $B \in M(l, \mathbb{C})$, in the above and take the derivatives at $t = 0$ of the both sides, we get

$$\sum_{i=1}^{k} \tilde{\varphi}(A_1, \ldots, [A_i, B], \ldots, A_k) = 0. \tag{8.6}$$

Remark 8.2. The polarization of a polynomial φ homogeneous of degree k is given by

$$\tilde{\varphi}(A_1, \ldots, A_k) = \frac{1}{k!} \frac{\partial^k}{\partial \lambda_1 \ldots \partial \lambda_k} \varphi(\lambda_1 A_1 + \cdots + \lambda_k A_k).$$

Note that the right-hand side does not depend on the λ_i's and is equal to the coefficient of $\lambda_1 \cdots \lambda_k$ in the expansion of $\varphi(\lambda_1 A_1 + \cdots + \lambda_k A_k)$ divided by $k!$.

In particular, for the q-th elementary polynomial σ_q, the polarization is given as follows. Let $\mathcal{I}$ be as in (8.5). Then

$$\tilde{\sigma}_q(A_1, \ldots, A_q) = \frac{1}{q!} \sum_{\rho \in \mathfrak{S}_q} \sum_{I \in \mathcal{I}} \det A_{\rho, I},$$

where $\mathfrak{S}_q$ denotes the symmetric group of degree q and $A_{\rho,I}$ the $q \times q$ matrix whose j-th column is that of $(A_{\rho(j)})_{I,I}$. For $\sigma_l = \det$,

$$\tilde{\sigma}_l(A_1, \ldots, A_l) = \frac{1}{l!} \sum_{\rho \in \mathfrak{S}_l} \det A_\rho, \tag{8.7}$$

where A_ρ denotes the $l \times l$ matrix whose j-th column is that of $A_{\rho(j)}$.

Characteristic forms

Let M be a C^∞ manifold and U an open set in M. We consider the space of $l \times l$ matrices whose entries are differential forms on U, which may be expressed as $M(l, \mathbb{C}) \otimes A^*(U)$. An invariant k-linear form $\tilde{\varphi}$ as above may be extended to an invariant k-linear form on $M(l, \mathbb{C}) \otimes A^*(U)$ as follows. For decomposable elements we set

$$\tilde{\varphi}(A_1 \otimes \omega_1, \ldots, A_k \otimes \omega_k) = \tilde{\varphi}(A_1, \ldots, A_k)\, \omega_1 \wedge \cdots \wedge \omega_k$$

and extend it linearly. Then we have, for $\xi_1, \ldots, \xi_k \in M(l, \mathbb{C}) \otimes A^*(U)$,

$$d\tilde{\varphi}(\xi_1, \ldots, \xi_k) = \sum_{i=1}^{k} (-1)^{\varepsilon(i)} \tilde{\varphi}(\xi_1, \ldots, d\xi_i, \ldots, \xi_k), \tag{8.8}$$

where $\varepsilon(i) = \sum_{j<i} \deg \xi_j$.

An invariant polynomial φ may also be extended to an invariant polynomial on $M(l,\mathbb{C})\otimes A^*(U)$ by $\varphi(\xi)=\tilde{\varphi}(\xi,\dots,\xi)$.

Example 8.2. Let $\Omega_\nu=(\omega_{ij}^{(\nu)})$, $\nu=1,2$, be 2×2 matrices with entries differential forms. Then we compute

$$\tilde{\sigma}_2(\Omega_1,\Omega_2)=\frac{1}{2}(\omega_{11}^{(1)}\wedge\omega_{22}^{(2)}-\omega_{12}^{(1)}\wedge\omega_{21}^{(2)}+\omega_{22}^{(1)}\wedge\omega_{11}^{(2)}-\omega_{21}^{(1)}\wedge\omega_{12}^{(2)}).$$

If E is a complex vector bundle of rank l on M and if Ξ is an element of $A^*(M;\mathrm{Hom}(E,E))$, choosing a frame of E, it is locally represented by a matrix ξ of forms and a different choice of the frame will give a matrix similar to ξ so that $\varphi(\Xi)$ is a well-defined differential form on M.

We also have a product, denoted by $\wedge$, in $M(l,\mathbb{C})\otimes A^*(U)$. It is defined for decomposable elements by

$$(A_1\otimes\omega_1)\wedge(A_2\otimes\omega_2)=(A_1A_2)\otimes(\omega_1\wedge\omega_2)$$

and then extended linearly. Note that the expressions $\theta\wedge\theta$ and $\kappa\wedge\theta$ appeared before are consistent with this. We also define the bracket product of ξ in $M(l,\mathbb{C})\otimes A^p(U)$ and η in $M(l,\mathbb{C})\otimes A^q(U)$ by

$$[\xi,\eta]=\xi\wedge\eta-(-1)^{pq}\eta\wedge\xi.$$

In particular, if ξ is of even degree, $[\xi,\eta]=\xi\wedge\eta-\eta\wedge\xi$ so that the Bianchi identity (8.3) may be written as

$$d\kappa=[\kappa,\theta]. \tag{8.9}$$

From (8.6), we have

$$\sum_{i=1}^{k}(-1)^{q\varepsilon'(i)}\tilde{\varphi}(\xi_1,\dots,[\xi_i,\eta],\dots,\xi_k)=0, \tag{8.10}$$

where $q=\deg\eta$ and $\varepsilon'(i)=\sum_{j>i}\deg\xi_j$.

Let E be a C^∞ vector bundle of rank l on M. Let ∇ be a connection for E and K its curvature. Since K is in $A^2(M;\mathrm{Hom}(E,E))$, for an invariant polynomial φ, we have a differential form $\varphi(K)$. We now show that the form is closed and its class in the de Rham cohomology depends only on E and not on the choice of the connection.

Proposition 8.2. *For every invariant polynomial φ, the form $\varphi(K)$ is closed.*

Proof. Since the problem is local, it is sufficient to show that $\varphi(\kappa)$ is closed, where κ is the curvature matrix of ∇ with respect to some (local) frame. By (8.8),

$$d\varphi(\kappa) = \sum_{i=1}^{k} \tilde{\varphi}(\kappa, \dots, \overset{i}{d\kappa}, \dots, \kappa).$$

Then by (8.9) and (8.10), this is zero. □

By a slight abuse of notation, we introduce the following:

Definition 8.3. For a connection ∇ and an invariant polynomial φ homogeneous of degree k, the *characteristic form* is defined by

$$\varphi(\nabla) = \left(\frac{\sqrt{-1}}{2\pi}\right)^k \varphi(K).$$

Difference forms

If we have several connections for a complex vector bundle E, we have the following:

Proposition 8.3. *Let $\nabla_0, \dots, \nabla_p$ be connections for E. For an invariant polynomial φ homogeneous of degree k, there exists a $(2k-p)$-form $\varphi(\nabla_0, \dots, \nabla_p)$ alternating in the $p+1$ entries and satisfying*

$$\sum_{\nu=0}^{p} (-1)^{\nu} \varphi(\nabla_0, \dots, \widehat{\nabla_\nu}, \dots, \nabla_p) + (-1)^p d\varphi(\nabla_0, \dots, \nabla_p) = 0. \qquad (8.11)$$

Proof. Consider the product $\mathbb{R}^p \times M$ with projection $\rho : \mathbb{R}^p \times M \to M$. Denoting by $(t_1, \dots, t_p)$ coordinates on $\mathbb{R}^p$, let $\tilde{\nabla}$ be the connection for $\rho^* E$ given by

$$\tilde{\nabla} = \left(1 - \sum_{\nu=1}^{p} t_\nu\right) \rho^* \nabla_0 + \sum_{\nu=1}^{p} t_\nu \rho^* \nabla_\nu.$$

Let $\Delta^p = (P_0, \dots, P_p)$ be the standard p-simplex in $\mathbb{R}^p$ (cf. Appendix B) and $\rho' : \Delta^p \times M \to M$ the restriction of ρ. Then we have the integration along fibers (cf. Section 6.3):

$$\rho'_* : A^{2k}(\Delta^p \times M) \longrightarrow A^{2k-p}(M).$$

We set

$$\varphi(\nabla_0, \dots, \nabla_p) = \rho'_* \varphi(\tilde{\nabla}).$$

Then by Propositions 6.2.1 and 6.3 and the identity

$$\partial\Delta^p = \sum_{\nu=0}^{p}(-1)^\nu (P_0, \dots, \widehat{P}_\nu, \dots, P_p),$$

we see that it has the desired property. □

Remark 8.3. If $p > k$, there are no terms in $\varphi(\tilde{\nabla})$ involving $dt_1 \wedge \cdots \wedge dt_p$. Thus

$$\varphi(\nabla_0, \dots, \nabla_p) = 0 \qquad \text{if } \ p > k.$$

The form $\varphi(\nabla_0, \dots, \nabla_p)$ as above is called a *difference form*. In particular, if ∇ and ∇' are two connections for E,

$$\varphi(\nabla') - \varphi(\nabla) = d\varphi(\nabla, \nabla').$$

From this we see that the class $[\varphi(\nabla)]$ of the closed form $\varphi(\nabla)$ in the de Rham cohomology $H_d^{2k}(M)$ depends only on E and not on the choice of the connection ∇.

Definition 8.4. We denote this class by $\varphi(E)$ and call it the *characteristic class* of E for the polynomial φ.

Definition 8.5. The q-th *Chern form* of ∇ is defined by

$$c^q(\nabla) = \Big(\frac{\sqrt{-1}}{2\pi}\Big)^q \sigma_q(K).$$

Its class in $H_d^{2q}(M)$ is denoted by $c^q_{\mathrm{diff}}(E)$ and is called the q-th *differential geometric Chern class* of E.

The class $c^q_{\mathrm{diff}}(E)$ is also denoted by $c^q(E)$, if there is no fear of confusion.

If κ is the curvature matrix of ∇ with respect to some frame of E on U, the total Chern form is given by

$$c^*(\nabla) = \det\Big(I + \frac{\sqrt{-1}}{2\pi}\kappa\Big). \tag{8.12}$$

We call

$$c^*_{\mathrm{diff}}(E) = 1 + c^1_{\mathrm{diff}}(E) + \cdots + c^l_{\mathrm{diff}}(E)$$

the total Chern class of E, which is considered as an element in the cohomology ring $H_d^*(M)$. Note that the class $c^*_{\mathrm{diff}}(E)$ is invertible in $H_d^*(M)$.

We give an alternative expression of the difference form in terms of polarization. Thus we consider the situation of Proposition 8.3. For $t = (t_1, \dots, t_p) \in \mathbb{R}^p$, we set

$$\nabla_t = \Big(1 - \sum_{\nu=1}^{p} t_\nu\Big)\nabla_0 + \sum_{\nu=1}^{p} t_\nu \nabla_\nu$$

and think of it as a connection for E. Let K_t denote its curvature. Recall that, for each $\nu = 1, \dots, p$, $\nabla_\nu - \nabla_0$ is in $A^1(M; \mathrm{Hom}(E, E))$ (cf. Remark 8.1). Also by Remark 8.3, we may assume that $p \leq k$. We set

$$\mathcal{I} = \{\, (i_1, \dots, i_p) \mid 1 \leq i_1 < \cdots < i_p \leq k \,\}$$

and

$$\varPhi_t = (-1)^{\frac{p(p-1)}{2}} p! \sum_{I \in \mathcal{I}} \tilde{\varphi}(K_t, \dots, \nabla_1 \overset{i_1}{-} \nabla_0, \dots, \nabla_p \overset{i_p}{-} \nabla_0, \dots, K_t).$$

Proposition 8.4. *The form $\varphi(\nabla_0, \dots, \nabla_p)$ as constructed in Proposition 8.3 is given by*

$$\varphi(\nabla_0, \dots, \nabla_p) = \Big(\frac{\sqrt{-1}}{2\pi}\Big)^k \int_{\Delta^p} \varPhi_t \, dt_1 \cdots dt_p.$$

Proof. For $\nu = 0, \dots, p$, let θ_ν be the connection matrix of ∇_ν with respect to some frame $e^{(l)}$ and let κ_t be the curvature matrix of ∇_t with respect to $e^{(l)}$. Then the curvature matrix $\tilde{\kappa}$ of the connection $\tilde{\nabla}$ in the proof of Proposition 8.3 with respect to the frame $\rho^* e^{(l)}$ is given by

$$\tilde{\kappa} = \sum_{\nu=1}^{p} dt_\nu \wedge (\theta_\nu - \theta_0) + \kappa_t$$

so that

$$\varphi(\tilde{\kappa}) = \tilde{\varphi}\Big(\sum_{\nu=1}^{p} dt_\nu \wedge (\theta_\nu - \theta_0) + \kappa_t, \dots, \sum_{\nu=1}^{p} dt_\nu \wedge (\theta_\nu - \theta_0) + \kappa_t\Big).$$

The term involving $dt_1 \wedge \cdots \wedge dt_p$ in the above is given by

$$(-1)^{\frac{p(p-1)}{2}} p! dt_1 \wedge \cdots \wedge dt_p \sum_{I \in \mathcal{I}} \tilde{\varphi}(\kappa_t, \dots, \theta_1 \overset{i_1}{-} \theta_0, \dots, \theta_p \overset{i_p}{-} \theta_0, \dots, \kappa_t).$$

Hence we have the proposition. $\square$

In particular, for two connections ∇ and ∇' for E, we have:

$$\varphi(\nabla,\nabla')=\Big(\frac{\sqrt{-1}}{2\pi}\Big)^k\int_0^1\sum_{i=1}^{k}\tilde{\varphi}(K_t,\dots,\overset{i}{\nabla'-\nabla},\dots,K_t)dt. \tag{8.13}$$

Remark 8.4. 1. As is shown below (cf. Theorem 10.12), the Chern class $c^q_{\text{diff}}(E)$ defined above is the image of the Chern class $c^q_{\text{top}}(E)$ defined by obstruction theory by the canonical morphism $H^{2q}(M;\mathbb{Z})\to H^{2q}(M;\mathbb{C})$.

In the rest of this chapter $c^q_{\text{diff}}(E)$ will simply be denoted by $c^q(E)$.

2. As noted above, for an invariant polynomial φ, there is a polynomial P such that $\varphi=P(\sigma_1,\sigma_2,\dots)$. We have $\varphi(\nabla)=P(c^1(\nabla),c^2(\nabla),\dots)$ and $\varphi(E)=P(c^1(E),c^2(E),\dots)$.

3. In general, suppose we have a C^∞ one parameter family of connections ∇_t and let K_t be its curvature. Then the assignment $s\mapsto \frac{d}{dt}(\nabla_t(s))$ is $A^0(M)$ linear so that we may think of $\dot{\nabla}_t=\frac{d}{dt}\nabla_t$ as being in $A^1(M;\text{Hom}(E,E))$ and we have

$$\varphi(\nabla_1)-\varphi(\nabla_0)=d\Big(\Big(\frac{\sqrt{-1}}{2\pi}\Big)^k\int_0^1\sum_{i=1}^{k}\tilde{\varphi}(K_t,\dots,\overset{i}{\dot{\nabla}_t},\dots,K_t)dt\Big).$$

The form in (8.13) is the special case where $\nabla_t=(1-t)\nabla_0+t\nabla_1$.

8.3 Further topics on connections

Pull-back: Let $f:M'\to M$ be a C^∞ map of C^∞ manifolds and let $\pi:E\to M$ be a complex vector bundle. For an open set U in M, f induces a morphism $f^*:A^p(U;E)\to A^p(f^{-1}(U);f^*E)$ (cf. (6.5)). If ∇ is a connection for E, we see that there is a unique connection for f^*E, denoted by $f^*\nabla$, making the following diagram commutative:

$$\begin{array}{ccc} A^0(M;E) & \xrightarrow{\ \nabla\ } & A^1(M;E)\\ \big\downarrow{\scriptstyle f^*} & & \big\downarrow{\scriptstyle f^*}\\ A^0(M';f^*E) & \xrightarrow{\ f^*\nabla\ } & A^1(M';f^*E). \end{array} \tag{8.14}$$

Let θ and κ be, respectively, the connection and curvature matrices of ∇ with respect to a frame $(e_1,\dots,e_l)$ of E on U. If $(f^*e_1,\dots,f^*e_l)$ form a frame of f^*E on $f^{-1}(U)$, the connection and curvature matrices of $f^*\nabla$ with respect to this frame are given, respectively, by $f^*\theta$ and $f^*\kappa$. This in particular shows that, for any invariant polynomial φ,

$$\varphi(f^*\nabla)=f^*\varphi(\nabla)\quad\text{and}\quad\varphi(f^*E)=f^*\varphi(E),$$

where f^* in the right-hand side of the second identity denotes the canonical morphism $H^*(M;\mathbb{C}) \to H^*(M';\mathbb{C})$.

Direct sum: Let E_1 and E_2 be complex vector bundles on M and ∇_1 and ∇_2 connections for E_1 and E_2, respectively. Every section s of the direct sum $E = E_1 \oplus E_2$ is written as $s = (s_1, s_2)$ with s_1 and s_2 sections of E_1 and E_2. If we define $\nabla : A^0(M;E) \to A^1(M;E)$ by $\nabla(s) = (\nabla_1(s_1), \nabla_2(s_2))$, it is a connection for E, which we call the direct sum of ∇_1 and ∇_2 and denote by $\nabla_1 \oplus \nabla_2$. If we take some frames of E_1 and E_2 and if we denote by θ_1, κ_1 and θ_2, κ_2 the connection and curvature matrices for ∇_1 and ∇_2, respectively, the connection and curvature matrices θ, κ for ∇ with respect to the combined frame are given by $\theta = \theta_1 \oplus \theta_2$, $\kappa = \kappa_1 \oplus \kappa_2$. This in particular shows that

$$c^*(\nabla_1 \oplus \nabla_2) = c^*(\nabla_1) \wedge c^*(\nabla_2) \quad \text{and} \quad c^*(E_1 \oplus E_2) = c^*(E_1) \smile c^*(E_2). \tag{8.15}$$

The latter is referred to as the *Whitney sum formula for Chern classes* (cf. Corollary 5.6 for the case of localized top Chern classes from the topological viewpoint).

Virtual bundles: Let E and F be vector bundles on M. We consider the formal difference $F - E$ and define its total Chern class by

$$c^*(F - E) = c^*(F)/c^*(E). \tag{8.16}$$

For connections ∇^E and ∇^F for E and F, we set $\nabla^\star = (\nabla^E, \nabla^F)$. Then the above class is represented by the form

$$c^*(\nabla^\star) = c^*(\nabla^F)/c^*(\nabla^E).$$

The q-th Chern class $c^q(F - E)$ of $F - E$ is the component of $c^*(F - E)$ in $H_d^{2q}(M)$. It is represented by $c^q(\nabla^\star)$, the term in $c^*(\nabla^\star)$ of degree $2q$.

If φ is an invariant polynomial, we write $\varphi = P(c^1, c^2, \dots)$ and define the characterstic class of $F - E$ for φ by $\varphi(F-E) = P(c^1(F-E), c^2(F-E), \dots)$. It is represented by the form $\varphi(\nabla^\star) = P(c^1(\nabla^\star), c^2(\nabla^\star), \dots)$.

Suppose we have $p+1$ pairs of connections $\nabla_\nu^\star = (\nabla_\nu^E, \nabla_\nu^F)$, $\nu = 0, \dots, p$. Then there exists a form $\varphi(\nabla_0^\star, \dots, \nabla_p^\star)$ satisfying

$$\sum_{\nu=0}^{p} (-1)^\nu \varphi(\nabla_0^\star, \dots, \widehat{\nabla_\nu^\star}, \dots, \nabla_p^\star) + (-1)^p d\varphi(\nabla_0^\star, \dots, \nabla_p^\star) = 0.$$

This is proved in more generality in Proposition 15.2 below, the proof being essentially the same as the one for vector bundles (cf. Proposition 8.3).

In particular, for two pairs of connections $\nabla_0^\star$ and $\nabla_1^\star$, we have

$$\varphi(\nabla_1^\star) - \varphi(\nabla_0^\star) = d\varphi(\nabla_0^\star, \nabla_1^\star).$$

Hence we reprove that the class $[\varphi(\nabla^\star)]$ of the closed form $\varphi(\nabla^\star)$, which is in fact $\varphi(F - E)$, depends only on $F - E$ and not on the choice of $\nabla^\star$.

We discuss more general cases in Section 15.2 below.

Exact sequences: Let

$$0 \longrightarrow E' \stackrel{\iota}{\longrightarrow} E \stackrel{\eta}{\longrightarrow} E'' \longrightarrow 0 \tag{8.17}$$

be an exact sequence of vector bundles on M and let ∇', ∇ and ∇'' be connections for E', E and E'', respectively. We say that the triple $(\nabla', \nabla, \nabla'')$ is *compatible* with (8.17) if the following diagram is commutative:

$$\begin{array}{ccccc} A^0(M;E') & \stackrel{\iota}{\longrightarrow} & A^0(M;E) & \stackrel{\eta}{\longrightarrow} & A^0(M;E'') \\ \big\downarrow{\scriptstyle \nabla'} & & \big\downarrow{\scriptstyle \nabla} & & \big\downarrow{\scriptstyle \nabla''} \\ A^1(M;E') & \stackrel{1\otimes\iota}{\longrightarrow} & A^1(M;E) & \stackrel{1\otimes\eta}{\longrightarrow} & A^1(M;E''). \end{array}$$

Proposition 8.5. 1. *Given a connection ∇' for E', there exist connections ∇ and ∇'' for E and E'' such that $(\nabla', \nabla, \nabla'')$ is* compatible *with* (8.17).

2. *Given a connection ∇'' for E'', there exist connections ∇' and ∇ for E' and E such that $(\nabla', \nabla, \nabla'')$ is* compatible *with* (8.17).

Proof. **1.** By Proposition 3.3 (see also Remark 3.4), $E \simeq E' \oplus E''$. Let ∇'' be an arbitrary connection for E'' and ∇ the connection for E corresponding to $\nabla' \oplus \nabla''$. Then $(\nabla', \nabla, \nabla'')$ has the desired property. Similarly for **2.** □

Proposition 8.6. *For a triple $(\nabla', \nabla, \nabla'')$ compatible with* (8.17),

$$c^*(\nabla) = c^*(\nabla') \wedge c^*(\nabla'').$$

Proof. We may find a (local) frame $e'^{(l')} = (e'_1, \dots, e'_{l'})$ of E' such that E has a frame $e^{(l)}$ of the form $e^{(l)} = (\iota(e'_1), \dots, \iota(e'_{l'}), e_{l'+1}, \dots, e_l)$. Then $e''^{(l'')} = (\eta(e_{l'+1}), \dots, \eta(e_l))$ is a frame of E''. Let θ', θ and θ'' be the connection matrices of ∇', ∇ and ∇'' with respect to $e'^{(l')}$, $e^{(l)}$ and $e''^{(l'')}$, respectively. Then we have

$$\theta = \begin{pmatrix} \theta' & * \\ 0 & \theta'' \end{pmatrix},$$

which implies the proposition. □

From this we have the following:

Proposition 8.7. *Let φ be an invariant polynomial and $(\nabla'_\nu, \nabla_\nu, \nabla''_\nu)$, $\nu = 0, \ldots, p$, triples of connections compatible with* (8.17).

1. *If we set $\nabla^\star_\nu = (\nabla''_\nu, \nabla_\nu)$, then*

$$\varphi(\nabla^\star_0, \ldots, \nabla^\star_p) = \varphi(\nabla'_0, \ldots, \nabla'_p), \quad \text{in particular} \quad \varphi(E - E'') = \varphi(E').$$

2. *If we set $\nabla^\star_\nu = (\nabla'_\nu, \nabla_\nu)$, then*

$$\varphi(\nabla^\star_0, \ldots, \nabla^\star_p) = \varphi(\nabla''_0, \ldots, \nabla''_p), \quad \text{in particular} \quad \varphi(E - E') = \varphi(E'').$$

Tensor product: Let E_1 and E_2 be complex vector bundles of ranks l_1 and l_2 on M and ∇_1 and ∇_2 connections for E_1 and E_2, respectively. Every section s of the tensor product $E = E_1 \otimes E_2$ is written locally as a finite sum of sections of the form $s_1 \otimes s_2$ with s_1 and s_2 sections of E_1 and E_2. If we define $\nabla : A^0(M; E_1 \otimes E_2) \to A^1(M; E_1 \otimes E_2)$ by

$$\nabla(s_1 \otimes s_2) = \nabla_1(s_1) \otimes s_2 + s_1 \otimes \nabla_2(s_2),$$

it is a connection for E. If we take some frames of E_1 and E_2 and if we denote by θ_1, κ_1 and θ_2, κ_2 the connection and curvature matrices of ∇_1 and ∇_2, respectively, the connection and curvature matrices θ and κ of ∇ with respect to the combined frame are given by

$$\theta = \theta_1 \otimes I_{l_2} + I_{l_1} \otimes \theta_2, \quad \kappa = \kappa_1 \otimes I_{l_2} + I_{l_1} \otimes \kappa_2.$$

Exterior product: Let E be a complex vector bundle and ∇ a connection for E. If we define $\bigwedge^r \nabla : A^0(M; \bigwedge^r E) \to A^1(M; \bigwedge^r E)$ by

$$\textstyle\bigwedge^r \nabla(s_1 \wedge \cdots \wedge s_r) = \sum_{i=1}^r s_1 \wedge \cdots \wedge \nabla(s_i) \wedge \cdots \wedge s_r,$$

it is a connection for $\bigwedge^r E$.

Dual: Let E be a complex vector bundle of rank l and ∇ a connection for E. The connection ∇^* for E^* *dual* to ∇ is determined by the relation

$$d\varphi(s) = \nabla^*(\varphi)(s) + \varphi(\nabla(s)) \quad \text{for} \quad s \in A^0(M; E), \quad \varphi \in A^0(M; E^*).$$

If θ and κ are the connection and curvature matrices of ∇ with respect to some frame, those matrices θ^* and κ^* of ∇^* with respect to the dual frame are given by $\theta^* = -{}^t\theta$ and $\kappa^* = -{}^t\kappa$ so that we have

$$c^q(\nabla^*) = (-1)^q c^q(\nabla) \quad \text{and} \quad c^q(E^*) = (-1)^q c^q(E).$$

8.4 Characteristic classes in Čech-de Rham cohomology

Let $\mathcal{U} = \{U_\alpha\}_{\alpha\in I}$ be an open covering of a C^∞ manifold M as in Section 7.3. Also, let $\pi : E \to M$ be a C^∞ complex vector bundle of rank l and φ an invariant polynomial homogeneous of degree k. For each α, we choose a connection ∇_α for E on U_α, and for the collection $\nabla_* = (\nabla_\alpha)_\alpha$, we define the element $\varphi(\nabla_*)$ in $A^{2k}(\mathcal{U})$ by

$$\varphi(\nabla_*)_{\alpha_0\dots\alpha_p} = \varphi(\nabla_{\alpha_0},\dots,\nabla_{\alpha_p}).$$

Then we have $D\varphi(\nabla_*) = 0$. Moreover, if we take another collection $\nabla'_* = (\nabla'_\alpha)_\alpha$,

$$\varphi(\nabla'_*) - \varphi(\nabla_*) = D\psi,$$

where ψ is the element in $A^{2k-1}(\mathcal{U})$ given by

$$\psi_{\alpha_0\dots\alpha_p} = \sum_{\nu=0}^{p}(-1)^\nu\varphi(\nabla_{\alpha_0},\dots,\nabla_{\alpha_\nu},\nabla'_{\alpha_\nu},\dots,\nabla'_{\alpha_p}).$$

Hence the form $\varphi(\nabla_*)$ defines a class in $H^{2k}_D(\mathcal{U})$ which depends only on E but not on the choice of the collection of connections ∇_*. Comparing with the class defined by a global connection, we have:

Theorem 8.1. *The class $[\varphi(\nabla_*)]$ in $H^{2k}_D(\mathcal{U})$ corresponds to the class $\varphi(E)$ in $H^{2k}_d(M)$ under the isomorphism of Theorem 7.1.*

For a virtual bundle $F-E$, we take a pair of connections $\nabla^\star_\alpha = (\nabla^E_\alpha, \nabla^F_\alpha)$ on each U_α, and for the collection $\nabla^\star_* = (\nabla^\star_\alpha)_\alpha$, we define the element $\varphi(\nabla^\star_*)$ in $A^{2k}(\mathcal{U})$ by

$$\varphi(\nabla^\star_*)_{\alpha_0\dots\alpha_p} = \varphi(\nabla^\star_{\alpha_0},\dots,\nabla^\star_{\alpha_p}).$$

Then we have $D\varphi(\nabla^\star_*) = 0$. Moreover, it is shown as above that the class of $\varphi(\nabla^\star_*)$ in $H^{2k}_D(\mathcal{U})$ does not depend on the choice of the collection of families of connections $\nabla^\star_*$ and the class $[\varphi(\nabla^\star_*)]$ corresponds to the class $\varphi(F-E)$ in $H^{2k}_d(M)$ under the isomorphism of Theorem 7.1.

Notes

For the Chern-Weil theory, we refer to [Baum and Bott (1972); Bott and Chern (1965); Milnor and Stasheff (1974)]. The construction of the difference form in Proposition 8.3 is due to [Bott (1972)]. Here the sign convention is altered so that it is compatible with the Čech-de Rham cocycle condition (cf. (7.4) and Section 8.4). As to Remark 8.4. 3, we refer to [Bott and Chern (1965)].

Chapter 9

Vector Bundles with Metrics and Related Topics

The notion of an inner product on a real vector space or a Hermitian inner product on a complex vector space is generalized to that of a Riemannian metric on a real vector bundle or a Hermitian metric on a complex vector bundle, respectively. A metric on a C^∞ or complex manifold is defined to be that on the real tangent bundle or the holomorphic tangent bundle, respectively. For each case we have the theory of harmonic forms, which we review briefly. On the way we introduce the Dolbeault cohomology, a complex counterpart of the de Rham cohomology. For a compact Riemannian manifold, the de Rham cohomology is isomorphic with the space of real harmonic forms and, for a compact Hermaitian manifold, the Dolbeault cohomology is isomorphic with the space of complex harmonic forms. The Poincaré duality is described in trems of real harmonic forms. We have the Kodaira-Serre duality as a complex counterpart of this.

For a complex manifold, the space of r-forms is the direct sum of the spaces of (p, q)-forms for (p, q) with $p+q = r$. In general, this relation does not directly descends to the cohomology level, i.e., the de Rham cohomology may not be the direct sum of the Dolbeault cohomologies. If this is the case, we say that the manifold admits a Hodge structure. We give some conditions for this to hold.

We study vector bundles on projective spaces. We see that there is a natural metric on the projective space, called the Fubini-Study metric.

For a Hermitian metric on a complex manifold, associated is a form of type $(1, 1)$. A Kähler manifold is a complex manifold that admits a Hermitian metric whose associated form is closed. For such a metric, a form is real harmonic if and only if it is complex harmonic. From this, we see that any compact Kähler manifold admits a Hodge structure. Every submanifold of the projective space is Kähler with the metric induced from the Fubini-Study metric.

We define the Atiyah classes of a holomorphic vector bundle in the Dolbeault cohomology, taking connections of type $(1,0)$ and developing a theory in parallel to the Chern-Weil theory. We also discuss the Bott-Chern cohomology of a complex manifold, which refines both the de Rham and Dolbeault cohomologies. Then we define the Bott-Chern classes of a Hermitian vector bundle, taking a metric connection for the bundle.

9.1 Metrics on vector bundles and harmonic forms

In this section we discuss metrics on vector bundles and briefly review the theory of harmonic forms.

Riemannian metrics

Inner products: Let V be a real vector space. An *inner product* on V is a bilinear form on V that is symmetric and positive definite, i.e., a bilinear form

$$g : \mathsf{V} \times \mathsf{V} \longrightarrow \mathbb{R}$$

such that $g(u,v) = g(v,u)$ for u and v in V and that $g(u,u) > 0$ for $u \neq 0$. An *inner product space* (V, g) is a vector space V together with an inner product g on V. For u in V, its *norm* is defined by $\|u\| = \sqrt{g(u,u)}$.

Example 9.1. Let $\mathsf{V} = \mathbb{R}^m$ be the vector space of column m-vectors. Then every bilinear form g is expressed as

$$g(u,v) = {}^t u G v = \sum_{i,j=1}^{m} x_i g_{ij} y_j \tag{9.1}$$

for some $m \times m$ matrix $G = (g_{ij})$, where $u = {}^t(x_1, \dots, x_m)$ and $v = {}^t(y_1, \dots, y_m)$. The form g is symmetric if and only if G is symmetric, in which case all the eigenvalues of G are real. Thus every inner product on $\mathbb{R}^m$ is expressed as (9.1) with G a positive symmetric matrix, i.e., a symmetric matrix whose eigenvalues are all positive. If G is the identity matrix, the corresponding inner product is the *standard inner product* (or *Euclidean inner product*) on $\mathbb{R}^m$.

In the following, we assume that V is finite dimensional of dimension m. In this case, fixing a basis of V, we have an isomorphism $\mathsf{V} \simeq \mathbb{R}^m$. Then the inner products on $\mathbb{R}^m$ are carried to those on V. In fact every inner product on V is obtained this way.

Let (V, g) be an inner product space. A basis $(e_1, \ldots, e_m)$ of V is said to be *g-orthonormal*, or simply *orthonormal*, if $g(e_i, e_j) = \delta_{ij}$. The standard basis of $\mathbb{R}^m$ is an orthonormal basis for the standard inner product. Starting with an arbitrary basis of V, we may construct an orthonomal basis by the Gram-Schmidt process. An *orthogonal transformation* of (V, g) is a linear transformation φ of V that preserves the inner product, i.e.,

$$g(\varphi(u), \varphi(v)) = g(u, v) \qquad \text{for all } u,\ v \text{ in } \mathsf{V}.$$

Recall that an $m \times m$ matrix P is *orthogonal* if ${}^tP\,P = I$. A linear transformation φ of V is orthogonal if and only if the matrix of φ with respect to an orthonormal basis is orthogonal.

Let $\mathsf{V}^* = \mathrm{Hom}(\mathsf{V}, \mathbb{R})$ be the dual space of V. Then an inner product g on V defines an isomorphism

$$\varphi_g : \mathsf{V} \xrightarrow{\sim} \mathsf{V}^* \qquad \text{by } \ \varphi_g(u)(v) = g(u, v).$$

Riemannian metrics on vector bundles: Let $\pi : E \to M$ be a C^∞ real vector bundle of rank l'.

Definition 9.1. A *Riemannian metric* g on E is a rule that assigns to each point x in M an inner product g_x on the fiber E_x and the assignment is C^∞ in the sense that for every C^∞ sections s and s' of E on an open set U, the function $g(s, s')$ defined by $g(s, s')(x) = g_x(s(x), s'(x))$ is C^∞ on U.

A *Riemannian vector bundle* (E, g) is a vector bundle E together with a Riemannian metric g on E.

Example 9.2. Let $E = \mathbb{R}^{l'} \times M$ be the product bundle. An inner product on $\mathbb{R}^{l'}$ defines a Riemannian metric on E.

Proposition 9.1. *Every vector bundle admits a Riemannian metric.*

Proof. Let $\{U_\alpha\}$ be a locally finite open covering of M such that E is trivial on each U_α and let g_α be a Riemannian metric on $E|_{U_\alpha}$. Taking a C^∞ partition of unity $\{\rho_\alpha\}$ subordinate to $\{U_\alpha\}$, set $g = \sum \rho_\alpha g_\alpha$. Then it is a Riemannian metric on E. □

Let (E, g) be a Riemannian vector bundle on M and $\{U_\alpha\}$ a covering of M such that E is trivial on each U_α. Letting $e_\alpha^{(l')} = (e_1^\alpha, \ldots, e_{l'}^\alpha)$ be a frame of E on U_α, set $G_\alpha = (g_{ij}^\alpha)$, $g_{ij}^\alpha = g(e_i^\alpha, e_j^\alpha)$. Then it is an $(l' \times l')$-matrix valued C^∞ function on U_α whose value at each point is a positive definite symmetric matrix. By a frame change $e_\beta^{(l')} = e_\alpha^{(l')} A_{\alpha\beta}$, we have

$$G_\beta = {}^t A_{\alpha\beta} G_\alpha A_{\alpha\beta}. \tag{9.2}$$

Note that $\{A_{\alpha\beta}\}$ is a system of transition matrices for E (cf. (3.5)). Conversely if we have a system of transition matrices $\{A_{\alpha\beta}\}$ for E and a system $\{G_\alpha\}$ satisfying (9.2), it defines a Riemannian metric on E.

A frame of a Riemannian vector bundle is *orthonormal* if it is orthonormal at each point. An orthonormal frame exists locally by the Gram-Schmidt process. If $e_\beta^{(l')}$ and $e_\alpha^{(l')}$ above are both orthonormal, then $A_{\alpha\beta}$ is an orthogonal matrix valued function. Thus in this case, (9.2) may be written $G_\beta = A_{\beta\alpha} G_\alpha A_{\alpha\beta}$.

Proposition 9.2. *The structure group of a real vector bundle E may be reduced to $O(l')$.*

Proof. Choose a Riemannian metric g on E and let $\{U_\alpha\}$ be a covering of M such that $E|_{U_\alpha}$ admits an orthonormal frame $e_\alpha^{(l')}$. Let ψ_α be a trivialization on U_α which takes $e_\alpha^{(l')}$ to the standard basis of $\mathbb{R}^{l'}$. Then $\psi_\alpha \circ \psi_\beta^{-1}$ is $O(l')$-valued. □

Remark 9.1. 1. Let $\pi : E \to M$ be a vector bundle and ε a C^∞ positive function on M. Choosing a Riemannian metric g on E, we may consider the open ε-ball bundle $\{\, e \in E \mid \|e\| < \varepsilon(\pi(e)) \,\}$ in E, $\|e\| = \sqrt{g(e,e)}$.

2. The bundle E is orientable if and only if its structure group can be reduced to $SO(l')$.

Let E be a real vector bundle and E^* its dual. If g is a Riemannian metric on E, it induces a morphism

$$\varphi_g : E \longrightarrow E^* \qquad \text{by } \ \varphi_g(\xi)(\eta) = g(\xi, \eta).$$

Exercise 9.1. Verify that φ_g is an isomorphism.

Riemannian metrics on C^∞ manifolds: Let M be a C^∞ manifold of dimension m. A *Riemannian metric* on M is a Riemannian metric g on its tangent bundle $T_{\mathbb{R}}M$. A *Riemannian manifold* (M, g) is a C^∞ manifold M together with a Riemannian metric g on M. Let $(x_1, \dots, x_m)$ be local coordinates on M and set

$$g_{ij} = g\Big(\frac{\partial}{\partial x_i}, \frac{\partial}{\partial x_j}\Big).$$

The Riemannian metric g defines a bilinear form

$$g : T_{\mathbb{R},x}M \times T_{\mathbb{R},x}M \longrightarrow \mathbb{R},$$

which defines a section of $(T_{\mathbb{R}}M \otimes T_{\mathbb{R}}M)^* \simeq T^*_{\mathbb{R}}M \otimes T^*_{\mathbb{R}}M$, given locally by

$$\sum_{i,j=1}^{m} g_{ij}(x)\, dx_i \otimes dx_j.$$

Suppose M is oriented and $(x_1, \ldots, x_m)$ a positive coordinate system on an open set U in M. Let $\tau^{(m)} = (\tau_1, \ldots, \tau_m)$ be a g-orthonormal frame of $T_{\mathbb{R}}M$ on U that defines the orientation of $T_{\mathbb{R}}(M)$, i.e., the orientation determined by $(\frac{\partial}{\partial x_1}, \ldots, \frac{\partial}{\partial x_m})$ (cf. Section 3.7). If we let $(\tau_1^*, \ldots, \tau_m^*)$ be the frame of $T^*_{\mathbb{R}}M$ dual to $\tau^{(m)}$, we see that

$$\tau_1^* \wedge \cdots \wedge \tau_m^* = \sqrt{\det(g_{ij})}\, dx_1 \wedge \cdots \wedge dx_m$$

and that it defines a global positive form, called the *volume form* of g and is denoted by $d\mu$.

Hermitian metrics

Hermitian inner products: Let V be a complex vector space.

Definition 9.2. A *Hermitian inner product* on V is a map

$$h : \mathsf{V} \times \mathsf{V} \longrightarrow \mathbb{C}$$

satisfying the following conditions:

(1) $h(u, v)$ is $\mathbb{C}$-linear in u,
(2) $h(v, u) = \overline{h(u, v)}$,
(3) if $u \neq 0$, then $h(u, u) > 0$.

A *Hermitian vector space* (V, h) is a vector space V together with a Hermitian inner product h on V. For u in V, its *norm* is defined by $\|u\| = \sqrt{h(u, u)}$.

Example 9.3. Let $\mathsf{V} = \mathbb{C}^n$ be the vector space of column n-vectors. Then every Hermitian inner product h is expressed as

$$h(u, v) = {}^t u H \bar{v} = \sum_{i,j=1}^{n} z_i h_{ij} \bar{w}_j,$$

for some positive Hermitian matrix $H = (h_{ij})$, where $u = {}^t(z_1, \ldots, z_n)$ and $v = {}^t(w_1, \ldots, w_n)$. Thus H is an $n \times n$ matrix such that ${}^tH = \bar{H}$ and that all of its eigenvalues are positive. If H is the identity matrix, the corresponding inner product is the *standard Hermitian inner product* on $\mathbb{C}^n$.

In the following, we assume that V is finite dimensional of dimension n. As in the real case, the Hermitian inner products on $\mathbb{C}^n$ are carried over to those on V. Let (V, h) be an n-dimensional Hermitian vector space. A basis $(e_1, \ldots, e_n)$ of V is said to be *h-orthonormal*, or simply *orthonormal* if

$$h(e_i, e_j) = \delta_{ij}.$$

The standard basis of $\mathbb{C}^n$ is an orthonormal basis for the standard inner product. Starting with an arbitrary basis of V, we may construct an orthonomal basis by the Gram-Schmidt process.

A *unitary transformation* of (V, h) is a linear transformation φ of V that preserves the inner product, i.e.,

$$h(\varphi(u), \varphi(v)) = h(u, v) \qquad \text{for all } u,\ v \text{ in } \mathsf{V}.$$

Recall that an $n \times n$ matrix U is unitary if ${}^tU\bar{U} = I$. A linear transformation φ of V is unitary if and only if the matrix U of φ with respect to an orthonormal basis is unitary.

Let $\mathsf{V}^* = \mathrm{Hom}(\mathsf{V}, \mathbb{C})$ be the dual space of V. Then a Hermitian inner product h on V defines an isomorphism (cf. Remark 2.11)

$$\varphi_h : \mathsf{V} \xrightarrow{\sim} \overline{\mathsf{V}^*} \qquad \text{by } \varphi_h(u)(v) = h(u, v).$$

Hermitian metrics on vector bundles: Let $\pi : E \to M$ be a C^∞ complex vector bundle of rank l on a C^∞ manifold M.

Definition 9.3. A *Hermitian metric* h on E is a rule that assigns to each point x in M a Hermitian inner product h_x on the fiber E_x and the assignment is C^∞ in the sense that for every C^∞ sections s and s' of E, $h(s, s')$ is a C^∞ function.

A *Hermitian vector bundle* (E, h) is a complex vector bundle E together with a Hermitian metric h on E.

Example 9.4. Let $E = \mathbb{C}^l \times M$ be the product vector bundle. Hermitian inner products on $\mathbb{C}^n$ define Hermitian metrics on E.

The following is proved as Proposition 9.1:

Proposition 9.3. *Every complex vector bundle admits a Hermitian metric.*

As in the Riemannian case, we have the following local representation of a Hermitian metric. Thus let (E, h) be a Hermitian vector bundle and $\{U_\alpha\}$ a covering of M such that $E|_{U_\alpha}$ is trivial for each α. Letting $e_\alpha^{(l)} = (e_1^\alpha, \ldots, e_l^\alpha)$ be a frame of E on U_α, set $H_\alpha = (h_{ij}^\alpha)$, $h_{ij}^\alpha = h(e_i^\alpha, e_j^\alpha)$.

Then it is an $(l \times l)$-matrix valued C^∞ function on U_α whose value at each point is a positive definite Hermitian matrix. By a frame change $e_\beta^{(l)} = e_\alpha^{(l)} A_{\alpha\beta}$, we have

$$H_\beta = {}^t A_{\alpha\beta} H_\alpha \overline{A_{\alpha\beta}}. \tag{9.3}$$

Conversely if we have a system $\{H_\alpha\}$ satisfying (9.3), it defines a Hermitian metric on E.

A frame of a Hermitian vector bundle is *orthonormal* if it is orthonormal at each point. An orthonormal frame exists locally by the Gram-Schmidt process. If $e_\beta^{(l)}$ and $e_\alpha^{(l)}$ above are both orthonormal, then $A_{\alpha\beta}$ is a unitary matrix valued function.

The following is proved as Proposition 9.2:

Proposition 9.4. *The structure group of a complex vector bundle may be reduced to $U(l)$.*

Let E be a complex vector bundle and E^* its dual. If h is a Hermitian metric on E, it defines an isomorphism (cf. Exercise 9.1)

$$\varphi_h : E \xrightarrow{\sim} \overline{E^*} \qquad \text{by } \varphi_h(\xi)(\eta) = h(\xi, \eta). \tag{9.4}$$

The metric h induces a Hermitian metric $\bar{h}$ on $\overline{E}$ by $\bar{h}(\xi,\eta) = \overline{h(\xi,\eta)}$. This defines an isomorphism $\varphi_{\bar{h}} : \overline{E} \xrightarrow{\sim} E^*$ by $\varphi_{\bar{h}}(\xi)(\eta) = \bar{h}(\xi,\eta)$.

Hermitian metrics on complex manifolds: Let M be a complex manifold of dimension n. A *Hermitian metric* on M is a Hermitian metric h on its holomorphic tangent bundle TM. A *Hermitian manifold* (M,h) is a complex manifold M together with a Hermitian metric h on M. Let $(z_1, \ldots, z_n)$ be local coordinates on M and set

$$h_{ij} = h\Big(\frac{\partial}{\partial z_i}, \frac{\partial}{\partial z_j}\Big).$$

The Hermitian metric h induces a bilinear form

$$h' : T_z M \times \overline{T}_z M \longrightarrow \mathbb{C} \qquad \text{given by} \quad h'(u,v) = h(u, \bar{v}).$$

This in turn induces a linear map

$$T_z M \otimes \overline{T}_z M \longrightarrow \mathbb{C},$$

which defines a section of $(TM \otimes \overline{T}M)^* \simeq T^*M \otimes \overline{T}^* M$, given locally by

$$ds^2 = \sum_{i,j=1}^{n} h_{ij}(z)\, dz_i \otimes d\bar{z}_j.$$

Recall that there is an isomorphism $TM \simeq T_{\mathbb{R}}M$ as real bundles (cf. Proposition 3.5) so that we have $\mathbb{R}$-bilinear forms as the real and imaginary parts of h:

$$\Re h,\ \Im h : T_{\mathbb{R}}M \times T_{\mathbb{R}}M \longrightarrow \mathbb{R}.$$

We see that $g = \Re h = \frac{1}{2}(h + \bar{h})$ is a positive definite symmetric bilinear form and defines a Riemannian metric on $T_{\mathbb{R}}M$. Locally it is given by

$$\frac{1}{2}\sum_{i,j=1}^{n} h_{ij}(z)(dz_i \otimes d\bar{z}_j + d\bar{z}_j \otimes dz_i).$$

On the other hand, $\Im h$ is a non-degenerate alternating form. We set

$$\omega = -\Im h = \frac{\sqrt{-1}}{2}(h - \bar{h})$$

and call it the *associated Hermitian form.* Locally it is given by

$$\omega = \frac{\sqrt{-1}}{2}\sum_{i,j=1}^{n} h_{ij}(z) dz_i \wedge d\bar{z}_j. \tag{9.5}$$

From the local expressions of g and ω, we see that the volume form $d\mu$ of g is given by

$$d\mu = \frac{\omega^n}{n!}. \tag{9.6}$$

From the local representations, we see that every one of the three forms h, g, and ω uniquely determine the other two (with $H = (h_{ij})$ positive definite Hermitian matrix).

Harmonic forms

Recall that, for a C^∞ manifold M, we denote by $A^r(M)$ the $\mathbb{C}$-vector space of complex valued r-forms on M (cf. Section 7.1).

Let (M, g) be an oriented Riemannian manifold of dimension m. The *star operator*

$$* : A^r(M) \longrightarrow A^{m-r}(M) \tag{9.7}$$

is defined as follows. Let $(\tau_1, \ldots, \tau_m)$ be a g-orthonormal frame of $T_{\mathbb{R}}M$ on some open set U of M and $(\tau_1^*, \ldots, \tau_m^*)$ the dual frame of T_M^*. For a subset $I = \{i_1, \ldots, i_r\}$ of $\{1, \ldots, m\}$ we let $J = \{j_1, \ldots, j_{m-r}\}$ denote its complement and $\varepsilon_{I,J}$ the sign of the permutation $\{1, \ldots, m\} \to \{I, J\}$. We set

$$* \tau_{i_1}^* \wedge \cdots \wedge \tau_{i_r}^* = \varepsilon_{I,J}\, \tau_{j_1}^* \wedge \cdots \wedge \tau_{j_{m-r}}^*$$

and extend it $A^0(U)$-linearly. Then we see that it does not depend on the choice of the orthonormal frame and defines an $A^0(M)$-module isomorphism. It is a local operator and has the following properties:

(1) $*1 = d\mu$, the volume form of g,
(2) $**\omega = (-1)^{r(m-r)}\omega$, for $\omega \in A^r(M)$,
(3) $\omega \wedge *\theta = \theta \wedge *\omega$, for $\omega,\ \theta \in A^r(M)$.

We set

$$d^* = (-1)^r *^{-1} d* : A^r(M) \longrightarrow A^{r-1}(M).$$

Note that $d^* d^* = 0$ and that, if m is even, then $d^* = - * d *$.

The *d-Laplacian* $\triangle_d : A^r(M) \to A^r(M)$ is defined by

$$\triangle_d = d d^* + d^* d.$$

It is what is called an elliptic differential operator.

Definition 9.4. An r-form ω is *d-harmonic* if $\triangle_d \omega = 0$.

We denote by $\mathcal{H}^r(M)$ the space of d-harmonic r-forms on M.

Now we assume that M is compact. If we define a pairing on $A^r(M)$ by

$$(\omega, \theta) = \int_M \omega \wedge *\bar{\theta}, \tag{9.8}$$

we see that it is a Hermitian inner product on $A^r(M)$. Then d^* is the adjoint of d; $(d\omega, \theta') = (\omega, d^*\theta')$, for $\omega \in A^r(M)$ and $\theta' \in A^{r+1}(M)$, and $\triangle_d$ is self-adjoint; $(\triangle_d \omega, \theta) = (\omega, \triangle_d \theta)$. From

$$(\triangle_d \omega, \omega) = \|d\omega\|^2 + \|d^*\omega\|^2,$$

we see that an r-form ω is d-harmonic if and only if $d\omega = 0$ and $d^*\omega = 0$. Since $\triangle_d = (d+d^*)(d+d^*)$, this is equivalent to saying that $(d+d^*)\omega = 0$. The above holds also for forms of mixed degree:

Proposition 9.5. *Let M be a compact Riemannian manifold of dimension m. For $\omega \in \bigoplus_{r=0}^m A^r(M)$, the following three conditions are equivalent:*

(1) $\triangle_d \omega = 0$, (2) $d\omega = d^*\omega = 0$, (3) $(d + d^*)\omega = 0$.

From general results for self-adjoint elliptic operators, we have:

Theorem 9.1. *Let M be a compact Riemannian manifold. Then the space $\mathcal{H}^r(M)$ is finite dimensional and there exist orthogonal decompositions:*

$$A^r(M) = \mathcal{H}^r(M) \oplus dA^{r-1}(M) \oplus d^* A^{r+1}(M),$$
$$\operatorname{Ker} d^r = \mathcal{H}^r(M) \oplus dA^{r-1}(M).$$

Corollary 9.1. *We have an isomorphism*

$$H_d^r(M) \simeq \mathcal{H}^r(M).$$

This reproves the finite dimensionality of $H_d^r(M)$ for compact M.

Dolbeault cohomology and its conjugate

Let M a complex manifold of dimension n and $A^{p,q}(M)$ the $\mathbb{C}$-vector space of C^∞ (p,q)-forms on M (cf. Section 6.1). By (6.9), we have a complex of $\mathbb{C}$-vector spaces for each fixed p:

$$0 \longrightarrow A^{p,0}(M) \xrightarrow{\bar{\partial}^{p,0}} A^{p,1}(M) \xrightarrow{\bar{\partial}^{p,1}} \cdots \xrightarrow{\bar{\partial}^{p,n-1}} A^{p,n}(M) \longrightarrow 0.$$

It is called the p-th *Dolbeault complex* of M and is denoted by $(A^{p,\bullet}(M), \bar{\partial})$.

Definition 9.5. The *Dolbeault cohomology* of M of type (p,q) is the q-th cohomology of $(A^{p,\bullet}(M), \bar{\partial})$:

$$H^{p,q}_{\bar{\partial}}(M) = \operatorname{Ker} \bar{\partial}^{p,q} / \operatorname{Im} \bar{\partial}^{p,q-1}.$$

For a $\bar{\partial}$-closed (p,q)-form ω, its class in $H^{p,q}_{\bar{\partial}}(M)$ is denoted by $[\omega]$. Since $\operatorname{Ker} \bar{\partial}^0 = \Gamma(M; \bigwedge^p T^*M)$, the space of holomorphic p-forms on M (cf. Section 6.1), we have:

$$H^{p,0}_{\bar{\partial}}(M) = \Gamma(M; \textstyle\bigwedge^p T^*M).$$

In general there is the Dolbealt theorem which states that $H^{p,q}_{\bar{\partial}}(M)$ is isomorphic with the cohomology $H^q(M; \Omega^p)$ of M with coefficients in the sheaf Ω^p of germs of holomorphic p-forms (cf. Section 11.4 below).

Denoting by $A^r(M)$ the $\mathbb{C}$-vector space of complex valued C^∞ r-forms on M, we have a canonical direct sum decomposition

$$A^r(M) = \bigoplus_{p+q=r} A^{p,q}(M)$$

and the de Rham complex $(A^\bullet(M), d)$ is the single complex associated with the double complex $(A^{\bullet,\bullet}(M), \partial, \bar{\partial})$ (cf. Section A.1). In general this does not lead to a decomposition on the cohomology level. However, there is a spectral sequence relating the Dolbeault cohomology $H^{p,q}_{\bar{\partial}}(M)$ and the de Rham cohomology $H^r_d(M)$ (cf. Section 9.2 below).

Exercise 9.2. 1. Show that the identity map of forms induces a canonical morphism

$$H^{n,q}_{\bar{\partial}}(M) \longrightarrow H^{n+q}_d(M),$$

which is surjective if $q = n$.

2. Show that the assignment of the $(0,q)$-component to each q-form induces a morphism

$$H^q_d(M) \longrightarrow H^{0,q}_{\bar{\partial}}(M).$$

Complex conjugate: We may as well consider, for each fixed q, the complex $(A^{\bullet,q}(M), \partial)$ and its p-th cohomology:

$$H_{\partial}^{p,q}(M) = \operatorname{Ker} \partial^{p,q} / \operatorname{Im} \partial^{p-1,q}.$$

From $A^{p,q}(M) = \overline{A^{q,p}(M)}$, we have a canonical identification (cf. Remark 2.11):

$$H_{\partial}^{p,q}(M) = \overline{H_{\bar{\partial}}^{q,p}(M)}, \qquad [\omega] \leftrightarrow [\bar{\omega}]. \tag{9.9}$$

Complex harmonic forms

Let (M, h) be a Hermitian manifold of dimension n. As we have seen, h defines a Riemannian metric g on M and we have the star operator (9.7). In fact it induces an $A^0(M)$-module isomorphism

$$* : A^{p,q}(M) \longrightarrow A^{n-q,n-p}(M). \tag{9.10}$$

We set

$$\bar{\partial}^* = (-1)^{p+q} *^{-1} \partial * = - * \partial * : A^{p,q}(M) \longrightarrow A^{p,q-1}(M).$$

The *$\bar{\partial}$-Laplacian* $\triangle_{\bar{\partial}} : A^{p,q}(M) \to A^{p,q}(M)$ is then defined by

$$\triangle_{\bar{\partial}} = \bar{\partial}\,\bar{\partial}^* + \bar{\partial}^*\,\bar{\partial},$$

which is shown to be an elliptic differential operator.

Definition 9.6. A (p,q)-form ω is *$\bar{\partial}$-harmonic* if $\triangle_{\bar{\partial}}\,\omega = 0$.

We denote by $\mathcal{H}^{p,q}(M)$ the space of $\bar{\partial}$-harmonic (p,q)-forms on M.

Now we assume that M is compact. Then the Hermitian inner product (9.8) restricts to that on $A^{p,q}(M)$. With this, $\bar{\partial}^*$ is the adjoint of $\bar{\partial}$ and $\triangle_{\bar{\partial}}$ is self-adjoint. We also see that a (p,q)-form ω is $\bar{\partial}$-harmonic if and only if $\bar{\partial}\omega = \bar{\partial}^*\omega = 0$. As in the case of Theorem 9.1, we have:

Theorem 9.2. *For a compact Hermitian manifold M, the space $\mathcal{H}^{p,q}(M)$ is finite dimensional and there exist orthogonal decompositions*:

$$A^{p,q}(M) = \mathcal{H}^{p,q}(M) \oplus \bar{\partial} A^{p,q-1}(M) \oplus \bar{\partial}^* A^{p,q+1}(M),$$
$$\operatorname{Ker} \bar{\partial}^{p,q} = \mathcal{H}^{p,q}(M) \oplus \bar{\partial} A^{p,q-1}(M).$$

Corollary 9.2. *We have an isomorphism*

$$H_{\bar{\partial}}^{p,q}(M) \simeq \mathcal{H}^{p,q}(M).$$

This proves the finite dimensionality of $H_{\bar{\partial}}^{p,q}(M)$ for compact M.

Remark 9.2. We may also define the adjoint of $\partial : A^{p,q}(M) \to A^{p+1,q}(M)$ by $\partial^* = - * \bar{\partial} *$. The ∂-Laplacian $\triangle_\partial : A^{p,q}(M) \to A^{p,q}(M)$ is then defined by

$$\triangle_\partial = \partial \partial^* + \partial^* \partial$$

and there is the corresponding harmonic theory. In particular, if M is compact, a (p,q)-form is ∂-harmonic if and only if it is ∂- and ∂^*-closed. We also have a decomposition theorem as Theorem 9.2.

Poincaré duality: Let M be a compact oriented Riemannian manifold of dimension m. The bilinear map

$$A^r(M) \times A^{m-r}(M) \xrightarrow{\wedge} A^m(M) \xrightarrow{\int_M} \mathbb{C}$$

is non-degenerate, since if $\int_M \omega \wedge \theta = 0$ for all $\theta \in A^{m-r}(M)$, setting $\theta = *\bar{\omega}$ we see that $\omega = 0$. By restricting the bilinear map to harmonic representatives and using the fact that the spaces of harmonic forms are finite dimensional, we see that

$$H^r_d(M) \simeq H^{m-r}_d(M)^*.$$

Thus we recover Corollary 7.9.

Kodaira-Serre duality: Let M be a compact Hermitian manifold of dimension n. The bilinear map

$$A^{p,q}(M) \times A^{n-p,n-q}(M) \xrightarrow{\wedge} A^{n,n}(M) = A^{2n}(M) \xrightarrow{\int_M} \mathbb{C}$$

is non-degenerate. By restricting the bilinear map to harmonic representatives and using the fact that the spaces of harmonic forms are finite dimensional, we see that

$$H^{p,q}_{\bar{\partial}}(M) \simeq H^{n-p,n-q}_{\bar{\partial}}(M)^*.$$

Cohomology with coefficients in a vector bundle: Let E be a holomorphic vector bundle of rank l on a complex manifold M of dimension n. For each p, we have a complex (cf. (6.10)):

$$0 \longrightarrow A^{p,0}(M;E) \xrightarrow{\bar{\partial}} A^{p,1}(M;E) \xrightarrow{\bar{\partial}} \cdots \xrightarrow{\bar{\partial}} A^{p,n}(M;E) \longrightarrow 0,$$

the q-th cohomology of which is denoted by $H^{p,q}_{\bar{\partial}}(M;E)$.

Let E^* be the dual bundle. Then we have a wedge product

$$A^{p,q}(M;E) \times A^{r,s}(M;E^*) \xrightarrow{\wedge} A^{p+r,q+s}(M)$$

defined by

$$(\omega \otimes s) \wedge (\theta \otimes \sigma) = \sigma(s) \cdot \omega \wedge \theta$$

for decomposable elements and then extended $A^0(M)$-bilinearly. It satisfies
$\bar\partial(\xi\wedge\psi)=\bar\partial\xi\wedge\psi+(-1)^{p+q}\xi\wedge\bar\partial\psi$ for $\xi\in A^{p,q}(M;E)$, $\psi\in A^{r,s}(M;E^*)$.
Thus we have a wedge product

$$H^{p,q}_{\bar\partial}(M;E)\times H^{r,s}_{\bar\partial}(M;E^*)\xrightarrow{\ \wedge\ }H^{p+r,q+s}_{\bar\partial}(M).$$

In general, for a holomorphic vector bundle E, if we set $A^{p,q}(M;\overline{E})=C^\infty(\bigwedge^p T^*M\otimes\bigwedge^q\overline{T}^*M\otimes\overline{E})$, the operator $\partial: A^{p,q}(M;\overline{E})\to A^{p+1,q}(M;\overline{E})$ is well-defined.

Now suppose M and E are endowed with Hermitian metrics. Then we have the star operator (9.10) and the isomorphism (9.4) for the metric h on E. We define an isomorphism

$$*_h: A^{p,q}(M;E)\longrightarrow A^{n-q,n-p}(M;\overline{E^*})$$

by $*_h(\omega\otimes s)=(*\omega)\otimes\varphi_h(s)$ for decomposable elements and then extending it $A^0(M)$-linearly. We set

$$\bar\partial_h^*=(-1)^{p+q}*_h^{-1}\circ\partial\circ *_h: A^{p,q}(M;E)\longrightarrow A^{p,q-1}(M;E). \qquad (9.11)$$

The *h-$\bar\partial$-Laplacian* $\triangle_{\bar\partial,h}: A^{p,q}(M;E)\to A^{p,q}(M;E)$ is then defined by

$$\triangle_{\bar\partial,h}=\bar\partial\,\bar\partial_h^*+\bar\partial_h^*\,\bar\partial,$$

which is shown to be an elliptic differential operator. The space of harmonic (p,q)-forms with coefficients in E is defined by

$$\mathcal{H}^{p,q}(M;E)=\{\,\xi\in A^{p,q}(M;E)\mid\triangle_{\bar\partial,h}\xi=0\,\}.$$

Note that, for $\xi\in A^{p,q}(M;E)$, we may think of $\bar\xi$ as being in $A^{q,p}(M;\overline{E})$ and that we also have $\varphi_{\bar h}:\overline{E}\xrightarrow{\sim}E^*$. Thus we have $*_{\bar h}: A^{q,p}(M;\overline{E})\xrightarrow{\sim}A^{n-p,n-q}(M;E^*)$.

Now we assume that M is compact. If we define a pairing on $A^{p,q}(M;E)$ by

$$(\xi,\eta)=\int_M\xi\wedge *_{\bar h}\,\bar\eta, \qquad (9.12)$$

we see that it is a Hermitian inner product on $A^{p,q}(M;E)$. Then $\bar\partial_h^*$ is the adjoint of $\bar\partial$ and $\triangle_{\bar\partial,h}$ is self-adjoint. Thus we have a decomposition theorem as Theorem 9.2 and an isomorphism (cf. Corollary 9.2)

$$H^{p,q}_{\bar\partial}(M;E)\simeq\mathcal{H}^{p,q}(M;E),$$

which shows the finite dimensionality of $H^{p,q}_{\bar\partial}(M;E)$ for compact M.

A statement similar to Proposition 9.5 is valid in this case as well so that, for $\xi\in\bigoplus_{q=0}^n A^{p,q}(M;E)$, the following three conditions are equivalent:

(1) $\triangle_{\bar\partial,h}\,\xi=0$, (2) $\bar\partial\,\xi=\bar\partial_h^*\,\xi=0$, (3) $(\bar\partial+\bar\partial_h^*)\,\xi=0$.

We also have a general form of the Kodaira-Serre duality, which reads

$$H^{p,q}_{\bar\partial}(M;E)\simeq H^{n-p,n-q}_{\bar\partial}(M;E^*)^*. \qquad (9.13)$$

9.2 Hodge structures

Let M be a complex manifold. The de Rham complex $(A^\bullet(M), d)$ of M is the single complex associated with the double complex $(A^{\bullet,\bullet}(M), \partial, \bar\partial)$, $d = \partial + \bar\partial$. We have:

$$A^{p,q}(M) = \overline{A^{q,p}(M)} \quad \text{and} \quad A^r(M) = \bigoplus_{p+q=r} A^{p,q}(M).$$

We try to find when the above relations carry on to the cohomologies.

Hodge filtrations: Note that $A^\bullet(M)$ has two natural filtrations (cf. Section A.1). The first filtration on $A^r(M)$ is given by

$$'F^p A^r(M) = \bigoplus_{i=p}^{r} A^{i,r-i}(M).$$

It induces a filtration on $H^r_d(M)$ by

$$'F^p H^r_d(M) = \operatorname{Ker} d^r \cap {'F^p} A^r(M) / \operatorname{Im} d^{r-1} \cap {'F^p} A^r(M),$$

for which we have

$$H^r_d(M) = {'F^0} H^r_d(M) \supset {'F^1} H^r_d(M) \supset \cdots \supset {'F^{r+1}} H^r_d(M) = 0. \tag{9.14}$$

The second filtration on $A^r(M)$ is given by

$$''F^q A^r(M) = \bigoplus_{j=q}^{r} A^{r-j,j}(M)$$

and it induces a filtration $(''F^q H^r_d(M))$ on $H^r_d(M)$.

As $\overline{A^{q,p}(M)} = A^{p,q}(M)$, we may identify the filtration $(\overline{'F^q A^r(M)})$ conjugate to $('F^q A^r(M))$ with the second filtration: $\overline{'F^q A^r(M)} = {''F^q} A^r(M)$, which leads to the identification

$$\overline{'F^q H^r_d(M)} = {''F^q} H^r_d(M).$$

Definition 9.7. The filtration $('F^p H^r_d(M))$ is a *Hodge filtration* of weight r if

$$H^r_d(M) = \bigoplus_{p+q=r} {'F^p} H^r_d(M) \cap \overline{'F^q H^r_d(M)}.$$

Lemma 9.1. *The filtration $('F^p H^r_d(M))$ is a Hodge filtration of weight r if and only if*

$$H^r_d(M) = {'F^p} H^r_d(M) \oplus \overline{'F^q H^r_d(M)} \qquad \textit{for every } (p,q) \textit{ with } p+q = r+1.$$

Moreover, if this is the case, there is a canonical isomorphism

$$'F^p H^r_d(M) \cap \overline{'F^q H^r_d(M)} \simeq {'G^p} H^r_d(M) \qquad \textit{for every } (p,q) \textit{ with } p+q = r,$$

where $'G^p H^r_d(M) = {'F^p} H^r_d(M) / {'F^{p+1}} H^r_d(M)$.

Proof. In the following, we denote $H^r_d(M)$ by H^r.

Suppose the filtration $('F^pH^r)$ is a Hodge filtration of weight r. Take $c \in H^r$. Then we may write uniquely $c = \sum_{i+j=r} c^{i,j}$, $c^{i,j} \in 'F^iH^r \cap \overline{'F^jH^r}$. For (p,q) with $p+q = r+1$, $c = c_1 + c_2$, where $c_1 = \sum_{i \geq p} c^{i,r-i} \in 'F^pH^r$ and $c_2 = \sum_{i \leq p-1} c^{i,r-i} = \sum_{j \geq q} c^{r-j,j} \in \overline{'F^qH^r}$.

For the converse, take $c \in H^r$. Then we may write uniquely $c = a_r + b_1$ with $a_r \in 'F^rH^r = 'F^rH^r \cap \overline{'F^0H^r}$ and $b_1 \in \overline{'F^1H^r}$. We may write uniquely $b_1 = a_{r-1} + b_2$ with $a_{r-1} \in 'F^{r-1}H^r$ and $b_2 \in \overline{'F^2H^r} \subset \overline{'F^1H^r}$. Thus $a_{r-1} \in 'F^rH^{r-1} \cap \overline{'F^1H^r}$. Continuing this, we have the desired direct sum decomposition.

For the last statement, for $c \in 'F^pH^r$ we denote by $[c]^p$ its class in $'G^pH^r$. We define a morphism

$$'F^pH^r \cap \overline{'F^qH^r} \longrightarrow 'G^pH^r \qquad \text{by } \; c \mapsto [c]^p$$

and show that it is an isomorphism. For the surjectivity, take $[c]^p \in 'G^pH^r$, $c \in 'F^pH^r$. Then we may write uniquely $c = \sum_{i=0}^r c^{i,r-i}$ with $c^{i,r-i} \in 'F^iH^r \cap \overline{'F^{r-i}H^r}$. We have $[c]^p = [c']^p$, $c' = \sum_{i=0}^p c^{i,r-i}$. Since $\sum_{i=p+1}^r c^{i,r-i} \in 'F^{p+1}H^r \subset 'F^pH^r$, we have $c' \in 'F^pH^r$. On the other hand, c' is also in $\overline{'F^qH^r}$ and $c' \mapsto [c]^p$. For the injectivity, take $c \in 'F^pH^r \cap \overline{'F^qH^r}$ such that $[c]^p = 0$. This means that $c \in 'F^{p+1}H^r \cap \overline{'F^qH^r} = 0$. □

Hodge structures: We first give the following:

Definition 9.8. 1. We say that M admits a *Hodge structure* of weight r if there exist isomorphisms

$$H^{p,q}_{\bar\partial}(M) \simeq \overline{H^{q,p}_{\bar\partial}(M)},\ p+q=r, \quad \text{and} \quad H^r_d(M) \simeq \bigoplus_{p+q=r} H^{p,q}_{\bar\partial}(M).$$

2. A Hodge structure as above is said to be *natural* if the following conditions hold:

(H1) Every class in $H^{p,q}_{\bar\partial}(M)$, $p+q=r$, has a representative ω with $\partial\omega = 0$ and $\bar\partial\omega = 0$, i.e., $d\omega = 0$, and the assignment $\omega \mapsto \overline{\omega}$ induces the first isomorphism above.

(H2) Every class in $H^r_d(M)$ has a representative ω which may be written $\omega = \sum_{p+q=r} \omega^{p,q}$, where $\omega^{p,q}$ is a (p,q)-form with $d\omega^{p,q} = 0$, and the assignment $\omega \mapsto (\omega^{p,q})_{p+q=r}$ induces the second isomorphism above.

Remark 9.3. 1. We may rephrase (H1) as (cf. (9.9)):

(H1)′ Every class in $H^{p,q}_{\bar\partial}(M)$, $p+q=r$, has a representative ω with $\partial\omega = 0$ and $\bar\partial\omega = 0$, i.e., $d\omega = 0$, and the assignment $\omega \mapsto \omega$ induces an isomorphism $H^{p,q}_{\bar\partial}(M) \simeq H^{p,q}_{\partial}(M)$.

2. Every compact Kähler manifold admits a natural Hodge structure of weight r for every r (cf. Corollary 9.5 below).

Note that every element in $'G^pH^r_d(M) = {}'F^pH^r_d(M)/{}'F^{p+1}H^r_d(M)$ is written $[[\omega]]^p$, where $\omega \in {}'F^pA^r(M)$ with $d\omega = 0$, $[\omega]$ is the class of ω in $'F^pH^r_d(M)$ and $[[\omega]]^p$ is the class of $[\omega]$ in $'G^pH^r_d(M)$. The condition $d\omega = 0$ implies $\bar\partial\omega^{p,r-p} = 0$ when we write $\omega = \sum_{i=p}^r \omega^{i,r-i}$.

Theorem 9.3. *A complex manifold M admits a natural Hodge structure of weight r if and only if the following conditions hold:*

(1) *the morphism* $\operatorname{Ker} d \cap {}'F^pA^r(M) \to A^{p,r-p}(M)$, $\omega \mapsto \omega^{p,r-p}$, *induces an isomorphism* $'G^pH^r_d(M) \simeq H^{p,r-p}_{\bar\partial}(M)$ *for every p,*
(2) $('F^pH^r_d(M))$ *is a Hodge filtration on $H^r_d(M)$ of weight r.*

Proof. Suppose M admits a natural Hodge structure of weight r. We claim that there is an isomorphism

$$'F^pH^r_d(M) \simeq \bigoplus_{i=p}^r H^{i,r-i}_{\bar\partial}(M) \tag{9.15}$$

compatible with the one in (H2) in the sense that the following is commutative:

$$\begin{array}{ccc} 'F^pH^r_d(M) & \xrightarrow{\sim} & \bigoplus_{i=p}^r H^{i,r-i}_{\bar\partial}(M) \\ \cap & & \cap \\ H^r_d(M) & \xrightarrow{\sim} & \bigoplus_{i=0}^r H^{i,r-i}_{\bar\partial}(M). \end{array}$$

For this, take $\theta \in \operatorname{Ker} d^r \cap {}'F^pA^r(M)$ and write $\theta = \sum_{i=p}^r \theta^{i,r-i}$ with $\theta^{i,r-i} \in A^{i,r-i}(M)$. From $d\theta = 0$ and (H1), we see that there exist $\omega^{i,r-i}$ and $\alpha^{i,r-i}$ in $A^{i,r-i}(M)$, $p \le i \le r$, such that $d\omega^{i,r-i} = 0$ and that

$$\theta^{i,r-i} = \omega^{i,r-i} + \partial\alpha^{i-1,r-i} + \bar\partial\alpha^{i,r-i-1},$$

where we set $\alpha^{p-1,r-p} = 0$. Then we have

$$\theta = \sum_{i=p}^r \omega^{i,r-i} + d\sum_{i=p}^r \alpha^{i,r-i-1}.$$

By (H2), the assignment $\theta \mapsto (\omega^{i,r-i})_{i=p}^r$ induces a well-defined morphism $'F^pH^r_d(M) \to \bigoplus_{i=p}^r H^{i,r-i}_{\bar\partial}(M)$ compatible with the isomorphism of (H2).

It is obviously injective. The surjectivity follows from (H1) and it is the desired isomorphism.

From (9.15), we have $'G^pH_d^r(M) \simeq H_{\bar\partial}^{p,r-p}(M)$ and the correspondence is the one as given in (1).

We also have an isomorphism $\overline{'F^qH_d^r(M)} \simeq \bigoplus_{j=q}^r H_{\bar\partial}^{r-j,j}(M)$ compatible with the one in (H2). Thus for (p,q) with $p+q = r+1$, $H_d^r(M) = {'F^pH_d^r(M)} \oplus \overline{'F^qH_d^r(M)}$ and we have (2).

Now we prove the converse. The condition (2) implies (cf. Lemma 9.1)

$$H_d^r(M) = \bigoplus_{p+q=r} {'F^pH_d^r(M)} \cap \overline{'F^qH_d^r(M)} \qquad \text{and} \tag{9.16}$$

$$'F^pH_d^r(M) \cap \overline{'F^qH_d^r(M)} \simeq {'G^pH_d^r(M)}, \quad r = p+q. \tag{9.17}$$

From the condition (1) and (9.17), we have

$$\begin{aligned} H_{\bar\partial}^{p,q}(M) &\simeq {'G^pH_d^r(M)} \simeq {'F^pH_d^r(M)} \cap \overline{'F^qH_d^r(M)} \\ &\simeq \overline{'G^qH_d^r(M)} \simeq \overline{H_{\bar\partial}^{q,p}(M)}, \quad r = p+q. \end{aligned}$$

We look at the correspondence above. Take $c \in {'F^pH_d^r(M)} \cap \overline{'F^qH_d^r(M)}$, we may write $c = [\omega_1] = [\omega_2]$, where $\omega_1 = \sum_{i=p}^r \omega_1^{i,h-i} \in {'F^pA^r(M)}$, $d\omega_1 = 0$ (thus $\bar\partial\omega_1^{p,q} = 0$) and $\omega_2 = \sum_{j=q}^r \omega_1^{h-j,j} \in \overline{'F^qA^r(M)}$, $d\omega_2 = 0$ (thus $\partial\omega_2^{p,q} = 0$). Then the correspondence is given by $[\omega_1^{p,q}] \leftrightarrow [\overline{\omega_2^{p,q}}]$. From $[\omega_1] = [\omega_2]$, we have $\omega_1 - \omega_2 = d\theta$ for some $\theta \in A^{r-1}(M)$. Denoting by $\theta^{p,q-1}$ and $\theta^{p-1,q}$ the $(p,q-1)$- and $(p-1,q)$-components of θ, we have

$$\omega_1^{p,q} - \omega_2^{p,q} = \partial\theta^{p-1,q} + \bar\partial\theta^{p,q-1}.$$

Then $\omega = \omega_1^{p,q} - \bar\partial\theta^{p,q-1} = \omega_2^{p,q} + \partial\theta^{p-1,q}$ is a representative as in (H1). From (9.16), (9.17) and the condition (1), we have (H2). □

Remark 9.4. If M admits a natural Hodge structure, the isomorphisms giving the Hodge structure are uniquely determined.

Frölicher spectral sequence: Recall that there are two spectral sequences associated with a double complex (cf. Theorem A.13). For the first filtration $('F^pA^\bullet(M))$, we have:

Proposition 9.6. *There is a spectral sequence* $(E_k^{p,q}, H^r)$ *with*

$$E_1^{p,q} \simeq H_{\bar\partial}^{p,q}(M) \quad \text{and} \quad H^r = H_d^r(M).$$

It is referred to as the *Frölicher spectral sequence* for M. We have

$$E_\infty^{p,q} \simeq {'G^pH_d^{p+q}(M)}. \tag{9.18}$$

Proposition 9.7. *The Frölicher spectral sequence for* M *degenerates at* E_1 *if and only if the condition* (1) *in Theorem 9.3 holds for every* r.

Proof. If the Frölicher spectral sequence degenerates at E_1,

$$'G^p H_d^r(M) \simeq E_\infty^{p,r-p} = E_1^{p,r-p} \simeq H_{\bar\partial}^{p,r-p}(M)$$

and by (A.9) and (A.10), we have the condition (1) in Theorem 9.3 for every r.

For the converse, note that $E_1^{p,q} \simeq H_{\bar\partial}^{p,q}(M)$ and $d_1 : E_1^{p,q} \to E_1^{p+1,q}$ corresponds to $\partial : H_{\bar\partial}^{p,q}(M) \to H_{\bar\partial}^{p+1,q}(M)$. By assumption, every class in $H_{\bar\partial}^{p,q}(M)$ has a representative $\omega^{p,q}$ which is the (p,q)-component of some $\omega = \sum_{i=p}^r \omega^{i,r-i} \in {'F^p} H_d^r(M)$ with $d\omega = 0$. The last condition implies $\bar\partial\omega^{p,q} = 0$ and $\partial\omega^{p,q} = -\bar\partial\omega^{p+1,q-1}$. Thus $d_1 = 0$ and by Proposition A.15, the Frölicher spectral sequence degenerates at E_1. □

From Theorem 9.3 and Proposition 9.7, we have:

Corollary 9.3. *A complex manifold M admits a natural Hodge structure of weight r for every r, if and only if the following conditions hold:*

(1) *the Frölicher spectral sequence for M degenerates at E_1,*
(2) *$({'F^p} H_d^r(M))$ is a Hodge filtration on $H_d^r(M)$ of weight r for every r.*

Numerical invariants: Suppose now that M is compact and let $n = \dim M$. By the de Rham theorem, $\dim H_d^r(M) = \dim H^r(M;\mathbb{C}) = b_r$, the r-th Betti number. The Euler-Poincaré characteristic is defined by $\chi(M) = \sum_{r=0}^{2n}(-1)^r b_r$ (cf. (5.14)). In this case $H_{\bar\partial}^{p,q}(M)$ is also finite dimensional (cf. Corollary 9.2) and we set $h^{p,q} = \dim H_{\bar\partial}^{p,q}(M)$.

Proposition 9.8. *Let M be a compact complex manifold of dimension n. Then we have:*
1. $b_r \le \sum_{p+q=r} h^{p,q}$, *where the equality holds if and only if the Frölicher spectral sequence degenerates at E_1.*
2. $\chi(M) = \sum_{p,q=0}^n (-1)^{p+q} h^{p,q}$.

Proof. Consider the Frölicher spectral sequence $(E_k^{p,q}, H^r)$. From (9.14) and (9.18), we have

$$b_r = \sum_{p=0}^r \dim {'G^p} H_d^r(M) = \sum_{p+q=r} \dim E_\infty^{p,q}.$$

Since $E_{k+1}^{p,q}$ is the cohomology of $E_k^{p,q}$ by the differential operator d_k, we have

$$\dim E_\infty^{p,q} \le \cdots \le \dim E_2^{p,q} \le \dim E_1^{p,q} = h^{p,q}$$

and the statement 1.

Also, since the alternating sum is invariant under taking cohomology, we have the equality in 2. □

Remark 9.5. A compact complex manifold M admits a Hodge structure (not necessarily the natural one) of weight r if and only if $h^{p,q} = h^{q,p}$, $p+q=r$, and $b_r = \sum_{p+q=r} h^{p,q}$.

9.3 Vector bundles on projective space

Tautological bundle

Since the projective space $\mathbb{P}^n$ is a particular Grassmann manifold, there is a natural line bundle T_n on $\mathbb{P}^n$, the *tautological bundle*, as we saw in Section 3.6. It is the bundle whose fiber at a point L in $\mathbb{P}^n$ is itself considered as a line and is described as follows. First, it can be expressed as

$$T_n = \{ (z, L) \in \mathbb{C}^{n+1} \times \mathbb{P}^n \mid z \in L \}.$$

Thus it is a subset of $\mathbb{C}^{n+1} \times \mathbb{P}^n$ together with a map $\pi : T_n \to \mathbb{P}^n$, the restriction of the projection onto the second factor. In terms of coordinates $(z, [\zeta]) = (z_1, \ldots, z_{n+1}, [\zeta_0, \ldots, \zeta_n])$ on $\mathbb{C}^{n+1} \times \mathbb{P}^n$, it is given by

$$T_n = \{ (z, [\zeta]) \in \mathbb{C}^{n+1} \times \mathbb{P}^n \mid z_{\alpha+1}\zeta_\beta - z_{\beta+1}\zeta_\alpha = 0, \quad \alpha,\ \beta = 0, \ldots, n \}.$$

For each $\alpha = 0, \ldots, n$, we set $U_\alpha = \{ [\zeta] \in \mathbb{P}^n \mid \zeta_\alpha \neq 0 \}$ and $\xi_\alpha = z_{\alpha+1}$. Then on $\pi^{-1}(U_\alpha)$, we have $z_{\beta+1} = \zeta_\beta/\zeta_\alpha \cdot \xi_\alpha$, $\beta = 0, \ldots, n$, and we have a trivialization

$$\psi_\alpha : \pi^{-1}(U_\alpha) \xrightarrow{\sim} \mathbb{C} \times U_\alpha, \qquad (z, [\zeta]) \mapsto (\xi_\alpha, [\zeta]).$$

We have $\psi_\alpha \circ \psi_\beta^{-1}(\xi_\beta, [\zeta]) = (\zeta_\alpha/\zeta_\beta \cdot \xi_\beta, [\zeta])$. Thus T_n is a line bundle with a system of transition functions $\{g^{\alpha\beta}\}$ given by $g^{\alpha\beta} = \zeta_\alpha/\zeta_\beta$, so that it is the bundle associated with the $\mathbb{C}^*$-principal bundle $\mathbb{C}^{n+1} \setminus \{0\} \to \mathbb{P}^n$ (cf. Example 3.4) and is dual to the hyperplane bundle H_n on $\mathbb{P}^n$ (cf. Example 3.6).

Let e_α be the frame of T_n on U_α given by $e_\alpha([\zeta]) = \psi_\alpha^{-1}(1, [\zeta])$. If we set

$$h^\alpha(e_\alpha, e_\alpha) = \frac{1}{\zeta_\alpha \overline{\zeta}_\alpha} \sum_{\gamma=0}^{n} \zeta_\gamma \overline{\zeta}_\gamma, \tag{9.19}$$

it defines a Hermitian metric h on T_n (cf (9.3), where $A_{\alpha\beta} = \zeta_\alpha/\zeta_\beta$).

There is another way of viewing T_n. Namely let $\varpi : T_n \to \mathbb{C}^{n+1}$ be the restriction of the projection to the first factor. Then it is in fact the blowing-up of 0 in $\mathbb{C}^{n+1}$ (cf. Sections 2.2 and 11.6) and the above metric is induced from the standard metric on $\mathbb{C}^{n+1}$.

As seen in the above, the bundle T_n is a subbundle of the product bundle $\mathbb{I}^{n+1} = \mathbb{C}^{n+1} \times \mathbb{P}^n$ and the quotient $Q_n = \mathbb{I}^{n+1}/T_n$ is the *universal bundle* on $\mathbb{P}^n$ (cf. Section 3.6). We record the exact sequence:

$$0 \longrightarrow T_n \longrightarrow \mathbb{I}^{n+1} \longrightarrow Q_n \longrightarrow 0. \tag{9.20}$$

Euler sequence

Now we study the tangent bundle of $\mathbb{P}^n$. We denote by $\zeta = (\zeta_0, \dots, \zeta_n)$ coordinates on $\mathbb{C}^{n+1} \setminus \{0\}$ and by $\varpi : \mathbb{C}^{n+1} \setminus \{0\} \to \mathbb{P}^n$ the canonical surjection so that $\varpi(\zeta) = [\zeta] = [\zeta_0, \dots, \zeta_n]$. We have the differential

$$\varpi_* : T_\zeta(\mathbb{C}^{n+1} \setminus \{0\}) \longrightarrow T_{\varpi(\zeta)}\mathbb{P}^n$$

at each point ζ in $\mathbb{C}^{n+1} \setminus \{0\}$.

Lemma 9.2. *The vector space $T_{\varpi(\zeta)}\mathbb{P}^n$ is spanned by the vectors $\varpi_*(\frac{\partial}{\partial\zeta_\alpha})$, $\alpha = 0, \dots, n$, with the relation*

$$\sum_{\alpha=0}^{n} \zeta_\alpha \, \varpi_*\Big(\frac{\partial}{\partial\zeta_\alpha}\Big) = 0.$$

Proof. Without loss of generality, we may assume $\zeta_0 \neq 0$. Then we compute

$$\varpi_*\Big(\frac{\partial}{\partial\zeta_\alpha}\Big) = \begin{cases} \frac{1}{\zeta_0} \frac{\partial}{\partial(\zeta_\alpha/\zeta_0)} & \alpha \neq 0, \\ -\frac{1}{\zeta_0} \sum_{\beta\neq 0} \frac{\zeta_\beta}{\zeta_0} \frac{\partial}{\partial(\zeta_\beta/\zeta_0)} & \alpha = 0, \end{cases} \tag{9.21}$$

which implies the lemma. □

From (9.20), we have the exact sequence

$$0 \longrightarrow Q_n^* \longrightarrow (\mathbb{I}^{n+1})^* \xrightarrow{\rho} H_n \longrightarrow 0,$$

where $(\mathbb{I}^{n+1})^* = \mathrm{Hom}(\mathbb{C}^{n+1}, \mathbb{C}) \times \mathbb{P}^n$. An element in $\mathrm{Hom}(\mathbb{C}^{n+1}, \mathbb{C})$ is given by a linear function ℓ on $\mathbb{C}^{n+1} = \{(z_1 \dots, z_{n+1})\}$:

$$\ell(z) = a_1 z_1 + \cdots + a_{n+1} z_{n+1}.$$

The fiber of H_n at L in $\mathbb{P}^n$ is L^* and ρ maps an element (ℓ, L) of $\mathbb{I}^{n+1*}$ to the restriction $\ell|_L$; $\rho(\ell, L) = \ell|_L$. Thus every element of H_n is expressed as $\rho(\ell, L)$ and two expressions $\rho(\ell, L)$ and $\rho(\ell', L)$ represent the same element if and only if $\ell' - \ell$ vanishes on L. From (9.21) and the above remark, we see that the map

$$\varepsilon : H_n^{\oplus n+1} \longrightarrow T\mathbb{P}^n$$

that assigns $\sum_{\alpha=0}^n \ell_\alpha(\zeta)\varpi_*(\frac{\partial}{\partial\zeta_\alpha})$ to $(\rho(\ell_0, L), \dots, \rho(\ell_n, L))$ is a well-defined vector bundle morphism, with $L = \varpi(\zeta)$.

Proposition 9.9 (Euler sequence). *There is an exact sequence of vector bundles on $\mathbb{P}^n$:*

$$0 \longrightarrow \mathbb{I} \longrightarrow H_n^{\oplus n+1} \xrightarrow{\varepsilon} T\mathbb{P}^n \longrightarrow 0.$$

Proof. By Lemma 9.2, ε is surjective. Moreover there is a non-vanishing section $(\rho(\ell_0, L), \dots, \rho(\ell_n, L))$ of $H_n^{\oplus n+1}$ given by $\ell_\alpha(z) = z_{\alpha+1}$, $\alpha = 0, \dots, n$, which is in the kernel of ε. $\square$

From the above, we see that the total Chern class of $T\mathbb{P}^n$ is given by (cf. (8.15))

$$c^*(T\mathbb{P}^n) = c^*(H_n^{\oplus n+1}) = (1 + h_n)^{n+1}, \tag{9.22}$$

where $h_n = c^1(H_n)$ is the first Chern class of H_n.

Remark 9.6. Let ℓ be a linear function on $\mathbb{C}^{n+1}$ as above. Then it defines a section of $H_n \to \mathbb{P}^n$ by $L \mapsto \rho(\ell, L)$. In terms of trivializations whose system of transition function is $\{\zeta_\beta/\zeta_\alpha\}$, the fiber coordinate ℓ_α of the section on U_α is given by

$$\ell_\alpha(z^\alpha) = a_1 z_1^\alpha + \cdots + a_\alpha z_\alpha^\alpha + a_{\alpha+1} + a_{\alpha+2} z_{\alpha+1}^\alpha + \cdots + a_{n+1} z_n^\alpha,$$

where $z^\alpha = (z_1^\alpha, \dots, z_n^\alpha)$ is the coordinate system on U_α as in (2.2).

Exercise 9.3. Show that the holomorphic tangent bundle of a Hopf manifold $\pi : M \to \mathbb{P}^{n-1}$ is isomorphic to $\pi^* H_{n-1}^{\oplus n}$ (cf. Exercise 3.5).

Fubini-Study metric

Denoting by $\zeta = (\zeta_0, \dots, \zeta_n)$ coordinates on $\mathbb{C}^{n+1} \setminus \{0\}$, we consider the differential form

$$\omega = \frac{1}{2\pi\sqrt{-1}} \bar{\partial}\partial \log \|\zeta\|^2.$$

We show that it defines a global $(1,1)$-form on $\mathbb{P}^n$. To see this, recall that the canonical surjection $\varpi : \mathbb{C}^{n+1} \setminus \{0\} \to \mathbb{P}^n$ has the structure of a principal $\mathbb{C}^*$-bundle (cf. Example 3.4). Thus if we have a section s on an open set U in $\mathbb{P}^n$ and another s' on U', there is a non-vanishing holomorphic function f such that $s' = fs$ on $U \cap U'$ and we see that $(s')^*\omega = s^*\omega$. For each $\alpha = 0, \dots, n$, let U_α be the open set of $\mathbb{P}^n$ given by $\zeta_\alpha \neq 0$ and $z^\alpha = (z_1^\alpha, \dots, z_n^\alpha)$ the coordinate system on U_α as in (2.2). Since $\mathbb{P}^n$ is covered by the U_α's and on U_α there is a section s^α given by

$$s^\alpha(z^\alpha) = (z_1^\alpha, \dots, z_\alpha^\alpha, 1, z_{\alpha+1}^\alpha, \dots, z_n^\alpha),$$

we see that ω defines a global form on $\mathbb{P}^n$, which we denote also by ω and call the *Fubini-Study form.* On U_α we compute

$$\begin{aligned}\omega &= \frac{1}{2\pi\sqrt{-1}} \bar{\partial}\partial \log(1 + \|z^\alpha\|^2) \\ &= \frac{\sqrt{-1}}{2\pi} \Big(\frac{\sum_{i=1}^n dz_i^\alpha \wedge d\bar{z}_i^\alpha}{1 + \|z^\alpha\|^2} - \frac{(\sum_{i=1}^n \bar{z}_i^\alpha dz_i^\alpha) \wedge (\sum_{i=1}^n z_i^\alpha d\bar{z}_i^\alpha)}{(1 + \|z^\alpha\|^2)^2} \Big).\end{aligned}$$

If we set $z_i = z_i^0$, at the point $[1, 0, \ldots, 0]$, we have

$$\omega = \frac{\sqrt{-1}}{2\pi} \sum_{i=1}^{n} dz_i \wedge d\bar{z}_i.$$

Since $U(n+1)$ acts transitively on $\mathbb{P}^n$ (cf. Section 3.6) and leaves the form ω invariant, ω is a Hermitian form. The Hermitian metric defined by ω is called the *Fubini-Study metric* on $\mathbb{P}^n$. In fact it is a Kähler metric (cf. Section 9.4 below).

Later (see Example 9.6) we show that ω represents the first Chern class of the hyperplane bundle so that

$$\int_{\mathbb{P}^n} \omega^n = 1. \tag{9.23}$$

Thus from (9.6), we see that the volume of $\mathbb{P}^n$ with respect to the Fubini-Study metric is equal to $\pi^n/n!$.

9.4 Kähler manifolds

Let M be a complex manifold of dimension n. Recall that a Hermitian metric on M is a Hermitian metric on its holomorphic tangent bundle TM. For such a metric, there is the associated $(1,1)$-form locally given as (9.5).

Definition 9.9. A Hermitian metric on M is called *Kähler*, if the associated form is closed. A complex manifold is called *Kähler* if it admits a Kähler metric.

Example 9.5. 1. If $\dim M = 1$, every Hermitian metric on M is Kähler by dimension reason. Thus M is Kähler.

2. $\mathbb{C}^n$ is Kähler with the standard metric $ds^2 = \sum_{i=1}^n dz_i \otimes d\bar{z}_i$, which also defines a Kähler metric on every complex torus $T = G\backslash\mathbb{C}^n$ (cf. Example 3.2).

3. Let V be a complex submanifold of M. Then a Hermitian metric h on M induces a Hermitian metric on V whose sccociated form is the pull-back of the associated form of h. Thus if M is Kähler, V is also Kähler.

4. $\mathbb{P}^n$ is Kähler with the Fubini-Study metric, thus every complex submanifold of $\mathbb{P}^n$ is Kähler.

5. For a Hopf manifold M, $b_1 = 1$ (cf. Exercise 3.2). Thus it is non-Kähler (cf. Corollary 9.6 below).

We quote the following fundamental fact:

Theorem 9.4. *Let M be a Kähler manifold. If the Laplacians $\triangle_d$ and $\triangle_{\bar\partial}$ are defined with respect to the Kähler metric on M, we have:*

$$\triangle_d = 2\,\triangle_{\bar\partial} = 2\,\triangle_{\partial}.$$

Corollary 9.4. *For a Kähler manifold M,*

$$\mathcal{H}^{p,q}(M) = \overline{\mathcal{H}^{q,p}(M)} \quad \text{and} \quad \mathcal{H}^r(M) = \bigoplus_{p+q=r} \mathcal{H}^{p,q}(M).$$

For a compact Kähler manifold M, from Corollaries 9.1, 9.2 and 9.4, we have isomorphisms

$$H^{p,q}_{\bar\partial}(M) \simeq \overline{H^{q,p}_{\bar\partial}(M)} \quad \text{and} \quad H^r_d(M) \simeq \bigoplus_{p+q=r} H^{p,q}_{\bar\partial}(M),$$

that is, M admits a Hodge structure of weight r for every r (cf. Definition 9.8). The isomorphisms above *a priori* depends on the chosen Kähler metric, however, noticing that harmonic representatives satisfy the conditions (H1) and (H2) of Definition 9.8. 2, we see that the Hodge structure above is natural and the isomorphisms do not depend on the metric (cf. Remark 9.4). Thus:

Corollary 9.5. *Every compact Kähler manifold admits a natural Hodge structure of weight r for every r.*

Corollary 9.6. *For a compact Kähler manifold M,*

$$h^{p,q} = h^{q,p} \quad \text{and} \quad b_r = \sum_{p+q=r} h^{p,q}, \qquad p, q, r \geq 0.$$

In particular, b_r is even for r odd.

Thus b_1 is even and $h^{1,0} = \dim H^{1,0}_{\bar\partial}(M) = \dim H^0(M;\Omega^1) = \frac{1}{2}b_1$ is a topological invariant. In particular, for a compact Riemann surface C, it is called the *genus* of C. If we denote it by g, as a C^∞ manifold, C is an orientable compact real surface with g "holes".

Kodaira embedding theorem: Let M be a complex manifold. A *Hodge metric* on M is a Kähler metric whose associated $(1,1)$-form represents a class in the image of $H^2(M;\mathbb{Z}) \to H^2(M;\mathbb{C}) \simeq H^2_d(M)$. For example, the Fubini-Study metric on $\mathbb{P}^n$ is Hodge, since its associated form represents the first Chern class of the hyperplane bundle (cf. Example 9.6 below). Consequently, every complex submanifold of $\mathbb{P}^n$ carries a Hodge metric.

The Kodaira embedding theorem assures that the converse is true for compact complex manifolds, i.e., if a compact complex manifold carries a Hodge metric, it can be embedded in $\mathbb{P}^n$ for some n. Combined with Chow's thorem (cf. Remark 2.5), we have:

Theorem 9.5. *A compact complex manifold is (biholomorphic with) a projective algebraic manifold if and only if it carries a Hodge metric.*

9.5 Atiyah classes

Let M be a complex manifold of dimension n and E a holomorphic vector bundle of rank l on M.

Definition 9.10. A connection ∇ for E is *of type* $(1,0)$ if each entry of the connection matrix with respect to a holomorphic frame is a form of type $(1,0)$.

Note that the above property of ∇ does not depend on the choice of the holomorphic frame. A connection of type $(1,0)$ is also called a $(1,0)$-connection.

Proposition 9.10. *Every holomorphic vector bundle admits a connection of type* $(1,0)$.

Proof. Let $\mathcal{U} = \{U_\alpha\}$ be an open covering of M such that E is holomorphically trivial on each U_α. Let $e_\alpha^{(l)}$ be a holomorphic frame of E on U_α and ∇_α the connection trivial with respect to $e_\alpha^{(l)}$. Take a partition of unity $\{\rho_\alpha\}$ subordinate to $\mathcal{U}$. Then $\nabla = \sum \rho_\alpha \nabla_\alpha$ is a $(1,0)$-connection for E. □

Let ∇ be a connection for E. Since $T_{\mathbb{R}}^{c} M = TM \oplus \overline{T}M$, we may express as $\nabla = \nabla' + \nabla''$ with

$$\nabla' : A^0(E) \longrightarrow A^0(T^*M \otimes E) \quad \text{and} \quad \nabla'' : A^0(E) \longrightarrow A^0(\overline{T}^*M \otimes E).$$

The following is not difficult to see:

Proposition 9.11. *A connection* ∇ *is of type* $(1,0)$ *if and only if* $\nabla'' = \bar{\partial}$.

Let ∇ be a $(1,0)$-connection for E and K its curvature. The curvature matrix κ is related to the connection matrix θ by (8.2). Thus we have a decomposition

$$\kappa = \kappa^{2,0} + \kappa^{1,1}, \qquad \kappa^{2,0} = \partial\theta + \theta \wedge \theta \ \text{ and } \ \kappa^{1,1} = \bar{\partial}\theta. \tag{9.24}$$

Accordingly we have a decomposition

$$K = K^{2,0} + K^{1,1},$$

where $K^{2,0}$ and $K^{1,1}$ are, respectively, a $(2,0)$-form and a $(1,1)$-form with coefficients in $\mathrm{Hom}(E,E)$.

Definition 9.11. For a $(1,0)$-connection ∇ and an invariant polynomial φ homogeneous of degree k, the (k,k)-*characteristic form* is defined by

$$\varphi^{k,k}(\nabla) = \left(\frac{\sqrt{-1}}{2\pi}\right)^k \varphi(K^{1,1}).$$

Note that the characteristic form $\varphi(\nabla)$ (cf. Definition 8.3) has no other components than those of types $(2k,0),\ldots,(k,k)$ and $\varphi^{k,k}(\nabla)$ is the component of type (k,k). By (9.24), $\varphi^{k,k}(\nabla)$ is locally $\bar\partial$-exact so that we have:

Proposition 9.12. *The form $\varphi^{k,k}(\nabla)$ is $\bar\partial$-closed:*

$$\bar\partial\varphi^{k,k}(\nabla) = 0.$$

The following is proved using Proposition 8.3, in fact the difference form $\varphi(\nabla_0,\ldots,\nabla_p)$ constructed there has no other components than those of types $(2k-p,0),\ldots,(k,k-p)$ and $\varphi^{k,k-p}(\nabla_0,\ldots,\nabla_p)$ is the $(k,k-p)$-component of $\varphi(\nabla_0,\ldots,\nabla_p)$:

Proposition 9.13. *Let $\nabla_0,\ldots,\nabla_p$ be $(1,0)$-connections for E. For an invariant polynomial φ homogeneous of degree k, there exists a $(k,k-p)$-form $\varphi^{k,k-p}(\nabla_0,\ldots,\nabla_p)$ alternating in the $p+1$ entries and satisfying*

$$\sum_{\nu=0}^{p}(-1)^\nu\varphi^{k,k-p+1}(\nabla_0,\ldots,\widehat{\nabla_\nu},\ldots,\nabla_p) + (-1)^p\bar\partial\varphi^{k,k-p}(\nabla_0,\ldots,\nabla_p) = 0.$$

In particular, if ∇ and ∇' are two $(1,0)$-connections for E,

$$\varphi^{k,k}(\nabla') - \varphi^{k,k}(\nabla) = \bar\partial\varphi^{k,k-1}(\nabla,\nabla').$$

From this we see that the class $[\varphi^{k,k}(\nabla)]$ of the $\bar\partial$-closed form $\varphi^{k,k}(\nabla)$ in the Dolbeault cohomology $H^{k,k}_{\bar\partial}(M)$ depends only on E and not on the choice of the $(1,0)$-connection ∇.

Definition 9.12. We denote this class by $\varphi^{k,k}(E)$ and call it the *characteristic class* of E of type (k,k) for the polynomial φ.

Definition 9.13. Let ∇ be a $(1,0)$-connection for E. The q-th *Atiyah form* of ∇ is defined by

$$a^q(\nabla) = \Big(\frac{\sqrt{-1}}{2\pi}\Big)^q \sigma_q(K^{1,1}).$$

Its class in $H^{q,q}_{\bar\partial}(M)$ is denoted by $a^q(E)$ and is called the q-th *Atiyah class* of E.

Remark 9.7. 1. For a $(1,0)$-connection ∇ for E, the Chern form $c^q(\nabla)$ has no other components than those of types $(2q,0),\dots,(q,q)$ and the Atiyah form $a^q(\nabla)$ is the (q,q)-component. In particular, $a^n(\nabla) = c^n(\nabla)$.

The q-th Chern class $c^q(E)$ of E is the class of $c^q(\nabla)$ in $H^{2q}_d(M)$. Thus, although we may not be able to compare the Atiyah and Chern classes directly on the cohomology level in general, we may do so on the form level.

2. If ∇ is a metric connection for E, i.e., a $(1,0)$-connection compatible with some Hermitian metric on E (cf. Section 9.6 below), then its curvature is of type $(1,1)$ and thus we have $a^q(\nabla) = c^q(\nabla)$ for all q (cf. Corollary 9.7).

3. If M admits a natural Hodge structure of weight $2q$ (cf. Definition 9.8. 2), for instance if M is compact Kähler, there is a canonical monomorphism $h : H^{q,q}_{\bar\partial}(M) \to H^{2q}_d(M)$ and, taking a metric connection for E, we see that $h(a^q(E)) = c^q(E)$.

Similar remarks as above apply for any general invariant polynomial φ.

9.6 Hermitian connections and Bott-Chern classes

Let M be a C^∞ manifold and E a complex vector bundle of rank l on M. If h is a Hermitian metric on E, it induces pairings

$$h : A^1(M;E) \times A^0(M;E) \longrightarrow A^1(M),$$
$$h : A^0(M;E) \times A^1(M;E) \longrightarrow A^1(M).$$

To see this, let $(e_1,\dots,e_l)$ be a frame of E on an open set U in M. An element σ in $A^1(M;E)$ can be written as $\sigma = \sum_{i=1}^l \omega_i \otimes e_i$ on U with ω_i uniquely determined 1-forms. Since the coefficients (ω_i) behave in the same way as coefficient functions of a section by frame changes, if we set

$$h(\sigma,s) = \sum_{i=1}^l h(e_i,s)\,\omega_i, \qquad h(s,\sigma) = \sum_{i=1}^l h(s,e_i)\,\overline{\omega}_i,$$

they define 1-forms on M.

Definition 9.14. Let (E, h) be a Hermitian vector bundle. A connection ∇ for E is an *h-connection* if it is compatible with h in the sense that

$$dh(s, s') = h(\nabla s, s') + h(s, \nabla s'), \qquad s, s' \in A^0(M; E).$$

A connection ∇ for E is a *Hermitian connection* if it is an h-connection for some Hermitian metric h on E.

Proposition 9.14. *Let (E, h) be a Hermitian vector bundle. A connection ∇ for E is an h-connection if and only if the connection matrix with respect to an arbitrary h-orthonormal frame is skew-Hermitian.*

Proof. Let ∇ be an h-connection. Also let $e^{(l)} = (e_1, \ldots, e_l)$ be an h-orthonormal frame and $\theta = (\theta_{ij})$ the connection matrix of ∇ with respect to $e^{(l)}$. From $dh(e_i, e_j) = h(\nabla e_i, e_j) + h(e_i, \nabla e_j)$, we have

$${}^t\theta + \overline{\theta} = 0,$$

i.e., θ is skew-Hermitian. The converse can easily be checked. □

Proposition 9.15. *Every Hermitian vector bundle (E, h) admits an h-connection. Thus every complex vector bundle admits a Hermitian connection.*

Proof. Let $\mathcal{U} = \{U_\alpha\}$ be an open covering of M such that E is trivial on each U_α. For each α, let $e_\alpha^{(l)}$ be an h-orthonormal frame of E on U_α and ∇_α the connection for E on U_α trivial with respect to $e_\alpha^{(l)}$. Let $\{\rho_\alpha\}$ be the partition of unity subordinate to $\mathcal{U}$ and set $\nabla = \sum \rho_\alpha \nabla_\alpha$. We claim that ∇ is an h-connection. To see this, let $A^{\alpha\beta}$ be the unitary matrix valued function on $U_\alpha \cap U_\beta$ determined by $e_\beta^{(l)} = e_\alpha^{(l)} A^{\alpha\beta}$. Then the connection matrix θ^α of ∇ with respect to $e_\alpha^{(l)}$ is given by (cf. (8.4))

$$\theta^\alpha = \sum_\beta \rho_\beta A^{\alpha\beta} \cdot dA^{\beta\alpha},$$

which is skew-Hermitian, since the $A^{\alpha\beta}$'s are unitary. □

Proposition 9.16. *If ∇ is a Hermitian connection for E, the Chern forms $c^q(\nabla)$ are real.*

Proof. Let h be a Hermitian metric such that ∇ is an h-connection. Let $e^{(l)}$ be an h-orthonormal frame of E and θ and κ the connection and curvature matrices of ∇ with respect to $e^{(l)}$. By (8.2) and Proposition 9.14, κ is also skew-Hermitian. Since the total Chern form is locally given by (8.12), we easily see that $c^q(\nabla)$ are real. □

Now suppose M is a complex manifold and E a holomorphic vector bundle. Recall that the notion of connection of type $(1,0)$ is introduced in Definition 9.10.

Theorem 9.6. *A holomorphic Hermitian vector bundle (E,h) admits a unique h-connection of type $(1,0)$.*

Proof. Suppose we have an h-connection ∇ of type $(1,0)$. Let $e^{(l)}$ be a (local) holomorphic frame and θ the connection matrix of ∇ with respect to $e^{(l)}$. We also set $H = (h_{ij})$, $h_{ij} = h(e_i, e_j)$. Since ∇ is Hermitian, we have

$$dH = {}^t\theta H + H\overline{\theta}.$$

As θ is of type $(1,0)$, this is equivalent to

$$\partial H = {}^t\theta H \quad \text{and} \quad \overline{\partial} H = H\overline{\theta}.$$

Since H is Hermitian, we have the unique solution

$$\theta = {}^tH^{-1} \cdot \partial\, {}^tH. \tag{9.25}$$

Moreover, since we have (9.3) and (8.4) by a frame change $e'^{(l)} = e^{(l)}A$, we see that this defines a global h-connection of type $(1,0)$. □

We call a Hermitian connection of type $(1,0)$ a *metric connection.* For a Hermitian metric h, the h-metric connection is the h-connection of type $(1,0)$. The following is a consequence of direct computation using (9.25):

Theorem 9.7. *Let E be a holomorphic vector bundle. If ∇ is a metric connection for E, its curvature matrix κ with respect to a holomorphic frame is of type $(1,1)$ and is given by*

$$\kappa = \kappa^{1,1} = \overline{\partial}\theta.$$

Corollary 9.7. *Let E be a holomorphic vector bundle. If ∇ is a metric connection for E, for an invariant polynomial φ homogeneous of degree k,*

$$\varphi(\nabla) = \varphi^{k,k}(\nabla).$$

Thus they are simultaneously d- and $\overline{\partial}$-closed.

Remark 9.8. In fact, for metric connections $\nabla_0, \ldots, \nabla_p$ for E, we have

$$\varphi(\nabla_0, \ldots, \nabla_p) = \varphi^{k,k-p}(\nabla_0, \ldots, \nabla_p).$$

In particular, let $L \to M$ be a line bundle and h a Hermitian metric on L. Let $\{U_\alpha\}$ be a covering on M trivializing L so that we have a frame e^α on U_α. By (9.25), we have $\theta^\alpha = \partial \log h^\alpha$, $h^\alpha = h(e^\alpha, e^\alpha)$. Thus we have:

Proposition 9.17. *In the above situation, the first Chern form of the h-metric connection ∇^h is given by*

$$c^1(\nabla^h) = \frac{\sqrt{-1}}{2\pi}\,\bar{\partial}\partial \log h^\alpha \qquad \textit{on } U_\alpha,$$

which is also equal to the first Atiyah form $a^1(\nabla^h)$.

Example 9.6. Let $H_n \to \mathbb{P}^n$ be the hyperplane bundle on $\mathbb{P}^n$ (cf. Example 3.6). For each $\alpha = 0, \ldots, n$, we set $U_\alpha = \{\,[\zeta] \in \mathbb{P}^n \mid \zeta_\alpha \neq 0\,\}$ and let $z^\alpha = (z_1^\alpha, \ldots, z_n^\alpha)$ be coordinates on U_α defined by (2.2). Then H_n is the bundle defined by $\{g^{\alpha\beta}\}$, $g^{\alpha\beta} = \zeta_\beta/\zeta_\alpha$. If we set

$$h^\alpha = \frac{1}{1 + \|z^\alpha\|^2}, \tag{9.26}$$

we see that it defines a Hermitian metric h on H_n (cf. (9.3), where $A_{\alpha\beta} = \zeta_\beta/\zeta_\alpha$). It is the "dual" of the metric given by (9.19) for the tautological bundle. Then on U_α, we have

$$c^1(\nabla^h) = \frac{1}{2\pi\sqrt{-1}}\,\bar{\partial}\partial \log(1 + \|z^\alpha\|^2).$$

Thus the first Chern form for this metric on the hyperplane bundle H_n coincides with the Fubini-Study form (cf. Section 9.3).

Now we reprove the identity (7.71).

Proposition 9.18. *For the Bochner-Martinelli form β_n on $\mathbb{C}^n$,*

$$\int_{\mathbb{S}^{2n-1}} \beta_n = 1$$

Proof. Recall that (cf. Section 9.3), under the projection $\varpi : \mathbb{C}^n \smallsetminus \{0\} \to \mathbb{P}^{n-1}$ given by $z = (z_1, \ldots, z_n) \mapsto [\zeta] = [\zeta_0, \ldots, \zeta_{n-1}] = [z_1, \ldots, z_n]$, the Kähler form ω of the Fubini-Study metric on $\mathbb{P}^{n-1}$ pulls back to

$$\varpi^*\omega = \frac{1}{2\pi\sqrt{-1}}\bar{\partial}\partial \log \|z\|^2.$$

Recall also that $\varpi : \mathbb{C}^n \smallsetminus \{0\} \to \mathbb{P}^{n-1}$ has the structure of a $\mathbb{C}^*$-bundle (cf. Example 3.4). On each $U_\alpha = \{\zeta_\alpha \neq 0\}$, $\alpha = 0, \ldots, n-1$, it has a trivialization $\varpi^{-1}(U_\alpha) \simeq \mathbb{C}^* \times U_\alpha$, under which $(z_1, \dots, z_n)$ corresponds to $(z_{\alpha+1}, [z])$. By Proposition 7.27, we have

$$\beta_n = \frac{1}{2\pi\sqrt{-1}} \partial \log \|z\|^2 \wedge \varpi^* \omega^{n-1}.$$

On $\varpi^{-1}(U_\alpha)$, the form $\partial \log \|z\|^2$ corresponds to

$$\partial(\log \|\zeta/\zeta_\alpha\|^2 + \log |z_{\alpha+1}|^2) = \partial \log \|\zeta/\zeta_\alpha\|^2 + 2\pi\sqrt{-1}\beta_1,$$

where β_1 is the Cauchy form in $z_{\alpha+1}$. Since $\partial \log \|\zeta/\zeta_\alpha\|^2$ may be thought of as a $(1,0)$-form on $\mathbb{P}^{n-1}$, $\partial \log \|\zeta/\zeta_\alpha\|^2 \wedge \varpi^* \omega^{n-1} = 0$. The restriction of ϖ to $\mathbb{S}^{2n-1}$ is an $\mathbb{S}^1$-bundle on $\mathbb{P}^{n-1}$. Thus by the projection formula, we have

$$\int_{\mathbb{S}^{2n-1}} \beta_n = \int_{\mathbb{P}^{n-1}} \varpi_* \beta_1 \cdot \omega^{n-1} = \int_{\mathbb{P}^{n-1}} \omega^{n-1} = 1. \qquad \square$$

Bott-Chern cohomology

Let M be a complex manifold. For each (p,q), we set

$$Z^{p,q}_{\mathrm{BC}}(M) = \{\, \omega \in A^{p,q}(M) \mid d\omega = 0 \,\},$$
$$B^{p,q}_{\mathrm{BC}}(M) = \{\, \omega \in A^{p,q}(M) \mid \omega = \bar{\partial}\partial\theta \ \text{ for some } \theta \in A^{p-1,q-1}(M) \,\}.$$

Then $B^{p,q}_{\mathrm{BC}}(M)$ is a subspace of $Z^{p,q}_{\mathrm{BC}}(M)$.

Definition 9.15. The *Bott-Chern cohomology* of M of type (p,q) is defined by

$$H^{p,q}_{\mathrm{BC}}(M) = Z^{p,q}_{\mathrm{BC}}(M)/B^{p,q}_{\mathrm{BC}}(M).$$

Note that the identity map of forms induces canonical morphisms

$$\begin{array}{ccccc} & & H^{p,q}_{\mathrm{BC}}(M) & & \\ & \swarrow^{\alpha_{\bar{\partial}}} & \downarrow{\scriptstyle \alpha_d} & \searrow^{\alpha_\partial} & \\ H^{p,q}_{\bar{\partial}}(M) & & H^{p+q}_d(M) & & H^{p,q}_{\partial}(M). \end{array} \tag{9.27}$$

Exercise 9.4. Verify the above.

Remark 9.9. It is known that the conditions (1) and (2) of Corollary 9.3 are satisfied, that is, M admits a natural Hodge structure of weight r for every r, if and only if α_d is injective for every (p,q). The last condition means that for every (p,q)-form ω with $d\omega = 0$ and $\omega = d\theta'$ for some $(p+q-1)$-form θ', there exists a $(p-1,q-1)$-form θ such that $\omega = \bar{\partial}\partial\theta$. This property is referred to as the $\bar{\partial}\partial$-Lemma. If this is the case, $\alpha_{\bar{\partial}}$, α_∂ and $\alpha_d : \bigoplus_{p+q=r} H^{p,q}_{\mathrm{BC}}(M) \to H^r_d(M)$ are shown to be isomorphisms and the isomorphisms in (H1)$'$ and (H2) in Definition 9.8. 2 (see also Remark 9.3. 1) are given by $\alpha_\partial \circ \alpha_{\bar{\partial}}^{-1}$ and $\alpha_{\bar{\partial}} \circ \alpha_d^{-1}$, respectively.

Bott-Chern classes

Let E be a holomorphic vector bundle of rank l on a complex manifold M. Also let h be a Hermitian metric on E and ∇^h the h-metric connection for E. If φ is an invariant homogeneous polynomial of degree k, $\varphi(\nabla^h)$ is a (k,k)-form which is d-closed. We denote it simply by $\varphi(h)$. We show that its class in the Bott-Chern cohomology $H^{k,k}_{\mathrm{BC}}(M)$ does not depend on the choice of the metric h.

Thus let h_0 and h_1 be two Hermitian metrics on E. If we set $h_t = (1-t)h_0 + th_1$, then it is a Hermitian metric on E for $t \in \mathbb{R}$. We define an element L_t in $A^0(M; \mathrm{Hom}(E,E))$ by

$$h_t(L_t s, s') = \frac{d}{dt} h_t(s, s')$$

and set

$$\varphi(h_1, h_0) = \left(\frac{\sqrt{-1}}{2\pi}\right)^k \int_0^1 \sum_{i=1}^{k} \tilde{\varphi}(K_t, \dots, \overset{i}{L_t}, \dots, K_t) dt,$$

where $\tilde{\varphi}$ is the polarization of φ (cf. Section 8.2). Then it is a $(k-1,k-1)$-form and we have:

Proposition 9.19. *In the above situation,*

$$\bar{\partial}\partial\varphi(h_1, h_0) = \varphi(h_1) - \varphi(h_0).$$

Proof. Let λ_t denote the matrix of L_t with respect to a holomorphic frame $(e_1, \dots, e_l)$:

$$L_t e_i = \sum_{j=1}^{l} (\lambda_t)_{ji}\, e_j.$$

By setting $H_t = (h_{ij})$, $h_{ij} = h_t(e_i, e_j)$, we compute

$$\lambda_t = {}^t H_t^{-1} \cdot \frac{d}{dt}\, {}^t H_t.$$

Let θ_t and κ_t denote the connection and curvature matrices of ∇^{h_t}. Then by (8.2), (8.3) and Theorem 9.7, we have

$$\kappa_t = \bar{\partial}\theta_t, \quad \partial\theta_t + \theta_t \wedge \theta_t = 0, \quad \partial\kappa_t = [\kappa_t, \theta_t].$$

We also have

$$\partial\lambda_t - [\lambda_t, \theta_t] = \frac{d}{dt}\theta_t. \tag{9.28}$$

We claim that

$$\partial\tilde{\varphi}(\kappa_t,\dots,\overset{i}{\lambda_t},\dots,\kappa_t)=\tilde{\varphi}(\kappa_t,\dots,\overset{i}{\frac{d}{dt}}\theta_t,\dots,\kappa_t). \tag{9.29}$$

First observe that the entries are all of even degree. The left-hand side is

$$\sum_{j\neq i}\tilde{\varphi}(\kappa_t,\dots,\overset{j}{[\kappa_t,\theta_t]},\dots,\overset{i}{\lambda_t},\dots,\kappa_t)+\tilde{\varphi}(\kappa_t,\dots,\overset{i}{\partial\lambda_t},\dots,\kappa_t).$$

On the other hand, by the invariance condition (8.10),

$$\sum_{j\neq i}\tilde{\varphi}(\kappa_t,\dots,\overset{j}{[\kappa_t,\theta_t]},\dots,\overset{i}{\lambda_t},\dots,\kappa_t)+\tilde{\varphi}(\kappa_t,\dots,\overset{i}{[\lambda_t,\theta_t]},\dots,\kappa_t)=0.$$

The equality (9.29) then follows from (9.28). Thus we have

$$\partial\tilde{\varphi}(K_t,\dots,\overset{i}{L_t},\dots,K_t)=\tilde{\varphi}(K_t,\dots,\overset{i}{\dot{\nabla}_t},\dots,K_t)$$

and the proposition follows from Remark 8.4. 3. □

From the above we see that the class of $\varphi(h)$ in $H^{k,k}_{\mathrm{BC}}(M)$ does not depend on the choice of the metric h.

Definition 9.16. The *Bott-Chern class* $\varphi_{\mathrm{BC}}(E)$ of E with respect to φ is the class in $H^{k,k}_{\mathrm{BC}}(M)$ of $\varphi(h)$ for some Hermitian metric h on E.

Remark 9.10. 1. The Bott-Chern Chern class and the Bott-Chern Atiyah class are the same: $c^q_{\mathrm{BC}}(E)=a^q_{\mathrm{BC}}(E)$ in $H^{q,q}_{\mathrm{BC}}(M)$. It goes to the Chern class $c^q(E)$ by the morphism α_d in (9.27) and to the Atiyah class $a^q(E)$ by $\alpha_{\bar{\partial}}$.

2. Suppose E is a line bundle with Hermitian metrics h_0 and h_1. For a non-vanishing section s, we set $N_i(s)=h_i(s,s)$, $i=0,1$. Then, for $\varphi=c^1$, we compute

$$c^1(h_0,h_1)=\frac{\sqrt{-1}}{2\pi}\log\frac{N_1(s)}{N_0(s)}.$$

Notes

As to harmonic forms and related topics, we refer to [Griffiths and Harris (1978); Kodaira (1986); Voisin (2007); Wells (2008)] and references therein. The Kodaira-Serre duality is proved by K. Kodaira using harmonic forms. It is then proved in [Serre (1955a)] in the context of sheaf cohomology including the case the base manifold M may not be compact (cf. the expression in Theorem 11.13 below for the case M is compact).

Some part of Section 9.2 is taken from [Angella, Suwa, Tardini and Tomassini (2020)]. The original reference for the Frölicher spectral sequence is [Frölicher (1955)]. Theorem 9.5 is due to [Kodaira (1954)].

The Atiyah class of a holomorphic vector bundle is originally introduced in [Atiyah (1957)] as an obstruction to the existence of a complex analytic connection for the bundle. For the description as given in Section 9.5, including its localization theory, we refer to [Abate, Bracci, Suwa and Tovena (2013); Suwa (2009)].

The Bott-Chern cohomology is introduced in [Bott and Chern (1965)]. The reference for Remark 9.9 is [Deligne, Griffiths, Morgan and Sullivan (1975)]. The Bott-Chern difference form $\varphi(h_0, h_1)$ as above has also been defined by different approaches (cf. [Bismut, Gillet and Soulé (1988); Burgos Gil, Kramer and Kühn (2005); Burgos Gil and Liţkanu (2010); Gillet and Soulé (1990)]). For the relative Bott-Chern cohomology and its applications, we refer to [Corrêa and Suwa (2020)].

Chapter 10

Localization

In general, if we have a geometric object on a space, there is some vanishing theorem away from the singular set of the object and certain characteristic classes are localized at the set. The localizations yield local invariants, called residues, through the Alexander duality and we have a residue theorem that expresses global invariants of the space as a sum of the residues. The Poincaré-Hopf theorem for a vector field with isolated singularities on a compact manifold is a prototype of this.

In this chapter, we explain general philosophy and procedure of localizing characteristic classes and of obtaining the associated residues.

We state the residue theorem in general (Theorem 10.1), which appears in various forms throughout this book, and give an explicit expression of residues (Theorem 10.4). Various types of localizations and the relevant vanishing theorem for each type are explained. In particular, the differential geometric localization of characteristic classes of complex vector bundles by frames is discussed in detail.

We prove a fundamental theorem of residues (Theorem 10.8), which says that the differential geometric residue of a section at an isolated singularity coincides with the topological residue, i.e., the mapping degree of a certain map. In a similar way, we show that the Thom class is a localized top Chern class (Theorem 10.9).

We then generalize Theorem 10.8 to the case of frames (Theorem 10.11) and prove that the topological and differential geometric Chern classes are essentially the same, including the localized case (Theorems 10.12 and 10.13).

10.1 Localization and associated residues

General philosophy

Let M be a C^∞ manifold of dimension m and S a closed set in M. For a complex vector bundle E on M and an invariant polynomial φ, it sometimes happens that we have the vanishing $\varphi(E) = 0$ on $M \smallsetminus S$ by some reason. Then, in view of the exact sequence

$$\cdots \longrightarrow H^*(M, M \smallsetminus S) \xrightarrow{j^*} H^*(M) \longrightarrow H^*(M \smallsetminus S) \longrightarrow \cdots,$$

there is a class $\varphi_S(E)$ in $H^*(M, M \smallsetminus S)$ which is sent to $\varphi(E)$ by j^*. We call $\varphi_S(E)$ a *localization of* $\varphi(E)$ *at* S. Notice that, since j^* is not injective in general, $\varphi_S(E)$ is not uniquely determined. However, in the cases we consider below, the vanishing of $\varphi(E)$ occurs on the level of cocycles and, using this fact, we may define a natural localization. If M is oriented and if S is a subcomplex of M with respect to some triangulation of M, the localization gives rise to the "residue" in the homology of each connected component of S through the Alexander duality $H^*(M, M \smallsetminus S) \simeq \check{H}_*(S)$ and we have the residue theorem (cf. Theorem 10.1 below).

We have already seen this kind of phenomenon as localization by frames from topological viewpoint (cf. Section 5.6). We explain the procedure in the framework of Chern-Weil theory adapted to Čech-de Rham cohomology, which is applicable in other settings as well.

Residue theorem

Let M be a C^∞ manifold of dimension m and S a closed set in M. Letting $U_0 = M \smallsetminus S$ and U_1 a neighborhood of S, we consider the covering $\mathcal{U} = \{U_0, U_1\}$ of M. In the following, we consider the reduced expression of Čech-de Rham cochains and we denote $\hat{A}^r(\mathcal{U})$ simply by $A^r(\mathcal{U})$ (cf. Remark 7.4, also Example 7.1).

Localized classes: Let E be a C^∞ complex vector bundle on M. For an invariant polynomial φ homogeneous of degree k, the characteristic class $\varphi(E)$ is represented by the cocycle $\varphi(\nabla_*)$ in $A^{2k}(\mathcal{U})$ given by

$$\varphi(\nabla_*) = (\varphi(\nabla_0), \varphi(\nabla_1), \varphi(\nabla_0, \nabla_1)), \tag{10.1}$$

where ∇_0 and ∇_1 are connections for E on U_0 and U_1 respectively (cf. Section 8.4). Sometimes it happens that we have a "vanishing theorem" on

U_0 for some polynomials φ. To be a little more precise, there is some "geometric object" γ on U_0, with which is associated a class $\mathcal{C}$ of connections for E on U_0 such that, for a certain polynomial φ, we have

$$\varphi(\nabla_0^{(0)}, \dots, \nabla_0^{(p)}) = 0 \qquad \text{if every } \nabla_0^{(i)} \text{ belongs to } \mathcal{C}.$$

We call a connection belonging to $\mathcal{C}$ *special* for γ and a polynomial φ as above *adapted* to γ. As we see below, this kind of vanishing defines a localization of $\varphi(E)$ uniquely.

To see this, first note that, if ∇_0 is special and if φ is adapted to γ, the cocycle $\varphi(\nabla_*)$ of (10.1) is in $A^{2k}(\mathcal{U}, U_0)$ and defines a class in $H_D^{2k}(\mathcal{U}, U_0)$, which is denoted by $\varphi_S(E,\gamma)$. It is represented by

$$\varphi(\nabla_*) = (\varphi(\nabla_1), \varphi(\nabla_0, \nabla_1)).$$

Proposition 10.1. *The class $\varphi_S(E,\gamma)$ does not depend on the choice of the special connection ∇_0 or the connection ∇_1.*

Proof. If ∇_0 and ∇_0' are both special, using $\varphi(\nabla_0, \nabla_0') = 0$, we have

$$(\varphi(\nabla_1), \varphi(\nabla_0', \nabla_1)) - (\varphi(\nabla_1), \varphi(\nabla_0, \nabla_1)) = D(0, \varphi(\nabla_0, \nabla_0', \nabla_1)).$$

Similarly, for connections ∇_1 and ∇_1' on U_1,

$$\begin{aligned}(\varphi(\nabla_1'), \varphi(\nabla_0, \nabla_1')) &- (\varphi(\nabla_1), \varphi(\nabla_0, \nabla_1)) \\ &= D(\varphi(\nabla_1', \nabla_1), \varphi(\nabla_0, \nabla_1, \nabla_1')).\end{aligned}$$

□

We call $\varphi_S(E,\gamma)$ the *localization of $\varphi(E)$ at S by γ*. Sometimes we denote it simply by $\varphi(E,\gamma)$.

From now on we assume that M is oriented. Before we state the residue theorem in general, we discuss a special case to explain the idea.

A special case: We consider the case $2k = m$ and S is compact. Let $(S_\lambda)_\lambda$ be the connected components of S and, for each λ, we take an open neighborhood U_λ of S_λ in U_1 so that $U_\lambda \cap U_\mu = \emptyset$, if $\lambda \neq \mu$. For each λ, let R_λ be an m-dimensional manifold with C^∞ boundary in U_λ containing S_λ in its interior and set $R_{0\lambda} = -\partial R_\lambda$. Let $\varphi_S(E,\gamma)$ be as above. We define (cf. (7.42))

$$\operatorname{Res}_\varphi(\gamma, E; S_\lambda) = \int_{R_\lambda} \varphi(\nabla_1) + \int_{R_{0\lambda}} \varphi(\nabla_0, \nabla_1). \tag{10.2}$$

Then it does not depend on the choice of R_λ.

Let R be an m-dimensional compact manifold possibly with boundary in M. Letting O be a neighborhood of ∂R in M, we apply the above localization procedure to the case $S = R \smallsetminus O$. Thus let $U_0 = M \smallsetminus S$. As to U_1, we take a neighborhood of R and consider the covering $\mathcal{U} = \{U_0, U_1\}$ of M. Suppose we have a geometric object γ on U_0. Let ∇_0 be a special connection for E on U_0 and ∇_1 a connection for E on U_1. Then, for a polynomial φ adapted to γ, we have the relative class in $H^m_D(\mathcal{U}, U_0)$ represented by $(\varphi(\nabla_1), \varphi(\nabla_0, \nabla_1))$. It is denoted by $\varphi_R(E,\gamma)$ with a little abuse of notation, as it should be denoted by $\varphi_{R \smallsetminus O}(E,\gamma)$. We have the integration (cf. (7.44))

$$\int \varphi_R(E,\gamma) = \int_R \varphi(\nabla_1) - \int_{\partial R} \varphi(\nabla_0, \nabla_1). \qquad (10.3)$$

Note that, while $\varphi_R(E,\gamma)$ depends on O, the value of the integration does not.

Remark 10.1. 1. The integration above makes sense if we have γ only on O. In this case, let $U = R \cup O$, $U_0 = U \smallsetminus S = O$ and U_1 a neighborhood of R in U and proceed as above for the covering $\mathcal{U} = \{U_0, U_1\}$.

2. For some particular choice of O, there is a canonical isomorphism $H^m_D(\mathcal{U}, U_0) \simeq H^m(R, \partial R; \mathbb{C})$. In this case we write the above integral $\int_{\underline{R}} \varphi_R(E,\gamma)$ (cf. (7.61)).

Coming back to the previous situation, let S be a compact set in M and suppose a geometric object γ is given on $U_0 = M \smallsetminus S$ so that we have the localization $\varphi_S(E,\gamma)$. If R is an m-dimensional compact manifold with boundary in M containing S in its interior, we also have the localization $\varphi_R(E,\gamma)$ as above.

The following is a direct consequence of Stokes' theorem:

Proposition 10.2. *In the above situation, we have*

$$\sum_\lambda \mathrm{Res}_\varphi(\gamma, E; S_\lambda) = \int \varphi_R(E,\gamma).$$

Suppose M is compact. If we take M as R, the right-hand side is written $\int_M \varphi(E)$.

General case: We assume that S is a subcomplex of M with respect to some triangulation K_0 of M. Then there is a canonical isomorphism $H^{2k}_D(\mathcal{U}, U_0) \simeq H^{2k}(M, M \smallsetminus S; \mathbb{C})$ (cf. Corollary 7.8) and the class $\varphi_S(E,\gamma)$ is sent to $\varphi(E)$ by the canonical morphism $j^* : H^{2k}(M, M \smallsetminus S; \mathbb{C}) \to H^{2k}(M; \mathbb{C})$. We have the Alexander duality (Theorems 4.2 and 7.13):

$$A : H^{2k}(M, M \smallsetminus S; \mathbb{C}) \xrightarrow{\sim} \check{H}_{m-2k}(S; \mathbb{C}).$$

Definition 10.1. The *residue* $\operatorname{Res}_\varphi(\gamma, E; S)$ *of* γ *at* S *for* $\varphi(E)$ is the image of $\varphi_S(E,\gamma)$ by A.

Suppose that S has only a finite number of connected components $(S_\lambda)_\lambda$. Then we have a decomposition

$$\check{H}_{m-2k}(S;\mathbb{C}) = \bigoplus_\lambda \check{H}_{m-2k}(S_\lambda;\mathbb{C})$$

and accordingly we have the residue $\operatorname{Res}_\varphi(\gamma, E; S_\lambda)$ in $\check{H}_{m-2k}(S_\lambda;\mathbb{C})$ for each λ. For each λ, we take an open neighborhood U_λ of S_λ in U_1 so that $U_\lambda \cap U_\mu = \emptyset$, if $\lambda \neq \mu$. Let K, K' and K^* be as in Section 4.1. We may assume that $O_{K'}(S_\lambda) \subset U_\lambda$. Also let $\{R_0, R_1\}$ be a honeycomb system adapted to $\mathcal{U}$, K' and S (cf. Definitions 7.9 and 7.12) and set $R_\lambda = R_1 \cap U_\lambda$ and $R_{0\lambda} = -\partial R_\lambda$. Then we have (cf. (7.59)):

Proposition 10.3. *In the above situation, the residue* $\operatorname{Res}_\varphi(\gamma, E; S_\lambda)$ *is represented by the cycle*

$$C_\lambda = \sum_{\boldsymbol{s}} c_{\boldsymbol{s}} \boldsymbol{s}, \qquad c_{\boldsymbol{s}} = \int_{\boldsymbol{s}^* \cap R_\lambda} \varphi(\nabla_1) + \int_{\boldsymbol{s}^* \cap R_{0\lambda}} \varphi(\nabla_0, \nabla_1), \tag{10.4}$$

where $\boldsymbol{s}$ *runs through the* $(m-2k)$*-simplices of* K *in* S_λ.

In particular, if $2k = m$ and if S_λ is compact, $\check{H}_0(S_\lambda;\mathbb{C}) = H_0(S_\lambda;\mathbb{C}) \simeq \mathbb{C}$, the isomorphism given by the augmentation. Thus we may regard $\operatorname{Res}_\varphi(\gamma, E; S_\lambda)$ as a complex number and we have (cf. Proposition 7.16):

$$\operatorname{Res}_\varphi(\gamma, E; S_\lambda) = \int_{R_\lambda} \varphi(\nabla_1) + \int_{R_{0\lambda}} \varphi(\nabla_0, \nabla_1), \tag{10.5}$$

which coincides with the expression (10.2).

Let R be an m-dimensional manifold possibly with boundary in M. We may assume that R and ∂R are K_0-subcomplexes of M. In the above considerations, we let S be R. Then, for a geometric object γ on $U_0 = M \smallsetminus R$ as above, we have the relative class $\varphi_R(E,\gamma)$ in $H^{2k}(M, M \smallsetminus R;\mathbb{C}) \simeq H^{2k}(R, \partial R;\mathbb{C})$.

The relative class $\varphi_R(E,\gamma)$ above is also defined if we have γ only in a neighborhood of ∂R (cf. Remark 5.8 for topological localizations).

Coming back to the previous situation, Let S be a K_0-subcomplex of M and γ a geometric object on $U_0 = M \smallsetminus S$ as above. Let R be as above and suppose $S \subset \operatorname{Int} R$. Then the restriction of γ to $M \smallsetminus R$, or to a neighborhood of ∂R, defines the relative class $\varphi_R(E,\gamma)$ in $H^{2k}(R, \partial R;\mathbb{C})$.

From Corollary 4.3 (see also Remark 7.19), we have:

Theorem 10.1 (Residue theorem). *In the above situation, it holds:*

1. *For each* λ, *we have the residue* $\mathrm{Res}_\varphi(\gamma, E; S_\lambda)$ *in* $\check{H}_{m-2k}(S_\lambda; \mathbb{C})$, *which is represented by the cycle* (10.4).

2. *We have*

$$\sum_\lambda (\imath_\lambda)_* \mathrm{Res}_\varphi(\gamma, E; S_\lambda) = [\underline{R}] \frown \varphi_R(E, \gamma) \qquad \text{in } \check{H}_{m-2k}(R; \mathbb{C}),$$

where $\imath_\lambda : S_\lambda \hookrightarrow R$ *denotes the inclusion.*

Remark 10.2. 1. If we take M as R in 2 above, the right-hand side is expressed as $[M] \frown \varphi(E)$ in $\check{H}_{m-2k}(M; \mathbb{C})$.

2. In the case $2k = m$, if R is compact, each S_λ is compact and the identity in 2 above is expressed as that in Proposition 10.2 (cf. (7.63)).

3. As we have seen, if $2k = m$ and if S is compact, we do not have to assume that S is a subcomplex of M. Simply define $\mathrm{Res}_\varphi(\gamma, E; S_\lambda)$ as in (10.2). Then we have Proposition 10.2, for compact R. That is the content of Theorem 10.1.

4. The above theorem becomes especially meaningful every time we have an explicit description of the residues.

5. The above arguments also work, if we replace E with a virtual bundle and M with a possibly singular variety, with some modifications (cf. Chapter 13 below).

Transverse residues

Let M, K, K^* and S be as above. Suppose the maximum dimension of the simplices of K in S is $m - 2k$ and let S' be an oriented submanifold of M of dimension $m - 2k$ which is contained in S. Let x be a point in S' and $\mathbb{D}$ a slice of S' in M at x (cf. Definition 3.24). Suppose

(*) the restriction $\gamma|_\mathbb{D}$ of γ to $\mathbb{D}$ makes sense and the restriction of any connection special for γ is special for $\gamma|_\mathbb{D}$.

Then, restricting the bundle E and the connections ∇_0 and ∇_1 to $\mathbb{D}$, we have the localization $\varphi_x(E|_\mathbb{D}, \gamma|_\mathbb{D})$ and the residue $\mathrm{Res}_\varphi(\gamma|_\mathbb{D}, E|_\mathbb{D}; x)$, which correspond to each other by the Alexander isomorphism

$$H^{2k}(\mathbb{D}, \mathbb{D} \smallsetminus \{x\}; \mathbb{C}) \simeq H_0(\{x\}; \mathbb{C}).$$

As $H_0(\{x\}; \mathbb{C}) \simeq \mathbb{C}$, we may think of $\mathrm{Res}_\varphi(\gamma|_\mathbb{D}, E|_\mathbb{D}; x)$ as a number, which is called the *transverse residue* at x. If we take a closed $2k$-ball $\mathbb{B}_1$ with center x in $\mathbb{D}$ and set $\mathbb{B}_{01} = -\partial \mathbb{B}_1$, which is a $(2k-1)$-sphere with the orientation opposite to the natural one, it is given by (cf. (10.5))

$$\mathrm{Res}_\varphi(\gamma|_{\mathbb{D}}, E|_{\mathbb{D}}; x) = \int_{\mathbb{B}_1} \varphi(\nabla_1) + \int_{\mathbb{B}_{01}} \varphi(\nabla_0, \nabla_1). \tag{10.6}$$

With these, we have:

Theorem 10.2. *As a function of x, $\mathrm{Res}_\varphi(\gamma|_{\mathbb{D}}, E|_{\mathbb{D}}; x)$ is locally constant. Thus, if S' is connected, it is constant.*

Proof. Let U be a tubular neighborhood of S' in U_1 with a C^∞ projection $\pi : U \to S'$, which gives U the structure of a bundle of open balls of dimension $2k$. Setting $U_0' = U \smallsetminus S'$ and $U_1' = U$, we consider the covering $\mathcal{U}' = \{U_0', U_1'\}$ of U. Let R_1 be a bundle on S' of closed balls of dimension $2k$ in U. Then $R_{01} = -\partial R_1$ is a bundle on S' of $(2k-1)$-spheres. We denote the restrictions of π to R_1 and R_{01} by π_1 and π_{01}, respectively. We have the fiber integration on ČdR cochains (cf. Definition 7.13):

$$\pi_* : A^p(\mathcal{U}', U_0') \longrightarrow A^{p-2k}(S'),$$

which is compatible with the differentials D and d (cf. (7.67)). The restriction of the localization $\varphi_S(E, \gamma)$ to U is represented by the cocycle $\varphi(\nabla_*)|_U = (\varphi(\nabla_1)|_{U_1'}, \varphi(\nabla_0, \nabla_1)|_{U_{01}'})$ in $A^{2k}(\mathcal{U}', U_0')$. We have

$$\pi_*(\varphi(\nabla_*)|_U) = (\pi_1)_*\varphi(\nabla_1) + (\pi_{01})_*\varphi(\nabla_0, \nabla_1),$$

which is in $A^0(S')$, i.e., a C^∞ function on S'. By definition it coincides with $\mathrm{Res}_\varphi(\gamma|_{\mathbb{D}}, E|_{\mathbb{D}}; x)$. As $\varphi(\nabla_*)|_U$ is D-closed, it is d-closed and is locally constant. □

Remark 10.3. 1. We may assume that x is the barycenter $b_{\boldsymbol{s}}$ of some $(m-2k)$-simplex $\boldsymbol{s}$ of K in S' and $\mathbb{B}_1 = \boldsymbol{s}^* \cap R_1$ (cf. Theorem 5.1). The transverse residue is then expressed as (cf. Proposition 10.3)

$$\mathrm{Res}_\varphi(\gamma|_{\mathbb{D}}, E|_{\mathbb{D}}; x) = \int_{\boldsymbol{s}^* \cap R_1} \varphi(\nabla_1) + \int_{\boldsymbol{s}^* \cap R_{01}} \varphi(\nabla_0, \nabla_1).$$

2. If we denote by $\iota : \mathbb{D} \hookrightarrow M$ the inclusion, we have the commutative diagram (Proposition 4.13 with $\mathbb{C}$-coefficients):

$$\begin{array}{ccc}
H^{2q}(M, M \smallsetminus S; \mathbb{C}) & \xrightarrow[A_{M,S}]{\sim} & \breve{H}_{m-2q}(S; \mathbb{C}) \\
\Big\downarrow \iota^* & & \Big\downarrow (\mathbb{D} \cdot\ )_x \\
H^{2q}(\mathbb{D}, \mathbb{D} \smallsetminus \{x\}; \mathbb{C}) & \xrightarrow[A_{\mathbb{D},x}]{\sim} & \breve{H}_0(\{x\}; \mathbb{C})
\end{array}$$

so that we may write (cf. Proposition 7.8)

$$\mathrm{Res}_\varphi(\gamma|_{\mathbb{D}}, E|_{\mathbb{D}}; x) = \big(\mathbb{D} \bullet \mathrm{Res}_\varphi(\gamma, E; S)\big)_x .$$

3. In the cases we consider below, the assumption (*) is always satisfied.

Expression of residues

Integral representation: Before we proceed to express the residues in a rather general setting, we give a formula not involving combinatorial topology, in the case S_λ is compact, still assuming that S is a subcomplex of M.

Thus suppose S_λ is compact. Note that $\check{H}_{m-2k}(S_\lambda; \mathbb{C}) = H_{m-2k}(S_\lambda; \mathbb{C})$ then. We have seen that, in the case $2k = m$, the residue is expressed by a global integral (cf. (10.5)). We consider the case $2k < m$.

Let U_λ be as before and R_λ a compact m-dimensional manifold with boundary in U_λ containing S_λ in its interior and we set $R_{0\lambda} = -\partial R_\lambda$. From Proposition 7.17 we have:

Proposition 10.4. *The residue* $\mathrm{Res}_\varphi(\gamma, E; S_\lambda)$ *in* $H_{m-2k}(S_\lambda; \mathbb{C})$ *is represented by an* $(m-2k)$*-cycle* C *such that*

$$\int_C \eta = \int_{R_\lambda} \varphi(\nabla_1) \wedge \eta + \int_{R_{0\lambda}} \varphi(\nabla_0, \nabla_1) \wedge \eta \tag{10.7}$$

for every closed $(m-2k)$*-form* η *on* U_λ.

In particular if $2k = m$, the above reduces to (10.5).

Case S_λ is a submanifold: We give a formula for the residue at S_λ when it is a compact oriented submanifold of M of dimension $m-2k$ with orientation compatible with that of M. In this case we have the transverse residue $\mathrm{Res}_\varphi(\gamma|_{\mathbb{D}}, E|_{\mathbb{D}}; x)$ at each point x in S_λ and it is in fact constant (cf. Theorem 10.2).

Theorem 10.3. *Suppose that* S_λ *is a compact oriented submanifold of* M *of dimension* $m-2k$. *Then we have:*

$$\mathrm{Res}_\varphi(\gamma, E; S_\lambda) = \mathrm{Res}_\varphi(\gamma|_{\mathbb{D}}, E|_{\mathbb{D}}; x) \cdot S_\lambda \qquad \text{in} \ \ H_{m-2k}(S_\lambda; \mathbb{C}).$$

Proof. We try to find $\mathrm{Res}_\varphi(\gamma, E; S_\lambda)$ by Proposition 10.4. As U_λ, we take a tubular neighborhood of S_λ with a C^∞ projection $\pi : U_\lambda \to S_\lambda$, which gives U_λ the structure of a bundle of open balls of dimension $2k$.

Setting $U_0 = U_\lambda \smallsetminus S_\lambda$, we consider the covering $\mathcal{U}_\lambda = \{U_0, U_\lambda\}$ of U_λ. As R_λ, we take a bundle on S_λ of closed balls of dimension $2k$ in U_λ. Then $R_{0\lambda}$ is a bundle on S_λ of $(2k-1)$-spheres. We denote the restrictions of π to R_λ and $R_{0\lambda}$ by π_λ and $\pi_{0\lambda}$, respectively. For a closed $(m-2k)$-form η on U_λ, we compute the integral

$$I = \int_{R_\lambda} \varphi(\nabla_1) \wedge \eta + \int_{R_{0\lambda}} \varphi(\nabla_0, \nabla_1) \wedge \eta.$$

Since π induces an isomorphism $\pi^* : H_d^{m-2k}(S_\lambda) \xrightarrow{\sim} H_d^{m-2k}(U_\lambda)$, there exist a closed $(m-2k)$-form θ on S_λ and an $(m-2k-1)$-form τ on U_λ such that

$$\eta = \pi^*\theta + d\tau.$$

Using the projection formula, the fact that $\varphi(\nabla_1)$ is d-closed and the Stokes theorem, we have

$$\int_{R_\lambda} \varphi(\nabla_1) \wedge \eta = \int_{S_\lambda} (\pi_\lambda)_*\varphi(\nabla_1) \cdot \theta + (-1)^{2d+1} \int_{R_{0\lambda}} \varphi(\nabla_1) \wedge \tau,$$

where $(\pi_\lambda)_*$ denotes the integration along the fiber of π_λ. Note that $(\pi_\lambda)_*\varphi(\nabla_1)$ is a C^∞ function on S_λ. Noting that $d\varphi(\nabla_0, \nabla_1) = \varphi(\nabla_1)$ on $U_{0\lambda}$ and $\partial R_{0\lambda} = \emptyset$, we also compute to get

$$\int_{R_{0\lambda}} \varphi(\nabla_0, \nabla_1) \wedge \eta = \int_{S_\lambda} (\pi_{0\lambda})_*\varphi(\nabla_0, \nabla_1) \cdot \theta + (-1)^{2k} \int_{R_{0\lambda}} \varphi(\nabla_1) \wedge \tau.$$

Thus we have

$$I = \int_{S_\lambda} ((\pi_\lambda)_*\varphi(\nabla_1) + (\pi_{0\lambda})_*\varphi(\nabla_0, \nabla_1)) \cdot \theta$$

Now we recall the fiber integration on ČdR cochains (cf. Definition 7.13):

$$\pi_* : A^p(\mathcal{U}_\lambda, U_0) \longrightarrow A^{p-2k}(S_\lambda).$$

It is compatible with the differentials D and d (cf. (7.67)). Since $(\varphi(\nabla_1)|_{U_\lambda}, \varphi(\nabla_0, \nabla_1)|_{U_{0\lambda}})$ is a cocycle in $A^p(\mathcal{U}_\lambda, U_0)$, the function $(\pi_\lambda)_*\varphi(\nabla_1) + (\pi_{0\lambda})_*\varphi(\nabla_0, \nabla_1)$ is d-closed so that it is a constant. By definition the constant is exactly $\mathrm{Res}_\varphi(\gamma|_{\mathbb{D}}, E|_{\mathbb{D}}; x)$ above. Finally from

$$\int_{S_\lambda} \theta = \int_{S_\lambda} \eta,$$

we have the theorem. □

Case S_λ is a pseudo-manifold: We give an explicit expression of $\mathrm{Res}_\varphi(\gamma, E; S_\lambda)$ in the case S_λ is a pseudo-manifold (cf. Definition 5.12).

Let S_λ be a connected component of S as above and suppose it is an oriented pseudo-manifold of dimension $m-2k$. If $\boldsymbol{s}$ is an $(m-2k)$-simplex of K in S_λ, $\boldsymbol{s}^*$ is a $2k$-cell such that $\boldsymbol{s}^* \cap S_\lambda = \{b_{\boldsymbol{s}}\}$. Thus we have the residue $\mathrm{Res}_\varphi(\gamma, E|_{\boldsymbol{s}^*}; b_{\boldsymbol{s}})$, which is given by (cf. Remark 10.3. 1)

$$\mathrm{Res}_\varphi(\gamma, E|_{\boldsymbol{s}^*}; b_{\boldsymbol{s}}) = \int_{\boldsymbol{s}^* \cap R_1} \varphi(\nabla_1) + \int_{\boldsymbol{s}^* \cap R_{01}} \varphi(\nabla_0, \nabla_1).$$

From Proposition 10.3, we have:

Theorem 10.4. *Suppose S_λ is an oriented pseudo-manifold of dimension $m-2k$. Then the residue $\mathrm{Res}_\varphi(\gamma, E; S_\lambda)$ in $\check{H}_{m-2k}(S_\lambda; \mathbb{C})$ is represented by the cycle*

$$\sum_{\boldsymbol{s}} \mathrm{Res}_\varphi(\gamma, E|_{\boldsymbol{s}^*}; b_{\boldsymbol{s}}) \cdot \boldsymbol{s},$$

where $\boldsymbol{s}$ runs through all the $(m-2k)$-simplices of K in S_λ.

Corollary 10.1. *In the above situation, suppose the set of general points of S_λ has the structure of a C^∞ submanifold of dimension $m-2q$ of M. Then, if $S_\lambda = \bigcup_i S_{\lambda,i}$ is the irreducible decomposition,*

$$\mathrm{Res}_\varphi(\gamma, E; S_\lambda) = \sum_i \mathrm{Res}_\varphi(\gamma|_{\mathbb{D}}, E|_{\mathbb{D}}; x_{\lambda,i}) \cdot [S_{\lambda,i}] \qquad \text{in } \check{H}_{m-2k}(S_\lambda; \mathbb{C}),$$

where $x_{\lambda,i}$ is a general point of $S_{\lambda,i}$ and $\mathbb{D}$ a slice of $S_{\lambda,i}$ at $x_{\lambda,i}$.

In the case S_λ is a compact submanifold, the above reduces to Theorem 10.3. Note that we do not need the compactness of M or of S_λ in Corollary 10.1.

Various types of localizations

In the following, we discuss some vanishing theorems and the corresponding localizations. In the case (I) below, both the obstruction theory and the Chern-Weil theory are effective and the comparison of the two approaches yields interesting results. In the case (II), it seems difficult to apply the obstruction theory directly. We use either the relative K-theory or the Chern-Weil theory which we adopt in this text. In the case (III), we have continuous invariants and the topological method seems to be non-applicable in general. The method using connections is effective.

(I) Localization by a frame

Suppose a complex vector bundle E of rank l admits an r-frame $s^{(r)}$ on $M \smallsetminus S$. If we use the obstruction theory, this means the vanishing of the cocyle representing the Chern class $c^q(E)$, $q = l - r + 1$, on $M \smallsetminus S$ and we have naturally the topological localization $c^q_{\mathrm{top}}(E, s^{(r)})$ (cf. Section 5.6).

In the above setting, the geometric object γ is $s^{(r)}$ and the connections special for γ are the $s^{(r)}$-trivial connections. The Chern polynomial c^q is adapted to γ (in fact c^i, $q \leq i \leq l$, are adapted, however the case $i = q$ is of particular interest). The corresponding localization of $c^q(E)$ is denoted by $c^q_S(E, s^{(r)})$. This will be discussed in detail in Section 10.3 below.

In fact, the assumption that there exists an r-frame on $M \smallsetminus S$ is rather strong. We take a triangulation of M compatible with S (or the cellular decomposition dual to it), and $c^q(E)$ is localized if we are given an r-frame on the $2q$-skeleton of $M \smallsetminus S$, $q = l - r + 1$.

(II) Localization by exactness

Suppose we have a complex of complex vector bundles on M

$$0 \longrightarrow E_q \longrightarrow \cdots \longrightarrow E_0 \longrightarrow 0,$$

which is exact on $M \smallsetminus S$. In this situation we are to consider the characteristic classes of the virtual bundle $\xi = \sum_{\nu=0}^{q}(-1)^j E_\nu$ (cf. Section 8.3, also Sections 15.2–15.4 below). The geometric object γ is the exact sequence and the connections special for γ are the families of connections compatible with the sequence. Any polynomial φ without constant term is adapted to γ (cf. Proposition 15.4 below) and we have the localization $\varphi_S(\xi)$ in $H^*(M, M \smallsetminus S)$.

In particular, if M is a complex manifold and if $\mathscr{S}$ is a coherent $\mathscr{O}_M$-module, we have the localized class $\varphi_S(\mathscr{S})$ with S the support of $\mathscr{S}$. This will be used in Chapter 15 below.

(III) Localization by Bott type vanishing theorems

This type of vanishing occurs on some characteristic forms of a vector bundle which admits an action of a holomorphic subbundle of the holomorphic tangent bundle. Thus the geometric object γ is such an action and the connections special for γ are the ones compatible with the action. This will be explained in detail in (III) of Section 10.2. It leads to a general residue theory for singular holomorphic foliations or distributions.

10.2 Vanishing theorems

(I) *Vanishing by frames*

Let M be a C^∞ manifold and E a C^∞ complex vector bundle of rank l on M. We begin with an obvious case of "vanishing by rank reason": for arbitrary k connections $\nabla_1,\dots,\nabla_k$ for E,

$$c^i(\nabla_1,\dots,\nabla_k)=0 \qquad \text{for } i\geq l+1.$$

The following is essentially a consequence of this:

Proposition 10.5. *Let $s^{(r)}=(s_1,\dots,s_r)$ be an r-frame of E on an open set U in M and let $\nabla_1,\dots,\nabla_k$ be $s^{(r)}$-trivial connections for E on U, then on U,*

$$c^i(\nabla_1,\dots,\nabla_k)=0 \qquad \textit{for } i\geq l-r+1.$$

Proof. Let E' be the subbundle of E on U spanned by $s^{(r)}$ and, for each $j=0,\dots,p$, let ∇'_j be the connection for E' obtained by restricting ∇_j to the sections of E'. Then ∇_j induces a connection ∇''_j for the quotient bundle $E''=E/E'$ on U so that the triple $(\nabla'_j,\nabla_j,\nabla''_j)$ is compatible with the exact sequence

$$0\longrightarrow E'\longrightarrow E\longrightarrow E''\longrightarrow 0.$$

We set $\nabla^\bullet_j=(\nabla'_j,\nabla_j)$. Then, since the connection matrix of ∇'_j with respect to the frame $s^{(r)}$ of E' is zero, we have $c_i(\nabla_1,\dots,\nabla_k)=c_i(\nabla^\bullet_1,\dots,\nabla^\bullet_k)$. By Proposition 8.4, it is equal to $c_i(\nabla''_1,\dots,\nabla''_k)$, which vanishes for $i\geq r-l+1$, since the rank of E'' is $r-l$. □

(II) *Vanishing by exactness*

This will be explained in detail in Section 15.4 below.

(III) *Bott vanishing theorem*

Let M be a C^∞ manifold of dimension m and $T_{\mathbb{R}}M$ its tangent bundle. We set $\tau=T_{\mathbb{R}}M\otimes\mathbb{C}$, the complexification of $T_{\mathbb{R}}M$. Let E be a complex vector bundle of rank l on M. For an open set U of M we denote by $A^r(U;E)$ the space of complex valued r-forms with coefficients in E on U, i.e., sections of $\bigwedge^r\tau^*\otimes E$ on U. We denote by $A^r(U)$ the space of C^∞ functions on U.

Let H be a complex subbundle of τ and $\rho : \tau^* \to H^*$ the canonical surjection. Then we have the exterior derivative along H:

$$d_H : A^0(M) \longrightarrow A^0(M; H^*),$$

which is defined by $d_H = \rho \circ d$ so that $d_H(f)(u) = u(f)$ for $f \in A^0(M)$ and $u \in A^0(M; H)$.

Definition 10.2. A *partial connection for E along H* is a $\mathbb{C}$-linear map

$$\delta : A^0(M; E) \longrightarrow A^0(M; H^* \otimes E)$$

satisfying the Leibniz rule

$$\delta(fs) = d_H(f) \otimes s + f\delta(s) \qquad \text{for } f \in A^0(M) \text{ and } s \in A^0(M; E).$$

Thus a connection for E is nothing but a partial connection along τ.

Example 10.1. The exterior derivative d_H along H is a partial connection for the product bundle $\mathbb{C} \times M$ along H.

Example 10.2. Let M be a complex manifold and TM its holomorphic tangent bundle. Then we have a natural decomposition $\tau = TM \oplus \overline{T}M$. If E is a holomorphic vector bundle on M,

$$\bar{\partial} : A^0(M; E) \longrightarrow A^0(M; \overline{T}^* M \otimes E)$$

is a partial connection for E along $\overline{T}M$.

The following is proved as Lemma 8.1:

Lemma 10.1. *A partial connection δ is a local operator, i.e., if a section s is identically* 0 *on an open set U, so is $\delta(s)$.*

Thus the restriction of δ to U is well-defined. We also have:

Lemma 10.2. *Let $\delta_1, \dots, \delta_k$ be partial connections for E along H and $f_1, \dots, f_k$ C^∞ functions on M with $\sum_{i=1}^k f_i = 1$. Then $\sum_{i=1}^k f_i\delta_i$ is a partial connection for E along H.*

Proposition 10.6. *For arbitrary H and E, E admits a partial connection along H.*

Proposition 10.7. *Let δ_1 and δ_2 be partial connections for E along H_1 and H_2, respectively. Suppose $H_{1,x} \cap H_{2,x} = \{0\}$ for all x in M. Then*

$$\delta_1 \oplus \delta_2 : A^0(M; E) \longrightarrow A^0(M; (H_1^* \oplus H_2^*) \otimes E)$$

given by $\delta_1 \oplus \delta_2(s) = (\delta_1(s), \delta_2(s))$ is a partial connection for E along $H_1 \oplus H_2$.

If δ is a partial connection for E along H, for each u in $A^0(M;H)$, we have a $\mathbb{C}$-linear map:

$$\delta_u : A^0(M;E) \longrightarrow A^0(M;E),$$

which is defined by $\delta_u(s) = \delta(s)(u)$. It satisfies

$$\delta_u(fs) = u(f)s + f\delta_u(s).$$

By Lemma 10.1, if $u \in A^0(U;H)$, $\delta_u : A^0(U;E) \to A^0(U;E)$ is a well-defined operator. Note that δ_u cannot be thought of as an element in $A^0(M;E^* \otimes E)$ so that a partial connection cannot be though of as a map $A^0(M;H) \to A^0(M;E^* \otimes E)$.

Note that a connection ∇ defines a partial connection for arbitrary H i.e., $(\rho \otimes 1) \circ \nabla$.

Definition 10.3. We say that a connection ∇ for E extends δ if the following diagram is commutative:

$$\begin{array}{ccc} A^0(M;E) & \xrightarrow{\nabla} & A^0(M;\tau^* \otimes E) \\ & \overset{\delta}{\searrow} & \downarrow{\scriptstyle \rho\otimes 1} \\ & & A^0(M;H^* \otimes E). \end{array}$$

Lemma 10.3. *A connection ∇ extends δ if and only if*

$$\nabla_u = \delta_u \qquad \textit{for } \ u \in A^0(M;H).$$

Proposition 10.8. *For arbitrary δ, there is a connection ∇ that extends δ.*

Proof. Take a C^∞ decomposition $\tau = H \oplus H'$ and let $\nabla = \delta \oplus \delta'$ with δ' a partial connection along H'. □

Suppose H is involutive. Then we may define

$$\begin{aligned} d_H :& A^0(M;H^*) \longrightarrow A^0(M;\textstyle\bigwedge^2 H^*) \qquad \text{by} \\ & d_H\,\eta(u,v) = u(\eta(v)) - v(\eta(u)) - \eta([u,v]). \end{aligned}$$

Then the following diagram is commutative:

$$\begin{array}{ccc} A^0(M,\tau^*) & \xrightarrow{d} & A^0(M,\bigwedge^2\tau^*) \\ \downarrow{\scriptstyle \rho} & & \downarrow{\scriptstyle \bigwedge^2\rho} \\ A^0(M,H^*) & \xrightarrow{d_H} & A^0(M,\bigwedge^2 H^*). \end{array}$$

If δ is a partial connection for E along H, it induces a $\mathbb{C}$-linear map

$$\hat{\delta} : A^0(M, H^* \otimes E) \longrightarrow A^0(M, \bigwedge^2 H^* \otimes E)$$

satisfying

$$\hat{\delta}(\eta \otimes s) = d_H\eta \otimes s - \eta \wedge \delta(s) \quad \text{for } \eta \in A^0(M, H^*) \text{ and } s \in A^0(M, E).$$

It also satisfies

$$\hat{\delta}(f\sigma) = d_H(f) \wedge \sigma + f\hat{\delta}(\sigma) \quad \text{for } f \in A^0(M) \text{ and } \sigma \in A^0(M, H^* \otimes E).$$

The composition

$$K_\delta = \hat{\delta} \circ \delta : A^0(M, E) \longrightarrow A^0(M, \bigwedge^2 H^* \otimes E)$$

is called the *curvature of* δ. The curvature is $A^0(M)$-linear:

$$K_\delta(fs) = fK_\delta(s)$$

so that we may think of K_δ as being in $A^0(M, \bigwedge^2 H^* \otimes \mathrm{Hom}(E, E))$.

Definition 10.4. Let H be involutive. A partial connection δ along H is called *flat* if $K_\delta = 0$.

Let $e^{(l)} = (e_1, \dots, e_l)$ be a frame of E on U. We write

$$\delta(e_i) = \sum_{j=1}^{l} \theta_{ji} \otimes e_j, \quad \theta_{ij} \in A^0(U; H^*).$$

The matrix $\theta_\delta = (\theta_{ij})$ is called the connection matrix of δ with respect to $e^{(l)}$. We also write

$$K_\delta(e_i) = \sum_{j=1}^{l} \kappa_{ji} \otimes e_j, \quad \kappa_{ij} \in A^0(U; \bigwedge^2 H^*).$$

The matrix $\kappa_\delta = (\kappa_{ij})$ is called the curvature matrix of δ with respect to $e^{(l)}$. We have

$$\kappa_\delta = d_H\theta_\delta + \theta_\delta \wedge \theta_\delta. \tag{10.8}$$

By the frame change $e'^{(l)} = e^{(l)}A$,

$$\theta'_\delta = A^{-1} \cdot d_H A + A^{-1}\theta_\delta A \quad \text{and} \quad \kappa'_\delta = A^{-1}\kappa_\delta A. \tag{10.9}$$

The identity (10.8) implies the following:

Proposition 10.9 (Ricci identity). *We have*

$$K_\delta(u, v) = \delta_u \circ \delta_v - \delta_v \circ \delta_u - \delta_{[u,v]}.$$

From now on let M be a complex manifold and E a holomorphic vector bundle on M. We call a holomorphic subbundle F of TM a (non-singular) *holomorphic distribution* on M. The dimension of the distribution is the rank of F. If F is involutive, it is a holomorphic foliation on M (cf. Section 6.4).

Definition 10.5. Let F be a holomorphic distribution on M. A partial connection δ for E along F is *holomorphic* if $\delta(s)(u)$ is holomorphic whenever s and u are.

We recall Definition 9.10 for connections of type $(1,0)$.

Proposition 10.10. *A connection ∇ for E is of type $(1,0)$ if and only if it is a connection extending $\bar\partial$.*

Proof. This follows from the fact that the map $\rho : \tau^* \to \overline{T}^* M$ assigns to each 1-form its $(0,1)$-component and that $\bar\partial$ kills holomorphic sections. □

With these we have the following:

Theorem 10.5. *Let F be a holomorphic distribution of dimension p on M, δ a holomorphic partial connection for E along F and φ an invariant polynomial homogeneous of degree k. Let $\nabla_0, \dots, \nabla_q$ be connections for E extending $\delta \oplus \bar\partial$.*
(1) *If $k > n - p + [\frac{p}{2}]$,*

$$\varphi(\nabla_0, \dots, \nabla_q) = 0.$$

(2) *If $k > n - p$,*

$$\varphi^{k,k}(\nabla_0, \dots, \nabla_q) = 0.$$

(3) *In the case F is involutive and δ is flat, if $k > n - p$,*

$$\varphi(\nabla_0, \dots, \nabla_q) = 0.$$

Proof. It suffices to prove the theorem for the case $q = 0$. The case for general q follows from the construction of the difference form.

We take a C^∞ decomposition $TM = F \oplus G$. Let $(u_1, \dots, u_p)$ be a local holomorphic frame of F. Let $(u_1^*, \dots, u_p^*)$ be the dual frame of F^* and $(v_1^*, \dots, v_{n-p}^*)$ a local frame of G^*. Since ∇ extands $\bar\partial$, each entry of the curvature matrix κ of ∇ with respect to a holomorphic frame of E has only $(2,0)$ and $(1,1)$ components, i.e., a linear combination (with function coefficients) of

$$u_i^* \wedge u_j^*, \quad u_i^* \wedge v_j^*, \quad v_i^* \wedge v_j^*, \quad u_i^* \wedge d\bar z_j, \quad v_i^* \wedge d\bar z_j.$$

For holomorphic sections u and s of F and E and a section $\bar{w}$ of $\overline{T}M$, we have

$$K(u,\bar{w})(s) = \nabla_u(\nabla_{\bar{w}}(s)) - \nabla_{\bar{w}}(\nabla_u(s)) - \nabla_{[u,\bar{w}]}(s).$$

In the above, $\nabla_{\bar{w}}(s) = \bar{\partial}_{\bar{w}}(s) = \bar{\partial}(s)(\bar{w}) = 0$, also $\nabla_{\bar{w}}(\nabla_u(s)) = 0$ as $\nabla_u(s) = \delta_u(s)$ is holomorphic. Finally $\nabla_{[u,\bar{w}]}(s) = 0$ as $[u,\bar{w}] = 0$. Since holomorphic sections locally generates the space of sections, we have $K(u,\bar{w}) = 0$, which shows we do not have $u_i^* \wedge d\bar{z}_j$ in the above list.

Finally if F is involutive and if δ is flat, we do not have $u_i^* \wedge u_j^*$ in the above list. □

The theories developed above may be done in parallel using the notion of an "action" of H on E.

Definition 10.6. An *action of* H *on* E is a $\mathbb{C}$-bilinear map

$$\alpha : A^0(M,H) \times A^0(M,E) \longrightarrow A^0(M,E)$$

satisfying, for $f \in A^0(M)$, $u \in A^0(M,H)$ and $s \in A^0(M,E)$,

(1) $\alpha(fu,s) = f\alpha(u,s)$,
(2) $\alpha(u,fs) = u(f)s + f\alpha(u,s)$.

The following is not difficult to see:

Proposition 10.11. *If δ is a partial connection δ for E along H,*

$$\alpha_\delta : A^0(M,H) \times A^0(M,E) \longrightarrow A^0(M,E),$$

defined by $\alpha_\delta(u,s) = \delta(s)(u)$, is an action of H on E. Conversely an action determines a partial connection.

If H is involutive, for an action α, the corresponding partial connection is flat if and only if it further satisfies

(3) $\alpha([u,v],s) = \alpha(u,\alpha(v,s)) - \alpha(v,\alpha(u,s))$.

In this case we say that the action is involutive.

If M is a complex manifold and if F is a holomorphic subbundle of TM, for an action α of F, the corresponding partial connection is holomorphic if and only if

(4) $\alpha(u,s)$ is holomorphic whenever u and s are.

In this case we say that the action is holomorphic.

We give some example of holomorphic actions. In the following we denote by M a complex manifold of dimension n and F a subbundle of TM of rank p.

(A) Action on the normal bundle of the foliation: Suppose F is involutive, i.e., a foliation. We call $N_F = TM/F$ the normal bundle of the foliation. We denote by $\eta : TM \to N_F$ the canonical surjection. Then

$$\alpha : A^0(M, F) \times A^0(M, N_F) \longrightarrow A^0(M, N_F)$$

defined by $\alpha(u, \eta(v)) = \eta([u, v])$, for $u \in A^0(M, F)$ and $v \in A^0(M, TM)$, is a well-defined action.

(B) Action on the normal bundle of an invariant submanifold: Let V be a complex submanifold of M. Let N_V be the normal bundle of V in M so that we have the exact sequence (3.13). Let F be a distribution of dimension p on M. We say that F leaves V invariant if $F|_V \subset TV$. In this case we set $F_V = F|_V$, which is a distribution of dimension p on V. In this situation, there is a natural holomorphic action of F_V on N_V.

The action is constructed as follows. Let u and ν be C^∞ sections of F_V and N_V, respectively. Take sections $\tilde{u}$ of F and $\tilde{v}$ of TM so that $\tilde{u}|_V = u$ and $\varpi(\tilde{v}|_V) = \nu$, where $|_V$ means the restriction as sections. Define

$$\alpha : A^0(V, F_V) \times A^0(V, N_V) \longrightarrow A^0(V, N_V) \qquad \text{by } \alpha(u, \nu) = \varpi([\tilde{u}, \tilde{v}]|_V).$$

Then it is a well-defined action.

(C) Action on the normal bundle of the ambient foliation: Let V be as in (B) and let F be a foliation on M leaving V in variant. Then the action in (A) induces an action of F_V on $N_F|_V$.

Remark 10.4. If F is a foliation on M leaving a submanifold V invariant, we have the commutative diagram with exact rows and columns:

$$\begin{array}{ccccccccc}
 & & 0 & & 0 & & & & \\
 & & \uparrow & & \uparrow & & & & \\
0 & \longrightarrow & N_{F_V} & \longrightarrow & N_F|_V & \longrightarrow & N_V & \longrightarrow & 0 \\
 & & \uparrow & & \uparrow & & \uparrow{\scriptstyle =} & & \\
0 & \longrightarrow & TV & \longrightarrow & TM|_V & \xrightarrow{\varpi} & N_V & \longrightarrow & 0 \\
 & & \uparrow & & \uparrow & & & & \\
 & & F_V & \xrightarrow{=} & F_V & & & & \\
 & & \uparrow & & \uparrow & & & & \\
 & & 0 & & 0. & & & &
\end{array}$$

The bundle F_V acts on each of the bundles in the first row. The action on N_{F_V} is the one in (A) for $M = V$, the action on $N_F|_V$ is the one in (C) and the action on N_V is the one in (B) (for this one the involutivity of F is not necessary).

10.3 Differential geometric localization by frames

Let M be a C^∞ manifold of dimension m and E a C^∞ complex vector bundle of rank l on M. Also let S be a closed set in M and suppose we have a C^∞ r-frame $s^{(r)}$ of E on $M \smallsetminus S$. Then we will see that there is a natural localization $c^i(E, s^{(r)})$ in $H^{2i}(M, M \smallsetminus S; \mathbb{C})$ of the Chern class $c^i(E)$ for $i \geq l-r+1$, letting γ and φ in Section 10.1 be $s^{(r)}$ and c^i, respectively.

Letting $U_0 = M \smallsetminus S$ and U_1 a neighborhood of S, we consider the covering $\mathcal{U} = \{U_0, U_1\}$ of M. Recall the class $c^i(E)$ is represented by the cocycle $c^i(\nabla_*)$ in $A^{2i}(\mathcal{U})$ given by

$$c^i(\nabla_*) = (c^i(\nabla_0), c^i(\nabla_1), c^i(\nabla_0, \nabla_1)), \tag{10.10}$$

where ∇_0 and ∇_1 denote connections for E on U_0 and U_1, respectively. If we take as ∇_0 an $s^{(r)}$-trivial connection, then $c^i(\nabla_0) = 0$ and the cocycle is in $A^{2i}(\mathcal{U}, U_0)$ for $i \geq l-r+1$ (cf. Proposition 10.5). Moreover, the class of $c^i(\nabla_*)$ in $H^{2i}(M, M \smallsetminus S; \mathbb{C})$ does not depend on the choice of the $s^{(r)}$-trivial connection ∇_0 or the connection ∇_1 (cf. Proposition 10.1). Thus the class is well-defined in $H^{2i}(M, M \smallsetminus S; \mathbb{C})$, which we denote by $c^i_{S,\mathrm{diff}}(E, s^{(r)})$ and call the *differential geometric localization* of $c^i(E)$ by $s^{(r)}$ at S. It will also be denoted by $c^i_S(E, s^{(r)})$, $c^i_{\mathrm{diff}}(E, s^{(r)})$ or $c^i(E, s^{(r)})$, if there is no fear of confusion. Its image by the canonical morphism $H^{2i}(M, M \smallsetminus S; \mathbb{C}) \to H^{2i}(M; \mathbb{C})$ is the class $c^i(E)$.

Remark 10.5. From the fact that $c^i_{\mathrm{diff}}(E, s^{(r)})$ does not depend on ∇_0, we see that, for $i = l-r+2, \dots, l$, we have $c^i_{\mathrm{diff}}(E, s^{(r)}) = c^i_{\mathrm{diff}}(E, s^{(r')})$, where $s^{(r')}$ denotes an r'-frame, $r' = r - i + 1$, made of r' arbitrary members of $s^{(r)}$. Thus the case $i = l-r+1$ will be of essential interest. In the following, we consider only this case and set $q = l - r + 1$.

Suppose M is oriented and S is a subcomplex of M with respect to some triangulation of M. Then we have the Alexander duality (Theorems 4.2 and 7.13):

$$A : H^{2q}(M, M \smallsetminus S; \mathbb{C}) \xrightarrow{\sim} \check{H}_{m-2q}(S; \mathbb{C}).$$

Definition 10.7. The *differential geometric residue* $\mathrm{Res}_{c^q}(s^{(r)}, E; S)$ *of* $s^{(r)}$ *for* $c^q(E)$ *at* S is the image of $c^q_{\mathrm{diff}}(E, s^{(r)})$ by A.

Suppose that S has a finite number of connected components $(S_\lambda)_\lambda$. Then we have the residue $\mathrm{Res}_{c^q}(s^{(r)}, E; S_\lambda)$ in $\check{H}_{m-2q}(S_\lambda; \mathbb{C})$ for each λ. It is represented by a cycle as in (10.4) with $\varphi(\nabla_1)$ and $\varphi(\nabla_0, \nabla_1)$ replaced by $c^q(\nabla_1)$ and $c^q(\nabla_0, \nabla_1)$, respectively.

In particular, if $2q = m$ and if S_λ is compact, we have (cf. (10.5)):

$$\operatorname{Res}_{c^q}(s^{(r)}, E; S_\lambda) = \int_{R_\lambda} c^q(\nabla_1) + \int_{R_{0\lambda}} c^q(\nabla_0, \nabla_1). \tag{10.11}$$

Let R be an m-dimensional manifold with boundary in M. If we have an r-frame $s^{(r)}$ on $M \smallsetminus R$, or on a neighborhood of ∂R, we have the relative class $c_R^q(E, s^{(r)})$ in $H^{2q}(R, \partial R; \mathbb{C})$.

Coming back to the previous situation, suppose S is contained in the interior of an m-dimensional manifold R with boundary in M. An r-frame $s^{(r)}$ as above restricts to an r-frame on $M \smallsetminus R$, or on a neighborhood of ∂R, and we have the relative class $c_R^q(E, s^{(r)})$ in $H^{2q}(R, \partial R; \mathbb{C})$.

From Theorem 10.1 we have the differential geometric counterpart of Theorem 5.2:

Theorem 10.6. *In the above situation, it holds:*

1. *For each λ, we have the residue* $\operatorname{Res}_{c^q}(s^{(r)}, E; S_\lambda)$ *in* $\check{H}_{m-2q}(S_\lambda; \mathbb{C})$.
2. *We have*

$$\sum_\lambda (\imath_\lambda)_* \operatorname{Res}_{c^q}(s^{(r)}, E; S_\lambda) = [\underline{R}] \frown c_R^q(E, s^{(r)}) \qquad \text{in } \check{H}_{m-2q}(R; \mathbb{C}).$$

Remark 10.6. Remarks as in Remark 10.2. 1, 2 and 3 apply to this case as well, replacing φ, k and γ with c^q, q and $s^{(r)}$, respectively. In particular, if $2q = m$ and if S is compact, it is not necessary to assume that S is a subcomplex of M to have the residue theorem.

Suppose S is compact. For each λ we take a neighborhood U_λ of S_λ in U_1, having the same homotopy type as S_λ, so that the U_λ's are mutually disjoint. Let R_λ be a compact m-dimensional manifold with C^∞ boundary in U_λ containing S_λ in its interior. We set $R_{0\lambda} = -\partial R_\lambda$. Then, by Proposition 10.4, the residue $\operatorname{Res}_{c^q}(s^{(r)}, E; S_\lambda)$ is represented by an $(m-2q)$-cycle C in S_λ such that

$$\int_C \eta = \int_{R_\lambda} c^q(\nabla_1) \wedge \eta + \int_{R_{0\lambda}} c^q(\nabla_0, \nabla_1) \wedge \eta \tag{10.12}$$

for every closed $(m-2q)$-form η on U_λ.

We also have the transverse residues and the statements corresponding to Theorems 10.3 and 10.4 and Corollary 10.1.

Remark 10.7. In Section 5.6, we defined the topological localization $c_{\mathrm{top}}^q(E, s^{(r)})$ and the topological residue $\operatorname{TRes}_{c^q}(s^{(r)}, E; S_\lambda)$. In Section 10.5 below, we prove that they are essentially the same as the ones defined using connections.

Before we proceed further we give some basic examples to illustrate the procedure explained above.

Example 10.3. Let U be a neighborhood of 0 in $\mathbb{R}^2$ and $E = \mathbb{C} \times U$ the product bundle on U. Suppose we have a non-vanishing C^∞ section s of E on $U_0 = U \smallsetminus \{0\}$. Set $U_1 = U$ and consider the covering $\mathcal{U} = \{U_0, U_1\}$ of U. Then we have the localization $c^1(E, s)$ in $H^2(\mathcal{U}, U_0) \simeq H^2(U, U \smallsetminus \{0\}; \mathbb{C})$ and the residue $\mathrm{Res}_{c^1}(s, E; 0)$ in $H_0(\{0\}; \mathbb{C}) = \mathbb{C}$, which we try to find. See Example 5.1 and (5.10) for the topological counterpart. The two residues are related by Proposition 10.13.

If we denote by e the frame of E given by $e(z) = (1, z)$, we may write $s = fe$ with f a non-vanishing C^∞ function on U_0. Let $\mathbb{B}^2$ be a closed 2-ball around 0 in U. Its boundary $\partial\mathbb{B}^2$ is a circle $\mathbb{S}^1$ oriented counterclockwise. In the expression (10.11) of the residue, we take as ∇_1 the e-trivial connection on U, thus $c^1(\nabla_1) = 0$ and

$$\mathrm{Res}_{c^1}(s, E; 0) = -\int_{\mathbb{S}^1} c^1(\nabla_0, \nabla_1)$$

with ∇_0 the s-trivial connection on U_0. Now we recall how the difference form $c^1(\nabla_0, \nabla_1)$ is defined (cf. Proposition 8.3). Consider the product $\mathbb{R} \times U$ with projection $\rho : \mathbb{R} \times U \to U$. Define a connection for $\rho^* E$ on $\mathbb{R} \times U_0$ by $\tilde{\nabla} = (1-t)\nabla_0 + t\nabla_1$ (we omit ρ^*). Then $c^1(\nabla_0, \nabla_1) = \rho'_* c^1(\tilde{\nabla})$, where $\rho' : [0,1] \times U_0 \to U_0$ is the restriction of ρ. Let θ_i be the connection matrix, in fact a form, of ∇_i with respect to the frame e, $i = 0, 1$. Then $\theta_1 = 0$. Noting that the connection form of ∇_0 with respect to s is zero, we get from (8.4),

$$\theta_0 = -\frac{df}{f}.$$

Hence $\tilde{\theta} = (1-t)\theta_0 = (t-1)\frac{df}{f}$ and the curvature form $\tilde{\kappa}$ is given by

$$\tilde{\kappa} = d\tilde{\theta} + \tilde{\theta} \wedge \tilde{\theta} = dt \wedge \frac{df}{f}.$$

Thus

$$c^1(\nabla_0, \nabla_1) = \rho'_* c^1(\tilde{\nabla}) = \frac{\sqrt{-1}}{2\pi} \rho'_* \left(dt \wedge \frac{df}{f} \right) = -\frac{1}{2\pi\sqrt{-1}} \frac{df}{f}$$

and

$$\mathrm{Res}_{c^1}(s, E; 0) = \frac{1}{2\pi\sqrt{-1}} \int_{\mathbb{S}^1} \frac{df}{f}. \tag{10.13}$$

If we denote by $p : E \to \mathbb{C}$ the projection in the fiber direction and by ζ a coordinate on $\mathbb{C}$, we may write $\frac{df}{f} = f^* \frac{d\zeta}{\zeta} (= s^* p^* \frac{d\zeta}{\zeta})$. Thus $\operatorname{Res}_{c^1}(s, E, 0)$ is equal to $\deg f$ (cf. Proposition 10.13 below).

We naturally identify $\mathbb{R}^2$ with $\mathbb{C}$ and suppose s, thus f, is holomorphic on U_0. If f extends to a holomorphic function on U with a zero at 0, then the residue (10.13) is equal to the order of the zero (cf. Section 1.2). It may also be expressed as either $\dim \mathscr{O}_1/(f)$ or $\operatorname{Res}_0 \begin{bmatrix} df \\ f \end{bmatrix}$ (cf. Example 12.2 below). If f has a pole at 0, it is the order of the pole.

Hyperplane bundle

We also take up the following example with non-isolated zero set and prove a statement corresponding to Proposition 5.6. As is seen in the proof below, the essential ingredient is the Cauchy integral formula.

Let $\mathbb{P}^n$ be the n-dimensional complex projective space with homogeneous coordinates $[\zeta_0, \dots, \zeta_n]$. We denote by U_α the open set in $\mathbb{P}^n$ defined by $\zeta_\alpha \neq 0$, $\alpha = 0, \dots, n$. Let H denote the hyperplane defined by $\zeta_0 = 0$ and L_H the line bundle associated with H. Recall that L_H is defined by the system of transition functions $\{\zeta_\beta/\zeta_\alpha\}$ (cf. Example 3.6). It admits a section s represented by the collection (s_α), where s_α is a holomorphic function on U_α given by $s_\alpha = \zeta_0/\zeta_\alpha$. Since the zero set of s is H, we have the localization $c^1(L_H, s)$ in $H^2(\mathbb{P}^n, \mathbb{P}^n \smallsetminus H; \mathbb{C})$ of $c^1(L_H)$ and the associated residue $\operatorname{Res}_{c^1}(s, L_H; H)$, which is the image of $c^1(L_H, s)$ by the Alexander isomorphism

$$H^2(\mathbb{P}^n, \mathbb{P}^n \smallsetminus H; \mathbb{C}) \xrightarrow{\sim} H_{2n-2}(H; \mathbb{C}).$$

Proposition 10.12. *We have:*

$$\operatorname{Res}_{c^1}(s, L_H; H) = [H] \qquad \textit{in} \;\; H_{2n-2}(H; \mathbb{C}).$$

Proof. By Theorem 10.3, it suffices to prove that the transverse residue at a point in H is equal to 1. Thus let p be a point in H and assume that it is in U_α, $\alpha \neq 0$. We fix a trivialization $L_H|_{U_\alpha} \simeq \mathbb{C} \times U_\alpha$ so that the section s corresponds to $[\zeta] \mapsto (\zeta_0/\zeta_\alpha, [\zeta])$ on U_α. The function ζ_0/ζ_α is a part of a coordinate system on U_α, which will be denoted by z. As a slice $\mathbb{D}$ of H at p, we may take a complex one-dimensional disk in U_α with coordinate z. If we let $\mathbb{B}$ be a closed 2-ball with center p in $\mathbb{D}$, by Example 10.3, the transverse residue is given by

$$\operatorname{Res}_{c^1}(s|_{\mathbb{D}}, L_H|_{\mathbb{D}}; p) = \frac{1}{2\pi\sqrt{-1}} \int_{\partial \mathbb{B}} \frac{dz}{z} = 1. \qquad \square$$

As noted in Remark 5.18. 2, the above is a special case of the situation we discuss in the next section.

Whitney sum formula

Let M be a C^∞ manifold of dimension m and

$$0 \longrightarrow E' \longrightarrow E \longrightarrow E'' \longrightarrow 0 \tag{10.14}$$

an exact sequence of C^∞ complex vector bundles on M. Let S be a closed set in M and s a non-vanishing section of E' on $M \smallsetminus S$. Then we have the localizations $c^{l'}(E', s)$ in $H^{2l'}(M, M \smallsetminus S; \mathbb{C})$ and $c^l(E, s)$ in $H^{2l}(M, M \smallsetminus S; \mathbb{C})$.

Theorem 10.7. *In the above situation,*

$$c^l(E, s) = c^{l'}(E', s) \smile c^{l''}(E'') \qquad \textit{in} \ \ H^{2l}(M, M \smallsetminus S).$$

Proof. Take a covering $\mathcal{U} = \{U_0, U_1\}$ with $U_0 = M \smallsetminus S$ and U_1 a neighborhood of S. Let ∇'_0 be an s-trivial connection for E' on U_0 and ∇'_1 an arbitrary connection for E' on U_1. Also let ∇'' be an arbitrary connection for E''. Take connections ∇_0 and ∇_1 for E on U_0 and U_1, respectively, so that $(\nabla'_0, \nabla_0, \nabla'')$ and $(\nabla'_1, \nabla_1, \nabla'')$ are compatible with (10.14). Then ∇_0 is s-trivial. The class $c^l(E, s)$ is represented by the Čech-de Rham cocycle $(0, c^l(\nabla_1), c^l(\nabla_0, \nabla_1))$ and the class $c^{l'}(E', s)$ by $(0, c^{l'}(\nabla_1), c^{l'}(\nabla'_0, \nabla'_1))$. Then we see (cf. the proof of Proposition 8.6) that the curvature matrices with respect to suitable frames have the relation

$$\kappa_i = \begin{pmatrix} \kappa'_i & * \\ 0 & \kappa'' \end{pmatrix}, \qquad i = 0, 1, \tag{10.15}$$

which shows $c^l(\nabla_1) = c^{l'}(\nabla'_1) \cdot c^{l''}(\nabla'')$.

To find difference forms, set $\tilde{\nabla} = (1-t)\nabla_0 + t\nabla_1$ and $\tilde{\nabla}' = (1-t)\nabla'_0 + t\nabla'_1$. Then, from (10.15) we have

$$\tilde{\kappa} = \begin{pmatrix} \tilde{\kappa}' & * \\ 0 & \kappa'' \end{pmatrix},$$

which shows $c^l(\nabla_0, \nabla_1) = c^{l'}(\nabla'_0, \nabla'_1) \cdot c^{l''}(\nabla'')$. □

Suppose that S is a subcomplex of M for some triangulation and that M is oriented. Then we have the residues $\mathrm{Res}_{c^{l'}}(s, E'; S)$ in $\breve{H}_{m-2l'}(S; \mathbb{C})$ and $\mathrm{Res}_{c^l}(s, E; S)$ in $\breve{H}_{m-2l}(S; \mathbb{C})$ as the images of $c^{l'}(E', s)$ and $c^l(E, s)$, respectively, by the Alexander isomorphism.

Letting $p = 2l'$, $r = 2l$ and $\gamma = c^{l''}(E'')$ in Proposition 4.3, we have:

Corollary 10.2. *If S is a subcomplex of M for some triangulation of M,*

$$\operatorname{Res}_{c^l}(s, E; S) = \operatorname{Res}_{c^{l'}}(s, E'; S) \frown i^* c^{l''}(E'') \qquad \text{in } \check{H}_{m-2l}(S; \mathbb{C}),$$

where $i : S \hookrightarrow M$ denotes the inclusion.

Remark 10.8. The above is a differential geometric counterpart of Proposition 5.10. Note that here we only need a non-vanishing section away from S. Also, compare Theorem 10.7 with Corollary 10.5.1 below.

10.4 Thom class of a complex vector bundle

We first try to generalize Example 10.3 to the higher-dimensional case, which will be of fundamental importance in the subsequent discussions.

Let U be a neighborhood of 0 in $\mathbb{R}^{2n}$ and $E = \mathbb{C}^n \times U$ the product bundle on U. Suppose we have a non-vanishing C^∞ section s of E on $U_0 = U \smallsetminus \{0\}$. Set $U_1 = U$ and consider the covering $\mathcal{U} = \{U_0, U_1\}$ of U. We have the localization $c^n(E, s)$ in $H^{2n}(\mathcal{U}, U_0) \simeq H^{2n}(U, U \smallsetminus \{0\}; \mathbb{C})$ and the associated residue $\operatorname{Res}_{c^n}(s, E, 0)$ in $H_0(\{0\}; \mathbb{C}) = \mathbb{C}$. If we denote by $e^{(n)} = (e_1, \dots, e_n)$ the frame of E given by $e_i(z) = ({}^t(0, \dots, \overset{i}{1}, \dots, 0), z)$, we may write $s = \sum_{i=1}^n f_i e_i$ with f_i C^∞ functions on U_0 such that, at every point of U_0, at least one of them is non-zero. Let $\mathbb{B}^{2n}$ be a closed $2n$-ball around 0 in U. In the expression (10.11) of the residue, we take as ∇_1 the $e^{(n)}$-trivial connection on U so that $c^n(\nabla_1) = 0$ and

$$\operatorname{Res}_{c^n}(s, E, 0) = -\int_{\partial \mathbb{B}^{2n}} c^n(\nabla_0, \nabla_1)$$

with ∇_0 an s-trivial connection on U_0. Let β_n denote the Bochner-Martinelli form on $\mathbb{C}^n$ (cf. Section 7.10).

Theorem 10.8 (Fundamental theorem of residues). *In the above situation, we have $c^n(\nabla_0, \nabla_1) = -f^* \beta_n$ for a suitable choice of an s-trivial connection ∇_0. Thus*

$$\operatorname{Res}_{c^n}(s, E, 0) = \int_{\partial \mathbb{B}^{2n}} f^* \beta_n.$$

Proof. For each $i = 1, \dots, n$, we set $V_i = \{\, z \in U_0 \mid f_i(z) \neq 0 \,\}$. Then $\mathcal{V} = \{V_i\}_{i=1}^n$ is an open covering of U_0. On V_i, $e_i^{(n)} = (e_1, \dots, \overset{i}{s}, \dots, e_n)$ is a frame of E. Let D_i be the connection for E on V_i trivial with respect to $e_i^{(n)}$.

Set $\rho_i = |f_i|^2/\|f\|^2$, $\|f\| = \sqrt{|f_1|^2 + \cdots + |f_n|^2}$, which is a C^∞ function on U_0. On V_i, $D_i\, e_k = 0$ for $k \neq i$ and $D_i\, e_i = -1/f_i \cdot \sum_{j=1}^n df_j \otimes e_j$. Thus we may define an operator $\rho_i D_i : A^0(U_0, E) \to A^1(U_0, E)$ by setting

$$(\rho_i D_i)e_k = \begin{cases} -\frac{\bar{f}_i}{\|f\|^2} \sum_{j=1}^n df_j \otimes e_j & \text{for} \quad k = i, \\ 0 & \text{for} \quad k \neq i \end{cases}$$

and $(\rho_i D_i)(\sum_{k=1}^n g_k e_k) = \rho_i \sum_{k=1}^n dg_k \otimes e_k + \sum_{k=1}^n g_k(\rho_i D_i)e_k$ for general sections. Then, from $\sum_{i=1}^n \rho_i = 1$, we see that $\nabla_0 = \sum_{i=1}^n \rho_i D_i$ is a connection for E on U_0. It is s-trivial, since each D_i is.

Now we compute $c^n(\nabla_0, \nabla_1)$ with ∇_0 as above. Let P_i be the matrix of frame change from $e^{(n)}$ to $e_i^{(n)}$, i.e., $e_i^{(n)} = e^{(n)} P_i$. Then P_i is obtained from the identity matrix by replacing the i-th column with ${}^t(f_1, \ldots, f_n)$. Since $e^{(n)} = e_i^{(n)} P_i^{-1}$ and D_i is trivial with respect to $e_i^{(n)}$, the connection matrix θ_0 of ∇_0 with respect to $e^{(n)}$ is given by (cf. (8.4))

$$\theta_0 = \sum_{i=1}^n \rho_i\, P_i \cdot dP_i^{-1} = -\sum_{i=1}^n \rho_i\, dP_i \cdot P_i^{-1}, \tag{10.16}$$

while the connection matrix θ_1 of ∇_1 with respect to $e^{(n)}$ is zero. In the subsequent computations we use the fact that, if the entries of a matrix are forms of even degrees possibly except for the ones in a single column, we may treat its determinant the same way as usual matrices of numbers. Consider the connection $\tilde{\nabla} = (1-t)\nabla_0 + t\nabla_1$ as in Proposition 8.3. Then by definition

$$c^n(\nabla_0, \nabla_1) = \Big(\frac{\sqrt{-1}}{2\pi}\Big)^n \rho'_* \det \tilde{\kappa},$$

where $\tilde{\kappa} = -dt \wedge \theta_0 + \kappa_t$, $\kappa_t = (1-t)d\theta_0 + (1-t)^2 \theta_0 \wedge \theta_0$. From definition of the integration along fibers, we have

$$c^n(\nabla_0, \nabla_1) = -\Big(\frac{\sqrt{-1}}{2\pi}\Big)^n \sum_{j=1}^n \int_0^1 \det Q_j\, dt, \tag{10.17}$$

where Q_j denotes the matrix obtained from κ_t by replacing the j-th column by that of θ_0. Note that this can be also obtained using Proposition 8.4 and (8.7).

From (10.16), we see that the (i,j)-entry of θ_0 is $-\|f\|^{-2} \bar{f}_j df_i$. Thus the (i,j)-entry of $d\theta_0$ is

$$-(\|f\|^{-2} d\bar{f}_j + \bar{f}_j d(\|f\|^{-2})) \wedge df_i, \tag{10.18}$$

while the (i,j)-entry of $\theta_0 \wedge \theta_0$ is $\|f\|^{-4} \sum_{k=1}^n \bar{f}_k \bar{f}_j df_i \wedge df_k$. Thus in $\det Q_j$, each term containing an entry of $\theta_0 \wedge \theta_0$ has more than n of the df_i's and is equal to 0. Hence $\det Q_j = \det Q_j'$, where Q_j' is the matrix obtained from $(1-t)d\theta_0$ by replacing the j-th column by that of θ_0. We claim that, with the notation in Section 7.10,

$$\det Q_j' = (-1)^{\frac{n(n-1)}{2}+n} n! \frac{\overline{\Phi_j(f)} \wedge \Phi(f)}{\|f\|^{2n}} (1-t)^{n-1} + \sum_{k\neq j} r_{jk} \tag{10.19}$$

for some forms r_{jk} with $r_{kj} = -r_{jk}$. The identity (10.19) together with (10.17) will prove the theorem.

Let R denote the matrix whose (i,j)-entry is $-d\bar{f}_j \wedge df_i$. We also denote by R_j the matrix obtained from R by replacing the j-th column by $\bar{f}_j\,{}^t(df_1, \dots, df_n)$ and by R_{jk}, $k \neq j$, the matrix obtained from R_j by replacing the k-th column by $-\bar{f}_k d(\|f\|^{-2}) \wedge {}^t(df_1, \dots, df_n)$. From (10.18) we have

$$\det Q_j' = \Big(\frac{1}{\|f\|^{2n}} \det R_j + \frac{1}{\|f\|^{2(n-1)}} \sum_{k \neq j} \det R_{jk} \Big) (1-t)^{n-1}.$$

We compute

$$\begin{aligned}\det R_j &= (-1)^n \sum_{\sigma \in \mathfrak{S}_n} \operatorname{sgn}\sigma \cdot d\bar{f}_1 \wedge df_{\sigma(1)} \wedge \cdots \wedge \bar{f}_j df_{\sigma(j)} \wedge \cdots \wedge d\bar{f}_n \wedge df_{\sigma(n)} \\ &= (-1)^{\frac{n(n-1)}{2}+n} n! \overline{\Phi_j(f)} \wedge \Phi(f).\end{aligned}$$

We also see that, from the expression of the determinant as above, $\det R_{kj} = -\det R_{jk}$. Thus setting $r_{jk} = \|f\|^{-2(n-1)} \det R_{jk}\,(1-t)^{n-1}$, we have (10.19). □

From the above we see that $\operatorname{Res}_{c^n}(s, E, 0)$ is equal to $\deg f$ (cf. Proposition 10.13 below). Moreover, if we identify $\mathbb{R}^{2n}$ with $\mathbb{C}^n$ and if s, thus f is defined and holomorphic on U, this may be expressed as either $\dim \mathscr{O}_n/(f_1, \dots, f_n)$ or $\operatorname{Res}_0 \begin{bmatrix} df_1 \wedge \cdots \wedge df_n \\ f_1, \dots, f_n \end{bmatrix}$.

Thom class as a localized top Chern class

Let M a C^∞ manifold of dimension m and $\pi : E \to M$ a complex C^∞ vector bundle of rank l. We apply the above consideration to the diagonal section s_Δ of the pull-back π^*E (cf. Section 5.6). The zero set of s_Δ is the image Σ of the zero section of $E \to M$, which is naturally diffeomorphic with M. In this situation, on the one hand there is the differential geometric localization $c^l_{\text{diff}}(\pi^*E, s_\Delta)$ in $H^{2l}(E, E \smallsetminus \Sigma; \mathbb{C})$ of $c^l(\pi^*E)$ by s_Δ. On the other hand, as an oriented real vector bundle of rank $2l$, there is the Thom class Ψ_E of E

in $H^{2l}(E, E \smallsetminus \Sigma; \mathbb{Z})$ (cf. Section 4.3). Its image by the canonical morphism $H^{2l}(E, E \smallsetminus \Sigma; \mathbb{Z}) \to H^{2l}(E, E \smallsetminus \Sigma; \mathbb{C})$ is the class $\Psi_E^{\mathbb{C}}$ (cf. Section 7.9).

If M is oriented, we have the residue $\mathrm{Res}_{c^l}(s_\Delta, \pi^*E; \Sigma)$ as the image of $c^l_{\mathrm{diff}}(\pi^*E, s_\Delta)$ by the Alexander isomorphism

$$A : H^{2l}(E, E \smallsetminus \Sigma; \mathbb{C}) \xrightarrow{\sim} \check{H}_m(\Sigma; \mathbb{C}).$$

In this case, $\Psi_E^{\mathbb{C}}$ corresponds to Σ by A and may be identified with Ψ_E (cf. Remark 7.20).

Theorem 10.9. *In the above situation,*

1. $c^l_{\mathrm{diff}}(\pi^*E, s_\Delta) = \Psi_E^{\mathbb{C}}$ *in* $H^{2l}(E, E \smallsetminus \Sigma; \mathbb{C})$.

2. *If M is oriented,*

$$\mathrm{Res}_{c^l}(s_\Delta, \pi^*E; \Sigma) = \Sigma \qquad \text{in } \check{H}_m(\Sigma; \mathbb{C}).$$

Proof. Since the statement 2 directly follows from 1, we prove 1. Let $W_0 = E \smallsetminus \Sigma$ and W_1 a neighborhood of Σ in E and consider the covering $\mathcal{W} = \{W_0, W_1\}$ of E. Also let $\{T_0, T_1\}$ be a honeycomb system adapted to $\mathcal{W}$ as in Section 7.9. Let D_0 be an s_Δ-trivial connection for π^*E on W_0 and D_1 a connection for π^*E on W_1. The class $c^l(\pi^*E, s_\Delta)$ is represented by the cocycle $c^l(D_*)$ in $A^{2l}(\mathcal{W}, W_0)$ given by

$$c^l(D_*) = (0, c^l(D_1), c^l(D_0, D_1)).$$

By Corollary 7.11, it suffices to prove that

$$(\pi_1)_* c^l(D_1) + (\pi_{01})_* c^l(D_0, D_1) = 1. \tag{10.20}$$

Note that the left-hand side of the above does not depend on the choices of D_0 and D_1 as above and that it suffices to integrate over a single fiber. Thus let x_0 be a point in M and U a neighborhood of x_0 in M such that $\pi^{-1}(U) \simeq \mathbb{C}^l \times U$. Let $e^{(l)} = (e_1, \dots, e_l)$ be a frame of E on U which gives this trivialization and $\zeta = (\zeta_1, \dots, \zeta_l)$ the fiber coordinates with respect to this frame. Hereafter, we identify $\pi^{-1}(U)$ with $\mathbb{C}^l \times U = \{(\zeta, x)\}$. Let $p : \pi^{-1}(U) = \mathbb{C}^l \times U \to \mathbb{C}^l$ be the projection. We also identify the last $\mathbb{C}^l$ with $\pi^{-1}(x_0)$. We may assume that $T_{01} \cap \pi^{-1}(U) = -(\mathbb{S}^{2l-1} \times U)$, where $\mathbb{S}^{2l-1}$ is the unit $(2l-1)$-sphere in $\mathbb{C}^l$, oriented as the boundary of the unit ball. We claim that by a suitable choice of D_0 and D_1, we have $c^l(D_1)|_{\pi^{-1}(U)} = 0$ and $c^l(D_0, D_1)|_{\pi^{-1}(U)} = -p^*\beta_l$ so that

$$c^l(D_*)|_{\pi^{-1}(U)} = (0, 0, -p^*\beta_l),$$

which will imply (10.20).

To define D_0 and D_1, it suffices to define these on $\pi^{-1}(x_0)(=\mathbb{C}^l)$, as they naturally define connections on $\pi^{-1}(U)$ by the product structure and then on the whole E by Lemma 8.3. Now $\pi^*e^{(l)} = (\pi^*e_1, \dots, \pi^*e_l)$ is a frame of π^*E on $\pi^{-1}(U)$, which gives a trivialization $\varpi^{-1}\pi^{-1}(U) = \mathbb{C}^l \times \pi^{-1}(U)$. Let ∇_1 be a connection for $\pi^*E|_{\pi^{-1}(x_0)}$ trivial with respect to $\pi^*e^{(l)}|_{\pi^{-1}(x_0)}$ and set $D_1 = p^*(\nabla_1)$. Thus D_1 is trivial with respect to $\pi^*e^{(l)}$ and we have $c^l(D_1) = p^*c^l(\nabla_1) = 0$. The restriction of the diagonal section

$$s_\Delta|_{\pi^{-1}(x_0)} : \pi^{-1}(x_0) = \mathbb{C}^l \longrightarrow \pi^*E|_{\pi^{-1}(x_0)} = \mathbb{C}^l \times \mathbb{C}^l$$

is given by $(\zeta, x_0) \mapsto (\zeta, \zeta, x_0)$. Let ∇_0 be the connection for $\pi^*E|_{\pi^{-1}(x_0)}$ as constructed in the proof of Theorem 10.8 and set $D_0 = p^*(\nabla_0)$. Then $c^l(D_0, D_1) = p^*c^l(\nabla_0, \nabla_1) = -p^*\beta_l$. □

Remark 10.9. 1. The above choice of D_0 and D_1 gives explicit local expressions of the forms in Proposition 7.22 for a complex vector bundle. In particular, for the product bundle $E = \mathbb{C}^l \times M$ with projection $p : E \to \mathbb{C}^l$ in the fiber direction, the Thom class Ψ_E is represented by the cocycle $(0, 0, -p^*\beta_l)$ in $A^{2l}(\mathcal{W}, W_0)$ for a covering $\mathcal{W}$ of E as above.

2. As in the case of topological counterparts (cf. Remark 5.18. 2), we may think of Proposition 10.12 as a special case of Theorem 10.9.

From Corollary 5.5 and Theorem 10.9, we have:

Corollary 10.3. *Let E be a complex vector bundle of rank l on a C^∞ manifold M.*

1. *$c^l_{\text{diff}}(\pi^*E, s_\Delta)$ is the image of $c^l_{\text{top}}(\pi^*E, s_\Delta)$ by the canonical morphism $H^{2l}(E, E \smallsetminus \Sigma; \mathbb{Z}) \to H^{2l}(E, E \smallsetminus \Sigma; \mathbb{C})$.*

2. *If M is oriented, $\check{H}_m(\Sigma; \mathbb{Z}) \to \check{H}_m(\Sigma; \mathbb{C})$ is injective so that we may identify $c^l_{\text{diff}}(\pi^*E, s_\Delta)$ with $c^l_{\text{top}}(\pi^*E, s_\Delta)$ and $\operatorname{Res}_{c^l}(s_\Delta, \pi^*E; \Sigma)$ with $\operatorname{TRes}_{c^l}(s_\Delta, \pi^*E; \Sigma)$.*

The above is a special case of Theorem 10.13 below.

Universality of the Thom class: Let $\pi : E \to M$ be as above. Suppose we have a section s of E on M with the zero set S. Then it induces a morphism

$$s^* : H^{2l}(E, E \smallsetminus \Sigma; \mathbb{C}) \longrightarrow H^{2l}(M, M \smallsetminus S; \mathbb{C}).$$

Theorem 10.10. *In the above situation,*

$$c^l_{\text{diff}}(E, s) = s^*\Psi^{\mathbb{C}}_E.$$

Proof. Let $\mathcal{W} = \{W_0, W_1\}$ be a covering of π^*E and let D_0 and D_1 be an s_Δ-trivial connection for π^*E on W_0 and a connection for π^*E on W_1, respectively, as above. We take a covering $\mathcal{U} = \{U_0, U_1\}$ of M with $U_0 = M \smallsetminus S$ and U_1 a neighborhood of S such that $s(U_1) \subset W_1$. Recall that there are natural isomorphisms:

$$s^*\pi^*E \simeq (\pi \circ s)^*E = (1_M)^*E \simeq E$$

and that the correspondence is given by $(\xi_1, \xi_2, x) \leftrightarrow (\xi_1, x) \leftrightarrow \xi_1$ with $x = \pi(\xi_1)$ and $\xi_2 = s(x)$ (cf. Proposition 3.2). For $i = 0, 1$, we denote by ∇_i the connection for E on U_i that corresponds to s^*D_i in the above isomorphism. For the pull-back s^*s_Δ of s_Δ by s, we have $s^*s_\Delta(x) = (s(x), s(x), x)$ (cf. (3.6)). Thus it corresponds to s in the above isomorphism. From $(s^*D_0)(s^*s_\Delta) = s^*(D_0(s_\Delta)) = 0$ (cf. (8.14)), we see that ∇_0 is s-trivial. Hence the class $c^l_S(E, s)$ is represented by the cocycle $(0, c^l(\nabla_0), c^l(\nabla_0, \nabla_1))$ and we have the theorem by Proposition 7.8. □

Remark 10.10. The above gives a differential geometric proof of the fact that the Thom class is a universal localization of the top Chern class (cf. Remarks 5.17.2 and 5.18.1). If M is oriented, we have a commutative diagram as (5.22) with $\mathbb{C}$-coefficients and l' replaced by $2l$. We may write

$$\mathrm{Res}_{c^l}(s, E; S) = (M \bullet_s \Sigma)_S.$$

Comparing with (5.23), we have:

Corollary 10.4. *Let E be a complex vector bundle of rank l on a C^∞ manifold M and s a section of E on M with the zero set S a subcomplex of M.*

1. *$c^l_{\mathrm{diff}}(E, s)$ is the image of $c^l_{\mathrm{top}}(E, s)$ by the canonical morphism $H^{2l}(M, M \smallsetminus S; \mathbb{Z}) \to H^{2l}(M, M \smallsetminus S; \mathbb{C})$.*

2. *If M is oriented, $\mathrm{Res}_{c^l}(s, E; S)$ is the image of $\mathrm{TRes}_{c^l}(s, E; S)$ by the canonical morphism $\check{H}_{m-2l}(S, \mathbb{Z}) \to \check{H}_{m-2l}(S; \mathbb{C})$.*

Compare the above with Theorem 10.13 and Corollary 10.6 below.

For $i = 1, 2$, let E_i be a complex vector bundle of rank l_i on M. Let s_i be a section of E_i on M with the zero set S_i. We set $E = E_1 \oplus E_2$, $s = s_1 \oplus s_2$, $l = l_1 + l_2$ and $S = S_1 \cap S_2$, which is the zero set of s. Then, from the identity $\Psi^c_E = (\tilde{d})^*(\Psi^c_{E_1} \times \Psi^c_{E_2})$ (cf. Corollary 4.5), we have:

Corollary 10.5 (Whitney sum formula). *In the above situation,*

1. $c^l_{\mathrm{diff}}(E, s) = c^{l_1}_{\mathrm{diff}}(E_1, s_1) \smile c^{l_2}_{\mathrm{diff}}(E_2, s_2)$.

2. *If M is oriented and if each S_i is a subcomplex of M,*

$$\operatorname{Res}_{c^l}(s,E;S) = (\operatorname{Res}_{c^{l_1}}(s_1,E_1;S_1)\cdot \operatorname{Res}_{c^{l_2}}(s_2,E_2;S_2))_S \quad \text{in } \breve{H}_{m-2l}(S;\mathbb{C}).$$

The above is a differential geometric counterpart of Corollary 5.6. Note that, in the item 1, it is not necessary to assume that S_i is a subcomplex of M. The Whitney sum formula is further refined in Section 14.1 below.

Intermediate Thom classes

We may also define the differential geometric intermediate Thom classes (cf. Section 5.6) in the relative Čech-de Rham cohomology.

Let M be a C^∞ manifold of dimension m and $\pi : E \to M$ a complex vector bundle of rank l. Let $E^r = E \oplus \cdots \oplus E$ be the direct sum of r copies of E with the projection $\rho : E^r \to M$. Consider the vector bundle ρ^*E of rank l on E^r (see (5.24)). It has the diagonal r-section $s_\Delta^{(r)}$ with singular set Σ given by (5.26). For $q = l - r + 1$, the *differential geometric q-th Thom class* Ψ_E^q of E is defined by

$$\Psi_E^q = c^q_{\text{diff}}(\rho^*E, s_\Delta^{(r)}),$$

which is in $H^{2q}(E^r, E^r \smallsetminus \Sigma; \mathbb{C})$. By Definition 5.16 and Theorem 10.13 below, the above class is the image of the topological q-th Thom class by the canonical morphism $H^{2q}(E^r, E^r \smallsetminus \Sigma; \mathbb{Z}) \to H^{2q}(E^r, E^r \smallsetminus \Sigma; \mathbb{C})$.

Let $\mathcal{W} = \{W_0, W_1\}$ be a covering of E^r with $W_0 = E^r \smallsetminus \Sigma = W(E,r)$ and W_1 a neighborhood of Σ in E^r. In $H^{2q}(E^r, E^r \smallsetminus \Sigma; \mathbb{C}) = H^{2q}(\mathcal{W}, W_0)$ the class Ψ_E^q is represented by a ČdR cocycle

$$c^q(D_*) = (c^q(D_1), c^q(D_0, D_1)),$$

where D_0 is an $s_\Delta^{(r)}$-trivial connection for ρ^*E on W_0 and D_1 a connection for ρ^*E on W_1. We may choose D_1 so that $D_1 = \rho^*\nabla$ with ∇ a connection for E on M.

In particular, if M is a point, then $E = \mathbb{C}^l$, $E^r = \mathbb{C}^{lr}$, $\Sigma = \Sigma_0$ (cf. (5.25)) and $\mathbb{C}^{lr} \smallsetminus \Sigma_0 = W(l,r)$. There exist isomorphisms

$$H^{2q}(\mathbb{C}^{lr}, \mathbb{C}^{lr} \smallsetminus \Sigma_0; \mathbb{C}) \simeq H^{2q-1}(W(l,r); \mathbb{C}) \simeq H_d^{2q-1}(W(l,r))$$

By a suitable choice of connections D_0 and D_1, we may make $c^q(D_1) = 0$ and $c^q(D_0, D_1)$ the negative of the explicit generator of $H^{2q-1}(W(l,r))$ given in Proposition 7.28. Also by the Alexander isomorphism

$$A : H^{2q}(\mathbb{C}^{lr}, \mathbb{C}^{lr} \smallsetminus \Sigma_0; \mathbb{C}) \xrightarrow{\sim} \breve{H}_{2(rl-q)}(\Sigma_0; \mathbb{C}) \simeq \mathbb{C},$$

the class $\Psi_{\mathbb{C}^l}^q$ corresponds to the generator $[\Sigma_0]$.

10.5 Coincidence of topological and differential geometric localizations

First we give an expression of the mapping degree in terms of differential forms.

Let $W(l, r)$ be the Stiefel manifold of r-frames in $\mathbb{C}^l$ and set $q = l - r + 1$ as before. We have $H^{2q-1}((W(l, r); \mathbb{C}) \simeq \mathbb{C}$ with a canonical generator. By the de Rham theorem (Theorem 7.5), the generator is represented by a closed $(2q - 1)$-form ω_{2q-1} on $W(l, r)$. By definition of the mapping degree and Proposition 7.5, we have:

Proposition 10.13. *For a C^∞ map $\varphi : \mathbb{S}^{2q-1} \to W(l, r)$,*

$$\deg \varphi = \int_{\mathbb{S}^{2q-1}} \varphi^* \omega_{2q-1}.$$

An explicit expression of ω_{2q-1} is given in Proposition 7.28. In particular, if $r = 1$, then $q = l$ and $W(l, 1) = \mathbb{C}^l \smallsetminus \{0\}$. In this case we may take as ω_{2q-1} the Bochner-Martinelli form β_l (cf. Section 7.10).

Now we give a topological interpretation of the differential geometric residue in a special but fundamental case.

Let $\mathbb{B}^{2l}$ be a closed ball of dimension $2l$ in $\mathbb{R}^{2l}$ and E a C^∞ complex vector bundle of rank l on a small neighborhood U of $\mathbb{B}^{2l}$. Suppose we have a non-vanishing C^∞ section s on a neighborhood of the boundary $\partial \mathbb{B}^{2l} = \mathbb{S}^{2l-1}$ of $\mathbb{B}^{2l}$. We may extend s to a non-vanishing section, denoted also by s, on $U \smallsetminus \{x\}$, where x is a point in the interior of $\mathbb{B}^{2l}$. Thus on the one hand we have the index $I(s, x)$ (cf. Definition 5.3), which is an integer, and on the other hand we have the residue $\mathrm{Res}_{c^l}(s, E; x)$, which is in $H_0(\{x\}; \mathbb{C}) = \mathbb{C}$ and is a complex number. From Theorem 10.8 and Proposition 10.13 we see that they are in fact the same:

Proposition 10.14. *In the above situation,*

$$\mathrm{Res}_{c^l}(s, E; x) = I(s, x).$$

Thus $\mathrm{Res}_{c^l}(s, E; x)$ *is an integer.*

As we will see, many significant results follow from this fact

Now we go on to a more general case. Let $\mathbb{B}^{2q}$ be a closed ball of dimension $2q$ in $\mathbb{R}^{2q}$ and E a complex vector bundle of rank l on a small neighborhood U of $\mathbb{B}^{2q}$. Suppose we have an r-frame $s^{(r)} = (s_1, \dots, s_r)$, $r = l - q + 1$, on a neighborhood of $\partial \mathbb{B}^{2q}$. We may extend $s^{(r)}$ to an r-frame on $U \smallsetminus \{x\}$, where x is a point in the interior of $\mathbb{B}^{2q}$. We set

$s^{(r-1)} = (s_1, \dots, s_{r-1})$. By dimension reason, we may assume that $s^{(r-1)}$ is extended to an $(r-1)$-frame on U. Then it generates a trivial subbundle $\mathbb{I}$ of E of rank $r-1$ so that we have an exact sequence

$$0 \longrightarrow \mathbb{I} \longrightarrow E \longrightarrow E' \longrightarrow 0, \tag{10.21}$$

where E' is a vector bundle of rank q on U. If we denote by s'_r the section of E' determined by s_r, it is non-vanishing on $U \setminus \{x\}$.

Lemma 10.4. *In the above situation, we have*

$$\mathrm{Res}_{c^q}(s^{(r)}, E; x) = \mathrm{Res}_{c^q}(s'_r, E'; x).$$

Proof. Letting $U_0 = U \setminus \{0\}$ and U_1 a neighborhood of x in U, we consider the covering $\mathcal{U} = \{U_0, U_1\}$ of U. Let ∇ be the $s^{(r-1)}$-trivial connection for $\mathbb{I}$. Also, let ∇'_0 be an s'_r-trivial connection for E' on U_0 and take an $s^{(r)}$-trivial connection ∇_0 for E on U_0 so that $(\nabla, \nabla_0, \nabla'_0)$ is compatible with (10.21). Also, let ∇'_1 be a connection for E' on U and take a connection ∇_1 for E so that $(\nabla, \nabla_1, \nabla'_1)$ is compatible with (10.21). Then we have

$$c^q(\nabla_1) = c^q(\nabla'_1) \quad \text{and} \quad c^q(\nabla_0, \nabla_1) = c^q(\nabla'_0, \nabla'_1).$$

The identity follows from (10.11). □

By Lemma 10.4, Propositions 5.1 and 10.14, we have:

Theorem 10.11. *In the above situation, we have*

$$\mathrm{Res}_{c^q}(s^{(r)}, E; x) = I(s^{(r)}, x).$$

Thus $\mathrm{Res}_{c^q}(s^{(r)}, E; x)$ *is an integer.*

Now we prove that the Chern class defined by obstruction theory and the one defined by Chern-Weil theory are essentially the same.

Let M be a C^∞ manifold of dimension m. We take a C^∞ triangulation K_0 of M and let K and K^* be as before. Let W_0 be a neighborhood of the $(p-1)$-skeleton $(K^*)^{p-1}$ of K^* such that the barycenter $b_{\boldsymbol{s}}$ of each p-cell $\boldsymbol{s}^*$ is not in W_0, and let W_1 be an open set so that $\mathcal{W} = \{W_0, W_1\}$ is a covering of M, in fact we may take M itself as W_1. Let $\{R_0, R_1\}$ be a system of honeycomb cells adapted to $\mathcal{W}$ such that R_0 contains $(K^*)^{p-1}$ in its interior. Recall that the isomorphism $\psi_D : H^p_D(\mathcal{W}) \xrightarrow{\sim} H^p_{K^*}(M; \mathbb{C})$ is induced by the linear map

$$\psi_D : A^p(\mathcal{W}) \longrightarrow C^p_{K^*}(M; \mathbb{C})$$

given by

$$\langle \boldsymbol{s}^*, \psi_D(\xi)\rangle = \int_{\boldsymbol{s}^*\cap R_0} \xi_0 + \int_{\boldsymbol{s}^*\cap R_1} \xi_1 + \int_{\boldsymbol{s}^*\cap R_{01}} \xi_{01},$$

for $\xi = (\xi_0, \xi_1, \xi_{01})$ in $A^p(\mathcal{W})$ (cf. (7.40)).

Let E be a C^∞ complex vector bundle of rank l on M. We have the q-th Chern class $c^q_{\text{diff}}(E)$ in $H^{2q}(M;\mathbb{C})$ defined via Chern-Weil theory and the q-th Chern class $c^q_{\text{top}}(E)$ in $H^{2q}(M,\mathbb{Z})$ defined via obstruction theory.

Theorem 10.12. *In the above situation, $c^q_{\text{diff}}(E)$ is the image of $c^q_{\text{top}}(E)$ by the canonical morphism $H^{2q}(M;\mathbb{Z}) \to H^{2q}(M;\mathbb{C})$.*

Proof. Suppose we have an r-frame $s^{(r)}$ on the $(2q-1)$-skeleton $(K^*)^{2q-1}$ of K^*, $r = l - q + 1$. Let W_0 be a neighborhood of $(K^*)^{2q-1}$ in M, so small that $s^{(r)}$ can be extended to W_0 without singularities, and set $W_1 = M$. Consider the covering $\mathcal{W} = \{W_0, W_1\}$ of M and let $\{R_0, R_1\}$ be a honeycomb system adapted to $\mathcal{W}$ such that R_0 contains $(K^*)^{2q-1}$ in its interior.

Let ∇_0 be an $s^{(r)}$-trivial connection for E on W_0 and ∇_1 an arbitrary connection for E on W_1. Then the class $c^q_{\text{diff}}(E)$ is represented by the cocycle

$$c^q(\nabla_*) = (c^q(\nabla_0), c^q(\nabla_1), c^q(\nabla_0,\nabla_1)) \qquad \text{in } A^{2q}(\mathcal{W}).$$

Note that, since ∇_0 is $s^{(r)}$-trivial, we have $c^q(\nabla_0) = 0$ and $c^q(\nabla_*)$ is in fact in $A^{2q}(\mathcal{W}, W_0)$. In the isomorphism $\psi_D : H^{2q}_D(\mathcal{W}) \xrightarrow{\sim} H^{2q}(M;\mathbb{C})$, the cocycle $c^q(\nabla_*)$ assigns to each $2q$-cell $\boldsymbol{s}^*$ the value

$$\int_{\boldsymbol{s}^*\cap R_1} c^q(\nabla_1) + \int_{\boldsymbol{s}^*\cap R_{01}} c^q(\nabla_0,\nabla_1),$$

which is equal to $\mathrm{Res}_{c^q}(s^{(r)}|_{\boldsymbol{s}^*}, E|_{\boldsymbol{s}^*}; b_{\boldsymbol{s}})$. The theorem follows from Theorem 10.11. □

We have a similar theorem for localized classes. Thus let S be a K_0-subcomplex of M and $s^{(r)}$ an r-frame of E on $M \smallsetminus S$, $r = l - q + 1$. Then we have the localized classes $c^q_{\text{diff}}(E, s^{(r)})$ in $H^{2q}(M, M \smallsetminus S;\mathbb{C})$ and $c^q_{\text{top}}(E, s^{(r)})$ in $H^{2q}(M, M \smallsetminus S;\mathbb{Z})$.

Theorem 10.13. *In the above situation, if the (real) codimension of S is greater than or equal to two, $c^q_{\text{diff}}(E, s^{(r)})$ is the image of $c^q_{\text{top}}(E, s^{(r)})$ by the canonical morphism $H^{2q}(M, M \smallsetminus S;\mathbb{Z}) \to H^{2q}(M, M \smallsetminus S;\mathbb{C})$.*

Proof. Let $U_0 = M \setminus S$ and U_1 a neighborhood of S. Also let W_0 be a neighborhood of $(K^*)^{2q-1}$ as in Theorem 10.12 and set $U_0' = U_0 \cup W_0$. Consider the coverings $\mathcal{U} = \{U_0, U_1\}$ and $\mathcal{U}' = \{U_0', U_1\}$ of M. Then we have a canonical morphism

$$H_D^{2q}(\mathcal{U}', U_0') \longrightarrow H_D^{2q}(\mathcal{U}, U_0) \tag{10.22}$$

We may think of U_0' as being the union of U_0 and a neighborhood U of $S \cap (K^*)^{2q-1}$. We may suppose that U retracts to $S \cap (K^*)^{2q-1}$, which is of dimension at most $2q-3$. Hence we see that the canonical morphism $H^{2q}(M, U_0') \to H^{2q}(M, U_0)$ is an isomorphism. Thus (10.22) is also an isomorphism.

We take a honeycomb system $\{R_0', R_1\}$ adapted to $\mathcal{U}'$ so that for $\boldsymbol{s}^*$ with $\boldsymbol{s}^* \cap S = \emptyset$, it holds that $\boldsymbol{s}^* \cap R_1 = \emptyset$ (see the proof of Theorem 7.10). We may extend $s^{(r)}$ to U_0' and let ∇_0 be an $s^{(r)}$-trivial connection for E on U_0'. The class $c^q_{\mathrm{diff}}(E, s^{(r)})$ is represented by a cocycle in $A^{2q}(\mathcal{U}', U_0')$ assigning to each $2q$-cell $\boldsymbol{s}^*$ the value (cf. (7.46))

$$\int_{\boldsymbol{s}^* \cap R_1} c^q(\nabla_1) - \int_{\partial(\boldsymbol{s}^* \cap R_1)} c^q(\nabla_0, \nabla_1),$$

which is zero for $\boldsymbol{s}^*$ not intersecting with S. □

Corollary 10.6. *Suppose M is oriented. Then $\mathrm{Res}_{c^q}(s^{(r)}, E; S)$ is the image of $\mathrm{TRes}_{c^q}(s^{(r)}, E; S)$ by the canonical morphism*

$$\check{H}_{m-2q}(S; \mathbb{Z}) \longrightarrow \check{H}_{m-2q}(S; \mathbb{C}).$$

In particular, if $m = 2q$ and if S is compact and connected, the both are the identical integers.

Remark 10.11. 1. From the above arguments, we see that we may define the localizations and associated residues for a frame $s^{(r)}$ given only on $(K^*)^{2q} \cap (M \setminus S)$ using the Chern-Weil theory in the integral cohomology and homology.

2. In the case $r = 1$, the above may be proved using the fact that the Thom class is universal both in topological and differential geometric senses.

Notes

In each of the types of localization discussed in Section 10.1, if we use the differential geometric methods, we take some special connection ∇ on $M \setminus S$ first. As to the treatment after this, there are several ways:

(1) Extend ∇ to the whole M by modifying it in a neighborhood of S. Then we get a characteristic form with compact support on M, for example [Baum and Bott (1972)].
(2) Consider ∇ as a singular connection on M and extend the characteristic form as a current on M, for example [Harvey and Lawson (1993)].
(3) Let ∇ as it is and take another connection on a neighborhood of S. In this case the difference in the intersection becomes important. The localization is naturally represented by a Čech-de Rham cocycle and this is the way used in this text.

The proof of Theorem 10.5 as given here is due to [Abate, Bracci and Tovena (2008)] and [Abate, Bracci, Suwa and Tovena (2013)].

As is mentioned in Notes of Chapter 5, the residue defined in Section 10.3 (likewise the one in Section 5.6), in the case $r = 1$, corresponds to the "localized top Chern class" defined in [Fulton (1984)] in the algebraic category. In particular, see Example 19.2.6 in there concerning Theorem 10.10 and Remark 10.10.

Parts of Chapters 4, 5, 7, 8 and 10 of this book are summarized in [Suwa (2022b)].

Chapter 11

Further Topics

We present various topics that are basic in complex analytic geometry and are needed in later chapters. We start by reviewing sheaves and sheaf cohomology. There is an important class of sheaves that are called coherent sheaves. We quote fundamental coherence theorems of K. Oka and discuss some of their consequences. The Čech-Dolbeault cohomology, a complex counterpart of the Čech-de Rham cohomology, is then introduced. It allows us to regard both Dolbeault cochains and Čech cochains naturally as Čech-Dolbeault cochains and leads to a canonical Dolbeault theorem (Corollary 11.6). After somewhat detailed discussion of complex analytic spaces, generalizations of varieties, presentation of divisors and local complete intersections follows.

11.1 Sheaves

In this section, we review fundamental materials on sheaf theory.

Let X be a topological space.

Definition 11.1. A *sheaf of Abelian groups* on X is a topological space $\mathscr{S}$ together with a surjective continuous map $\pi : \mathscr{S} \to X$ such that

(1) π is a local homeomorphism,
(2) for each x in X, $\mathscr{S}_x = \pi^{-1}(x)$, called the *stalk* of $\mathscr{S}$ on x, has the structure of an Abelian group,
(3) the operations $a \mapsto -a$ of $\mathscr{S}$ to $\mathscr{S}$ and $(a, b) \mapsto a + b$ of $\mathscr{S} \times_X \mathscr{S}$ to $\mathscr{S}$ are continuous, where we set

$$\mathscr{S} \times_X \mathscr{S} = \{\, (a, b) \in \mathscr{S} \times \mathscr{S} \mid \pi(a) = \pi(b) \,\}.$$

Example 11.1. 1. Let $\pi : \mathscr{S} \to X$ be a sheaf on X and A a subspace of X. Then $\pi^{-1}(A)$ is a sheaf on A, which is called the *restriction* of $\mathscr{S}$ to A and is denoted by $\mathscr{S}|_A$.

2. Let G be an Abelian group and set $G_X = G \times X$ with the discrete topology on G. Then it is a sheaf with π the projection onto the second factor. We call it the *constant sheaf* with stalk G. It will also be denoted by G, if there is no fear of confusion.

Definition 11.2. Let $\pi : \mathscr{S} \to X$ be a sheaf and U an open set in X. A *section* of $\mathscr{S}$ on U is a continuous map $s : U \to \mathscr{S}$ such that $\pi \circ s = 1_U$.

From the condition (1) in Definition 11.1, we have:

Proposition 11.1. *Let $\pi : \mathscr{S} \to X$ be a sheaf.*

1. *For every point a in $\mathscr{S}$, there exists a section s on a neighborhood of $x = \pi(a)$ such that $s(x) = a$.*

2. *Let s and t be sections on neighborhoods U and V of x, respectively. If $s(x) = t(x)$, there exists a neighborhood W of x with $W \subset U \cap V$ such that $s = t$ on W.*

Exercise 11.1. Let $\mathscr{S}$ be a sheaf on X.

1. Show that the map $X \to \mathscr{S}$ defined by $x \mapsto 0 \in \mathscr{S}_x$ is continuous. Thus it is a section of $\mathscr{S}$ on X, called the *zero section.*

2. Let s be a section of $\mathscr{S}$ on an open set U in X. Show that the set $\{x \in U \mid s(x) \neq 0\}$ is closed in U.

We denote the set of sections of $\mathscr{S}$ on U by $\Gamma(U; \mathscr{S})$. This has naturally the structure of an Abelian group. By convention we set $\Gamma(\emptyset; \mathscr{S}) = 0$. If (U, V) is a pair of open sets with $V \subset U$, the restriction defines a morphism $r_{VU} : \Gamma(U; \mathscr{S}) \to \Gamma(V; \mathscr{S})$. We have $r_{UU} = 1$ and $r_{WV} \circ r_{VU} = r_{WU}$ for every triple (U, V, W) of open sets with $W \subset V \subset U$.

Thus for each x in X, the collection $(\Gamma(U; \mathscr{S}), r_{VU})$ forms a direct system on the set of open neighborhoods of x and from Proposition 11.1 we see that

$$\mathscr{S}_x = \varinjlim_{U \ni x} \Gamma(U; \mathscr{S}),$$

the direct limit (cf. Section A.1, also Remark 1.6).

In view of the above, we introduce the following:

Definition 11.3. A *presheaf* on X is a system (S_U, ρ_{VU}), where, for every open set U in X, S_U is an Abelian group and, for every pair of open sets

(U, V) with $V \subset U$, $\rho_{VU} : S_U \to S_V$ is a morphism of Abelian groups, called the "restriction", such that

(1) $S_\emptyset = 0$,
(2) $\rho_{UU} = 1$ and $\rho_{WV} \circ \rho_{VU} = \rho_{WU}$ for every triple (U, V, W) of open sets with $W \subset V \subset U$.

If (S_U, ρ_{VU}) is a presheaf, we may construct a sheaf as follows. Let

$$\mathscr{S}_x = \varinjlim_{U \ni x} S_U$$

and $\mathscr{S} = \bigsqcup_{x \in X} \mathscr{S}_x$, the disjoint union with the natural map $\pi : \mathscr{S} \to X$. Then, for every open set U and a point x in U, there is a canonical morphism $\rho_{U,x} : S_U \to \mathscr{S}_x$. We topologize $\mathscr{S}$ so that for every open set U and σ in S_U, the set $\{ \rho_{U,x}(\sigma) \mid x \in U \}$ is an open set in $\mathscr{S}$. This way $\mathscr{S}$ becomes a sheaf.

In the following, if the restriction morphisms ρ_{VU} of the presheaf (S_U, ρ_{VU}) are obvious, we abbreviate the above procedure by saying that $\mathscr{S}$ is the sheaf defined by the presheaf $U \mapsto S_U$.

For a presheaf (S_U, ρ_{VU}) and the sheaf $\mathscr{S}$ constructed as above, we have a morphism

$$\varphi_U : S_U \longrightarrow \Gamma(U; \mathscr{S}) \qquad \text{defined by } \varphi(\sigma)(x) = \rho_{U,x}(\sigma).$$

Proposition 11.2. *The morphisms φ_U are isomorphisms if and only if the following conditions are satisfied:*

(1) *for $\sigma \in S_U$, the condition $\rho_{U_i,U}(\sigma) = 0$, $i = 1, 2$, implies $\sigma = 0$,*
(2) *for $\sigma_i \in S_{U_i}$ with $\rho_{U_1 \cap U_2, U_1}(\sigma_1) = \rho_{U_1 \cap U_2, U_2}(\sigma_2)$, there exists $\sigma \in S_U$ with $\rho_{U_i,U}(\sigma) = \sigma_i$.*

In the above, U_1 and U_2 are arbitrary open sets and $U = U_1 \cup U_2$.

Exercise 11.2. Verify Proposition 11.2.

Sheaf morphisms: Let $\pi : \mathscr{S} \to X$ and $\varpi : \mathscr{T} \to X$ be sheaves on X.

Definition 11.4. A *morphism* $\varphi : \mathscr{S} \to \mathscr{T}$ is a continuous map such that

(1) $\varpi \circ \varphi = \pi$,
(2) the restriction $\varphi_x : \mathscr{S}_x \to \mathscr{T}_x$ of φ to each stalk is a morphism of Abelian groups.

A morphism $\varphi : \mathscr{S} \to \mathscr{T}$ is said to be an *isomorphism* if there exists a morphism $\psi : \mathscr{T} \to \mathscr{S}$ such that $\psi \circ \varphi = 1_{\mathscr{S}}$ and $\varphi \circ \psi = 1_{\mathscr{T}}$.

Let (S_U, ρ_{VU}) and (T_U, ρ'_{VU}) be presheaves. A *morphism* $(S_U) \to (T_U)$ is a collection (φ_U) of morphisms $\varphi_U : S_U \to T_U$ compatible with the restrictions. Such a morphism defines a morphism $\varphi : \mathscr{S} \to \mathscr{T}$ of sheaves defined by the presheaves.

A morphism $\varphi : \mathscr{S} \to \mathscr{T}$ defines a morphism $\varphi_U : \Gamma(U; \mathscr{S}) \to \Gamma(U; \mathscr{T})$ for each open set U and the collection (φ_U) is a morphism of presheaves.

Sub and quotient sheaves: Let $\mathscr{S} \to X$ be a sheaf. A subset $\mathscr{S}'$ of $\mathscr{S}$ is said to be a *subsheaf* of $\mathscr{S}$ if it is a sheaf with respect to the topology and algebraic structure induced from those of $\mathscr{S}$. In this case the disjoint union $\bigsqcup_{x \in X} \mathscr{S}_x / \mathscr{S}'_x$ has naturally the structure of a sheaf on X. We call this the *quotient sheaf* of $\mathscr{S}$ by $\mathscr{S}'$ and denote by $\mathscr{S}/\mathscr{S}'$.

Let (S'_U, ρ'_{VU}) and (S_U, ρ_{VU}) be presheaves such that, for each open set U, S'_U is a subgroup of S_U and that the inclusions $S'_U \hookrightarrow S_U$ are compatible with the restrictions. Then the sheaf $\mathscr{S}'$ defined by (S'_U) is a subsheaf of the sheaf $\mathscr{S}$ defined by (S_U). The quotient sheaf $\mathscr{S}/\mathscr{S}'$ is defined by the presheaf (S_U/S'_U).

Exercise 11.3. Show that a section of a quotient sheaf $\mathscr{S}/\mathscr{S}'$ on an open set U is represented by a collection $\{(U_\alpha, s_\alpha)\}_{\alpha \in I}$, where $\{U_\alpha\}$ is a covering of U, s_α is a section of $\mathscr{S}$ on U_α, for each α, and $s_\beta - s_\alpha$ is a section of $\mathscr{S}'$ on $U_\alpha \cap U_\beta$, for each pair (α, β).

Let $\varphi : \mathscr{S} \to \mathscr{T}$ be a morphism. Then, $\operatorname{Ker} \varphi = \bigsqcup_{x \in X} \operatorname{Ker} \varphi_x$ has the structure of a subsheaf of $\mathscr{S}$ and $\operatorname{Im} \varphi = \bigsqcup_{x \in X} \operatorname{Im} \varphi_x$ that of $\mathscr{T}$. We set $\operatorname{Coker} \varphi = \mathscr{T}/\operatorname{Im} \varphi$.

A sequence of sheaf morphisms

$$\cdots \longrightarrow \mathscr{S}^{p-1} \xrightarrow{\varphi^{p-1}} \mathscr{S}^{p} \xrightarrow{\varphi^{p}} \mathscr{S}^{p-1} \longrightarrow \cdots$$

is said to be *exact* if $\operatorname{Im} \varphi^{p-1} = \operatorname{Ker} \varphi^p$ for all p.

Thus if

$$0 \longrightarrow \mathscr{S}' \xrightarrow{\iota} \mathscr{S} \xrightarrow{\varphi} \mathscr{S}'' \longrightarrow 0$$

is an exact sequence of sheaves, ι is an isomorphism of $\mathscr{S}'$ onto $\operatorname{Im} \iota$ and φ induces an isomorphism $\operatorname{Coker} \iota \simeq \mathscr{S}''$.

Direct and inverse images: Let $f : Y \to X$ be a continuous map of topological spaces.

For a sheaf $\mathscr{T}$ on Y, the *direct image* $f_*\mathscr{T}$ is the sheaf on X defined by the presheaf that assigns to each open set U in X the group $\Gamma(f^{-1}U;\mathscr{T})$. We have $\Gamma(U;f_*\mathscr{T}) = \Gamma(f^{-1}U;\mathscr{T})$ for every open set U in X.

In particular, if S is a closed subset of X with the inclusion $i : S \hookrightarrow X$, then, for a sheaf $\mathscr{T}$ on S, $i_*\mathscr{T}$ is the sheaf on X such that

$$\begin{cases} i_*\mathscr{T}|_S = \mathscr{T}, \\ i_*\mathscr{T}|_{X\smallsetminus S} = 0. \end{cases} \tag{11.1}$$

For a sheaf $\mathscr{S}$ on X, the *inverse image* $f^{-1}\mathscr{S}$ is the sheaf on Y defined by the presheaf

$$W \mapsto \varinjlim_{U\supset f(W)} \Gamma(U;\mathscr{S}),$$

where the limit is taken over all open sets U of X containing $f(W)$. From definition we see that, for every point y in Y,

$$(f^{-1}\mathscr{S})_y = \mathscr{S}_{f(y)}. \tag{11.2}$$

If $\mathscr{S}$ is a sheaf on X and A a subspace of X with the inclusion $i : A \hookrightarrow X$, $i^{-1}\mathscr{S}$ coincides with the restriction $\mathscr{S}|_A$ (cf. Example 11.1. 1).

Proposition 11.3. *Let $f : Y \to X$ be a continuous map of topological spaces and let $\mathscr{S}$ and $\mathscr{T}$ be sheaves on X and Y, respectively. Then there exist canonical morphisms*

$$\mathscr{S} \longrightarrow f_*f^{-1}\mathscr{S} \qquad \text{and} \qquad f^{-1}f_*\mathscr{T} \longrightarrow \mathscr{T}.$$

Thus giving a morphism $\mathscr{S} \to f_*\mathscr{T}$ is equivalent to giving a morphism $f^{-1}\mathscr{S} \to \mathscr{T}$.

Exercise 11.4. Verify the above.

Remark 11.1. Let $\mathcal{S}h(X)$ and $\mathcal{S}h(Y)$ be the categories of sheaves of Abelian groups on X and Y, respectively (cf. Example A.1. 2). Then, for a continuous map $f : Y \to X$, we may think of $f_* : \mathcal{S}h(Y) \to \mathcal{S}h(X)$ and $f^{-1} : \mathcal{S}h(X) \to \mathcal{S}h(Y)$ as covariant functors. The functor f^{-1} is exact by (11.2), while f_* is only left exact in general (cf. Remark 11.10 below). In the particular case where S is a closed subset of X with the inclusion $i : S \hookrightarrow X$, $i_* : \mathcal{S}h(S) \to \mathcal{S}h(X)$ is exact (cf. (11.1)).

Morphism of sheaves on distinct spaces: Let X and Y be topological spaces and $\mathscr{S}$ and $\mathscr{T}$ sheaves on X and Y, respectively.

Definition 11.5. A *morphism* $(Y, \mathscr{T}) \to (X, \mathscr{S})$ is a pair (f, φ) of a continuous map $f : Y \to X$ and a morphism $\varphi : \mathscr{S} \to f_* \mathscr{T}$ of sheaves on X.

In the case $Y = X$ and $f = 1_X$, φ is a morphism as in Definition 11.4.

A morphism (f, φ) as above defines a morphism $f^{-1}\mathscr{S} \to f^{-1}f_*\mathscr{T} \to \mathscr{T}$ so that, for each $y \in Y$, it defines a morphism of Abelian groups

$$\varphi_y : \mathscr{S}_{f(y)} \longrightarrow \mathscr{T}_y.$$

For a sequence $(Z, \mathscr{U}) \overset{(g,\psi)}{\longrightarrow} (Y, \mathscr{T}) \overset{(f,\varphi)}{\longrightarrow} (X, \mathscr{S})$ of morphisms, the *composition* $(f, \varphi) \circ (g, \psi) : (Z, \mathscr{U}) \to (X, \mathscr{S})$ is defined as $(f \circ g, f_*\psi \circ \varphi)$. The identity morphism of $(X, \mathscr{S})$ is the pair $(1_X, 1_{\mathscr{S}})$ of the identities. A morphism $(f, \varphi) : (Y, \mathscr{T}) \to (X, \mathscr{S})$ is an *isomorphism* if it admits an inverse, i.e., if there exist a morphism $(g, \psi) : (X, \mathscr{S}) \to (Y, \mathscr{T})$ with $(f, \varphi) \circ (g, \psi) = (1_X, 1_{\mathscr{S}})$ and $(g, \psi) \circ (f, \varphi) = (1_Y, 1_{\mathscr{T}})$.

Support of a sheaf: Let $\mathscr{S}$ be a sheaf on X. The *support* of $\mathscr{S}$ is defined by

$$\mathrm{Supp}(\mathscr{S}) = \{ x \in X \mid \mathscr{S}_x \neq 0 \}.$$

Exercise 11.5. In the above situation, suppose that $\mathrm{Supp}(\mathscr{S})$ is contained in a closed set S in X and denote by $i : S \hookrightarrow X$ the inclusion. Show that $i_*i^{-1}\mathscr{S} = \mathscr{S}$.

Sheaves with other algebraic structures: We may similarly define sheaves of rings, modules and so forth. Thus a *sheaf of commutative rings with unity* is a sheaf $\mathscr{R}$ of Abelian groups such that each stalk $\mathscr{R}_x$ has furthermore the structure of a commutative ring with unity and that the operation $(a, b) \mapsto ab$ of $\mathscr{R} \times_X \mathscr{R}$ to $\mathscr{R}$ and the assignment $x \mapsto 1 \in \mathscr{R}_x$ of X to $\mathscr{R}$ are continuous. A *sheaf of $\mathscr{R}$-modules*, an *$\mathscr{R}$-module* for short, is a sheaf $\mathscr{S}$ of Abelian groups such that each stalk $\mathscr{S}_x$ has furthermore the structure of an $\mathscr{R}_x$-module and that the operation $(a, f) \mapsto af$ of $\mathscr{R} \times_X \mathscr{S}$ to $\mathscr{S}$ is continuous. A sheaf of Abelian groups may be thought of as a $\mathbb{Z}_X$-module with $\mathbb{Z}_X = \mathbb{Z} \times X$ the constant sheaf with stalk $\mathbb{Z}$.

In the following, a sheaf of rings means a sheaf of commutative rings with unity.

Algebraic operations: As in the case of vector bundles, we may consider algebraic operations on sheaves. Thus, for two $\mathscr{R}$-modules $\mathscr{S}$ and $\mathscr{T}$, we have the direct sum $\mathscr{S} \oplus \mathscr{T}$, the tensor product $\mathscr{S} \otimes_{\mathscr{R}} \mathscr{T}$ and the sheaf of morphisms $\mathscr{H}om_{\mathscr{R}}(\mathscr{S}, \mathscr{T})$. The direct sum of p copies of $\mathscr{S}$ will be denoted by $\mathscr{S}^p$. They are all $\mathscr{R}$-modules.

Some fundamental sheaves

In each of the following, the sheaf $\mathscr{S}$ we introduce is defined by a presheaf (S_U, ρ_{VU}) for which the restrictions ρ_{VU} are the obvious ones. Also, in each case, the conditions in Proposition 11.2 are satisfied so that there is a canonical isomorphism $S_U \xrightarrow{\sim} \Gamma(U; \mathscr{S})$.

Sheaf of continuous functions: Let X be a topological space. The *sheaf of germs of continuous functions* on X, denoted by $\mathscr{C}_X$ or $\mathscr{C}$, is the sheaf defined by the presheaf $U \mapsto C_U$, the ring of complex valued continuous functions on U. It is a sheaf of rings and will also be denoted by $\mathscr{C}^0_X$.

Sheaf of C^∞ functions and forms: Let M be a C^∞ manifold. The *sheaf of germs of C^∞ functions* on M, denoted by $\mathscr{A}_M$ or $\mathscr{A}$, is the sheaf defined by the presheaf $U \mapsto A_U = A^0(U)$, the ring of complex valued C^∞ functions on U. It is a sheaf of rings and will also be denoted by $\mathscr{C}^\infty_M$.

Let E be a C^∞ vector bundle on M. The *sheaf of germs of C^∞ sections of E*, denoted by $\mathscr{A}(E)$, is the sheaf defined by the presheaf $U \mapsto A^0(U; E)$, the A_U-module of C^∞ sections of E on U. It is an $\mathscr{A}_M$-module.

In particular, for $E = \bigwedge^p T^*_{\mathbb{R}} M$, $\mathscr{A}^p_M = \mathscr{A}(\bigwedge^p T^*_{\mathbb{R}} M)$ is the *sheaf of germs of C^∞ p-forms* on M. The exterior derivative induces a morphism $d : \mathscr{A}^p \to \mathscr{A}^{p+1}$ of sheaves of $\mathbb{C}$-vector spaces.

Sheaf of real analytic functions: We may likewise consider the sheaf of real analytic functions on a real analytic manifold M. It is a sheaf of rings and is denoted by $\mathscr{C}^\omega_M$.

Sheaf of holomorphic functions and forms: Let M be a complex manifold. The *sheaf of germs of holomorphic functions* on M, denoted by $\mathscr{O}_M$ or $\mathscr{O}$, is the sheaf defined by the presheaf $U \mapsto O_U$, the ring of holomorphic functions on U. It is a sheaf of rings. Note that for each point z in M, $\mathscr{O}_{M,z}$ is naturally isomorphic with the ring $\mathscr{O}_n$ as constructed in Section 1.4.

Let E be a holomorphic vector bundle on M. The sheaf of germs of holomorphic sections of E, denoted by $\mathscr{O}(E)$, is the sheaf defined by the presheaf $U \mapsto O(U; E)$, the O_U-module of holomorphic sections of E on U. It is an $\mathscr{O}_M$-module. In particular, for $E = \bigwedge^p T^*M$, $\Omega_M^p = \mathscr{O}(\bigwedge^p T^*M)$ is the *sheaf of germs of holomorphic p-forms* on M. We have $\Omega_M^0 = \mathscr{O}_M$. Also for $E = TM$, $\Theta_M = \mathscr{O}(TM)$ is the *sheaf of germs of holomorphic vector fields* on M. Sometimes it is referred to as the *tangent sheaf* of M.

We may also consider $\mathscr{A}_M^{p,q} = \mathscr{A}(\bigwedge^p T^*M \otimes \bigwedge^q \overline{T}^* M)$, the *sheaf of germs of* C^∞ *(p,q)-forms* on M (cf. Definition 6.2). It is an $\mathscr{A}_M$-module. We have the morphism $\bar{\partial} : \mathscr{A}^{p,q} \to \mathscr{A}^{p,q+1}$ of sheaves of $\mathbb{C}$-vector spaces.

Sheaf of algebraic functions: Let M be a quasi-projective manifold and M_Z the manifold M with Zarisky topology (cf. Section 2.2). Starting from the rings $O_{U_Z}^{\mathrm{r}}$ of regular functions on Zariski open sets U_Z in M_Z, we may construct the sheaf $\mathscr{O}^{\mathrm{r}}$ of germs of regular functions on M_Z. The sheaf $\mathscr{O}_M^{\mathrm{a}}$ of germs of algebraic functions on M is defined to be the inverse image $\mathscr{O}_M^{\mathrm{a}} = 1_M^{-1}\mathscr{O}_{M_Z}^{\mathrm{r}}$ by the identity map $1_M : M \to M_Z$.

11.2 Sheaf cohomology

There are several ways to define cohomology of sheaves. Here we adopt the one used in Section 7.2, i.e., the Čech method.

Let $\mathscr{S}$ be a sheaf on X and $\mathcal{U} = \{U_\alpha\}_{\alpha \in I}$ an open covering of X. For $\alpha_0, \ldots, \alpha_p$ in I, we set $U_{\alpha_0 \ldots \alpha_p} = \bigcap_{\nu=0}^p U_{\alpha_\nu}$, as before. We define $C^p(\mathcal{U}; \mathscr{S})$ to be the direct product:

$$C^p(\mathcal{U}; \mathscr{S}) = \prod_{(\alpha_0, \ldots, \alpha_p) \in I^{p+1}} \Gamma(U_{\alpha_0 \ldots \alpha_p}; \mathscr{S}).$$

The coboundary operator

$$\delta = \delta^p : C^p(\mathcal{U}; \mathscr{S}) \longrightarrow C^{p+1}(\mathcal{U}; \mathscr{S})$$

is defined as before. The operators have the property $\delta^{p+1} \circ \delta^p = 0$.

Definition 11.6. The p-th *cohomology of* $\mathcal{U}$ *with coefficients in* $\mathscr{S}$ is the p-th cohomology of $(C^\bullet(\mathcal{U}; \mathscr{S}), \delta)$:

$$H^p(\mathcal{U}; \mathscr{S}) = \operatorname{Ker} \delta^p / \operatorname{Im} \delta^{p-1}.$$

It is also called the p-th *cohomology of* $\mathscr{S}$ *on* $\mathcal{U}$.

Remark 11.2. 1. We may perform a similar construction using only alternating cochains and the resulting cohomology turns out to be isomorphic with the one above (cf. Section 7.2). Thus, if the covering $\mathcal{U}$ has order $m+1$ (cf. Section 2.5), then $H^p(\mathcal{U};\mathscr{S}) = 0$ for $p > m$.

Also, if I is properly ordered (cf. Definition 7.4), we may use the reduced expression to the same effect.

2. Let M be a C^∞ manifold. If $\mathscr{S} = \mathsf{R}_M$, the constant sheaf with stalk a commutative ring with unity R, the above cohomology coincides with $H^p(\mathcal{U};\mathsf{R})$ in Definition 7.2. Also if $\mathscr{S} = \mathscr{A}^q_M$, it coincides with $H^p(\mathcal{U};A^q)$ in Definition 7.3.

Let $\mathcal{V} = \{V_\lambda\}_{\lambda\in J}$ be a refinement of $\mathcal{U}$ and choose a map $\iota : J \to I$ with $V_\lambda \subset U_{\iota(\lambda)}$ for all λ. Then we have a morphism

$$\iota^* : C^p(\mathcal{U};\mathscr{S}) \longrightarrow C^p(\mathcal{V};\mathscr{S}),$$

which is defined by $\iota^*(\sigma)_{\lambda_0\dots\lambda_p} = \sigma_{\iota(\lambda_0)\dots\iota(\lambda_p)}|_{V_{\lambda_0\dots\lambda_p}}$. It is compatible with the coboundary operators and induces a morphism

$$h_{\mathcal{V}\mathcal{U}} : H^p(\mathcal{U};\mathscr{S}) \longrightarrow H^p(\mathcal{V};\mathscr{S}),$$

which can be shown not to depend on the choice of ι. The set of open coverings of X is a directed set (cf. Remark 2.1. 1) and the system $(H^p(\mathcal{U};\mathscr{S}), h_{\mathcal{V}\mathcal{U}})$ forms a direct system (cf. Definition A.8). The p-th *cohomology of X with coefficients in $\mathscr{S}$*, or *cohomology of $\mathscr{S}$ on X*, is defined as the direct limit:

$$H^p(X;\mathscr{S}) = \varinjlim H^p(\mathcal{U};\mathscr{S}).$$

The 0-th cohomology is easy to find; since, for every $\mathcal{U}$, $\operatorname{Ker}\delta^0 = \Gamma(X;\mathscr{S})$, we have $H^0(\mathcal{U};\mathscr{S}) = \Gamma(X;\mathscr{S})$ so that

$$H^0(X;\mathscr{S}) = \Gamma(X;\mathscr{S}).$$

Remark 11.3. 1. If X is of finite dimension m, then $H^p(X;\mathscr{S}) = 0$ for $p > m$ (cf. Definition 2.15 and Remark 11.2. 1).

2. If it is necessary to distinguish the above cohomology from the one defined some other way, we denote it by $\check{H}^p(X;\mathscr{S})$ and call it the p-th *Čech cohomology of X with coefficients in $\mathscr{S}$*.

3. In the case $X = M$, a C^∞ manifold, and $\mathscr{S} = \mathsf{R}_M$, the constant sheaf with stalk a commutative ring R with unity, then there is a canonical isomorphism

$$\check{H}^p(M;\mathsf{R}_M) \simeq H^p(M;\mathsf{R}),$$

the singular or simplicial cohomology of M, as the set of simplicial coverings is cofinal in the set of coverings of M (cf. Remark 11.2. 2 and Theorem 7.3).

Exercise 11.6. Let S be a closed set in a topological space X, with inclusion $i : S \hookrightarrow X$, and $\mathscr{T}$ a sheaf on S. Show that, there is a canonical isomorphism $H^p(X; i_*\mathscr{T}) \simeq H^p(S; \mathscr{T})$ for all $p \geq 0$.

Exercise 11.7. Let X be a topological space and $\mathscr{C}^*$ the sheaf of germs of non-vanishing complex valued continuous functions on X with multiplicative group structure. Show that $H^1(X; \mathscr{C}^*)$ may be thought of as the set of isomorphism classes of continuous complex line bundles on X, the group structure of $H^1(X; \mathscr{C}^*)$ corresponding to the tensor product of line bundles (cf. Definition 3.5 and Exercise 3.3).

Likewise, if M is a C^∞ manifold and if $\mathscr{A}^*$ is the sheaf of germs of non-vanishing complex valued C^∞ functions, $H^1(M; \mathscr{A}^*)$ may be thought of as the set of isomorphism classes of C^∞ complex line bundles on M. Also, if we denote by $\mathscr{O}^*$ the sheaf of non-vanishing holomorphic functions on a complex manifold M, $H^1(M; \mathscr{O}^*)$ is the set of isomorphism classes of holomorphic line bundles on M, which is referred to as the *Picard group* of M and is denoted by $\mathrm{Pic}(M)$.

Let $\varphi : \mathscr{S} \to \mathscr{T}$ be a morphism of sheaves on X. For any covering $\mathcal{U}$ of X, it induces a morphism $\varphi^\bullet : C^\bullet(\mathcal{U}; \mathscr{S}) \to C^\bullet(\mathcal{U}; \mathscr{T})$ of complexes. It then induces a morphism $\varphi^* : H^p(\mathcal{U}; \mathscr{S}) \to H^p(\mathcal{U}; \mathscr{T})$, which is compatible with the morphisms $h_{\mathcal{V}\mathcal{U}}$, and, passing to the direct limit, induces a morphism

$$\varphi^* : H^p(X; \mathscr{S}) \longrightarrow H^p(X; \mathscr{T}).$$

In the case X is paracompact (cf. Remark 2.1. 3), we have:

Theorem 11.1. *If X is paracompact, a short exact sequence of sheaves*

$$0 \longrightarrow \mathscr{S}' \xrightarrow{\iota} \mathscr{S} \xrightarrow{\varphi} \mathscr{S}'' \longrightarrow 0 \tag{11.3}$$

on X yields a long exact sequence

$$\begin{aligned} 0 \to H^0(X; \mathscr{S}') &\xrightarrow{\iota^*} H^0(X; \mathscr{S}) \xrightarrow{\varphi^*} H^0(X; \mathscr{S}'') \xrightarrow{\delta^*} H^1(X; \mathscr{S}') \xrightarrow{\iota^*} \cdots \\ \xrightarrow{\delta^*} H^p(X; \mathscr{S}') &\xrightarrow{\iota^*} H^p(X; \mathscr{S}) \xrightarrow{\varphi^*} H^p(X; \mathscr{S}'') \xrightarrow{\delta^*} H^{p+1}(X; \mathscr{S}') \xrightarrow{\iota^*} \cdots . \end{aligned}$$

Here is an outline of the proof. From (11.3), we have the exact sequence of complexes

$$0 \longrightarrow C^\bullet(\mathcal{U}; \mathscr{S}') \xrightarrow{\iota^\bullet} C^\bullet(\mathcal{U}; \mathscr{S}) \xrightarrow{\varphi^\bullet} C^\bullet(\mathcal{U}; \mathscr{S}''),$$

where $\varphi^\bullet$ is not surjective in general. Thus we denote by $C_0^\bullet(\mathcal{U};\mathscr{S}'')$ the image of $\varphi^\bullet$, which is a subcomplex of $C^\bullet(\mathcal{U};\mathscr{S}'')$, and by $H_0^p(\mathcal{U};\mathscr{S}'')$ its cohomology. Then the exact sequence of complexes

$$0 \longrightarrow C^\bullet(\mathcal{U};\mathscr{S}') \xrightarrow{\iota^\bullet} C^\bullet(\mathcal{U};\mathscr{S}) \xrightarrow{\varphi^\bullet} C_0^\bullet(\mathcal{U};\mathscr{S}'') \longrightarrow 0$$

yields a long exact sequences of cohomologies on $\mathcal{U}$ (cf. Proposition A.11). Since taking direct limite is an exact functor, we have a long exact sequences of cohomologies as above with $H^p(X;\mathscr{S}'')$ replaced by $H_0^p(X;\mathscr{S}'')$. The inclusions $C_0^\bullet(\mathcal{U};\mathscr{S}'') \hookrightarrow C^\bullet(\mathcal{U};\mathscr{S}'')$ induce a canonical morphism $H_0^p(X;\mathscr{S}'') \to H^p(X;\mathscr{S}'')$, which is shown to be an isomorphism, if X is paracompact.

Fine sheaves

Let X be a paracompact topological space and $\mathscr{S}$ a sheaf on X.

Definition 11.7. We say that $\mathscr{S}$ is *fine* if, for every locally finite covering $\mathcal{U} = \{U_\alpha\}$ of X, it admits a partition of unity subordinate to $\mathcal{U}$, i.e., for each α, there is a sheaf morphism $\eta_\alpha : \mathscr{S} \to \mathscr{S}$ such that

(1) $\overline{\{x \in X \mid \eta_\alpha(x) \neq 0\}} \subset U_\alpha$,
(2) $\sum_\alpha \eta_\alpha = 1$.

The following is proved as Corollary 7.2:

Proposition 11.4. *If $\mathscr{S}$ is a fine sheaf, for every locally finite covering $\mathcal{U}$, the complex $C^\bullet(\mathcal{U},\mathscr{S})$ is acyclic* (cf. Appendix A), *i.e.,*

$$H^p(\mathcal{U};\mathscr{S}) = 0 \qquad \textit{for } p \geq 1.$$

Since the locally finite coverings are cofinal in the set of coverings of a paracompact space, we have:

Corollary 11.1. *If $\mathscr{S}$ is a fine sheaf on a paracompact space X,*

$$H^p(X;\mathscr{S}) = 0 \qquad \textit{for } p \geq 1.$$

Example 11.2. The sheaf $\mathscr{C}_X$ of germs of continuous functions is fine. If M is a C^∞ manifold, the sheaves $\mathscr{A}_M^r$ are fine. Also for a complex manifold M, the sheaves $\mathscr{A}_M^{p,q}$ are fine.

Exponential sequence

Let X be a paracompact topological space and consider the exact sequence of sheaves on X:

$$0 \longrightarrow \mathbb{Z}_X \longrightarrow \mathscr{C}_X \xrightarrow{\exp} \mathscr{C}_X^* \longrightarrow 0,$$

where exp assigns to a germ f the germ $e^{2\pi\sqrt{-1}f}$. There is the associated long exact sequence (subscript X being omitted):

$$\cdots \to H^1(X;\mathscr{C}) \to H^1(X;\mathscr{C}^*) \xrightarrow{\delta^*} H^2(X;\mathbb{Z}) \to H^2(X;\mathscr{C}) \to \cdots. \quad (11.4)$$

We examine the connecting morphism δ^* more closely. Note that every class $h \in H^1(X;\mathscr{C}^*)$ is represented by a cocycle $\{h^{\alpha\beta}\} \in C^1(\mathcal{U};\mathscr{C}^*)$ for some covering $\mathcal{U}$ of X. Then $\delta^*(h)$ is represented by the cocycle $c = \delta g$, $g^{\alpha\beta} = \frac{1}{2\pi\sqrt{-1}} \log h^{\alpha\beta}$. Fixing a branch of $\log h^{\alpha\beta}$ for each pair (α,β), we have:

$$c_{\alpha\beta\gamma} = \frac{1}{2\pi\sqrt{-1}}(\log h^{\beta\gamma} - \log h^{\alpha\gamma} + \log h^{\alpha\beta}), \quad (11.5)$$

which is an integer, as $h^{\alpha\beta}h^{\beta\gamma} = h^{\alpha\gamma}$.

Note that the above $H^2(X;\mathbb{Z})$ is the Čech cohomology $\check{H}^2(X;\mathbb{Z}_X)$ (cf. Remark 11.3.2). Also, since the sheaf $\mathscr{C}$ is fine, $\delta^* : H^1(X;\mathscr{C}^*) \xrightarrow{\sim} \check{H}^2(X;\mathbb{Z}_X)$ is an isomorphism (cf. Corollary 11.1).

If M is a C^∞ manifold, there are similar sequences, replacing $\mathscr{C}$ and $\mathscr{C}^*$ with $\mathscr{A}$ and $\mathscr{A}^*$. In this case, $\delta^* : H^1(M;\mathscr{A}^*) \xrightarrow{\sim} \check{H}^2(M;\mathbb{Z}_M)$ is also an isomorphism. There is a canonical isomorphism $\check{H}^2(M;\mathbb{Z}_M) \xrightarrow{\sim} H^2(M;\mathbb{Z})$, the singular or simplicial cohomology. Let K be a C^∞ triangulation of M and $\mathcal{U} = \{U_\alpha\}$ the associated simplicial covering of M. Then the isomorphism is induced from the morphism (cf. Lemma 7.6)

$$\psi_\delta : C^2(\mathcal{U};\mathbb{Z}) \longrightarrow C_K^2(M;\mathbb{Z}) \qquad \text{given by } \langle(\alpha,\beta,\gamma),\psi_\delta(c)\rangle = c_{\alpha\beta\gamma}.$$

Example 11.3. We try to compute $\delta^*(h)$ for the class h of the hyperplane bundle H_1 on $\mathbb{P}^1$ explicitly. Let $[\zeta_0,\zeta_1]$ be homogeneous coordinates on $\mathbb{P}^1$ and $z = \zeta_1/\zeta_0$ a coordinate on the open set $\{\zeta_0 \neq 0\}$, which is identified with the z-plane $\mathbb{C}$. We may triangulate $\mathbb{P}^1$ by the boundary of a tetrahedron $|K|$ with vertices v_0, v_1, v_2, v_3. We assume that the vertex v_0 is the point $[1,0]$ and that the fundamental cycle of $\mathbb{P}^1$ which is given as the sum of oriented simplices (cf. Section B.2)

$$\partial(v_0,v_1,v_2,v_3) = (v_1,v_2,v_3) - (v_0,v_2,v_3) + (v_0,v_1,v_3) - (v_0,v_1,v_2)$$

has the same orientation as the z-plane $\mathbb{C}$. Let $\mathcal{U} = \{U_\alpha\}_{0\le\alpha\le3}$ be the simplicial covering of $\mathbb{P}^1$ associated with the triangulation (cf. Section 7.5).

The hyperplane bundle H_1 is given by the system $\{h^{\alpha\beta}\}$ of transition functions given by (here we use the reduced expression (cf. Remark 11.2.1) and we describe the case $\alpha < \beta$)

$$h^{\alpha\beta} = \begin{cases} z & \alpha = 0,\ \beta \neq 0, \\ 1 & \text{otherwise.} \end{cases}$$

The class $\delta^*(h)$ is represented by the cocycle $(c_{\alpha\beta\gamma})$ given by (11.5). Thus

$$c_{\alpha\beta\gamma} = \begin{cases} \frac{1}{2\pi\sqrt{-1}}(\log h^{\alpha\beta} - \log h^{\alpha\gamma}) & \alpha = 0, \\ 0 & \text{otherwise.} \end{cases}$$

We fix a branch of $\log h^{01}$ and take as $\log h^{02}$ its analytic continuation. Then $c_{012} = 0$. We also take as $\log h^{03}$ the analytic continuation of $\log h^{02}$. Then $c_{023} = 0$ and $c_{013} = \frac{1}{2\pi\sqrt{-1}}(\log h^{01} - \log h^{03}) = 1$, since we turn around v_0 clockwise as we proceed $v_1 \to v_2 \to v_3$.

We have the Poincaré duality

$$H^2(\mathbb{P}^1;\mathbb{Z}) \xrightarrow{\sim} H_0(\mathbb{P}^1;\mathbb{Z}) \simeq \mathbb{Z}.$$

given by the evaluation on the fundamental cycle. In our case, from the above computation,

$$[\mathbb{P}^1] \frown \delta^*(h) = 1.$$

In fact we have the following:

Proposition 11.5. *For a continuous line bundle L on a C^∞ manifold M,*

$$\delta^*(L) = c^1(L).$$

Proof. Let $\{(U_\alpha, \psi_\alpha)\}$ be a system of trivializations of L and $\{h^{\alpha\beta}\}$ the associated system of transition functions. Let e_α be the frame of L on U_α given by $e_\alpha(x) = \psi_\alpha^{-1}(1, x)$. On $U_{\alpha\beta}$ we have $e_\beta = h^{\alpha\beta} e_\alpha$.

Suppose we have a frame s on the 1-skeleton K^1 of K. The first Chern class $c^1(L)$ is represented by a cocycle γ which assigns to each 2-simplex $\boldsymbol{s} = (\alpha, \beta, \gamma)$ the index $I(s, b_{\boldsymbol{s}})$ of s at the barycenter $b_{\boldsymbol{s}}$ (cf. Sections 5.1 and 5.2). Let U_0 be an open set containing $\boldsymbol{s}$ such that L is trivial on U_0. Let e_0 be a frame of L on U_0 and write $s = f e_0$. Then (cf. Example 5.1):

$$I(s, b_{\boldsymbol{s}}) = \frac{1}{2\pi\sqrt{-1}} \int_{\partial \boldsymbol{s}} d \log f.$$

To compute the above, we fix a branch $(\log f)_\alpha$ of $\log f$ on $\partial\boldsymbol{s} \cap U_\alpha$. As a branch $(\log f)_\beta$ of $\log f$ on $\partial\boldsymbol{s} \cap U_\beta$, we take the analytic continuation of $(\log f)_\alpha$ and as $(\log f)_\gamma$, we take the analytic continuation of $(\log f)_\beta$. Then, letting p denote the barycenter of the 1-simplex (γ, α), we have

$$I(s, b_{\boldsymbol{s}}) = \frac{1}{2\pi\sqrt{-1}}((\log f)_\gamma(p) - (\log f)_\alpha(p)).$$

For $\mu = \alpha, \beta, \gamma$, we define $h^{0\mu}$ and f_μ by $e_\mu = h^{0\mu} e_0$ and $s = f_\mu e_\mu$ so that $f = h^{0\mu} f_\mu$ on $U_{0\mu} = U_0 \cap U_\mu$. We choose branches of $\log h^{0\mu}$ and $\log f_\mu$ so that

$$(\log f)_\mu = \log h^{0\mu} + \log f_\mu \qquad \text{on} \quad U_{0\mu}.$$

Finally, for $\mu, \nu = \alpha, \beta, \gamma$, we choose the branch of $\log h^{\mu\nu}$ so that

$$\log f_\mu = \log h^{\mu\nu} + \log f_\nu.$$

Then we see that

$$c_{\alpha\beta\gamma} = I(s, b_{\boldsymbol{s}}). \qquad \square$$

For C^∞ line bundles, we also have:

Proof. [Proof of Proposition 11.5 via Chern-Weil theory] Let $\{(U_\alpha, \psi_\alpha)\}$ be a system of trivializations of L with $\{U_\alpha\}$ a good covering and $\{h^{\alpha\beta}\}$ the associated system of transition functions. Take a connection ∇ for L. Let e_α be the frame of L on U_α given by $e_\alpha(x) = \psi_\alpha^{-1}(1, x)$ and θ_α the connection form of ∇ with respect to e_α. On $U_{\alpha\beta}$ we have $e_\beta = h^{\alpha\beta} e_\alpha$. Hence by (8.4),

$$\theta_\beta = d \log h^{\alpha\beta} + \theta_\alpha$$

The curvature form is given by

$$\kappa = d\theta_\alpha.$$

Recall that $c^1(L)$ is represented by the form $\frac{\sqrt{-1}}{2\pi}\kappa$. In (7.10), we may set $\chi_\alpha^{(0)} = \frac{\sqrt{-1}}{2\pi}\theta_\alpha$ and $\chi_{\alpha\beta}^{(1)} = \frac{\sqrt{-1}}{2\pi} \log h^{\alpha\beta}$. Thus it correspond to the cocycle c in $C^2(\mathcal{U}; \mathbb{C})$ given by $c = -\delta\chi^{(1)}$.

On the other hand, the image of the class of $h = \{h^{\alpha\beta}\}$ by δ^* is represented by the cochain $c' = \delta g$, $g^{\alpha\beta} = \frac{1}{2\pi\sqrt{-1}} \log h^{\alpha\beta}$. Thus $c = c'$. $\square$

In view of Exercise 11.7 (see also the paragraph right after), we have:

Corollary 11.2. *On a C^∞ manifold, the set of isomorphism classes of continuous line bundles and the set of isomorphism classes of C^∞ line bundles are the same. Moreover they are classified by their Chern classes.*

For a complex manifold M, we have a similar exact sequence

$$0 \longrightarrow \mathbb{Z} \xrightarrow{\iota} \mathscr{O} \xrightarrow{\exp} \mathscr{O}^* \longrightarrow 0,$$

which yields the exact sequence

$$\cdots \to H^1(M;\mathbb{Z}) \xrightarrow{\iota} H^1(M;\mathscr{O}) \to H^1(M;\mathscr{O}^*) \xrightarrow{\delta^*} H^2(M;\mathbb{Z}) \to \cdots. \quad (11.6)$$

Also, for a holomorphic line bundle L on M, we have

$$\delta^*(L) = c^1(L).$$

Note that $\operatorname{Ker}\delta^* \simeq H^1(M;\mathscr{O})/\iota H^1(M;\mathbb{Z})$ is the group of holomorphic line bundles that are trivial as continuous or C^∞ bundles.

Complexes of sheaves

A *complex of sheaves* is a sequence of sheaf morphisms

$$0 \longrightarrow \mathscr{K}^0 \xrightarrow{d^0} \mathscr{K}^1 \xrightarrow{d^1} \cdots \xrightarrow{d^{q-1}} \mathscr{K}^q \xrightarrow{d^q} \cdots$$

with $d^{q+1} \circ d^q = 0$ for all $q \geq 0$. It is also denoted by $(\mathscr{K}^\bullet, d)$. We say that $(\mathscr{K}^\bullet, d)$ is a *resolution* of a sheaf $\mathscr{S}$ if there is a morphism $\iota : \mathscr{S} \to \mathscr{K}^0$ such that

$$0 \longrightarrow \mathscr{S} \xrightarrow{\iota} \mathscr{K}^0 \xrightarrow{d^0} \mathscr{K}^1 \xrightarrow{d^1} \cdots \xrightarrow{d^{q-1}} \mathscr{K}^q \xrightarrow{d^q} \cdots$$

is exact. This is equivalent to saying that $\mathscr{K}^\bullet$ is acyclic and $\operatorname{Ker} d_0$ is isomorphic with $\mathscr{S}$. We abbreviate it by saying that $0 \to \mathscr{S} \to \mathscr{K}^\bullet$ is a resolution.

de Rham complex: Let M be a C^∞ manifold of dimension m. Then we have the complex

$$0 \longrightarrow \mathscr{A}^0 \xrightarrow{d^0} \mathscr{A}^1 \xrightarrow{d^1} \cdots \xrightarrow{d^{m-1}} \mathscr{A}^m \longrightarrow 0,$$

also called the *de Rham complex* on M (cf. Section 7.1). By the Poincaré lemma (Lemma 7.1), it is acyclic and gives a resolution $0 \to \mathbb{C} \to \mathscr{A}^\bullet$ of the constant sheaf $\mathbb{C}$.

Dolbeault complex: Let M be a complex manifold of dimension n. Then, for each p, we have the complex

$$0 \longrightarrow \mathscr{A}^{p,0} \xrightarrow{\bar\partial^{p,0}} \mathscr{A}^{p,1} \xrightarrow{\bar\partial^{p,1}} \cdots \xrightarrow{\bar\partial^{p,n-1}} \mathscr{A}^{p,n} \longrightarrow 0, \quad (11.7)$$

also called the p-th *Dolbeault complex* on M (cf Section 9.1). By the Dolbeault-Grothendieck lemma (Lemma 11.4 below), it is acyclic and gives a resolution $0 \to \Omega^p \to \mathscr{A}^{p,\bullet}$ of the sheaf Ω^p of holomorphic p-forms.

Hypercohomology

Let $(\mathscr{K}^\bullet, d)$ be a complex of sheaves on a topological space X and let $\mathcal{U} = \{U_\alpha\}_{\alpha \in I}$ be an open covering of X. Then we have a double complex $(C^\bullet(\mathcal{U}; \mathscr{K}^\bullet), \delta, (-1)^\bullet d)$. We consider the associated single complex $(K^\bullet(\mathcal{U}), D)$ as in Section 7.3. Thus

$$K^r(\mathcal{U}) = \bigoplus_{p+q=r} C^p(\mathcal{U}; \mathscr{K}^q), \qquad D = \delta + (-1)^p d.$$

Definition 11.8. The *hypercohomology of $\mathscr{K}^\bullet$ on $\mathcal{U}$* is the cohomology of $(K^\bullet(\mathcal{U}), D)$:

$$\boldsymbol{H}^r(\mathcal{U}; \mathscr{K}^\bullet) = \operatorname{Ker} D^r / \operatorname{Im} D^{r-1}.$$

Remark 11.4. Taking the inductive limit over the open coverings of X of the above cohomology, we obtain the hypercohomology $\boldsymbol{H}^r(X; \mathscr{K}^\bullet)$ of $\mathscr{K}^\bullet$ on X. Sometimes it is referred to as the *Čech hypercohomology.*

In the above situation, we have the complex $(\Gamma(X; \mathscr{K}^\bullet), d)$, whose cohomology is denoted by $H^r(\Gamma(X; \mathscr{K}^\bullet))$. Note that the inclusion $\Gamma(X; \mathscr{K}^r) \hookrightarrow C^0(\mathcal{U}; \mathscr{K}^r) \subset K^r(\mathcal{U})$ is compatible with the differentials induces a morphism

$$\varphi_{\mathcal{U}} : H^r(\Gamma(X; \mathscr{K}^\bullet)) \longrightarrow \boldsymbol{H}^r(\mathcal{U}; \mathscr{K}^\bullet).$$

Proposition 11.6. *If the complex $C^\bullet(\mathcal{U}; \mathscr{K}^p)$ is acyclic for every $p \geq 0$, then $\varphi_{\mathcal{U}}$ is an isomorphism.*

Proof. We consider the first spectral sequence $({}'E_k^{p,q}, H^r)$ associated with the double complex $C^\bullet(\mathcal{U}; \mathscr{K}^\bullet)$ (cf. Theorem A.13), where

$${}'E_2^{p,q} = H_d^p H_\delta^q(C^\bullet(\mathcal{U}; \mathscr{K}^\bullet)) \quad \text{and} \quad H^r = \boldsymbol{H}^r(\mathcal{U}; \mathscr{K}^\bullet).$$

By assumption, $H_\delta^q(C^\bullet(\mathcal{U}; \mathscr{K}^\bullet)) = 0$ for $q \geq 1$, while $H_\delta^0(C^\bullet(\mathcal{U}; \mathscr{K}^\bullet)) = \Gamma(X; \mathscr{K}^\bullet)$. The proposition follows then from Proposition A.13. □

Let $\mathscr{K}^\bullet$ and $\mathcal{U}$ be as before.

Definition 11.9. We say that $\mathcal{U}$ is *good for $\mathscr{K}^\bullet$*, if the complex $\Gamma(U_{\alpha_0 \dots \alpha_p}; \mathscr{K}^\bullet)$ is acyclic for every $(\alpha_0, \dots, \alpha_p) \in I^{p+1}$, $p \geq 0$.

If we set $\mathscr{S} = \operatorname{Ker} d^0$, the inclusion $C^r(\mathcal{U}; \mathscr{S}) \hookrightarrow C^r(\mathcal{U}; \mathscr{K}^0) \subset K^r(\mathcal{U})$ is compatible with the differentials and induces a morphism

$$\psi_{\mathcal{U}} : H^r(\mathcal{U}; \mathscr{S}) \longrightarrow \boldsymbol{H}^r(\mathcal{U}; \mathscr{K}^\bullet).$$

Proposition 11.7. *If $\mathcal{U}$ is good for $\mathscr{K}^\bullet$, then $\psi_{\mathcal{U}}$ is an isomorphism.*

Proof. We consider the second spectral sequence $({}''E_k^{p,q}, H^r)$ associated with the double complex $C^\bullet(\mathcal{U};\mathscr{S}^\bullet)$, where

$$''E_2^{p,q} = H_\delta^p H_d^q(C^\bullet(\mathcal{U};\mathscr{S}^\bullet)) \quad \text{and} \quad H^r = \boldsymbol{H}^r(\mathcal{U};\mathscr{S}^\bullet).$$

By assumption, $H_d^q(C^\bullet(\mathcal{U};\mathscr{S}^\bullet)) = 0$ for $q \geq 1$, while $H_d^0(C^\bullet(\mathcal{U};\mathscr{S}^\bullet)) = C^\bullet(\mathcal{U};\mathscr{S})$. □

From Propositions 11.6 and 11.7, we have:

Theorem 11.2. *Let $(\mathscr{K}^\bullet, d)$ be a complex of sheaves on X, $\mathscr{S}$ the kernel of $d^0 : \mathscr{K}^0 \to \mathscr{K}^1$ and $\mathcal{U}$ a covering of X. If $C^\bullet(\mathcal{U};\mathscr{K}^p)$ is acyclic for $p \geq 0$ and if $\mathcal{U}$ is good for $\mathscr{K}^\bullet$, there exist canonical isomorphisms*

$$H^r(\Gamma(X;\mathscr{K}^\bullet)) \xrightarrow{\sim} \boldsymbol{H}^r(\mathcal{U};\mathscr{K}^\bullet) \xleftarrow{\sim} H^r(\mathcal{U};\mathscr{S}).$$

de Rham case: If $X = M$ is a C^∞ manifold and $\mathscr{K}^\bullet = \mathscr{A}^\bullet$, the de Rham complex, $\boldsymbol{H}^r(\mathcal{U};\mathscr{A}^\bullet)$ is the Čech-de Rham cohomology $H_D^r(\mathcal{U})$ introduced in Section 7.3. The cohomology $H^r(\Gamma(M;\mathscr{A}^\bullet))$ is the de Rham cohomology $H_d^r(M)$. The assumption of Proposition 11.6 is satisfied if $\mathcal{U}$ is locally finite (cf. Proposition 11.4). Thus Proposition 11.6 gives an alternative proof of Theorem 7.1. Also $\mathcal{U}$ is good for $\mathscr{A}^\bullet$ if it is good in the sense of Definition 7.6 and $0 \to \mathbb{C} \to \mathscr{A}^\bullet$ is a resolution. Thus Proposition 11.7 gives an alternative proof of Theorem 7.2. Finally Theorem 11.2 reduces to Corollary 7.4; for a locally finite good covering $\mathcal{U}$, there exist canonical isomorphisms

$$H_d^r(M) \xrightarrow{\sim} H_D^r(\mathcal{U}) \xleftarrow{\sim} H^r(\mathcal{U};\mathbb{C}). \tag{11.8}$$

In particular, snce the locally finite good coverings are cofinal in the set of coverings of M, we have:

$$\check{H}^r(M;\mathbb{C}) \simeq H^r(\mathcal{U};\mathbb{C}), \quad \text{for every locally finite good covering } \mathcal{U}. \tag{11.9}$$

Dolbeault case: If $X = M$ is a complex manifold and $\mathscr{K}^\bullet = \mathscr{A}^{p,\bullet}$, the Dolbeault complex, $\boldsymbol{H}^q(\mathcal{U};\mathscr{A}^{p,\bullet})$ is the Čech-Dolbeault cohomology $H_{\bar\vartheta}^{p,q}(\mathcal{U})$ we discuss in Section 11.4 below.

11.3 Coherent sheaves

Let X be a topological space and $\mathscr{R}$ a sheaf of rings on X. Also let $\mathscr{S}$ be an $\mathscr{R}$-module. For sections $s_1, \ldots, s_p$ of $\mathscr{S}$ on an open set U in X, we may define a morphism

$$\varphi : \mathscr{R}^p|_U \longrightarrow \mathscr{S}|_U \tag{11.10}$$

by assigning to $(a_1, \ldots, a_p) \in \mathscr{R}_x^p$ the element $\sum_{i=1}^{p} a_i s_i(x)$ in $\mathscr{S}_x$, $x \in U$. Conversely a morphism as above defines sections $s_1, \ldots, s_p$ of $\mathscr{S}$ on U by $s_i(x) = \varphi_x(0, \ldots, \overset{i}{1}, \ldots, 0)$.

In the following, by convention, we set $\mathscr{R}^0 = 0$.

Sheaves of finite type

Definition 11.10. An $\mathscr{R}$-module $\mathscr{S}$ is *of finite type*, *$\mathscr{R}$-finite* for short, if every point x in X admits a neighborhood U such that $\mathscr{S}|_U$ is generated by a finite number of sections, i.e., there exists a morphism as in (11.10) that is surjective.

From definition we have (cf. Exercise 11.1. 2):

Proposition 11.8. *If $\mathscr{S}$ is $\mathscr{R}$-finite, its support* Supp($\mathscr{S}$) *is a closed set in X*

Exercise 11.8. Suppose $\mathscr{S}$ is of finite type. Let $s_1, \ldots, s_p$ be sections of $\mathscr{S}$ on a neighborhood of x in X. Show that, if $s_1(x), \ldots, s_p(x)$ generate the $\mathscr{R}_x$-module $\mathscr{S}_x$, then there is a neighborhood U of x such that $s_1(y), \ldots, s_p(y)$ generate $\mathscr{S}_y$ for all y in U.

Locally free sheaves

Definition 11.11. An $\mathscr{R}$-module $\mathscr{S}$ is said to be *locally free* if every point in X admits a neighborhood U such that $\mathscr{S}|_U$ is isomorphic with $\mathscr{R}^p|_U$ for some non-negative integer p.

The integer p above is determined uniquely on each connected component of X and is called the *rank* of $\mathscr{S}$ on the component.

Example 11.4. Let M be a complex manifold of dimension n. It is not difficult to see that an $\mathscr{O}_M$-module $\mathscr{S}$ is locally free if and only if $\mathscr{S} = \mathscr{O}_M(E)$ for some holomorphic vector bundle E on M. In this case the rank of $\mathscr{S}$ is the rank of E.

For example, the tangent sheaf Θ_M is locally free of rank n; if U is a coordinate neighborhood with coordinates $(z_1, \ldots, z_n)$, then $\Theta_M|_U$ is isomorphic with $\mathscr{O}_M^n|_U$, a germ $v = \sum_{i=1}^{n} f_i \cdot \partial/\partial z_i$ corresponding to $(f_1, \ldots, f_n)$. Also the sheaf Ω_M^p of holomorphic p-forms is locally free of rank $\binom{n}{p}$.

Let M be a complex manifold of dimension n and $\mathscr{O} = \mathscr{O}_M$ the sheaf of holomorphic functions on M. Recall that, for each point z in M, $\mathscr{O}_z$ is canonically isomorphic with the ring $\mathscr{O}_n$ of germs of holomorphic functions of n-variables and it admits a unique maximal ideal $\mathfrak{m}_z$, i.e., the ideal of germs of functions whose value at z is zero. There is a canonical isomorphism $\mathscr{O}_z/\mathfrak{m}_z \simeq \mathbb{C}$ (cf. Section 1.4). For an $\mathscr{O}$-module $\mathscr{S}$, there is a canonical isomorphism $\mathscr{S}_z/\mathfrak{m}_z\mathscr{S}_z \simeq \mathscr{O}_z/\mathfrak{m}_z \otimes_{\mathscr{O}_z} \mathscr{S}_z$ and it may be thought of as a $\mathbb{C}$-vector space. If $\mathscr{S}$ is of finite type, then $\mathscr{S}_z/\mathfrak{m}_z\mathscr{S}_z$ is finite dimensional.

Proposition 11.9. *Let $\mathscr{S}$ be an $\mathscr{O}$-module of finite type and set $r(z) = \dim \mathscr{S}_z/\mathfrak{m}_z\mathscr{S}_z$. If $r(z)$ is locally constant as a function of $z \in M$, then $\mathscr{S}$ is locally free.*

Proof. Let z_0 be an arbitrary point in M and set $r = r(z_0)$. By Proposition A.20 (see also Proposition 11.1.1 and Exercise 11.8), there exist a neighborhood U of z_0 and an epimorphism

$$\mathscr{O}^r|_U \xrightarrow{\varphi} \mathscr{S}|_U \longrightarrow 0.$$

By assumption, taking a smaller U if necessary, we may assume that $r(z) = r$ for all $z \in U$. Note that, by Proposition A.20, r is the minimum number of generators of $\mathscr{S}_z$, $z \in U$. We claim that φ is injective, which will prove the proposition.

Define sections $s_1, \ldots, s_r$ of $\mathscr{S}$ on U by $s_i(z) = \varphi_z(0, \ldots, \overset{i}{1}, \ldots, 0)$. Then $s_1(z), \ldots, s_r(z)$ generate $\mathscr{S}_z$ for all $z \in U$. Take $(f_1, \ldots, f_r) \in \operatorname{Ker} \varphi_z$, $z \in U$. Suppose the f_i's have representatives on a neighborhood U_z of z in U. From $0 = \varphi_z(f_1, \ldots, f_r) = \sum_{i=1}^r f_i \cdot s_i(z) = 0$, we have that $\sum_{i=1}^r f_{i,z'} \cdot s_i(z') = 0$, for all $z' \in U_z$. Since the minimum number of generators of $\mathscr{S}_{z'}$ is r, we see that the $f_{i,z'}$'s are non-unit for all $z' \in U_z$, which shows that $f_i = 0$, $i = 1, \ldots, r$, as germs at z. □

Coherent sheaves

For a morphism φ as in (11.10), its image is called the *sheaf generated* by the s_i's and the kernel is called the *sheaf of relations* among the s_i's.

Definition 11.12. An $\mathscr{R}$-module $\mathscr{S}$ is *coherent* if it is of finite type and if the sheaf of relations among every finite set of its (local) sections is of finite type.

In this case, we say that $\mathscr{S}$ is $\mathscr{R}$-coherent. If $\mathscr{R}$ is $\mathscr{R}$-coherent, we say that it is a coherent sheaf of rings.

Proposition 11.10. *If $\mathscr{S}$ is locally free, then for every x in X, $\mathscr{S}_x$ is $\mathscr{R}_x$-free. The converse is true if $\mathscr{S}$ is coherent.*

Exercise 11.9. Verify Proposition 11.10.

We quote some of the fundamental properties of coherent sheaves:

Proposition 11.11. *Let $\mathscr{R}$ be a sheaf of rings on a topological space X.*

1. *Let $\mathscr{S}$ be a coherent $\mathscr{R}$-module. Then a sub-$\mathscr{R}$-module $\mathscr{S}'$ of $\mathscr{S}$ is coherent if and only if it is finite.*

2. *Suppose we have an exact sequence of $\mathscr{R}$-modules*

$$0 \longrightarrow \mathscr{S}' \longrightarrow \mathscr{S} \longrightarrow \mathscr{S}'' \longrightarrow 0.$$

If any two of the above are coherent, so is the rest.

3. *If $\varphi : \mathscr{S} \to \mathscr{T}$ is a morphism of $\mathscr{R}$-modules, then* $\operatorname{Ker}\varphi$ *and* $\operatorname{Coker}\varphi$ *are coherent.*

4. *If $\mathscr{S}$ and $\mathscr{T}$ are two coherent $\mathscr{R}$-modules, then $\mathscr{S} \oplus \mathscr{T}$, $\mathscr{S} \otimes_{\mathscr{R}} \mathscr{T}$ and $\mathscr{H}om_{\mathscr{R}}(\mathscr{S}, \mathscr{T})$ are coherent.*

As a consequence we have:

Proposition 11.12. *Let $\mathscr{S}$ be an $\mathscr{R}$-module.*

1. *If $\mathscr{S}$ is $\mathscr{R}$-coherent, every point in X admits a neighborhood U such that there exists an exact sequence of the form*

$$\mathscr{R}^q|_U \longrightarrow \mathscr{R}^p|_U \longrightarrow \mathscr{S}|_U \longrightarrow 0 \tag{11.11}$$

with p and q positive integers.

2. *Suppose $\mathscr{R}$ itself is coherent as an $\mathscr{R}$-module. If every point in X admits a neighborhood U and an exact sequence as* (11.11)*, then $\mathscr{S}$ is coherent.*

Proof. The first statement is immediate from the definition. The second follows from Proposition 11.11. 3 and 4. □

We prepare a lemma for later use. Thus let $\mathscr{R}$ be a coherent sheaf of rings on X and $\mathscr{I}$ an ideal sheaf of finite type in $\mathscr{R}$. Then $\mathscr{R}/\mathscr{I}$ is $\mathscr{R}$-coherent and its support S is a closed set in X (cf. Propositions 11.11 and 11.8). Denoting by $i : S \hookrightarrow X$ the inclusion, we set $\mathscr{R}_S = i^{-1}(\mathscr{R}/\mathscr{I})$. If $\mathscr{S}$ is an $\mathscr{R}_S$-module, then $i_*\mathscr{S}$ is an $i_*\mathscr{R}_S$-module. Since $i_*\mathscr{R}_S = \mathscr{R}/\mathscr{I}$

(cf. Exercise 11.5), we may think of $i_*\mathscr{S}$ as an $\mathscr{R}$-module via the canonical epimorphism $\mathscr{R} \to \mathscr{R}/\mathscr{I} = i_*\mathscr{R}_S$.

Lemma 11.1. *In the above situation, it holds:*

1. *$\mathscr{S}$ is $\mathscr{R}_S$-finite if and only if $i_*\mathscr{S}$ is $\mathscr{R}$-finite.*

2. *$\mathscr{S}$ is $\mathscr{R}_S$-coherent if and only if $i_*\mathscr{S}$ is $\mathscr{R}$-coherent.*

3. *$\mathscr{R}_S$ is a coherent sheaf of rings.*

Proof. **1.** Suppose $\mathscr{S}$ is $\mathscr{R}_S$-finite. Then each point in S admits a neighborhood W in S such that there exists an epimorphism $\varphi : (\mathscr{R}_S)^p|_W \to \mathscr{S}|_W$. Note that $W = U \cap S$ for some open set U in X. We have an epimorphism $\mathscr{R}^p|_U \to (i_*\mathscr{S})|_U$ as the composition of $\mathscr{R}^p|_U \to (\mathscr{R}/\mathscr{I})^p|_U = (i_*\mathscr{R}_S)^p|_U$ and $i_*\varphi : (i_*\mathscr{R}_S)^p|_U \to (i_*\mathscr{S})|_U$ (cf. Remark 11.1).

Suppose $i_*\mathscr{S}$ is $\mathscr{R}$-finite. Then each point in X admits a neighborhood U such that there exists an epimorphism of the form $\mathscr{R}^p|_U \to (i_*\mathscr{S})|_U$. The fact that $i_*\mathscr{S}$ is an $i_*\mathscr{R}_S = \mathscr{R}/\mathscr{I}$-module implies that the morphism induces an epimorphism $(\mathscr{R}_S)^p|_W \to \mathscr{S}|_W$, $W = U \cap S$.

2. Suppose $\mathscr{S}$ is $\mathscr{R}_S$-coherent. Then each point in S admits a neighborhood W in S such that there exists an exact sequence of the form

$$(\mathscr{R}_S)^q|_W \longrightarrow (\mathscr{R}_S)^p|_W \longrightarrow \mathscr{S}|_W \longrightarrow 0.$$

Noting that $W = U \cap S$ for some open set U in X, we have the exact sequence (cf. Remark 11.1)

$$(i_*\mathscr{R}_S)^q|_U \longrightarrow (i_*\mathscr{R}_S)^p|_U \longrightarrow (i_*\mathscr{S})|_U \longrightarrow 0.$$

Noting that $i_*\mathscr{R}_S = \mathscr{R}/\mathscr{I}$, which is $\mathscr{R}$-coherent, we see that $i_*\mathscr{S}$ is $\mathscr{R}$-coherent, as it is locally the cokernel of a morphism of $\mathscr{R}$-coherent modules.

Suppose $i_*\mathscr{S}$ is $\mathscr{R}$-coherent. Then $i_*\mathscr{S}$ is $\mathscr{R}$-finite and by the above, $\mathscr{S}$ is $\mathscr{R}_S$-finite. We show that every sheaf of relations among sections of $\mathscr{S}$ is $\mathscr{R}_S$-finite. Suppose an $\mathscr{R}_S$-morphism $\varphi : (\mathscr{R}_S)^p|_W \to \mathscr{S}|_W$ is given on an open set W in S so that we have the exact sequence

$$0 \longrightarrow \operatorname{Ker}\varphi \longrightarrow (\mathscr{R}_S)^p|_W \xrightarrow{\varphi} \mathscr{S}|_W.$$

We may write $W = U \cap S$ for some open set U in X and we have the exact sequence (cf. Remark 11.1)

$$0 \longrightarrow i_* \operatorname{Ker}\varphi \longrightarrow (i_*\mathscr{R}_S)^p|_U \xrightarrow{i_*\varphi} (i_*\mathscr{S})|_U,$$

which shows that $i_* \operatorname{Ker}\varphi = \operatorname{Ker} i_*\varphi$. Since $(i_*\mathscr{R}_S)^p = (\mathscr{R}/\mathscr{I})^p$ is $\mathscr{R}$-coherent and $i_*\mathscr{S}$ is $\mathscr{R}$-coherent by assumption, $\operatorname{Ker} i_*\varphi$ is $\mathscr{R}$-coherent (cf. Proposition 11.11. 3, 4), in particular it is $\mathscr{R}$-finite. Therefore $\operatorname{Ker}\varphi$ is $\mathscr{R}_S$-finite.

The statement 3 is a direct consequence of 2. □

Fundamental theorems of coherence

In the following we denote by M a complex manifold and $\mathscr{O}_M$ the sheaf of germs of holomorphic functions on M.

We start by quoting the two fundamental theorems of coherence due to K. Oka.

Theorem 11.3. *The sheaf $\mathscr{O}_M$ is a coherent sheaf of rings.*

Let U be an open set in M. We may identify the ring O_U of holomorphic functions on U with the ring $\Gamma(U; \mathscr{O}_M)$ of sections of $\mathscr{O}_M$ on U.

For a variety V in M, we denote by $\mathscr{I}_V$ the sheaf defined by the presheaf

$$U \mapsto \{f \in O_U \mid f(z) = 0 \text{ for all } z \in V \cap U\}.$$

It is a sub-$\mathscr{O}_M$-module of $\mathscr{O}_M$ (an ideal sheaf in $\mathscr{O}_M$) and is called the ideal sheaf of holomorphic functions vanishing on V, or simply the *ideal sheaf of* V. For each point z in M, we have $\mathscr{I}_{V,z} = I(V_z)$, where V_z is the germ of V at z. Note that, if $z \notin V$, then $V_z = \emptyset$ and $\mathscr{I}_{V,z} = \mathscr{O}_{M,z}$.

Theorem 11.4. *The ideal sheaf $\mathscr{I}_V$ is $\mathscr{O}_M$-coherent.*

The essential part of the above is that $\mathscr{I}_V$ is $\mathscr{O}_M$-finite (cf. Proposition 11.11.1).

We discuss some of the applications of the above theorems.

Reduced germs of holomorphic functions: Let $\mathscr{O}_n$ be the ring of germs of holomorphic functions at 0 in $\mathbb{C}^n$.

Definition 11.13. A germ f in $\mathscr{O}_n$ is *reduced* if $\sqrt{(f)} = (f)$.

By Corollary 2.1, it is equivalent to saying that $I(V(f)) = (f)$. For a germ f to be reduced there are three cases; (1) $f = 0$, (2) f is a unit, (3) the irreducible decomposition of f has no multiple factors, i.e., in the irreducible decomposition $f = p_1^{m_1} \cdots p_r^{m_r}$, we have $m_i = 1$ for all i.

We represent f by a holomorphic function, denoted also by f, in a neighborhood U of 0 and denote by f_z the germ of f at $z \in U$.

Proposition 11.13. *If f is reduced at* 0, *then the germ f_z in $\mathscr{O}_{U,z}$ is reduced for all z sufficiently close to* 0.

Proof. Let $V = V(f)$ be the variety in U defined by f. Then $\mathscr{I}_{V,z} = I(V_z) = I(V(f_z))$ for all $z \in U$. Since f generates $\mathscr{I}_{V,0}$, by assumption, f_z generates $\mathscr{I}_{V,z}$, i.e., $I(V(f_z)) = (f_z)$, for all z sufficiently close to 0 (cf. Exercise 11.8). □

Note that an irreducible germ p in $\mathscr{O}_n$ is reduced. Thus p_z is reduced for z close to 0. However, p_z may not be irreducible even for z in $V(p)$ and close to 0 (cf. Remark 2.6).

Singular set of a variety: The singular set of a variety is expressed explicitly as follows:

Theorem 11.5. *Let V be a variety in M and z_0 a point in* $\mathrm{Sing}(V)$. *Suppose that V_{z_0} is of pure codimension k. Then for every set of generators $\{f_1, \dots, f_r\}$ of $\mathscr{I}_{V,z_0}$, there exists a neighborhood U of z_0 in M such that:*

(1) *$f_1, \dots, f_r$ have representatives on U,*
(2) $V \cap U = \{\, z \in U \mid f_1(z) = \cdots = f_r(z) = 0 \,\}$,
(3) $\mathrm{Sing}(V) \cap U$
$= \{\, z \in V \cap U \mid df_{i_1} \wedge \cdots \wedge df_{i_k}(z) = 0, \quad 1 \le i_1 < \cdots < i_k \le r \,\}$.

Proof. The statements (1) and (2) follow from the coherence of $\mathscr{I}_V$ (cf. Exercise 11.8). Let z be a regular point of $V \cap U$. Then there exist defining functions $h_1, \dots, h_k$ for V near z such that

$$dh_1 \wedge \cdots \wedge dh_k(z) \neq 0. \tag{11.12}$$

Since the germs of $f_1, \dots, f_r$ at z generate $\mathscr{I}_{V,z}$, we may write $h_{j,z} = \sum_{i=1}^{r} a_{ij} f_{i,z}$, $a_{ij} \in \mathscr{O}_{M,z}$, $j = 1, \dots, k$. Then from (11.12), we see that

$$df_{i_1} \wedge \cdots \wedge df_{i_k}(z) \neq 0 \qquad \text{for some } \ 1 \le i_1 < \cdots < i_k \le r. \tag{11.13}$$

Conversely, let z be a point in $V \cap U$ with the property (11.13). Then, in a neighborhood of z, $V(f_{i_1}, \dots, f_{i_k})$ is a submanifold of codimension k that contains V. We have $\mathscr{O}_{M,z}/(f_{i_1}, \dots, f_{i_k})_z \simeq \mathscr{O}_d$, $d = n - k$. By Theorem 2.10, the class of $f_{i,z}$ in $\mathscr{O}_{M,z}/(f_{i_1}, \dots, f_{i_k})_z$ is 0, i.e., $f_{i,z}$ is in $(f_{i_1}, \dots, f_{i_k})_z$, for each $i = 1, \dots, r$. Thus $V_z = V(f_{i_1}, \dots, f_{i_k})_z$, which shows that z is a regular point of V. □

From the above we see that $\mathrm{Sing}(V)$ is a variety and by Theorem 2.7 (2), $\dim \mathrm{Sing}(V) \le \dim V - 1$.

Singular set of a coherent sheaf: We introduce the following:

Definition 11.14. For a coherent $\mathscr{O}_M$-module $\mathscr{S}$, we set

$$\mathrm{Sing}(\mathscr{S}) = \{\, z \in M \mid \mathscr{S}_z \text{ is not } \mathscr{O}_{M,z}\text{-free} \,\}$$

and call it the *singular set* of $\mathscr{S}$.

Let $\mathscr{S}$ be a coherent $\mathscr{O}_M$-module. Then from definition, $\mathscr{S}$ is locally free on $M \setminus \mathrm{Sing}(\mathscr{S})$. Its rank is called the *rank* of $\mathscr{S}$.

We show that $\mathrm{Sing}(\mathscr{S})$ is a variety in M. Since $\mathscr{O}$ itself is coherent, each point in M admits a neighborhood U such that there exists an exact sequence of the form

$$\mathscr{O}^q|_U \xrightarrow{\varphi} \mathscr{O}^p|_U \longrightarrow \mathscr{S}|_U \longrightarrow 0.$$

For $z \in U$, taking the tensor products with $\mathscr{O}_z/\mathfrak{m}_z \simeq \mathbb{C}$ over $\mathscr{O}_z$, we have the exact sequence

$$\mathbb{C}^q \xrightarrow{\varphi(z)} \mathbb{C}^p \longrightarrow \mathscr{S}_z/\mathfrak{m}_z\mathscr{S}_z \longrightarrow 0,$$

where $\varphi(z)$ is the $\mathbb{C}$-linear map obtained by evaluating φ at z. If we set $r(z) = \dim \mathscr{S}_z/\mathfrak{m}_z\mathscr{S}_z$, we have $\operatorname{rank}\varphi(z) = p - r(z)$.

Now we assume that U is connected and set $r = \min_{z\in U} r(z)$.

Proposition 11.14. *In the above situation,*

$$\mathrm{Sing}(\mathscr{S}) \cap U = \{\, z \in U \mid \operatorname{rank}\varphi(z) < p - r \,\}.$$

Thus $\mathrm{Sing}(\mathscr{S})$ *is an analytic variety in* M.

Proof. Let S denote the right-hand side. On $U \setminus S$, $r(z)$ is constantly equal to r and $U \setminus S \subset U \setminus \mathrm{Sing}(\mathscr{S})$ by Proposition 11.9. On the other hand, let z be a point in S. Since S is a variety in U, which is assumed to be connected, and $S \neq U$, every neighborhood of z contains points not in S (cf. Theorem 2.1). Thus $\mathscr{S}_z$ cannot be $\mathscr{O}_z$-free and $S \subset \mathrm{Sing}(\mathscr{S})$. □

Support of a coherent sheaf: As a related topic, we prove the following:

Proposition 11.15. *The support* $\mathrm{Supp}(\mathscr{S}) = \{\, z \in M \mid \mathscr{S}_z \neq 0\}$ *of a coherent* $\mathscr{O}_M$*-module* $\mathscr{S}$ *is a variety in* M.

Proof. Let U and φ be as above. Note that $\mathrm{Supp}(\mathscr{S}) \cap U$ is the set of points $z \in U$ where φ_z is not surjective. If φ_z is surjective, so is $\varphi(z)$. Conversely, if $\varphi(z)$ is surjective, so is φ_z by Theorem A.18. Thus we have $\mathrm{Supp}(\mathscr{S}) \cap U = \{\, z \in U \mid \operatorname{rank}\varphi(z) < p \}$. □

Holomorphically convex domains

A *domain* in $\mathbb{C}^n$ is a connected open set in $\mathbb{C}^n$.

Definition 11.15. A domain D in $\mathbb{C}^n$ is *holomorphically convex* if for every compact set K in D, the set

$$\{ z \in D \mid |f(z)| \le \|f\|_K \quad \text{for all } f \in O_D \}$$

is compact, where O_D denotes the ring of holomorphic functions on D.

Theorem 11.6. 1. *Let D and D' be domains in $\mathbb{C}^n$ and $\varphi : D \to D'$ a biholomorphic map. If D is holomorphically convex, so is D'.*

2. *The intersection of a finite number of holomorphically convex domains is holomorphically convex.*

Example 11.5. 1. Every domain in $\mathbb{C}$ is holomorphically convex.

2. Every convex domain is holomorphically convex. Although the image of a convex domain by a biholomorphic map may not be convex, it is holomorphically convex, which is one of the advantages of holomorphical convexity.

Theorem 11.7 (Oka-Cartan). *Let D be a holomorphically convex domain. For every coherent $\mathscr{O}_D$-module $\mathscr{S}$,*

$$H^q(D; \mathscr{S}) = 0 \quad \text{for } q \ge 1.$$

11.4 Dolbeault theorem

Let M be a complex manifold of dimension n. In Definition 9.5, we introduced the Dolbeault cohomology $H^{p,q}_{\bar\partial}(M)$ for each (p,q). We show that this is canonically isomorphic with $H^q(M; \Omega^p)$.

We start with the one variable case. Let U be a relatively compact open set in $\mathbb{C}$ such that $\overline{U}$ is a real two-dimensional submanifold with piecewise C^∞ boundary. For a C^∞ function φ in a neighborhood of $\overline{U}$ and a point z in U, we set

$$\int_{\overline{U}} \varphi(\zeta) \frac{d\zeta \wedge d\bar\zeta}{\zeta - z} = \lim_{\varepsilon \to 0} \int_{\overline{U} \setminus \Delta_\varepsilon} \varphi(\zeta) \frac{d\zeta \wedge d\bar\zeta}{\zeta - z},$$

where $\Delta_\varepsilon = \Delta(z, \varepsilon)$ is the disk with center z and radius ε.

Lemma 11.2. *The above limit exists for every point z in U.*

Proof. We take a disk Δ_ρ with center z and radius ρ so that its closure is contained in U. Then for $\varepsilon < \rho$,

$$\int_{\overline{U} \setminus \Delta_\varepsilon} \varphi(\zeta) \frac{d\zeta \wedge d\bar\zeta}{\zeta - z} = \int_{\overline{U} \setminus \Delta_\rho} \varphi(\zeta) \frac{d\zeta \wedge d\bar\zeta}{\zeta - z} + \int_{\overline{\Delta}_\rho \setminus \Delta_\varepsilon} \varphi(\zeta) \frac{d\zeta \wedge d\bar\zeta}{\zeta - z}.$$

By the parametrization $\zeta = z + re^{i\theta}$, $i = \sqrt{-1}$, we have

$$\frac{d\zeta \wedge d\bar{\zeta}}{\zeta - z} = -2ie^{-i\theta}\, dr \wedge d\theta.$$

Thus

$$\int_{\overline{\Delta}_\rho \setminus \Delta_\varepsilon} \varphi(\zeta)\, \frac{d\zeta \wedge d\bar{\zeta}}{\zeta - z} = -2i \int_0^{2\pi} \int_\varepsilon^\rho \varphi(z + re^{i\theta}) e^{-i\theta}\, dr\, d\theta,$$

which proves the lemma. □

Note that the above proof also shows that $\int_{\overline{U}} \varphi(\zeta)\, \frac{d\zeta \wedge d\bar{\zeta}}{\zeta - z}$ is C^∞ as a function of $z \in U$ (cf. Lemma 11.3 below).

In the following, let U be as above and γ its boundary (cf. Remark 1.2).

Theorem 11.8 (Generalized Cauchy integral formula). *Let f be a C^∞ function in a neighborhood of $\overline{U}$. Then we have, for $z \in U$,*

$$f(z) = \frac{1}{2\pi\sqrt{-1}} \Big(\int_\gamma \frac{f(\zeta)\, d\zeta}{\zeta - z} + \int_{\overline{U}} \frac{\partial f}{\partial \bar{\zeta}}(\zeta)\, \frac{d\zeta \wedge d\bar{\zeta}}{\zeta - z} \Big).$$

Proof. Let Δ_ε be a disk with center z and radius ε whose closure is in U and $\gamma_\varepsilon = \partial \Delta_\varepsilon$. From

$$\frac{\partial f}{\partial \bar{\zeta}}(\zeta)\, \frac{d\zeta \wedge d\bar{\zeta}}{\zeta - z} = -d\Big(\frac{f(\zeta) d\zeta}{\zeta - z}\Big)$$

and Stokes' theorem, we have

$$\int_{\overline{U} \setminus \Delta_\varepsilon} \frac{\partial f}{\partial \bar{\zeta}}(\zeta)\, \frac{d\zeta \wedge d\bar{\zeta}}{\zeta - z} = -\int_\gamma \frac{f(\zeta) d\zeta}{\zeta - z} + \int_{\gamma_\varepsilon} \frac{f(\zeta) d\zeta}{\zeta - z}.$$

If we make ε tend to 0, the left-hand side approaches the integral over $\overline{U}$, by Lemma 11.2. Taking the parametrization $\zeta = z + \varepsilon e^{\sqrt{-1}\,\theta}$ of γ_ε, we see that the second term in the right-hand side approaches $2\pi\sqrt{-1} f(z)$. □

Lemma 11.3. *Let U and f be as in Theorem 11.8. If we set*

$$g(z) = \frac{1}{2\pi\sqrt{-1}} \int_{\overline{U}} f(\zeta)\, \frac{d\zeta \wedge d\bar{\zeta}}{\zeta - z},$$

it is a C^∞ function on U and

$$\frac{\partial g}{\partial \bar{z}} = f.$$

Proof. Let z be a point in U and take two open sets V_1 and V_2 so that $z \in V_1 \subset V_2 \subset \overline{V}_2 \subset U$. Let ψ be a C^∞ function on $\mathbb{C}$ such that $\psi \equiv 1$ on $\overline{V}_1$ and $\psi \equiv 0$ on $\mathbb{C} \setminus V_2$ and set $f_1 = (1-\psi)f$ and $f_2 = \psi f$. Then we have a decomposition $f = f_1 + f_2$ and the corresponding decomposition $g = g_1 + g_2$. Since $f_1 \equiv 0$ on $\overline{V}_1$, we have

$$g_1(z) = \frac{1}{2\pi\sqrt{-1}} \int_{\overline{U} \setminus V_1} f_1(\zeta) \frac{d\zeta \wedge d\bar{\zeta}}{\zeta - z},$$

which is holomorphic in z so that $\frac{\partial g_1}{\partial \bar{z}} = 0$. On the other hand, since f_2 may be thought of as a C^∞ function on $\mathbb{C}$ with $f_2 \equiv 0$ on $\mathbb{C} \setminus V_2$, by the change of variable $\zeta' = \zeta - z$ we have

$$g_2(z) = \frac{1}{2\pi\sqrt{-1}} \int_{\mathbb{C}} f_2(\zeta) \frac{d\zeta \wedge d\bar{\zeta}}{\zeta - z} = \frac{1}{2\pi\sqrt{-1}} \int_{\mathbb{C}} f_2(\zeta' + z) \frac{d\zeta' \wedge d\bar{\zeta}'}{\zeta'},$$

which shows that $g_2(z)$ is C^∞ and

$$\frac{\partial g_2}{\partial \bar{z}} = \frac{1}{2\pi\sqrt{-1}} \int_{\mathbb{C}} \frac{\partial f_2}{\partial \bar{z}}(\zeta' + z) \frac{d\zeta' \wedge d\bar{\zeta}'}{\zeta'} = \frac{1}{2\pi\sqrt{-1}} \int_{\overline{U}} \frac{\partial f_2}{\partial \bar{\zeta}}(\zeta) \frac{d\zeta \wedge d\bar{\zeta}}{\zeta - z}.$$

By Theorem 11.8, this is equal to

$$f_2(z) - \frac{1}{2\pi\sqrt{-1}} \int_{\gamma} \frac{f_2(\zeta) d\zeta}{\zeta - z},$$

where the first term is equal to $f(z)$ and the second 0, as $f_2 \equiv f$ on $\overline{V}_1$ and $f_2 \equiv 0$ outside of V_2. □

Lemma 11.4 (Dolbeault-Grothendieck lemma). *Let U be an open set in $\mathbb{C}^n$ which is the product $U = U_1 \times \cdots \times U_n$ of open sets U_i, $i = 1, \ldots, n$, in $\mathbb{C}$ as considered above. If ω a (p,q)-form on a neighborhood of $\overline{U}$, $q \geq 1$, with $\bar{\partial}\omega = 0$, then there exists a $(p, q-1)$-form η on U such that $\omega = \bar{\partial}\eta$.*

Proof. First we note that it suffices to prove the case $p = 0$. Indeed, if we write

$$\omega = \sum_{|I|=p, |J|=q} f_{IJ} dz_I \wedge d\bar{z}_J,$$

for each I, the $(0,q)$-form $\omega_I = \sum_{|J|=q} f_{IJ} d\bar{z}_J$ is also $\bar{\partial}$-closed. Thus, if $\omega_I = \bar{\partial}\eta_I$, setting $\eta = (-1)^p \sum_I dz_I \wedge \eta_I$, we have $\omega = \bar{\partial}\eta$.

Letting k be the least integer such that the expression

$$\omega = \sum_{|J|=q} f_J d\bar{z}_J$$

does not involve $d\bar{z}_{k'}$, $k' > k$, we prove the lemma by induction on k. If $k = 0$, then $\omega = 0$ and the lemma holds by setting $\eta = 0$. Suppose it holds for $k = k_0$ and we prove it for $k = k_0 + 1$. We may write

$$\omega = \omega_1 + \omega_2 \wedge d\bar{z}_{k_0+1},$$

where ω_1 and ω_2 are forms not involving $d\bar{z}_{k'}$, $k' > k_0$. We may write

$$\omega_2 = \sum_{J \ni k_0+1} f_J d\bar{z}_{J \setminus \{k_0+1\}}.$$

Since $\bar{\partial}\omega_1$ does not have terms involving $d\bar{z}_{k_0+1} \wedge d\bar{z}_j$ for $j > k_0 + 1$ and $0 = \bar{\partial}\omega = \bar{\partial}\omega_1 + \bar{\partial}\omega_2 \wedge d\bar{z}_{k_0+1}$, we see that

$$\frac{\partial f_J}{\partial \bar{z}_j} = 0$$

for J containing $k_0 + 1$ and $j > k_0 + 1$. Thus if we set

$$g_J(z) = \frac{1}{2\pi\sqrt{-1}} \int_{\bar{U}_{k_0+1}} f_J(z_1, \ldots, \zeta_{k_0+1}, \ldots, z_n) \frac{d\zeta_{k_0+1} \wedge d\bar{\zeta}_{k_0+1}}{\zeta_{k_0+1} - z_{k_0+1}},$$

by Lemma 11.3, we have

$$\frac{\partial g_J}{\partial \bar{z}_{k_0+1}} = f_J \quad \text{and} \quad \frac{\partial g_J}{\partial \bar{z}_j} = 0, \quad \text{for } j > k_0 + 1.$$

If we set

$$\eta_0 = \sum_{J \ni k} g_J d\bar{z}_{J \setminus \{k_0+1\}},$$

the form $\omega - (-1)^{q-1}\bar{\partial}\eta_0$ does not involve $d\bar{z}_{k'}$, $k' > k_0$ and $\bar{\partial}$-closed. By induction hypothesis, there exists a $(0, q-1)$-form η_1 such that $\omega - (-1)^{q-1}\bar{\partial}\eta_0 = \bar{\partial}\eta_1$ and $\omega = \bar{\partial}((-1)^{q-1}\eta_0 + \eta_1)$. □

Corollary 11.3. *The Dolbeault complex* (11.7) *is acyclic and gives a resolution* $0 \to \Omega^p \to \mathscr{A}^{p,\bullet}$ *of the sheaf* Ω^p *of holomorphic p-forms.*

Remark 11.5. The above is a weak form in the sense that the domain of definition of η as given in the proof is smaller than that of ω. This is sufficient to have Corollary 11.3 and the following theorem.

For the full version we need further assumption on U. For example, if each U_i is convex, we may proceed further to prove that if ω is $\bar{\partial}$-closed on U, it is $\bar{\partial}$-exact on U. This is one of the essential ingredients in the proof of Theorem 11.7.

Theorem 11.9 (Dolbeault theorem). *There is an isomorphism*

$$H^{p,q}_{\bar{\partial}}(M) \simeq H^q(M; \Omega^p).$$

Proof. We set $\mathscr{A}^{p,n+1} = 0$ and, for $r = 0, \ldots, n$, we denote by $\mathscr{K}^{p,r}$ the kernel of $\bar{\partial}^{p,r} : \mathscr{A}^{p,r} \to \mathscr{A}^{p,r+1}$. Then we have $\mathscr{K}^{p,0} = \Omega^p$ and $\mathscr{K}^{p,n} = \mathscr{A}^{p,n}$. We have the short exact sequences:

$$0 \longrightarrow \mathscr{K}^{p,r} \longrightarrow \mathscr{A}^{p,r} \xrightarrow{\bar{\partial}} \mathscr{K}^{p,r+1} \longrightarrow 0, \qquad 0 \le r \le n,$$

where we set $\mathscr{K}^{p,n+1} = 0$. We have the associated long exact sequence (cf. Theorem 11.1):

$$\begin{aligned}\cdots \longrightarrow H^q(M; \mathscr{K}^{p,r}) \longrightarrow H^q(M; \mathscr{A}^{p,r}) &\xrightarrow{\bar{\partial}^*} H^q(M; \mathscr{K}^{p,r+1}) \\ \xrightarrow{\delta^*} H^{q+1}(M; \mathscr{K}^{p,r}) &\longrightarrow H^{q+1}(M; \mathscr{A}^{p,r}) \longrightarrow \cdots . \end{aligned} \tag{11.14}$$

First, letting $q = r = 0$, we have the exact sequence

$$0 \longrightarrow \Gamma(M; \Omega^p) \longrightarrow \Gamma(M; \mathscr{A}^{p,0}) \xrightarrow{\bar{\partial}^*} \Gamma(M; \mathscr{K}^{p,1}),$$

which gives the theorem for $q = 0$.

Suppose $q \ge 1$. Since the sheaf $\mathscr{A}^{p,r}$ is fine, by Proposition 11.4,

$$\delta^* : H^q(M; \mathscr{K}^{p,r+1}) \xrightarrow{\sim} H^{q+1}(M; \mathscr{K}^{p,r}).$$

Thus by a repeated use of the above

$$H^q(M; \Omega^p) = H^q(M; \mathscr{K}^{p,0}) \simeq H^1(M; \mathscr{K}^{p,q-1}).$$

From (11.14),

$$\Gamma(M; \mathscr{A}^{p,q-1}) \xrightarrow{\bar{\partial}^*} \Gamma(M; \mathscr{K}^{p,q}) \xrightarrow{\delta^*} H^1(M; \mathscr{K}^{p,q-1}) \longrightarrow 0.$$

Thus we have the theorem for $q \ge 1$. □

From Theorems 11.9 and 11.7, we have:

Corollary 11.4. *For a holomorphically convex domain D, the Dolbeault complex $(\Gamma(D; \mathscr{A}^{p,\bullet}), \bar{\partial})$ is acyclic:*

$$H^{p,q}_{\bar{\partial}}(D) = 0 \qquad \text{for } q \ge 1.$$

Remark 11.6. 1. We use the above theorem for M holomorphically convex domains, i.e., the case the right-hand side vanishes for $q \ge 1$, to obtain a canonical Dolbeault theorem (cf. Corollary 11.6 below). The difference between the above isomorphism and the canonical one is a sign of $(-1)^{\frac{q(q+1)}{2}}$.

2. We could also prove, for a C^∞ manifold M, that $\check{H}^r(M; \mathbb{C})$ is isomorphic with $H^r_d(M)$ in a similar manner, using the resolution $0 \to \mathbb{C} \to \mathscr{A}^\bullet$. However the isomorphism obtained this way differs from the canonical one (cf. Corollary 7.4, also (11.8) and (11.9)) by a sign of $(-1)^{\frac{r(r+1)}{2}}$.

Čech-Dolbeault cohomology

Let M be a complex manifold of dimension n. We have the Dolbeault complex $(\mathscr{A}^{p,\bullet}, \bar\partial)$. Let $\mathcal{U} = \{U_\alpha\}_{\alpha\in I}$ be an open covering of M. Then we have the double complex $(C^\bullet(\mathcal{U}; \mathscr{A}^{p,\bullet}), \delta, (-1)^\bullet\bar\partial)$. We denote the associated single complex by $(A^{p,\bullet}(\mathcal{U}), \bar\vartheta)$. Thus

$$A^{p,q}(\mathcal{U}) = \bigoplus_{q_1+q_2=q} C^{q_1}(\mathcal{U}; \mathscr{A}^{p,q_2}), \qquad \bar\vartheta = \delta + (-1)^{q_1}\bar\partial.$$

We could define the Čech-Dolbeault cohomology as in the case of Čech-de Rham cohomology. Here we define simply as

Definition 11.16. The *Čech-Dolbeault cohomology* of type (p,q) of $\mathcal{U}$ is the q-th hypercohomology of $\mathscr{A}^{p,\bullet}$ on $\mathcal{U}$:

$$H^{p,q}_{\bar\vartheta}(\mathcal{U}) = \boldsymbol{H}^q(\mathcal{U}; \mathscr{A}^{p,\bullet}).$$

From Proposition 11.6 we have the following, which can also be proved as Theorem 7.1:

Theorem 11.10. *If $\mathcal{U}$ is locally finite, then the inclusion $A^{p,q}(M) \hookrightarrow C^0(\mathcal{U}, A^{p,q}) \subset A^{p,q}(\mathcal{U})$ induces an isomorphism*

$$H^{p,q}_{\bar\partial}(M) \xrightarrow{\sim} H^{p,q}_{\bar\vartheta}(\mathcal{U}).$$

We consider only locally finite coverings hereafter.

Definition 11.17. A covering $\mathcal{U} = \{U_\alpha\}$ of M is *analytically good*, if every non-empty intersection $U_{\alpha_0\dots\alpha_r}$ is biholomorphic to a holomorphically convex domain in $\mathbb{C}^n$.

By Corollary 11.4, if $\mathcal{U}$ is analytically good, it is good for $\mathscr{A}^{p,\bullet}$ (cf. Definition 11.9). From Proposition 11.7, we have the following, which can also be proved as Theorem 7.2:

Theorem 11.11. *If $\mathcal{U}$ is analytically good, the inclusion $C^q(\mathcal{U}; \Omega^p) \hookrightarrow C^q(\mathcal{U}; A^{p,0}) \subset A^{p,q}(\mathcal{U})$ induces an isomorphism*

$$H^q(\mathcal{U}; \Omega^p) \xrightarrow{\sim} H^{p,q}_{\bar\vartheta}(\mathcal{U}).$$

From Theorems 11.10 and 11.11, we have:

Corollary 11.5. *If $\mathcal{U}$ is analytically good, there is a canonical isomorphism*

$$H^{p,q}_{\bar\partial}(M) \simeq H^q(\mathcal{U}; \Omega^p).$$

Note that every complex manifold admits an analytically good covering and the set of analytically good coverings is cofinal in the set of coverings of M. Since $H^q(\mathcal{U};\Omega^p)$ is independent of the analytically good covering $\mathcal{U}$ by Corollary 11.5, we have:

Theorem 11.12. *If $\mathcal{U}$ is an analytically good covering of a complex manifold M, the canonical morphism is an isomorphism:*

$$H^q(\mathcal{U};\Omega^p) \xrightarrow{\sim} H^q(M;\Omega^p).$$

Combining with Corollary 11.5, we have:

Corollary 11.6 (Canonical Dolbeault theorem). *There is a canonical isomorphism*

$$H^{p,q}_{\bar\partial}(M) \simeq H^q(M;\Omega^p).$$

An explicit correspondence is given as (7.10) with D replaced by $\bar\vartheta$. In fact we used the Dolbeault theorem (Theorem 11.9) for Corollary 11.6, however the advantage of this is that it gives a canonical correspondence.

In the following, we consider the covering $\mathcal{U}=\{U_i\}_{i=1}^n$ of $\mathbb{C}^n\smallsetminus\{0\}$ given by $U_i=\{z_i\neq 0\}$. It is an analytically good covering. The n-form, the *Cauchy form in n-variables,*

$$\kappa_n=\Big(\frac{1}{2\pi\sqrt{-1}}\Big)^n\frac{dz_1\wedge\cdots\wedge dz_n}{z_1\cdots z_n}$$

may be thought of as a cocycle c in $C^{n-1}(\mathcal{U};\Omega^n)$ given by $c_{1\dots n}=\kappa_n$. We also have the Bochner-Martinelli form β_n, which is a $\bar\partial$-closed $(n,n-1)$-form on $\mathbb{C}^n\smallsetminus\{0\}$ (cf. Section 7.10).

Proposition 11.16. *In the isomorphism*

$$H^{n,n-1}_{\bar\partial}(\mathbb{C}^n\smallsetminus 0)\simeq H^{n-1}(\mathcal{U};\Omega^n)$$

of Corollary 11.6, the class of β_n corresponds to the class of $(-1)^{\frac{n(n-1)}{2}}\kappa_n$.

Proof. If $n=1$, the cohomologies are the same and $\beta_1=\kappa_1$. Thus we assume $n\geq 2$. We may think of β_n as being in $C^0(\mathcal{U};\mathscr{A}^{n,n-1})\subset\mathscr{A}^{n,n-1}(\mathcal{U})$ and κ_n in $C^{n-1}(\mathcal{U};\mathscr{A}^{n,0})\subset\mathscr{A}^{n,n-1}(\mathcal{U})$. We construct a cochain χ in $\mathscr{A}^{n,n-2}(\mathcal{U})=\bigoplus_{p=0}^{n-2}C^p(\mathcal{U};\mathscr{A}^{n,q})$, $q=n-p-2$, so that

$$\beta_n-(-1)^{\frac{n(n-1)}{2}}\kappa_n=\bar\vartheta\chi \qquad \text{in } \mathscr{A}^{n,n-1}(\mathcal{U}).$$

Writing $\chi = \sum_{p=0}^{n-2} \chi^p$, $\chi^p \in C^p(\mathcal{U}; \mathscr{A}^{n,q})$, this is expressed as

$$\begin{cases} \beta_n = \bar\partial \chi^0, \\ 0 = \delta\chi^{p-1} + (-1)^p \bar\partial \chi^p, \qquad 1 \le p \le n-2, \\ -(-1)^{\frac{n(n-1)}{2}} \kappa_n = \delta \chi^{n-2}. \end{cases} \tag{11.15}$$

Note that the condition in the middle is vacuous if $n = 2$.

Let $0 \le p \le n-2$ so that $0 \le q \le n-2$. For a $(p+1)$-tuple of integers $I = (i_0, \dots, i_p)$ with $1 \le i_0 < \cdots < i_p \le n$, let $I^* = (j_0, \dots, j_q)$ denote the complement of $\{i_0, \dots, i_p\}$ in $\{1, \dots, n\}$ with $1 \le j_0 < \cdots < j_q \le n$. Setting $\Phi(z) = dz_1 \wedge \cdots \wedge dz_n$, $z_I = z_{i_0} \cdots z_{i_p}$, $|I^*| = j_0 + \cdots + j_q$ and

$$\bar\Phi_{I^*}(z) = \sum_{\mu=0}^{q} (-1)^\mu \bar z_{j_\mu} d\bar z_{j_0} \wedge \cdots \wedge \widehat{d\bar z_{j_\mu}} \wedge \cdots \wedge d\bar z_{j_q},$$

we define a cochain χ^p by

$$\chi^p_I = (-1)^{\varepsilon_I} \frac{q!\, C_n}{(n-1)!} \frac{\bar\Phi_{I^*}(z) \wedge \Phi(z)}{z_I \, \|z\|^{2(q+1)}}, \qquad \varepsilon_I = |I^*| + \frac{q(n+p-1)}{2}.$$

Then it can be shown that it satisfies (11.15). □

Remark 11.7. Let W be a holomorphically convex neighborhood of 0 in $\mathbb{C}^n$ and $\mathcal{W} = \{W_i\}_{i=1}^n$ the covering of $W \smallsetminus 0$ given by $W_i = W \cap U_i$. Then we have a canonical isomorphism $H^{n,n-1}_{\bar\partial}(W \smallsetminus 0) \simeq H^{n-1}(\mathcal{W}; \Omega^n)$, under which the class of β_n (restricted to $W \smallsetminus 0$) corresponds to the class of $(-1)^{\frac{n(n-1)}{2}} \kappa_n$ (restricted to $\mathcal{W}$). Suppose the class of θ corresponds to the class of γ under the above isomorphism. If h is a holomorphic function on V, since $\bar\partial h = 0$, we see that the class of $h\theta$ corresponds to the class of $h\gamma$ (cf. the relation similar to (7.10) for Čech-Dolbeault cocycles and (11.15)).

In the above situation set

$$R_1 = \{\, z \in \mathbb{C}^n \mid \|z\|^2 \le n\varepsilon^2 \,\}.$$

The boundary ∂R_1 is a usually oriented $(2n-1)$-sphere $\mathbb{S}^{2n-1}$. We also set

$$\Gamma = \{\, z \in \mathbb{C}^n \mid |z_i| = \varepsilon,\ i = 1, \dots, n \,\},$$

which is an n-cycle oriented so that $\arg z_1 \wedge \cdots \wedge \arg z_n$ is positive.

Proposition 11.17. *Let θ be a $\bar\partial$-closed $(n, n-1)$-form on $\mathbb{C}^n \smallsetminus 0$ and γ a cocycle in $C^{n-1}(\mathcal{U}; \Omega^n)$. If the class of θ corresponds to the class of γ by the canonical isomorphism*

$$H^{n,n-1}_{\bar\partial}(\mathbb{C}^n \smallsetminus 0) \simeq H^{n-1}(\mathcal{U}; \Omega^n),$$

then

$$\int_{\mathbb{S}^{2n-1}} \theta = (-1)^{\frac{n(n-1)}{2}} \int_{\Gamma} \gamma.$$

Proof. Recall that we have canonical isomorphisms

$$H^{n,n-1}_{\bar\partial}(\mathbb{C}^n\smallsetminus 0)\xrightarrow{\sim} H^{n,n-1}_{\bar\vartheta}(\mathcal{U})\xleftarrow{\sim} H^{n-1}(\mathcal{U};\Omega^n).$$

The assumption implies that there exists a cochain χ in $\mathscr{A}^{n,n-2}(\mathcal{U})$ such that $\theta-\gamma=\bar\vartheta\chi$. Consider the commutative diagram

$$\begin{array}{ccccccc}
\mathscr{A}^{n,n-2}(\mathcal{U}) & \longrightarrow & \mathscr{A}^{2n-2}(\mathcal{U}) & \longrightarrow & \mathscr{A}^{2n-2}(\mathcal{U}\cap\mathbb{S}^{2n-1}) & & \\
\downarrow{\scriptstyle\bar\vartheta} & & \downarrow{\scriptstyle D} & & \downarrow{\scriptstyle D} & & \\
\mathscr{A}^{n,n-1}(\mathcal{U}) & \longrightarrow & \mathscr{A}^{2n-1}(\mathcal{U}) & \longrightarrow & \mathscr{A}^{2n-1}(\mathcal{U}\cap\mathbb{S}^{2n-1}) & \xrightarrow{\int_{\mathbb{S}^{2n-1}}} & \mathbb{C},
\end{array}$$

where $\mathcal{U}\cap\mathbb{S}^{2n-1}$ denotes the covering of $\mathbb{S}^{2n-1}$ consisting of the $U_i\cap\mathbb{S}^{2n-1}$'s. For each $i=1,\dots,n$, we set

$$Q_i=\{\, z\in\mathbb{S}^{2n-1}\mid |z_i|\geq|z_j|\ \text{ for all } j\neq i\,\}.$$

Then $\{Q_i\}$ is a honeycomb system adapted to $\mathcal{U}\cap\mathbb{S}^{2n-1}$ and, by the Stokes formula for Čech-de Rham cochains,

$$0=\int_{\mathbb{S}^{2n-1}}D\chi=\sum_{p=0}^{n-1}\sum_{i_0<\cdots<i_p}\int_{Q_{i_0\dots i_p}}(D\chi)_{i_0\dots i_p}=\sum_{i=1}^{n}\int_{Q_i}\theta-\int_{Q_{1\dots n}}\gamma.$$

Noting that $Q_{1\dots n}=(-1)^{\frac{n(n-1)}{2}}\Gamma$, we have the proposition. □

Note that the above is consistent with Proposition 11.16:

$$\int_{\mathbb{S}^{2n-1}}\beta_n=1=\int_\Gamma\kappa_n.$$

Dolbeault theorem with coefficients in a vector bundle: Let M be a complex manifold of dimension n. For a holomorphic vector bundle E on M, we set $\mathscr{A}^{p,q}(E)=\mathscr{A}(\bigwedge^p T^*M\otimes\bigwedge^q \overline{T}^* M\otimes E)$ and $\Omega^p(E)=\mathscr{O}(\bigwedge^p T^*M\otimes E)$. In particular, $\Omega^0(E)=\mathscr{O}(E)$. The sheaves $\mathscr{A}^{p,q}(E)$ are fine and, for each p, give a fine resolution of $\Omega^p(E)$:

$$0\longrightarrow\Omega^p(E)\longrightarrow\mathscr{A}^{p,0}(E)\xrightarrow{\bar\partial}\mathscr{A}^{p,1}(E)\xrightarrow{\bar\partial}\cdots\xrightarrow{\bar\partial}\mathscr{A}^{p,n}(E)\longrightarrow 0.$$

Since $\Gamma(M;\mathscr{A}^{p,q}(E))=A^{p,q}(M;E)$, as Theorem 11.9, we have an isomorphism

$$H^q(M;\Omega^p(E))\simeq H^{p,q}_{\bar\partial}(M;E).$$

Considering the double complex $(C^\bullet(\mathcal{U};\mathscr{A}^{p,\bullet}(E)),\bar\vartheta)$, we have a canonical isomorphism as Corollary 11.6.

Thus, if M is compact, $H^q(M;\Omega^p(E))$ is finite dimensional for all q and is equal to 0 for $q>n$ (cf. Section 9.1).

Definition 11.18. For a holomorphic vector bundle E on a compact complex manifold M of dimension n, the *Euler-Poincaré characteristic* is defined by

$$\chi(M; \mathscr{O}(E)) = \sum_{q=0}^{n} (-1)^q \dim H^q(M; \mathscr{O}(E)).$$

This is used in the expression of the Hirzebruch-Riemann-Roch theorem in Chapter 15 below.

Remark 11.8. In fact, for a complex manifold M of dimension n and a coherent $\mathscr{O}_M$-module $\mathscr{S}$, the following is known:

(1) $H^q(M; \mathscr{S}) = 0$ for $q > n$.
(2) If M is compact, $H^q(M; \mathscr{S})$ are finite dimensional for all $q \geq 0$.

We also have (cf. (9.13)):

Theorem 11.13 (Kodaira-Serre duality). *For a holomorphic vector bundle E on a compact complex manifold M of dimension n, there is a canonical isomorphism*

$$H^q(M; \Omega^p(E)) \simeq H^{n-q}(M; \Omega^{n-p}(E^*))^*.$$

11.5 Complex analytic spaces

In this section we introduce complex analytic spaces as generalizations of analytic varieties. We discuss some fundamental cases before we go into generalities.

$\mathbb{C}$-algebras: A $\mathbb{C}$-*algebra* is a commutative ring with unity R together with a ring morphism $\mathbb{C} \to \mathsf{R}$, called a *structural morphism.* A *morphism of* $\mathbb{C}$-*algebras* is a ring morphism compatible with the structural morphisms. As a commutative ring with unity, $\mathbb{C}$ has a unique structural morphism $\mathbb{C} \to \mathbb{C}$, i.e., the identity, and the identity is the only $\mathbb{C}$-algebra morphism of $\mathbb{C}$ to $\mathbb{C}$ itself.

If R is a $\mathbb{C}$-algebra, any R-module is endowed with a $\mathbb{C}$-vector space structure via the structural morphism of R. Thus any ideal I in R is a sub-$\mathbb{C}$-vector space of R and the quotient R/I is a $\mathbb{C}$-algebra with the structural morphism $\mathbb{C} \to \mathsf{R} \to \mathsf{R}/\mathsf{I}$.

A *local* $\mathbb{C}$-*algebra* is a $\mathbb{C}$-algebra R such that it is a local ring and that $\mathsf{R}/\mathfrak{m}$ is isomorphic with $\mathbb{C}$ as a $\mathbb{C}$-algebra, $\mathfrak{m}$ being the maximal ideal in R

(cf. Section A.2). In this case, $\mathsf{R} \simeq \mathbb{C} \oplus \mathfrak{m}$ as $\mathbb{C}$-vector spaces. A $\mathbb{C}$-algebra morphism $\varphi : \mathsf{R} \to \mathsf{S}$ of local $\mathbb{C}$-algebras is *local* if $\varphi(\mathfrak{m}) \subset \mathfrak{n}$, where $\mathfrak{m}$ and $\mathfrak{n}$ denote the maximal ideals of R and S, respectively. In fact, in this case, it holds that $\mathfrak{m} = \varphi^{-1}(\mathfrak{n})$.

Let X be a topological space and $\mathbb{C}_X = \mathbb{C} \times X$ the constant sheaf on X with stalk $\mathbb{C}$. A sheaf of $\mathbb{C}$-algebras is a sheaf $\mathscr{R}$ of rings on X together with a sheaf morphism $\mathbb{C}_X \to \mathscr{R}$, called also a structural morphism. In this case, for each x in X, $\mathscr{R}_x$ is a $\mathbb{C}$-algebra. A morphism of sheaves of $\mathbb{C}$-algebras is a morphism of sheaves of rings compatible with the structural morphisms.

Complex manifolds: Let M be a complex manifold of dimension n. Then it is equipped with the sheaf of rings $\mathscr{O}_M$ of germs of holomorphic functions. For each point z in M, the ring $\mathscr{O}_{M,z}$ is isomorphism with the ring $\mathscr{O}_n$ of germs of holomorphic functions at 0 in $\mathbb{C}^n$. Thus it is a Noetherian local ring with the ideal $\mathfrak{m}_z$ of germs of holomorphic functions vanishing at z as its unique maximal ideal. Note that $\mathscr{O}_M$ is in fact a sheaf of $\mathbb{C}$-algebras and, for each $z \in M$, $\mathscr{O}_{M,z}$ is a local $\mathbb{C}$-algebra. There is a canonical morphism $e_z : \mathscr{O}_{M,z} \to \mathscr{O}_{M,z}/\mathfrak{m}_z \simeq \mathbb{C}$, which we call the *evaluation morphism* at z. Recall that, if we denote by O_U the ring of holomorphic functions on an open set U in M, there is a canonical isomorphism

$$O_U \xrightarrow{\sim} \Gamma(U, \mathscr{O}_M)$$

that assigns to f in O_U the section s given by $s(z) = f_z$, the germ of f at z in U. The inverse assigns to a section s the function f given by $f(z) = e_z(s(z))$. We have been and will be identifying O_U and $\Gamma(U, \mathscr{O}_M)$ via the above isomorphism.

If $f : M' \to M$ is a holomorphic map of complex manifolds, it induces a morphism $\Gamma(U; \mathscr{O}_M) \to \Gamma(f^{-1}U; \mathscr{O}_{M'})$ by the pull-back of holomorphic functions, for each open set U in M. Thus it induces in turn a sheaf morphism $f^* : \mathscr{O}_M \to f_*\mathscr{O}_{M'}$. It induces a sheaf morphism

$$f^{-1}\mathscr{O}_M \xrightarrow{f^{-1}(f^*)} f^{-1}f_*\mathscr{O}_{M'} \longrightarrow \mathscr{O}_{M'}.$$

For each point z' in M' it restricts to a morphism $f^*_{z'} : \mathscr{O}_{M,f(z')} \to \mathscr{O}_{M',z'}$ of local $\mathbb{C}$-algebras.

Varieties: Let V be a variety in a complex manifold M with the inclusion $i : V \hookrightarrow M$. A complex valued function f on an open set W in V is defined to be *holomorphic*, if each point z in W admits a neighborhood U_z in M such that $f|_{W \cap U_z}$ may be extended to a holomorphic function on U_z. Let

$\mathscr{I}_V$ be the ideal sheaf of V. Then the quotient $\mathscr{O}_M/\mathscr{I}_V$ is supported on V. If we set $\mathscr{O}_V = i^{-1}(\mathscr{O}_M/\mathscr{I}_V)$, i.e., the restriction $(\mathscr{O}_M/\mathscr{I}_V)|_V$, the sections of $\mathscr{O}_V$ on an open set W in V may be identified with holomorphic functions on W (cf. Exercise 11.3). Thus $\mathscr{O}_V$ may be thought of as the sheaf of germs of holomorphic functions on V. For each point z in V, $\mathscr{O}_{V,z} = \mathscr{O}_{M,z}/\mathscr{I}_{V,z}$ is a Noetherian local $\mathbb{C}$-algebra with the maximal ideal given by $\mathfrak{m}_{V,z} = \mathfrak{m}_z/\mathscr{I}_{V,z}$. Note that there is the evaluation morphism $e_z : \mathscr{O}_{V,z} \to \mathscr{O}_{V,z}/\mathfrak{m}_{V,z} \simeq \mathbb{C}$. By Theorem 11.4 and Lemma 11.1, $\mathscr{O}_V$ is a coherent sheaf of rings. As in the case of manifolds, we may identify a section of $\mathscr{O}_V$ with a holomorphic function via evaluation morphisms.

Let V and V' be varieties in complex manifolds M and M', respectively. A map $f : V' \to V$ is defined to be *holomorphic*, if each point z' in V' admits a neighborhood U' in M' such that $f|_{V'\cap U'}$ may be extended to a holomorphic map $\tilde{f} : U' \to M$. Such an extension $\tilde{f}$, referred to as a *local extension* of f, induces a morphism $(\tilde{f})^*_{z'} : \mathscr{O}_{M,f(z')} \to \mathscr{O}_{M',z'}$ with $(\tilde{f})^*_{z'}\mathscr{I}_{V,f(z')} \subset \mathscr{I}_{V',z'}$. Hence we have a morphism $(\tilde{f})^*_{z'} : (\mathscr{O}_M/\mathscr{I}_V)_{f(z')} \to (\mathscr{O}_{M'}/\mathscr{I}_{V'})_{z'}$, which does not depend on the choice of the local extension $\tilde{f}$. Thus a holomorphic map f is continuous and induces a sheaf morphism $f^* : \mathscr{O}_V \to f_*\mathscr{O}_{V'}$. For each point z' in V', it restricts to a morphism $f^*_{z'} : \mathscr{O}_{V,f(z')} \to \mathscr{O}_{V',z'}$ of local $\mathbb{C}$-algebras.

$\mathbb{C}$-ringed spaces: More generally we introduce the following:

Definition 11.19. A *$\mathbb{C}$-ringed space* $\boldsymbol{X} = (X, \mathscr{O}_{\boldsymbol{X}})$ is a topological space X together with a sheaf of $\mathbb{C}$-algebras $\mathscr{O}_{\boldsymbol{X}}$ on X.

In the above, X is referred to as the *support* of $\boldsymbol{X}$ and $\mathscr{O}_{\boldsymbol{X}}$ the *structure sheaf* of $\boldsymbol{X}$.

Let $\boldsymbol{X} = (X, \mathscr{O}_{\boldsymbol{X}})$ and $\boldsymbol{Y} = (Y, \mathscr{O}_{\boldsymbol{Y}})$ be $\mathbb{C}$-ringed spaces. A *morphism* $\boldsymbol{Y} \to \boldsymbol{X}$ is defined as in Definition 11.5. Thus it is a pair (f, φ) of a continuous map $f : Y \to X$ and a morphism $\varphi : \mathscr{O}_{\boldsymbol{X}} \to f_*\mathscr{O}_{\boldsymbol{Y}}$ of sheaves of $\mathbb{C}$-algebras. A morphism (f, φ) defines, for each $y \in Y$, a morphism of $\mathbb{C}$-algebras

$$\varphi_y : \mathscr{O}_{\boldsymbol{X},f(y)} \longrightarrow \mathscr{O}_{\boldsymbol{Y},y}.$$

The composition of morphisms is defined as in the paragraph after Definition 11.5. The identity morphism of $\boldsymbol{X} = (X, \mathscr{O}_{\boldsymbol{X}})$ is the pair $1_{\boldsymbol{X}} = (1_X, 1_{\mathscr{O}_{\boldsymbol{X}}})$ of the identities. A morphism $(f, \varphi) : \boldsymbol{Y} \to \boldsymbol{X}$ is an *isomorphism* if there exist a morphism $(g, \psi) : \boldsymbol{X} \to \boldsymbol{Y}$ with $(f, \varphi) \circ (g, \psi) = 1_{\boldsymbol{X}}$ and $(g, \psi) \circ (f, \varphi) = 1_{\boldsymbol{Y}}$.

Example 11.6. 1. If $(X, \mathscr{O}_{\mathbf{X}})$ is a $\mathbb{C}$-ringed space, for every open set W in X, $(W, \mathscr{O}_{\mathbf{X}}|_W)$ is a $\mathbb{C}$-ringed space.

2. If V is a variety and $\mathscr{O}_V$ the sheaf as defined above, then $(V, \mathscr{O}_V)$ is a $\mathbb{C}$-ringed space. If $f : V' \to V$ is a holomorphic map of varieties, then $(f, f^*) : (V', \mathscr{O}_{V'}) \to (V, \mathscr{O}_V)$ is a morphism of $\mathbb{C}$-ringed spaces. In fact, any morphism of varieties as $\mathbb{C}$-ringed spaces is of this form (cf. Proposition 11.20 below).

Model complex spaces: Let U be an open set in $\mathbb{C}^n$ and $\mathscr{I}$ an ideal sheaf of finite type in $\mathscr{O}_U$. Then $\mathscr{I}$ is a coherent $\mathscr{O}_U$-module and $\mathscr{O}_U/\mathscr{I}$ is also $\mathscr{O}_U$-coherent (cf. Proposition 11.11.1 and 2). Thus the support V of $\mathscr{O}_U/\mathscr{I}$ is a variety in U (cf. Proposition 11.15) and $\mathscr{I} \subset \mathscr{I}_V$. Let $i : V \hookrightarrow U$ be the inclusion and set $\mathscr{O}_{\mathbf{V}} = i^{-1}(\mathscr{O}_U/\mathscr{I})$. For each point z in V, $\mathscr{O}_{\mathbf{V},z} = \mathscr{O}_{U,z}/\mathscr{I}_z$ is a Noetherian local ring with the maximal ideal given by $\mathfrak{m}_{\mathbf{V},z} = \mathfrak{m}_z/\mathscr{I}_z$. Note that there is the evaluation morphism $\mathscr{O}_{\mathbf{V},z} \to \mathscr{O}_{\mathbf{V},z}/\mathfrak{m}_{\mathbf{V},z} \simeq \mathbb{C}$. We call a $\mathbb{C}$-ringed space as $\boldsymbol{V} = (V, \mathscr{O}_{\mathbf{V}})$ a *model complex space.* By Lemma 11.1, $\mathscr{O}_{\mathbf{V}}$ is a coherent sheaf of rings.

See Proposition 11.20 below for morphisms of model complex spaces as $\mathbb{C}$-ringed spaces.

Complex analytic spaces: Now we introduce the following:

Definition 11.20. A *complex analytic space*, or *complex space* for short, is a $\mathbb{C}$-ringed space $\boldsymbol{X} = (X, \mathscr{O}_{\mathbf{X}})$, where X is a Hausdorff topological space and every point in X admits a neighborhood W such that $(W, \mathscr{O}_{\mathbf{X}}|_W)$ is isomorphic with a model complex space.

Note that, if $\boldsymbol{X} = (X, \mathscr{O}_{\mathbf{X}})$ is a complex space, then $\mathscr{O}_{\mathbf{X}}$ is a coherent sheaf of rings. Also for each point x in X, $\mathscr{O}_{\mathbf{X},x}$ is a Noetherian local $\mathbb{C}$-algebra whose residual field is isomorphism with $\mathbb{C}$.

A *morphism of complex spaces* is, by definition, a morphism as $\mathbb{C}$-ringed spaces.

Proposition 11.18. *If* $(f, \varphi) : (Y, \mathscr{O}_{\mathbf{Y}}) \to (X, \mathscr{O}_{\mathbf{X}})$ *is a morphism of complex spaces, then, for each* y *in* Y, $\varphi_y : \mathscr{O}_{\mathbf{X},f(y)} \to \mathscr{O}_{\mathbf{Y},y}$ *is a morphism of local* $\mathbb{C}$*-algebras.*

Proof. Let $\mathfrak{m}_{f(y)}$ and $\mathfrak{n}_y$ denote the maximal ideals in $\mathscr{O}_{\mathbf{X},f(y)}$ and $\mathscr{O}_{\mathbf{Y},y}$, respectively. Consider the composition ψ_y

$$\mathscr{O}_{\mathbf{X},f(y)} \xrightarrow{\varphi_y} \mathscr{O}_{\mathbf{Y},y} \longrightarrow \mathscr{O}_{\mathbf{Y},y}/\mathfrak{n}_y \simeq \mathbb{C}.$$

of φ_y and the evaluation morphism. Since ψ_y is a $\mathbb{C}$-algebra morphism, its image is $\mathbb{C}$ and $\mathscr{O}_{\boldsymbol{X},f(y)}/\operatorname{Ker}\psi_y \simeq \mathbb{C}$. Thus $\operatorname{Ker}\psi_y$ is a maximal ideal in $\mathscr{O}_{\boldsymbol{X},f(y)}$. Since $\mathscr{O}_{\boldsymbol{X},f(y)}$ has the unique maximal ideal $\mathfrak{m}_{f(y)}$, $\operatorname{Ker}\psi_y = \mathfrak{m}_{f(y)}$. On the other hand, $\operatorname{Ker}\psi_y = \varphi_y^{-1}(\mathfrak{n}_y)$. □

Let $\boldsymbol{X} = (X, \mathscr{O}_{\boldsymbol{X}})$ be a complex space. A section s of $\mathscr{O}_{\boldsymbol{X}}$ on an open set W in X defines a complex valued function on W by $x \mapsto e_x(s(x))$, where $e_x : \mathscr{O}_{\boldsymbol{X},x} \to \mathscr{O}_{\boldsymbol{X},x}/\mathfrak{m}_x \simeq \mathbb{C}$ is the evaluation morphism. Since $\boldsymbol{X}$ is locally isomorphic with a model complex space, we see that this function is continuous and we have a canonical morphism

$$\gamma_W : \Gamma(W; \mathscr{O}_{\boldsymbol{X}}) \longrightarrow C_W \simeq \Gamma(W; \mathscr{C}_X),$$

where C_W denotes the ring of continuous functions on W and $\mathscr{C}_X$ the sheaf of continuous functions on X. In general this morphism is not injective and we may not think of s as a usual function (cf. Example 11.7 below). Note that the morphisms γ_W induces a sheaf morphism

$$\gamma : \mathscr{O}_{\boldsymbol{X}} \longrightarrow \mathscr{C}_X$$

and that γ is injective if and only if γ_W is injective for all open set W in X.

A complex space $\boldsymbol{X}$ is said to be *reduced* if, for every x in X, $\mathscr{O}_{\boldsymbol{X},x}$ has no non-zero nilpotent elements, i.e., if $a^k = 0$ for $a \in \mathscr{O}_{\boldsymbol{X},x}$ and a positive integer k, then $a = 0$.

A model complex space $\boldsymbol{V} = (V, \mathscr{O}_{\boldsymbol{V}})$, $\mathscr{O}_{\boldsymbol{V}} = i^{-1}(\mathscr{O}_U/\mathscr{I})$, is reduced if and only if $\sqrt{\mathscr{I}_z} = \mathscr{I}_z$ for all $z \in V$ (in fact for all $z \in U$). By the Nullstellensatz (Theorem 2.8), this is equivalent to saying that $\mathscr{I} = \mathscr{I}_V$, i.e., $\boldsymbol{V}$ is a variety in U.

Proposition 11.19. *For a complex space $\boldsymbol{X}$, the canonical morphism γ is injective if and only if $\boldsymbol{X}$ is reduced.*

Proof. If γ is injective, then $\mathscr{O}_{\boldsymbol{X},x}$ has no non-zero nilpotent elements, as $\mathscr{C}_{X,x}$ has this property. Conversely, if $\boldsymbol{X}$ is reduced, then γ is injective, as $\boldsymbol{X}$ is locally isomorphic with a model complex space that is a variety. □

In general, for a complex space $\boldsymbol{X}$, we may construct a reduced analytic space $\boldsymbol{X}_{\text{red}}$, called the reduction of $\boldsymbol{X}$ as follows.

Let $\mathscr{N}_{\boldsymbol{X}}$ be the ideal sheaf in $\mathscr{O}_{\boldsymbol{X}}$ defined by the presheaf

$$W \mapsto \{\, s \in \Gamma(W; \mathscr{O}_{\boldsymbol{X}}) \mid s^k = 0 \text{ for some positive integer } k \,\}.$$

For a model complex space $\boldsymbol{V} = (V, \mathscr{O}_{\boldsymbol{V}})$, $\mathscr{O}_{\boldsymbol{V}} = i^{-1}(\mathscr{O}_U/\mathscr{I})$, we have $\mathscr{N}_{\boldsymbol{V}} = i^{-1}(\sqrt{\mathscr{I}}/\mathscr{I})$, which is equal to $i^{-1}(\mathscr{I}_V/\mathscr{I})$, by the Nullstellensatz. Thus $\mathscr{O}_{\boldsymbol{V}}/\mathscr{N}_{\boldsymbol{V}} \simeq \mathscr{O}_V$.

The above shows that, for a complex space $\boldsymbol{X}$, $\operatorname{Supp}(\mathscr{O}_{\boldsymbol{X}}/\mathscr{N}_{\boldsymbol{X}}) = X$ and that $(X, \mathscr{O}_{\boldsymbol{X}}/\mathscr{N}_{\boldsymbol{X}})$ is a reduced complex space. We call it the *reduction* of $\boldsymbol{X}$ and denote it by $\boldsymbol{X}_{\text{red}}$. There is a canonical morphism $(1_X, \pi) : \boldsymbol{X}_{\text{red}} \to \boldsymbol{X}$, where $\pi : \mathscr{O}_{\boldsymbol{X}} \to \mathscr{O}_{\boldsymbol{X}}/\mathscr{N}_{\boldsymbol{X}}$ is the canonical epimorphism. Note that $\mathscr{N}_{\boldsymbol{X}}$ is coherent and that $\mathscr{N}_{\boldsymbol{X}} = \operatorname{Ker}\gamma$.

For a model complex space $\boldsymbol{V} = (V, \mathscr{O}_{\boldsymbol{V}})$, $\mathscr{O}_{\boldsymbol{V}} = i^{-1}(\mathscr{O}_U/\mathscr{I})$, the variety $V = (V, \mathscr{O}_V)$, $\mathscr{O}_V = i^{-1}(\mathscr{O}_U/\mathscr{I}_V)$, in U is the reduction of $\boldsymbol{V}$.

If $(f, \varphi) : \boldsymbol{Y} \to \boldsymbol{X}$ is a morphism of complex spaces, then $\varphi_y(\mathscr{N}_{\boldsymbol{X}, f(y)}) \subset \mathscr{N}_{\boldsymbol{Y}, y}$ for every y in Y and φ induces a morphism $\varphi_{\text{red}} : \mathscr{O}_{\boldsymbol{X}}/\mathscr{N}_{\boldsymbol{X}} \to f_*(\mathscr{O}_{\boldsymbol{Y}}/\mathscr{N}_{\boldsymbol{Y}})$ (here we note that f_* is left exact). Thus we have a morphism $(f, \varphi_{\text{red}}) : \boldsymbol{Y}_{\text{red}} \to \boldsymbol{X}_{\text{red}}$ making the following diagram commutative:

$$\begin{array}{ccc} \boldsymbol{Y}_{\text{red}} & \xrightarrow{(f,\varphi_{\text{red}})} & \boldsymbol{X}_{\text{red}} \\ \downarrow & & \downarrow \\ \boldsymbol{Y} & \xrightarrow{(f,\varphi)} & \boldsymbol{X}\,, \end{array}$$

where the vertical arrows are the canonical morphisms.

Proposition 11.20. *Let $\boldsymbol{V}' = (V', \mathscr{O}_{\boldsymbol{V}'})$ and $\boldsymbol{V} = (V, \mathscr{O}_{\boldsymbol{V}})$ be model complex sapces and $(f, \varphi) : \boldsymbol{V}' \to \boldsymbol{V}$ a morphism of complex spaces. Then:*

1. *$f : V' \to V$ is holomorphic. Thus we have $f^* : \mathscr{O}_V \to f_*\mathscr{O}_{V'}$.*

2. *If $\boldsymbol{V}'$ is reduced, then $\varphi = f^* \circ \pi$, where $\pi : \mathscr{O}_{\boldsymbol{V}} \to \mathscr{O}_V$ is the canonical morphism. In particular, if $\boldsymbol{V}$ is also reduced, then $\varphi = f^*$.*

Proof. For each open set W in V, we have the diagram

$$\begin{array}{ccccc} \Gamma(W; \mathscr{O}_{\boldsymbol{V}}) & \xrightarrow{\pi} & \Gamma(W; \mathscr{O}_V) & \overset{f^*}{\searrow} & \\ {\scriptstyle\varphi}\downarrow & & {\scriptstyle\varphi_{\text{red}}}\downarrow & & \\ \Gamma(f^{-1}W; \mathscr{O}_{\boldsymbol{V}'}) & \xrightarrow{\pi'} & \Gamma(f^{-1}W; \mathscr{O}_{V'}) & \xrightarrow{\gamma'} & \Gamma(f^{-1}W; \mathscr{C}_{V'}), \end{array}$$

where the rectangle is commutative. Since γ' is injective, we identify $\Gamma(f^{-1}W; \mathscr{O}_{V'})$ with a subspace of $\Gamma(f^{-1}W; \mathscr{C}_{V'})$ via γ'. The proposition is prove if we show that $f^* = \varphi_{\text{red}}$.

For every point z' in V', we have the commutative diagram

$$\begin{array}{ccccc} \mathscr{O}_{V, f(z')} & \longrightarrow & \mathscr{O}_{V, f(z')}/\mathfrak{m}_{f(z')} & \xrightarrow{\sim} & \mathbb{C} \\ {\scriptstyle\varphi_{\text{red}, z'}}\downarrow & & \downarrow & & \downarrow{\scriptstyle =} \\ \mathscr{O}_{V', z'} & \longrightarrow & \mathscr{O}_{V', z'}/\mathfrak{m}_{z'} & \xrightarrow{\sim} & \mathbb{C}. \end{array}$$

The composition of the morphisms in the first row is the evaluation morphism $e_{f(z')}$ and that in the second row is $e_{z'}$. For $h \in \Gamma(W; \mathscr{O}_V)$, $\varphi_{\rm red}(h)$ is the function that assigns to $z' \in f^{-1}W$ the value $e_{z'}(\varphi_{{\rm red},z'}(h_{f(z')})) = e_{f(z')}(h_{f(z')}) = h(f(z')) = f^*(h)(z')$. □

Remark 11.9. 1. Let $\boldsymbol{X} = (X, \mathscr{O}_{\boldsymbol{X}})$ be a complex space. If W is an open set in X such that $(W, \mathscr{O}_{\boldsymbol{X}}|_W)$ is isomorphic to a model space, W may be thought of as a variety and a section of $\mathscr{O}_{\boldsymbol{X}}|_W$ defines a holomorphic function on W, however the correspondence is not injective, unless $\boldsymbol{X}$ is reduced.

2. Proposition 11.20 shows that, if $(f, \varphi) : \boldsymbol{Y} \to \boldsymbol{X}$ is a morphism of complex spaces, f may locally be thought of as a holomorphic map of varieties and that, if $\boldsymbol{Y}$ is reduced, φ is determined uniquely by f.

A complex space $\boldsymbol{X} = (X, \mathscr{O}_{\boldsymbol{X}})$ is said to be *non-singular* if every point in X admits a neighborhood W such that $(W, \mathscr{O}_{\boldsymbol{X}}|_W)$ is isomorphic with a model complex space of the form $(U, \mathscr{O}_U)$ with U an open set in $\mathbb{C}^n$. By Proposition 11.20, a complex manifold as defined in Definition 2.3 is nothing but a non-singular complex space whose support has a countable basis.

A *germ of a complex space* is defined to be a pair $\boldsymbol{X} = (X, \mathscr{O}_n/I)$, where I is an ideal in $\mathscr{O}_n$ and X the germ of the variety defined by I. The germ is reduced if and only if $\sqrt{I} = I$.

Example 11.7. Let $M = \mathbb{C} = \{z\}$ be the complex plane. For a positive integer m, let $\mathscr{I}_m$ be the ideal sheaf in $\mathscr{O}_{\mathbb{C}}$ generated by z^m. Then the support of $\mathscr{O}_{\mathbb{C}}/\mathscr{I}_m$ is $\{0\}$ and $\boldsymbol{V}_m = (\{0\}, i^{-1}(\mathscr{O}_{\mathbb{C}}/\mathscr{I}_m))$, $i : \{0\} \hookrightarrow \mathbb{C}$ being the inclusion, is a model complex space, called an *m-fold point*. In the case $m = 1$, it is called a *simple point*. The identity map of $\mathbb{C}$ induces a morphism $\boldsymbol{V}_1 \to \boldsymbol{V}_m$. The simple point $\boldsymbol{V}_1$ is the reduction of $\boldsymbol{V}_m$.

Complex subspaces

Let V be a variety in a complex manifold M. A subset V' of V is said to be a variety in V, or a subvariety of V, if each point z in V admits a neighborhood W of z in V such that $V' \cap W$ is the common set of zeros of a finite number of holomorphic functions on W. Thus V' is a closed set in V and is in fact a variety in M. In particular, if X is a variety in M, $V \cap X$ is a variety in V.

Let $\boldsymbol{V} = (V, \mathscr{O}_{\boldsymbol{V}})$, $\mathscr{O}_{\boldsymbol{V}} = i^{-1}(\mathscr{O}_U/\mathscr{I})$, be a model complex space. Let $\mathscr{I}'$ be an ideal sheaf in $\mathscr{O}_{\boldsymbol{V}}$ of finite type. Then $\mathscr{I}'$ is $\mathscr{O}_{\boldsymbol{V}}$-coherent and $\mathscr{O}_{\boldsymbol{V}}/\mathscr{I}'$ is also $\mathscr{O}_{\boldsymbol{V}}$-coherent. Let V' denote the support of $\mathscr{O}_{\boldsymbol{V}}/\mathscr{I}'$ with the inclusion $i' : V' \hookrightarrow V$.

Proposition 11.21. *In the above situation, the* $\mathbb{C}$*-ringed space* $\boldsymbol{V}' = (V', \mathscr{O}_{\boldsymbol{V}'})$, $\mathscr{O}_{\boldsymbol{V}'} = (i')^{-1}(\mathscr{O}_{\boldsymbol{V}}/\mathscr{I}')$, *is isomorphic with a model complex space.*

Proof. Let $\pi : i^{-1}\mathscr{O}_U \to i^{-1}\mathscr{O}_U/i^{-1}\mathscr{I} = \mathscr{O}_{\boldsymbol{V}}$ denote the canonical epimorphism. Then π induces an isomorphism $i^{-1}\mathscr{O}_U/\pi^{-1}\mathscr{I}' \xrightarrow{\sim} \mathscr{O}_{\boldsymbol{V}}/\mathscr{I}'$ of $i^{-1}\mathscr{O}_U$-modules. Since V is a closed set in U, we may write $\pi^{-1}\mathscr{I}' = i^{-1}i_*(\pi^{-1}\mathscr{I}')$ (cf. (11.1)). Here $i_*(\pi^{-1}\mathscr{I}')$ is *a priori* an $i_*i^{-1}\mathscr{O}_U$-module, however it may canonically be thought of as an $\mathscr{O}_U$-module via the canonical morphism $\mathscr{O}_U \to i_*i^{-1}\mathscr{O}_U$ (cf. Proposition 11.3). Thus $\mathscr{O}_{\boldsymbol{V}}/\mathscr{I}' \simeq i^{-1}(\mathscr{O}_U/i_*(\pi^{-1}\mathscr{I}'))$ as $i^{-1}\mathscr{O}_U$-modules. Since $\mathscr{I}'$ is $\mathscr{O}_{\boldsymbol{V}}$-finite and $\mathscr{I}$ is $\mathscr{O}_U$-finite, we see that $\pi^{-1}\mathscr{I}'$ is $i^{-1}\mathscr{O}_U$-finite. Thus $i_*(\pi^{-1}\mathscr{I}')$ is $i_*i^{-1}\mathscr{O}_U$-finite and, in turn, $\mathscr{O}_U$-finite. $\square$

Note that, in the above, V' is a subvariety of V.

In general, let $\boldsymbol{X} = (X, \mathscr{O}_{\boldsymbol{X}})$ be a complex space.

Definition 11.21. A *complex subspace* of $\boldsymbol{X}$ is a $\mathbb{C}$-ringed space $\boldsymbol{X}' = (X', \mathscr{O}_{\boldsymbol{X}'})$ for which there is an ideal sheaf $\mathscr{I}$ of finite type in $\mathscr{O}_{\boldsymbol{X}}$ such that

(1) X' is the support of $\mathscr{O}_{\boldsymbol{X}}/\mathscr{I}$.
(2) $\mathscr{O}_{\boldsymbol{X}'} = i^{-1}(\mathscr{O}_{\boldsymbol{X}}/\mathscr{I})$, $i : X' \hookrightarrow X$ being the inclusion.

By Proposition 11.21, a complex subspace of a complex space is a complex space. Also by Propositions 11.8 and 11.11, X' is a closed set in X.

There is a canonical morphism $(i, \pi) : \boldsymbol{X}' \to \boldsymbol{X}$, where $\pi : \mathscr{O}_{\boldsymbol{X}} \to \mathscr{O}_{\boldsymbol{X}}/\mathscr{I} \to i_*i^{-1}(\mathscr{O}_{\boldsymbol{X}}/\mathscr{I}) = i_*\mathscr{O}_{\boldsymbol{X}'}$ is the composition of the canonical epimorphism and the first morphism in Proposition 11.3. For each x in X', $\pi_x : \mathscr{O}_{\boldsymbol{X},x} \to \mathscr{O}_{\boldsymbol{X}',x} = \mathscr{O}_{\boldsymbol{X},x}/\mathscr{I}_x$ is the canonical epimorphism.

In particular, a model complex space $\boldsymbol{V} = (V, \mathscr{O}_{\boldsymbol{V}})$, $\mathscr{O}_{\boldsymbol{V}} = i^{-1}(\mathscr{O}_U/\mathscr{I})$, is a complex subspace of $(U, \mathscr{O}_U)$.

Let $\boldsymbol{X}_1$ and $\boldsymbol{X}_2$ be complex subspaces of $\boldsymbol{X}$ defined by ideal sheaves $\mathscr{I}_1$ and $\mathscr{I}_2$ of $\mathscr{O}_{\boldsymbol{X}}$, respectively. The subspace defined by $\mathscr{I}_1 + \mathscr{I}_2$ is called the *intersection* of $\boldsymbol{X}_1$ and $\boldsymbol{X}_2$. It has $X_1 \cap X_2$ as its support. It is a subspace of $\boldsymbol{X}_i$ defined by the ideal $(\mathscr{I}_1 + \mathscr{I}_2)/\mathscr{I}_i$, $i = 1, 2$.

Analytic inverse image: Let $(f, \varphi) : \boldsymbol{Y} = (Y, \mathscr{O}_{\boldsymbol{Y}}) \to \boldsymbol{X} = (X, \mathscr{O}_{\boldsymbol{X}})$ be a morphism of complex spaces. The morphism φ induces a morphism

$f^{-1}\mathscr{O}_{\boldsymbol{X}} \xrightarrow{f^{-1}\varphi} f^{-1}f_*\mathscr{O}_{\boldsymbol{Y}} \longrightarrow \mathscr{O}_{\boldsymbol{Y}}$, which makes $\mathscr{O}_{\boldsymbol{Y}}$ an $f^{-1}\mathscr{O}_{\boldsymbol{X}}$-module. For an $\mathscr{O}_{\boldsymbol{X}}$-module $\mathscr{S}$, we set

$$(f,\varphi)^*\mathscr{S} = \mathscr{O}_{\boldsymbol{Y}} \otimes_{f^{-1}\mathscr{O}_{\boldsymbol{X}}} f^{-1}\mathscr{S}$$

and call it the *analytic inverse image* of $\mathscr{S}$ by (f,φ). It is an $\mathscr{O}_{\boldsymbol{Y}}$-module. In particular, $(f,\varphi)^*\mathscr{O}_{\boldsymbol{X}} = \mathscr{O}_{\boldsymbol{Y}}$. When it is apparent what is φ, as in the case $\boldsymbol{Y}$ is reduced (cf. Proposition 11.20) or the case of canonical morphism of a subspace, which we discuss right after, we simply denote $(f,\varphi)^*$ by f^*.

If $\boldsymbol{X}'$ is a subspace of $\boldsymbol{X}$ with defining ideal $\mathscr{I}$, for an $\mathscr{O}_{\boldsymbol{X}}$-module $\mathscr{S}$, the sheaf

$$i^*\mathscr{S} = \mathscr{O}_{\boldsymbol{X}'} \otimes_{i^{-1}\mathscr{O}_{\boldsymbol{X}}} i^{-1}\mathscr{S} = i^{-1}((\mathscr{O}_{\boldsymbol{X}}/\mathscr{I}) \otimes_{\mathscr{O}_{\boldsymbol{X}}} \mathscr{S})$$

is called the *analytic restriction* of $\mathscr{S}$ to $\boldsymbol{X}'$.

Let $(f,\varphi) : \boldsymbol{Y} = (Y,\mathscr{O}_{\boldsymbol{Y}}) \to \boldsymbol{X} = (X,\mathscr{O}_{\boldsymbol{X}})$ be a morphism of complex spaces and $\boldsymbol{X}'$ a subspace of $\boldsymbol{X}$ with defining ideal $\mathscr{I}$. The canonical monomorphism $\mathscr{I} \to \mathscr{O}_{\boldsymbol{X}}$ induces a morphism

$$(f,\varphi)^*\mathscr{I} \longrightarrow (f,\varphi)^*\mathscr{O}_{\boldsymbol{X}} = \mathscr{O}_{\boldsymbol{Y}},$$

which is not injective in general. We denote the image by $\mathscr{O}_{\boldsymbol{Y}} \cdot (f,\varphi)^*\mathscr{I}$. We also denote it by $\mathscr{O}_{\boldsymbol{Y}} \cdot \mathscr{I}$, since $(\mathscr{O}_{\boldsymbol{Y}} \cdot (f,\varphi)^*\mathscr{I})_y = \mathscr{O}_{\boldsymbol{Y},y} \cdot \mathscr{I}_{f(y)}$, where $\mathscr{O}_{\boldsymbol{Y},y}$ is considered as an $\mathscr{O}_{\boldsymbol{X},f(y)}$-module via $\varphi_y : \mathscr{O}_{\boldsymbol{X},f(y)} \to \mathscr{O}_{\boldsymbol{Y},y}$. In this situation, the subspace $\boldsymbol{Y}'$ of $\boldsymbol{Y}$ defined by the ideal $\mathscr{O}_{\boldsymbol{Y}} \cdot \mathscr{I}$ is called the inverse image of $\boldsymbol{X}'$ by (f,φ) and is denoted by $(f,\varphi)^*\boldsymbol{X}'$, or by $f^*\boldsymbol{X}'$ when φ is apparent. Let $(j,\varpi) : \boldsymbol{Y}' \to \boldsymbol{Y}$ be the canonical morphism. Then (f,φ) induces a morphism $(f',\varphi') : \boldsymbol{Y}' \to \boldsymbol{X}'$ and there is a commutative diagram

$$\begin{array}{ccc} \boldsymbol{Y}' & \xrightarrow{(f',\varphi')} & \boldsymbol{X}' \\ {\scriptstyle (j,\varpi)}\downarrow & & \downarrow{\scriptstyle (i,\pi)} \\ \boldsymbol{Y} & \xrightarrow{(f,\varphi)} & \boldsymbol{X}. \end{array}$$

Exercise 11.10. Show that, if $\boldsymbol{X}_1$ and $\boldsymbol{X}_2$ are complex subspaces of $\boldsymbol{X}$, then the inverse image of $\boldsymbol{X}_2$ by the canonical morphism $\boldsymbol{X}_1 \to \boldsymbol{X}$ coincides with the intersection of $\boldsymbol{X}_1$ and $\boldsymbol{X}_2$.

Complex spaces in complex manifolds: Let M be a complex manifold and $\mathscr{I}$ an ideal sheaf of finite type in $\mathscr{O}_M$. Then $\mathscr{O}_M/\mathscr{I}$ is coherent and its support, denoted by X, is a variety in M. Let $i : X \hookrightarrow M$ be the inclusion and set $\mathscr{O}_{\boldsymbol{X}} = i^{-1}(\mathscr{O}_M/\mathscr{I})$. Then $\boldsymbol{X} = (X,\mathscr{O}_{\boldsymbol{X}})$ is a complex subspace of M. Such a complex space is referred to as a complex space in M.

The local and global dimensions of $\boldsymbol{X}$ are defined to be those of X (cf. Section 2.4).

Higher direct images of sheaves

Let $f : Y \to X$ be a continuous map of topological spaces and $\mathscr{S}$ a sheaf on Y. The p-th *direct image* $R^p f_* \mathscr{S}$ of $\mathscr{S}$ by f is the sheaf on X defined by the presheaf $U \mapsto H^p(f^{-1}U; \mathscr{S})$. For an exact sequence as (11.3), we have an exact sequence

$$\begin{aligned} 0 \to f_*\mathscr{S}' \to f_*\mathscr{S} \to f_*\mathscr{S}'' \to R^1 f_*\mathscr{S}' \to \cdots \\ \to R^p f_*\mathscr{S}' \to R^p f_*\mathscr{S} \to R^p f_*\mathscr{S}'' \to R^{p+1} f_*\mathscr{S}' \to \cdots . \end{aligned}$$

Remark 11.10. If we denote by $\mathcal{S}h(X)$ and $\mathcal{S}h(Y)$ categories of sheaves of Abelian groups on X and Y (cf. Example A.1. 2), we may think of $f_* : \mathcal{S}h(Y) \to \mathcal{S}h(X)$ as a covariant functor and $R^p f_* : \mathcal{S}h(Y) \to \mathcal{S}h(X)$ as its p-th derived functor.

From Theorem A.14, we have:

Proposition 11.22. *If $f : Y \to X$ and $g : Z \to Y$ are continuous maps, for a sheaf $\mathscr{S}$ on Z, there exists a spectral sequence $(E_k^{p,q}, H^r)$ with*

$$E_2^{p,q} = R^p f_* R^q g_* \mathscr{S} \quad \textit{and} \quad H^r = R^r (f \circ g)_* \mathscr{S}.$$

Let $(f, \varphi) : \boldsymbol{Y} = (Y, \mathscr{O}_{\boldsymbol{Y}}) \to \boldsymbol{X} = (X, \mathscr{O}_{\boldsymbol{X}})$ be a morphism of complex spaces. If $\mathscr{S}$ is an $\mathscr{O}_{\boldsymbol{Y}}$-module, then for any open set U in X, $H^p(f^{-1}U; \mathscr{S})$ is a $\Gamma(f^{-1}U; \mathscr{O}_{\boldsymbol{Y}})$-module. On the other hand, φ induces a morphism $\Gamma(U; \mathscr{O}_{\boldsymbol{X}}) \to \Gamma(f^{-1}U; \mathscr{O}_{\boldsymbol{Y}})$. Thus $R^p f_* \mathscr{S}$ may be thought of as an $\mathscr{O}_{\boldsymbol{X}}$-module.

For coherent sheaves on complex spaces, there is the following fundamental theorem of H. Grauert:

Theorem 11.14. *Let $(f, \varphi) : \boldsymbol{Y} \to \boldsymbol{X}$ be a morphism of complex spaces with f proper. If $\mathscr{S}$ is a coherent $\mathscr{O}_{\boldsymbol{Y}}$-module, then $R^p f_* \mathscr{S}$ are coherent $\mathscr{O}_{\boldsymbol{X}}$-modules for all $p \geq 0$.*

For a complex space $\boldsymbol{X}$, let $K_h(\boldsymbol{X})$ denote the K-group of coherent $\mathscr{O}_{\boldsymbol{X}}$-modules. Then for a morphism $(f, \varphi) : \boldsymbol{Y} \to \boldsymbol{X}$ as above, we have a morphism

$$(f, \varphi)_! : K_h(\boldsymbol{Y}) \longrightarrow K_h(\boldsymbol{X}), \tag{11.16}$$

which is induced from $\mathscr{S} \mapsto \sum_{p \geq 0} (-1)^p R^p f_* \mathscr{S}$. In the case itis apparent what is φ, we denote $(f, \varphi)_!$ simply by $f_!$.

In particular, consider the case $\boldsymbol{X}$ is a point with reduced structure. Then, f being proper means that Y is compact. We have that, if Y is compact, for any coherent $\mathscr{O}_{\boldsymbol{Y}}$-module $\mathscr{S}$, $H^p(Y;\mathscr{S})$ are finite dimensional for all $p \geq 0$. This generalizes the statement (2) in Remark 11.8.

Proposition 11.23. *Let $\boldsymbol{X} = (X, \mathscr{O}_{\boldsymbol{X}})$ be a complex space in a complex manifold M with the embedding $(i, \pi) : \boldsymbol{X} \to M$. Then, for any coherent $\mathscr{O}_{\boldsymbol{X}}$-module $\mathscr{S}$, $R^p i_* \mathscr{S} = 0$ for $p \geq 1$ so that*

$$i_! \mathscr{S} = i_* \mathscr{S}.$$

Proof. First note that, since X is a closed set in M, i is proper. Let z be a point in M. For a holomorphically convex neighborhood U of z, $H^p(i^{-1}U;\mathscr{S}) \simeq H^p(U, i_*\mathscr{S})$ (cf. Exercise 11.6), which is 0 for $p \geq 1$ as $i_*\mathscr{S}$ is $\mathscr{O}_M$-coherent (cf. Theorem 11.7). □

We come back to this subject in Chapter 15 below.

11.6 Divisors

In this section, we let M denote a complex manifold of dimension n and $\mathscr{O}$ the sheaf of germs of holomorphic functions on M.

Meromorphic functions

Let U be an open set in M. Recall that we may identify the ring O_U of holomorphic functions on U with the ring $\Gamma(U;\mathscr{O})$ of sections of $\mathscr{O}$ on U.

We let S_U denote the set

$$\{ g \in O_U \mid \text{for any open set } V \subset U \text{ and } h \in O_V,\ g|_V \cdot h = 0 \text{ implies } h = 0 \}.$$

It is multiplicative in O_U so that we may form the fraction ring $S_U^{-1}O_U$ (cf. Section A.2). Note that, if U is non-empty and connected, then O_U is an integral domain and $S_U = O_U \smallsetminus \{0\}$ (cf. Theorem 1.7) so that $S_U^{-1}O_U$ is the fraction field $F(O_U)$ of O_U. If V is an open set with $V \subset U$, there is the restriction map $\rho_{VU} : S_U^{-1}O_U \to S_V^{-1}O_V$. From the system $\{S_U^{-1}O_U, \rho_{VU}\}$, we may construct a sheaf $\mathscr{M}$ of rings (cf. Section 11.1). Noting that, for each point z in M, the set of connected neighborhoods of z is cofinal in the set of neighborhoods of z, we see that the stalk $\mathscr{M}_z$ coincides with the fraction field of $\mathscr{O}_z$ (cf. Section 1.4). The sheaf $\mathscr{M}$ is called the *sheaf of germs of meromorphic functions* on M. It naturally contains $\mathscr{O}$ as a subsheaf.

There is a canonical morphism $S_U^{-1}O_U \to \Gamma(U;\mathscr{M})$, which is not surjective in general, as the condition (2) in Proposition 11.2 does not hold.

A *meromorphic function* φ on an open set U in M is a section of $\mathscr{M}$ on U. Such a function is represented by a system $\{(U_\alpha, f^\alpha, g^\alpha)\}$, where $\{U_\alpha\}$ is a covering of U and f^α, g^α are holomorphic functions on U_α such that the germ g_z^α at z is non-zero for all z in U_α, the germs f_z^α and g_z^α are relatively prime for all z in U_α (cf. Proposition 1.8) and that $f^\alpha g^\beta = f^\beta g^\alpha$ in $U_\alpha \cap U_\beta$. We write $\varphi = f^\alpha / g^\alpha$ on U_α. A *pole* of φ is a zero of g_α for some α.

Divisors

Let $\mathscr{M}^*$ denote the sheaf of multiplicative groups of non-zero germs of meromorphic functions, which contains the sheaf $\mathscr{O}^*$ of germs of non-vanishing holomorphic functions as a subsheaf. Here we note that $\mathscr{M}^*$ and $\mathscr{O}^*$ are the sets of invertible elements in $\mathscr{M}$ and $\mathscr{O}$, respectively. Thus, while a germ in $\mathscr{M}^*$ may be represented by a function with zeros, every germ in $\mathscr{O}^*$ is represented by a function without zeros.

The sheaf $\mathscr{D}iv$ of Cartier divisors on M is defined by

$$0 \longrightarrow \mathscr{O}^* \longrightarrow \mathscr{M}^* \longrightarrow \mathscr{D}iv \longrightarrow 0. \tag{11.17}$$

Definition 11.22. A *Cartier divisor* on M is a section of $\mathscr{D}iv$ on M.

A Cartier divisor D is represented by a system $\{(U_\alpha, \varphi^\alpha)\}$, where $\{U_\alpha\}$ is a covering of M, φ^α is in $\Gamma(U_\alpha; \mathscr{M}^*)$ for each α, and $f^{\alpha\beta} = \varphi^\alpha/\varphi^\beta$ is in $\Gamma(U_\alpha \cap U_\beta; \mathscr{O}^*)$ for each pair (α, β). While we use the multiplicative notation for $\mathscr{O}^*$ and $\mathscr{M}^*$, we use the additive notation for Cartier divisors. Thus if divisors D_i, $i = 1, 2$, are represented by $\{(U_\alpha, \varphi_i^\alpha)\}$, then $D_1 + D_2$ is represented by $\{(U_\alpha, \varphi_1^\alpha \cdot \varphi_2^\alpha)\}$. If D is represented by $\{(U_\alpha, \varphi^\alpha)\}$, then $-D$ is represented by $\{(U_\alpha, (\varphi^\alpha)^{-1})\}$. Also, $D = 0$ if and only if it is represented by $\{(U_\alpha, \varphi^\alpha)\}$ with φ^α in $\Gamma(U_\alpha; \mathscr{O}^*)$ for every α.

From (11.17) we have the exact sequence:

$$\cdots \longrightarrow \Gamma(M; \mathscr{M}^*) \longrightarrow \Gamma(M; \mathscr{D}iv) \xrightarrow{\delta^*} H^1(M; \mathscr{O}^*) \longrightarrow \cdots. \tag{11.18}$$

Definition 11.23. For a Cartier divisor D, the *line bundle* L_D *associated with* D is defined to be $\delta^*(D)^{-1}$.

If D is represented by the system as above, L_D is defined by the transition functions $\{f^{\alpha\beta}\}$ (cf. Definition 3.15). The line bundle L_D has a natural meromorphic section s_D, i.e., the section represented by φ^α on each U_α.

If φ is a section of $\mathscr{M}^*$, i.e., a meromorphic function not identically zero on any connected component of M, then it defines a Cartier divisor (cf. (11.18)), which we call the divisor of φ and denote by (φ). Such a divisor is called *principal*. If φ is represented by $\{(U_\alpha, f^\alpha, g^\alpha)\}$, we may write $(\varphi) = D_0 - D_\infty$, where D_0 and D_∞ are represented, respectively, by $\{(U_\alpha, f^\alpha)\}$ and $\{(U_\alpha, g^\alpha)\}$.

Two Cartier divisors D_1 and D_2 are said to be *linearly equivalent* if $D_1 - D_2$ is principal. By the exactness of (11.18), for a Cartier divisor D, L_D is trivial if and only if D is principal. Thus D_1 and D_2 are linearly equivalent if and only if $L_{D_1} = L_{D_2}$.

Definition 11.24. A *Weil divisor* on M is a locally finite formal sum $\sum m_i V_i$, where the V_i's are (globally) irreducible varieties of pure codimension one in M and the m_i's are integers.

In the above, the sum being locally finite means that each point in M has a neighborhood that intersects with only a finite number of V_i. The Weil divisors on M form an Abelian group, which we denote by $\mathrm{W}(M)$.

We show that a Cartier divisor naturally defines a Weil divisor. For this, we define the notion of the *order* of a meromorphic function, or section, along an irreducible variety of codimension one.

First we recall that there is a one-to-one correspondence between the set of irreducible germs of varieties of codimension one at 0 in $\mathbb{C}^n$ and the set of principal prime ideals in $\mathscr{O}_n$ (cf. Remark 2.9). Let V be an irreducible germ of a variety of codimension one at 0 and f a non-zero germ in $\mathscr{O}_n$. We define the order $\mathrm{ord}_V(f)$ of f along V as follows. If f is a unit, we set $\mathrm{ord}_V(f) = 0$. If f is not a unit, let $f = p_1^{m_1} \cdots p_r^{m_r}$ be the irreducible decomposition. If there is an i such that $V(p_i) = V$, we set $\mathrm{ord}_V(f) = m_i$, otherwise, $\mathrm{ord}_V(f) = 0$. If $\varphi = f/g$ is a germ in $\mathscr{M}^*$, we set $\mathrm{ord}_V(\varphi) = \mathrm{ord}_V(f) - \mathrm{ord}_V(g)$.

Next, let V be a globally irreducible variety of codimension one in M and V_{reg} the set of regular points in V. Also let L be a holomorphic line bundle on M and s a meromorphic section of L which is not identically zero on any connected component of M. Let $\{U_\alpha\}$ be a covering of M trivializing L and $\{f^{\alpha\beta}\}$ the system of transition functions with respect to some frames of L. Then s is represented by a collection $\{\varphi^\alpha\}$, where $\varphi^\alpha \in \Gamma(U_\alpha; \mathscr{M}^*)$, such that $\varphi^\alpha = f^{\alpha\beta}\varphi^\beta$ on $U_\alpha \cap U_\beta$. Let z be a point in V_{reg} (here we could take a point z where V_z is irreducible). Thus the prime ideal corresponding to V_z is generated by the germ of a holomorphic function without singular point. If z is in U_α, we have $\mathrm{ord}_{V_z}(\varphi_z^\alpha)$ as defined above. It does not depend on the choice of U_α containing z and is a locally constant function of z. Thus it does not depend on the choice of the point z, since $V_{\rm reg}$ is connected. We denote it by $\operatorname{ord}_V(s)$.

Proposition 11.24. *There is a natural Abelian group morphism*

$$\Phi : \Gamma(M; \mathscr{D}iv) \longrightarrow \mathrm{W}(M).$$

Proof. For a Cartier divisor D, associated is a line bundle L_D together with the natural meromorphic section s_D. Then

$$\operatorname{ord}_V : \Gamma(M; \mathscr{D}iv) \longrightarrow \mathbb{Z}, \qquad D \mapsto \operatorname{ord}_V(s_D)$$

is an Abelian group morphism. The formal sum

$$\sum \operatorname{ord}_V(s_D) \cdot V \tag{11.19}$$

over the irreducible varieties of codimension one in M is locally finite and is a Weil divisor. If we define $\Phi(D)$ to be the above sum, Φ is an Abelian group morphism. □

For a Cartier divisor D we call $\Phi(D)$ the Weil divisor associate with D. The union of varieties appearing in the sum (11.19) is called the *support* of D and is denoted by $|D|$. Note that, if $D = 0$, then $|D| = \emptyset$.

Remark 11.11. Let D be a Cartier divisor. Then the natural meromorphic section s_D of the associated bundle L_D is a non-vanishing section away from $|D|$. Thus there is the localization $c^1(L_D, s_D)$ in $H^2(M, M \smallsetminus |D|; \mathbb{Z})$ (cf. Section 5.6) and the associated residue $\operatorname{TRes}_{c^1}(s_D, L_D; |D|)$ as its image by the Alexander isomorphism

$$H^2(M, M \smallsetminus |D|; \mathbb{Z}) \xrightarrow{\sim} \check{H}_{2(n-1)}(|D|; \mathbb{Z}).$$

We see later that the Weil divisor associated with D naturally defines a class in $\check{H}_{2(n-1)}(|D|; \mathbb{Z})$ that coincides with $\operatorname{TRes}_{c^1}(s_D, L_D; |D|)$ (cf. Theorem 12.10 below).

On complex manifolds, a Weil divisor naturally defines a Cartier divisor as well:

Proposition 11.25. *On a complex manifold M, the morphism Φ is an isomorphism.*

Proof. We define a morphism

$$\Psi : \mathrm{W}(M) \longrightarrow \Gamma(M; \mathscr{D}iv),$$

which will be the inverse of Φ. For this, let $\sum m_i V_i$ be a non-zero Weil divisor. Then there exists a covering $\{U_\alpha\}$ of M such that each V_i with

$V_i \cap U_\alpha \neq \emptyset$ is defined by a holomorphic function f_i^α on U_α whose germ is reduced at every point in U_α (cf. Theorem 2.12 and Proposition 11.13). We set $I_\alpha = \{i \mid V_i \cap U_\alpha \neq \emptyset\}$. Taking a finer covering, if necessary, we may assume that it is a finite set. If we set $\varphi^\alpha = \prod_{i \in I_\alpha}(f_i^\alpha)^{m_i}$ on U_α, the system $\{(U_\alpha, \varphi^\alpha)\}$ defines a Cartier divisor, which is the image of $\sum m_i V_i$ by Ψ.

To show that $\Phi \circ \Psi = 1$, recall that, for a globally irreducible variety V, the regular part V_{reg} is connected and its closure in M is V (cf. Section 2.4). Thus in the sum (11.19) for the Cartier divisor $D = \Psi(\sum m_i V_i)$, only the V's with $V = V_i$ for some i appear. Moreover, for such a V, $\mathrm{ord}_V(s_D) = m_i$ by definition. The identity $\Psi \circ \Phi = 1$ also follows from definition. □

Thus as long as we consider divisors on complex manifolds, it is not necessary to distinguish Cartier and Weil divisors and we make identification via Φ, or Ψ, hereafter.

We may likewise define Cartier and Weil divisors on singular varieties (cf. Sections 14.3 and 14.4 below). In general, a Cartier divisor defines a Weil divisor, but not vice versa, as the following example shows.

Example 11.8. Let V the variety defined by $z_1 z_2 = z_3^2$ in $\mathbb{C}^3 = \{(z_1, z_2, z_3)\}$. Let V_1' and V_2' be defined by $z_1 = z_3 = 0$ and $z_2 = z_3 = 0$, respectively. Then they are irreducible varieties of codimension one in V and are Weil divisors, but not Cartier. The union $V_1' \cup V_2'$ is defined by $z_3 = 0$ and is a Cartier divisor.

A divisor $D = \sum m_i V_i$ is said to be *effective* if $m_i \geq 0$ for all i. In this case we write $D \geq 0$. It is equivalent to saying that it has a representative $(U_\alpha, \varphi_\alpha)$ with φ_α a holomorphic function on U_α for every α, i.e., the natural section s_D of the associated bundle L_D is holomorphic.

Let $D = \sum m_i V_i$ be a non-zero effective divisor. Then we may think of it as a complex space of pure dimension $n-1$ in M as follows. Let $\{(U_\alpha, \varphi_\alpha)\}$ be a representative with φ_α a holomorphic function on U_α. As $\varphi_\alpha/\varphi_\beta$ is a non-vanishing holomorphic function on $U_\alpha \cap U_\beta$, we may consider the ideal sheaf $\mathscr{I}$ in $\mathscr{O}_M$ generated by φ_α on U_α. The support of $\mathscr{O}_M/\mathscr{I}$ is $\bigcup_i V_i = |D|$ and D is the complex space $(|D|, i^{-1}(\mathscr{O}_M/\mathscr{I}))$, where $i : |D| \hookrightarrow M$ is the inclusion.

Let $f : M' \to M$ be a holomorphic map of complex manifolds and D a divisor on M. Taking the inverse images by f of meromorphic functions locally defining D, we get a divisor on M', which we call the inverse image of D by f and denote by f^*D. If D is effective, it is the inverse image of D as a complex subspace of M.

For a divisor $D = \sum_{i=1}^{r} m_i V_i$ which is a finite sum, we may write $L_D = \bigotimes_{i=1}^{r} L_{V_i}^{m_i}$, where $L_{V_i}^{m_i}$ denotes the tensor product of m_i copies of L_{V_i}, for $m_i > 0$, and the tensor product of $-m_i$ copies of $L_{V_i}^*$, for $m_i < 0$.

Let K_M be the canonical bundle of M (cf. Definition 3.13). The divisor defined by a meromorphic section of K_M is called the *canonical divisor*. Note that it is determined up to linear equivalence. The line bundle defined by the canonical divisor is K_M. The canonical divisor is also denoted by K_M, if there is no fear of confusion.

We discuss more on divisors in Section 12.6 and Chapter 14 below.

Blowing-up

Let U be a neighborhood of 0 in $\mathbb{C}^n = \{(z_1, \ldots, z_n)\}$ and $\tilde{U}$ the variety in $U \times \mathbb{P}^{n-1}$ defined by

$$\tilde{U} = \{\, (z, [\zeta]) \in U \times \mathbb{P}^{n-1} \mid z_{\alpha+1}\zeta_\beta - z_{\beta+1}\zeta_\alpha = 0, \quad 0 \le \alpha < \beta \le n-1 \}.$$

If we set $W_\alpha = \{\, [\zeta] \in \mathbb{P}^{n-1} \mid \zeta_\alpha \neq 0 \}$, $\alpha = 0, \ldots, n-1$, the manifold $U \times \mathbb{P}^{n-1}$ is covered by n coordinate neighborhoods $U \times W_\alpha$ with coordinates $(z_1, \ldots, z_n, \xi_1^\alpha, \ldots, \xi_{n-1}^\alpha)$, where (cf. (2.2))

$$\xi_i^\alpha = \begin{cases} \zeta_{i-1}/\zeta_\alpha & i = 1, \ldots, \alpha, \\ \zeta_i/\zeta_\alpha & i = \alpha+1, \ldots, n-1. \end{cases}$$

In $U \times W_\alpha$, $\tilde{U}$ is given by

$$\begin{cases} z_{\alpha+1}\xi_i^\alpha - z_i = 0 & i = 1, \ldots, \alpha \quad \text{and} \\ z_{\alpha+1}\xi_i^\alpha - z_{i+1} = 0 & i = \alpha+1, \ldots, n-1. \end{cases}$$

Thus $\tilde{U}$ is an n-dimensional complex manifold, which is covered by n coordinate neighborhoods $\tilde{U}_\alpha = \tilde{U} \cap (U \times W_\alpha)$ with coordinates $(z_{\alpha+1}, \xi_1^\alpha, \ldots, \xi_{n-1}^\alpha)$. If we denote by $\pi : \tilde{U} \to U$ the restriction of the projection $U \times \mathbb{P}^{n-1} \to U$ to $\tilde{U}$, it is a holomorphic map such that $D = \pi^{-1}(0) \simeq \mathbb{P}^{n-1}$ and that its restriction to $\tilde{U} \setminus D$ is a biholomorphic map onto $U \setminus \{0\}$. This procedure is called the *blowing-up* of U at 0. It replaces 0 with the projective space $\mathbb{P}^{n-1}$ to separate the lines through 0 in U. Note that D is defined by $z_{\alpha+1} = 0$ in $\tilde{U}_\alpha$ and is a divisor in $\tilde{U}$, which is called the *exceptional divisor*. The associated line bundle L_D is defined by the system of transition functions $\{z_{\alpha+1}/z_{\beta+1}\}$. Since $z_{\alpha+1}/z_{\beta+1} = \zeta_\alpha/\zeta_\beta$ on $\tilde{U}$, we see that $L_D|_D = -H_{n-1}$, where H_{n-1} is the hyperplane bundle on

$D = \mathbb{P}^{n-1}$ (cf. Example 3.6). Note that $\tilde{U}$ may be viewed as a neighborhood of the zero section of the tautological bundle on $\mathbb{P}^{n-1}$ (cf. Section 9.3).

Let $\boldsymbol{X} = (X, \mathscr{O}_{\boldsymbol{X}})$ be a complex space in U defined by an ideal sheaf $\mathscr{I}$ of finite type in $\mathscr{O}_U$ so that $\mathscr{O}_{\boldsymbol{X}} = i^{-1}(\mathscr{O}_U/\mathscr{I})$ (cf. Section 11.5). Suppose the underlying space X, which is a variety in U, contains 0. The *total transform* of $\boldsymbol{X}$ is the inverse image $\pi^*\boldsymbol{X}$ of $\boldsymbol{X}$ by π. It is the complex space in $\tilde{U}$ defined by the ideal sheaf $\mathscr{O}_{\tilde{U}} \cdot \pi^*\mathscr{I}$. If $f(z_1, \ldots, z_n)$ is a function in $\mathscr{I}$ and if m is the order of f at 0 (cf. Definition 1.10), then we may write, on $\tilde{U}_\alpha$,

$$\pi^* f(z_{\alpha+1}, \xi_1^\alpha, \ldots, \xi_{n-1}^\alpha) = z_{\alpha+1}^m \cdot \tilde{f}_\alpha(z_{\alpha+1}, \xi_1^\alpha, \ldots, \xi_{n-1}^\alpha) \tag{11.20}$$

with $\tilde{f}_\alpha$ a holomorphic function on $\tilde{U}_\alpha$. In $\tilde{U}_\alpha \cap \tilde{U}_\beta$, the difference of $\tilde{f}_\alpha$ and $\tilde{f}_\beta$ is a multiple by a non-vanishing holomorphic function. Thus we may consider the ideal sheaf $\tilde{\mathscr{I}}$ in $\mathscr{O}_{\tilde{U}}$ generated by the $\tilde{f}_\alpha$'s. The *proper transform* of $\boldsymbol{X}$, denoted by $\tilde{\boldsymbol{X}}$, is the complex space in $\tilde{U}$ defined by $\tilde{\mathscr{I}}$. The underlying space of the total transform $\pi^*\boldsymbol{X}$ is the union of D and $\tilde{X}$, the underlying space of $\tilde{\boldsymbol{X}}$. The holomorphic map π restricts to an isomorphism $\tilde{\boldsymbol{X}}|_{\tilde{X}\setminus\pi^{-1}(0)} \xrightarrow{\sim} \boldsymbol{X}|_{X\setminus\{0\}}$ of complex spaces.

Example 11.9. Let U be a neighborhood of 0 in $\mathbb{C}^2$ and $\pi : \tilde{U} \to U$ the blowing-up of U at 0. Let C be a curve (one-dimensional variety) in U through 0 defined by a (reduced) holomorphic function f. Then the total transform π^*C is the pull-back of the divisor C and is written $\pi^*C = \tilde{C} + mD$, where $\tilde{C}$ is the proper transform of C, $D = \pi^{-1}(0) \simeq \mathbb{P}^1$ and m is the order of f at 0. The restriction ϖ of π to $\tilde{C}$ is a surjective holomorphic map $\varpi : \tilde{C} \to C$ such that $\varpi^{-1}(0)$ consists of a finite number of points and that ϖ induces a biholomorphic map from $\tilde{C} \setminus \varpi^{-1}(0)$ onto $C \setminus \{0\}$.

It is known that, continuing the above process, we may "resolve" the singularity of a plane curve. Namely, let C be a curve in U with singularity only at 0 and let $C = \bigcup_{i=1}^r C_i$ be the irreducible decomposition at 0. Then, after a finite number of suitable blowing-ups, we obtain a two-dimensional complex manifold $\tilde{U}$ together with a proper surjective holomorphic map $\pi : \tilde{U} \to U$ such that we may write

$$\pi^*C = \sum_{i=1}^r \tilde{C}_i + \sum_{j=1}^s m_j D_j,$$

where

(1) the $\tilde{C}_i$'s are mutually disjoint connected non-singular curves,
(2) for each j, D_j is a non-singular curve biholomorphic with $\mathbb{P}^1$ and m_j is a positive integer,

(3) if we set $D = \bigcup_{j=1}^{s} D_j$, then $\pi^{-1}(0) = D$ and $\pi|_{\tilde{U} \setminus D}$ is biholomorphic onto $U \setminus \{0\}$,
(4) for each i, $\tilde{C}_i$ intersects D at one point q_i and $\pi(\tilde{C}_i) = C_i$. Thus, if we set $\varpi_i = \pi|_{\tilde{C}_i}$, then $\varpi_i^{-1}(0) = q_i$ and $\varpi|_{\tilde{C}_i \setminus \{q_i\}}$ is biholomorphic onto $C_i \setminus \{0\}$ and
(5) every singular point of $\tilde{C} \cup D$, $\tilde{C} = \bigcup_{i=1}^{r} \tilde{C}_i$, is equivalent to the one in Example 2.3. 1, i.e., every intersection of $\tilde{C} \cup D$ is a "normal crossing".

Let C be a (possibly singular) compact curve in a (non-singular) complex surface (complex manifold of dimension two) X. The singular set $\mathrm{Sing}(C)$ consists of a finite number of points. Using the above, we may construct a non-singular compact curve $\tilde{C}$ together with a holomorphic map $\varpi : \tilde{C} \to X$ with $\varpi(\tilde{C}) = C$ such that

(1) for each p in $\mathrm{Sing}(C)$, $\varpi^{-1}(p)$ consists of a finite number of points, and
(2) $\varpi|_{\tilde{C} \setminus \varphi^{-1}(S)}$ is a biholomorphic map onto $C \setminus S$.

The above $\tilde{C}$ is determined uniquely up to biholomorphic equivalence. If C is (globally) irreducible, $\tilde{C}$ is connected. We call $\tilde{C}$ the *non-singular model* of C.

Exercise 11.11. Let V be the variety defined by $z_1 z_2 - z_3^2 = 0$ in $\mathbb{C}^3 = \{(z_1, z_2, z_3)\}$. Find the total transform and the proper transform of V under the blowing-up of $\mathbb{C}^3$ at 0.

11.7 Local complete intersections

Complete intersections

Let $\mathscr{O}_n$ denote the ring of germs of holomorphic functions at 0 in $\mathbb{C}^n$. For a germ V of a variety at 0 in $\mathbb{C}^n$, let $I(V)$ denote the ideal in $\mathscr{O}_n$ of germs vanishing on V. Also, for an ideal I in $\mathscr{O}_n$, let $V(I)$ denote the germ of the variety defined by I. Recall that $V(I(V)) = V$ and $I(V(I)) = \sqrt{I}$ (cf. Sections 1.4, 2.3 and 2.4).

Note that, if an ideal I is generated by k germs, then $\dim V(I) \geq n - k$ (cf. Theorem 2.11).

Definition 11.25. Let V be a germ of a variety of pure dimension d at 0 in $\mathbb{C}^n$. Set $k = n - d$.

1. We call V a *set-theoretic complete intersection germ*, if $V = V(I)$ for some ideal I generated by k germs in $\mathscr{O}_n$.
2. We call V a *complete intersection germ*, a *CI germ* for short, if the ideal $I(V)$ is generated by k germs in $\mathscr{O}_n$.

A CI germ is a set-theoretic one as $V = V(I(V))$, however the converse is not true in general (cf. Example 11.10. 2 below). The germ of a non-singular variety V is a CI. Indeed, by a suitable change of coordinates $(z_1, \ldots, z_n)$ of $\mathbb{C}^n$, we may assume that $I(V) = (z_1, \ldots, z_k)$. Every germ of a hypersurface, i.e., the case $k = 1$, is a CI germ by Theorem 2.12.

Example 11.10. 1. The germ at 0 of the variety in Exercise 2.5. 4 is not a set-theoretical complete intersection.
2. Let V be the germ at 0 of the union of the coordinate axes in $\mathbb{C}^3 = \{(z_1, z_2, z_3)\}$. Then $V = V(I)$ with $I = (z_1z_2, (z_1 + z_2)z_3)$ and thus V is a set-theoretic complete intersection. On the other hand, it can be shown that $I(V) = (z_1z_2, z_2z_3, z_3z_1)$ and that the number of generators cannot be reduced.

We quote the following:

Theorem 11.15. *Let $f_1, \ldots, f_k$ be germs in $\mathscr{O}_n$ and $V = V(I)$, $I = (f_1, \ldots, f_k)$. If $C(f) \cap V$ is nowhere dense in V, V is a CI of dimension $d = n - k$ and $I(V) = I$.*

In the above, we take a neighborhood U of 0 in $\mathbb{C}^n$ where the germs $f_1, \ldots, f_k$ have representatives and denote by $C(f)$ the *critical set* of $f = (f_1, \ldots, f_k)$, i.e., the set of critical points of f, in U (cf. Definition 1.9). The converse of the above statement is proved below (cf. (11.22)), i.e., it holds that, if V is a CI with $I(V) = (f_1, \ldots, f_k)$, then $C(f) \cap V$ is nowhere dense in V, $f = (f_1, \ldots, f_k)$.

Exercise 11.12. Show that, if we set $f = (z_1z_2, (z_1 + z_2)z_3)$ in Example 11.10. 2, then $C(f) \cap V$ is not nowhere dense in V.
Singular set: For complete intersections we give a simpler proof of Theorem 11.5, although we still use the coherece of ideal sheaves (cf. Theorem 11.4).

In general, let $f_1, \ldots, f_r$ be germs in $\mathscr{O}_n$ and set $g_i = \sum_{j=1}^r a_{ij} f_j$ with $a_{ij} \in \mathscr{O}_n$, for $i = 1, \ldots, s$. Then we see that

$$\operatorname{rank} \frac{\partial(g_1, \ldots, g_s)}{\partial(z_1, \ldots, z_n)}(0) \leq \operatorname{rank} \frac{\partial(f_1, \ldots, f_r)}{\partial(z_1, \ldots, z_n)}(0).$$

From this we have:

Lemma 11.5. *Let V be a germ of a variety at 0 in $\mathbb{C}^n$. Then the rank of the Jacobian matrix $\partial(f_1, \dots, f_r)/\partial(z_1, \dots, z_n)$ at 0 does not depend on the choice of the generators $f_1, \dots, f_r$ of $I(V)$.*

We denote the above rank by $\rho(0)$.

Now let V be a CI germ of dimension d and $f_1, \dots, f_k$ generators of $I(V)$. We take a neighborhood U of 0 in $\mathbb{C}^n$ such that V and $f_1, \dots, f_k$ have representatives on U. We may assume that the germs $f_{1,z}, \dots, f_{k,z}$ generate $\mathscr{I}_{V,z} = I(V_z)$ for every z in U (cf. Theorem 11.4 and Exercise 11.8). This shows that V_z is a CI of dimension d for every $z \in U \cap V$. We may also write

$$V = \{ z \in U \mid f_1(z) = \cdots = f_k(z) = 0 \}. \tag{11.21}$$

Proposition 11.26. *In the above situation,*

$$\operatorname{Sing}(V) = \left\{ z \in V \;\middle|\; \operatorname{rank} \frac{\partial(f_1, \dots, f_k)}{\partial(z_1, \dots, z_n)}(z) < k \right\}.$$

Proof. If z is a regular point of V, then $\rho(z) = k$ (cf. Exercise 2.7.1). Conversely, if $\rho(z) = k$, then z is a regular point of V by definition (cf. Definition 2.8). □

In the above situation, we think of $f = (f_1, \dots, f_k) : U \to \mathbb{C}^k$ as being a holomorphic map and denote by $C(f)$ its critical set. Then Proposition 11.26 implies:

$$\operatorname{Sing}(V) = C(f) \cap V. \tag{11.22}$$

For the hypersurface case, we have:

Proposition 11.27. *In the case $k = 1$, $C(f) \subset V$ so that*

$$\operatorname{Sing}(V) = C(f), \qquad f = f_1.$$

Proof. Let z be a point in $C(f)$. There exist a connected neighborhood D of 0 in $\mathbb{C} = \{t\}$ and a holomorphic map $\varphi = (\varphi_1, \dots, \varphi_n) : D \to U$ such that $\varphi(0) = 0$, $\varphi(t_0) = z$ for some t_0 and $\varphi(t) \in C(f)$ for all t in D. Then we have

$$\frac{d}{dt}(f \circ \varphi) = \sum_{i=1}^{n} \Big(\frac{\partial f}{\partial z_i} \circ \varphi \Big) \frac{d\varphi_i}{dt} = 0,$$

which shows that $f \circ \varphi$ is constantly equal to 0 and thus z is in V. □

Isolated singularity case: In the situation of Proposition 11.26, we think of $f = (f_1, \dots, f_k) : U \to \mathbb{C}^k$ as being a holomorphic map and let $C(f)$ denote the critical set of f, as before. We also set $D(f) = f(C(f))$, the set of critical values.

Proposition 11.28. *Suppose* $\mathrm{Sing}(V) = \{0\}$. *Then there exist a neighborhood* U *of* 0 *in* $\mathbb{C}^n$ *as above and a neighborhood* U' *of* 0 *in* $\mathbb{C}^k$ *such that*

(1) $C(f)$ *is a variety of pure dimension* $k-1$ *in* U,
(2) $f(C(f)) \subset U'$ *and* $f|_{C(f)} : C(f) \to U'$ *is a finite map,*
(3) $D(f)$ *is a variety of pure dimension* $k-1$, *i.e., a hypersurface in* U'.

Proof. Since $C(f)$ is the set of common zeros of the $k \times k$ minors of the Jacobian matrix $\partial(f_1, \dots, f_k)/\partial(z_1, \dots, z_n)$, it is a variety in U. By the assumption $\mathrm{Sing}(V) = \{0\}$, it is non-empty (cf. (11.22)) and, for every $z \in C(f)$, $\dim C(f)_z \geq k-1$ (cf. Proposition A.23.1 and (A.16)). The assumption $\mathrm{Sing}(V) = \{0\}$ also implies that there exists a neighborhood U' of 0 in $\mathbb{C}^k$ such that $f|_{C(f)\cap f^{-1}U'} : C(f) \cap f^{-1}U' \to U'$ is proper and thus finite (cf. Proposition 2.7). By Proposition 2.8, $D(f)$ is a variety in U'. By Sard's theorem (cf. Remark 2.10), $\dim D(f)_{z'} \leq k-1$ for all $z' \in D(f)$. Therefore, again by Proposition 2.8, $C(f)$ and $D(f)$ are pure $(k-1)$-dimensional. Replacing U with $f^{-1}U'$, we have the proposition. □

Global complete intersections: Let M be a complex manifold of dimension n.

Definition 11.26. Let V be a variety of pure dimension d in M and set $k = n - d$. We call V a *complete intersection*, a *CI* for short, in M if there exist k holomorphic functions $f_1, \dots, f_k$ on M such that

(1) $V = \{ z \in M \mid f_1(z) = \cdots = f_k(z) = 0 \}$,
(2) $C(f) \cap V$ is nowhere dense in V, $f = (f_1, \dots, f_k)$.

If V is as above, for every $z \in V$, V_z is a CI germ of dimension d by Theorem 11.15. We have $\mathrm{Sing}(V) = C(f) \cap V$.

If V is a CI germ, there exists a neighborhood U of 0 in $\mathbb{C}^n$ such that V has a representative that is a complete intersection in U (cf. (11.21)). In this situation, we set $V_t = f^{-1}(t)$ for $t \in \mathbb{C}^k$. Then we have:

Proposition 11.29. *If* V *is a CI germ, there exists a neighborhood* U *of* 0 *in* $\mathbb{C}^n$ *such that* V_t *is a complete intersection in* U *for all* t *near* $0 \in \mathbb{C}^k$. *Moreover,* $\mathrm{Sing}(V_t) = C(f) \cap V_t$.

Milnor fibration

In general, let V be a variety of pure dimension d in a neighborhood U of 0 in $\mathbb{C}^n$ such that V contains 0 and that $V \setminus \{0\}$ is non-singular. Let

$$\mathbb{B}_\varepsilon = \{ (z_1, \dots, z_n) \mid |z_1|^2 + \cdots + |z_n|^2 \leq \varepsilon^2 \}$$

be the closed ball of radius ε and $\mathbb{S}_\varepsilon = \partial\mathbb{B}_\varepsilon$ the $(2n-1)$-sphere of radius ε.

We quote the following two theorems of J. Milnor.

Theorem 11.16. *For sufficiently small ε, V intersects $\mathbb{S}_\varepsilon$ transversely and the pair $(\mathbb{B}_\varepsilon, V \cap \mathbb{B}_\varepsilon)$ is homeomorphic with the cone over $(\mathbb{S}_\varepsilon, V \cap \mathbb{S}_\varepsilon)$.*

Thus, in the above situation, $L = V \cap \mathbb{S}_\varepsilon$ is a $(2d-1)$-dimensional C^∞ manifold, which is called the *link* of V at 0.

Remark 11.12. The above theorem is proved by constructing a C^∞ vector field v on $O \setminus \{0\}$, where O is an open set in U containing $\mathbb{B}_\varepsilon$, such that $v(x)$ is pointing away from 0 for all $x \in O \setminus \{0\}$ and that $v(x)$ is tangent to V for all $x \in V \cap (O \setminus \{0\})$.

Let V be a hypersuface and f a generator of $I(V)$. We suppose f has a representative on U and V has at most an isolated singularity at 0.

Theorem 11.17. *Let V be a hypersurface as above.*

1. *For sufficiently small ε,*

$$\varphi = \frac{f}{|f|} : \mathbb{S}_\varepsilon \setminus L \longrightarrow \mathbb{S}^1$$

is a C^∞ locally trivial fibration.

2. *The fiber of φ has the homotopy type of a bouquet of $(n-1)$-spheres.*

In the first step of the proof it is shown that φ is a submersion. However we cannnot directly apply the Ehresmann fibration theorem (Theorem 3.12) as φ is not proper. It is done by constructing some special horizontal vector field on $\mathbb{S}_\varepsilon \setminus L$. Since $\mathbb{S}^1$ is connected, each fiber $F_t = \varphi^{-1}(t)$ is diffeomorphic to a fixed C^∞ manifold F, which is called the *Milnor fiber*. The closure of each fiber F_t is equal to $F_t \cup L$. In fact F has a natural complex structure.

Now let V be a CI of dimension d, having at most an isolated singularity at 0, and $f_1, \dots, f_k$ generators of $I(V)$, $k = n - d$. Let $C(f)$ be the critical set of $f = (f_1, \dots, f_k)$ and $D(f) = f(C(f))$ the set of critical values. If V is non-singular at 0, $C(f) = \emptyset$ and $D(f) = \emptyset$. If V has a singularity at

0, we take neighborhoods U and U' of 0 in $\mathbb{C}^n$ and $\mathbb{C}^k$, respectively, as in Proposition 11.28.

We have the following:

Theorem 11.18. *Let V be a CI of dimension d as above.*

1. *For sufficiently small balls $\mathbb{B}_\varepsilon$ around 0 in U and $\mathbb{B}'_\delta$ around 0 in U', f induces the structure of a C^∞ locally trivial fibration on*

$$\mathbb{B}_\varepsilon \cap f^{-1}(\mathbb{B}'_\delta \setminus D(f)) \longrightarrow \mathbb{B}'_\delta \setminus D(f).$$

2. *The fiber of the above fibration has the homotopy type of a bouquet of d-spheres.*

The statement 1 follows from the Ehresmann theorem with boundary. The fiber of the map makes sense, since $\mathbb{B}'_\delta \setminus D(f)$ is connected (cf. Remark 3.9). The interior F of the fiber has a natural complex structure and, in the case $k = 1$, it is diffeomorphic to the Milnor fiber above. Thus it is also referred to as the Milnor fiber and the number of spheres appearing in the above is called the *Milnor number* of V at 0 and is denoted by $\mu(V, 0)$. The number $\mu(V, 0)$ does not depend on the choice of generators of $I(V)$. It is positive if and only if 0 is a singular point of V.

There is an algebraic formula for this number. We set, for $i = 1, \ldots, k$,

$$a_i = \dim_{\mathbb{C}} \mathscr{O}_n/(J(f_1, \ldots, f_i), f_1, \ldots, f_{i-1}),$$

where the denominator in the right-hand side is the ideal generated by the Jacobians $\det(\partial(f_1, \ldots, f_i)/\partial(z_{\nu_1}, \ldots, z_{\nu_i}))$, $1 \leq \nu_1 < \cdots < \nu_i \leq n$, and $f_1, \ldots, f_{i-1}$. Then

$$\mu(V, 0) = \sum_{i=1}^{k} (-1)^{k-i} a_i. \tag{11.23}$$

In particular if $k = 1$,

$$\mu(V, 0) = \dim_{\mathbb{C}} \mathscr{O}_n/J(f), \tag{11.24}$$

where $J(f) = (\partial f/\partial z_1, \ldots, \partial f/\partial z_n)$, $f = f_1$. The right-hand side of (11.24) is referred to as the *multiplicity* of the function f at 0 and is denoted by $m(f, 0)$.

Tangent and cotangent sheaves on a variety

Let M be a complex manifold and V an analytic variety in M with the inclusion $i : V \hookrightarrow M$. We denote by $\mathscr{I}_V$ the ideal sheaf of holomorphic functions vanishing on V (cf. Sections 11.3 and 11.5). Then the sheaf $\mathscr{O}_V = i^{-1}(\mathscr{O}_M/\mathscr{I}_V)$ is the sheaf of holomorphic functions on V. We wish to define the sheaf of holomorphic 1-forms on V. For this we "eliminate", from the $\mathscr{O}_M$-module Ω_M of holomorphic 1-forms on M, the differentials of functions in $\mathscr{I}_V$ in the following manner. We consider the sheaf $\mathscr{I}_V/\mathscr{I}_V^2$, which is an $(\mathscr{O}_M/\mathscr{I}_V)$-module, and define

$$d : \mathscr{I}_V/\mathscr{I}_V^2 \longrightarrow (\mathscr{O}_M/\mathscr{I}_V) \otimes_{\mathscr{O}_M} \Omega_M \qquad \text{by } [f] \mapsto [1] \otimes df.$$

It is a well-defined morphism of $(\mathscr{O}_M/\mathscr{I}_V)$-modules and induces a morphism $i^{-1}d : i^{-1}(\mathscr{I}_V/\mathscr{I}_V^2) \to \mathscr{O}_V \otimes_{i^{-1}\mathscr{O}_M} i^{-1}\Omega_M = i^*\Omega_M$ of $\mathscr{O}_V$-modules.

Definition 11.27. We define the *sheaf of holomorphic* 1*-forms* Ω_V on V, or the *cotangent sheaf* of V, by the exact sequence

$$i^{-1}(\mathscr{I}_V/\mathscr{I}_V^2) \xrightarrow{i^{-1}d} i^*\Omega_M \longrightarrow \Omega_V \longrightarrow 0.$$

If we denote by V_{reg} the regular part $V \smallsetminus \mathrm{Sing}(V)$ of V, the image of the restriction of $i^{-1}d$ to V_{reg} is identified with the sheaf $\mathscr{O}_{V_{\mathrm{reg}}}(N^*_{V_{\mathrm{reg}}})$ of germs of holomorphic sections of the "conomal bundle" $N^*_{V_{\mathrm{reg}}}$. We also have $\Omega_{V_{\mathrm{reg}}} = \mathscr{O}_{V_{\mathrm{reg}}}(T^*V_{\mathrm{reg}})$.

We call $\Theta_V = \mathscr{H}om_{\mathscr{O}_V}(\Omega_V, \mathscr{O}_V)$ the *tangent sheaf* of V and $\mathscr{N}_V = \mathscr{H}om_{\mathscr{O}_V}(i^{-1}(\mathscr{I}_V/\mathscr{I}_V^2), \mathscr{O}_V)$ the *normal sheaf* of V in M. Since $\Theta_{V_{\mathrm{reg}}} = \mathscr{O}_{V_{\mathrm{reg}}}(TV_{\mathrm{reg}})$ and $\mathscr{N}_{V_{\mathrm{reg}}} = \mathscr{O}_{V_{\mathrm{reg}}}(N_{V_{\mathrm{reg}}})$, we have the following commutative diagram with exact rows:

$$\begin{array}{ccccccccc} 0 & \longrightarrow & \Theta_{V_{\mathrm{reg}}} & \longrightarrow & (i^*\Theta_M)|_{V_{\mathrm{reg}}} & \longrightarrow & \mathscr{N}_{V_{\mathrm{reg}}} & \longrightarrow & 0 \\ & & \downarrow & & \downarrow & & \downarrow & & \\ 0 & \longrightarrow & \Theta_V & \longrightarrow & i^*\Theta_M & \longrightarrow & \mathscr{N}_V. & & \end{array} \tag{11.25}$$

Local complete intersections

Let M be a complex manifold of dimension n and V a variety of pure d dimension in M. Set $k = n - d$.

Definition 11.28. We call V a *local complete intersection*, an *LCI* for short, *in* M, if $\mathscr{I}_{V,z}$ is generated by k germs for each z in V.

If V is as above, V_z is a CI germ for each z in V, as $\mathscr{I}_{V,z} = I(V_z)$. If $f_1, \dots, f_k$ are generators of $\mathscr{I}_{V,z}$, we have an expression as (11.21) for a neighborhood U of z. We call $f_1, \dots, f_k$ local generators of $\mathscr{I}_V$. Their classes $[f_1], \dots, [f_k]$ locally generate $i^{-1}(\mathscr{I}_V/\mathscr{I}_V^2)$ and are linearly independent over $\mathscr{O}_V$. Hence $i^{-1}(\mathscr{I}_V/\mathscr{I}_V^2)$ is locally $\mathscr{O}_V$-free of rank k and $\mathscr{N}_V$ is also locally $\mathscr{O}_V$-free of rank k. Thus there is a vector bundle N_V on V of rank k extending $N_{V_{\mathrm{reg}}}$. We have a commutative diagram with an exact row:

$$\begin{array}{ccccccccc} 0 & \longrightarrow & TV_{\mathrm{reg}} & \longrightarrow & TM|_{V_{\mathrm{reg}}} & \xrightarrow{\varpi} & N_{V_{\mathrm{reg}}} & \longrightarrow & 0 \\ & & & & \downarrow & & \downarrow & & \\ & & & & TM|_V & \xrightarrow{\varpi} & N_V. & & \end{array} \tag{11.26}$$

If we take the sheaves of germs of holomorphic sections of the bundles in (11.26), we recover (11.25) except for the term Θ_V. If $f_1, \dots, f_k$ are local generators of $\mathscr{I}_V$, then the trivialization of N_V dual to $([f_1], \dots, [f_k])$ extends the trivialization $(\varpi(\partial/\partial f_1), \dots, \varpi(\partial/\partial f_k))$ of $N_{V_{\mathrm{reg}}}$. (Note that near a regular point of $f = (f_1, \dots, f_k)$, we may take $(f_1, \dots, f_k)$ as a part of a local coordinate system on M.) We call it the *trivialization of N_V associated with $f = (f_1, \dots, f_k)$*.

For an LCI V in M, we set $\tau_V = TM|_V - N_V$ and call it the *virtual tangent bundle* of V. Note that, if V is non-singular, $TM|_V - N_V = TV$ in the K-group $K(V)$ of V (cf. Section 15.2 below).

Sometimes we impose a stronger condition:

Definition 11.29. Let V be a variety of pure codimension k in a complex manifold M. We say that V is a *local complete intersection defined by a section*, an *SLCI* for short, *in M*, if there exist a holomorphic vector bundle N on M of rank k and a holomorphic section s of N such that $\mathscr{I}_V$ is locally generated by the components of s (with respect to some local holomorphic frame of N).

Thus V is the zero set of s and is an LCI. We may express the virtual tangent bundle as $\tau_V = (TM - N)|_V$.

Example 11.11. As examples of SLCI's, we have the following:

1. A hypersurface V in M $(k = 1)$. In this case, we may take as N the line bundle L_V associated with V and as s the natural section (cf. Definition 3.15 and the subsequent notes).

2. A global complete intersection V in M. In this case we may take as N the product bundle $\mathbb{C}^k \times M$ and as s a family of defining functions of V (cf. Definition 11.26).

3. A (projective algebraic) complete intersection V in the projective space $\mathbb{P}^n$. This means that the ideal $I(V)$ of homogeneous polynomials vanishing on V is generated by k homogeneous polynomials $P_1, \dots, P_k$. Let p_i denote the degree of P_i for $i = 1, \dots, k$ and U_α the affine coordinate $\zeta_\alpha \neq 0$ for $\alpha = 0, \dots, n$. Then, in U_α, V is defined by $f_1 = \cdots = f_k = 0$, $f_i = P_i/\zeta_\alpha^{p_i}$. Note that it is only locally a complete intersection. In this case, we may take as N the bundle $H_n^{\otimes p_1} \oplus \cdots \oplus H_n^{\otimes p_k}$, where H_n denotes the hyperplane bundle (cf. Example 3.6).

4. Let M be a complex manifold of dimension n and N a holomorphic vector bundle of rank k on M. Let V be the zero set of a holomorphic section s of N generically transverse to the zero section. This means that, if $(f_1, \dots, f_k)$ denote the components of s with respect to a holomorphic frame on an open set U in M, V is given by $f_1 = \cdots = f_k = 0$ in U and $df_1 \wedge \cdots \wedge df_k \not\equiv 0$ on $V \cap U$. In this case the singular set $\mathrm{Sing}(V)$ of V is given by $df_1 \wedge \cdots \wedge df_k = 0$ in $V \cap U$ and the restriction of N to the regular part $V_{\mathrm{reg}} = V \smallsetminus \mathrm{Sing}(V)$ coincides with the normal bundle of V_{reg} in M. By Theorem 11.15, V is a local complete intersection of dimension $d = n - k$.

Complex spaces defined by a regular section

We refer to Section A.2 for the definition and fundamental properties of regular sequences. In the following, an $\mathcal{O}_n$-regular sequence in the maximal ideal $\mathfrak{m}$ of $\mathcal{O}_n$ is simply called a regular sequence.

Let I be the ideal generated by k germs $f_1, \dots, f_k$ in $\mathcal{O}_n$ and $V = V(I)$. Recall that V is pure d-dimensional, $d = n - k$, if and only if $f_1, \dots, f_k$ form a regular sequence (cf. Proposition A.23).

Definition 11.30. A germ $\boldsymbol{X} = (X, \mathcal{O}_{\boldsymbol{X}})$ of a complex space is said to be defined by a regular sequence if $\mathcal{O}_{\boldsymbol{X}} = \mathcal{O}_n/I$ for some ideal $I = (f_1, \dots, f_k)$ with $f_1, \dots, f_k$ a regular sequence.

If $\boldsymbol{X} = (X, \mathcal{O}_{\boldsymbol{X}})$ is as above, X is a set-theoretical complete intersection. If V is a set-theoretical complete intersection germ defined by I, then $(V, \mathcal{O}_n/I)$ is a germ of a complex space defined by a regular sequence.

Remark 11.13. If I is as in Definition 11.30, then $\mathcal{O}_n/I$ is a Cohen-Macaulay ring (cf. Proposition A.22).

Let M be a complex manifold of dimension n and E a holomorphic vector bundle of rank l on M. Also let s be a holomorphic section of E and $\boldsymbol{S}$ the complex space defined by (the ideal generated by the local components of) s. Thus the support of $\boldsymbol{S}$ is the zero set S of s.

Definition 11.31. The section s is called a *regular section* if, at each point p in S, the germs at p of local components of s form a regular sequence in the ring $\mathscr{O}_{M,p}$ of germs holomorphic functions at p on M.

Note that this is equivalent to saying that $\dim S = n - l$.

Let $\boldsymbol{X} = (X, \mathscr{O}_{\boldsymbol{X}})$ be a complex space of pure dimension d in M so that we may write $\mathscr{O}_{\boldsymbol{X}} = i^{-1}(\mathscr{O}_M/\mathscr{I})$ for some finitely generated ideal sheaf $\mathscr{I}$ in $\mathscr{O}_M$, $i : X \hookrightarrow M$ being the inclusion. Set $k = n - d$.

Definition 11.32. We say that $\boldsymbol{X}$ as above is a *complex space defined by a regular section* if there exist a holomorphic vector bundle E on M of rank k and a regular section s of E such that $\mathscr{I}$ is locally generated by the components of s (with respect to some local holomorphic frame of E).

If $\boldsymbol{X}$ is as in Definition 11.32, the class $c^k(E, s)$ in $H^{2l}(M, M \smallsetminus X)$ is Alexander dual to $[\boldsymbol{X}]$ in $\breve{H}_{2d}(X)$ (cf. Theorem 12.9 below).

Notes

There are a number of references on sheaves and sheaf cohomology. Here we list [Godement (1958); Hirzebruch (1966); Serre (1955b)].

Here is a word about the exponetial sequence. In some textbooks, the first Chern class of a C^∞ or holomorphic line bundle L is defined as $c^1(L) = \delta^*(L)$ with δ^* the connecting morphism defined the same way as in (11.4) with $\mathscr{C}$ and $\mathscr{C}^*$ replaced by $\mathscr{A}$ and $\mathscr{A}^*$ or as in (11.6). Then they try to represent it by a differential form via the canonical morphism $H^2(M; \mathbb{Z}) \to H^2(M; \mathbb{C})$ followed by the de Rham isomorphism $H^2(M; \mathbb{C}) \simeq H^2_d(M)$ that is proved sheaf theoretically as Theorem 11.9 (cf. Remark 11.6. 2). Suppose L is defined by a system of transition functions $\{f^{\alpha\beta}\}$ on some covering and let $\{h_\alpha\}$ be a Hermitian metric on L, $h_\beta = |f^{\alpha\beta}|^2 h_\alpha$. If one proceeds this way correctly, one gets the differential form $\omega(L) = -\frac{1}{2\pi\sqrt{-1}}\bar{\partial}\partial \log h_\alpha$, with a sign different from the expected one. For example if L is the hyperplane bundle H, $\omega(H)$ is the negatve of the Fubini-Study form. For this reason, in some books, the first Chern class is defined by $c^1(L) = -\delta^*(L)$. However, this is not correct from the combinatorial viewpoint (cf. Example 11.3, see also the proof of Theorem 4.3.1 in [Hirzebruch (1966)], where a similar computation is done). The point is that the proof of the de Rham theorem as above gives one of the isomorphisms and the correspondence it provides does not have much significance. If one uses the correspondence (7.10) which gives the canonical isomorphism in Corollary 7.4, as in the proof of Proposition 11.5 via Chern-Wel theory, one obtains a correct form. A similar remark applies to the Dolbeault case, see for example Proposition 11.16.

For the materials in Section 11.3, we refer to [Gunning and Rossi (1965); Narasimhan (1966); Noguchi (2016)]. In some literatures Theorem 11.4 is referred to as a theorem of Oka-Cartan. See [Noguchi (2016)] for a historical account.

The theory of relative Čech-Dolbeault cohomology may be developed as in the case of de Rham (cf. [Suwa (2009, 2022a, 2023)]). Applications of the theory include, besides the localization of Atiyah classes mentioned in Notes of Chapter 9, the behavior of Hodge structure under blowing-up [Angella, Suwa, Tardini and Tomassini (2020)] and explicit expressions of Sato hyperfunctions and related operations [Honda, Izawa and Suwa (2023); Suwa (2022a)]. See [Suwa (2022a)] for the detailed proof of Proposition 11.16.

In Remark 11.8, the statement (1) is due to [Malgrange (1957)] and (2) to [Cartan et Serre (1953)].

We list [de Jong and Pfister (2000); Fischer (1976); Grauert and Remmert (1984); Grothendieck (1960-61)] as references for Section 11.5. Theorem 11.14 is due to [Grauert (1960)].

The existence of resolution of singularities for algebraic varieties in caracteristic zero was proved in [Hironaka (1964)]. See also [Hauser (2003)] for an expository account on this.

We list [Looijenga (1984); Milnor (1968); Oka (1997); Seade (2019)] as references for Section 11.7. Theorems 11.16 and 11.17 are due to [Milnor (1968)]. The fibration theorem (Theorem 11.18) is proved in [Milnor (1968)] for the case $k = 1$ and in [Hamm (1971)] for general k. The formula (11.23) is due to [Greuel (1975); Lê (1974)].

We refer to [Tsikh (1988)] for Theorem 11.15, see also Chapter VI. 1. 6 in [Łojasiewicz (1991)].

For Definition 11.31, see Section B.3 in [Fulton (1984)].

Chapter 12

Residues of Chern Classes on Manifolds

We describe explicitly the residues determined by holomorphic r-sections of holomorphic vector bundles on complex manifolds. A fundamental theorem (Theorem 12.2) says that, if the singular set has an expected dimension (the proper case), the residue is determined by the residues at isolated singularities of the r-section that appear on the transverse slices of irreducible components of the singular set. We give various expressions for the residue at an isolated singularity (Theorems 12.3–12.5) as well as some fundamental examples.

As a related topic, we discuss the local duality concerning complex spaces defined by an r-section. We introduce the notion of the multiplicity of an irreducible component of such a space as the transverse residue (Definition 12.5), which leads to the natural definition of the class of the space in the homology of its support. The local duality theorem (Theorem 12.8), which says that the homology class of the space is the Alexander dual of the Chern class localized by the r-section, then immediately follows from Theorem 12.2. We also generalize the local duality theorem for the case $r = 1$ to the case of divisors, introducing the notion of localization by meromorphic sections (Theorems 12.9 and 12.10). These will be used in Chapter 14 below for the localized intersection theory.

12.1 Triangulation of subanalytic sets

Let M be a real analytic manifold.

Definition 12.1. A subset X of M is *semianalytic* if every point a in M has a neighborhood U such that $X \cap U = \bigcup_{i=1}^{p} \bigcap_{j=1}^{q} S_{ij}$, where each S_{ij} is a set of the form $\{f_{ij}(x) = 0\}$ or $\{f_{ij}(x) > 0\}$ for some real valued C^ω function f_{ij} on U.

For example, a variety in a complex manifold is semianalytic. This can be seen by taking the real and imaginary parts of defining functions at each point in M (cf. Definition 2.7).

Definition 12.2. A subset X in M is *subanalytic* if every point a in M has a neighborhood U such that $X \cap U = \pi(A)$, where A is a relatively compact semianalytic set in $M \times M'$, for some C^ω manifold M', and $\pi : M \times M' \to M$ the projection.

Semianalytic sets are subanalytic. Also, a polyhedron P in $\mathbb{R}^N$ (cf. Definition 5.6) is subanalytic in an open set in $\mathbb{R}^N$ containing P as a closed set.

Let $X \subset M$ and $X' \subset M'$ be subanalytic sets. A continuous map $f : X \to X'$ is said to be *subanalytic* if the graph of f is subanalytic in $M \times M'$.

Let X be a subanalytic set in M and K a simplicial complex in $\mathbb{R}^N$ (cf. Definition B.13). Thus $|K|$ is subanalytic in some open set in $\mathbb{R}^N$.

Definition 12.3. A *subanalytic triangulation* of X is a triangulation (K, h) of X such that $h : |K| \to X$ is a subanalytic homeomorphism.

We quote the following:

Theorem 12.1. *Let X be a subanalytic set in a C^ω manifold M. Then:*
1. *There exists a subanalytic triangulation (K, h) of M compatible with X such that h is C^1.*

We call such a triangulation a C^1 triangulation of (M, X) for short.

2. *If (K_1, h_1) and (K_2, h_2) are C^1 triangulations of (M, X), there exists a common refinement (K, h), i.e., (K, h) is a C^1 triangulation of (M, X) and there is a commutative diagram*

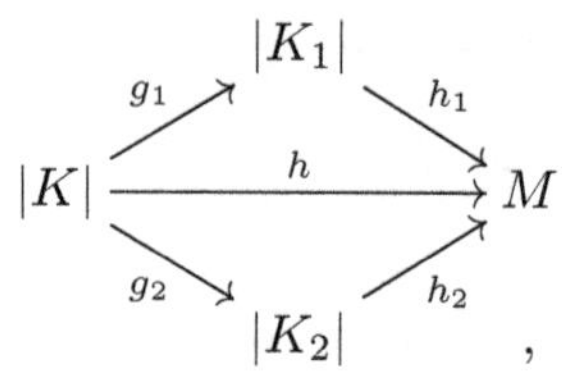

,

where, for each $i = 1, 2$,

(1) *(K, g_i) is a C^1 triangulation of $(|K_i|, |L_i|)$, L_i being the subcomplex of K_i with $h_i(|L_i|) = X$,*

(2) *every simplex of* K_i *is the union of the images by* g_i *of a finite number of simplices of* K.

In the statement 1 above, h being C^1 means the following. For each simplex $\boldsymbol{s}$ of K, we think of it as being in the affine space spanned by $\boldsymbol{s}$ and require that $h|_{\boldsymbol{s}}$ be C^1 in the sense of Definition 3.19 with C^∞ replaced by C^1. Contrary to the case of C^∞ triangulations as in Definition 4.1, we do not require that the rank of $h|_{\boldsymbol{s}}$ at each point of $\boldsymbol{s}$ be equal to $\dim \boldsymbol{s}$. In the theorem above, the set of points in $\boldsymbol{s}$ where the rank of $h|_{\boldsymbol{s}}$ is smaller than $\dim \boldsymbol{s}$ has measure zero in $\boldsymbol{s}$.

12.2 Residues of Chern classes on manifolds

Let M be a complex manifold of dimension n and E a holomorphic vector bundle of rank l on M. Let $s^{(r)} = (s_1, \dots, s_r)$ be a holomorphic r-section of E. Then it defines a complex space $\boldsymbol{S} = (S, \mathscr{O}_{\boldsymbol{S}})$ in M (cf. Section 11.5). It is given as follows. Let $\mathscr{F}$ be the ideal sheaf in $\mathscr{O}_M$ locally generated by the $(r \times r)$-minors of the matrix of the local components of the s_i's. The support S of $\boldsymbol{S}$ is the support of $\mathscr{O}_M/\mathscr{F}$, which is the singular set $S(s^{(r)})$ of $s^{(r)}$ and $\mathscr{O}_{\boldsymbol{S}} = \iota^{-1}(\mathscr{O}_M/\mathscr{F})$, where $\iota : S \hookrightarrow M$ is the inclusion. By (A.16), We have $\dim S \geq n - q$, $q = l - r + 1$ (cf. (A.16), in the case $r = 1$ this also follows from Theorem 2.11).

We take a C^1 triangulation K_0 of M compatible with S as in Theorem 12.1 and let K be its barycentric subdivision. Also, let K^* denote the cellular decomposition dual to K, which is constructed from the barycentric subdivision K' of K (cf. Section 4.1).

In this situation, we have the topological localization $c^q_{\text{top}}(E, s^{(r)})$ in $H^{2q}(M, M \smallsetminus S; \mathbb{Z})$ (cf. Section 5.6) and the differential geometric localization $c^q_{\text{diff}}(E, s^{(r)})$ in $H^{2q}(M, M \smallsetminus S; \mathbb{C})$ (cf. Section 10.3). We also have the associated residues $\mathrm{TRes}_{c^q}(s^{(r)}, E; S)$ in $\breve{H}_{2(n-q)}(S; \mathbb{Z})$ and $\mathrm{Res}_{c^q}(s^{(r)}, E; S)$ in $\breve{H}_{2(n-q)}(S; \mathbb{C})$. We have seen (cf. Corollary 10.6) that $\mathrm{Res}_{c^q}(s^{(r)}, E; S)$ is the image of $\mathrm{TRes}_{c^q}(s^{(r)}, E; S)$ by the canonical morphism

$$\breve{H}_{2(n-q)}(S; \mathbb{Z}) \longrightarrow \breve{H}_{2(n-q)}(S; \mathbb{C}). \tag{12.1}$$

Now we consider the case where S is of pure dimension $n - q$. In the following, we refer this situation as the *proper case*. Let $(S_i)_i$ be the (global) irreducible components of S. Then the set $\{S_i\}_i$ is locally finite. Each S_i defines a class $[S_i]$ in $\breve{H}_{2(n-q)}(S; \mathbb{Z})$ and it is the free Abelian group generated by the $[S_i]$'s. Thus the morphism (12.1) is injective so that we may identify the two residues:

$$\operatorname{Res}_{c^q}(s^{(r)}, E; S) = \operatorname{TRes}_{c^q}(s^{(r)}, E; S). \tag{12.2}$$

Let p_i be a non-singular point of $S_i^\circ = S_i \smallsetminus \bigcup_{j \neq i} S_j$ and $\mathbb{D}_i$ a complex slice of S_i in M at p_i (cf. Definition 3.24). Recall that $\mathbb{D}_i$ is a locally closed complex sudmanifold of dimension q in M through p_i and transverse to S_i at p_i. The r-section $s_i^{(r)} = s^{(r)}|_{\mathbb{D}_i}$ of $E_i = E|_{\mathbb{D}_i}$ has an isolated singularity at p_i so that we have the residue $\operatorname{Res}_{c^q}(s_i^{(r)}, E_i; p_i)$, which is an integer (cf. Theorem 10.11).

Considering a triangulation as above of a neighborhood of S that contains $\mathbb{D}_i$ as a closed set and is compatible with $S \cup \mathbb{D}_i$, we may assume that p_i is the barycenter $b_{\boldsymbol{s}}$ of a $2(n-q)$-simplex $\boldsymbol{s}$ in the non-singular part of S_i° and that the $2q$-cell $\boldsymbol{s}^*$ dual to $\boldsymbol{s}$ is in $\mathbb{D}_i$. Then by Corollary 10.6, or as a special case of (12.2),

$$\operatorname{Res}_{c^q}(s_i^{(r)}, E_i; p_i) = \operatorname{TRes}_{c^q}(s_i^{(r)}, E_i; p_i), \quad p_i = b_{\boldsymbol{s}}.$$

Note that this number does not depend on the choice of p_i on the non-singular part of S_i°, since the residue is locally constant in p_i (topologically, this follows from the homotopy invariance of the indices and differential geometrically by continuity of the residue with respect to p_i) and the non-singular part of S_i° is connected.

We restate Corollary 5.1 (see also Corollary 10.1) in the above situation:

Theorem 12.2. *Suppose S is pure $(n-q)$-dimensional. Then we have:*

$$\operatorname{TRes}_{c^q}(s^{(r)}, E; S) = \sum_i \operatorname{TRes}_{c^q}(s_i^{(r)}, E_i; p_i) \cdot [S_i] \qquad \text{in} \quad \breve{H}_{2(n-q)}(S; \mathbb{Z}).$$

Thus we see that, in order to find the residue in the proper case, it suffices to know the residue at an isolated singularity. We discuss this situation in Section 12.4 below.

12.3 Grothendieck residues

In this section, we briefly review Grothendieck residues, which are used for analytic expressions of the residues.

Let U be a neighborhood of the origin 0 in $\mathbb{C}^n$ and $f_1, \dots, f_n$ holomorphic functions on U such that their common set of zeros $V(f_1, \dots, f_n)$ consists only of 0. For small positive numbers ε_i, $i = 1, \dots, n$, we set

$$\Gamma = \{\, z \in U \mid |f_i(z)| = \varepsilon_i,\ i = 1, \dots, n \,\},$$

which is an n-cycle in U. It is oriented so that the form $d\theta_1 \wedge \cdots \wedge d\theta_n$ is positive, $\theta_i = \arg f_i$. For a holomorphic n-form ω on U, we set

$$\mathrm{Res}_0 \begin{bmatrix} \omega \\ f_1, \ldots, f_n \end{bmatrix} = \left(\frac{1}{2\pi\sqrt{-1}}\right)^n \int_\Gamma \frac{\omega}{f_1 \cdots f_n} \tag{12.3}$$

and call it the *Grothendieck residue* of $\omega / f_1 \cdots f_n$ at 0.

Note that this residue is alternating in $(f_1, \ldots, f_n)$. In general, this is computed as follows. From the condition $V(f_1, \ldots, f_n) = \{0\}$, we see that, for each i, z_i is in the radical $\sqrt{(f_1, \ldots, f_n)}$ of the ideal $(f_1, \ldots, f_n)$ (cf. Theorem 2.8 (Nullstellensatz)). Hence there is a positive integer k_i such that $z_i^{k_i}$ is in $(f_1, \ldots, f_n)$ and we may write $z_i^{k_i} = \sum_{j=1}^n c_{ij} f_j$ with $c_{ij} \in \mathscr{O}_n$. Then

$$\mathrm{Res}_0 \begin{bmatrix} \omega \\ f_1, \ldots, f_n \end{bmatrix} = \mathrm{Res}_0 \begin{bmatrix} \det(c_{ij})\omega \\ z_1^{k_1}, \ldots, z_n^{k_n} \end{bmatrix}.$$

If we write $\omega = f\, dz_1 \wedge \cdots \wedge dz_n$ with f in $\mathscr{O}_n$, the right-hand side of the above is, by the Cauchy integral formula, the coefficient of $z_1^{k_1 - 1} \cdots z_n^{k_n - 1}$ in the power series expansion of $f \det(c_{ij})$.

In particular, if $f = (f_1, \ldots, f_n)$ is non-degenerate, i.e., if the Jacobian $J_f = \det(\partial f_i / \partial z_j)$ is non-zero at 0, then we have

$$\mathrm{Res}_0 \begin{bmatrix} \omega \\ f_1, \ldots, f_n \end{bmatrix} = \frac{f(0)}{J_f(0)}.$$

Example 12.1. In the case $n = 1$, the residue (12.3) is the usual Cauchy residue at 0 of the meromorphic 1-form ω / f_1.

Example 12.2. If $\omega = df_1 \wedge \cdots \wedge df_n$, then

$$\mathrm{Res}_0 \begin{bmatrix} df_1 \wedge \cdots \wedge df_n \\ f_1, \ldots, f_n \end{bmatrix}$$

is a positive integer which is simultaneously equal to (cf. Sections 12.4 and 12.5 below):

(1) the mapping degree of $f = (f_1, \ldots, f_n)$, thus the Poincaré-Hopf index at 0 in $\mathbb{C}^n$ of the vector field $v = \sum_{i=1}^n f_i \cdot \partial / \partial z_i$,
(2) $\dim_{\mathbb{C}} \mathscr{O} / (f_1, \ldots, f_n)$, where $\mathscr{O}$ denotes the ring of germs at 0 of holomorphic functions on $\mathbb{C}^n$ and $(f_1, \ldots, f_n)$ the ideal generated by the germs of $f_1, \ldots, f_n$ at 0.

It can also be interpreted as the intersection number $(D_1 \bullet \cdots \bullet D_n)_0$ at 0 of the divisors D_i defined by f_i (cf. (14.9) below).

Example 12.3. In particular, if $f_i = \partial f/\partial z_i$ for some holomorphic function f on U, then it is the *Milnor number* $\mu(V,0)$ of the hypersurface V defined by f at 0:

$$\operatorname{Res}_0\begin{bmatrix} d(\frac{\partial f}{\partial z_1}) \wedge \cdots \wedge d(\frac{\partial f}{\partial z_n}) \\ \frac{\partial f}{\partial z_1}, \dots, \frac{\partial f}{\partial z_n} \end{bmatrix} = \mu(V,0),$$

which is also the *multiplicity* $m(f,0)$ of f at 0 (cf. Section 11.7 and Section 12.5 below).

12.4 Residues at an isolated singularity

Let M be a complex manifold of dimension n and E a holomorphic vector bundle of rank l on M, $l \geq n$. Let $r = l - n + 1$ and suppose we have a holomorphic r-section $s^{(r)}$ with an isolated singularity at p in M. Thus this is a proper case. In this situation, we have $\operatorname{TRes}_{c^n}(s^{(r)}, E; p)$ and $\operatorname{Res}_{c^n}(s^{(r)}, E; p)$, which are identified. It is an integer, in fact we will see that it is positive in the holomorphic case.

Recall that $\operatorname{Res}_{c^n}(s^{(r)}, E; p)$ is expressed as follows (cf. (10.11)). Let U be a neighborhood of p where the bundle E is trivial. We may assume that U is a coordinate neighborhood and sometimes we identify p with 0 in $\mathbb{C}^n$ and U with a neighborhood of 0. Letting $U_0 = U \smallsetminus \{p\}$ and $U_1 = U$, we consider the covering $\{U_0, U_1\}$ of U. We take an $s^{(r)}$-trivial connection ∇_0 for E on U_0 and a connection ∇_1 for E on U_1 trivial with respect to some holomorphic frame $e^{(l)} = (e_1, \dots, e_l)$ of E. Thus $c^n(\nabla_0) = 0$ and $c^n(\nabla_1) = 0$. Let R be a compact real $2n$-dimensional manifold with C^∞ boundary in U containing p in its interior. Then we have

$$\operatorname{Res}_{c^n}(s^{(r)}, E; p) = -\int_{\partial R} c^n(\nabla_0, \nabla_1). \qquad (12.4)$$

We give various expressions of this number.

Topological expression

We have already seen that $\operatorname{Res}_{c^n}(s^{(r)}, E; p)$ may be expressed as a mapping degree, even if E and $s^{(r)}$ are not holomorphic, in fact $s^{(r)}$ may not be defined at p (cf. Definition 5.3, (5.10) and Theorem 10.11). Let $\mathbb{S}^{2n-1}$ denote a small $(2n-1)$-sphere in U with center p. Then we have the mapping as given in (5.1):

$$\varphi : \mathbb{S}^{2n-1} \longrightarrow W(l,r).$$

Theorem 12.3. *In the above situation, we have:*

$$\mathrm{Res}_{c^n}(s^{(r)}, E; p) = \deg \varphi.$$

Remark 12.1. If E and $s^{(r)}$ are holomorphic, then $\mathrm{Res}_{c^n}(s^{(r)}, E; p)$ is a positive integer, by our orientation convention.

Residue at a non-degenerate zero: Let us assume that $l = n$. Thus $r = 1$ and we have only one section s. We write $s = \sum_{i=1}^n f_i\, e_i$ with f_i holomorphic functions on U. We have $W(l, r) = W(n, 1) = \mathbb{C}^n \smallsetminus \{0\}$, which deformation retracts to $\mathbb{S}^{2n-1}$.

We say that p is a *non-degenerate zero* of s if

$$\det \frac{\partial(f_1, \ldots, f_n)}{\partial(z_1, \ldots, z_n)}(p) \neq 0.$$

In this case, $(f_1, \ldots, f_n)$ form a coordinate system around p. Hence from Theorem 12.3 we have:

Proposition 12.1. *If p is a non-degenerate zero of s,*

$$\mathrm{Res}_{c^n}(s, E; p) = 1.$$

Analytic expression

First we prepare some preliminary materials.

Lemma 12.1. *Let V be a germ of a variety of dimension d at 0 in $\mathbb{C}^n$ and $g_1, \ldots, g_N$ germs in $\mathcal{O}_n$. Suppose $V(g_1 \ldots, g_N) \cap V = \{0\}$. Then there exists an $N \times d$ matrix $C = (c_{ij})$ of complex numbers such that, for germs $f_j = \sum_{i=1}^N c_{ij} g_i$, $j = 1, \ldots, d$, we have $V(f_1, \ldots, f_d) \cap V = \{0\}$.*

Proof. It suffices to show the following for $\nu = 1, \ldots, d$:

(*) If there is an $N \times (\nu - 1)$ matrix (c_{ij}) such that $\dim V(f_1, \ldots, f_{\nu-1}) \cap V = d - \nu + 1$, for $f_j = \sum_{i=1}^N c_{ij} g_j$, $j = 1, \ldots, \nu - 1$, then there exist complex numbers $c_{i\nu}$, $i = 1, \ldots, N$, such that $\dim V(f_1, \ldots, f_\nu) \cap V = d - \nu$, for $f_\nu = \sum_{i=1}^N c_{i\nu} g_i$.

In the above, if $\nu = 1$, $V(f_1, \ldots, f_{\nu-1})$ is understood to be (the germ at 0 of) $\mathbb{C}^n$. To show (*), let $V(f_1, \ldots, f_{\nu-1}) \cap V = \bigcup_q V_q$ be the irreducible decomposition (cf. Theorem 2.3). Since $V(g_1 \ldots, g_N) \cap V = \{0\}$, for each q, there exist a point x_q in a representative of V_q and g_i with $g_i(x_q) \neq 0$. Let H_q denote the hyperplane in $\mathbb{C}^N = \{(\xi_1, \ldots, \xi_N)\}$ defined by $\sum_{i=1}^{N} g_i(x_q)\xi_i = 0$. Let $(c_{1\nu}, \ldots, c_{N\nu})$ be a point in $\mathbb{C}^N \setminus \bigcup_q H_q$ and set $f_\nu = \sum_{i=1}^{N} c_{i\nu} g_i$. Then $V_q \not\subset V(f_\nu)$ for each q. We have

$$V(f_1, \ldots, f_\nu) \cap V = \bigcup_q (V_q \cap V(f_\nu)).$$

Since each V_q is irreducible and $V_q \not\subset V(f_\nu)$, $\dim V_q \cap V(f_\nu) = \dim V_q - 1$ (cf. Theorem 2.11). Therefore, we have (*) and the lemma. □

Note that the above is also valid if $d = n$, in which case V is the germ of $\mathbb{C}^n$. In fact it is this case that we consider in this chapter and the general case of $d < n$ will be used in the next chapter.

Let r and l be integers with $1 \leq r \leq l$ and set $N = \binom{l}{r}$. We denote by $M(l, \mathbb{C})$ the space of $l \times l$ complex matrices, which is naturally identified with $\mathbb{C}^{l^2}$. For a matrix A in $M(l, \mathbb{C})$, its r-th exterior power $\bigwedge^r A$ is an $N \times N$ matrix defined as in Section 8.2. The following is proved by linear algebra:

Lemma 12.2. *For every matrix A in $M(l, \mathbb{C})$ and every neighborhood W of A, there exists a matrix A' in W such that the matrix consisting of the first d columns of $\bigwedge^r A'$ satisfies the condition of Lemma 12.1.*

Let E, U, $e^{(l)}$ and $s^{(r)}$ be as in the beginning of this section. We write $s_i = \sum_{j=1}^{l} f_{ji}\, e_j$, $i = 1, \ldots, r$, with f_{ji} holomorphic functions on U. Let F be the $l \times r$ matrix whose (i,j)-entry is f_{ij}. We set

$$\mathcal{I} = \{ (i_1, \ldots, i_r) \mid 1 \leq i_1 < \cdots < i_r \leq l \}. \tag{12.5}$$

For an element $I = (i_1, \ldots, i_r)$ in $\mathcal{I}$, let F_I denote the $r \times r$ matrix whose k-th row is the i_k-th row of F, $k = 1, \ldots, r$, and set $\varphi_I = \det F_I$. Note that $\mathscr{F}_p$ is the ideal generated by the germs of the φ_I's and $S(s^{(r)})$ is the set of common zeros of the φ_I's. From the assumption $S(s^{(r)}) = \{p\}$, we have the following as a consequence of Lemmas 12.1 and 12.2 with $d = n$:

Lemma 12.3. *We may choose a holomorphic frame $e^{(l)} = (e_1, \ldots, e_l)$ of E so that there exist n elements $I^{(1)}, \ldots, I^{(n)}$ in $\mathcal{I}$ with the property $V(\varphi_{I^{(1)}}, \ldots, \varphi_{I^{(n)}}) = \{p\}$, where $V(\varphi_{I^{(1)}}, \ldots, \varphi_{I^{(n)}})$ denotes the common zero set of $\varphi_{I^{(1)}}, \ldots, \varphi_{I^{(n)}}$.*

Let $e^{(l)}$ be a frame of E as in Lemma 12.3. We write $I^{(\alpha)} = (i_1^{(\alpha)}, \ldots, i_r^{(\alpha)})$, $\alpha = 1, \ldots, n$, and let $F^{(\alpha)}$ be the $l \times l$ matrix obtained from the $l \times l$ identity matrix by replacing the $i_j^{(\alpha)}$-th column with the j-th column of F, $j = 1, \ldots, r$. Note that $\det F^{(\alpha)} = \varphi_{I^{(\alpha)}}$. Let $\check{F}^{(\alpha)}$ denote the adjoint matrix of $F^{(\alpha)}$ and set

$$\Theta^{(\alpha)} = dF^{(\alpha)} \cdot \check{F}^{(\alpha)},$$

which is an $l \times l$-matrix with entries holomorphic 1-forms. The significance of this matrix will be clarified in the proof of the following theorem. Denoting by $\tilde{\sigma}_n$ the polarization of the n-th elementary invariant polynomial σ_n (cf. Section 8.2), we have the holomorphic n-form $\tilde{\sigma}_n(\Theta^{(1)},\dots,\Theta^{(n)})$ on U.

Theorem 12.4. *In the above notation,*

$$\mathrm{Res}_{c^n}(s^{(r)},E;p)=\mathrm{Res}_p\begin{bmatrix}\tilde{\sigma}_n(\Theta^{(1)},\dots,\Theta^{(n)})\\ \varphi_{I^{(1)}},\dots,\varphi_{I^{(n)}}\end{bmatrix}.$$

Proof. Recall that the residue is given by (12.4), where ∇_0 is an $s^{(r)}$-trivial connection for E on $U_0=U\smallsetminus\{p\}$ and ∇_1 the $e^{(l)}$-trivial connection for E on $U_1=U$. We compute this by setting

$$R=\{\,q\in U\mid |\varphi_{I^{(1)}}(q)|^2+\cdots+|\varphi_{I^{(n)}}(q)|^2\le n\varepsilon^2\,\}$$

with ε a small positive number. The idea is to introduce a suitable ČdR-cochain that constitute a bridge between the $(2n-1)$-form $c^n(\nabla_0,\nabla_1)$ and the n-form appearing in the Grothendieck residue in the statement.

We consider the covering $\mathcal{U}=\{U^{(\alpha)}\}_{\alpha=1}^n$ of U_0 given by

$$U^{(\alpha)}=\{\,q\in U_0\mid \varphi_{I^{(\alpha)}}(q)\neq 0\,\}$$

and work on the Čech-de Rham cohomology of $\mathcal{U}$. On $U^{(\alpha)}$, we may replace, in the frame $e^{(l)}$, $(e_{i_1^{(\alpha)}},\dots,e_{i_r^{(\alpha)}})$ with $(s_1,\dots,s_r)$ to obtain a frame $e_\alpha^{(l)}$ for E. The matrix of frame change from $e^{(l)}$ to $e_\alpha^{(l)}$ is given by $F^{(\alpha)}$. We denote by $\nabla^{(\alpha)}$ the connection for E on $U^{(\alpha)}$ trivial with respect to $e_\alpha^{(l)}$. The connection matrix $\theta^{(\alpha)}$ of $\nabla^{(\alpha)}$ with respect to the frame $e^{(l)}$ is obtained by setting $P=(F^{(\alpha)})^{-1}$ and $\theta=0$ in (8.4):

$$\theta^{(\alpha)}=F^{(\alpha)}\cdot d(F^{(\alpha)})^{-1}=-dF^{(\alpha)}\cdot(F^{(\alpha)})^{-1}=-\frac{1}{\varphi_{I^{(\alpha)}}}\Theta^{(\alpha)}.$$

We define an alternating cochain τ in $A^{2n-2}(\mathcal{U})$ by

$$\tau_{\alpha_1\dots\alpha_q}=c^n(\nabla_0,\nabla_1,\nabla^{(\alpha_1)},\dots,\nabla^{(\alpha_q)}),$$

which is a $(2n-q-1)$-form on $U^{(\alpha_1\dots\alpha_q)}=U^{(\alpha_1)}\cap\cdots\cap U^{(\alpha_q)}$. We claim that its coboundary is given, for $(\alpha_1,\dots,\alpha_q)$ with $1\le\alpha_1<\cdots<\alpha_q\le n$, as follows:

$$(D\tau)_{\alpha_1\dots\alpha_q}=\begin{cases}-c^n(\nabla_0,\nabla_1) & q=1,\\ 0 & q=2,\dots,n-1,\\ -c^n(\nabla_1,\nabla^{(1)},\dots,\nabla^{(n)}) & q=n.\end{cases}\qquad(12.6)$$

To prove this, first notice that, since ∇_0 and $\nabla^{(\alpha)}$ are all $s^{(r)}$-trivial,

$$c^n(\nabla_0, \nabla^{(\alpha_1)}, \dots, \nabla^{(\alpha_q)}) = 0 \qquad \text{for } q \geq 1. \tag{12.7}$$

Then we compute $c^n(\nabla_1, \nabla^{(\alpha_1)}, \dots, \nabla^{(\alpha_q)})$ via Proposition 8.4. Thus let ∇_t denote the connection for E on $U^{(\alpha_1 \dots \alpha_q)}$ given by

$$\nabla_t = \Big(1 - \sum_{\nu=1}^{q} t_\nu\Big)\nabla_1 + \sum_{\nu=1}^{q} t_\nu \nabla^{(\alpha_\nu)}.$$

Since the connection matrix θ_1 of ∇_1 with respect to $e^{(l)}$ is zero, the connection matrix θ_t of ∇_t with respect to $e^{(l)}$ is given by

$$\theta_t = \Big(1 - \sum_{\nu=1}^{q} t_\nu\Big)\theta_1 + \sum_{\nu=1}^{q} t_\nu \theta^{(\alpha_\nu)} = \sum_{\nu=1}^{q} t_\nu \theta^{(\alpha_\nu)}.$$

The curvature matrix κ_t of ∇_t is given by $\kappa_t = d\theta_t + \theta_t \wedge \theta_t$, which is holomorphic for each fixed t. The form $c^n(\nabla_1, \nabla^{(\alpha_1)}, \dots, \nabla^{(\alpha_q)})$ is a constant multiple of the integral on Δ^n of the sum of the forms

$$\tilde{\sigma}_n(\kappa_t, \dots, \overset{\alpha_1}{\theta_1}, \dots, \overset{\alpha_q}{\theta_q}, \dots, \kappa_t),$$

which are holomorphic $(2n-q)$-forms on $U^{(\alpha_1 \dots \alpha_q)}$. Thus we have

$$c^n(\nabla_1, \nabla^{(\alpha_1)}, \dots, \nabla^{(\alpha_q)}) = 0 \qquad \text{for } q = 1, \dots, n-1. \tag{12.8}$$

Now we compute $D\tau$. For $q = 1$, we have, by (12.7) and (12.8),

$$(D\tau)_\alpha = dc^n(\nabla_0, \nabla_1, \nabla^{(\alpha)}) = -c^n(\nabla_0, \nabla_1).$$

For $q = 2, \dots, n$, we have, by (12.7),

$$\begin{aligned}
(D\tau)_{\alpha_1 \dots \alpha_q} &= \sum_{\nu=1}^{q} (-1)^\nu c^n(\nabla_0, \nabla_1, \nabla^{(\alpha_1)}, \dots, \widehat{\nabla^{(\alpha_\nu)}}, \dots, \nabla^{(\alpha_q)}) \\
&\quad + (-1)^{q-1} dc^n(\nabla_0, \nabla_1, \nabla^{(\alpha_1)}, \dots, \nabla^{(\alpha_q)}) \\
&= -c^n(\nabla_1, \nabla^{(\alpha_1)}, \dots, \nabla^{(\alpha_q)}) + c^n(\nabla_0, \nabla^{(\alpha_1)}, \dots, \nabla^{(\alpha_q)}) \\
&= -c^n(\nabla_1, \nabla^{(\alpha_1)}, \dots, \nabla^{(\alpha_q)}),
\end{aligned}$$

which proves (12.6) in view of (12.8).

Denoting by ι the inclusion map $\partial R \hookrightarrow U_0$, we let $\iota^*\mathcal{U}$ be the covering of ∂R by the open sets $\partial R \cap U^{(\alpha)}$. Then, as a honeycomb system $\{R^{(\alpha)}\}_{\alpha=1}^n$ adapted to $\iota^*\mathcal{U}$, we take

$$R^{(\alpha)} = \{\, q \in \partial R \mid \ |\varphi_{I^{(\alpha)}}(q)| \geq |\varphi_{I^{(\beta)}}(q)| \quad \text{for all } \beta \,\}$$

and, for a q-tuple $(\alpha_1, \dots, \alpha_q)$ with $1 \leq \alpha_1 < \dots < \alpha_q \leq n$, we set $R^{(\alpha_1 \dots \alpha_q)} = R^{(\alpha_1)} \cap \dots \cap R^{(\alpha_q)}$, which is a $(2n-q)$-dimensional manifold with boundary oriented as an intersection of honeycomb cells. Considering the integration

$$\int_{\partial R} : A^{2n-1}(\iota^* \mathcal{U}) \longrightarrow \mathbb{C},$$

from the identity $\int_{\partial R} D\tau = 0$, we obtain

$$\mathrm{Res}_{c^n}(s^{(r)}, E; p) = \int_{R^{(1\dots n)}} c^n(\nabla_1, \nabla^{(1)}, \dots, \nabla^{(n)}).$$

By Proposition 8.4, setting

$$\Phi = (-1)^{\frac{n(n-1)}{2}} n!\, \tilde{\sigma}_n(\theta^{(1)}, \dots, \theta^{(n)}),$$

we have

$$\begin{aligned} c^n(\nabla_1, \nabla^{(1)}, \dots, \nabla^{(n)}) &= \left(\frac{\sqrt{-1}}{2\pi}\right)^n \int_{\Delta^n} \Phi\, dt_1 \cdots dt_n \\ &= (-1)^{\frac{n(n-1)}{2}} \left(\frac{\sqrt{-1}}{2\pi}\right)^n \tilde{\sigma}_n(\theta^{(1)}, \dots, \theta^{(n)}) \\ &= (-1)^{\frac{n(n-1)}{2}} \left(\frac{1}{2\pi\sqrt{-1}}\right)^n \frac{\tilde{\sigma}_n(\Theta^{(1)}, \dots, \Theta^{(n)})}{\varphi_{I(1)} \cdots \varphi_{I(n)}}. \end{aligned}$$

Noting that the n-cycle Γ appearing in the Grothendieck residue with respect to the functions $\varphi_{I(1)}, \dots, \varphi_{I(n)}$ is given by $\Gamma = (-1)^{\frac{n(n-1)}{2}} R^{(1\dots n)}$, we obtain the formula. □

Special cases: **1.** The case $l = n$ and $r = 1$. We may write $s^{(1)} = (s)$, $s = \sum_{i=1}^n f_i e_i$. Then we may set $\varphi_{I(i)} = f_i$, $i = 1, \dots, n$. We see that $\Theta^{(i)}$ is the $l \times l$ matrix whose i-th column is ${}^t(df_1, \dots, df_n)$ with all the other entries equal to 0. Thus we have $\tilde{\sigma}_n(\Theta^{(1)}, \dots, \Theta^{(n)}) = df_1 \wedge \cdots \wedge df_n$ and

$$\mathrm{Res}_{c^n}(s, E; p) = \mathrm{Res}_p \begin{bmatrix} df_1 \wedge \cdots \wedge df_n \\ f_1, \dots, f_n \end{bmatrix}.$$

In the case $r = 1$ and $n = 1$, the proof is directly done by computing the difference form $c^1(\nabla_0, \nabla_1)$, which turns out to be (cf. Example 10.3)

$$c^1(\nabla_0, \nabla_1) = -\frac{1}{2\pi\sqrt{-1}} \frac{df}{f}.$$

For general n, the above can be also obtained from Theorem 10.8, Propositions 11.16 and 11.17.

2. The case $n = 1$ and $r = l$. Let $e^{(l)} = (e_1, \dots, e_l)$ be an arbitrary frame of E and write $s_i = \sum_{j=1}^l f_{ij}\, e_j$, $i = 1, \dots, l$. Let $F = (f_{ij})$ and set $\varphi = \det F$. Then we may set $\varphi_{I(1)} = \varphi$ and we have $\sigma_1(\Theta) = d\varphi$ so that we have

$$\mathrm{Res}_{c^1}(s^{(r)}, E; p) = \mathrm{Res}_p \begin{bmatrix} d\varphi \\ \varphi \end{bmatrix}.$$

Note that $\mathrm{Res}_{c^1}(s^{(r)}, E; p)$ in this case coincides with the residue $\mathrm{Res}_{c^1}(s, \det E; p)$ of the section $s = s_1 \wedge \cdots \wedge s_l$ of the line bundle $\det E = \bigwedge^l E$ at p.

Fundamental properties of the residues

We consider the situation in the beginning of this section.

Letting T be a neighborhood of 0 in $\mathbb{C} = \{t\}$, we consider the bundle $\tilde{E} = E \times T$ on $\tilde{U} = U \times T$. Suppose we have an r-tuple of holomorphic sections $\tilde{s}^{(r)}$ of $\tilde{E}$ on $\tilde{U}$ such that $\tilde{s}^{(r)}(z,0) = s^{(r)}(z)$. For t in T, we set $E_t = \tilde{E}|_{U\times\{t\}}$ and $s_t^{(r)}(z) = \tilde{s}^{(r)}(z,t)$. We call such an $\tilde{s}^{(r)}$ (or $s_t^{(r)}$) a *perturbation* of $s^{(r)}$. Sometimes we identify $U \times \{t\}$ with U and $E|_t$ with E.

Lemma 12.4. *We have* $\dim S(\tilde{s}^{(r)}) = 1$ *and* $S(s_t^{(r)})$ *is a non-empty finite set.*

Proof. By the upper semicontinuity of the dimension of $S(s_t^{(r)})$ (cf. Theorem 2.13), we have $\dim S(\tilde{s}^{(r)}) \le 1$. On the other hand, we have $\operatorname{codim} S(\tilde{s}^{(r)}) \le l - r + 1 = n$ by (A.16). □

Lemma 12.5. *The sum* $\sum_{q\in S(s_t^{(r)})} \operatorname{Res}_{c^n}(s_t^{(r)}, E_t; q)$ *is continuous in* t.

Proof. Let $\tilde{\nabla}_0$ be an $\tilde{s}^{(r)}$-trivial connection for $\tilde{E}$ on $\tilde{U}_0 = \tilde{U} \smallsetminus S(\tilde{s}^{(r)})$ and $\tilde{\nabla}_1$ a connection for $\tilde{E}$ on $\tilde{U}$. The statement follows from Theorem 10.6 by computing the residues taking the restrictions of $\tilde{\nabla}_0$ and $\tilde{\nabla}_1$ and using (10.11). □

Since the residues are integers, from the above, we have:

Proposition 12.2. *The sum* $\sum_{q\in S(s_t^{(r)})} \operatorname{Res}_{c^n}(s_t^{(r)}, E_t; q)$ *is constant. In particular,*

$$\operatorname{Res}_{c^n}(s^{(r)}, E; p) = \sum_{q\in S(s_t^{(r)})} \operatorname{Res}_{c^n}(s_t^{(r)}, E_t; q).$$

Note that this also follows from Theorems 5.2 and the homotopy invariance of topological residues.

We prepare a proposition for later use. Suppose one of the sections in $s^{(r)}$ is non-zero at p. For example we assume that $s_1(p) \neq 0$ so that we have an exact sequence of vector bundles on a neighborhood of p:

$$0 \longrightarrow \mathbb{I} \longrightarrow E \longrightarrow E' \longrightarrow 0, \tag{12.9}$$

where $\mathbb{I}$ denotes the trivial line bundle determined by s_1 and E' is a vector bundle (still trivial) of rank $l-1$. Let $s'^{(r-1)} = (s'_2, \dots, s'_r)$ denote the $(r-1)$-sections of E' determined by $(s_2, \dots, s_r)$. The following is proved as Lemma 10.4:

Proposition 12.3. *In the above situation, we have*

$$\mathrm{Res}_{c^n}(s^{(r)}, E; p) = \mathrm{Res}_{c^n}({s'}^{(r-1)}, E'; p).$$

We also need the following:

Lemma 12.6. *If $r = 1$, there exists always a "good perturbation" of s, i.e., a holomorphic sections $\tilde{s}$ of $\tilde{E}$ near 0 such that $\tilde{s}(z,0) = s(z)$ and that s_t has only non-degenerate zeros, for $t \neq 0$.*

Proof. We may write $s = \sum_{i=1}^n f_i e_i$ and think of $f = (f_1, \dots, f_n)$ as a holomorphic map of U into $\mathbb{C}^n$. We assume that 0 is a degenerate zero of s and consider the hypersurface C in U defined by

$$\det \frac{\partial(f_1, \dots, f_n)}{\partial(z_1, \dots, z_n)} = 0.$$

By Theorem 2.14, $D = f(C)$ is a hypersurface in a neighborhood of 0 in $\mathbb{C}^n$. Let T be a one-dimensional disk intersection with D only at 0. Then $s_t = \sum_{i=1}^n (f_i - t)e_i$ is a desired perturbation. □

Algebraic expression

We refer to Section A.2 for relevant materials in algebra.

Let E, U, $e^{(l)}$ and $s^{(r)}$ be as in the beginning of this section. Also let F, F_I and $\varphi_I = \det F_I$ be as before. We denote by $\mathcal{O}_U$ the sheaf of germs of holomorphic functions on U and by $\mathscr{F}$ the ideal sheaf in $\mathcal{O}_U$ generated by the (germs of) φ_I's. Note that $\mathscr{F}$ does not depend on the choice of the frame $e^{(l)}$ of E. We have the complex space $\boldsymbol{S} = (S, \mathcal{O}_{\boldsymbol{S}})$ in U with support $S = S(s^r) = \{p\}$ and structure sheaf $\mathcal{O}_{\boldsymbol{S}} = \iota^{-1}(\mathcal{O}_U/\mathscr{F})$, $\iota : S \hookrightarrow U$ being the inclusion, as above.

First we consider the case $l = n$ and $r = 1$. Thus we have one section $s = \sum_{i=1}^n f_i e_i$ and $\mathscr{F}_p = (f_1, \dots, f_n)$. Obviously we have:

Proposition 12.4. *If p is a non-degenerate zero of s,*

$$\dim_{\mathbb{C}} \mathcal{O}_n/(f_1, \dots, f_n) = 1.$$

Now we come back to the general case and let $\tilde{s}^{(r)} = (\tilde{s}_1, \dots, \tilde{s}_r)$ be a perturbation of $s^{(r)}$ as before. We define $\tilde{F}_I$ and $\tilde{\varphi}_I$ as above, using the $\tilde{s}_i$'s. Let T be a small neighborhood of 0 in $\mathbb{C}$ and $\tilde{\mathscr{F}}$ the ideal sheaf generated by the $\tilde{\varphi}_I$'s in $\mathcal{O}_{\tilde{U}}$, $\tilde{U} = U \times T$. Let $\tilde{\boldsymbol{S}}$ be the complex space in $\tilde{U}$ with support $\tilde{S} = S(\tilde{s}^{(r)})$ and structure sheaf $\mathcal{O}_{\tilde{\boldsymbol{S}}} = (\tilde{\iota})^{-1}(\mathcal{O}_{\tilde{U}}/\tilde{\mathscr{F}})$, where $\tilde{\iota} : \tilde{S} \hookrightarrow \tilde{U}$ is the inclusion. We denote by $p : \tilde{U} = U \times T \to T$ the projection onto the second factor and by π the restriction of p to $\tilde{S}$ so that $\pi = p \circ \tilde{\iota}$. We have a morphism

$$(\pi, \varpi) : \tilde{\boldsymbol{S}} \longrightarrow (T, \mathscr{O}_T) \tag{12.10}$$

as the composition of the canonical morphism $\tilde{\boldsymbol{S}} \to (\tilde{U}, \mathscr{O}_{\tilde{U}})$ and $(p, p^*) : (\tilde{U}, \mathscr{O}_{\tilde{U}}) \to (T, \mathscr{O}_T)$. For each x in $\tilde{S}$,

$$\varpi_x : \mathscr{O}_{T,\pi(x)} \to \mathscr{O}_{\tilde{\boldsymbol{S}},x}$$

is the composition of $p_x^* : \mathscr{O}_{T,\pi(x)} \to \mathscr{O}_{\tilde{U},x}$ and the canonical epimorphism $\mathscr{O}_{\tilde{U},x} \to \mathscr{O}_{\tilde{U},x}/\tilde{\mathscr{F}}_x = \mathscr{O}_{\tilde{\boldsymbol{S}},x}$.

Let $\mathscr{F}_t$ be the ideal sheaf generated by the $\varphi_{I,t}$'s in $\mathscr{O}_{U_t}$, $U_t = U \times \{t\}$. The fiber of (π, ϖ) over $t \in T$ is the complex space $\boldsymbol{S}_t$ in $\tilde{\boldsymbol{S}}$ with support $S_t = S(s_t^{(r)})$ and structure sheaf $\mathscr{O}_{\boldsymbol{S}_t} = \iota_t^{-1}(\mathscr{O}_{U_t}/\mathscr{F}_t)$, where $\iota_t : S_t \hookrightarrow U_t$ is the inclusion.

By Lemma 12.4, $\pi : \tilde{S} \to T$ is a finite map. Thus each point x in $\tilde{S}$ is isolated in S_t, $t = \pi(x)$, and by the Nullstellensatz, $\dim_{\mathbb{C}} \mathscr{O}_{\boldsymbol{S}_t,x} = \dim_{\mathbb{C}} \mathscr{O}_{U_t,x}/\mathscr{F}_{t,x}$ is finite (cf. (A.18)).

Proposition 12.5. *In the above situation,*

$$\dim_{\mathbb{C}} \mathscr{O}_{U,p}/\mathscr{F}_p = \sum_{q \in S(s_t^{(r)})} \dim_{\mathbb{C}} \mathscr{O}_{U_t,q}/\mathscr{F}_{t,q}.$$

Proof. We claim that the morphism (π, ϖ) in (12.10) is flat (cf. Section A.2). Then the proposition follows from (A.21).

Let x be a point in $\tilde{S}$ and set $t = \pi(x)$. In the following, we set $\mathscr{O}'_x = \mathscr{O}_{\tilde{U},x}$, $\mathscr{O}_x = \mathscr{O}_{\tilde{\boldsymbol{S}},x}$ and $\mathscr{O}_t = \mathscr{O}_{T,t}$. Note that $\mathscr{O}'_x$ and $\mathscr{O}_t$ are regular local rings of dimensions $n+1$ and 1, respectively. We have ht $\tilde{\mathscr{F}}_x = n = l-r+1$. Hence by (A.17), the ring $\mathscr{O}_x$ is CM. Since the morphism $\varpi_x : \mathscr{O}_t \to \mathscr{O}_x$ is finite (cf. (A.20)), $\mathscr{O}_x$ is a CM $\mathscr{O}_t$-module by (A.14). By (A.15), denoting by $\mathfrak{m}_t$ the maximal ideal in $\mathscr{O}_t$,

$$\operatorname{depth}(\mathfrak{m}_t; \mathscr{O}_x) + \operatorname{pd}_{\mathscr{O}_t} \mathscr{O}_x = \operatorname{depth} \mathfrak{m}_t.$$

We have $\operatorname{depth}(\mathfrak{m}_t; \mathscr{O}_x) = \dim_{\mathscr{O}_t} \mathscr{O}_x = \dim_{\mathscr{O}_x} \mathscr{O}_x = 1$ and $\operatorname{depth} \mathfrak{m}_t = \dim \mathscr{O}_t = 1$. Therefore, $\operatorname{pd}_{\mathscr{O}_t} \mathscr{O}_x = 0$ and (π, ϖ) is flat. □

From Propositions 12.1, 12.2, 12.4 and 12.5 and Lemma 12.6, we have:

Corollary 12.1. *In the case $l = n$ and $r = 1$,*

$$\operatorname{Res}_{c^n}(s, E; p) = \dim_{\mathbb{C}} \mathscr{O}_{U,p}/\mathscr{F}_p = \dim_{\mathbb{C}} \mathscr{O}_n/(f_1, \dots, f_n).$$

We continue with the general situation where $l \geq n$. We assume that $s_1(p) \neq 0$ as in the situation of Proposition 12.3. Then we may write $s'_i = \sum_{j=2}^{l} f'_{ij}\, e'_j$, $i = 2, \dots, r$, with f'_{ij} holomorphic functions on U and $e'^{(l-1)} = (e'_2, \dots, e'_l)$ a frame of E' (cf. (12.9)). Let F' be the $(r-1)\times(l-1)$ matrix whose (i,j)-entry is f'_{ij}. We set

$$\mathcal{I}' = \{\, (i_2, \dots, i_r) \mid 2 \leq i_2 < \cdots < i_r \leq l \,\}.$$

For an element $I' = (i_2, \dots, i_r)$ in $\mathcal{I}'$, let $F'_{I'}$ denote the $(r-1)\times(r-1)$ matrix consisting of the columns of F' corresponding to I' and set $\varphi'_{I'} = \det F'_{I'}$.

Note that the set of common zeros of the $\varphi'_{I'}$'s consists only of p. Let $\mathscr{F}'_p$ denote the ideal of $\mathscr{O}_{U,p}$ generated by the $\varphi'_{I'}$'s.

Proposition 12.6. *In the above situation, we have*

$$\mathscr{F}_p = \mathscr{F}'_p \quad \textit{and thus} \quad \dim_{\mathbb{C}} \mathscr{O}_{U,p}/\mathscr{F}_p = \dim_{\mathbb{C}} \mathscr{O}_{U,p}/\mathscr{F}'_p.$$

Proof. We may assume, without loss of generality, that $f_{11}(p) \neq 0$. Then, we may take as $(e'_2, \dots, e'_l)$ the sections determined by $(e_2, \dots, e_l)$. For $i \geq 2$, we have

$$s_i = \frac{1}{f_{11}}\left(f_{i1}s_1 + \sum_{j=2}^{l} \begin{vmatrix} f_{11} & f_{1j} \\ f_{i1} & f_{ij} \end{vmatrix} e_j \right).$$

Hence

$$f'_{ij} = \frac{1}{f_{11}} \begin{vmatrix} f_{11} & f_{1j} \\ f_{i1} & f_{ij} \end{vmatrix}.$$

For $I' = (i_2, \dots, i_l)$, we compute $\varphi_{(1,I')} = f_{11} \cdot \varphi'_{I'}$. Thus the ideal $\mathscr{F}'_p$ is generated by $\{\, \varphi_{(1,I')} \mid I' \in \mathcal{I}' \,\}$. On the other hand, for every $I = (i_1, \dots, i_r)$, considering the determinant of the $(r+1)\times(r+1)$ matrix whose first and second rows are $(f_{11}, f_{1i_1}, \dots, f_{1i_r})$ and whose k-th row is $(f_{k-1,1}, f_{k-1,i_1}, \dots, f_{k-1,i_r})$, $k \geq 3$, we have

$$f_{11} \cdot \varphi_I = \sum_{j=1}^{r} (-1)^{j-1} f_{1i_j} \cdot \varphi_{(1,i_1,\dots,\widehat{i_j},\dots,i_r)}.$$

Hence we have $\mathscr{F}'_p = \mathscr{F}_p$. □

Theorem 12.5. *We have*

$$\operatorname{Res}_{c^n}(s^{(r)}, E; p) = \dim_{\mathbb{C}} \mathscr{O}_{U,p}/\mathscr{F}_p.$$

Proof. We prove this by induction on r. The case $r = 1$ is Corollary 12.1. Suppose that the statement is true for $(r-1)$-sections (with isolated singularity). If there is a section in $s^{(r)}$ which does not vanish at p, we are done by Propositions 12.3 and 12.6. Otherwise, we take a perturbation $s_{1,t}$ of s_1 so that $s_{1,t}(p) \neq 0$, for $t \neq 0$, and set $s_t^{(r)} = (s_{1,t}, s_2, \dots, s_r)$. Recalling that none of the s_i's vanishes on $U \smallsetminus \{0\}$, we see that, for $t \neq 0$ and every point $q \in S(s_t^{(r)})$, there is a section in $s_t^{(r)}$ which does not vanish at q. The theorem follows from Propositions 12.2, 12.3, 12.5 and 12.6 and the induction hypothesis. □

Special cases: **1.** The case $l = n$ and $r = 1$. In this case, we restate Corollary 12.1. We may write $s^{(1)} = (s)$, $s = \sum_{i=1}^n f_i e_i$. Then we have $\mathscr{F}_p = (f_1, \dots, f_n)$ and

$$\mathrm{Res}_{c^n}(s, E; p) = \dim_{\mathbb{C}} \mathscr{O}_n/(f_1, \dots, f_n).$$

2. The case $n = 1$ and $r = l$. Let $e^{(l)} = (e_1, \dots, e_l)$ be an arbitrary frame of E and write $s_i = \sum_{j=1}^l f_{ij}\, e_j$, $i = 1, \dots, r$. Let $F = (f_{ij})$ and set $\varphi = \det F$. Then we have $\mathscr{F}_p = (\varphi)$ and

$$\mathrm{Res}_{c^1}(s, E; p) = \dim_{\mathbb{C}} \mathscr{O}_1/(\varphi).$$

.

12.5 Examples

In the following, we denote by M a complex manifold of dimension n.

Poincaré-Hopf index of a vector field

We take as E the holomorphic tangent bundle TM. A section of TM may be thought of either as a complex vector field or as a real vector field (cf. Proposition 3.6). Let S be a compact set in M and S_λ a connected component of S. For a non-vanishing C^∞ vector field v on $U \smallsetminus S_\lambda$, where U is a neighborhood of S_λ, we may define the *Poincaré-Hopf index* of v at S_λ by (cf. Remark 10.6)

$$\mathrm{PH}(v, S_\lambda) = \mathrm{Res}_{c^n}(v, TM; S_\lambda).$$

If S_λ is a subcomplex with respect to some triangulation of M, this definition coincides with the one in (5.15) by Proposition 5.8 and Corollary 10.6. If S_λ consists of a point p and if v is defined and holomorphic in a neighborhood of p, it has algebraic and analytic expressions (cf. Theorems 12.5 and 12.4, also Example 12.2).

Let R be a compact manifold of dimension $2n$ with boundary in M. For a non-vanishing vector field v on a neighborhood of ∂R, we may also define (cf. (10.3))

$$\mathrm{PH}(v, R) = \int_R c^n_R(TM, v),$$

which coincides with the one in (5.16) (cf. Proposition 7.18 and Remark 7.18).

Suppose S has only a finite number of connected components $(S_\lambda)_\lambda$. Then if v is a non-vanishing vector field on $M \smallsetminus S$ and if S is in the interior of R, we have (cf. Proposition 10.2 and Theorem 10.6)

$$\sum_\lambda \mathrm{PH}(v, S_\lambda) = \mathrm{PH}(v, R).$$

If M is compact, taking M as R, we have

$$\sum_\lambda \mathrm{PH}(v, S_\lambda) = \int_M c^n(M),$$

where $c^n(M) = c^n(TM)$ and the right-hand side equals the Euler-Poincaré characteristic $\chi(M)$ of M (cf. (5.20), Gauss-Bonnet formula).

Index of a $(1,0)$*-form*

We take as E the holomorphic cotangent bundle T^*M of M. Then a section of T^*M is a $(1,0)$-form. Let θ be a C^∞ $(1,0)$-form. For a compact connected component S_λ of the zero set or the set of points where it is not defined $S(\theta)$ of θ having a neighborhood disjoint from the other components, we define the *index* $\mathrm{Ind}(\theta, S_\lambda)$ of θ at S_λ by

$$\mathrm{Ind}(\theta, S_\lambda) = \mathrm{Res}_{c^n}(\theta, T^*M; S_\lambda).$$

If M is compact and if $S(\theta)$ admits only a finite number of connected components (S_λ), we have (cf. Proposition 10.2, Theorem 10.6, (5.20))

$$\sum_\lambda \mathrm{Ind}(\theta, S_\lambda) = (-1)^n \chi(M).$$

If S_λ consists of a point p and if θ is defined and holomorphic in a neighborhood of p, $\mathrm{Ind}(\theta, p)$ has the topological, algebraic and analytic expressions as given above.

Multiplicity of a function

For a holomorphic function f on M, its differential df is a section of T^*M. The zero set $S(df)$ of df coincides with the critical set $C(f)$ of f.

Definition 12.4. The *multiplicity* $m(f, S_\lambda)$ of f at a compact connected component S_λ of $C(f)$ as above is defined by

$$m(f, S_\lambda) = \mathrm{Res}_{c^n}(df, T^*M; S_\lambda).$$

Note that, if S_λ consists of a point p, it coincides with the multiplicity $m(f,p)$ of f at p described in Example 12.3.

Now we consider the global situation. Let $f\colon M \to C$ be a holomorphic map of M onto a complex curve (Riemann surface) C. The differential

$$df\colon TM \to f^*TC$$

of f determines a section of $T^*M \otimes f^*TC$, which is also denoted by df. The set of zeros of df is the critical set $C(f)$ of f. Suppose $C(f)$ is a compact set with a finite number of connected components $(S_\lambda)_\lambda$. Then we have the residue $\mathrm{Res}_{c^n}(df, T^*M \otimes f^*TC; S_\lambda)$ for each λ. If M is compact, we have (cf. Proposition 10.2, Theorem 10.6)

$$\sum_\lambda \mathrm{Res}_{c^n}(df, T^*M \otimes f^*TC; S_\lambda) = \int_M c^n(T^*M \otimes f^*TC).$$

We look at the both sides of the above more closely. In the following, we set $D(f) = f(C(f))$, the set of critical values. Then, if M is compact, f defines the structure of a C^∞ locally trivial fibration on $M \smallsetminus C(f) \to C \smallsetminus D(f)$ (cf. Theorem 3.12). Denoting by F a general fiber of f, we have:

Lemma 12.7. *If M is compact and if $D(f)$ consists of isolated points,*

$$\int_M c^n(T^*M \otimes f^*TC) = (-1)^n(\chi(M) - \chi(F) \cdot \chi(C)).$$

Suppose that $f(S_\lambda)$ consists of a point. Taking a coordinate on C around $f(S_\lambda)$, we think of f as a holomorphic function near S_λ. Then we may write

$$\mathrm{Res}_{c^n}(df, T^*M \otimes f^*TC; S_\lambda) = \mathrm{Res}_{c^n}(df, T^*M; S_\lambda) = m(f, S_\lambda),$$

the multiplicity of f at S_λ. Thus we have

Theorem 12.6. *Let $f\colon M \to C$ be a holomorphic map of a compact complex manifold M of dimension n onto a complex curve C. If $D(f)$ consists of isolated points, then we have*

$$\sum_\lambda m(f, S_\lambda) = (-1)^n(\chi(M) - \chi(F) \cdot \chi(C)).$$

In particular, we have:

Corollary 12.2. *In the above situation, if the critical set $C(f)$ of f consists of only isolated points,*

$$\sum_{p\in C(f)} m(f,p) = (-1)^n(\chi(M) - \chi(F)\cdot\chi(C)).$$

The above notion of multiplicity and the formula for these multiplicities will be generalized to the case of functions on singular varieties in Section 13.7 below.

12.6 Dual class of a complex subspace

We come back to the situation of the beginning of Section 12.2 and let M, E, $s^{(r)}$, $\boldsymbol{S}$, S and so forth be as given there. We adopt Convention 4.2 for notation of the class of a cycle.

We assume that $\dim S = n - q$ so that we may identify the topological residue $\mathrm{TRes}_{c^q}(s^{(r)}, E; S)$ and the differential geometric residue $\mathrm{Res}_{c^q}(s^{(r)}, E; S)$ (cf. (12.2)). In this case, we may also identify $c^q_{\mathrm{top}}(E, s^{(r)})$ and $c^q_{\mathrm{diff}}(E, s^{(r)})$, which we denote by $c^q(E, s^{(r)})$. Let (S_i) be the irreducible components of S. Also let p_i, $\mathbb{D}_i$, E_i and $s_i^{(r)}$ be as before.

Definition 12.5. The *multiplicity* m_i of S_i in $\boldsymbol{S}$ is defined by

$$m_i = \mathrm{Res}_{c^q}(s_i^{(r)}, E_i; p_i).$$

Note that m_i has topological, analytic and algebraic expressions as given in Theorems 12.3–12.5.

Definition 12.6. The homology class of $\boldsymbol{S}$ is defined by $[\boldsymbol{S}] = \sum m_i\,[S_i]$ in $\breve{H}_{2(n-q)}(M;\mathbb{Z})$ or by $\boldsymbol{S} = \sum m_i\,[S_i]$ in $\breve{H}_{2(n-q)}(S;\mathbb{Z})$.

In this situation, we have:

Theorem 12.7. *The class $c^q(E)$ corresponds to $[\boldsymbol{S}]$ under the Poincaré duality $H^{2q}(M;\mathbb{Z}) \xrightarrow{\sim} \breve{H}_{2(n-q)}(M;\mathbb{Z})$.*

In fact, we have the following more "precise" theorem, which is a direct consequence of Theorem 12.2:

Theorem 12.8. *Let $\boldsymbol{S}$ be the complex space of dimension $n-q$ in M as above. Then the class of $\boldsymbol{S}$ in $\breve{H}_{2(n-q)}(S;\mathbb{Z})$ is equal to* $\mathrm{TRes}_{c^q}(s^{(r)}, E; S)$, *which may be identified with* $\mathrm{Res}_{c^q}(s^{(r)}, E; S)$. *Thus the class* $c^q(E, s^{(r)})$ *corresponds to the class* $\boldsymbol{S}$ *under the Alexander duality*

$$H^{2q}(M, M \setminus S; \mathbb{Z}) \xrightarrow{\sim} \check{H}_{2(n-q)}(S; \mathbb{Z}).$$

We specialize the above to the case $r = 1$. In this case, we have $q = l$ and we are to consider the top Chern class.

Let M and E be as above. Also let s be a holomorphic section of E and $\boldsymbol{S}$ the complex space defined by (the ideal generated by the local components of) s. Thus the support of $\boldsymbol{S}$ is the zero set S of s. In this situation, we have the localization $c^l(E, s)$ in $H^{2l}(M, M \setminus S)$ and the associated residue $\mathrm{TRes}_{c^l}(s, E; S)$ in $\check{H}_{2(n-l)}(S)$. In general, $\dim S \geq n - l$ and the equality holds if and only if s is a regular section (cf. Definition 11.31).

The following is a special case of Theorem 12.8:

Theorem 12.9. *Let* $\boldsymbol{S}$ *be the complex space of dimension* $n - l$ *in* M *as above. Then the class of* $\boldsymbol{S}$ *in* $\check{H}_{2(n-l)}(S; \mathbb{Z})$ *is equal to* $\mathrm{TRes}_{c^l}(s, E; S)$, *which may be identified with* $\mathrm{Res}_{c^l}(s, E; S)$. *Thus the class* $c^l(E, s)$ *corresponds to the class* $\boldsymbol{S}$ *under the Alexander duality*

$$H^{2l}(M, M \setminus S; \mathbb{Z}) \xrightarrow{\sim} \check{H}_{2(n-l)}(S; \mathbb{Z}).$$

Remark 12.2. 1. Corollary 5.5 (see also Theorem 10.9) may be understood in this context.

2. In the above situation, we call $c^l(E, s)$ the Thom class of $\boldsymbol{S}$ in M and denote it by $\Psi_{\boldsymbol{S}}$ (cf. Definition 15.4 and the subsequent paragraph). This is used to express the localized Riemann-Roch theorem for embeddings in a certain case in Section 15.6 below (cf. Theorem 15.5).

Localization by meromorphic sections

Let M be a complex manifold of dimension n and D a divisor on M (cf. Section 11.6). Recall that D is represented by a system $\{(U_\alpha, \varphi^\alpha)\}$, where $\{U_\alpha\}$ is a covering of M, φ^α is in $\Gamma(U_\alpha; \mathscr{M}^*)$, for each α, and $f^{\alpha\beta} = \varphi^\alpha/\varphi^\beta$ is in $\Gamma(U_\alpha \cap U_\beta; \mathscr{O}^*)$, for each pair (α, β). The system $\{f^{\alpha\beta}\}$ defines the associated line bundle L_D and the collection $\{\varphi_\alpha\}$ the natural meromorphic section s_D of L_D.

Conversely, let L be a holomorphic line bundle on M and s a meromorphic section of L. Let $\{U_\alpha\}$ be a covering of M trivializing L and $\{f^{\alpha\beta}\}$ a system of transition functions. Then s is represented by a collection $\{\varphi^\alpha\}$ of meromorphic functions such that $\varphi^\alpha = f^{\alpha\beta}\varphi^\beta$ on $U_\alpha \cap U_\beta$. If φ_α is in

$\Gamma(U_\alpha;\mathscr{M}^*)$ for each α, then s defines a divisor D. Moreover L is the line bundle L_D associated with D and s is the natural section s_D of L_D.

In the above situation, we have the topological localization $c^1_{\text{top}}(L_D, s_D)$ in $H^2(M, M \smallsetminus |D|;\mathbb{Z})$ (cf. Section 5.6) and the differential geometric localization $c^1_{\text{diff}}(L_D, s_D)$ in $H^2(M, M\smallsetminus|D|;\mathbb{C})$ (cf. Section 10.3). We also have the associated residues $\mathrm{TRes}_{c^1}(s_D, L_D; |D|)$ in $\breve{H}_{2(n-1)}(|D|;\mathbb{Z})$ and $\mathrm{Res}_{c^1}(s_D, L_D; |D|)$ in $\breve{H}_{2(n-1)}(|D|;\mathbb{C})$. We have seen (cf. Corollary 10.6) that $\mathrm{Res}_{c^1}(s_D, L_D; |D|)$ is the image of $\mathrm{TRes}_{c^1}(s_D, L_D; |D|)$ by the canonical morphism

$$\breve{H}_{2(n-1)}(|D|;\mathbb{Z}) \longrightarrow \breve{H}_{2(n-1)}(|D|;\mathbb{C}). \tag{12.11}$$

Suppose D has only a finite number of connected components $(S_\lambda)_\lambda$. Then we have a decomposition $\breve{H}_{2(n-1)}(|D|;\mathbb{Z}) = \bigoplus_\lambda \breve{H}_{2(n-1)}(S_\lambda;\mathbb{Z})$ and accordingly we have the residue $\mathrm{TRes}_{c^1}(s_D, L_D; S_\lambda)$ in $\breve{H}_{2(n-1)}(S_\lambda;\mathbb{Z})$ for each λ. We have (cf. Theorem 5.2):

$$\sum_\lambda (\imath_\lambda)_* \mathrm{TRes}_{c^1}(s_D, L_D; S_\lambda) = M \frown c^1(L_D) \qquad \text{in } \breve{H}_{2(n-1)}(M;\mathbb{Z}),$$

where $\imath_\lambda : S_\lambda \hookrightarrow M$ denotes the inclusion. Passing to the homology with $\mathbb{C}$-coefficient, we have a similar statement for $\mathrm{Res}_{c^1}(s_D, S_\lambda; |D|)$ (cf. Theorem 10.6).

Example 12.4. Let C be a compact Riemann surface and L a holomorphic line bundle on C. Suppose we have a non-zero meromorphic section s of L. Then the support S of the divisor defined by s is the set of zeros and poles of s. The previous construction gives us the localization $c^1(L, s)$ in $H^2(C, C\smallsetminus S)$ of $c^1(L)$ in $H^2(C)$. Note that S consists of a finite number of points and, for each p in S, $\mathrm{TRes}_{c^1}(s, L; p)$ and $\mathrm{Res}_{c^1}(s, L; p)$ are the same integer. We have

$$\sum_{p\in S} \mathrm{Res}_{c^1}(s, L; p) = \langle C, c^1(L)\rangle.$$

The residue $\mathrm{Res}_{c^1}(s, L; p)$ is given as follows. We choose an open neighborhood U of p not containing any other points of S and trivializing L. Let e be a holomorphic frame of L on U, and write $s = fe$ with f a meromorphic function on U. Let $\mathbb{S}^1$ be a circle around p in U oriented counterclockwise. Then we have (cf. Example 10.3)

$$\mathrm{Res}_{c^1}(s, L; p) = \frac{1}{2\pi\sqrt{-1}} \int_{\mathbb{S}^1} \frac{df}{f},$$

which is the order of zero or pole according as f has a zero or a pole at p.

Let D a non-zero divisor on M and write $D = \sum m_i V_i$, $m_i \neq 0$. Then each V_i defines a class $[V_i]$ in $\check{H}_{2(n-1)}(|D|;\mathbb{Z})$ and it is the free Abelian group generated by the $[V_i]$'s. Thus the morphism (12.11) is injective so that we may identify the two residues:

$$\mathrm{Res}_{c^1}(s_D, L_D; |D|) = \mathrm{TRes}_{c^1}(s_D, L_D; |D|). \tag{12.12}$$

Let p_i be a non-singular point of $V_i \smallsetminus \bigcup_{j \neq i} V_j$ and $\mathbb{D}_i$ a complex slice of V_i in M at p_i, similarly as in the beginning of Section 12.2. Note that in this case $\mathbb{D}_i$ is complex one dimensional. The section $s_i = s_D|_{\mathbb{D}_i}$ of $L_i = L_D|_{\mathbb{D}_i}$ is non-vanishing and holomorphic away from p_i so that we have the residues $\mathrm{TRes}_{c^1}(s_i, L_i; p_i)$ and $\mathrm{Res}_{c^1}(s_i, L_i; p_i)$, both of which are the same integer $\mathrm{ord}_{V_i}(s_D)$ (cf. Example 10.3). It is the coefficient m_i of V_i in D (cf. (11.19)).

Definition 12.7. For a divisor $D = \sum m_i V_i$, we define its homology class by $[D] = \sum m_i [V_i]$ in $\check{H}_{2n-2}(M;\mathbb{Z})$ or by $D = \sum m_i [V_i]$ in $\check{H}_{2n-2}(|D|,\mathbb{Z})$.

If D is non-zero and effective, we have seen that it may be thought of as a complex space of pure dimension $n-1$ in M (cf. Section 11.6). In this case, the natural section s_D of the associated line bundle L_D is holomorphic and D is the complex space defined by s_D. The multiplicity of V_i in D is m_i so that the above definition coincides with the one in Definition 12.6.

Note that, if $D = 0$, then $|D| = \emptyset$ and the both sides of (12.12) are zero. The statement of Theorem 12.2 holds in the case of divisors as well:

$$\mathrm{TRes}_{c^1}(s_D, L_D; |D|) = \mathrm{TRes}_{c^1}(s_i, L_i; p_i) \cdot [V_i] \qquad \text{in} \quad \check{H}_{2(n-1)}(|D|;\mathbb{Z})$$

so that we have:

Theorem 12.10. *Let D be a divisor on M. Then the class of D in $\check{H}_{2(n-1)}(|D|;\mathbb{Z})$ is equal to $\mathrm{TRes}_{c^1}(s_D, L_D; |D|)$, which may be identified with $\mathrm{Res}_{c^1}(s_D, L_D; |D|)$. Thus the class $c^1(L_D, s_D)$ corresponds to the class D under the Alexander duality*

$$H^2(M, M \smallsetminus |D|;\mathbb{Z}) \xrightarrow{\sim} \check{H}_{2(n-1)}(|D|;\mathbb{Z}).$$

In particular, the class $[D]$ in $\check{H}_{2(n-1)}(M;\mathbb{Z})$ is the Poincaré dual of $c_1(L_D)$ in $H^2(M;\mathbb{Z})$. Thus if D_1 and D_2 are two linearly equivalent divisors, then $[D_1] = [D_2]$.

Also, if D_1 and D_2 are two divisors, the intersection product $[D_1] \bullet [D_2]$ in $\check{H}_{2(n-2)}(M;\mathbb{Z})$ is the Poincaré dual of $c_1(L_{D_1}) \smile c_1(L_{D_2})$ in $H^4(M;\mathbb{Z})$ (cf. Theorem 4.7). If M is compact, for n divisors $D_1, \ldots, D_n$ on M, the "global intersection number" $\sharp([D_1] \bullet \cdots \bullet [D_n])$ is defined and is given by

$\langle M, c_1(L_{D_1}) \cdots c_1(L_{D_n}) \rangle$, where the product is the cup product (cf (14.8) below). If the intersection $\bigcap_{i=1}^n |D_i|$ consists of isolated points, then this number is the sum of the intersection numbers at the points of intersection (cf. Theorem 14.2 below). See Example 12.2 for this "local intersection number" when the divisors are effective. We descuss these more in detail in Section 14 below.

In the case D is a non-zero effective divisor, Theorem 12.10 is a special case of Theorem 12.9, where $l = 1$.

Example 12.5. Let V be the projective algebraic variety in $\mathbb{P}^n = \{[\zeta_0, \dots, \zeta_n]\}$ defined by a homogeneous polynomial P of degree p (cf. Section 2.2). The function $\varphi = P/\zeta_0^p$ is a well-defined meromorphic function on $\mathbb{P}^n$, which is given as the quotient of P/ζ_α^p by ζ_0^p/ζ_α^p on each affine open set $U_\alpha = \{\zeta_\alpha \neq 0\}$. Thus, if we denote by D_∞ the hyperplane defined by $\zeta_0 = 0$, then V is linearly equivalent to $p\,D_\infty$ and $[V] = p\,[D_\infty]$ in $H_{2n-2}(\mathbb{P}^n; \mathbb{Z})$. Recall that $[D_\infty] = [\mathbb{P}^{n-1}]$ is a generator of $H_{2n-2}(\mathbb{P}^n; \mathbb{Z}) \simeq \mathbb{Z}$. Also, the intersection product of k copies of $[\mathbb{P}^{n-1}]$ generates $H_{2n-2k}(\mathbb{P}^n; \mathbb{Z}) \simeq \mathbb{Z}$, for $k = 1, \dots, n$ (cf. Example 2.4. 2).

Notes

The presentation of this chapter is mainly based on Section 4 of [Suwa (2008)] and the references therein.

For subanalytic sets, we refer to [Bierstone and Milman (1988); Shiota (1997)] and the references therein. Theorem 12.1 is due to [Ohmoto and Shiota (2017)].

We list [Griffiths and Harris (1978); Hartshorne (1966)] as references for Section 12.3. In Section 12.4, see [Suwa (2005)] for algebraic expression and [Suwa (2003a)] for analytic expression with more examples. The formula in Corollary 12.2 is proved in [Iversen (1971)], see also Example 14.1.5 in [Fulton (1984)]. See [Izawa and Suwa (2003)] for a precise proof of Lemma 12.7.

See [Suwa (2000a)] for a proof of Theorem 12.9 via Čech-de Rham cohomology and some more materials in this direction, including the above type of duality for non-compact varieties, which is effectively used in [Brasselet, Lehmann, Seade and Suwa (2001)].

Chapter 13

Residues of Chern Classes on Singular Varieties

We describe explicitly the residues determined by holomorphic r-sections of vector bundles on singular varieties.

For this, we first look into the neighborhood structure of a Whitney stratification, introducing the notion of a controlled tube system. We then discuss the Poincaré, Alexander, Lefschetz and Thom morphisms for singular varieties, along the line of Sections 4.2 and 4.3. In order to represent these in terms of differential forms, we discuss the de Rham and Čech-de Rham theories for varieties. We define the Chern classes of complex vector bundles on varieties and their localizations by r-frames from both topological and differential geometric viewpoints. The classes defined by two ways are shown to be essentially the same.

As in the case of r-sections on complex manifolds, we have a fundamental theorem (Theorem 13.9) saying that the residue in the proper case is determined by the residues at isolated singularities of the r-section that appear on the transverse slices of irreducible components of the singular set. We give various expressions for the residue at an isolated singularity as well as some fundamental examples.

We also discuss the local duality concerning complex spaces defined by an r-section, as in the case of complex manifolds. It will be used in Chapter 14 below for the localized intersection theory.

13.1 Controlled tube systems for Whitney stratifications

Let M be a C^∞ manifold and X a submanifold of M. A *tube* around X is a triple $\boldsymbol{T} = (U, r, \rho)$, where U is an open neighborhood of X in M, $r : U \to X$ a C^∞ submersive retraction and ρ a non-negative C^∞ function on U such that $\rho^{-1}(0) = X$ and that each point $x \in X$ is a unique non-degenerate critical point of $\rho|_{r^{-1}(x)}$. For example, if $r : U \to X$ is a tubular neighborhood (cf. Theorem 3.10) and ρ the function corresponding to the square of distance from the zero section of the normal bundle with respect to some Riemannian metric, then (U, r, ρ) is a tube. It can be shown that every tube is essentially given as above.

Let S be a closed set in M and $\{X_\alpha\}$ a stratification of S (cf. Definition 2.13). A tube system for $\{X_\alpha\}$ is a collection $\{\boldsymbol{T}_\alpha\}$, where, for each α, $\boldsymbol{T}_\alpha = (U_\alpha, r_\alpha, \rho_\alpha)$ is a tube around X_α. The tube system $\{\boldsymbol{T}_\alpha\}$ is *controlled* if, for every pair (α, β) with $X_\alpha \cap (\overline{X_\beta} \smallsetminus X_\beta) \neq \emptyset$, the following holds:

$$r_\alpha \circ r_\beta(x) = r_\alpha(x) \quad \text{and} \quad \rho_\alpha \circ r_\beta(x) = \rho_\alpha(x)$$

for all $x \in U_\alpha \cap U_\beta$ with $r_\beta(x) \in U_\alpha$.

We quote:

Theorem 13.1. *Every Whitney stratification admits a controlled tube system.*

Here are some aspects of controlled tube systems. Let $\{\boldsymbol{T}_\alpha\}$ be a tube system for $\{X_\alpha\}$. For each α, we take a positive function ε_α on X_α so that $T_\alpha = \{\, x \in U_\alpha \mid \rho_\alpha(x) \leq \varepsilon_\alpha(r_\alpha(x)) \,\}$ is a closed ball bundle on X_α. Thus $O_\alpha = \{\, x \in U_\alpha \mid \rho_\alpha(x) < \varepsilon_\alpha(r_\alpha(x)) \,\}$ is an open ball bundle on X_α. Then $r_\alpha|_{T_\alpha} : T_\alpha \to X_\alpha$ is a proper deformation retraction and $r_\alpha|_{O_\alpha} : O_\alpha \to X_\alpha$ a deformation retraction. If $\{\boldsymbol{T}_\alpha\}$ is controlled, then:

(1) For every pair (α, β) with $X_\alpha \cap (\overline{X_\beta} \smallsetminus X_\beta) \neq \emptyset$, ∂T_α intersects both X_β and ∂T_β transversally.
(2) The inclusion $S \hookrightarrow \bigcup_\alpha O_\alpha$ is a homotopy equivalence.

13.2 Poincaré, Alexander, Lefschetz and Thom morphisms

Let M be a complex manifold of dimension n and V an analytic variety of pure dimension d in M. We denote by $\operatorname{Sing} V$ the singular set of V and by $V_{\text{reg}} = V \smallsetminus \operatorname{Sing} V$ the regular part, which is naturally oriented as a complex manifold. We take a C^1 triangulation (K_0, h) of M compatible with V and $\operatorname{Sing} V$ as in Theorem 12.1. Let K be the barycentric subdivision of K_0 and K^* the cellular decomposition dual to K, which is constructed from the barycentric subdivision K' of K (cf. Section 4.1). We denote by K_V the triangulation of V induced from K and by K'_V its barycentric subdivision. Then K'_V is the set of simplices of K' that are in V (cf. Section 4.3). Note that if $\boldsymbol{s}$ is a p-simplex of K in V, then $\boldsymbol{s}^*$ is a $(2n-p)$-cell of K^* and $\boldsymbol{s}^* \cap V$

is a $(2d-p)$-chain of K'_V. The chain $\boldsymbol{s}^* \cap V$ is a sum of K'_V-simplices each of which is oriented so that its orientation followed by that of $\boldsymbol{s}$ gives the orientation of V_{reg}.

In this section, homology and cohomology are with $\mathbb{Z}$-coefficient. We also use the notation in Section 4.1 for various chain and cochain groups. Thus, denoting by $C_\bullet^{K_V}(V)$ and $S_\bullet(V)$ the complexes of (finite) chains of K_V and of singular chains of V, there is a natural chain morphism

$$\eta_\bullet^{K_V} : C_\bullet^{K_V}(V) \longrightarrow S_\bullet(V),$$

which is defined as (4.1). It induces isomorphisms

$$\eta_*^{K_V} : H_p^{K_V}(V) \xrightarrow{\sim} H_p(V) \quad \text{and} \quad \eta_{K_V}^* : H^p(V) \xrightarrow{\sim} H_{K_V}^p(V). \qquad (13.1)$$

Also, denoting by $\breve{C}_\bullet^{K_V}(V)$ and $\breve{S}_\bullet(V)$ the complexes of locally finite chains of K_V and of locally finite singular chains of V, respectively, there is a natural chain morphism $\breve{\eta}_\bullet^{K_V} : \breve{C}_\bullet^{K_V}(V) \to \breve{S}_\bullet(V)$, which induces an isomorphism

$$\breve{\eta}_*^{K_V} : \breve{H}_p^{K_V}(V) \xrightarrow{\sim} \breve{H}_p(V). \qquad (13.2)$$

In the following, the suffix V in K_V or in K'_V will be omitted if there is no fear of confusion.

Poincaré morphism

We define a morphism

$$P_V : C_{K'}^p(V) \longrightarrow \breve{C}_{2d-p}^K(V) \qquad \text{by} \quad P(u_V) = \sum_{\boldsymbol{s}} \langle \boldsymbol{s}^* \cap V, u_V \rangle\, \boldsymbol{s}, \qquad (13.3)$$

where the sum is taken over all $(2d-p)$-simplices $\boldsymbol{s}$ of K_V. We may prove that it is compatible with boundary and coboundary operators along the same line as in the case of manifolds (cf. Section 4.1). Thus let $V_{K'}$ denote the fundamental cycle of V in K', i.e., the sum of $2d$-simplices of K' in V. The following is proved as Lemma 4.1:

Lemma 13.1. *For every p-simplex $\boldsymbol{t}$ of K' in V, we have:*

$$V_{K'} \frown \vartheta(\boldsymbol{t}) = \begin{cases} \boldsymbol{s} & \text{if } \boldsymbol{t} \subset \boldsymbol{s}^* \cap V \text{ for some } (2d-p)\text{-simplex } \boldsymbol{s} \text{ of } K_V, \\ 0 & \text{otherwise}, \end{cases}$$

as K'_V-chains.

From this we have the commutative diagram

$$\begin{array}{ccc} C^p_{K'}(V) & \xrightarrow{\ P\ } & \check{C}^K_{2d-p}(V) \\ & \searrow^{V_{K'}\frown} & \downarrow{\scriptstyle \kappa} \\ & & \check{C}^{K'}_{2d-p}(V), \end{array}$$

where κ is the natural monomorphism, and we have a relation as the one in Lemma 4.2. Thus, combining with the isomorphisms $\eta^*_{K'_V}$ in (13.1) for K'_V and $\check{\eta}^{K_V}_*$ in (13.2), we have:

Theorem 13.2. *The morphism P in* (13.3) *induces a morphism*

$$P_V : H^p(V) \longrightarrow \check{H}_{2d-p}(V).$$

We call P_V the *Poincaré morphism.* It is given by the cap product with the fundamental class $[V]$ and is independent of the triangulation. If V is non-singular, this is the Poincaré isomorphism in Theorem 4.1.

Alexander morphism

Let Z be a K_0-subcomplex of V. We denote by $S_{K'}(Z)$ and $O_{K'}(Z)$ the star and the open star of Z in K'. We may also consider $S_{K'_V}(Z) = S_{K'}(Z) \cap V$ and $O_{K'_V}(Z) = O_{K'}(Z) \cap V$. Both $S_{K'}(Z)$ and $S_{K'_V}(Z)$ proper deformation retract to Z and both $O_{K'}(Z)$ and $O_{K'_V}(Z)$ deformation retract to Z. Note that $V \smallsetminus O_{K'_V}(Z) = V \smallsetminus O_{K'}(Z)$.

We may write

$$C^p_{K'}(V, V \smallsetminus O_{K'}(Z)) = \{\, u_V \in C^p_{K'}(V) \mid \langle \boldsymbol{t}, u_V \rangle = 0 \text{ for } \boldsymbol{t} \text{ with } \boldsymbol{t} \cap Z = \emptyset \,\},$$

where $\boldsymbol{t}$ denotes a p-simplex of K'_V. Denoting by $H^p_{K'}(V, V \smallsetminus O_{K'}(Z))$ the cohomology of $C^*_{K'}(V, V \smallsetminus O_{K'}(Z))$, we have a natural isomorphism

$$\eta^*_{K'} : H^p(V, V \smallsetminus O_{K'}(Z)) \xrightarrow{\ \sim\ } H^p_{K'}(V, V \smallsetminus O_{K'}(Z)).$$

There is also a natural isomorphism

$$H^p(V, V \smallsetminus O_{K'}(Z)) \simeq H^p(V, V \smallsetminus Z).$$

If u_V is in $C^p_{K'}(V, V \smallsetminus O_{K'}(Z))$, in the sum in (13.3), only the simplices of K in Z appear. Hence the morphism P_V induces a morphism

$$A_V : C^p_{K'}(V, V \smallsetminus O_{K'}(Z)) \longrightarrow \check{C}^K_{2d-p}(Z). \tag{13.4}$$

It is compatible with the boundary and coboundary operators so that we have:

Theorem 13.3. *For a subcomplex Z of V, the morphism A_V in* (13.4) *induces a morphism*

$$A_{V,Z} : H^p(V, V \smallsetminus Z) \longrightarrow \breve{H}_{2d-p}(Z).$$

It is called the *Alexander morphism.* If V is non-singular, this is the Alexander isomorphism in Theorem 4.2.

Remark 13.1. As in the case of manifolds, we may write (cf. (4.13))

$$A(\alpha) = [\underline{S_{K'_V}(Z)}] \frown \alpha \qquad \text{for } \alpha \in H^p(V, V \smallsetminus Z),$$

where $[\underline{S_{K'_V}(Z)}]$ denotes the class in $\breve{H}_{2d}(S_{K'_V}(Z), \partial S_{K'_V}(Z))$ of the relative cycle $\underline{S_{K'_V}(Z)}$ in $\breve{C}^{K'}_{2d}(S_{K'_V}(Z), \partial S_{K'_V}(Z))(= \breve{C}^{K'}_{2d}(S_{K'_V}(Z))$.

Also, if Z is compact, we may write (cf. Remark 4.5. 3)

$$A(\alpha) = [\underline{V}] \frown \alpha \qquad \text{for } \alpha \in H^p(V, V \smallsetminus Z).$$

Lefschetz morphism

Let R be a real $2n$-dimensional manifold with boundary in M such that R contains $\operatorname{Sing} V$ in its interior and that ∂R is transverse to V_{reg}. We may assume that the triangulation K_0 is also compatible with R and ∂R. We set $Q = R \cap V$ and apply the above considerations for $Z = Q$. Since we have natural isomorphisms

$$H^p(V, V \smallsetminus O_{K'}(Q)) \simeq H^p(S_{K'_V}(Q), \partial S_{K'_V}(Q)) \simeq H^p(Q, \partial Q), \qquad (13.5)$$

where the boundaries are taken in V, we obtain a morphism

$$L : H^p(Q, \partial Q) \longrightarrow \breve{H}_{2d-p}(Q), \qquad (13.6)$$

which we call the *Lefschetz morphism.* Note that the above morphism is given by the cap product with the relative class $[\underline{Q}]$ of Q in $\breve{H}_{2d}(Q, \partial Q)$. It is the class corresponding to $[\underline{S_{K'_V}(Q)}]$ in the isomorphism $\breve{H}_{2d}(Q, \partial Q) \simeq \breve{H}_{2d}(S_{K'_V}(Q), \partial S_{K'_V}(Q))$.

If V is non-singular, (13.6) is the Lefschetz isomorphism in Theorem 4.3.

From the construction, we have (cf. Proposition 4.4):

Proposition 13.1. *Let R be a real $2n$-dimensional manifold possibly with boundary in M as above. For a subcomplex Z in the interior of $Q = R \cap V$, we have the commutative diagram:*

$$\begin{array}{ccccc}
H^p(V, V \smallsetminus Z) & \xrightarrow{j^*} & H^p(Q, \partial Q) & \xrightarrow{j'^*} & H^p(V) \\
\downarrow A & & \downarrow L & & \downarrow P \\
\breve{H}_{2d-p}(Z) & \xrightarrow{j_*} & \breve{H}_{2d-p}(Q) & \xrightarrow{j'_*} & \breve{H}_{2d-p}(V),
\end{array}$$

where $j : (Q, \partial Q) \hookrightarrow (V, V \smallsetminus Z)$, $\jmath : Z \hookrightarrow Q$ *and* $\jmath' : Q \hookrightarrow V$ *denote the inclusions and* j'^* *denotes the isomorphism in* (13.5) *followed by the canonical morphism* $H^p(V, V \smallsetminus O_{K'}(Q)) \to H^p(V)$.

Suppose Z has a finite number of connected components $(Z_\lambda)_\lambda$. Then $\breve{H}_{2d-p}(Z) = \bigoplus_\lambda \breve{H}_{2d-p}(Z_\lambda)$ and each class α in $H^p(V, V \smallsetminus Z)$ determines a class a_λ in $\breve{H}_{2d-p}(Z_\lambda)$ via the Alexander morphism.

Corollary 13.1. *In the above situation,*

$$\sum_\lambda (\jmath_\lambda)_* a_\lambda = [Q] \frown j^*\alpha \qquad \text{in } \breve{H}_{2d-p}(Q),$$

where $\jmath_\lambda : Z_\lambda \hookrightarrow Q$ *denotes the inclusion.*

In some situations, we do not consider the auxiliary manifold R and take M itself as R. In this case, we have the commutative diagram

$$\begin{array}{ccc} H^p(V, V \smallsetminus Z) & \xrightarrow{j^*} & H^p(V) \\ {\scriptstyle A}\downarrow & & \downarrow{\scriptstyle P} \\ \breve{H}_{2d-p}(Z) & \xrightarrow{\jmath_*} & \breve{H}_{2d-p}(V) \end{array} \tag{13.7}$$

and the right-hand side of the identity in Corollary 13.1 is written $V \frown j^*\alpha$ in $\breve{H}_{2d-p}(V)$.

The above is the basis of the residue theorem we discuss in Section 13.4 below (cf. Theorems 13.5 and 13.6).

Thom morphism

We consider the morphism

$$T_V : C^p_{K'}(V) \longrightarrow C^{2k+p}_{K^*}(M, M \smallsetminus O_{K'}(V)), \qquad u_V \mapsto u,$$

where u is given by, for each $(2d-p)$-simplex s of K,

$$\langle s^*, u \rangle = \begin{cases} \langle s^* \cap V, u_V \rangle & \text{if } s \subset V, \\ 0 & \text{otherwise.} \end{cases}$$

Then it is compatible with coboundary operators and induces a morphism

$$T_V : H^p(V) \longrightarrow H^{2k+p}(M, M \smallsetminus V), \tag{13.8}$$

called the *Thom morphism*. If V is non-singular, it is the isomorphism introduced in Section 4.3.

Definition 13.1. The *Thom class* of V, denoted by Ψ_V, is the image of [1] by $T_V : H^0(V) \to H^{2k}(M, M \smallsetminus V)$.

If V is non-singular, it coincides with the one in Definition 4.2.

Remark 13.2. A remark similar to Remark 4.9.1 applies to this case. Namely, If ψ is a cocycle in $C^{2k}_{K'}(M, M \smallsetminus O_{K'}(V))$ representing the Thom class, we have

$$\langle \iota \boldsymbol{s}^*, \psi \rangle = \langle \boldsymbol{s}^*, \iota^* \psi \rangle = \begin{cases} 1 & \text{if } \boldsymbol{s} \subset V, \\ 0 & \text{otherwise,} \end{cases}$$

where ι and ι^* are defined similarly as in Remark 4.9.1 with $k' = 2k$.

From definition we see that the following diagram is commutative:

$$\begin{array}{ccc} H^p(V) & \xrightarrow{T} & H^{2k+p}(M, M \smallsetminus V) \\ {\scriptstyle P}\downarrow & \swarrow {\scriptstyle A}\ \sim & \\ \breve{H}_{2d-p}(V). & & \end{array} \tag{13.9}$$

Thus the Thom class Ψ_V in $H^{2k}(M, M \smallsetminus V)$ correspons to the fundamental class V in $\breve{H}_{2d}(V)$ by the Alexander isomorphism $A_{M,V}$.

Note that there is the cup product

$$H^{2k}(M, M \smallsetminus V) \times H^p(M) \xrightarrow{\smile} H^{2k+p}(M, M \smallsetminus V).$$

Denoting by $i : V \hookrightarrow M$ the inclusion, the following is proved as Proposition 4.10:

Proposition 13.2. *The following diagram is commutative:*

$$\begin{array}{ccc} H^p(M) & \xrightarrow{\Psi_V \smile (\ )} & H^{2k+p}(M, M \smallsetminus V) \\ {\scriptstyle i^*}\downarrow & & \wr\downarrow {\scriptstyle A_{M,V}} \\ H^p(V) & \xrightarrow{P_V} & \breve{H}_{2d-p}(V). \end{array}$$

In particular, combined with (13.9), we have:

$$T_V(i^* \alpha) = \Psi_V \smile \alpha \qquad \text{for } \alpha \in H^p(M).$$

As in (4.25), if we use the isomorphisms $r^* : H^p(V) \xrightarrow{\sim} H^p(O_{K'}(V))$ and $H^{2k+p}(M, M \smallsetminus V) \xrightarrow{\sim} H^{2k+p}(O_{K'}(V), O_{K'}(V) \smallsetminus V)$, the above identity may be expressed as

$$T_V(\alpha_V) = \Psi_V \smile r^*(\alpha_V) \qquad \text{for } \alpha_V \in H^p(V). \tag{13.10}$$

If S is a K_0-subcomplex of M, setting $Z = S \cap V$, we have the cup product

$$H^{2k}(M, M \smallsetminus V) \times H^p(M, M \smallsetminus S) \xrightarrow{\smile} H^{2k+p}(M, M \smallsetminus Z).$$

The following is proved as Proposition 4.11 (cf. Remark 13.2):

Proposition 13.3. *The following diagram is commutative*:

$$\begin{array}{ccc} H^p(M, M \smallsetminus S) & \xrightarrow{\Psi_V \smile (\ )} & H^{2k+p}(M, M \smallsetminus Z) \\ {\scriptstyle i^*}\big\downarrow & & \wr\big\downarrow{\scriptstyle A_{M,Z}} \\ H^p(V, V \smallsetminus Z) & \xrightarrow{A_{V,Z}} & \breve{H}_{2d-p}(Z). \end{array}$$

Gysin morphism: In the above situation, let $i : V \hookrightarrow M$ denote the inclusion. As in Section 4.3, the composition

$$H^p(V) \xrightarrow{T_V} H^{2k+p}(M, M \smallsetminus V) \xrightarrow{j^*} H^{2k+p}(M)$$

is referred to as the *Gysin morphism* and denoted by i_*. We have the commutative diagram (cf. (4.18))

$$\begin{array}{ccc} H^p(V) & \xrightarrow{i_*} & H^{2k+p}(M) \\ \big\downarrow{\scriptstyle P_V} & & \wr\big\downarrow{\scriptstyle P_M} \\ \breve{H}_{2d-p}(V) & \xrightarrow{i_*} & \breve{H}_{2d-p}(M). \end{array} \tag{13.11}$$

If V is non-singular, this coincides with (4.27).

13.3 de Rham and Čech-de Rham theories for singular varieties

Let M be a complex manifold of dimension n and V a variety of dimension d in M. Also, let K_0, K, K' and K^* be as in the previous section.

Integration

Since the triangulation $h : |K_0| \to M$ is C^1, for a p-simplex $\boldsymbol{s}$ of K, or of K', and a C^∞ p-form ω on a neighborhood of $h(\boldsymbol{s})$ in M, we may define the integration as (6.15):

$$\int_{\boldsymbol{s}} \omega = \int_{\boldsymbol{s}} (h|_{\boldsymbol{s}})^* \omega,$$

which extends to integrations on finite chains.

Suppose $\operatorname{Sing} V$ is compact. Let R be a real $2n$-dimensional compact manifold possibly with boundary in M containing $\operatorname{Sing} V$ in its interior such that ∂R is transverse to V_{reg}. We set $Q = R \cap V$, which may be assumed to be a finite K-chain. For a neighborhood U of Q in M, we have the integration

$$\int_Q : A^{2d}(U) \longrightarrow \mathbb{C}.$$

We have the Stokes theorem (cf. Theorem 6.2):

$$\int_Q d\omega = \int_{\partial Q} \omega \qquad \text{for } \omega \in A^{2d-1}(U).$$

If V is compact, we have the integration

$$\int_V : H_d^{2d}(U) \longrightarrow \mathbb{C}, \tag{13.12}$$

where U is a neighborhood of V in U.

Let Z be a compact set in V (which may not be compact) containing $\operatorname{Sing} V$. Let U_0 be a neighborhood of $V \setminus Z$ in M and U_1 a neighborhood of Z in M and consider the covering $\mathcal{U} = \{U_0, U_1\}$ of $U = U_0 \cup U_1$. Let R_1 be a compact $2n$-dimensional manifold with boundary in U_1 containing Z in its interior such that ∂R_1 is transverse to V_{reg}. We set $Q_1 = R_1 \cap V$ and $Q_{01} = -\partial Q_1$. Then we have the integration

$$\int_V : A^{2d}(\mathcal{U}, U_0) \longrightarrow \mathbb{C} \qquad \text{given by } \xi \mapsto \int_{Q_1} \xi_1 + \int_{Q_{01}} \xi_{01}, \tag{13.13}$$

which induces

$$\int : H_D^{2d}(\mathcal{U}, U_0) \longrightarrow \mathbb{C}.$$

Let R be a real $2n$-dimensional compact manifold with boundary in M such that R contains $\operatorname{Sing} V$ in its interior and that ∂R is transverse to V_{reg}. We set $Q = R \cap V$ and let O be a neighborhood of ∂Q in M. Letting $U_0 = O$ and U_1 a neighborhood of Q in M, we consider the covering $\mathcal{U} = \{U_0, U_1\}$ of $U = U_0 \cup U_1$. We have the integration

$$\int : H_D^{2d}(\mathcal{U}, U_0) \longrightarrow \mathbb{C} \qquad \text{given by } [\xi] \mapsto \int_Q \xi_1 - \int_{\partial Q} \xi_{01}. \tag{13.14}$$

de Rham isomorphism

Let U be a neighborhood of V in M such that the inclusion $i : V \hookrightarrow U$ is a homotopy equivalence. For example, letting $\{X_\alpha\}$ be a Whitney stratification of V, we may take as U the union of open ball bundles $\bigcup O_\alpha$ with respect to a controlled tube system for $\{X_\alpha\}$. Then we have an isomorphism

$$i^* \circ \chi_d : H_d^p(U) \xrightarrow{\sim} H^p(V; \mathbb{C}) \tag{13.15}$$

as the composition of $\chi_d : H^p_d(U) \xrightarrow{\sim} H^p(U;\mathbb{C})$ (cf. Theorem 7.5) and $i^* : H^p(U;\mathbb{C}) \xrightarrow{\sim} H^p(V;\mathbb{C})$.

To give a simplicial expression of the above, we assume that U contains $S_K(V)$ and modify K away from a neighborhood of V to obtain a triangulation K_U of U as in Theorem 12.1, which is always possible. Then the isomorphism $i^* \circ \chi_d$ corresponds, via the isomorphism η^*_K of (13.1), to the isomorphism

$$i^* \circ \psi_d : H^p_d(U) \xrightarrow{\sim} H^p_K(V;\mathbb{C}), \tag{13.16}$$

which is the composition of $\psi_d : H^p_d(U) \xrightarrow{\sim} H^p_{K_U}(U;\mathbb{C})$ (cf. (7.30)) and $i^* : H^p_{K_U}(U;\mathbb{C}) \xrightarrow{\sim} H^p_K(V;\mathbb{C})$. The isomorphism $i^* \circ \psi_d$ is induced from the morphism

$$i^* \circ \psi_d : A^p(U) \longrightarrow C^p_K(V;\mathbb{C}) \tag{13.17}$$

that assigns to a p-form ω the cochain u given by

$$\langle \boldsymbol{s}, u \rangle = \int_{\boldsymbol{s}} \omega \qquad \text{for every } p\text{-simplex } \boldsymbol{s} \text{ of } K \text{ in } V.$$

In the following, we identify the singular homology and the simplicial homology, via the isomorphism η_* in (13.1) or $\breve{\eta}_*$ in (13.2), and the singular cohomology and the simplicial cohomology, via the isomorphism η^* in (13.1). Also, we use the integration of forms on simplicial chains, rather than that on singular chains, and use ψ instead of χ to denote the morphism from the space of forms to that of topological cochains (cf. Section 7.5).

If V is compact, we have the integration $\int_V : H^{2d}_d(U) \to \mathbb{C}$ (cf. 13.12). The following is proved as Proposition 7.11:

Proposition 13.4. *Suppose V is compact and let U be a neighborhood of V in M such that the inclusion $i : V \hookrightarrow U$ is a homotopy equivalence. Then, for a class α in $H^{2d}(V;\mathbb{C})$ and the class $[\omega]$ in $H^{2d}_d(U)$ corresponding to α under the isomorphism* (13.15), *we have*

$$\langle V, \alpha \rangle = \int_V [\omega].$$

Poincaré morphism: Let U be a neighborhood of V such that the inclusion $i : V \hookrightarrow U$ is a homotopy equivalence, as above. In terms of differential forms, the Poincaré morphism (cf. Theorem 13.2) with $\mathbb{C}$-coefficient

$$\tilde{P} : H^p_d(U) \xrightarrow{\sim} H^p(V;\mathbb{C}) \longrightarrow \breve{H}_{2d-p}(V;\mathbb{C}) \tag{13.18}$$

is induced from the morphism

$$A^p(U) \overset{i^* \circ \psi'_d}{\longrightarrow} C^p_{K'}(V;\mathbb{C}) \overset{P}{\longrightarrow} \breve{C}^K_{2d-p}(V;\mathbb{C}), \quad \omega \mapsto \sum_{\boldsymbol{s}} \Big(\int_{\boldsymbol{s}^* \cap V} \omega \Big) \boldsymbol{s},$$

where $i^* \circ \psi'_d$ denotes the morphism (13.17) for K' and the sum is taken over the $(2d-p)$-simplices $\boldsymbol{s}$ of K in V.

In particular consider the case $p = 2d$. If V is compact, $\breve{H}_0(V;\mathbb{C}) = H_0(V;\mathbb{C})$ and there is the augmentation $\varepsilon_* : H_0(V;\mathbb{C}) \to \mathbb{C}$. In this case, we have

$$\varepsilon_* \tilde{P}([\omega]) = \int_V \omega,$$

for any closed $2d$-form ω on U, which restates the equality in Proposition 13.4. If, moreover, V is connected, ε_* is an isomorphism so that (13.18) may be thought of as given by the integration

$$\int_V : H^{2d}_d(U) \longrightarrow \mathbb{C}.$$

Čech-de Rham isomorphism

Let U be a neighborhood of V in M such that the inclusion $i : V \hookrightarrow U$ is a homotopy equivalence and $\mathcal{U}$ a covering of U. Then we have a canonical isomorphism

$$H^p_D(\mathcal{U}) \xrightarrow{\sim} H^p(V;\mathbb{C}) \tag{13.19}$$

as the composition of $H^p_D(\mathcal{U}) \xrightarrow{\sim} H^p(U;\mathbb{C})$ and $i^* : H^p(U;\mathbb{C}) \xrightarrow{\sim} H^p(V;\mathbb{C})$. If there is a honeycomb system adapted to $\mathcal{U}$, the former is given by χ_D as in (7.24) (cf. Theorem 7.6).

We may give a simplicial expression of the above isomorphism as in the de Rham case. We do this in the case $\mathcal{U}$ consists of two open sets; $\mathcal{U} = \{U_0, U_1\}$, for simplicity. Thus let K_U be a triangulation of U as before. Suppose there exists a honeycomb system $\{R_0, R_1\}$ adapted to $\mathcal{U}$ that is also adapted to K in V, i.e., for each p-simplex $\boldsymbol{s}$ of K in V, $\boldsymbol{s} \cap R_i$ is a p-chain, $i = 0, 1$, and $\boldsymbol{s} \cap R_{01}$ is a $(p-1)$-chain with respect to some subdivision of K (cf. Definition 7.9). Then the isomorphism (13.19) corresponds, via the isomorphism η^*_K of (13.1), to

$$i^* \circ \psi_D : H^p_D(\mathcal{U}) \xrightarrow{\sim} H^p_K(V;\mathbb{C}), \tag{13.20}$$

which is induced from the morphism

$$i^* \circ \psi_D : A^p(\mathcal{U}) \longrightarrow C^p_K(V;\mathbb{C}) \tag{13.21}$$

that assigns to a ČdR cochain $\xi = (\xi_0, \xi_1, \xi_{01})$ the cochain u given by

$$\langle \boldsymbol{s}, u \rangle = \int_{\boldsymbol{s}\cap R_0} \xi_0 + \int_{\boldsymbol{s}\cap R_1} \xi_1 + \int_{\boldsymbol{s}\cap R_{01}} \xi_{01}$$

for every p-simplex $\boldsymbol{s}$ of K in V (cf. Remark 7.10).

Relative Čech-de Rham isomorphism

Let Z be a K_0-subcomplex of V containing $\mathrm{Sing}(V)$. Let U be a neighborhood of V in M such that the inclusion $i : V \hookrightarrow U$ is a homotopy equivalence. We take a covering $\mathcal{U} = \{U_0, U_1\}$ of U as follows. For U_0, we take a neighborhood of $V \smallsetminus Z$ in U such that the inclusion $i_0 : V \smallsetminus Z \hookrightarrow U_0$ is a homotopy equivalence. For U_1, we take a neighborhood of Z such that $U_0 \cup U_1 = U$. For example, we may take a covering defined as follows:

(C) Let $\{X_\alpha\}$ be a Whitney stratification of V. We may assume that the stratum X_{α_0} of the largest dimension is equal to $V \smallsetminus Z$ and that $\{X_\alpha\}_{\alpha \neq \alpha_0}$ is a Whitney stratification of Z. Let $\{O_\alpha\}$ be a system of open ball bundles with respect to a controlled tube system for $\{X_\alpha\}$. With these, we set $U_0 = O_{\alpha_0}$, $U_1 = \bigcup_{\alpha \neq \alpha_0} O_\alpha$ and $U = U_0 \cup U_1$. Then there is a C^∞ deformation retraction $r_0 : U_0 \to V \smallsetminus Z$ and the inclusion $i : V \hookrightarrow U$ is a homotopy equivalence. In this case, $i_1 : Z \hookrightarrow U_1$ is also a homotopy equivalence.

We may assume that U contains $S_K(V)$ and modify K away from a neighborhood of V to obtain a triangulation K_U of U, as before.

Suppose $\{R_0, R_1\}$ is a honeycomb system subordinate to $\mathcal{U}$, i.e., $R_i \subset U_i$, $i = 0, 1$, (cf. Remark 7.5) satisfying the following:

Condition (V, Z)**:**
(1) $V \subset R_0 \cup R_1$, (2) adapted to K' in V, (3) adapted to Z.

Recall that (2) means that, for each p-simplex $\boldsymbol{t}$ of K' in V, $\boldsymbol{t} \cap R_i$ is a p-chain, $i = 0, 1$, and $\boldsymbol{t} \cap R_{01}$ is a $(p-1)$-chain with respect to some subdivision of K' (cf. Definition 7.9) and (3) means that $R_1 \subset O_{K'}(Z)$ (cf. Definition 7.12). For example, take the barycentric subdivision K'' of K' and set $R_0 = S_{K''}(V \smallsetminus O_{K'}(Z))$ and $R_1 = S_{K''}(Z)$, the stars in K''. Then $\{R_0, R_1\}$ is a desired honeycomb system.

We have a morphism

$$i^* \circ \psi'_D : A^p(\mathcal{U}) \longrightarrow C^p_{K'}(V; \mathbb{C}), \tag{13.22}$$

which is defined as (13.21) for K'. It induces a morphism

$$i^* \circ \psi'_D : A^p(\mathcal{U}, U_0) \longrightarrow C^p_{K'}(V, V \smallsetminus O_{K'}(Z); \mathbb{C}) \tag{13.23}$$

that assigns to a relative p-ČdR cochain $\xi = (\xi_1, \xi_{01})$ the cocycle u given by

$$\langle \boldsymbol{t}, u \rangle = \int_{\boldsymbol{t} \cap R_1} \xi_1 + \int_{\boldsymbol{t} \cap R_{01}} \xi_{01}$$

for a p-simplex $\boldsymbol{t}$ of K' in V.

Theorem 13.4. *The morphism $i^* \circ \psi'_D$ in* (13.23) *induces an isomorphism*

$$i^* \circ \psi'_D : H^p_D(\mathcal{U}, U_0) \xrightarrow{\sim} H^p(V, V \smallsetminus Z; \mathbb{C}).$$

Proof. The morphism in (13.22) also induces a morphism

$$A^p(U_0) \longrightarrow C^p_{K'}(V \smallsetminus O_{K'}(Z); \mathbb{C}), \quad \text{given by } \xi_0 \mapsto \Big(\boldsymbol{t} \mapsto \int_{\boldsymbol{t} \cap R_0} \xi_0\Big), \tag{13.24}$$

and we have a commutative diagram of complexes with exact rows, omitting the coefficient $\mathbb{C}$ in the bottom row:

$$\begin{array}{ccccccccc}
0 & \longrightarrow & A^\bullet(\mathcal{U}, U_0) & \longrightarrow & A^\bullet(\mathcal{U}) & \longrightarrow & A^\bullet(U_0) & \longrightarrow & 0 \\
 & & \downarrow & & \downarrow & & \downarrow & & \\
0 & \longrightarrow & C^\bullet_{K'}(V, V \smallsetminus O_{K'}(Z)) & \longrightarrow & C^\bullet_{K'}(V) & \longrightarrow & C^\bullet_{K'}(V \smallsetminus O_{K'}(Z)) & \longrightarrow & 0.
\end{array}$$

We denote by j_0 the composition of the inclusions that are homotopy equivalences

$$V \smallsetminus O_{K'}(Z) \hookrightarrow V \smallsetminus Z \overset{i_0}{\hookrightarrow} U_0.$$

Then the morphism (13.24) may be thought of as the composition

$$A^p(U_0) \xrightarrow{\psi'_d} C^p_{K'_{U_0}}(U_0; \mathbb{C}) \xrightarrow{j_0^*} C^p_{K'}(V \smallsetminus O_{K'}(Z); \mathbb{C}),$$

where K'_{U_0} denotes a triangulation of U_0 obtained by extending the triangulation K' on $V \smallsetminus O_{K'}(Z)$. Thus it induces an isomorphism $H^p_d(U_0) \xrightarrow{\sim} H^p_{K'}(V \smallsetminus O_{K'}(Z); \mathbb{C})$ (cf. (13.16)).

In conclusion, we have the following commutative diagram with exact rows:

$$\begin{array}{ccccccccc}
\cdots \longrightarrow & H^{p-1}_d(U_0) & \longrightarrow & H^p_D(\mathcal{U}, U_0) & \longrightarrow & H^p_D(\mathcal{U}) & \longrightarrow & H^p_d(U_0) & \longrightarrow \cdots \\
 & \wr\downarrow & & \downarrow & & \wr\downarrow & & \wr\downarrow & \\
\cdots \longrightarrow & H^{p-1}(V \smallsetminus Z) & \longrightarrow & H^p(V, V \smallsetminus Z) & \longrightarrow & H^p(V) & \longrightarrow & H^p(V \smallsetminus Z) & \longrightarrow \cdots.
\end{array}$$

We then have the theorem by the five lemma. □

Remark 13.3. In the absolute case, we assumed the existence of a honeycomb system that is also adapted with the triangulation to have the isomorphism in (13.20). In the relative case, we need a honeycomb system that is adapted to the triangulation and the set Z. As noted above, such a honeycomb system exists so that we always have the isomorphism in Theorem 13.4.

Alexander morphism: Let V, Z, $\mathcal{U} = \{U_0, U_1\}$ and $\{R_0, R_1\}$ be as above. In terms of ČdR cochains, the Alexander morphism (cf. Theorem 13.3) with $\mathbb{C}$-coefficient

$$\tilde{A} : H^p_D(\mathcal{U}, U_0) \xrightarrow{\sim} H^p(V, V \smallsetminus Z; \mathbb{C}) \longrightarrow \breve{H}_{2d-p}(Z, \mathbb{C}) \tag{13.25}$$

is induced from the morphism

$$A^p(\mathcal{U}, U_0) \xrightarrow{i^* \circ \psi'_D} C^p_{K'}(V, V \smallsetminus O_{K'}(Z); \mathbb{C}) \longrightarrow \breve{C}^K_{2d-p}(Z; \mathbb{C}),$$

$$\xi \mapsto \sum_{\boldsymbol{s}} \Big(\int_{\boldsymbol{s}^* \cap Q_1} \xi_1 + \int_{\boldsymbol{s}^* \cap Q_{01}} \xi_{01} \Big) \boldsymbol{s}, \tag{13.26}$$

where $Q_1 = R_1 \cap V$, $Q_{01} = R_{01} \cap V$ and the sum is taken over the $(2d-p)$-simplices of K in Z.

In particular consider the case $p = 2d$. If Z is compact, $\breve{H}_0(Z; \mathbb{C}) = H_0(Z; \mathbb{C})$ and there is the augmentation $\varepsilon_* : H_0(Z; \mathbb{C}) \to \mathbb{C}$. In this case, since the $2n$-cells $\boldsymbol{s}^*$ appearing in (13.26) cover $S_{K'}(Z)$, we have

$$\varepsilon_* \tilde{A}([\xi]) = \int_{Q_1} \xi_1 + \int_{Q_{01}} \xi_{01}, \tag{13.27}$$

which may as well be written $\int_V \xi$ (cf. (13.13)). If, moreover, Z is connected, ε_* is an isomorphism and we have:

Proposition 13.5. *If Z is compact and connected, in the Alexander morphism*

$$H^{2d}_D(\mathcal{U}, U_0) \longrightarrow H_0(Z; \mathbb{C}) \simeq \mathbb{C},$$

the class of $\xi = (\xi_1, \xi_{01})$ is sent to the complex number $\int_V \xi$.

The following is proved as Proposition 7.17:

Proposition 13.6. *Suppose Z is compact. Then in (13.25), a class $[\xi]$ in $H^p_D(\mathcal{U}, U_0)$ is sent to the class of a $(2d-p)$-cycle c in Z such that*

$$\int_c \eta_1 = \int_{Q_1} \xi_1 \wedge \eta_1 + \int_{Q_{01}} \xi_{01} \wedge \eta_1$$

for every closed $(2d-p)$-form η_1 on U_1.

In the case $p = 2d$, we recover (13.27).

Before we proceed to the Lefschetz morphism, we consider the following situation. Let R be a real $2n$-dimensional compact manifold with boundary in M such that R contains $\operatorname{Sing} V$ in its interior and that ∂R is transverse to V_{reg}. We set $Q = R \cap V$. By setting $Z = Q$ in the above, we have the integration, which we denote by $\int_{\underline{Q}}$:

$$\int_{\underline{Q}} : H_D^{2d}(\mathcal{U}, U_0) \longrightarrow \mathbb{C}.$$

As noted above, it coincides with the integration (13.13) for $Z = Q$.

We also have a canonical isomorphism (cf. Theorem 13.4) :

$$H_D^p(\mathcal{U}, U_0) \simeq H^p(Q, \partial Q; \mathbb{C}).$$

In this situation, we have a generalization of Proposition 13.4:

Proposition 13.7. *In the above situation, for a class α in $H^{2d}(Q, \partial Q; \mathbb{C})$ and the class $[\xi]$ in $H_D^{2d}(\mathcal{U}, U_0)$ corresponding to α under the above isomorphism, we have*

$$\langle [\underline{Q}], \alpha \rangle = \int_{\underline{Q}} [\xi].$$

Remark 13.4. In the equality above, the right-hand side is given by $\int_Q \xi = \int_{Q_1} \xi_1 + \int_{Q_{01}} \xi_{01}$ with Q_1 and Q_{01} as in (13.13) for $Z = Q$. If ξ_{01} is defined on a neighborhood of ∂Q, we may write (cf. (13.14) and Remark 7.18)

$$\int_{\underline{Q}} [\xi] = \int_Q \xi_1 - \int_{\partial Q} \xi_{01}.$$

Lefschetz morphism: Let R be a real $2n$-dimensional manifold with boundary in M containing $\operatorname{Sing} V$ in its interior such that ∂R is transverse to V_{reg}. We set $Q = R \cap V$. In terms of ČdR cochains, the Lefschetz morphism (cf. (13.6)) with $\mathbb{C}$-coefficient

$$\tilde{L} : H_D^p(\mathcal{U}, U_0) \xrightarrow{\sim} H^p(Q, \partial Q; \mathbb{C}) \longrightarrow \breve{H}_{2d-p}(Q; \mathbb{C}) \tag{13.28}$$

is obtained by setting $Z = Q$ in (13.25). Thus it is induced from the morphism assigning to $\xi = (\xi_1, \xi_{01})$, the cycle as given in (13.26).

In particular, in the case $p = 2d$ and Q is compact, we have an expression as (13.27):

$$\varepsilon_* L([\xi]) = \int_Q \xi$$

for any $2d$-ČdR cocycle ξ, which restates the equality in Proposition 13.7. If, moreover, Q is connected, we have a statement as in Proposition 13.5 for the Lefschetz duality:

Proposition 13.8. *If Q is compact and connected, in the Lefschetz morphism*

$$H_D^{2d}(\mathcal{U}, U_0) \longrightarrow H_0(Q; \mathbb{C}) \simeq \mathbb{C},$$

the class of $\xi = (\xi_1, \xi_{01})$ is sent to $\int_Q \xi$.

If Q is compact, we have a statement corresponding to Proposition 13.6, setting $Z = Q$.

Remark 13.5. With the expressions as above of the duality morphisms, we have a diagram similar to the one in Proposition 13.1 with $\mathbb{C}$-coefficient.

13.4 Chern classes and localization

Let M, V, K_0, K and K' be as before.

Chern classes from topological viewpoint

If we have a continuous complex vector bundle E_V of rank l on V, we may define the Chern class $c^q(E_V)$ in $H^{2q}(V;\mathbb{Z})$ by obstruction theory using the triangulation K or K' as in Section 5.2. If we use K', it is the primary obstruction to constructing an r-frame on the $2q$-skeleton $(K')^{2q}$ of K', $r = l - q + 1$. The image of $c^q(E_V)$ by the Poincaré morphism is represented in $\breve{H}_{2(d-q)}(V;\mathbb{Z})$ by the cycle

$$\sum_{\boldsymbol{s}} \sum_{\boldsymbol{t}} I(s^{(r)}, b_{\boldsymbol{t}})\, \boldsymbol{s}, \tag{13.29}$$

where the first sum is taken over all the $2(d-q)$-simplices $\boldsymbol{s}$ of K in V, the second sum is taken over all the $2q$-simplices $\boldsymbol{t}$ of K' in $\boldsymbol{s}^* \cap V$ and $b_{\boldsymbol{t}}$ denotes the barycenter of $\boldsymbol{t}$. Recall that we may construct an r-frame $s^{(r)}$ on the $(2q-1)$-skeleton of K' and $I(s^{(r)}, b_{\boldsymbol{t}})$ is the index of $s^{(r)}$ on $\partial \boldsymbol{t}$ at $b_{\boldsymbol{t}}$.

Topological localization

Let Z be a K_0-subcomplex of V containing $\mathrm{Sing}(V)$. Suppose we are already given an r-frame $s_V^{(r)}$ of E_V on the $2q$-skeleton of $V \smallsetminus O_{K'}(Z)$. Then, for a $2q$-simplex $\boldsymbol{t}$ of K' not intersecting with Z, we have $I(s_V^{(r)}, b_{\boldsymbol{t}}) = 0$. Thus the cocycle representing the Chern class $c^q(E_V)$ is in fact in $C^{2q}_{K'}(V, V \smallsetminus O_{K'}(Z))$. It represents a class in $H^{2q}(V, V \smallsetminus Z;\mathbb{Z})$, which is denoted by $c^q_{Z,\mathrm{top}}(E_V, s_V^{(r)})$ and is called the *topological localization* of $c^q(E_V)$ by $s_V^{(r)}$ at Z. It will also be denoted by $c^q_Z(E_V, s_V^{(r)})$, $c^q_{\mathrm{top}}(E_V, s_V^{(r)})$ or $c^q(E_V, s_V^{(r)})$.

Its image by the canonical morphism $H^{2q}(V, V \smallsetminus Z;\mathbb{Z}) \to H^{2q}(V;\mathbb{Z})$ is the Chern class $c^q(E_V)$, however as a relative class it depends on the frame $s_V^{(r)}$.

Definition 13.2. The *topological residue* $\mathrm{TRes}_{c^q}(s_V^{(r)}, E_V; Z)$ of $s_V^{(r)}$ for $c^q(E_V)$ at Z is the image of $c^q_{\mathrm{top}}(E_V, s_V^{(r)})$ by the Alexander morphism

$$A_{V,Z} : H^{2q}(V, V \smallsetminus Z; \mathbb{Z}) \longrightarrow \check{H}_{2(d-q)}(Z; \mathbb{Z}).$$

Note that $\mathrm{TRes}_{c^q}(s_V^{(r)}, E_V; Z)$ is represented by a cycle as (13.29), where the first sum is taken over the $2(d-q)$-simplices of K in Z. Note also that the possible singular point $b_{\boldsymbol{t}}$ of $s^{(r)}$ in $\boldsymbol{t}$ may not be in Z.

Suppose now that Z has only a finite number of connected components $(Z_\lambda)_\lambda$. Then we have a decomposition

$$\check{H}_{2(d-q)}(Z; \mathbb{Z}) = \bigoplus_\lambda \check{H}_{2(d-q)}(Z_\lambda; \mathbb{Z})$$

and accordingly, we have the residue $\mathrm{TRes}_{c^q}(s_V^{(r)}, E_V; Z_\lambda)$ in $\check{H}_{2(d-q)}(Z_\lambda; \mathbb{Z})$ for each λ. It is represented by a cycle as (13.29), where the sum is taken over the $2(d-q)$-simplices in Z_λ.

In the case $d = q$ and Z_λ is compact, the residue $\mathrm{TRes}_{c^q}(s_V^{(r)}, E_V; Z_\lambda)$ is in $H_0(Z_\lambda; \mathbb{Z}) \simeq \mathbb{Z}$ and may be thought of as an integer. It is given by

$$\mathrm{TRes}_{c^q}(s_V^{(r)}, E_V; Z_\lambda) = \sum_{\boldsymbol{s}} I(s^{(r)}, b_{\boldsymbol{s}}), \tag{13.30}$$

where the sum is taken over all the 0-simplices $\boldsymbol{s}$ of K in Z_λ, in fact $b_{\boldsymbol{s}} = \boldsymbol{s}$.

Let R be a real $2n$-dimensional manifold with boundary in M containing $\mathrm{Sing}\, V$ in its interior such that ∂R is transverse to V_{reg}. We set $Q = R \cap V$. In the above considerations, we let Z be Q. Thus suppose we are given an r-frame $s_V^{(r)}$ of E_V on the $2q$-skeleton of $V \smallsetminus O_{K'}(Q)$. Then we have the relative class $c^q_Q(E_V, s_V^{(r)})$ in $H^{2q}(V, V \smallsetminus O_{K'}(Q); \mathbb{Z}) \simeq H^{2q}(Q, \partial Q; \mathbb{Z})$.

To have the above relative class $c^q_Q(E_V, s_V^{(r)})$, it suffices to have $s_V^{(r)}$ on the $2q$-skeleton of $\partial S_{K'_V}(Q)$ (cf. Remark 5.8).

Coming back to the previous situation, let Z be a K_0-subcomplex of V containing $\mathrm{Sing}\, V$ and suppose we have an r-frame $s_V^{(r)}$ of E_V on the $2q$-skeleton of $V \smallsetminus O_{K'}(Z)$. Let R be as above and suppose that $Z \subset \mathrm{Int}\, R$. The r-frame $s_V^{(r)}$ restricts to an r-frame on the $2q$-skeleton of $V \smallsetminus O_{K'}(Q)$. Thus we have the relative class $c^q_Q(E_V, s_V^{(r)})$ in $H^{2q}(Q, \partial Q; \mathbb{Z})$ as above.

From Proposition 13.1, we have the following "residue theorem":

Theorem 13.5. *In the above situation it holds:*

1. *For each λ, we have the residue* $\mathrm{TRes}_{c^q}(s_V^{(r)}, E_V; Z_\lambda)$ *in* $\check{H}_{2(d-q)}(Z_\lambda; \mathbb{Z})$.

2. *We have*

$$\sum_{\lambda}(\jmath_\lambda)_* \operatorname{TRes}_{c^q}(s_V^{(r)}, E_V; Z_\lambda) = [Q] \frown c_Q^q(E_V, s_V^{(r)}) \qquad \text{in } \breve{H}_{2(d-q)}(Q;\mathbb{Z}),$$

where $\jmath_\lambda : Z_\lambda \hookrightarrow Q$ *denotes the inclusion.*

Note that if we take M as R, then $Q = V$ in the statement 2 above and the right-hand side is expressed as $V \frown c^q(E_V)$ in $\breve{H}_{2(d-q)}(V;\mathbb{Z})$.

We now treat the same problem from the differential geometric viewpoint. In fact these two are essentially the same (cf. Theorem 13.8 below).

Chern classes from differential geometric viewpoint

Let V be a variety in a complex manifold M as before. Also let U be a neighborhood of V in M such that the inclusion $i : V \hookrightarrow U$ is a homotopy equivalence. Suppose we have a continuous complex vector bundle E_V on V. Then we have a vector bundle E on U by pulling back E_V by a homotopy inverse φ of i; $E = \varphi^* E_V$. Note that E admits a C^∞ structure, which is essentially unique (cf. Remark 3.3). Thus choosing a connection for E, we may define the Chern classes $c^q(E)$ of E in $H_d^{2q}(U)$. Recall the isomorphism $i^* \circ \psi_d : H_d^{2q}(U) \xrightarrow{\sim} H^{2q}(V;\mathbb{C})$ (cf. (13.16)).

Definition 13.3. The differential geometric Chern classes $c^q(E_V)$ of E_V is defined to be the image in $H^{2q}(V;\mathbb{C})$ of $c^q(E)$ by $i^* \circ \psi_d$.

Remark 13.6. Suppose we have a C^∞ vector bundle E on U and set $E_V = E|_V (= i^* E)$. Then $\varphi^* E_V \simeq E$ for a homotopy inverse φ of i, as $i \circ \varphi \simeq 1_U$.

We may as well define the Chern classes of E_V using $i^* \circ \psi_D$ in (13.20). Consider in particular the case the covering $\mathcal{U}$ of U consists of two open sets ; $\mathcal{U} = \{U_0, U_1\}$, as before. If E is as above, its q-th Chern class $c^q(E)$ in $H_D^{2q}(\mathcal{U})$ is represented by the cocycle $c^q(\nabla_*)$ in $A^{2q}(\mathcal{U})$ given by

$$c^q(\nabla_*) = (c^q(\nabla_0), c^q(\nabla_1), c^q(\nabla_0, \nabla_1)), \tag{13.31}$$

where ∇_0 and ∇_1 are connections for E on U_0 and U_1, respectively (cf. (10.10)). If $\mathcal{U}$ admits a suitable honeycomb system so that the isomorphism $i^* \circ \psi_D : H_D^{2q}(\mathcal{U}) \xrightarrow{\sim} H^{2q}(V;\mathbb{C})$ in (13.20) is defined, we may define the Chern class $c^q(E_V)$ as the image of $c^q(E)$ by $i^* \circ \psi_D$.

Remark 13.7. We may also define the Chern class of a virtual bundle on V using (8.16) and the above arguments.

Differential geometric localization

For the relative case, let Z be a K_0-subcomplex of V containing $\mathrm{Sing}(V)$. We take a covering $\mathcal{U}$ as defined in (C) above. We may take a homotopy inverse φ of $i : V \hookrightarrow U$ so that we have the commutative diagram:

$$\begin{array}{ccccc} U_0 & \xrightarrow{\ \imath\ } & U & \xrightarrow{\ \varphi\ } & V \\ {\scriptstyle r_0}\downarrow & & \downarrow & \nearrow{\scriptstyle \varphi'} & \\ V \smallsetminus Z & \xrightarrow{\ \jmath\ } & U', & & \end{array}$$

where U' is the space obtained from U by retracting U_0 in U to $V \smallsetminus Z$, φ' is a homotopy inverse of the inclusion $i' : V \hookrightarrow U'$ such that $\varphi' \circ i' = \varphi \circ i$, $\imath$ and $\jmath$ are the inclusions.

Let E_V be a continuous vector bundle on V and E the pullback $\varphi^* E$, as above. Recall that E has a C^∞ structure. Note that $E_V|_{V \smallsetminus Z}$ also has a C^∞ structure (cf. Remark 3.3). Suppose we have a C^∞ r-frame $s_V^{(r)}$ of E_V on $V \smallsetminus Z$. Then we will see that there is a "localization" $c^q(E_V, s_V^{(r)})$ in $H^{2q}(V, V \smallsetminus Z; \mathbb{C})$ of the Chern class $c^q(E_V)$ in $H^{2q}(V;\mathbb{C})$, $q = l - r + 1$. In fact we have the localization $c^i(E_V, s_V^{(r)})$ for $i \geq l - r + 1$, however, the case $i = l - r + 1$ is of essential interest (cf. Remark 10.5).

First we show that $E|_{U_0}$ is isomorphic with $r_0^*(E_V|_{V \smallsetminus Z})$ as a C^∞ bundle. To see this, noting that $\varphi' \circ i' \simeq 1_V$, we have

$$E|_{U_0} = \imath^* \varphi^* E_V = r_0^*(\varphi' \circ \jmath)^* E_V = r_0^*((\varphi' \circ i')^* E_V)|_{V \smallsetminus Z}) \simeq r_0^*(E_V|_{V \smallsetminus Z}).$$

Let ∇ be an $s_V^{(r)}$-trivial connection for E_V on $V \smallsetminus Z$. We take as ∇_0 in (13.31) the connection corresponding to the pull-back $r_0^* \nabla$ (cf. Section 8.3). Then $c^q(\nabla_0) = r_0^* c^q(\nabla) = 0$ by Proposition 10.5. Thus the cocycle is in $A^{2q}(\mathcal{U}, U_0)$ and it defines a class in $H_D^{2q}(\mathcal{U}, U_0)$, which we denote by $c^q(E, s_V^{(r)})$. It does not depend on the choice of the $s_V^{(r)}$-trivial connection ∇ or on the choice of the connection ∇_1 (cf. Proposition 10.1). Recall that in this situation there is always an isomorphism (cf. Theorem 13.4)

$$i^* \circ \psi_D' : H_D^{2q}(\mathcal{U}, U_0) \xrightarrow{\ \sim\ } H^{2q}(V, V \smallsetminus Z; \mathbb{C}).$$

We denote the image of $c^q(E, s_V^{(r)})$ by $c^q_{\mathrm{diff}}(E_V, s_V^{(r)})$ and call it the *differential geometric localization* of $c^q(E_V)$ at Z by $s_V^{(r)}$. It is sent to the class $c^q(E_V)$ by the canonical morphism $H^{2q}(V, V \smallsetminus Z; \mathbb{C}) \to H^{2q}(V; \mathbb{C})$.

Suppose now that Z has only a finite number of connected components $(Z_\lambda)_\lambda$. We have the Alexander morphism

$$A : H^{2q}(V, V \smallsetminus Z; \mathbb{C}) \longrightarrow \check{H}_{2(d-q)}(Z; \mathbb{C}) = \bigoplus_\lambda \check{H}_{2(d-q)}(Z_\lambda; \mathbb{C}).$$

Thus, for each λ, $c^q_{\text{diff}}(E_V, s^{(r)}_V)$ defines a class in $\breve{H}_{2(d-q)}(Z_\lambda; \mathbb{C})$, which we denote by $\text{Res}_{c^q}(s^{(r)}_V, E_V; Z_\lambda)$ and call the residue of $s^{(r)}_V$ at Z_λ with respect to c^q. The residues are expressed in the following way (cf. (13.25)). For each λ, we choose a neighborhood U_λ of Z_λ in U_1, so that the U_λ's are mutually disjoint. We may assume that $O_{K'}(Z_\lambda) \subset U_\lambda$. Let $\{R_0, R_1\}$ be a honeycomb system subordinate to $\mathcal{U}$ and satisfying Condition (V, Z) above. We set $Q_\lambda = R_1 \cap V \cap U_\lambda$ and $Q_{0\lambda} = -\partial Q_\lambda$. Then the residue $\text{Res}_{c^q}(s^{(r)}_V, E_V; Z_\lambda)$ is represented by the $2(d-q)$-cycle

$$C_\lambda = \sum_{\boldsymbol{s}} \left(\int_{\boldsymbol{s}^* \cap Q_\lambda} c^q(\nabla_1) + \int_{\boldsymbol{s}^* \cap Q_{0\lambda}} c^q(\nabla_0, \nabla_1) \right) \boldsymbol{s}, \tag{13.32}$$

where the sum is taken over the $2(d-q)$-simplices of K in Z_λ.

In particular, if $d = q$ and if Z_λ is compact, $\breve{H}_0(Z_\lambda; \mathbb{C}) = H_0(Z_\lambda; \mathbb{C}) \simeq \mathbb{C}$. Thus we may regard $\text{Res}_{c^q}(s^{(r)}_V, E_V; Z_\lambda)$ as a complex number and we have

$$\text{Res}_{c^q}(s^{(r)}_V, E_V; Z_\lambda) = \int_{Q_\lambda} c^q(\nabla_1) + \int_{Q_{0\lambda}} c^q(\nabla_0, \nabla_1). \tag{13.33}$$

Let R be a real $2n$-dimensional manifold with boundary in M containing $\text{Sing}\, V$ in its interior such that ∂R is transverse to V_{reg}. We set $Q = R \cap V$. If we have an r-frame $s^{(r)}_V$ on $V \setminus Q$, or on a neighborhood of ∂Q, we have the relative class $c^q_Q(E, s^{(r)}_V)$ in $H^{2q}(Q, \partial Q; \mathbb{C})$.

Coming back to the previous situation, suppose Z is contained in the interior of a $2n$-dimensional manifold R as above. An r-frame $s^{(r)}_V$ as above restricts to an r-frame on $V \setminus Q$, or on a neighborhood of ∂Q, and we have the relative class $c^q_Q(E_V, s^{(r)}_V)$ in $H^{2q}(Q, \partial Q; \mathbb{C})$.

We have the differential geometric counterpart of Theorem 13.5:

Theorem 13.6. *In the above situation it holds:*

1. *For each λ, we have the residue* $\text{Res}_{c^q}(s^{(r)}_V, E_V; Z_\lambda)$ *in* $\breve{H}_{2(d-q)}(Z_\lambda; \mathbb{C})$.

2. *We have*

$$\sum_\lambda (\jmath_\lambda)_* \text{Res}_{c^q}(s^{(r)}_V, E_V; Z_\lambda) = [\underline{Q}] \frown c^q_Q(E_V, s^{(r)}_V) \qquad \textit{in } \breve{H}_{2(d-q)}(Q; \mathbb{C}).$$

If we take M as R, then $Q = V$ in the statement 2 above and the right-hand side is expressed as $V \frown c^q(E_V)$ in $\breve{H}_{2(d-q)}(V; \mathbb{C})$.

Remark 13.8. 1. In (13.32) and (13.33), the second integration is performed on $V \setminus Z$ and we may replace ∇_0 with the connection ∇ originally given on $V \smallsetminus Z$ (precisely speaking, the connection for $E|_{V \smallsetminus Z}$ corresponding to ∇ by the isomorphism $E|_{V \smallsetminus Z} \simeq E_V|_{V \smallsetminus Z}$).

2. Suppose we have a C^∞ vector bundle E on U and an r-section $s^{(r)}$ of E on U such that $S(s^{(r)}) \cap V$ is a K_0 subcomplex of V containing $\mathrm{Sing}(V)$. Then the above procudure works by setting $E_V = E|_V$, $s_V^{(r)} = s^{(r)}|_V$ and $Z = S \cap V$, $S = S(s^{(r)})$ (cf. Remark 13.6). In this setting, the localization $c^q(E_V, s_V^{(r)})$ is the image of $c^q(E, s^{(r)})$ by the canonical morphism

$$i^* : H^{2q}(U, U \smallsetminus S; \mathbb{C}) \longrightarrow H^{2q}(V, V \smallsetminus Z; \mathbb{C}).$$

3. In the case $d = q$ and $Z = \{p\}$ is an isolated point, the residue $\mathrm{Res}_{c^d}(s_V^{(r)}, E_V; \{p\})$ is a complex number given as follows (cf. (13.33) and 1, 2 above). Take a neighborhood O of p in U such that $E|_O$ is trivial and let $\mathbb{B}$ be a closed $2n$-ball with center p in O such that $\mathbb{S} = \partial\mathbb{B}$ intersects transversely with V (cf. Theorem 11.16). Then letting $L = V \cap \mathbb{S}$ be the link of V oriented as the boundary of $V \cap \mathbb{B}$, we have

$$\mathrm{Res}_{c^d}(s_V^{(r)}, E_V; \{p\}) = -\int_L c^d(\nabla, \nabla_1),$$

where ∇ is an $s_V^{(r)}$-trivial connection for E_V on $V \smallsetminus \{p\}$ and ∇_1 a connection for E on O trivial with respect to some frame.

Example 13.1. Let V be a one-dimensional variety in a neighborhood U of 0 with an isolated singularity at 0. Let $E = \mathbb{C} \times U$ be the product bundle on U. Suppose we have a C^∞ section s of E on U such that $S(s) \cap V = \{0\}$. Set $E_V = E|_V$ and $s_V = s|_V$. Then we have the localization $c^1(E_V, s_V)$ in $H^2(V, V \smallsetminus \{0\}; \mathbb{C})$ and the residue $\mathrm{Res}_{c^1}(s_V, E_V; 0)$ in $H_0(\{0\}; \mathbb{C}) = \mathbb{C}$, which we try to find.

If we denote by e the frame of E given by $e(z) = (1, z)$, we may write $s = fe$ with f a C^∞ function on U. Let ∇_0 be an s-trivial connection for E on $U \smallsetminus S(s)$ and ∇_1 a connection for E on U trivial with respect to e. Then, letting L be the link of V at 0, we have (cf. Remark 13.8):

$$\mathrm{Res}_{c^1}(s_V, E_V; 0) = -\int_L c^1(\nabla_0, \nabla_1).$$

The difference form $c^1(\nabla_0, \nabla_1)$ is computed as in Example 10.3 and we have $c^1(\nabla_0, \nabla_1) = -\frac{1}{2\pi\sqrt{-1}} \frac{df}{f}$ so that

$$\mathrm{Res}_{c^1}(s_V, E_V; 0) = \frac{1}{2\pi\sqrt{-1}} \int_L \frac{df}{f}. \tag{13.34}$$

If f is holomorphic on U, then the residue (13.34) may also be expressed as the Grothendieck residue $\operatorname{Res}_0\begin{bmatrix} df \\ f \end{bmatrix}_V$ relative to V (cf. Section 13.6, Theorem 13.11 below). If V is a complete intersection defined by $h_1 = \cdots = h_{n-1} = 0$ in U, it is equal to

$$\operatorname{Res}_0\begin{bmatrix} df \wedge dh_1 \wedge \cdots \wedge dh_{n-1} \\ f, h_1, \ldots, h_{n-1} \end{bmatrix}.$$

In this case, it is also equal to $\dim \mathscr{O}_n/(f, h_1, \ldots, h_{n-1})$ (cf. Example 12.2, Theorem 13.12 below).

Coincidence of topological and differential geometric localizations

Let M and V be as above and let E_V be a continuous complex vector bundle of rank l on V. Then we have the Chern class $c^q_{\mathrm{top}}(E_V)$ in $H^{2q}(V;\mathbb{Z})$ defined via obstruction theory and the Chern class $c^q_{\mathrm{diff}}(E_V)$ in $H^{2q}(V;\mathbb{C})$ defined via Chern-Weil theory. As in the case of manifolds, they are essentially the same as stated below.

The following can be proved similarly as Theorem 10.12:

Theorem 13.7. *In the above situation, $c^q_{\mathrm{diff}}(E_V)$ is the image of $c^q_{\mathrm{top}}(E_V)$ by the canonical morphism $H^{2q}(V;\mathbb{Z}) \to H^{2q}(V;\mathbb{C})$.*

We have the following theorem for localized classes, which can be proved the same way as Theorem 10.13. Thus let Z be a subvariety of V of dimension smaller than $d = \dim V$ and containing $\operatorname{Sing} V$ and $s^{(r)}_V$ an r-frame of E_V on $V \smallsetminus Z$, $r = l - q + 1$. Then we have the localized classes $c^q_{\mathrm{top}}(E_V, s^{(r)}_V)$ in $H^{2q}(V, V \smallsetminus Z;\mathbb{Z})$ and $c^q_{\mathrm{diff}}(E_V, s^{(r)}_V)$ in $H^{2q}(V, V \smallsetminus Z;\mathbb{C})$.

Theorem 13.8. *In the above situation, $c^q_{\mathrm{diff}}(E_V, s^{(r)}_V)$ is the image of $c^q_{\mathrm{top}}(E_V, s^{(r)}_V)$ by the canonical morphism*

$$H^{2q}(V, V \smallsetminus Z;\mathbb{Z}) \to H^{2q}(V, V \smallsetminus Z;\mathbb{C}).$$

Corollary 13.2. *In the above situation, the residue $\operatorname{Res}_{c^q}(s^{(r)}_V, E_V; Z)$ is the image of the topological residue $\operatorname{TRes}_{c^q}(s^{(r)}_V, E_V; Z)$ by the canonical morphism $\breve{H}_{2(d-q)}(Z;\mathbb{Z}) \to \breve{H}_{2(d-q)}(Z;\mathbb{C})$.*

In particular, if $d = q$ and if Z is compact and connected, the both may be thought of as an identical integer.

13.5 Residues of Chern classes on singular varieties

Let V be a variety of dimension d in a complex manifold M of dimension n. Also, let E_V be a continuous vector bundle of rank l on V. Let Z be a subvariety of V containing $\operatorname{Sing} V$ and suppose an r-frame $s_V^{(r)}$ of E_V is given on $V \smallsetminus Z$.

In this situation, we have the topological localization $c_{\mathrm{top}}^q(E_V, s_V^{(r)})$ in $H^{2q}(V, V \smallsetminus Z; \mathbb{Z})$ and the residue $\mathrm{TRes}_{c^q}(s_V^{(r)}, E_V; Z)$ in $\check{H}_{2(d-q)}(Z; \mathbb{Z})$, $q = l - r + 1$. If $s_V^{(r)}$ is C^∞, we also have the differential geometric localization $c_{\mathrm{diff}}^q(E_V, s_V^{(r)})$ in $H^{2q}(V, V \smallsetminus Z; \mathbb{C})$ and the residue $\mathrm{Res}_{c^q}(s_V^{(r)}, E_V; Z)$ in $\check{H}_{2(d-q)}(Z; \mathbb{C})$. We have seen (cf. Corollary 13.2) that $\mathrm{Res}_{c^q}(s_V^{(r)}, E_V; Z)$ is the image of $\mathrm{TRes}_{c^q}(s_V^{(r)}, E_V; Z)$ by the canonical morphism

$$\check{H}_{2(d-q)}(Z; \mathbb{Z}) \longrightarrow \check{H}_{2(d-q)}(Z; \mathbb{C}). \tag{13.35}$$

Now we consider the case where Z is of pure dimension $d - q$. We refer this situation as the *proper case*, as in Chapter 12. Let $(Z_i)_i$ be the irreducible components of Z. Each Z_i defines a class $[Z_i]$ in $\check{H}_{2(d-q)}(Z; \mathbb{Z})$ and it is the free Abelian group generated by the $[Z_i]$'s. Thus the morphism (13.35) is injective so that we may identify the two residues:

$$\mathrm{Res}_{c^q}(s_V^{(r)}, E_V; Z) = \mathrm{TRes}_{c^q}(s_V^{(r)}, E_V; Z). \tag{13.36}$$

Let p_i be a non-singular point of $Z_i^\circ = Z_i \smallsetminus \bigcup_{j \neq i} Z_j$ and $\mathbb{D}_i$ a complex slice of Z_i° in M at p_i (cf. Definition 3.24). Recall that $\mathbb{D}_i$ is a locally closed complex sudmanifold of dimension $q + k$ in M through p_i and transverse to Z_i at p_i. Let U be a neighborhood of Z in M that contains $\mathbb{D}_i$ as a closed submanifold and take a triangulation K_0 of U as in Theorem 12.1, compatible with $V \cap U$ and $Z \cup \mathbb{D}_i$, and let K be its barycentric subdivision. Let K' and K^* be as before. We may assume that p_i is the barycenter $b_{\boldsymbol{s}}$ of a $2(d-q)$-simplex $\boldsymbol{s}$ in the non-singular part of Z_i° and that the $2(q+k)$-cell $\boldsymbol{s}^*$ dual to $\boldsymbol{s}$ is in $\mathbb{D}_i$. Thus $\mathbb{D}_i$ is also transverse to V. The r-section $s_i^{(r)} = s_V^{(r)}|_{\mathbb{D}_i \cap V}$ of $E_i = E_V|_{\mathbb{D}_i \cap V}$ has an isolated singularity at p_i so that we have the residue $\mathrm{TRes}_{c^q}(s_i^{(r)}, E_i; p_i)$, which is an integer. If $s_V^{(r)}$ is C^∞, we also have $\mathrm{Res}_{c^q}(s_i^{(r)}, E_i; p_i)$, which may be identified with $\mathrm{TRes}_{c^q}(s_i^{(r)}, E_i; p_i)$. Note that this number does not depend on the choice of p_i as in the case of residues on complex manifolds.

Theorem 13.9. *In the above situation, we have*

$$\mathrm{TRes}_{c^q}(s_V^{(r)}, E_V; Z) = \sum_i \mathrm{TRes}_{c^q}(s_i^{(r)}, E_i; p_i) \cdot [Z_i] \qquad \text{in} \quad \check{H}_{2(d-q)}(Z; \mathbb{Z}).$$

Proof. By definition, the residue $\mathrm{TRes}_{c^q}(s_V^{(r)}, E_V; Z)$ is represented by the cycle

$$\sum_{\boldsymbol{s}} \sum_{\boldsymbol{t}} I(s^{(r)}, b_{\boldsymbol{t}})\, \boldsymbol{s}, \tag{13.37}$$

where the first sum is taken over all the $2(d-q)$-simplices $\boldsymbol{s}$ of K in Z, the second sum is taken over all the $2q$-simplices $\boldsymbol{t}$ of K' in $\boldsymbol{s}^* \cap V$ and $I(s^{(r)}, b_{\boldsymbol{t}})$ denotes the index of $s^{(r)}$ at the barycenter $b_{\boldsymbol{t}}$ of $\boldsymbol{t}$.

Since $\boldsymbol{s}$ is a simplex of dimension $2(d-q)$, its interior is in the non-singular part of Z and we may express the cycle (13.37) as

$$\sum_i \sum_{\boldsymbol{s}_{(i)}} \sum_{\boldsymbol{t}_{(i)}} I(s^{(r)}, b_{\boldsymbol{t}_{(i)}})\, \boldsymbol{s}_{(i)},$$

where the second sum is taken over all the $2(d-q)$-simplices $\boldsymbol{s}_{(i)}$ of K in Z_i and the third sum is taken over all the $2q$-simplices $\boldsymbol{t}_{(i)}$ of K' in $\boldsymbol{s}^*_{(i)} \cap V$. Moreover, we have $\sum_{\boldsymbol{t}_{(i)}} I(s^{(r)}, b_{\boldsymbol{t}_{(i)}}) = \mathrm{TRes}_{c^q}(s_i^{(r)}, E_i; p_i)$, which is independent of $\boldsymbol{s}_{(i)}$. □

Remark 13.9. If V is Cohen-Macaulay, in particular if it is a local complete intersection, then $\dim Z \geq d-q$ (cf. Remark 11.13, Proposition A.23 and (A.16)).

13.6 Residues at an isolated singularity

Let V be a variety of dimension d in a complex manifold M of dimension $n = d+k$ and let E be a holomorphic vector bundle of rank l ($\geq d$) on a neighborhood of V in M. Suppose that V has at most an isolated singularity at p. Let $r = l-d+1$ and suppose we have a holomorphic r-section $s^{(r)}$ of E with $S(s^{(r)}) \cap V = \{p\}$. Then, setting $E_V = E|_V$ and $s_V^{(r)} = s^{(r)}|_V$, we have $\mathrm{TRes}_{c^d}(s_V^{(r)}, E_V; p)$ and $\mathrm{Res}_{c^d}(s_V^{(r)}, E_V; p)$, which are identified. It is an integer, in fact we will see that it is positive.

Recall that $\mathrm{Res}_{c^d}(s_V^{(r)}, E_V; p)$ is expressed as follows (cf. (13.33) and Remark 13.8). Let U be a neighborhood of p in M where the bundle E is trivial and set $W = U \cap V$. We may assume that U is a coordinate neighborhood and sometimes we identify p with 0 in $\mathbb{C}^n$ and U with a neighborhood of 0. Let ∇_0 be an $s^{(r)}$-trivial connection for E on $U \smallsetminus S(s^{(r)})$ and ∇_1 a connection for E on U trivial with respect to some holomorphic frame $e^{(l)} = (e_1, \dots, e_l)$ of E. Then, letting L be the link of V at p, we have

$$\operatorname{Res}_{c^d}(s_V^{(r)}, E_V; p) = -\int_L c^d(\nabla_0, \nabla_1). \tag{13.38}$$

Note that we may replace ∇_0 with an $s_V^{(r)}$-trivial connection ∇ for E_V on $W \smallsetminus \{p\}$.

We give various expressions of this number.

Topological expression

We take a small closed $2n$-ball $\mathbb{B}$ with center p and let $\mathbb{S}$ be the boundary $\partial\mathbb{B}$, which is a sphere of dimension $2n-1$. Recall that the link L of V at p is the intersection $V \cap \mathbb{S}$, which is a naturally oriented C^∞ manifold of dimension $2d-1$ (cf. Section 11.7). Let $(L_i)_i$ be the connected components of L. Since E_V is trivial on W, $s_V^{(r)}$ defines, for each i, a map (cf. (5.1))

$$\varphi_i : L_i \longrightarrow W(l, r).$$

Since $W(l,r)$ is $2l - 2r = 2d-2$ connected and $H_{2d-1}(W(l,r), \mathbb{Z}) = \mathbb{Z}$, we may consider the mapping degree $\deg \varphi_i$ of φ_i.

Theorem 13.10. *In the above situation,*

$$\operatorname{Res}_{c^d}(s_V^{(r)}, E_V; p) = \sum_i \deg \varphi_i.$$

Proof. We try to compute $\operatorname{TRes}_{c^d}(s_V^{(r)}, E_V; p)$, as it is the same as $\operatorname{Res}_{c^d}(s_V^{(r)}, E_V; p)$. We may assume that $\mathbb{B} = \boldsymbol{s}^*$ is the $2n$-cell dual to p. We may also extend $s_V^{(r)}|_{\partial\mathbb{B}}$ to $K'^{(2d-1)} \cap (\boldsymbol{s}^* \cap V)$, which includes p. Note that, in this process, we only keep $s^{(r)}|_{\partial\mathbb{B}\cap V}$ unchanged and modify $s_V^{(r)}$ in $(\operatorname{Int}\mathbb{B}) \cap (V \smallsetminus \{p\})$ to obtain a frame on $K'^{(2d-1)} \cap (\boldsymbol{s}^* \cap V)$. If we denote this frame by $s'^{(r)}$, by definition, we have (cf. (13.29))

$$\operatorname{TRes}_{c^d}(s_V^{(r)}, E_V; p) = \sum_{\boldsymbol{t}} I(s'^{(r)}, b_{\boldsymbol{t}}),$$

where the sum is taken over all the $2d$-simplices $\boldsymbol{t}$ of K' in $\boldsymbol{s}^* \cap V$ and $b_{\boldsymbol{t}}$ denotes the barycenter of $\boldsymbol{t}$. By definition of $I(s'^{(r)}, b_{\boldsymbol{t}})$ and properties of mapping degree, we have the theorem. □

For the above formula, we only need to have a continuous frame $s_V^{(r)}$ on $W \smallsetminus \{p\}$.

Analytic expression

Grothendieck residues relative to a subvariety: Let U be a neighborhood of 0 in $\mathbb{C}^n$ and V a variety of dimension d in U. We assume that V contains 0 and that $V \smallsetminus \{0\}$ is non-singular. Let $f_1, \dots, f_d$ be holomorphic functions on U and $V(f_1, \dots, f_d)$ the variety defined by them. Suppose that $V(f_1, \dots, f_d) \cap V = \{0\}$. For small positive numbers ε_i, $i = 1, \dots, d$, we set

$$\Gamma = \{ z \in U \cap V \mid |f_i(z)| = \varepsilon_i,\ i = 1, \dots, d \},$$

which is a d-cycle in $V \smallsetminus \{0\}$. It is oriented so that the form $d\theta_1 \wedge \cdots \wedge d\theta_d$ is positive, $\theta_i = \arg f_i$. For a holomorphic d-from ω on U, we set

$$\operatorname{Res}_0 \begin{bmatrix} \omega \\ f_1, \dots, f_d \end{bmatrix}_V = \Big(\frac{1}{2\pi\sqrt{-1}}\Big)^d \int_\Gamma \frac{\omega}{f_1 \cdots f_d}$$

and call it the *Grothendieck residue of* $\omega / f_1 \cdots f_d$ *at* 0 *relative to* V. In the case $d = n$, it is the usual Grothendieck residue in Section 12.3.

If V is a complete intersection defined by $h_1 = \cdots = h_k = 0$ in U, we have

$$\operatorname{Res}_0 \begin{bmatrix} \omega \\ f_1, \dots, f_d \end{bmatrix}_V = \operatorname{Res}_0 \begin{bmatrix} \omega \wedge dh_1 \wedge \cdots \wedge dh_k \\ f_1, \dots, f_d, h_1, \dots, h_k \end{bmatrix}.$$

Exercise 13.1. Verify the above.

Analytic expression: We consider the situation in the beginning of this section. We write $s_i = \sum_{j=1}^{l} f_{ji}\, e_j$, $i = 1, \dots, r$, with f_{ji} holomorphic functions on U. Let F be the $l \times r$ matrix whose (i, j)-entry is f_{ij}. Let $\mathcal{I}$ be the set as given in (12.5) and let F_I and φ_I be as defined before. If we write $e_I = e_{i_1} \wedge \cdots \wedge e_{i_r}$, we have

$$s_1 \wedge \cdots \wedge s_r = \sum_{I \in \mathcal{I}} \varphi_I\, e_I.$$

and $S(s^{(r)})$ is the set of common zeros of the φ_I's. From the assumption $S(s^{(r)}) \cap V = \{p\}$, we have the following by Lemmas 12.1 and 12.2:

Lemma 13.2. *We may choose a holomorphic frame* $e^{(l)} = (e_1, \dots, e_l)$ *of* E *so that there exist* d *elements* $I^{(1)}, \dots, I^{(d)}$ *in* $\mathcal{I}$ *with*

$$V(\varphi_{I^{(1)}}, \dots, \varphi_{I^{(d)}}) \cap V = \{p\}.$$

Let $e^{(l)}$ be a frame of E as in Lemma 13.2. We write $I^{(\alpha)} = (i_1^{(\alpha)}, \dots, i_r^{(\alpha)})$, $\alpha = 1, \dots, d$, and let $F^{(\alpha)}$ and $\Theta^{(\alpha)}$ be defined as in the paragraph after Lemma 12.3. Thus $F^{(\alpha)}$ is an $l\times l$ matrix of holomorphic functions and $\Theta^{(\alpha)}$ is an $l\times l$ matrix of holomorphic 1-forms.

With these we have:

Theorem 13.11. *In the above notation,*

$$\operatorname{Res}_{c^d}(s_V^{(r)},E_V;p)=\operatorname{Res}_p\begin{bmatrix}\tilde\sigma_d(\Theta^{(1)},\dots,\Theta^{(d)})\\ \varphi_{I^{(1)}},\dots,\varphi_{I^{(d)}}\end{bmatrix}_V.$$

The above theorem is proved exactly the same way as Theorem 12.4. The only difference is that, in Theorem 12.4, U is a neighborhood of p in a manifold, while in Theorem13.11, it is replaced by a neighborhood W of p in a possibly singular variety. However everything is performed on $W\smallsetminus\{p\}$, which is non-singular.

Special cases: **1.** The case $l=d$ and $r=1$. Let $e^{(l)}=(e_1,\dots,e_d)$ be an arbitrary frame of E and write $s=\sum_{i=1}^d f_ie_i$. Then we may set $\varphi_{I^{(i)}}=f_i$, $i=1,\dots,d$, and we have $\tilde\sigma_d(\Theta^{(1)},\dots,\Theta^{(d)})=df_1\wedge\cdots\wedge df_d$. Thus

$$\operatorname{Res}_{c^d}(s_V,E_V;p)=\operatorname{Res}_p\begin{bmatrix}df_1\wedge\cdots\wedge df_d\\ f_1,\dots,f_d\end{bmatrix}_V.$$

2. The case $d=1$ and $r=l$. Let $e^{(l)}=(e_1,\dots,e_l)$ be an arbitrary frame of E and write $s_i=\sum_{j=1}^r f_{ij}\,e_j$, $i=1,\dots,l$. Let $F=(f_{ij})$ and set $\varphi=\det F$. Then we may set $\varphi_{I^{(1)}}=\varphi$ and we have $\sigma_1(\Theta)=d\varphi$ so that we have

$$\operatorname{Res}_{c^1}(s_V^{(r)},E_V;p)=\operatorname{Res}_p\begin{bmatrix}d\varphi\\ \varphi\end{bmatrix}_V.$$

Fundamental properties of the residues

Let p be an isolated singular point in V and suppose that V is a complete intersection defined by $h=(h_1,\dots,h_k):(U,p)\to(\mathbb{C}^k,0)$. Let T be a small neighborhood of 0 in $\mathbb{C}^k$. For a point t in T, we set $V_t=h^{-1}(t)$. Let $C(h)$ denote the critical set of h and $D(h)=h(C(h))$ the discriminant, which is a hypersurface in T (cf. Proposition 11.28). We have $\operatorname{Sing}(V_t)=C(h)\cap V_t$, which consists of at most a finite number of points. We set $E_t=E|_{V_t}$ and $s_t^{(r)}=s^{(r)}|_{V_t}$ so that $S(s_t^{(r)})=S(s^{(r)})\cap V_t$. By the assumption $S(s^{(r)})\cap V=\{p\}$, we have $\dim S(s^{(r)})\le k$. Hence $S(s_t^{(r)})$ also consists of at most a finite number of points. Note that even if q is in $\operatorname{Sing}(V_t)$, if $q\notin S(s_t^{(r)})$, then $\operatorname{Res}_{c^d}(s_t^{(r)},E_t;q)=0$. We set $V_0=V$.

Lemma 13.3. *We have* $\dim S(s^{(r)}) = k$ *and* $S(s_t^{(r)})$ *is a non-empty finite set.*

Proof. By assumption, we have $\dim S(s^{(r)}) \leq k$. On the other hand, $\operatorname{codim} S(s^{(r)}) \leq l - r + 1 = d$, by (A.16). □

Lemma 13.4. *The sum* $\sum_{q \in S(s_t^{(r)})} \operatorname{Res}_{c^d}(s_t^{(r)}, E_t; q)$ *is continuous in* t.

Proof. Let ∇_0 be an $s^{(r)}$-trivial connection for E on $U \smallsetminus S(s^{(r)})$ and ∇_1 a connection for E on U. Then by Theorem 13.6, the above sum is equal to an integral over $Q_t = R \cap V_t$, which is continuous in t. □

Since $T \smallsetminus D(h)$ is dense in T, by Lemma 2.7, we have

Proposition 13.9. *The sum* $\sum_{q \in S(s_t^{(r)})} \operatorname{Res}_{c^n}(s_t^{(r)}, E_t; q)$ *is constant. In particular,*

$$\operatorname{Res}_{c^n}(s_V^{(r)}, E_V; p) = \sum_{q \in S(s_t^{(r)})} \operatorname{Res}_{c^n}(s_t^{(r)}, E_t; q),$$

which is a non-negative integer.

Remark 13.10. By Lemma 13.3, $\operatorname{Res}_{c^n}(s_V^{(r)}, E_V; p)$ is in fact a positive integer.

Algebraic expression

We refer to Section A.2 for relevant materials in algebra.

Let E, U, and $s^{(r)}$ be as in the beginning of this section. Let $e^{(l)}$ be a holomorphic frame of E on U. Also let F, F_I and $\varphi_I = \det F_I$ be as before. We denote by $\mathscr{O}_U$ the sheaf of germs of holomorphic functions on U and by $\mathscr{F}$ the ideal sheaf in $\mathscr{O}_U$ generated by the (germs of) φ_I's. Note that $\mathscr{F}$ does not depend on the choice of the frame $e^{(l)}$ of E. We have the complex space $\boldsymbol{S} = (S, \mathscr{O}_{\boldsymbol{S}})$ in U with support $S = S(s^r)$ and structure sheaf $\mathscr{O}_{\boldsymbol{S}} = \iota^{-1}(\mathscr{O}_U/\mathscr{F})$, where $\iota : S \hookrightarrow U$ is the inclusion (cf. Section 11.5). We suppose that V is a complete intersection defined by $h : (U, p) \to (\mathbb{C}^k, 0)$ as above. Let T be a small neighborhood of 0 in $\mathbb{C}^k$. Denoting by π the restriction of h to S, we have a morphism

$$(\pi, \varpi) : \boldsymbol{S} \longrightarrow (T, \mathscr{O}_T) \tag{13.39}$$

as the composition of the canonical morphism $\boldsymbol{S} \to (U, \mathscr{O}_U)$ and $(h, h^*) : (U, \mathscr{O}_U) \to (T, \mathscr{O}_T)$. For each x in S,

$$\varpi_x : \mathscr{O}_{T,\pi(x)} \to \mathscr{O}_{\boldsymbol{S},x}$$

is the composition of $h_x^* : \mathscr{O}_{T,\pi(x)} \to \mathscr{O}_{U,x}$ and the canonical epimorphism $\mathscr{O}_{U,x} \to \mathscr{O}_{U,x}/\mathscr{F}_x = \mathscr{O}_{\boldsymbol{S},x}$.

We suppose that V is a complete intersection defined by $h : (U,p) \to (\mathbb{C}^k, 0)$ as above. Let T be a small neighborhood of 0 in $\mathbb{C}^k$ and, for a point t in T, we set $V_t = h^{-1}(t)$. Also let $S(s_t^{(r)}) = S(s^{(r)}) \cap V_t$, as before. Let $\mathscr{I}_V = (h_1, \dots, h_k)$ be the ideal sheaf of V in $\mathscr{O}_U$ and $\mathscr{F}_V$ the ideal sheaf generated by $\mathscr{F}$ and $\mathscr{I}_V$ in $\mathscr{O}_U$. Also, for $t = (t_1, \dots, t_k) \in T$, we denote by $\mathscr{F}_{V_t}$ the ideal sheaf generated by $\mathscr{F}$ and $\mathscr{I}_{V_t} = (h_1 - t_1, \dots, h_k - t_k)$. The intersection $\boldsymbol{Z}$ of V and $\boldsymbol{S}$ in U is the complex space with support $Z = S \cap V = \{p\}$ and structure sheaf $\mathscr{O}_{\boldsymbol{Z}} = \iota^{-1}(\mathscr{O}_U/\mathscr{F}_V)$, where $\iota : Z \hookrightarrow U$ is the inclusion (cf. Section 11.5). Also, the intersection $\boldsymbol{Z}_t$ of V_t and $\boldsymbol{S}$ in U is the complex space with support $Z_t = S \cap V_t$ and structure sheaf $\mathscr{O}_{\boldsymbol{Z}_t} = \iota_t^{-1}(\mathscr{O}_U/\mathscr{F}_{V_t})$, where $\iota_t : Z_t \hookrightarrow U$ is the inclusion. It is the fiber of (π, ϖ) over t.

By Lemma 13.3, $\pi : S \to T$ is a finite map. Thus each point x in S is isolated in Z_t, $t = \pi(x)$, and $\dim_{\mathbb{C}} \mathscr{O}_{\boldsymbol{Z}_t,x} = \dim_{\mathbb{C}} \mathscr{O}_{U,x}/\mathscr{F}_{V_t,x}$ is finite.

Proposition 13.10. *In the above situation,*

$$\dim_{\mathbb{C}} \mathscr{O}_{U,p}/\mathscr{F}_{V,p} = \sum_{q \in S(s_t^{(r)})} \dim_{\mathbb{C}} \mathscr{O}_{U,q}/\mathscr{F}_{V_t,q}.$$

Proof. This is proved as Proposition 12.5 with a little modification, replacing $\tilde{\boldsymbol{S}}$, $\boldsymbol{S}$ and $\boldsymbol{S}_t$ with $\boldsymbol{S}$, $\boldsymbol{Z}$ and $\boldsymbol{Z}_t$, respectively.

We claim that the morphism (π, ϖ) in (13.39) is flat, which will prove the proposition. Let x be a point in S and set $t = \pi(x)$. In the following, we set $\mathscr{O}'_x = \mathscr{O}_{U,x}$, $\mathscr{O}_x = \mathscr{O}_{\boldsymbol{S},x}$ and $\mathscr{O}_t = \mathscr{O}_{T,t}$. Note that $\mathscr{O}'_x$ and $\mathscr{O}_t$ are regular local rings of dimensions n and k, respectively. We have $\operatorname{ht} \mathscr{F}_x = n - k = d = l - r + 1$. Hence by (A.17), the ring $\mathscr{O}_x$ is CM. Since the morphism $\varpi_x : \mathscr{O}_t \to \mathscr{O}_x$ is finite, $\mathscr{O}_x$ is a CM $\mathscr{O}_t$-module. By (A.15), denoting by $\mathfrak{m}_t$ the maximal ideal in $\mathscr{O}_t$,

$$\operatorname{depth}(\mathfrak{m}_t; \mathscr{O}_x) + \operatorname{pd}_{\mathscr{O}_t} \mathscr{O}_x = \operatorname{depth} \mathfrak{m}_t.$$

We have $\operatorname{depth}(\mathfrak{m}_t; \mathscr{O}_x) = \dim_{\mathscr{O}_t} \mathscr{O}_x = \dim_{\mathscr{O}_x} \mathscr{O}_x = k$ and $\operatorname{depth} \mathfrak{m}_t = \dim \mathscr{O}_t = k$. Therefore, $\operatorname{pd}_{\mathscr{O}_t} \mathscr{O}_x = 0$ and (π, ϖ) is flat. □

Since the regular values of h are dense, by Proposition 13.9, Theorem 12.5 and Proposition 13.10, we have the following theorem.

Theorem 13.12. *We have*

$$\operatorname{Res}_{c^d}(s_V^{(r)}, E_V; p) = \dim_{\mathbb{C}} \mathscr{O}_{U,p}/\mathscr{F}_{V,p}.$$

Remark 13.11. As we can see from the above proofs, the assumption that V is a complete intersection is necessary only to ensure that V admits a "smoothing" in U.

13.7 Examples and related topics

In the following, we denote by M a complex manifold of dimension n and by V a variety of dimension d in M.

Virtual index of a vector field

Suppose V is an SLCI, i.e., a local complete intersection defined by a section of a holomorphic vector bundle N of rank k on M (cf. Definition 11.29). We have the virtual tangent bundle $\tau_V = (TM - N)|_V$ of V. We also have its Chern class $c^*(\tau_V)$ in $H^*(V;\mathbb{C})$ (cf. Remark 13.7).

As V_{reg} is a complex manifold, it has the holomorphic tangent bundle TV_{reg}, which is canonically isomorphic with the real tangent bundle $T_{\mathbb{R}}V_{\mathrm{reg}}$ of V_{reg} as a real bundle. Let Z be a subvariety of V containing $\mathrm{Sing}(V)$ and v a C^∞ non-vanishing vector field on $V \smallsetminus Z$. Then we may define the localization $c^d_Z(\tau_V, v)$ by v of the class $c^d(\tau_V)$ as follows.

Let U be a neighborhood of V in M such that the inclusion $i : V \hookrightarrow U$ is a homotopy equivalence. Also let $\mathcal{U} = \{U_0, U_1\}$ be a covering of U such that U_0 is a neighborhood of $V \smallsetminus Z$ with a C^∞ deformation retraction $r_0 : U_0 \to V \smallsetminus Z$ and U_1 is a neighborhood of Z in U. For $i = 0, 1$, let ∇_i^M and ∇_i^N be connections for TM and N on U_i, respectively, and set $\nabla_i^\star = (\nabla_i^M, \nabla_i^N)$. Then the d-th Chern class in $H^{2d}_D(\mathcal{U})$ of the virtual bundle $(TM - N)|_U$ is represented by the cocycle (cf. Sections 8.3 and 8.4)

$$c^d(\nabla_*^\star) = (c^d(\nabla_0^\star), c^d(\nabla_1^\star), c^d(\nabla_0^\star, \nabla_1^\star)). \tag{13.40}$$

Suppose a non-vanishing C^∞ vector field v is given on $W_0 = V \smallsetminus Z$. Let ∇ be a v-trivial connection for TV_{reg} on W_0. We take connections ∇^M and ∇^N for $TM|_{W_0}$ and $N|_{W_0}$ so that $(\nabla, \nabla^M, \nabla^N)$ is compatible with the first row of (11.26) (cf. Proposition 8.5). Noting that $TM|_{U_0} \simeq r_0^*(TM|_{W_0})$ and $N|_{U_0} \simeq r_0^*(N|_{W_0})$, we take as ∇_0^M and ∇_0^N the connections for $TM|_{U_0}$ and $N|_{U_0}$ corresponding to $r_0^*\nabla^M$ and $r_0^*\nabla^N$, respectively. Then we have the vanishing $c^d(\nabla_0^\star) = r_0^* c^d(\nabla) = 0$ (cf. Proposition 8.7) and the cocycle in (13.40) defines a class in $H^{2d}_D(\mathcal{U}, U_0)$, whose image in $H^{2d}(V, V \smallsetminus Z; \mathbb{C})$ by the isomorphism of Theorem 13.4 is the localization $c^d_Z(\tau_V, v)$. We recall the Alexander morphism (cf. (13.25))

$$H^{2d}_D(\mathcal{U}, U_0) \longrightarrow \check{H}_0(Z;\mathbb{C}).$$

If Z has a finite number of connected components (Z_λ), for each λ, we have the residue $\mathrm{Res}_{c^d}(v, \tau_V; Z_\lambda)$ in $\check{H}_0(Z_\lambda;\mathbb{C})$. If Z_λ is a compact component, we have $\check{H}_0(Z_\lambda;\mathbb{C}) = H_0(Z_\lambda;\mathbb{C}) \simeq \mathbb{C}$, the isomorphism being given by the augmentation, so that we may think of $\mathrm{Res}_{c^d}(v, \tau_V; Z_\lambda)$ as a number.

Definition 13.4. For a compact connected component Z_λ, we define the *virtual index* $\mathrm{Vir}(v, Z_\lambda)$ of v at Z_λ as the residue $\mathrm{Res}_{c^d}(v, \tau_V; Z_\lambda)$.

The virtual index is expressed as (cf. (13.33))

$$\mathrm{Vir}(v, Z_\lambda) = \int_{Q_\lambda} c^d(\nabla_1^\star) + \int_{Q_{0\lambda}} c^d(\nabla_0^\star, \nabla_1^\star). \tag{13.41}$$

Note that we may replace $\nabla_0^\star$ with $\nabla^\star = (\nabla^M, \nabla^N)$ (cf. Remark 13.8. 1).

Proposition 13.11. *If Z_λ is in V_{reg}, then* $\mathrm{Vir}(v, Z_\lambda) = \mathrm{PH}(v, Z_\lambda)$.

Proof. Note that, in this case, the integrations in (13.41) are performed on V_{reg} and it suffices to define various connections involved on a neighborhood W of Z_λ in V_{reg}. Let ∇, ∇^M and ∇^N be as above and set $\nabla^\star = (\nabla^M, \nabla^N)$. We take a connections ∇' for TV_{reg} on W and connections ∇'^M and ∇'^N for TM and N on W so that $(\nabla', \nabla'^M, \nabla'^N)$ is compatible with the first row of (11.26) on W and set $\nabla'^\star = (\nabla'^M, \nabla'^N)$. Then, by Proposition 8.7, $c^d(\nabla'^\star) = c^d(\nabla')$ on W and $c^d(\nabla^\star, \nabla'^\star) = c^d(\nabla, \nabla')$ on $W \smallsetminus Z$. □

Let R be a compact manifold of dimension $2n$ with boundary in M containing $\mathrm{Sing}(V)$ in its interior such that ∂R is transverse to V. We set $Q = R \cap V$. For a non-vanishing vector field v on a neighborhood of ∂Q in V, we may also define $\mathrm{Vir}(v, Q)$ in $H_0(Q;\mathbb{C}) \simeq \mathbb{C}$.

If Z is contained in the interior of Q, for a non-vanishing vector field v on $V \smallsetminus Z$ (cf. Theorem 13.6),

$$\sum_\lambda \mathrm{Vir}(v, Z_\lambda) = \mathrm{Vir}(v, Q).$$

In particular, we have (cf. Proposition 13.4):

Proposition 13.12. *In the above situation, if V is compact, we have*

$$\sum_\lambda \mathrm{Vir}(v, Z_\lambda) = \int_V c^d(\tau_V).$$

Now we consider the case Z_λ consists of an isolated point p. In this case, we take as U a neighborhood of 0 in $\mathbb{C}^n$, p being identified with 0, and set $W = U \cap V$ (cf. the beginning of Section 13.6). Suppose W is a CI defined by $(h_1, \dots, h_k) : U \to \mathbb{C}^k$. Let v be a non-vanishing C^∞ vector field on $W_0 = W \smallsetminus \{0\}$. Then the $(k+1)$-tuple

$$v^{(k+1)} = \left(v, \frac{\partial}{\partial h_1}\bigg|_{W_0}, \dots, \frac{\partial}{\partial h_k}\bigg|_{W_0}\right)$$

of sections of $TM|_{W_0}$ form a frame so that we have the residue $\mathrm{Res}_{c^d}(v^{(k+1)}, TM|_V; 0)$.

Theorem 13.13. *In the above situation,*

$$\mathrm{Vir}(v, 0) = \mathrm{Res}_{c^d}(v^{(k+1)}, TM|_V; 0).$$

Proof. We denote by $\partial^{(k)}$ the frame $(\varpi(\partial/\partial h_1), \dots, \varpi(\partial/\partial h_k))$ of $N|_{W_0}$ and by $\nu^{(k)}$ the frame of N_U extending $\partial^{(k)}$. We take a v-trivial connection ∇ for TV_{reg} on W_0, a $v^{(k+1)}$-trivial connection ∇^M for $TM|_{W_0}$ and a $\partial^{(k)}$-trivial connection ∇^N for N_{W_0} so that $(\nabla, \nabla^M, \nabla^N)$ is compatible with the first row of (11.26) and set $\nabla^\star = (\nabla^M, \nabla^N)$. Also, let ∇_1^M be an arbitrary connection for TU and let ∇_1^N be the $\nu^{(k)}$-trivial connection for N and set $\nabla_1^\star = (\nabla_1^M, \nabla_1^N)$. Let R be a $2n$-dimensional manifold in U containing 0 in its interior such that ∂R is transverse to $V \smallsetminus \{0\}$ and set $Q = R \cap V$. Then (cf. (13.33))

$$\mathrm{Res}_{c^d}(v^{(k+1)}, TM|_V; 0) = \int_Q c^d(\nabla_1^M) - \int_{\partial Q} c^d(\nabla^M, \nabla_1^M).$$

From $c^*(\nabla_1^\star) = c^*(\nabla_1^M)/c^*(\nabla_1^N)$ and $c^*(\nabla_1^N) = 1$, we have

$$c^d(\nabla_1^\star) = c^d(\nabla_1^M). \tag{13.42}$$

In fact we may take ∇_1^M so that it is 0. To find $c^d(\nabla^\star, \nabla_1^\star)$, recall that it is given by integrating $c^d(\tilde{\nabla}^\star)$ over the 1-simplex $[0,1]$, where $\tilde{\nabla}^\star = (\tilde{\nabla}^M, \tilde{\nabla}^N)$ with $\tilde{\nabla}^M = (1-t)\nabla^M + t\nabla_1^M$ and $\tilde{\nabla}^N = (1-t)\nabla^N + t\nabla_1^N$. Since $\nabla^N = \nabla_1^N$ on W_0, we have $\tilde{\nabla}^N = \nabla^N$ and moreover, $c^*(\tilde{\nabla}^N) = 1$, as ∇^N is $\partial^{(k)}$-trivial. Thus we have $c^d(\tilde{\nabla}^\star) = c^d(\tilde{\nabla}^M)$ exactly as above. Therefore we have $c^d(\nabla^\star, \nabla_1^\star) = c^d(\nabla^M, \nabla_1^M)$, which together with (13.42) implies the equality. □

Remark 13.12. 1. Thus it has a topological expression as in Theorem 13.10. However, even if v is defined and holomorphic in a neighborhood of p, we cannot use the algebraic or analytic expression in Theorems 13.12

or 13.11, as the vector fields $\partial/\partial h_i$ cannot be extended holomorphically through 0.

2. As is seen in the proof, we could use an arbitrary frame of N on U_0 instead of $\partial^{(k)}$. In particular, if we take the frame $(\overline{\text{grad } h_1}, \dots, \overline{\text{grad } h_k})$, we get essentially what is called the GSV-index of v.

Now we consider the situation of Theorem 11.18, with f replaced by h. Thus suppose we have a holomorphic map $h : U \to U'$ from a neighborhood U of 0 in $\mathbb{C}^n$ onto a neighborhood U' of 0 in $\mathbb{C}^k$ with $h(0) = 0$ such that $V = h^{-1}(0)$ is a complete intersection of dimension $d = n - k$ with an isolated singularity at 0. In this case we may assume that, for each $t \in U'$, $V_t = h^{-1}(t)$ is a complete intersection of dimension d in U. Denoting by $C(h)$ the set of critical points of h, we have $\text{Sing}(V_t) = C(h) \cap V_t$. Let $\mathbb{B}_\varepsilon$ and $\mathbb{B}'_\delta$ be as in Theorem 11.18. Set $\mathbb{S}_\varepsilon = \partial\mathbb{B}_\varepsilon$ and suppose we are given a non-vanishing vector field $\tilde{v}$ on a neighborhood U_0 (disjoint from $C(h)$) of $\mathbb{S}_\varepsilon \cap h^{-1}(\mathbb{B}'_\delta)$ in U, which is tangent to $U_0 \cap V$. Setting $h_0 = h|_{U_0}$, we have the exact sequence of vector bundles

$$0 \longrightarrow Th_0 \longrightarrow TU_0 \longrightarrow h_0^*TU' \longrightarrow 0, \qquad (13.43)$$

where Th_0 is the bundle of vectors tangent to the fibers of h_0. Considering a C^∞ splitting of the above, we see that, taking a smaller U_0 if necessary, $\tilde{v}$ defines a non-vanishing C^∞ section $\tilde{v}'$ of Th_0. We set $v = \tilde{v}'|_V$ and $v_t = \tilde{v}'|_{V_t}$. Set $Q = \mathbb{B}_\varepsilon \cap V$ and, for each $t \in \mathbb{B}'_\delta$, $Q_t = \mathbb{B}_\varepsilon \cap V_t$. Then we have the virtual index $\text{Vir}(v_t, Q_t)$. If $t \in \mathbb{B}'_\delta \setminus D(h)$, then V_t is non-singular and $Q_t = \overline{F_t}$, where F_t is a Milnor fiber.

Theorem 13.14. *In the above situation,* $\text{Vir}(v_t, Q_t)$ *does not depend on* t. *In particular,*

$$\text{Vir}(v, Q) = \text{PH}(v_t, \overline{F}_t) \qquad \textit{for } t \notin D(h).$$

Proof. We compute $\text{Vir}(v_t, Q_t)$ using (the restriction to V_t of) connections as follows. We take connections ∇''_0, ∇_0 and ∇'_0 for $Th|_{U_0}$, $TU|_{U_0}$ and $h^*TU'|_{U_0}$, respectively, so that

(1) ∇''_0 is $\tilde{v}'$-trivial,
(2) the triple $(\nabla''_0, \nabla_0, \nabla'_0)$ is compatible with (13.43).

We set $\nabla^\star_0 = (\nabla_0, \nabla'_0)$.

There is a (trivial) vector bundle N with a frame $(\nu_1, \dots, \nu_k)$ on U and a surjective bundle morphism

$$TU \xrightarrow{\varpi} N \longrightarrow 0$$

such that, on $U \smallsetminus C(h)$, $\varpi(\frac{\partial}{\partial h_i}) = \nu_i$, $i = 1, \dots, k$. The restriction of the above to $V_{t,\mathrm{reg}}$ gives the sequence

$$0 \longrightarrow TV_{t,\mathrm{reg}} \longrightarrow TU|_{V_{t,\mathrm{reg}}} \xrightarrow{\varpi} N_{V_{t,\mathrm{reg}}} \longrightarrow 0, \tag{13.44}$$

where $N_{V_{t,\mathrm{reg}}}$ is the normal bundle of $V_{t,\mathrm{reg}}$. Thus $\tau_{V_t} = (TU - N)|_{V_t}$, the virtual tangent bundle of V_t. The bundle TU' is also trivial of rank k and we may identify $h_0^* TU'$ with $N|_{U_0}$ so that the restriction of (13.43) to V_t coincides with the restriction of (13.44) to U_0.

Let ∇_1 and ∇_1' be arbitrary connections for TU and N, respectively, on U and set $\nabla_1^\star = (\nabla_1, \nabla_1')$. Then we have

$$\mathrm{Vir}(v_t, Q_t) = \int_{Q_t} c^d(\nabla_1^\star) - \int_{\partial Q_t} c^d(\nabla_0^\star, \nabla_1^\star),$$

which varies continuously on t. For $t \in \mathbb{B}_\varepsilon \smallsetminus D(h)$, this is equal to $\mathrm{PH}(v_t, \overline{F_t})$, which is an integer. Therefore $\mathrm{Vir}(v_t, Q_t)$ is constant, as $\mathbb{B}_\varepsilon \smallsetminus D(h)$ is connected and dense in $\mathbb{B}_\varepsilon$. □

Recall that there exists a vector field v on $W_0 = U_0 \cap V$ pointing outward on Q (cf. Remark 11.12).

Corollary 13.3. *For a vector field v as above, we have*

$$\mathrm{Vir}(v, 0) = 1 + (-1)^d \mu(V, 0).$$

Euler-Poincaré characteristic of singular varieties

Let V be a local complete intersection defined by a section of a holomorphic vector bundle N of rank k on M and $\tau_V = (TM - N)|_V$ as before.

If V is non-singular, from (3.13) we have $c^d(\tau_V) = c^d(V)$ and moreover if V is compact, by the "Gauss-Bonnet formula" (cf. (5.20))

$$\chi(V) = \int_V c^d(\tau_V).$$

More generally, we have:

Theorem 13.15. *Let V be an SLCI of dimension d in M. If V is compact and has only isolated singularities $p_1, \dots, p_r$,*

$$\chi(V) = \int_V c^d(\tau_V) + (-1)^{d+1} \sum_{i=1}^{r} \mu(V, p_i).$$

Proof. For each $i = 1, \ldots, r$, we take a closed ball $\mathbb{B}_i$ in M around p_i so that they are mutually disjoint. Then $V^* = V \smallsetminus \bigcup_{i=1}^r \operatorname{Int} \mathbb{B}_i$ is a C^∞ manifold with boundary. Since ∂V^* is an odd-dimensional manifold, $\chi(\partial V^*) = 0$ and we have

$$\chi(V) = \chi(V^*) + \sum_{i=1}^{r} \chi(\mathbb{B}_i \cap V).$$

The space $\mathbb{B}_i \cap V$ is the cone with vertex p_i and base $\mathbb{S}_i \cap V$ and is contractible, where $\mathbb{S}_i = \partial \mathbb{B}_i$. Thus $\chi(\mathbb{B}_i \cap V) = 1$.

For each i, we take a neighborhood O_i of p_i containing $\mathbb{B}_i$ and a vector field $\tilde{v}_i$ on $O_i \smallsetminus \{p_i\}$ as in Remark 11.12. We may assume that the O_i's are mutually disjoint. Let v be the vector field on $\bigcup_{i=1}^r (O_i \smallsetminus \{p_i\}) \cap V$ which is the restriction of $\tilde{v}_i$ on $(O_i \smallsetminus \{p_i\}) \cap V$. By Propositions 13.12,

$$\int_V c^d(\tau_V) = \operatorname{Vir}(v, V^*) + \sum_{i=1}^{r} \operatorname{Vir}(v, p_i).$$

Since V^* is in V_{reg}, $\operatorname{Vir}(v, V^*) = \operatorname{PH}(v, V^*)$ by Proposition 13.11. We have $\operatorname{PH}(v, V^*) = \chi(V^*)$, since the vector field v is pointing inward on ∂V^* and $\chi(\partial V^*) = 0$ (cf. Remark 5.16). By Corollary 13.3, $\operatorname{Vir}(v, p_i) = 1 + (-1)^d \mu(V, p_i)$. □

Note that we need only the compactness of V but not of M.

Example 13.2. Let V be a complete intersection in $M = \mathbb{P}^n$ (cf. Example 11.11. 3). Then, from (8.16), (9.22) and $N = H_n^{\otimes p_1} \oplus \cdots \oplus H_n^{\otimes p_k}$, where H_n denotes the hyperplane bundle, we have

$$c^*(\tau_V) = i^* c^*(T\mathbb{P}^n - N), \quad c^*(T\mathbb{P}^n - N) = (1 + h_n)^{n+1} \cdot \prod_{\nu=1}^{k} \frac{1}{1 + p_\nu h_n}.$$

In the above, $i : V \hookrightarrow \mathbb{P}^n$ denotes the inclusion, $h_n = c^1(H_n)$ and the product is the cup product. On the other hand, V is the Alexander dual of a localization of $c^k(N)$ (cf. Theorem 12.9), in particular $[V]$ is the Poincaré dual of $c^k(N)$ and may be written

$$i_* V = [V] = \mathbb{P}^n \frown c^k(N) \qquad \text{in } H_{2d}(\mathbb{P}^n).$$

We have $c^k(N) = \prod_{\nu=1}^k p_\nu \cdot h_n^k$. Since h_n is the Poincaré dual of $[\mathbb{P}^{n-1}]$, we have $\mathbb{P}^n \frown h_n = [\mathbb{P}^{n-1}] = j_* \mathbb{P}^{n-1}$, where $j : \mathbb{P}^{n-1} \hookrightarrow \mathbb{P}^n$ is the inclusion. Thus, by (B.18) and (B.23), $\mathbb{P}^n \frown h_n^2 = j_* \mathbb{P}^{n-1} \frown h_n = j_*(\mathbb{P}^{n-1} \frown j^* h_n)$. Since $j^* h_n = h_{n-1}$, the first Chern class of the hyperplane bundle on $\mathbb{P}^{n-1}$, we have $\mathbb{P}^{n-1} \frown j^* h_n = j'_* \mathbb{P}^{n-2}$, where $j' : \mathbb{P}^{n-2} \hookrightarrow \mathbb{P}^{n-1}$ is the inclusion. Hence $\mathbb{P}^n \frown h_n^2 = [\mathbb{P}^{n-2}]$. Continuing this, we have

$$\mathbb{P}^n \frown c^k(N) = \prod_{\nu=1}^{k} p_\nu \cdot [\mathbb{P}^d].$$

Thus we compute, using (B.8) and noting that $\langle [\mathbb{P}^d], h_n^d \rangle = 1$,

$$\begin{aligned} \int_V c^d(\tau_V) &= \langle V, i^* c^d(T\mathbb{P}^n - N) \rangle = \langle i_* V, c^d(T\mathbb{P}^n - N) \rangle \\ &= \left[(1 + h_n)^{n+1} \cdot \prod_{\nu=1}^{k} \frac{p_\nu}{1 + p_\nu h_n} \right]_d, \end{aligned} \tag{13.45}$$

where $[\ ]_d$ denotes the coefficient of h_n^d in $[\ ]$. Thus we may compute $\chi(V)$ from Theorem 13.15.

Let V_0 be a non-singular complete intersection in $\mathbb{P}^n$ of dimension d with the same multi-degree as V. Then we have $\chi(V_0) = \int_{V_0} c^d(V_0)$, which is also given by (13.45). Hence we have the following formula, which is readily proved by a direct argument as well:

$$\chi(V) = \chi(V_0) + (-1)^{d+1} \sum_{i=1}^{r} \mu(V, p_i). \tag{13.46}$$

See Theorem 14.3 below for an application of Theorem 13.15 to the case V is a curve in a surface.

Virtual index of a (1,0)-form

Let V be a LCI defined by a section of a holomorphic vector bundle N of rank k on M as before. This time we consider the *virtual cotangent bundle* $\tau_V^* = (T^*M - N^*)|_V$. Let Z be a closed subset of V containing $\mathrm{Sing}(V)$ and θ a C^∞ non-vanishing (1,0)-form on $V \setminus Z$. Then we may define the localization $c_Z^d(\tau_V^*, \theta)$ of the class $c^d(\tau_V^*)$ as in the case of vector fields.

If Z has a finite number of connected components (Z_λ), for each λ we have the residue $\mathrm{Res}_{c^d}(\theta, \tau_V^*; Z_\lambda)$ and the residue formula. For a compact connected component Z_λ, we define the *virtual index* $\mathrm{Vir}(\theta, Z_\lambda)$ of θ at Z_λ as the residue $\mathrm{Res}_{c^d}(\theta, \tau_V^*; Z_\lambda)$.

Now we consider the case Z_λ consists of an isolated point p. In this case, we take as U a neighborhood of 0 in $\mathbb{C}^n$, p being identified with 0, and set $W = U \cap V$, as in the case of the virtual index of a vector field. Suppose W is a CI defined by $(h_1, \ldots, h_k) : U \to \mathbb{C}^k$. For a (1,0)-form θ on $W_0 = W \setminus \{0\}$, we consider the $(k+1)$-tuple $\theta^{(k+1)} = (\theta, dh_1|_{W_0}, \ldots, dh_k|_{W_0})$ of sections of $T^*M|_{W_0}$. It should be emphasized that here we take the restrictions as sections and not as differential forms.

The following is proved as Theorem 13.13:

Theorem 13.16. *In the above situation,*

$$\mathrm{Vir}(\theta, 0) = \mathrm{Res}_{c^d}(\theta^{(k+1)}, T^*M|_V; 0).$$

Thus it has a topological expression as in Theorem 13.10. Moreover, if θ is defined and holomorphic in a neighborhood of p, we have the analytic and algebraic expressions in Theorems 13.11 and 13.12.

Multiplicity of a function on a local complete intersection

Let V be an LCI defined by a section of N and consider the virtual cotangent bundle $\tau_V^* = (T^*M - N^*)|_V$ as above. Let $\tilde{f}$ be a holomorphic function on M and let f and f' be its restrictions to V and V_{reg}, respectively. We define the singular set $S(f)$ of f by $S(f) = \mathrm{Sing}(V) \cup C(f')$. As in the case of vector bundles, we may define the localization of the d-th Chern class of τ_V^* by df, which in turn defines the residue $\mathrm{Res}_{c^d}(df, \tau_V^*; Z)$ at each compact connected component Z of $S(f)$. For simplicity we consider the case Z consists of a point p. We define the *virtual multiplicity* $\widetilde{m}(f,p)$ of f at p by

$$\widetilde{m}(f,p) = \mathrm{Vir}(df, p) = \mathrm{Res}_{c^d}(df, \tau_V^*; p). \tag{13.47}$$

The *multiplicity* of f at p is then defined by

$$m(f,p) = \widetilde{m}(f,p) - \mu(V,p). \tag{13.48}$$

Note that, if p is in V_{reg}, we have $\mathrm{Res}_{c^d}(df, \tau_V^*; p) = \mathrm{Res}_{c^d}(df, T^*V_{\mathrm{reg}}; p)$. On the other hand, in this case we have $\mu(V,p) = 0$ so that $m(f,p)$ coincides with the one in Definition 12.4, see also Example 12.3.

Let $\tilde{f} : M \to C$ be a holomorphic map onto a non-singular complex curve C and set $f = \tilde{f}|_V$, $f' = \tilde{f}|_{V_{\mathrm{reg}}}$ and $S(f) = \mathrm{Sing}(V) \cup C(f')$. We further set $V_0 = V \setminus S(f)$ and $f_0 = \tilde{f}|_{V_0}$. Thus df_0 is a non-vanishing section of the bundle $T^*V_0 \otimes f_0^*TC$, which is of rank d. If we look at $c^d(\varepsilon)$, $\varepsilon = \tau_V^* \otimes f^*TC$, we see that there is a canonical localization $c_Z^d(\varepsilon, df)$ in $H^{2d}(V, V \setminus Z; \mathbb{C})$ of $c^d(\varepsilon)$, $Z = S(f)$.

Let $(Z_\lambda)_\lambda$ be the connected components of Z. Then $c_Z^d(\varepsilon, df)$ defines, for each λ, the residue $\mathrm{Res}_{c^d}(df, \tau_V^* \otimes f^*TC; Z_\lambda)$. If Z_λ is compact, it is given by a formula similar to (13.33), see also (13.41). Note that, if Z_λ is in the non-singular part $V_{\rm reg}$, it coincides with the one in Section 12.5. If V is compact, we have (cf. Proposition 13.12)

$$\sum_{\lambda} \mathrm{Res}_{c^d}(df, \tau_V^* \otimes f^*TC; Z_\lambda) = \int_V c^d(\tau_V^* \otimes f^*TC).$$

Now we assume that $S(f)$ consists of a finite number of isolated points and examine the both sides more closely. First, we may write, for a point p in $S(f)$,

$$\mathrm{Res}_{c^d}(df, \tau_V^* \otimes f^*TC; p) = \widetilde{m}(f,p) = m(f,p) - \mu(V,p).$$

If we set $C_0 = C \smallsetminus f(S(f))$, $V_0' = V \smallsetminus f^{-1}C_0$ and $f_0' = f|_{V_0'}$, then $f_0' : V_0' \to C_0$ is a C^∞ fiber bundle (cf. Theorem 3.12). Let F denote the fiber.

Lemma 13.5. *If V is compact, then we have*

$$\int_V c^d(\tau_V^* \otimes f^*TC) = (-1)^d\,(\chi(V) - \chi(F)\,\chi(C)) + \sum_{p\in S(f)} \mu(V,p).$$

Proof. Using the properties of Chern classes and noting that C is complex one-dimensional, we have

$$c^d(\tau_V^* \otimes f^*TC) = (-1)^d(c^d(\tau_V) - c^{d-1}(\tau_V)\cdot f^*c^1(C)).$$

Thus, in view of Theorem 13.15, it suffices to prove

$$\int_V c^{d-1}(\tau_V)\cdot f^*c^1(C)) = \chi(F)\chi(C). \tag{13.49}$$

For this, let $f(S(f)) = \{t_1, \dots, t_k\}$ and, for each t_i, take an open disk C_i around t_i in C so that they are mutually disjoint and that $TC|_{C_i}$ is trivial. Set $C_0 = C \smallsetminus S(f)$ and consider the covering $\mathcal{C} = \{C_0, \{C_i\}_{i=1}^k\}$ of C. We set $V_0' = V \smallsetminus f^{-1}(C_0)$ as above. Let $U_i = (\tilde{f})^{-1}C_i$ and let U_0 be a tubular neighborhood of V_0' in M with a deformation retraction $r_0 : U_0 \to V_0'$. We may take U_0 small enough so that $\tilde{f}$ does not have critical points in U_0. Set $\tilde{f}_0 = \tilde{f}|_{U_0}$. Let $\mathcal{U}$ denote the covering $\{U_0, \{U_i\}_{i=1}^k\}$ of $U_0 \cup \bigcup U_i$. In the following, we try to prove (13.49) taking connections for various bundles. As to the bundles on V, it is sufficient to consider connections on V_0' and U_i.

We take a connection ∇_0^C for TC on C_0. For each $i = 1, \dots, k$, let ∇_i^C be a connection for TC on C_i trivial with respect to some frame of TC. Noting that $c^1(\nabla_i^C) = 0$, $c^1(C)$ is represented, in $A^2(\mathcal{C})$, by the cocycle

$$c^1(\nabla_*^C) = (c^1(\nabla_0^C), 0, (c^1(\nabla_0^C, \nabla_i^C))_i).$$

We have an exact sequence of vector bundles on V_0':

$$0 \longrightarrow Tf_0' \longrightarrow TV_0' \longrightarrow f_0'^* TC \longrightarrow 0.$$

Take connections ∇_0^f and ∇_0^V for Tf_0' and TV_0', respectively, so that $(\nabla_0^f, \nabla_0^V, f^*\nabla_0^C)$ is compatible with the above sequence. Then we have $c^*(\nabla_0^V) = c^*(\nabla_0^f) \cdot f^*c^*(\nabla_0^C)$, in particular,

$$c^{d-1}(\nabla_0^V) = c^{d-1}(\nabla_0^f) + c^{d-2}(\nabla_0^f) \cdot f^*c^1(\nabla_0^C). \tag{13.50}$$

Take connections ∇_0 and ∇_0' for $TM|_{V_0'}$ and $N_{V_0'}$ so that $(\nabla_0^V, \nabla_0, \nabla_0')$ is compatible with

$$0 \longrightarrow TV_0' \longrightarrow TM|_{V_0'} \longrightarrow N_{V_0'} \longrightarrow 0$$

and set $\nabla_0^\star = (\nabla_0, \nabla_0')$. Then we have

$$c^{d-1}(\nabla_0^\star) = c^{d-1}(\nabla_0^V). \tag{13.51}$$

Also let ∇_i and ∇_i' be connections for TM and N, respectively, on U_i and set $\nabla_i^\star = (\nabla_i, \nabla_i')$. Then $c^{d-1}(\tau_V)$ is represented, in $A^{2(d-1)}(\mathcal{U})$, by the cocycle

$$c^{d-1}(\nabla_*^\star) = (c^{d-1}(\nabla_0^\star), (c^{d-1}(\nabla_i^\star))_i, (c^{d-1}(\nabla_0^\star, \nabla_i^\star))_i).$$

The class $c^{d-1}(\tau_V) \cdot f^*c^1(C)$ is then represented by the cup product

$$\begin{aligned} &c^{d-1}(\nabla_*^\star) \cdot f^*c^1(\nabla_*^C) \\ &\quad = (c^{d-1}(\nabla_0^\star) \cdot f^*c^1(\nabla_0^C), 0, (c^{d-1}(\nabla_0^\star) \cdot f^*c^1(\nabla_0^C, \nabla_i^C))_i). \end{aligned}$$

From (13.51) and (13.50) and by the dimension reason, we have

$$\begin{aligned} c^{d-1}(\nabla_0^\star) \cdot f^*c^1(\nabla_0^C) &= c^{d-1}(\nabla_0^f) \cdot f^*c^1(\nabla_0^C), \\ c^{d-1}(\nabla_0^\star) \cdot f^*c^1(\nabla_0^C, \nabla_i^C) &= c^{d-1}(\nabla_0^f) \cdot f^*c^1(\nabla_0^C, \nabla_i^C). \end{aligned}$$

For each $i = 1, \dots, k$, let D_i be a closed disk in C_i centered at t_i, and let $D_0 = C \smallsetminus \bigcup_{i=1}^k \operatorname{Int} D_i$. Then $\{D_0, \{D_i\}_{i=1}^k\}$ is a honeycomb system adapted to $\mathcal{C}$. If we set $Q_i = f^{-1}D_i$ and $Q_{0i} = f^{-1}D_{0i}$, $D_{0i} = -\partial D_i$, using the projection formula, the left-hand side of (13.49) becomes

$$\begin{aligned} &\int_{Q_0} c^{d-1}(\nabla_0^f) \cdot f^*c^1(\nabla_0^C) + \sum_{i=1}^k \int_{Q_{0i}} c^{d-1}(\nabla_0^f) \cdot f^*c^1(\nabla_0^C, \nabla_i^C) \\ &\quad = \int_{D_0} f_*c^{d-1}(\nabla_0^f) \cdot c^1(\nabla_0^C) + \sum_{i=1}^k \int_{D_{0i}} f_*c^{d-1}(\nabla_0^f) \cdot c^1(\nabla_0^C, \nabla_i^C). \end{aligned}$$

In the above, $f_* c^{d-1}(\nabla_0^f)$ denotes the integration along the fibers of f and it is equal to the constant $\chi(F)$. On the other hand,

$$\int_{D_0} c^1(\nabla_0^C) + \sum_{i=1}^{k} \int_{D_{0i}} c^1(\nabla_0^C, \nabla_i^C) = \int_C c^1(C) = \chi(C),$$

which proves (13.49). □

Thus we have:

Theorem 13.17. *In the above situation, we have*

$$\sum_{p \in S(f)} m(f,p) = (-1)^d \left(\chi(V) - \chi(F)\,\chi(C)\right).$$

The above generalizes the formula in Corollary 12.2 for singular varieties.

The virtual multiplicity $\tilde{m}(f,p) = \mathrm{Res}_{c^d}(df, \tau_V^*; p)$ is given as follows. Let U be a neighborhood of p in M so that the bundle N admits a frame $(\nu_1, \dots, \nu_k)$ on U. We write $\sigma = \sum_{i=1}^{k} h_i \, \nu_i$ with h_i holomorphic functions on U. Then V is defined by $(h_1, \dots, h_k)$ in U. Consider the $(k+1)$-tuple of sections

$$s^{(k+1)} = (d\tilde{f}, dh_1, \dots, dh_k)$$

of T^*U. By assumption we have $S(s^{(k+1)}) \cap V = \{p\}$. Since the rank of T^*U is n, we have the residue $\mathrm{Res}_{c^d}(s_V^{(k+1)}, T^*U|_V; p)$, $s_V^{(k+1)} = s^{(k+1)}|_V$. The following is proved as Theorem 13.13:

$$\widetilde{m}(f,p) = \mathrm{Res}_{c^d}(s_V^{(k+1)}, T^*U|_V; p). \tag{13.52}$$

The virtual multiplicity $\widetilde{m}(f,p)$ was defined as the residue of df on the virtual bundle τ_V^* (cf. (13.47)) and this definition led us to a global formula as in Theorem 13.17. The identity (13.52) shows that it coincides with the residue of $s_V^{(k+1)} = (d\tilde{f}|_V, dh_1|_V, \dots, dh_k|_V)$ on the vector bundle $T^*U|_V$. Thus we have various expressions for $\widetilde{m}(f,p)$ as given in the previous section; by Theorem 13.10, we have a topological expression, by Theorem 13.11 we have a way to compute $\widetilde{m}(f,p)$ explicitly and by Theorem 13.12 we may express

$$\widetilde{m}(f,p) = \dim_{\mathbb{C}} \mathcal{O}_n / (J(\tilde{f}, h_1, \dots, h_k), h_1, \dots, h_k), \tag{13.53}$$

where $J(\tilde{f}, h_1, \dots, h_k)$ denotes the Jacobian ideal of the map $(\tilde{f}, h_1, \dots, h_k)$, i.e., the ideal generated by the $(k+1) \times (k+1)$ minors of the Jacobian matrix $\frac{\partial(\tilde{f}, h_1, \dots, h_k)}{\partial(z_1, \dots, z_n)}$.

The multiplicity $m(f,p)$ is also interpreted as follows. Thus let $V_{\tilde{f}}$ denote the complete intersection defined by $(\tilde{f}, h_1, \dots, h_k)$, assuming $\tilde{f}(p) = 0$. Then, from (13.48), (13.53) and the identity

$$\mu(V,p) + \mu(V_{\tilde{f}}, p) = \dim_{\mathbb{C}} \mathcal{O}_{n+k}/(J(\tilde{f}, h_1, \dots, h_k), h_1, \dots, h_k),$$

which follows from (11.23), we get

$$m(f,p) = \mu(V_{\tilde{f}}, p).$$

Dual classes

We come back to the general situation as in the beginning of Section 13.5. Thus let V be a variety of dimension d in M. Here we let E_V be a holomorphic vector bundle of rank l on V. Note that it makes sense to talk about holomorphic vector bundles on varieties as we have the notion of a holomorphic function on a variety (cf. Section 11.5). Suppose we are given a holomorphic r-section $s_V^{(r)}$ of E_V. Let $\boldsymbol{Z}$ be the analytic space in V defined by $s_V^{(r)}$, i.e., the analytic space locally defined by the ideal generated by the $r \times r$ minors of the matrix whose entries are the local components of the $s_{V,i}$'s. The support Z of $\boldsymbol{Z}$ is the singular set of $s_V^{(r)}$. We assume that Z contains $\operatorname{Sing} V$. If we set $q = l - r + 1$, we have the topological and differential geometric localizations of $c^q(E_V)$ by $s_V^{(r)}$ and the associated residues $\operatorname{TRes}_{c^q}(s_V^{(r)}, E_V; Z)$ and $\operatorname{Res}_{c^q}(s_V^{(r)}, E_V; Z)$ as their images by the Alexander morphism.

In the following, we assume that $\dim Z = d - q$ (proper case) and let $(Z_i)_i$ be the irreducible components of Z. Note that in this case we may identify $\operatorname{TRes}_{c^q}(s_V^{(r)}, E_V; Z)$ and $\operatorname{Res}_{c^q}(s_V^{(r)}, E_V; Z)$ (cf. (13.36)).

Let p_i, E_i and $s_i^{(r)}$ be as in the paragraph following (13.36). Then we have the residues $\operatorname{TRes}_{c^q}(s_i^{(r)}, E_i; p_i)$ and $\operatorname{Res}_{c^q}(s_i^{(r)}, E_i; p_i)$, which may be identified. It is a positive integer.

Definition 13.5. The *multiplicity* m_i of Z_i in $\boldsymbol{Z}$ is defined by

$$m_i = \operatorname{Res}_{c^q}(s_i^{(r)}, E_i; p_i).$$

Note that, the multiplicity m_i has a topological expression as in Theorems 13.10. If E_V and $s_V^{(r)}$ are the restrictions of the ones on M, it has an analytic expression as in Theorem 13.11. If, moreover, V is a local complete intersection, it has an algebraic expression as in Theorem 13.12.

Definition 13.6. The homology class of $\boldsymbol{Z}$ is defined by $[\boldsymbol{Z}] = \sum m_i [Z_i]$ in $\check{H}_{2(d-q)}(V; \mathbb{Z})$ or by $\boldsymbol{Z} = \sum m_i [Z_i]$ in $\check{H}_{2(d-q)}(Z; \mathbb{Z})$.

From Theorem 13.9, we have:

Theorem 13.18. *Let* $\boldsymbol{Z}$ *be the complex subspace of dimension* $d-q$ *of* V *as above. Then the class of* $\boldsymbol{Z}$ *in* $\check{H}_{2(d-q)}(Z;\mathbb{Z})$ *is equal to* $\mathrm{TRes}_{c^q}(s_V^{(r)}, E_V; Z)$, *which may be identified with* $\mathrm{Res}_{c^q}(s_V^{(r)}, E_V; Z)$. *Thus the class* $c^q(E_V, s_V^{(r)})$ *is sent to the class* $\boldsymbol{Z}$ *by the Alexander morphism*

$$H^{2q}(V, V \smallsetminus Z; \mathbb{Z}) \longrightarrow \check{H}_{2(d-q)}(Z; \mathbb{Z}).$$

In particular, the class $[\boldsymbol{Z}]$ in $\check{H}_{2(d-q)}(V;\mathbb{Z})$ is the image of $c^q(L_D)$ in $H^{2q}(V;\mathbb{Z})$ by the Poincaré morphism.

Notes

The presentation of this chapter is mainly based on Sections 5 and 6 of [Suwa (2008)].

For controlled tube systems and Theorem 13.1, we refer to [Gibson, Wirthmüller, du Plessis and Looijenga (1976); Mather (2012); Shiota (1997)].

As to the Poincaré, Alexander and Thom morphisms, we followed the description of [Brasselet (1981)], except for the orientation convention (cf. Notes in Chapter 4).

In Section 13.6, see [Suwa (2003a)] for the analytic expression and [Suwa (2005)] for the algebraic expression with some more examples. As to the Grothendieck residue relative to a subvariety, see Ch.IV, 8 in [Suwa (1998)]. In Section 13.7, the virtual index is introduced in [Lehmann, Soares and Suwa (1995)]. The GSV-index of a vector field at an isolated singularity is introduced in [Gómez-Mont, Seade and Verjovsky (1991)] for hypersurfaces and in [Seade and Suwa (1996)] for complete intersections. As to the material there, we refer to the above literatures as well as [Brasselet, Seade and Suwa (2009); Suwa (2014)] and references therein.

In [Brasselet, Lehmann, Seade and Suwa (2001)], a generalized Milnor number $\mu(V,S)$ is defined for each compact connected component S of the singular set of an SLCI V and, if V is compact, a formula as in Theorem 13.15 is proved. In the case V is a hypersurface, the number and the formula coincide with the ones in [Parusiński (1988)], see also [Parusiński and Pragacz (1995)]. As to the formula (13.46), see also [Dimca (1992)].

The index of a 1-form is introduced and the algebraic expression is given in [Ebeling and Gusein-Zade (2001)]. For details on the multiplicity of a function, we refer to [Izawa and Suwa (2003)]. It is possible to define the multiplicity of a function at a non-isolated component of $S(f)$ using the number $\mu(V, S)$ above. See [Suwa (2005)] for more examples of related local invariants.

As another topic along the line above, there is the theory of characteristic classes of singular varieties. For the treatment in the framework of this book, we refer to [Brasselet, Lehmann, Seade and Suwa (2001); Brasselet, Seade and Suwa (2009); Suwa (1996)] and a survey paper [Suwa (2003b)]. We list, as related works, [Aluffi (1999); Brasselet, Schürmann and Yokura (2010); Brasselet and Schwartz (1981); Fulton and Johnson (1980); MacPherson (1974); Parusiński and Pragacz (2001); Schwartz (1965)] and survey papers [Schürmann and Yokura (2007); Yokura (2020)]. We also list [Brasselet (2022)] as a survey paper of characterictic classes in general.

Chapter 14

Intersection Product of Complex Subspaces

In this chapter, we discuss intersection products of the homology classes of complex spaces in complex manifolds or in singular varieties. The formulas we prove concerning intersection products are localized at the set of intersection. The corresponding global formulas are consequences of these.

We first generalize the Whitney sum formula for localized Chern classes (cf. Corollaries 5.6 and 10.5) so that it fits our purposes (Corollary 14.1).

In the case of manifolds, the intersection product in homology is defined combinatorially in Section 4.4. It is an operation dual to the cup product in cohomology. In general, the intersection product of the classes of subspaces is different from the class of the intersection of them as complex subspaces. The intersection products we consider involve, in some form or another, complex spaces defined by a family of sections of a vector bundle. In this situation, we have the localization of the Chern class of the bundle by the family of sections as well as the associated residue. The class of the complex space defined by the family of sections coincides with the residue if it has an expected dimension (proper case). We always have expected formulas for residues. Thus in the proper case, we have expected formulas for spaces as defined above.

We start by considering the intersection product of subspaces defined by a holomorphic section of a vector bundle. The aforementioned Whitney sum formula for localized classes and the duality in Theorem 12.9 are key ingredients for this. As a special case, we discuss intersection products of divisors and give Bézout's theorem as an example (Example 14.1). Intersection product with the class of a (fixed) subvariety is then studied. It leads to the intersection theory of the classes of complex spaces in complex manifolds in general. In particular, we consider the intersection product of the class of a variety and the class of the complex space defined by a family of holomorphic sections. Here we use the duality as in Theorem 12.8. As a related topic, we prove the adjunction formula for singular curves (Theorem 14.3).

As to the intersection product in sigular varieties, we do not have general combinatorial ways of defining it. Thus we consider intersection products of homology classes that come naturally as the images of the Poincaré or Alexander morphism and use the cup product in cohomology. For this, as in the case of manifolds, we consider intersection product involving subspaces defined by a family of sections and develop the theory in parallel with that case. We also discuss intersection products of divisors. Here we need to be a little careful, as unlike the case of complex manifolds, we have to distinguish between Cartier and Weil divisors. We also discuss the multiplicity of a variety at a point, giving its various expressions.

We then specialize the above considerations to the case of intersection products in singular surfaces. In particular, we study the behavior of intersection products under blowing-up, which is applied to define intersection product of Weil curves.

Finally we study the case where a variety and a complex space defined by a section intersect in a dimension greater than expected.

14.1 Refined Whitney sum formula

We recall the construction of the Euler class of an oriented real vector bundle and its localization in Sections 5.3, 5.4 and 5.6. Here we do it for bundles on simplicial complexes. In the following, a map means a continuous map and $\mathbb{R}^{l'} \smallsetminus \{0\}$ is denoted by $\mathbb{R}^{l'} \smallsetminus 0$.

Let l' be an integer with $l' > 1$. Recall that there is a canonical isomorphism (cf. Example B.6)

$$\delta^* : H^{l'-1}(\mathbb{S}^{l'-1}) \xrightarrow{\sim} H^{l'}(\mathbb{B}^{l'}, \mathbb{S}^{l'-1}) \simeq \mathbb{Z}.$$

There is also a canonical isomorphism $H^{l'}(\mathbb{R}^{l'}, \mathbb{R}^{l'} \smallsetminus 0) \simeq H^{l'}(\mathbb{B}^{l'}, \mathbb{S}^{l'-1})$. The group $H^{l'}(\mathbb{R}^{l'}, \mathbb{R}^{l'} \smallsetminus 0)$ has the canonical generator $\Psi_{\mathbb{R}^{l'}}$, the Thom class, which is equal to the l'-fold cross product $e^{l'}$ of the canonical generator e of $H^1(\mathbb{R}, \mathbb{R} \smallsetminus 0)$ (cf. Corollary 4.4.1). The generator of $H^{l'}(\mathbb{B}^{l'}, \mathbb{S}^{l'-1})$ corresponding to $\Psi_{\mathbb{R}^{l'}}$ is denoted by $u^{l'}$, which coincides with $\delta^*(\nu^*_{l'-1})$ (cf. Remark 5.1.2).

Thus we may define the degree of a map $(\mathbb{B}^{l'}, \mathbb{S}^{l'-1}) \to (\mathbb{R}^{l'}, \mathbb{R}^{l'} \smallsetminus 0)$ as well.

Proposition 14.1. *Giving a homotopy class of a map* $\varphi : \mathbb{S}^{l'-1} \to \mathbb{R}^{l'} \smallsetminus 0$ *is equivalent to giving that of a map* $\psi : (\mathbb{B}^{l'}, \mathbb{S}^{l'-1}) \to (\mathbb{R}^{l'}, \mathbb{R}^{l'} \smallsetminus 0)$, *in fact that of a map* $(\mathbb{B}^{l'}, \mathbb{B}^{l'} \smallsetminus 0) \to (\mathbb{R}^{l'}, \mathbb{R}^{l'} \smallsetminus 0)$.

Proof. Let φ be as above. Consider the product $I \times \mathbb{S}^{l'-1}$ with $I = [0,1]$, the unit interval, and define $\tilde{\varphi} : I \times \mathbb{S}^{l'-1} \to \mathbb{R}^{l'}$ by $(t,x) \mapsto t\varphi(x)$. The ball $\mathbb{B}^{l'}$ is obtained from $I \times \mathbb{S}^{l'-1}$ by shrinking $\{0\} \times \mathbb{S}^{l'-1}$ to a point and $\tilde{\varphi}$ induces a map $\psi : (\mathbb{B}^{l'}, \mathbb{S}^{l'-1}) \to (\mathbb{R}^{l'}, \mathbb{R}^{l'} \smallsetminus \{0\})$. The homotopy class of φ determines that of ψ as the homotopy class of such a map is uniquely determined by its degree. The converse is obvious. □

Let $E = \mathbb{R}^{l'} \times \mathbb{B}^{l'}$ be the product bundle of rank l' on $\mathbb{B}^{l'}$. Suppose we have a non-vanishing section s of E on $\mathbb{S}^{l'-1}$. We extend it to a section on $\mathbb{B}^{l'}$ non-vanishing section on $\mathbb{B}^{l'} \smallsetminus 0$. We have a composition as in (5.4):

$$\psi : (\mathbb{B}^{l'}, \mathbb{S}^{l'-1}) \xrightarrow{s} (\mathbb{R}^{l'}, \mathbb{R}^{l'} \smallsetminus 0) \times \mathbb{B}^{l'} \xrightarrow{p} (\mathbb{R}^{l'}, \mathbb{R}^{l'} \smallsetminus 0), \tag{14.1}$$

where p denotes the projection. Then the index $I(s,0)$ of s at 0 is the mapping degree of ψ:

$$I(s,0) = \deg \psi.$$

Suppose we have two product bundles $E_i = \mathbb{R}^{l'_i} \times \mathbb{B}^{l'_i}$, $i = 1, 2$. For each i, let s_i be a non-vanishing section of E_i on $\mathbb{S}^{l'_i-1}$. The product $(\mathbb{B}^{l'_1}, \mathbb{S}^{l'_1-1}) \times (\mathbb{B}^{l'_2}, \mathbb{S}^{l'_2-1})$ is homeomorphic with $(\mathbb{B}^{l'}, \mathbb{S}^{l'-1})$, $l' = l'_1 + l'_2$. If we identify the two pairs of spaces, the product $E = E_1 \times E_2$ is the product bundle $\mathbb{R}^{l'} \times \mathbb{B}^{l'}$ on $\mathbb{B}^{l'}$ and $s = s_1 \times s_2$ is a non vanishing section of E on $\mathbb{S}^{l'-1}$.

Lemma 14.1. *In the above situation, we have*

$$I(s,0) = I(s_1,0) \cdot I(s_2,0).$$

Proof. Let $\psi_i : (\mathbb{B}^{l'_i}, \mathbb{S}^{l'_i-1}) \to (\mathbb{R}^{l'_i}, \mathbb{R}^{l'_i} \smallsetminus 0)$ be the map as defined in (14.1) for s_i. Then we have $\psi_i^* : H^{l'_i}(\mathbb{R}^{l'_i}, \mathbb{R}^{l'_i} \smallsetminus 0) \to H^{l'_i}(\mathbb{B}^{l'_i}, \mathbb{S}^{l'_i-1})$ and by definition, $\psi_i^*(e^{l'_i}) = \deg \psi_i \cdot u^{l'_i}$. Also if we let ψ denote the map for s, we have $\psi^* : H^{l'}(\mathbb{R}^{l'}, \mathbb{R}^{l'} \smallsetminus 0) \to H^{l'}(\mathbb{B}^{l'}, \mathbb{S}^{l'-1})$. The lemma follows from the commutativity of the diagram

$$\begin{array}{ccc}
H^{l'_1}(\mathbb{R}^{l'_1}, \mathbb{R}^{l'_1} \smallsetminus 0) \times H^{l'_2}(\mathbb{R}^{l'_2}, \mathbb{R}^{l'_2} \smallsetminus 0) & \xrightarrow{\times} & H^{l'}(\mathbb{R}^{l'}, \mathbb{R}^{l'} \smallsetminus 0) \\
{\scriptstyle \psi_1^* \times \psi_2^*} \big\downarrow & & \big\downarrow {\scriptstyle \psi^*} \\
H^{l'_1}(\mathbb{B}^{l'_1}, \mathbb{S}^{l'_1-1}) \times H^{l'_2}(\mathbb{B}^{l'_2}, \mathbb{S}^{l'_2-1}) & \xrightarrow{\times} & H^{l'}(\mathbb{B}^{l'}, \mathbb{S}^{l'-1})
\end{array}$$

and Corollary 4.4.2. □

Let X be a space with a triangulation K_0. We denote by K the barycentric subdivision of K_0 and by K' that of K, as before. Also let $E \to X$ be an oriented real vector bundle of rank l' on X. We endow each simplex of K' with an orientation and define the Euler class of E and its localization using the triangulation K'. We also fix a simplicial ordering on the vertices of K' (cf. Definition B.14).

Let Z be a K_0-subcomplex of X and suppose we have a non-vanishing section s of E on the l'-skeleton of K' in $X \smallsetminus O_{K'}(Z)$. Then we have the Euler class $e(E, s)$ of E localized by s in $H^{l'}(X, X \smallsetminus Z)$. Recall that it is represented by the cocycle constructed as follows. First we extend s to a non-vanishing section $\tilde{s}$ of E on $X^{l'} \smallsetminus \{\text{isolated points}\}$, these isolated points being the barycenter of some K'-simplices in $S_{K'}(Z)$. The class $e(E, s)$ is represented by the cocycle γ that assigns to each l'-simplex $\boldsymbol{t}$ of K' the index $I(\tilde{s}, b_{\boldsymbol{t}})$ of $\tilde{s}$ at the barycenter $b_{\boldsymbol{t}}$ of $\boldsymbol{t}$.

For $i = 1, 2$, let E_i be an oriented real vector bundle of rank l'_i on X. Also let Z_i be a K_0-subcomplex of X and s_i a non-vanishing section of E_i on the l'_i-skeleton of K' in $X \smallsetminus O_{K'}(Z_i)$. We set $E = E_1 \oplus E_2$, $l' = l'_1 + l'_2$, $Z = Z_1 \cap Z_2$ and $s = s_1 \oplus s_2$. Note that $O_{K'}(Z) = O_{K'}(Z_1) \cap O_{K'}(Z_2)$. In this situation, s is a non-vanishing section of E on the l'-skeleton of K' in $X \smallsetminus O_{K'}(Z)$. Indeed, let $\boldsymbol{t} = \varepsilon(v_0, \cdots, v_{l'})$, $v_0 < \cdots < v_{l'}$, be an oriented l'-simplex of K' and set $\boldsymbol{t}_1 = \varepsilon_1(v_0, \cdots, v_{l'_1})$, $\boldsymbol{t}_2 = \varepsilon_2(v_{l'_1}, \cdots, v_{l'})$, where ε, ε_1 and ε_2 are ± 1, to be determined according to the prescribed orientations. If $\boldsymbol{t}$ is in $X \smallsetminus O_{K'}(Z)$, then either $\boldsymbol{t}_1$ is in $X \smallsetminus O_{K'}(Z_1)$ or $\boldsymbol{t}_2$ is in $X \smallsetminus O_{K'}(Z_2)$. There is a homeomorphism $\boldsymbol{t} \simeq \boldsymbol{t}_1 \times \boldsymbol{t}_2$, by which s corresponds to $s_1 \times s_2$. Hence s is non-vanishing on $\boldsymbol{t}$.

Thus we have the localizations $e(E_i, s_i)$ in $H^{l'_i}(X, X \smallsetminus Z_i)$, $i = 1, 2$, and $e(E, s)$ in $H^{l'}(X, X \smallsetminus Z)$. We also have the cup product

$$H^{l'_1}(X, X \smallsetminus Z_1) \times H^{l'_2}(X, X \smallsetminus Z_2) \overset{\smile}{\longrightarrow} H^{l'}(X, X \smallsetminus Z).$$

Theorem 14.1. *In the above situation,*

$$e(E, s) = e(E_1, s_1) \smile e(E_2, s_2).$$

Proof. For $i = 1, 2$, let $\tilde{s}_i$ be an extension of s_i and γ_i the cocycle representing the class $e(E_i, s_i)$ as above. Let $\boldsymbol{t}$ be an oriented l'-simplex of K' and $\boldsymbol{t}_1$, $\boldsymbol{t}_2$ be as above. Then $\gamma_1 \smile \gamma_2$ is the cocycle that assigns $I(\tilde{s}_1, b_{\boldsymbol{t}_1}) \cdot I(\tilde{s}_2, b_{\boldsymbol{t}_2})$ to $\boldsymbol{t}$. By Lemma 14.1, $I(\tilde{s}_1, b_{\boldsymbol{t}_1}) \cdot I(\tilde{s}_2, b_{\boldsymbol{t}_2}) = I(\tilde{s}_1 \times \tilde{s}_2, b_{\boldsymbol{t}_1} \times b_{\boldsymbol{t}_2})$. There is a homeomorphism $(\boldsymbol{t}_1 \times \boldsymbol{t}_2, b_{\boldsymbol{t}_1} \times b_{\boldsymbol{t}_2}) \simeq (\boldsymbol{t}, b_{\boldsymbol{t}})$, by which $\tilde{s}_1 \times \tilde{s}_2$ corresponds to an extension $\tilde{s}$ of s. The class $e(E, s)$

is represented by the cocycle γ that assigns $I(\tilde{s}, b_{\boldsymbol{t}})$, which is equal to $I(\tilde{s}_1, b_{\boldsymbol{t}_1}) \cdot I(\tilde{s}_2, b_{\boldsymbol{t}_2})$ by the above argument. $\square$

Corollary 14.1. *In the above situation, suppose E_i is a complex vector bundle of rank l_i, $i = 1, 2$, and set $l = l_1 + l_2$. Then we have*

$$c^l(E, s) = c^{l_1}(E_1, s_1) \smile c^{l_2}(E_2, s_2).$$

Remark 14.1. 1. In fact, if we use the cross product

$$H^{l'_1}(X, X \smallsetminus Z_1) \times H^{l'_2}(X, X \smallsetminus Z_2) \xrightarrow{\times} H^{l'}((X, X \smallsetminus Z_1) \times (X, X \smallsetminus Z_2)),$$

we have

$$e(E_1 \times E_2, s_1 \times s_2) = e(E_1, s_1) \times e(E_2, s_2).$$

The formula in Theorem 14.1 also follows from this by applying the morphism $d^* : H^{l'}((X, X \smallsetminus Z_1) \times (X, X \smallsetminus Z_2)) \to H^{l'}(X, X \smallsetminus Z)$ induced by the diagonal embedding $d : X \hookrightarrow X \times X$.

2. The advantage of the formula in Corollary 14.1 over that in Corollary 5.6 or in Corollary 10.5 is in two ways: (1) it can be applied to non-vanishing sections that are defined only away from the "singular set", for example, in the case of divisors, and (2) it can be applied to vector bundles on triangulable spaces, for example, singular varieties.

14.2 Intersection product in complex manifolds

Let M be a complex manifold of dimension n. We have the intersection product in $\check{H}_*(M; \mathbb{Z})$, which is denoted by $\bullet$ (cf. Section 4.4).

Let E be a holomorphic vector bundle of rank l on M. If $s^{(r)}$ is a holomorphic r-section of E with singular set S, we have the topological localization $c^q_{\mathrm{top}}(E, s^{(r)})$ in $H^{2q}(M, M \smallsetminus S; \mathbb{Z})$ and the differential geometric localization $c^q_{\mathrm{diff}}(E, s^{(r)})$ in $H^{2q}(M, M \smallsetminus S; \mathbb{C})$, $q = l - r + 1$. We also have the associated residues $\mathrm{TRes}_{c^q}(s^{(r)}, E; S)$ in $\check{H}_{2(n-q)}(S; \mathbb{Z})$ and $\mathrm{Res}_{c^q}(s^{(r)}, E; S)$ in $\check{H}_{2(n-q)}(S; \mathbb{C})$. Let $\boldsymbol{S}$ be the complex space defined by $s^{(r)}$. It has S as its support and, in general, $\dim S \geq n - q$. In the case $\dim S = n - q$ (proper case), we may identify $\mathrm{TRes}_{c^q}(s^{(r)}, E; S)$ and $\mathrm{Res}_{c^q}(s^{(r)}, E; S)$ (cf. (12.2)). Moreover, in this case, we may define the homology class of $\boldsymbol{S}$; $[\boldsymbol{S}]$ in $\check{H}_{2(n-q)}(M; \mathbb{Z})$ or $\boldsymbol{S}$ in $\check{H}_{2(n-q)}(S; \mathbb{Z})$ (cf. Definition 12.6). The class $\boldsymbol{S}$ coincides with $\mathrm{TRes}_{c^q}(s^{(r)}, E; S) = \mathrm{Res}_{c^q}(s^{(r)}, E; S)$ (cf. Theorems 12.8 and 12.9).

In the following, through the end of this chapter, the homology and cohomology are with $\mathbb{Z}$-coefficient, unless otherwise stated, and the coefficient is omitted.

Intersection product of subspaces defined by a section

For each $j = 1, \dots, \nu$, let E_j be a holomorphic vector bundle of rank l_j on M. Let s_j be a holomorphic section of E_j and $\boldsymbol{S}_j$ the complex space in M defined by s_j. The support S_j of $\boldsymbol{S}_j$ is the zero set of s_j. We have the (topological) localization $c^{l_j}(E_j, s_j)$ in $H^{2l_j}(M, M \smallsetminus S_j)$ of $c^{l_j}(E_j)$ by s_j. We also have the residue $\mathrm{TRes}_{c^{l_j}}(s_j, E_j; S_j)$ as the image of $c^{l_j}(E_j, s_j)$ by the Alexander isomorphism

$$H^{2l_j}(M, M \smallsetminus S_j) \xrightarrow{\sim} \breve{H}_{2(n-l_j)}(S_j).$$

Note that $\dim S_j \geq n - l_j$ and that the equality holds if and only if s_j is a regular section. If this is the case (proper case), we may define the class of $\boldsymbol{S}_j$, denoted also by $\boldsymbol{S}_j$, in $\breve{H}_{2(n-l_j)}(S_j)$. It coincides with $\mathrm{TRes}_{c^{l_j}}(s_j, E_j; S_j)$, which may be identified with $\mathrm{Res}_{c^{l_j}}(s_j, E_j; S_j)$ (cf. Theorem 12.9).

Let $E = E_1 \oplus \cdots \oplus E_\nu$ be the direct sum and s the section of E given by $s = s_1 \oplus \cdots \oplus s_\nu$. Let $\boldsymbol{S}$ be the complex space in M defined by s. As a complex subspace of M, it is the intersection of the $\boldsymbol{S}_j$'s. The support S of $\boldsymbol{S}$ is the zero set of s and is given by $S = \bigcap_{j=1}^{\nu} S_j$. We have the localization $c^l(E, s)$ in $H^{2l}(M, M \smallsetminus S)$ by s, $l = \sum_{j=1}^{\nu} l_j$. We also have the residue $\mathrm{TRes}_{c^l}(s, E; S)$ as the image of $c^l(E, s)$ by the Alexander isomorphism

$$H^{2l}(M, M \smallsetminus S) \xrightarrow{\sim} \breve{H}_{2(n-l)}(S).$$

Note that $\dim S \geq n - l$ and that the equality holds if and only if s is a regular section. In this case, we may define the class of $\boldsymbol{S}$, denoted also by $\boldsymbol{S}$, in $\breve{H}_{2(n-l)}(S)$. It coincides with $\mathrm{TRes}_{c^l}(s, E; S)$, which may be identified with $\mathrm{Res}_{c^l}(s, E; S)$.

We have the cup product

$$H^{2l_1}(M, M \smallsetminus S_1) \times \cdots \times H^{2l_\nu}(M, M \smallsetminus S_\nu) \longrightarrow H^{2l}(M, M \smallsetminus S)$$

and we have (cf. Corollary 14.1)

$$c^{l_1}(E_1, s_1) \smile \cdots \smile c^{l_\nu}(E_\nu, s_\nu) = c^l(E, s). \tag{14.2}$$

From (14.2) and Theorem 4.8, we have:

$$\begin{aligned}(\mathrm{TRes}_{c^{l_1}}(s_1, E_1; S_1) \bullet \cdots \bullet \mathrm{TRes}_{c^{l_\nu}}(s_\nu, E_\nu; S_\nu))_S \\ = \mathrm{TRes}_{c^l}(s, E; S) \qquad \text{in } \breve{H}_{2(n-l)}(S).\end{aligned} \tag{14.3}$$

Proposition 14.2. *In the above situation, we have:*

1. *If* $\dim S_j = n - l_j$, *for each* j,

$$(\boldsymbol{S}_1 \bullet \cdots \bullet \boldsymbol{S}_\nu)_S = \mathrm{TRes}_{c^l}(s, E; S) \qquad \text{in } \breve{H}_{2(n-l)}(S).$$

2. *Moreover, if* $\dim S = n - l$,

$$(\boldsymbol{S}_1 \bullet \cdots \bullet \boldsymbol{S}_\nu)_S = \boldsymbol{S} \qquad \text{in } \breve{H}_{2(n-l)}(S).$$

Proof. The proposition follows from Theorem 12.9. □

Remark 14.2. If $\dim S_j = n - l_j$, for each j, and if S is compact, we may write (cf. Remark 4.5. 3)

$$(\boldsymbol{S}_1 \bullet \cdots \bullet \boldsymbol{S}_\nu)_S = \underline{M} \frown c^l(E, s) \qquad \text{in } H_{2(n-l)}(S),$$

where $\underline{M}$ denotes the class of M in $H_{2n}(M, M \smallsetminus S)$.

Note that, if $l = n$ and if S is compact and connected, $\mathrm{TRes}_{c^n}(s, E; S)$ is in $H_0(S) \simeq \mathbb{Z}$ and may be regarded as an integer.

Suppose that $\dim S_j = n - l_j$, for each j, and that S has only a finite number of connected components $(S_\lambda)_\lambda$. Then the λ component of $(\boldsymbol{S}_1 \bullet \cdots \bullet \boldsymbol{S}_\nu)_S$ is given by $(\boldsymbol{S}_1 \bullet \cdots \bullet \boldsymbol{S}_\nu)_{S,\lambda} = \mathrm{TRes}_{c^l}(s, E; S_\lambda)$ in $\breve{H}_{2(n-l)}(S_\lambda)$ and we have

$$\sum_\lambda (i_\lambda)_*(\boldsymbol{S}_1 \bullet \cdots \bullet \boldsymbol{S}_\nu)_{S,\lambda} = (\boldsymbol{S}_1 \bullet \cdots \bullet \boldsymbol{S}_\nu)_S \quad \text{in} \quad \breve{H}_{2(n-l)}(S), \qquad (14.4)$$

where $i_\lambda : S_\lambda \hookrightarrow S$ denotes the inclusion.

Passing from the homologies of the S_j's and S to that of M, we have from (14.3) (cf. Remark 4.13),

$$[\mathrm{TRes}_{c^{l_1}}(s_1, E_1; S_1)] \bullet \cdots \bullet [\mathrm{TRes}_{c^{l_\nu}}(s_\nu, E_\nu; S_\nu)] \\ = [\mathrm{TRes}_{c^l}(s, E; S)] \qquad \text{in } \breve{H}_{2(n-l)}(M).$$

Since $[\mathrm{TRes}_{c^l}(s, E; S)] = P_M^{-1}(c^l(E)) = M \frown c^l(E)$, we have:

Proposition 14.3. *In the above situation, we have:*

1. *If* $\dim S_j = n - l_j$, *for each* j,

$$[\boldsymbol{S}_1] \bullet \cdots \bullet [\boldsymbol{S}_\nu] = M \frown c^l(E) \qquad \text{in } \breve{H}_{2(n-l)}(M).$$

2. *Moreover, if* $\dim S = n - l$,

$$[\boldsymbol{S}_1] \bullet \cdots \bullet [\boldsymbol{S}_\nu] = [\boldsymbol{S}] \qquad \text{in } \breve{H}_{2(n-l)}(M).$$

Intersection product of divisors

The divisors we consider in the following are assumed to be non-zero.

Recall that, for a divisor D on M, there is the associated line bundle L_D together with the natural meromorphic section s_D (cf. Section 11.6). We have the topological residue $\mathrm{TRes}_{c^1}(s_D, L_D; |D|)$ in $\check{H}_{2(n-1)}(|D|)$, which may be identified with the differential geometric residue $\mathrm{Res}_{c^1}(s_D, L_D; |D|)$ (cf. (12.12)). We have the homology class of D in $\check{H}_{2(n-1)}(|D|)$, denoted also by D, which coincides with $\mathrm{TRes}_{c^1}(s_D, L_D; |D|)$ (cf. Theorem 12.10). The results in the previous paragraph hold in this case with a little modification, in fact they include the case of effective divisors as a special case.

Thus, for divisors $D_1, \dots, D_\nu$ on M, we have the intersection product $(D_1 \bullet \cdots \bullet D_\nu)_S$ in $\check{H}_{2(n-\nu)}(S)$, $S = \bigcap_{j=1}^{\nu} |D_j|$ with $|D_j|$ the support of D_j. We set $E = L_{D_1} \oplus \cdots \oplus L_{D_\nu}$ and $s = s_{D_1} \oplus \cdots \oplus s_{D_\nu}$. Then we have the localization $c^\nu(E, s)$ in $H^{2\nu}(M, M \smallsetminus S)$ and the associated residue $\mathrm{TRes}_{c^\nu}(s, E; S)$ in $\check{H}_{2(n-\nu)}(S)$. If the D_j's are effective, the section s is holomorphic and defines a complex space $\boldsymbol{S}$ whose support is S. Moreover, if $\dim S = n-\nu$, then the class of $\boldsymbol{S}$ in $\check{H}_{2(n-\nu)}(S)$ is equal to $\mathrm{TRes}_{c^\nu}(s, E; S)$.

We have, as Proposition 14.2, the following:

Proposition 14.4. *In the above situation,*

1. $(D_1 \bullet \cdots \bullet D_\nu)_S = \mathrm{TRes}_{c^\nu}(s, E; S)$ *in* $\check{H}_{2(n-\nu)}(S)$.

2. *Moreover, if the* D_j*'s are effective and if* $\dim S = n - \nu$,

$$(D_1 \bullet \cdots \bullet D_\nu)_S = \boldsymbol{S} \quad \text{in } \check{H}_{2(n-\nu)}(S).$$

Note that, if S is compact, we may write (cf. Remark 14.2)

$$(D_1 \bullet \cdots \bullet D_\nu)_S = \underline{M} \frown c^\nu(E, s) \qquad \text{in } H_{2(n-\nu)}(S).$$

In the case $\nu = n$ and S is compact, $(D_1 \bullet \cdots \bullet D_n)_S$ is in $H_0(S)$ and we define the intersection number of the D_j's by

$$\sharp(D_1 \bullet \cdots \bullet D_n) = \varepsilon_*((D_1 \bullet \cdots \bullet D_n)_S), \tag{14.5}$$

where $\varepsilon_* : H_0(S) \to \mathbb{Z}$ is the augmentation.

We also have a formula as (14.4). Thus suppose that S has only a finite number of connected components $(S_\lambda)_\lambda$. Then the λ component of $(D_1 \bullet \cdots \bullet D_\nu)_S$ is given by $(D_1 \bullet \cdots \bullet D_\nu)_{S,\lambda} = \mathrm{TRes}_{c^\nu}(s, E; S_\lambda)$ in $\check{H}_{2(n-\nu)}(S_\lambda)$ and we have

$$\sum_\lambda (i_\lambda)_*(D_1 \bullet \cdots \bullet D_\nu)_{S,\lambda} = (D_1 \bullet \cdots \bullet D_\nu)_S \quad \text{in } \check{H}_{2(n-\nu)}(S). \tag{14.6}$$

In the global situation, we have (cf. Proposition 14.3.1)

$$[D_1] \bullet \cdots \bullet [D_\nu] = M \frown c^\nu(E) \qquad \text{in } \check{H}_{2(n-\nu)}(M), \tag{14.7}$$

where $c^\nu(E) = c^1(L_{D_1}) \smile \cdots \smile c^1(L_{D_\nu})$. In particular, if $\nu = n$ and if M is compact, we may define the intersection number of the $[D_j]$'s by

$$\sharp([D_1] \bullet \cdots \bullet [D_n]) = \varepsilon_*([D_1] \bullet \cdots \bullet [D_n]),$$

where $\varepsilon_* : H_0(M) \to \mathbb{Z}$ is the augmentation. By (14.7), we have

$$\sharp([D_1] \bullet \cdots \bullet [D_n]) = \langle M, c^1(L_{D_1}) \smile \cdots \smile c^1(L_{D_n})\rangle. \tag{14.8}$$

It coincides with the one in (14.5).

Let $D_1, \ldots, D_n$ be n divisors on M. Suppose that $S = \bigcap_{j=1}^n |D_j|$ consists of isolated points. For each point p in S, we have the localized intersection product $(D_1 \bullet \cdots \bullet D_n)_p$ in $H_0(\{p\}) \simeq \mathbb{Z}$. Thus it may be regarded as an integer.

From (14.6), we have the following:

Theorem 14.2. *Let M be a complex manifold of dimension n. For divisors $D_1, \ldots, D_n$ on M such that $S = \bigcap_{j=1}^n |D_j|$ consists of a finite number of isolated points, we have*

$$\sum_{p \in S} (D_1 \bullet \cdots \bullet D_n)_p = \sharp(D_1 \bullet \cdots \bullet D_n),$$

which coincides with $\sharp([D_1] \bullet \cdots \bullet [D_n])$, if M is compact.

In the case each D_j is defined by a holomorphic function f_j near p, the local intersection number $(D_1 \bullet \cdots \bullet D_n)_p$ is simultaneously equal to

(1) the mapping degree of $f = (f_1, \ldots, f_n)$, at p,
(2) the Grothendieck residue

$$\operatorname{Res}_p \begin{bmatrix} df_1 \wedge \cdots \wedge df_n \\ f_1, \ldots, f_n \end{bmatrix}, \tag{14.9}$$

(3) $\dim_{\mathbb{C}} \mathscr{O}/(f_1, \ldots, f_n)$.

(cf. Example 12.2, Theorems 12.3–12.5).

Example 14.1. Let $P_j(\zeta_0, \ldots, \zeta_n)$ be homogeneous polynomials of degree d_j, $j = 1, \ldots, n$. For each j, P_j defines an effective divisor D_j in the projective space $\mathbb{P}^n$. If $[\zeta_0, \cdots, \zeta_n]$ denote homogeneous coordinates on $\mathbb{P}^n$, the divisor D_j is defined by the holomorphic function $P_j(\zeta_0, \ldots, \zeta_n)/\zeta_\alpha^{d_j}$ on

$U_\alpha = \{\zeta_\alpha \neq 0\}$. Note that $L_{D_j} = H_n^{\otimes d_j}$, where H_n denotes the hyperplane bundle on $\mathbb{P}^n$.

In this situation, if $S = \bigcap_{j=1}^n |D_j|$ consists of isolated points, we have (Bézout's theorem):

$$\sum_{p \in S} (D_1 \bullet \cdots \bullet D_n)_p = d_1 \cdots d_n.$$

This follows from Theorem 14.2 and (14.8).

Intersection product with a subvariety

Let V be a variety of dimension d in M with the embedding $i : V \hookrightarrow M$. Also let K_0 be a triangulation of M as in Section 13.2. We may define the intersection product with V as in the case of submanifolds (cf. Section 4.4).

We consider the intersection of V and a homology class in M or in a K_0-subcomplex S of M.

First localization: Letting S_1 and S_2 be V and M, respectively, in (4.34), we have the intersection product localized at V:

$$\breve{H}_{2d}(V) \times \breve{H}_{2n-p}(M) \xrightarrow{(\ \bullet\ )_V} \breve{H}_{2d-p}(V).$$

For a class a in $\breve{H}_{2n-p}(M)$, we have the product $(V \bullet a)_V$. It is sent to $[V] \bullet a$ by $i_* : \breve{H}_{2d-p}(V) \to \breve{H}_{2d-p}(M)$.

We have the Alexander duality

$$A_{M,V} : H^{2k}(M, M \smallsetminus V) \xrightarrow{\sim} \breve{H}_{2d}(V),$$

under which the Thom class Ψ_V of V corresponds to the class V (cf. Definition 13.1 and (13.9)). Letting $S_1 = V$ and $S_2 = \emptyset$ in Theorem 4.8, we may write

$$(V \bullet a)_V = A_{M,V}(\Psi_V \smile P_M^{-1} a).$$

Second localization: Let S be a K_0-subcomplex of M and set $Z = V \cap S$. Letting S_1 and S_2 be V and S, respectively, in (4.34), we have the intersection product localized at Z:

$$\breve{H}_{2d}(V) \times \breve{H}_{2n-p}(S) \xrightarrow{(\ \bullet\ )_Z} \breve{H}_{2d-p}(Z). \tag{14.10}$$

For a class a in $\breve{H}_{2n-p}(S)$, we have the product $(V \bullet a)_Z$ in $\breve{H}_{2d-p}(Z)$.

Letting $S_1 = V$ and $S_2 = S$ in Theorem 4.8, we may write

$$(V \bullet a)_Z = A_{M,Z}(\Psi_V \smile A_{M,S}^{-1} a). \tag{14.11}$$

The following proposition follows from Propositions 13.2, 13.3, Theorems 4.7 and 4.8:

Proposition 14.5. *The following diagrams are commutative:*

$$\begin{array}{ccc} H^p(M) & \xrightarrow[P_M]{\sim} & \check{H}_{2n-p}(M) \\ {\scriptstyle i^*}\downarrow & & \downarrow{\scriptstyle (V\,\bullet\ )_V} \\ H^p(V) & \xrightarrow[P_V]{} & \check{H}_{2d-p}(V), \end{array} \qquad \begin{array}{ccc} H^p(M, M \smallsetminus S) & \xrightarrow[A_{M,S}]{\sim} & \check{H}_{2n-p}(S) \\ {\scriptstyle i^*}\downarrow & & \downarrow{\scriptstyle (V\,\bullet\ )_Z} \\ H^p(V, V \smallsetminus Z) & \xrightarrow[A_{V,Z}]{} & \check{H}_{2d-p}(Z). \end{array}$$

Thus, for a class α in $H^p(M)$, we have

$$(V \bullet P_M(\alpha))_V = V \frown i^*(\alpha). \tag{14.12}$$

Also, if Z is compact, for a class α in $H^p(M, M \smallsetminus S)$, we have (cf. Remark 13.1)

$$(V \bullet A_{M,S}(\alpha))_Z = \underline{V} \frown i^*(\alpha). \tag{14.13}$$

Remark 14.3. 1. The two diagrams in Proposition 14.5 are compatible through the diagrams (4.18) and (13.7).

2. Likewise, for a holomorphic map $f : V \to M$, we may define the intersection product with f as in Definition 4.4 and, if $Z = f^{-1}(S)$ has a finite number of connected components, we have the product componentwise and a formula as (4.35).

Intersection product of varieties

Let V and W be varieties of dimensions d and h, respectively, in M. We set $Z = V \cap W$, which is a variety in M. Letting $S = W$ in (14.10), we have the intersection product

$$\check{H}_{2d}(V) \times \check{H}_{2h}(W) \xrightarrow{(\ \bullet\ )_Z} \check{H}_{2(d+h-n)}(Z).$$

We have the intersection product $(V \bullet W)_Z$ in $\check{H}_{2(d+h-n)}(Z)$. It is sent to $[V] \bullet [W]$ by the canonical morphism $\check{H}_{2(d+h-n)}(Z) \to \check{H}_{2(d+h-n)}(M)$.

Now we assume that Z is of pure $(d+h-n)$-dimensional and let (Z_i) be irreducible components, which is locally finite. In this case, we have an explicit expression of the above intersection. Since $\check{H}_{2(d+h-n)}(Z)$ is a free Abelian group generated by the classes $[Z_i]$ in this case, we may write

$$(V \bullet W)_Z = \sum_i d_i \, [Z_i],$$

for some integers d_i. For each i, we take a point p_i in the non-singular part of $Z_i^\circ = Z_i \smallsetminus \bigcup_{j\neq i} Z_j$ and let $\mathbb{D}_i$ be a complex slice of Z_i at p_i (cf. Definition 3.24). Thus it is a $(2n-d-h)$-dimensional complex submanifold of M transverse to Z_i° at p_i. Let U be a neighborhood of Z in M that contains $\mathbb{D}_i$ as a closed submanifold and take a triangulation K_0 of U as in Theorem 12.1, compatible with $V \cap U$, $W \cap U$ and $Z \cup \mathbb{D}_i$, and let K be its barycentric subdivision. Let K' and K^* be as before. We may assume that p_i is the barycenter $b_{\boldsymbol{s}}$ of a $2(d+h-n)$-simplex $\boldsymbol{s}$ in the non-singular part of Z_i° and that the $2(2n-d-h)$-cell $\boldsymbol{s}^*$ dual to $\boldsymbol{s}$ is in $\mathbb{D}_i$. Thus $\mathbb{D}_i$ is also transverse to both V and W so that $\mathbb{D}_i \cap V$ and $\mathbb{D}_i \cap W$ are complex submanifolds of dimensions $(n-h)$ and $(n-d)$, respectively, of M. Note that $(\mathbb{D}_i \cap V) \cap (\mathbb{D}_i \cap W) = \mathbb{D}_i \cap V \cap W = \{p_i\}$. The following shows that d_i is the "transverse intersection number" of V and W at p_i:

Proposition 14.6. *We have*

$$d_i = ((\mathbb{D}_i \cap V) \cdot (\mathbb{D}_i \cap W))_{p_i}.$$

Proof. We claim that

$$((\mathbb{D}_i \cap V) \cdot (\mathbb{D}_i \cap W))_{p_i} = (\mathbb{D}_i \cdot (V \cdot W)_Z)_{p_i}. \tag{14.14}$$

Since the right-hand side above is d_i (cf. Exercise 4.2), the proposition then follows.

To show (14.14), let $\iota : \mathbb{D}_i \hookrightarrow M$ denote the inclusion. Then we have the following diagram (cf. Proposition 4.13):

$$\begin{array}{ccc} H^{2(n-d)}(M, M \smallsetminus V) & \xrightarrow[A_{M,V}]{\sim} & \breve{H}_{2d}(V) \\ \downarrow{\scriptstyle \iota^*} & & \downarrow{\scriptstyle (\mathbb{D}_i \,\cdot\;)_{\mathbb{D}_i \cap V}} \\ H^{2(n-d)}(\mathbb{D}_i, \mathbb{D}_i \smallsetminus (\mathbb{D}_i \cap V)) & \xrightarrow[A_{\mathbb{D}_i, \mathbb{D}_i \cap V}]{\sim} & H_{2(n-h)}(\mathbb{D}_i \cap V). \end{array} \tag{14.15}$$

Thus $(\mathbb{D}_i \cdot V)_{\mathbb{D}_i \cap V} = A\iota^* A^{-1} V$ and likewise $(\mathbb{D}_i \cdot W)_{\mathbb{D}_i \cap W} = A\iota^* A^{-1} W$, with A denoting appropriate Alexander isomorphisms. Considering the diagram in Theorem 4.8 for the case $(M, S_1) = (\mathbb{D}_i, \mathbb{D}_i \cap V)$, $(M, S_2) = (\mathbb{D}_i, \mathbb{D}_i \cap W)$ and using the functoriality of the cup product, we have

$$((\mathbb{D}_i \cap V) \cdot (\mathbb{D}_i \cap W))_{p_i} = A_{\mathbb{D}_i, p_i} \iota^* (A^{-1} V \smile A^{-1} W),$$

where

$$A_{\mathbb{D}_i, p_i} : H^r(\mathbb{D}_i, \mathbb{D}_i \smallsetminus \{p_i\}) \xrightarrow{\sim} H_0(\{p_i\}) \simeq \mathbb{Z}$$

denotes the Alexander isomorphism with $r = 2(2n-d-h)$. Considering again the diagram in Theorem 4.8 for the case $(M, S_1) = (M, V)$,

$(M, S_2) = (M, W)$, we see that the right-hand side above is equal to $A_{\mathbb{D}_i, p_i}\iota^* A^{-1}(V \bullet W)_Z$, which in turn equals the right-hand side of (14.14) by the commutativity of the diagram obtained by replacing V with Z in (14.15). □

Subspaces defined by a family of sections

Let E be a holomorphic vector bundle of rank l on M. Also, let $s^{(r)}$ be a holomorphic r-section of E and $\boldsymbol{S}$ the complex space defined by $s^{(r)}$, as in the beginning of this section.

Let V be a variety of dimension d in M and set $E_V = E|_V$ and $s_V^{(r)} = s^{(r)}|_V$. Let $\boldsymbol{Z}$ be the complex space in V defined by $s_V^{(r)}$. It is the intersection of V and $\boldsymbol{S}$ as a complex subspace of M and the support Z of $\boldsymbol{Z}$ is given by $Z = V \cap S$. In this situation, we have the (topological) localization $c^q(E_V, s_V^{(r)})$ in $H^{2q}(V, V \smallsetminus Z)$. Note that it is the image of $c^q(E, s^{(r)})$ by the pull-back

$$i^* : H^{2q}(M, M \smallsetminus S) \longrightarrow H^{2q}(V, V \smallsetminus Z).$$

We also have the residue $\mathrm{TRes}_{c^q}(s_V^{(r)}, E_V; Z)$ as its images by the Alexander morphism

$$H^{2q}(V, V \smallsetminus Z) \longrightarrow \breve{H}_{2(d-q)}(Z).$$

Recall that, if $\dim Z = d - q$, then $\mathrm{TRes}_{c^q}(s_V^{(r)}, E_V; Z)$ may be identified with the differential geometric residue $\mathrm{Res}_{c^q}(s_V^{(r)}, E_V; Z)$ (cf. (13.36)). Moreover, in this case, we may define the homology class of $\boldsymbol{Z}$; $[\boldsymbol{Z}]$ in $\breve{H}_{2(d-q)}(V)$ or $\boldsymbol{Z}$ in $\breve{H}_{2(d-q)}(Z)$ (cf. Definition 13.6). The class $\boldsymbol{Z}$ coincides with $\mathrm{TRes}_{c^q}(s_V^{(r)}, E_V; Z) = \mathrm{Res}_{c^q}(s_V^{(r)}, E_V; Z)$ (cf. Theorem 13.18).

By Proposition 14.5,

$$(V \bullet \mathrm{TRes}_{c^q}(s^{(r)}, E; S))_Z = \mathrm{TRes}_{c^q}(s_V^{(r)}, E_V; Z) \quad \text{in } \breve{H}_{2(d-q)}(Z). \quad (14.16)$$

Proposition 14.7. *In the above situation, we have:*

1. *If* $\dim S = n - q$,
$$(V \bullet \boldsymbol{S})_Z = \mathrm{TRes}_{c^q}(s_V^{(r)}, E_V; Z) \qquad \text{in } \breve{H}_{2(d-q)}(Z).$$

2. *Moreover, if* $\dim Z = d - q$,
$$(V \bullet \boldsymbol{S})_Z = \boldsymbol{Z} \qquad \text{in } \breve{H}_{2(d-q)}(Z).$$

Proof. The first statement follows from (14.16) and Theorem 12.8 and the second from Theorem 13.18. □

Remark 14.4. 1. If $\dim S = n - q$ and if Z is compact, we may write (cf. (14.13))

$$(V \cdot \boldsymbol{S})_Z = \underline{V} \frown c^q(E_V, s_V) \qquad \text{in } H_{2(d-q)}(Z).$$

2. See Section 14.5 below for the case $r = 1$ (thus $q = l$) and $\dim S > n - l$ or $\dim Z > d - l$.

Passing from the homology of Z to that of V, we have from (14.16) (cf. Remark 14.3. 1),

$$(V \cdot [\mathrm{TRes}_{c^q}(s^{(r)}, E; S)])_V = [\mathrm{TRes}_{c^q}(s_V^{(r)}, E_V; Z)]_V \qquad \text{in } \check{H}_{2(d-q)}(V),$$

where $[\ ]_V$ denotes the class in the homology of V. Since

$$[\mathrm{TRes}_{c^q}(s_V^{(r)}, E_V; Z)]_V = P_V(c^q(E_V)) = V \frown c^q(E_V),$$

we have:

Proposition 14.8. *In the above situation, we have:*

1. *If* $\dim S = n - q$,

$$(V \cdot [\boldsymbol{S}])_V = V \frown c^q(E_V) \qquad \text{in } \check{H}_{2(d-q)}(V).$$

2. *Moreover, if* $\dim Z = d - q$,

$$(V \cdot [\boldsymbol{S}])_V = [\boldsymbol{Z}]_V \qquad \text{in } \check{H}_{2(d-q)}(V).$$

Adjunction formula for singular curves

The classical "adjunction formula" says that, if C is a compact non-singular complex curve in a complex surface S (complex manifold of dimension two), then we have

$$\chi(C) = -C \cdot (K_S + C). \tag{14.17}$$

In the above, $\chi(C)$ denotes the Euler-Poincaré characteristic of C, which is equal to $2 - 2g$ with g the genus of C. Also, K_S is the canonical divisor of S and the $\cdot$ means intersection product. In the classical case, it is usually assumed that S is compact and the intersection product is in S. In this case, the intersection product is in $H_0(S)$ and the right-hand side of (14.17) is understood to be the image by the augmentation $\varepsilon_* : H_0(S) \to \mathbb{Z}$, which is an isomorphism, if S is connected. This formula follows from (6.11) using various relations among characteristic classes.

We prove a generalization of (14.17) as an application of Theorem 13.15 and the previous considerations. Here S is not assumed to be compact and the intersection product is in C.

Theorem 14.3. *Let C be a compact singular curve with singularities $p_1, \dots, p_r$ in a complex surface S. Then we have*

$$\chi(C) = -\sharp(C \cdot (K_S + C)) + \sum_{i=1}^{r} \mu(C, p_i),$$

where $\mu(C, p_i)$ is the Milnor number of C at p_i.

Proof. First note that $C \cap |K_M + C| = C$ and, since C is compact, $\sharp(C \cdot (K_S + C))$ makes sense (cf. (14.5)). Noting that C is an SLCI defined by the canonical section s_C of L_C, it suffices to show that (cf. Theorem 13.15)

$$\int_C c^1(\tau_C) = -\sharp(C \cdot (K_S + C)), \tag{14.18}$$

where $\tau_C = i^*(TS - L_C)$ with $i : C \hookrightarrow S$ the inclusion.

We have $(C \cdot (K_S + C))_C = (C \cdot ([K_S] + [C]))_C$ in $H_0(C)$. The classes $[K_S]$ and $[C]$ are the images of $c^1(\bigwedge^2 T^*S) = c^1(T^*S) = -c^1(TS)$ and $c^1(L_C)$, respectively, by the Poincaré isomorphism $P : H^2(S) \xrightarrow{\sim} \breve{H}_2(S)$. Thus by (14.12),

$$-(C \cdot ([K_S] + [C])_C = C \frown i^* c^1(TS - L_C) = C \frown c^1(\tau_V).$$

Taking the images of the both sides by the augmentation $\varepsilon_* : H_0(C) \to \mathbb{Z}$, we have (14.18). □

In the case C is connected, the number $\frac{1}{2}\sharp(C \cdot (K_S + C)) + 1$ is referred to as the *virtual genus* of C.

14.3 Intersection product in singular varieties

In this section, we let M denote a complex manifold of dimension n.

Intersection product in a variety

Let V be a variety of dimension d in M and take a triangulation of M as in Section 13.2. We wish to define the intersection product in V, denoted by $\bullet_V$, so that the following diagram is commutative:

$$\begin{array}{ccc} H^p(V) \times H^q(V) & \xrightarrow{\smile} & H^{p+q}(V) \\ {\scriptstyle P\times P}\downarrow & & \downarrow{\scriptstyle P} \\ \breve{H}_r(V) \times \breve{H}_s(V) & \xrightarrow{\bullet_V} & \breve{H}_{r+s-2d}(V), \end{array}$$

where P denotes the Poincaré morphism, $p + r = 2d$ and $q + s = 2d$. However, since P is not an isomorphism in general, we may define the product only for classes that come naturally from cohomology.

The situation is similar for localized intersection product. Let Z_1 and Z_2 be subcomplexes of V and set $Z = Z_1 \cap Z_2$. We wish to define the intersection product in V localized at Z, denoted by $(\ \bullet_V\ )_Z$, so that the following diagram is commutative:

$$\begin{array}{ccc} H^p(V, V \smallsetminus Z_1) \times H^q(V, V \smallsetminus Z_2) & \xrightarrow{\smile} & H^{p+q}(V, V \smallsetminus Z) \\ {\scriptstyle A_1\times A_2}\downarrow & & \downarrow{\scriptstyle A} \\ \breve{H}_r(Z_1) \times \breve{H}_s(Z_2) & \xrightarrow{(\ \bullet_V\ )_Z} & \breve{H}_{r+s-2d}(Z), \end{array}$$

where A, A_1 and A_2 denote the Alexander morphisms for Z, Z_1 and Z_2, $p + r = 2d$ and $q + s = 2d$.

Intersection product of subspaces defined by a section

Let V be a variety of dimension d in M. For each $j = 1, \ldots, \nu$, let $E_{V,j}$ be a holomorphic vector bundle of rank l_j on V. Let $s_{V,j}$ be a holomorphic section of $E_{V,j}$ and $\boldsymbol{Z}_j$ the complex space in V defined by $s_{V,j}$. We assume that the support Z_j of $\boldsymbol{Z}_j$ contains $\mathrm{Sing}(V)$. Then we have the localization $c^{l_j}(E_{V,j}, s_{V,j})$ in $H^{2l_j}(V, V \smallsetminus Z_j)$ by $s_{V,j}$, as in Section 13.4. We also have the residue $\mathrm{TRes}_{c^{l_j}}(s_{V,j}, E_{V,j}; Z_j)$ as the image of $c^{l_j}(E_{V,j}, s_{V,j})$ by the Alexander morphism

$$A_j : H^{2l_j}(V, V \smallsetminus Z_j) \longrightarrow \breve{H}_{2(d-l_j)}(Z_j).$$

If $\dim Z_j = d - l_j$, then $\mathrm{TRes}_{c^{l_j}}(s_{V,j}, E_{V,j}; Z_j)$ may be identified with the differential geometric residue $\mathrm{Res}_{c^{l_j}}(s_{V,j}, E_{V,j}; Z_j)$ (cf. (13.36)). Moreover, the class of $\boldsymbol{Z}_j$ in $\breve{H}_{2(d-l_j)}(Z_j)$ is equal to $\mathrm{TRes}_{c^{l_j}}(s_{V,j}, E_{V,j}; Z_j)$ (cf. Theorem 13.18).

Let $E_V = E_{V,1} \oplus \cdots \oplus E_{V,\nu}$ be the direct sum and s_V the section of E_V given by $s_V = s_{V,1} \oplus \cdots \oplus s_{V,\nu}$. Let $\boldsymbol{Z}$ be the complex space in V defined by s_V. As a complex subspace of V, it is the intersection of the $\boldsymbol{Z}_j$'s. The support Z of $\boldsymbol{Z}$ is the zero set of s_V and is given by $Z = \bigcap_{j=1}^{\nu} Z_j$. We have the localization $c^l(E_V, s_V)$ in $H^{2l}(V, V \smallsetminus Z)$ by s_V, $l = \sum_{j=1}^{\nu} l_j$. We also have the residue $\mathrm{TRes}_{c^l}(s_V, E_V; Z)$ as the image of $c^l(E_V, s_V)$ by the Alexander morphism

$$A : H^{2l}(V, V \smallsetminus Z) \longrightarrow \breve{H}_{2(d-l)}(Z).$$

If $\dim Z = d-l$, $\mathrm{TRes}_{c^l}(s_V, E_V; Z)$ may be identified with $\mathrm{Res}_{c^l}(s_V, E_V; Z)$. Moreover, the class of $\boldsymbol{Z}$ in $\breve{H}_{2(d-l)}(Z)$ is equal to $\mathrm{TRes}_{c^l}(s_V, E_V; Z)$.

In view of (14.3), we give the following:

Definition 14.1. The intersection product of the $\mathrm{TRes}_{c^{l_j}}(s_{V,j}, E_{V,j}; Z_j)$'s in V localized at Z is defined by

$$\begin{aligned}(\mathrm{TRes}_{c^{l_1}}(s_{V,1}, E_{V,1}; Z_1) \bullet_V \cdots \bullet_V \mathrm{TRes}_{c^{l_\nu}}(s_{V,\nu}, E_{V,\nu}; Z_\nu))_Z \\ = \mathrm{TRes}_{c^l}(s_V, E_V; Z) \quad \text{in } \breve{H}_{2(d-l)}(Z). \quad (14.19)\end{aligned}$$

We have the cup product

$$H^{2l_1}(V, V \smallsetminus Z_1) \times \cdots \times H^{2l_\nu}(V, V \smallsetminus Z_\nu) \longrightarrow H^{2l}(V, V \smallsetminus Z)$$

and we have (cf. Corollary 14.1)

$$c^{l_1}(E_{V,1}, s_{V,1}) \smile \cdots \smile c^{l_\nu}(E_{V,\nu}, s_{V,\nu}) = c^l(E_V, s_V).$$

Thus (14.19) is compatible with the Alexander morphisms A_j and A.

In the case $E_{V,j}$ and $s_{V,j}$ are the restriction of the ones E_j and s_j on M, by (14.16),

$$\begin{aligned}&\mathrm{TRes}_{c^{l_j}}(s_{V,j}, E_{V,j}; Z_j) = (V \bullet \mathrm{TRes}_{c^{l_j}}(s_j, E_j; S_j))_{Z_j} \quad \text{and} \\ &\mathrm{TRes}_{c^l}(s_V, E_V; Z) = (V \bullet \mathrm{TRes}_{c^l}(s, E; S))_Z\end{aligned}$$

so that (14.19) may be written

$$\begin{aligned}((V \bullet \mathrm{TRes}_{c^{l_1}}(s_1, E_1; S_1))_{Z_1} \bullet_V \cdots \bullet_V (V \bullet \mathrm{TRes}_{c^{l_\nu}}(s_\nu, E_\nu; S_\nu))_{Z_\nu})_Z \\ = (V \bullet \mathrm{TRes}_{c^l}(s, E; S))_Z \quad \text{in } \breve{H}_{2(d-l)}(Z).\end{aligned}$$

Here we recall the identity (14.3). See Propositions 14.15 and 14.16 below for related identities.

Coming back to the situation of the beginning of this paragraph, we have:

Proposition 14.9. *In the above situation, we have:*

1. *If* $\dim Z_j = d - l_j$*, for* $j = 1, \dots, \nu$*, then the intersection product of the* $\boldsymbol{Z}_j$*'s in* V *localized at* Z *is defined and is given by*

$$(\boldsymbol{Z}_1 \cdot_V \cdots \cdot_V \boldsymbol{Z}_\nu)_Z = \mathrm{TRes}_{c^l}(s_V, E_V; Z) \qquad \textit{in } \breve{H}_{2(d-l)}(Z).$$

2. *Moreover, if* $\dim Z = d - l$*,*

$$(\boldsymbol{Z}_1 \cdot_V \cdots \cdot_V \boldsymbol{Z}_\nu)_Z = \boldsymbol{Z} \qquad \textit{in } \breve{H}_{2(d-l)}(Z).$$

Proof. This follows from Theorem 13.18. □

Remark 14.5. If $\dim Z_j = d - l_j$, for $j = 1, \dots, \nu$, and if Z is compact, we may write

$$(\boldsymbol{Z}_1 \cdot_V \cdots \cdot_V \boldsymbol{Z}_\nu)_Z = \underline{V} \frown c^l(E_V, s_V) \qquad \text{in } H_{2(d-l)}(Z).$$

Note that, if $l = d$ and if Z is compact and connected, then $\mathrm{TRes}_{c^d}(s_V, E_V; Z)$ is in $H_0(Z) \simeq \mathbb{Z}$ and may be regarded as an integer.

Suppose that $\dim Z_j = d - l_j$, for $j = 1, \dots, \nu$, and that Z has only a finite number of connected components $(Z_\lambda)_\lambda$. Then the λ component of $(\boldsymbol{Z}_1 \cdot_V \cdots \cdot_V \boldsymbol{Z}_\nu)_Z$ is given by $(\boldsymbol{Z}_1 \cdot_V \cdots \cdot_V \boldsymbol{Z}_\nu)_{Z,\lambda} = \mathrm{TRes}_{c^l}(s_V, E_V; Z_\lambda)$ in $\breve{H}_{2(d-l)}(Z_\lambda)$ and we have

$$\sum_\lambda (i_\lambda)_*(\boldsymbol{Z}_1 \cdot_V \cdots \cdot_V \boldsymbol{Z}_\nu)_{Z,\lambda} = (\boldsymbol{Z}_1 \cdot_V \cdots \cdot_V \boldsymbol{Z}_\nu)_Z \qquad \text{in } \breve{H}_{2(d-l)}(Z), \tag{14.20}$$

where $i_\lambda : Z_\lambda \hookrightarrow Z$ denotes the inclusion.

Passing from the homologies of the Z_j's and Z to that of V, we define the intersection product of the $[\mathrm{TRes}_{c^{l_j}}(s_{V,j}, E_{V,j}; Z_j)]_V$'s in V by

$$[\mathrm{TRes}_{c^{l_1}}(s_{V,1}, E_{V,1}; Z_1)]_V \cdot_V \cdots \cdot_V [\mathrm{TRes}_{c^{l_\nu}}(s_{V,\nu}, E_{V,\nu}; Z_\nu)]_V = [\mathrm{TRes}_{c^l}(s_V, E_V; Z)]_V \qquad \text{in } \breve{H}_{2(d-l)}(V).$$

Since

$$[\mathrm{TRes}_{c^l}(s_V, E_V; Z)]_V = P_V(c^l(E_V)) = V \frown c^l(E_V),$$

we have:

Proposition 14.10. *In the above situation, we have:*

1. *If* $\dim Z_j = d - l_j$*, for* $j = 1, \dots, \nu$*, then the intersection product of the* $[\boldsymbol{Z}_j]_V$*'s in* V *is defined and is given by*

$$[\boldsymbol{Z}_1]_V \cdot_V \cdots \cdot_V [\boldsymbol{Z}_\nu]_V = V \frown c^l(E_V) \qquad \textit{in } \breve{H}_{2(d-l)}(V).$$

2. *Moreover, if* $\dim Z = d - l$*,*

$$[\boldsymbol{Z}_1]_V \cdot_V \cdots \cdot_V [\boldsymbol{Z}_\nu]_V = [\boldsymbol{Z}]_V \qquad \textit{in } \breve{H}_{2(d-l)}(V).$$

Intersection product with a subvariety

Let V' be a variety of dimension d' in V with the embedding $\iota : V' \hookrightarrow V$. Also let Z be a subcomplex of V and set $Z' = V' \cap Z$. We wish to define the intersection product in V with V', denoted by $(V' \bullet_V\ )_{V'}$, and the intersection product in V with V' localized at Z', denoted by $(V' \bullet_V\ )_{Z'}$, so that the following diagrams are commutative:

$$\begin{array}{ccc} H^p(V) & \xrightarrow[P_V]{} & \breve{H}_{2d-p}(V) \\ {\scriptstyle \iota^*}\big\downarrow & & \big\downarrow{\scriptstyle (V' \bullet_V\ )_{V'}} \\ H^p(V') & \xrightarrow[P_{V'}]{} & \breve{H}_{2d'-p}(V'), \end{array} \qquad \begin{array}{ccc} H^p(V, V \smallsetminus Z) & \xrightarrow[A_{V,Z}]{} & \breve{H}_{2d-p}(Z) \\ {\scriptstyle \iota^*}\big\downarrow & & \big\downarrow{\scriptstyle (V' \bullet_V\ )_{Z'}} \\ H^p(V', V' \smallsetminus Z') & \xrightarrow[A_{V',Z'}]{} & \breve{H}_{2d'-p}(Z'). \end{array}$$

Subspaces defined by a family of sections

Let V be a variety of dimension d in M. Also, let E_V be a holomorphic vector bundle of rank l on V, $s_V^{(r)}$ a holomorphic r-section of E_V and $\boldsymbol{Z}$ the complex space in V defined by $s_V^{(r)}$. We assume that the support Z of $\boldsymbol{Z}$ contains $\mathrm{Sing}(V)$. Then we have the localization $c^q(E_V, s_V^{(r)})$ in $H^{2q}(V, V \smallsetminus Z)$, $q = l - r + 1$, and the associated residue $\mathrm{TRes}_{c^q}(s_V^{(r)}, E_V; Z)$ in $\breve{H}_{2(d-q)}(Z)$ as its image by the Alexander morphism

$$H^{2q}(V, V \smallsetminus Z) \longrightarrow \breve{H}_{2(d-q)}(Z).$$

Let V' be a subvariety of V of dimension d' and set $E_{V'} = E_V|_{V'}$ and $s_{V'}^{(r)} = s_V^{(r)}|_{V'}$. Let $\boldsymbol{Z}'$ be the complex space in V' defined by $s_{V'}^{(r)}$. As a complex subspace of V, it is the intersection of V' and $\boldsymbol{Z}$ and the support Z' of $\boldsymbol{Z}'$ is given by $Z' = V' \cap Z$. We assume that Z' contains $\mathrm{Sing}(V')$. Then we have the topological localization $c^q(E_{V'}, s_{V'}^{(r)})$ in $H^{2q}(V', V' \smallsetminus Z')$ and the topological residue $\mathrm{TRes}_{c^q}(s_{V'}^{(r)}, E_{V'}; Z')$ as its image by the Alexander morphism

$$H^{2q}(V', V' \smallsetminus Z') \longrightarrow \breve{H}_{2(d'-q)}(Z').$$

In view of (14.16), we give the following:

Definition 14.2. The intersection pruduct of V' and $\mathrm{TRes}_{c^q}(s_V^{(r)}, E_V; Z)$ in V localized at Z' is defined by

$$(V' \bullet_V \mathrm{TRes}_{c^q}(s_V^{(r)}, E_V; Z))_{Z'} = \mathrm{TRes}_{c^q}(s_{V'}^{(r)}, E_{V'}; Z') \qquad \text{in } \breve{H}_{2(d'-q)}(Z'). \tag{14.21}$$

Proposition 14.11. *Definitions 14.1 and 14.2 are compatible in the case the both make sense.*

Proof. Suppose there exist a holomorphic vector bundle F of rank $k' = d - d'$ on V and a holomorphic section t of F such that V' is defined by t. We assume V' contains $\mathrm{Sing}(V)$. Then we have the localization $c^{k'}(F,t)$ in $H^{2k'}(V, V \smallsetminus V')$ and $V' = \mathrm{TRes}_{c^{k'}}(t, F; V')$ (cf. Theorem 13.18). Also let E_V be a holomorphic vector bundle of rank l on V and s_V a holomorphic section of E_V. Let $\boldsymbol{Z}$ be the complex space in V defined by s_V. We assume that its support Z containes the singular set $\mathrm{Sing}(V)$ of V. We set $E_{V'} = E_V|_{V'}$ and $s_{V'} = s_V|_{V'}$. Let $\boldsymbol{Z}'$ be the complex space in V' defined by $s_{V'}$. The support Z' of $\boldsymbol{Z}'$ is given by $Z' = Z \cap V'$. We assume that Z' contains $\mathrm{Sing}(V')$.

In this situation, we may define $(V' \bullet_V \mathrm{TRes}_{c^l}(s_V, E_V; Z))_{Z'}$ in two ways, i.e., by Definition 14.1 and by Definition 14.2, and we are to show that they are the same.

On the one hand, by Definition 14.2,

$$(V' \bullet_V \mathrm{TRes}_{c^l}(s_V, E_V; Z))_{Z'} = \mathrm{TRes}_{c^l}(s_{V'}, E_{V'}; Z') = A_{V',Z'}(c^l(E_{V'}, s_{V'})).$$

On the other hand, by Definition 14.1, the left-hand side is equal to

$$(\mathrm{TRes}_{c^{k'}}(t, F; V') \bullet_V \mathrm{TRes}_{c^l}(s_V, E_V; Z))_{Z'} = \mathrm{TRes}_{c^{k'+l}}(t \oplus s_V, F \oplus E_V; Z'),$$

which is equal to, by Corollary 14.1,

$$A_{V,Z'}(c^{k'+l}(F \oplus E_V, t \oplus s_V) = A_{V,Z'}((c^{k'}(F,t) \smile c^l(E_V, s_V)).$$

Thus the proposition follows from the commutativity of the diagram

$$\begin{array}{ccc} H^p(V, V \smallsetminus Z) & \xrightarrow{c^{k'}(F,t)\smile(\ )} & H^{2k'+p}(V, V \smallsetminus Z') \\ \downarrow{\scriptstyle \iota^*} & & \downarrow{\scriptstyle A_{V,Z'}} \\ H^p(V', V' \smallsetminus Z') & \xrightarrow{A_{V',Z'}} & \breve{H}_{2d'-p}(Z'), \end{array}$$

where $\iota : V' \hookrightarrow V$ is the embedding. For this, let γ be a cocycle in $C^{2k'}_{K'_V}(V, V \smallsetminus V')$ representing $c^{k'}(F,t)$. Then, for a $2d'$-simplex $\boldsymbol{s}$ of K, $\boldsymbol{s}^* \cap V$ is a $2k'$-chain of K'_V and we have (cf. (13.37)):

$$\langle \boldsymbol{s}^* \cap V, \gamma \rangle = \begin{cases} 1 & \text{if } \boldsymbol{s} \subset V', \\ 0 & \text{otherwise.} \end{cases}$$

The commutativity is then proved as Proposition 4.10, see also Proposition 13.2. □

If E_V and $s_V^{(r)}$ are the restrictions of E and $s^{(r)}$ on M, we have, by (14.16),

$$\mathrm{TRes}_{c^q}(s_V^{(r)}, E_V; Z) = (V \bullet \mathrm{TRes}_{c^q}(s^{(r)}, E; S))_Z \qquad \text{and}$$

$$\mathrm{TRes}_{c^q}(s_{V'}^{(r)}, E_{V'}; Z') = (V' \bullet \mathrm{TRes}_{c^q}(s^{(r)}, E; S))_{Z'}$$

so that (14.21) is written (contraction of V)

$$(V' \bullet_V (V \bullet \mathrm{TRes}_{c^q}(s^{(r)}, E; S))_Z)_{Z'} = (V' \bullet \mathrm{TRes}_{c^q}(s^{(r)}, E; S))_{Z'}.$$

Proposition 14.12. *Suppose that E_V and $s_V^{(r)}$ are the restrictions of E and $s^{(r)}$ on M. Then, if $\dim S = n - q$,*

$$(V' \bullet_V (V \bullet \boldsymbol{S})_Z)_{Z'} = (V' \bullet \boldsymbol{S})_{Z'} \qquad \text{in } \check{H}_{2(d'-q)}(Z').$$

Coming back to the situation of the beginning of this paragraph, we have:

Proposition 14.13. *In the above situation, we have:*

1. *If $\dim Z = d-q$, then the intersection product $(V' \bullet_V \boldsymbol{Z})_{Z'}$ in V localized at Z' is defined and is given by*

$$(V' \bullet_V \boldsymbol{Z})_{Z'} = \mathrm{TRes}_{c^q}(s_{V'}^{(r)}, E_{V'}; Z') \qquad \text{in } \check{H}_{2(d'-q)}(Z').$$

2. *Moreover, if $\dim Z' = d' - q$,*

$$(V' \bullet_V \boldsymbol{Z})_{Z'} = \boldsymbol{Z}' \qquad \text{in } \check{H}_{2(d'-q)}(Z').$$

Proof. The statements follow from Theorem 13.18. □

Remark 14.6. If Z' is compact and if $\dim Z = d - q$, we may write

$$(V' \bullet_V \boldsymbol{Z})_{Z'} = \underline{V'} \frown c^q(E_{V'}, s_{V'}) \qquad \text{in } H_{2(d'-q)}(Z').$$

Passing from the homology of Z' to that of V', we define the intersection pruduct of V' and $[\mathrm{TRes}_{c^q}(s_V^{(r)}, E_V; Z)]_V$ in V localized at V' by

$$\begin{aligned}(V' \bullet_V &[\mathrm{TRes}_{c^q}(s_V^{(r)}, E_V; Z)]_V)_{V'} \\ &= [\mathrm{TRes}_{c^q}(s_{V'}^{(r)}, E_{V'}; Z')]_{V'} \qquad \text{in } \check{H}_{2(d'-q)}(V'). \qquad (14.22)\end{aligned}$$

Since

$$[\mathrm{TRes}_{c^q}(s_{V'}^{(r)}, E_{V'}; Z')]_{V'} = P_{V'}(c^q(E_{V'})) = V' \frown c^q(E_{V'}), \qquad (14.23)$$

we have:

Proposition 14.14. *In the above situation, we have:*

1. *If $\dim Z = d - q$, then the intersection product $(V' \bullet_V [\boldsymbol{Z}]_V)_{V'}$ in V localized at V' is defined and is given by*

$$(V' \bullet_V [\boldsymbol{Z}]_V)_{V'} = V' \frown c^q(E_{V'}) \qquad \text{in } \check{H}_{2(d'-q)}(V').$$

2. *Moreover, if $\dim Z' = d' - q$,*

$$(V' \bullet_V [\boldsymbol{Z}]_V)_{V'} = [\boldsymbol{Z}']_{V'} \qquad \text{in } \check{H}_{2(d'-q)}(V').$$

Distributive property

For each $j = 1, \ldots, \nu$, let E_j be a holomorphic vector bundle of rank l_j on M and s_j a holomorphic section of E_j, as in Section 14.2. Also, let $E = E_1 \oplus \cdots \oplus E_\nu$, $l = \sum_{j=1}^{\nu} l_j$ and $s = s_1 \oplus \cdots \oplus s_\nu$. We denote by $\boldsymbol{S}_j$ and $\boldsymbol{S}$ the complex spaces in M defined by s_j and s, as before.

Let V be a variety of dimension d in M and set $E_{V,j} = E_j|_V$, $s_{V,j} = s_j|_V$, $E_V = E|_V$ and $s_V = s|_V$. We denote by $\boldsymbol{Z}_j$ and $\boldsymbol{Z}$ the complex spaces in V defined by $s_{V,j}$ and s_V. The support Z_j of $\boldsymbol{Z}_j$ is given by $Z_j = V \cap S_j$ and that of $\boldsymbol{Z}$ by $Z = V \cap S$. We assume that each Z_j contains $\mathrm{Sing}(V)$.

Proposition 14.15. *In the above situation, we have:*

1. *If each s_j, $j = 1, \ldots, \nu$, and s are regular sections, we have*

$$(V \bullet (\boldsymbol{S}_1 \bullet \cdots \bullet \boldsymbol{S}_\nu)_S)_Z$$
$$= ((V \bullet \boldsymbol{S}_1)_{Z_1} \bullet_V \cdots \bullet_V (V \bullet \boldsymbol{S}_\nu)_{Z_\nu})_Z \quad \text{in } \check{H}_{2(d-l)}(Z),$$

where the both sides are well-defined under the condition.

2. *Moreover, if $\dim Z_j = d - l_j$, for each j, then $(V \bullet \boldsymbol{S}_j)_{Z_j} = \boldsymbol{Z}_j$ so that*

$$(V \bullet (\boldsymbol{S}_1 \bullet \cdots \bullet \boldsymbol{S}_\nu)_S)_Z = (\boldsymbol{Z}_1 \bullet_V \cdots \bullet_V \boldsymbol{Z}_\nu)_Z \qquad \text{in } \check{H}_{2(d-l)}(Z).$$

Proof. If s_j is a regular section, the product $(V \bullet \boldsymbol{S}_j)_{Z_j}$ is equal to $\mathrm{TRes}_{c^{l_j}}(s_{V,j}, E_{V,j}; Z_j)$ by Proposition 14.7.1. If, moreover, s is a regular section, then $(\boldsymbol{S}_1 \bullet \cdots \bullet \boldsymbol{S}_\nu)_S = \boldsymbol{S}$ by Proposition 14.2. Thus $(V \bullet (\boldsymbol{S}_1 \bullet \cdots \bullet \boldsymbol{S}_\nu)_S)_Z = (V \bullet \boldsymbol{S})_Z$, which is equal to $\mathrm{TRes}_{c^l}(s_V, E_V; Z)$ by Proposition 14.7.1. Hence, by Definition 14.1, we have the identity in the first statement. The second statement follows from Proposition 14.7.2 and the first. $\square$

As noted in the proof, $(\boldsymbol{S}_1 \bullet \cdots \bullet \boldsymbol{S}_\nu)_S = \boldsymbol{S}$. Also, if $\dim Z = d - l$, $(\boldsymbol{Z}_1 \bullet_V \cdots \bullet_V \boldsymbol{Z}_\nu)_Z = \boldsymbol{Z}$ (cf. Proposition 14.9.2).

Let furthermore V' be a subvariety of V of dimension d' and set $E_{V',j} = E_j|_{V'}$, $s_{V',j} = s_j|_{V'}$, $E_{V'} = E|_{V'}$ and $s_{V'} = s|_{V'}$. We denote by $\boldsymbol{Z}'_j$ and $\boldsymbol{Z}'$ the complex spaces in V' defined by $s_{V',j}$ and $s_{V'}$. The support Z'_j of $\boldsymbol{Z}'_j$ is given by $Z'_j = V' \cap Z_j$ and that of $\boldsymbol{Z}'$ by $Z' = V' \cap Z$. We assume that each Z'_j contains $\mathrm{Sing}(V')$.

Proposition 14.16. *In the above situation we have:*

1. *If $\dim Z_j = d - l_j$, for each j, and $\dim Z = d - l$,*

$$(V' \bullet_V (\boldsymbol{Z}_1 \bullet_V \cdots \bullet_V \boldsymbol{Z}_\nu)_Z)_{Z'}$$
$$= ((V' \bullet_V \boldsymbol{Z}_1)_{Z'_1} \bullet_{V'} \cdots \bullet_{V'} (V' \bullet_V \boldsymbol{Z}_\nu)_{Z'_\nu})_{Z'} \qquad \text{in } \check{H}_{2(d'-l)}(Z'),$$

where the both sides are well-defined under the condition.

2. *Moreover, if* $\dim Z'_j = d' - l_j$, *for each* j, *then* $(V' \bullet_V \boldsymbol{Z}_j)_{Z'_j} = \boldsymbol{Z}'_j$ *so that*

$$(V' \bullet_V (\boldsymbol{Z}_1 \bullet_V \cdots \bullet_V \boldsymbol{Z}_\nu)_Z)_{Z'} = (\boldsymbol{Z}'_1 \bullet_{V'} \cdots \bullet_{V'} \boldsymbol{Z}'_\nu)_{Z'} \qquad \textit{in } \breve{H}_{2(d'-l)}(Z').$$

Proof. If $\dim Z_j = d - l_j$, for each j, the product $(V' \bullet_V \boldsymbol{Z}_j)_{Z'_j}$ makes sense and is equal to $\mathrm{TRes}_{c^{l_j}}(s_{V',j}, E_{V',j}; Z'_j)$ by Proposition 14.13. 1. If, moreover, $\dim Z = d - l$, then $(\boldsymbol{Z}_1 \bullet_V \cdots \bullet_V \boldsymbol{Z}_\nu)_Z = \boldsymbol{Z}$ by Proposition 14.9. Thus the product $(V' \bullet_V (\boldsymbol{Z}_1 \bullet_V \cdots \bullet_V \boldsymbol{Z}_\nu)_Z)_{Z'} = (V' \bullet_V \boldsymbol{Z})_{Z'}$ makes sense and is equal to $\mathrm{TRes}_{c^l}(s_{V'}, E_{V'}; Z')$ by Proposition 14.13. 1. Hence, by Definition 14.1 for the case V is V', we have the identity in the first statement. The second statement follows from Proposition 14.13. 2 and the first. □

As noted in the proof, $(\boldsymbol{Z}_1 \bullet_V \cdots \bullet_V \boldsymbol{Z}_\nu)_Z = \boldsymbol{Z}$. Also, if $\dim Z' = d' - l$, $(\boldsymbol{Z}'_1 \bullet_{V'} \cdots \bullet_{V'} \boldsymbol{Z}'_\nu)_{Z'} = \boldsymbol{Z}'$ (cf. Proposition 14.9. 2).

Intersection product of divisors

Meromorphic functions and divisors on singular varieties are defined as in the case of complex manifolds with a little modification (cf. Section 11.6).

First, let V be a non-empty germ of a variety at 0 in $\mathbb{C}^n$ and $V = \bigcup_{i=1}^r V(\mathfrak{p}_i)$ the irreducible decomposition with $\mathfrak{p}_i$ prime ideals in $\mathscr{O} = \mathscr{O}_n$. Thus $V = V(I)$, $I = \bigcap_{i=1}^r \mathfrak{p}_i$ (cf. Section 2.3). The quotient $\mathscr{O}_V = \mathscr{O}/I$ is the ring of germs of holomorphic functions on V (cf. Section 11.5). The set $S = \bigcap_{i=1}^r (\mathscr{O} \smallsetminus \mathfrak{p}_i) = \mathscr{O} \smallsetminus \bigcup_{i=1}^r \mathfrak{p}_i$ is mutiplicative in $\mathscr{O}$ and its image S' in $\mathscr{O}_V$ is multiplicative in $\mathscr{O}_V$ (cf. Section A.2). The fraction ring $\mathscr{M}_V = (S')^{-1}\mathscr{O}_V$ is the ring of germs of meromorphic functions on V. It is canonically isomorphic with $S^{-1}\mathscr{O}/I \cdot S^{-1}\mathscr{O}$ (cf. Proposition A.21). Thus every meromorphic function germ on V is represented by a germ of the form f/g with $f \in \mathscr{O}$ and $g \in S$. Note that S' does not contain any zero divisor in $\mathscr{O}_V$ so that $\mathscr{M}_V$ naturally contains $\mathscr{O}_V$.

Next, let V be a variety of dimension d in a complex manifold M. If we let $\mathscr{I}_V$ be the ideal sheaf of V, $\mathscr{O}_V = i^{-1}(\mathscr{O}/\mathscr{I}_V)$ is the sheaf of germs of holomorphic functions on V, where $\mathscr{O} = \mathscr{O}_M$ and $i : V \hookrightarrow M$ is the inclusion (cf. Section 11.5). Note that, for an open set W in V and an open set U in M with $W \subset U$, there is a canonical morphism $\Gamma(U;\mathscr{O}) \to \Gamma(W;\mathscr{O}_V)$. For f in $\Gamma(U;\mathscr{O})$, we denote by $[f]$ its image in $\Gamma(W;\mathscr{O}_V)$. We define the sheaf $\mathscr{M}_V$ of germs of meromorphic functions on V as in the case of complex manifolds, relpacing $\mathscr{O}$ with $\mathscr{O}_V$. For each point z in V, $\mathscr{M}_{V,z}$

is isomorphic with the ring of germs of meromorphic functions defined as above, i.e., if we define S_z in $\mathcal{O}_z$ and S'_z in $\mathcal{O}_{V,z} = \mathcal{O}_z/\mathcal{I}_{V,z}$ as above, then $\mathcal{M}_{V,z} \simeq (S'_z)^{-1}\mathcal{O}_{V,z} \simeq S_z^{-1}\mathcal{O}_z/(\mathcal{I}_{V,z} \cdot S_z^{-1}\mathcal{O}_z)$.

A *meromorphic function* ψ on an open set W in V is a section of $\mathcal{M}_V$ on W. Taking a suitable open set U in M with $W = U \cap V$, such a function is represented by a system $\{(U_\alpha, f^\alpha, g^\alpha)\}$, where $\{U_\alpha\}$ is a covering of U and f^α, g^α are holomorphic functions on U_α such that the germ g_z^α at z is in S_z for all z in $W_\alpha = U_\alpha \cap V$, the germs f_z^α and g_z^α are relatively prime for all z in U_α and that $[f^\alpha] \cdot [g^\beta] = [f^\beta] \cdot [g^\alpha]$ in $W_\alpha \cap W_\beta$. We write $\psi = [f^\alpha]/[g^\alpha]$ on W_α. A *pole* of ψ is a zero of $[g^\alpha]$ for some α.

Denoting by $\mathcal{O}_V^*$ and $\mathcal{M}_V^*$ the sets of invertible elements in $\mathcal{O}_V$ and $\mathcal{M}_V$, respectively, we define the sheaf $\mathcal{D}iv_V$ of Cartier divisors on V by a sequence as (11.17). A *Cartier divisor* on V is then defined to be a section of $\mathcal{D}iv_V$ on V. A Cartier divisor D is represented by a system $\{(W_\alpha, \psi^\alpha)\}$, where $\{W_\alpha\}$ is an open covering of V, ψ^α is in $\Gamma(W_\alpha; \mathcal{M}_V^*)$, for each α, and ψ^α/ψ^β is in $\Gamma(W_\alpha \cap W_\beta; \mathcal{O}_V^*)$, for each pair (α, β). Associated with D is a line bundle L_D on V together with a natural meromorphic section s_D. The bundle L_D is defined by the system of transition functions $\{\psi^\alpha/\psi^\beta\}$ and the section s_D is represented by the collection $\{\psi^\alpha\}$.

A *Weil divisor* on V is a locally finite formal sum $\sum m_i V'_i$ of varieties V'_i of codimension one in V with integral coefficients m_i. For a Weil divisor $D^{\mathrm{W}} = \sum m_i V'_i$, we define its support by $|D^{\mathrm{W}}| = \bigcup V'_i$.

Definition 14.3. For a non-zero Weil divisor $D^{\mathrm{W}} = \sum m_i V'_i$, we define its homology class by $[D^{\mathrm{W}}] = \sum m_i [V'_i]$ in $\breve{H}_{2d-2}(V;\mathbb{Z})$ or by $D^{\mathrm{W}} = \sum m_i [V'_i]$ in $\breve{H}_{2d-2}(|D^{\mathrm{W}}|, \mathbb{Z})$.

We would like to show that a Cartier divisor naturally defines a Weil divisor. Since the notion of the order of a meromorphic function along a subvariety of codimension one is not defined as easily as in the case of complex manifolds, we proceed as follows. Thus let D be a non-zero Cartier divisor. The set of zeros and poles of s_D is a variety of codimension one in V (cf. Theorem 2.11), which we denote by $|D|$ and call the support of D.

In the following we always asssume that $|D|$ contains $\mathrm{Sing}(V)$. Then we have the topological and differential geometric localizations of $c^1(L_D)$ by s_D and the associated residues $\mathrm{TRes}_{c^1}(s_D, L_D; |D|)$ in $\breve{H}_{2(d-1)}(|D|;\mathbb{Z})$ and $\mathrm{Res}_{c^1}(s_D, L_D; |D|)$ in $\breve{H}_{2(d-1)}(|D|;\mathbb{C})$ as their images by the Alexander morphism

$$H^2(V, V \smallsetminus |D|) \longrightarrow \breve{H}_{2(d-1)}(|D|).$$

If we let $|D| = \bigcup V_i'$ be the irreducible decomposition, each V_i' defines a class $[V_i']$ in $\check{H}_{2(d-1)}(|D|)$ and it is the free Abelian group generated by the $[V_i']$'s. Thus we may identify the two residues; $\mathrm{TRes}_{c^1}(s_D, L_D; |D|) = \mathrm{Res}_{c^1}(s_D, L_D; |D|)$.

Let p_i be a non-singular point of $V_i' \smallsetminus \bigcup_{j \neq i} V_j'$ and $\mathbb{D}_i$ a complex slice of V_i' in M at p_i, as in Section 13.5. The intersection $\mathbb{D}_i \cap V$ is a one-dimensional variety in $\mathbb{D}_i$. The section $s_i = s_D|_{\mathbb{D}_i \cap V}$ of $L_i = L_D|_{\mathbb{D}_i \cap V}$ is non-vanishing away from p_i so that we have the residues $\mathrm{TRes}_{c^1}(s_i, L_i; p_i)$ and $\mathrm{Res}_{c^1}(s_i, L_i; p_i)$, both of which may be thought of as the same integer.

Definition 14.4. The *order* of s_D along V_i' is defined by

$$\mathrm{ord}_{V_i'}(s_D) = \mathrm{TRes}_{c^1}(s_i, L_i; p_i).$$

Note that the definition does not depend on the choice of the point p_i as above. We then make the following:

Definition 14.5. We define the Weil divisor associated with D to be the locally finite formal sum $\sum \mathrm{ord}_{V_i'}(s_D) \cdot V_i'$.

The support of the Weil divisor defined by a Cartier divisor D coincides with $|D|$. Unlike the case of complex manifolds, a Weil divisor does not define a Cartier divisor in general (cf. Example 14.4 below).

For a non-zero Cartier divisor D, the statement as in Theorem 13.9 holds as well:

$$\mathrm{TRes}_{c^1}(s_D, L_D; |D|) = \sum_i \mathrm{TRes}_{c^1}(s_i, L_i; p_i) \cdot [V_i'] \qquad \text{in} \quad \check{H}_{2(d-1)}(|D|).$$

Thus we have:

Theorem 14.4. *Let D be a Cartier divisor on V. Then the class in $\check{H}_{2(d-1)}(|D|; \mathbb{Z})$ of the Weil divisor D^{W} associated with D is equal to $\mathrm{TRes}_{c^1}(s_D, L_D; |D|)$, which may be identified with $\mathrm{Res}_{c^1}(s_D, L_D; |D|)$. Thus the class $c^1(L_D, s_D)$ is sent to the class D^{W} by the Alexander morphism*

$$H^2(V, V \smallsetminus |D|; \mathbb{Z}) \longrightarrow \check{H}_{2(d-1)}(|D|; \mathbb{Z}).$$

In particular, the class $[D^{\mathrm{W}}]$ in $\check{H}_{2(d-1)}(V; \mathbb{Z})$ is the image of $c^1(L_D)$ in $H^2(V; \mathbb{Z})$ by the Poincaré morphism.

We say that a Weil divisor $D^{\mathrm{W}} = \sum m_i V_i'$ is *effective* if $m_i \geq 0$ for all i. Also, a Cartier divisor D is effective if its associated Weil divisor is effective. This is equivalent to saying that the natural section s_D of the associated bundle L_D is holomorphic.

If D is a non-zero effective Cartier divisor, s_D defines a complex space of dimension $d-1$ in V whose support is $|D|$. In this case, the order $\operatorname{ord}_{V_i'}(s_D)$ in Definition 14.4 of s_D along V_i' coincides with the multipicity m_i in Definiton 13.5 of V_i' in the complex space. Thus the homology class of the associated Weil divisor D^{W} coincides with that of the complex space (cf. Definitions 13.6, 14.3 and 14.5) so that Theorem 14.4 is a special case of Theorem 13.18, where $r=1$ and $q=l=1$.

The divisors we consider in the following are assumed to be non-zero. For a Cartier divisor D, we denote the associated Weil divisor also by D to avoid heavy notation.

Let $D_1,\dots,D_\nu$ be Cartier divisors on V and set $E_V = L_{D_1}\oplus\cdots\oplus L_{D_\nu}$, $s_V = s_{D_1}\oplus\cdots\oplus s_{D_\nu}$ and $Z=\bigcap_{j=1}^{\nu}|D_j|$. We assume that each $|D_j|$ contains $\operatorname{Sing}(V)$. As s_V is a non-vanishing holomorphic section of E_V away from Z, we have the topological localization $c^\nu(E_V,s_V)$ in $H^{2\nu}(V,V\smallsetminus Z)$ and the associated residue $\operatorname{TRes}_{c^\nu}(s_V,E_V;Z)$ in $\breve{H}_{2(d-\nu)}(Z)$. In view of Definition 14.1 and Theorem 14.4, we make the following

Definition 14.6. The intersection product of (the classes of the Weil divisors associated with) the D_j's in V localized at Z is defined by

$$(D_1\bullet_V\cdots\bullet_V D_\nu)_Z=\operatorname{TRes}_{c^\nu}(s_V,E_V;Z)\qquad\text{in }\breve{H}_{2(d-\nu)}(Z).$$

In the case the D_j's are effective, the above is compatible with Proposition 14.9. 1. If $\dim Z = d-\nu$, then $\operatorname{TRes}_{c^\nu}(s_V,E_V;Z)$ is equal to the class of the complex space $\boldsymbol{Z}$ defined by s_V in V (cf. Theorem 13.18).

If Z is compact, we may write (cf. Remark 14.5)

$$(D_1\bullet_V\cdots\bullet_V D_\nu)_Z=\underline{V}\frown c^\nu(E_V,s_V)\qquad\text{in } H_{2(d-\nu)}(Z).$$

In particular, if $\nu=d$ and if Z is compact, $(D_1\bullet_V\cdots\bullet_V D_\nu)_Z$ is in $H_0(Z)$ and we define the intersection number of the D_j's by

$$\sharp(D_1\bullet_V\cdots\bullet_V D_d)=\varepsilon_*((D_1\bullet_V\cdots\bullet_V D_d)_Z),\tag{14.24}$$

where $\varepsilon_*: H_0(Z)\to\mathbb{Z}$ is the augmentation.

In the global situation, we have (cf. Proposition 14.10. 1)

$$[D_1]_V\bullet_V\cdots\bullet_V[D_\nu]_V=V\frown c^\nu(E_V)\qquad\text{in }\breve{H}_{2(d-\nu)}(V),$$

where $c^\nu(E_V)=c^1(L_{D_1})\smile\cdots\smile c^1(L_{D_\nu})$. In particular, if $\nu=d$ and if V is compact, we may define the intersection number $\sharp([D_1]_V\bullet_V\cdots\bullet_V[D_d]_V)$, which coincides with the one in (14.24).

Note that the above product makes sense, if one of the divisors is only Weil. Indeed, suppose $D_1 = \sum m_i V_i'$ is Weil and the others are Cartier. We set $E_V^* = L_{D_2} \oplus \cdots \oplus L_{D_\nu}$, $s_V^* = s_{D_2} \oplus \cdots \oplus s_{D_\nu}$ and $Z^* = \bigcap_{j=2}^{\nu} |D_j|$. Then the product $(D_2 \bullet_V \cdots \bullet_V D_\nu)_{Z^*}$ is defined and is given by (cf. Definition 14.6)

$$(D_2 \bullet_V \cdots \bullet_V D_\nu)_{Z^*} = \mathrm{TRes}_{c^{\nu-1}}(s_V^*, E_V^*; Z^*) \qquad \text{in } \breve{H}_{2(d-\nu+1)}(Z^*).$$

Thus $(V_i' \bullet_V (D_2 \bullet_V \cdots \bullet_V D_\nu)_{Z^*})_{V_i' \cap Z^*}$ is defined in $\breve{H}_{2(d-\nu)}(V_i' \cap Z^*)$ and is equal to $\mathrm{TRes}_{c^{\nu-1}}(s_V^*|_{V_i'}, E_V^*|_{V_i'}; V_i' \cap Z^*)$ by Definition 14.2, for each i.

Definition 14.7. Let $D_1 = \sum m_i V_i'$ be a Weil divisor and $D_2, \ldots, D_\nu$ Cartier divisors on V. The intersection product of $D_1, \ldots, D_\nu$ in V localized at $Z = \bigcap_{j=1}^{\nu} |D_j|$ is defined by

$$\begin{aligned}(D_1 \bullet_V D_2 \bullet_V \cdots \bullet_V D_\nu)_Z \\ = \sum m_i \cdot (\jmath_i)_*(V_i' \bullet_V (D_2 \bullet_V \cdots \bullet_V D_\nu)_{Z^*})_{V_i' \cap Z^*} \qquad \text{in } \breve{H}_{2(d-\nu)}(Z),\end{aligned}$$

where $Z^* = \bigcap_{j=2}^{\nu} |D_j|$ and $\jmath_i : V_i' \cap Z^* \hookrightarrow Z$ is the inclusion.

Remark 14.7. In the case D_1 is Cartier and $m_i V_i'$ is the associated Weil divisor, the above definition coincides with the one in Definition 14.6 (cf. Proposition 14.11).

In particular, if $\nu = d$ and if Z is compact, we may define the intersection number $\sharp(D_1 \bullet_V \cdots \bullet_V D_d)$ as in (14.24).

Letting $D_1, D_2, \ldots, D_\nu$, Z, Z^*, E_V^* and s_V^* be as above, we have:

Proposition 14.17. *If Z is compact and if $\dim Z^* = d - \nu + 1$, we may write*

$$(D_1 \bullet_V D_2 \bullet_V \cdots \bullet_V D_\nu)_Z = \underline{D_1} \frown c^{\nu-1}(E_V^*|_{|D_1|}, s_V^*|_{|D_1|}) \qquad \text{in } H_{2(d-\nu)}(Z),$$

where $\underline{D_1}$ denotes the relative class of D_1 in $H_{2(d-1)}(|D_1|, |D_1| \smallsetminus Z)$.

Proof. Since $V_i' \cap Z^*$ is compact and $\dim Z^* = d - \nu + 1$, we may write (cf. Remark 14.6)

$$(V_i' \bullet_V (D_2 \bullet_V \cdots \bullet_V D_\nu)_{Z^*})_{V_i' \cap Z^*} = \underline{V_i'} \frown c^{\nu-1}(E_V^*|_{V_i'}, s_V^*|_{V_i'}).$$

Denoting by $\imath_i : V_i' \hookrightarrow |D_1|$ the inclusion, we have $c^{\nu-1}(E_V^*|_{V_i'}, s_V^*|_{V_i'}) = (\imath_i)^* c^{\nu-1}(E_V^*|_{|D_1|}, s_V^*|_{|D_1|})$. Thus

$$(\jmath_i)_*(V_i' \bullet_V (D_2 \bullet_V \cdots \bullet_V D_\nu)_{Z^*})_{V_i' \cap Z^*} = (\imath_i)_* \underline{V_i'} \frown c^{\nu-1}(E_V^*|_{V_i'}, s_V^*|_{V_i'}).$$

□

Passing from the homology of Z to that of $|D_1|$, from (14.22) with Z^* as Z and (14.23),

$$(D_1 \bullet_V [D_2 \bullet_V \cdots \bullet_V D_\nu]_V)_{|D_1|} \\ = D_1 \frown c^{\nu-1}(E_V^*|_{|D_1|}) \quad \text{in } \breve{H}_{2(d-\nu)}(|D_1|). \quad (14.25)$$

In the above situation, we have a formula as (14.20) and we have:

Theorem 14.5. *Let V be a variety of dimension d and let $D_1, \ldots, D_d$ be divisors, that are Cartier (possibly except for one) on V. If $Z = \bigcap_{j=1}^d |D_j|$ consists of a finite number of isolated points, we have*

$$\sum_{p \in Z} (D_1 \bullet_V \cdots \bullet_V D_d)_p = \sharp(D_1 \bullet_V \cdots \bullet_V D_d),$$

where $(D_1 \bullet_V \cdots \bullet_V D_d)_p$ is the intersection number at p.

In some cases we may express the intersection number at a point explicitly.

Thus let V be a variety of pure dimension d in a neighborhood U of 0 in $\mathbb{C}^n$ possibly with an isolated singularity at 0. Let $D_1, \ldots, D_d$ be effective Cartier divisors on V. Suppose $\bigcap_{j=1}^d |D_j| = \{0\}$ so that we have the intersection product $(D_1 \bullet_V \cdots \bullet_V D_d)_0$ in $H_0(\{0\}) \simeq \mathbb{Z}$. If U is sufficiently small, each D_j is defined by the restriction of a holomorphic function f_j on U. Note that $V(f_1, \ldots, f_d) \cap V = \{0\}$. By Definition 14.6, $(D_1 \bullet_V \cdots \bullet_V D_d)_0$ has various expressions as given in Section 13.6. Namely, we have a topological expression as in Theorem 13.10. Also, by Theorem 13.11, we have

$$(D_1 \bullet_V \cdots \bullet_V D_d)_0 = \operatorname{Res}_0 \begin{bmatrix} df_1 \wedge \cdots \wedge df_d \\ f_1, \ldots, f_d \end{bmatrix}_V. \quad (14.26)$$

Moreover, if V is a complete intersection defined by $h_1, \ldots, h_k$ in U, by Theorem 13.12, we have

$$(D_1 \bullet_V \cdots \bullet_V D_d)_0 = \dim \mathscr{O}_n/(f_1, \ldots, f_d, h_1, \ldots, h_k).$$

In the case $D_1 = \sum m_i V_i'$ is a Weil divisor, we have expressions as above on the V_i''s. For example,

$$(D_1 \bullet_V \cdots \bullet_V D_d)_0 = \sum m_i \operatorname{Res}_0 \begin{bmatrix} df_2 \wedge \cdots \wedge df_d \\ f_2, \ldots, f_d \end{bmatrix}_{V_i'}. \quad (14.27)$$

Multiplicity at a point

Let U be a neighborhood of 0 in $\mathbb{C}^n$ and V a variety of dimension d in U which contains 0.

The *tangent cone* $\boldsymbol{C}_0(V)$ of V at 0 is the complex space in U defined by the ideal sheaf generated by the leading homogeneous polynomials of germs in the ideal of V at 0. It is known that its support $C_0(V)$ has the same dimension as V.

Example 14.2. 1. Let V be the variety in $\mathbb{C}^2 = \{(z_1, z_2)\}$ defined by $z_1^3 - z_2^2 = 0$ (cf. Example 2.3. 2). Then $\boldsymbol{C}_0(V)$ is the complex space defined by the ideal (z_2^2) (a "double line").

2. Let V be the variety in $\mathbb{C}^3 = \{(z_1, z_2, z_3)\}$ defined by $z_1 z_2 - z_3^2 = 0$ (cf. Exercise 11.11). Then $\boldsymbol{C}_0(V)$ is V itself.

3. Let V be the variety in $\mathbb{C}^3 = \{(z_1, z_2, z_3)\}$ defined by $z_1 z_2^2 - z_3^2 = 0$ (cf. Example 2.3. 3). Then $\boldsymbol{C}_0(V)$ is the complex space defined by the ideal (z_3^2) (a "double plane").

A *hyperplane* H through 0 is a variety of dimension $n-1$ in U defined by a linear function ℓ. We may think of H as the complex space defined by a section s_H of the (trivial) line bundle L_H and ℓ the component of s_H with respect to some frame of L_H. Note that s_H is a regular section.

Let $\{H_1, \dots, H_i\}$ be a collection of hyperplanes, $1 \leq i \leq d$. We say that it is *excellent* for V if $\dim C_0(V) \cap H_1 \cap \cdots \cap H_i = d - i$. In this case, $s_{H_1} \oplus \cdots \oplus s_{H_i}$ is a regular section of $L_{H_1} \oplus \cdots \oplus L_{H_i}$.

If $\{H_1, \dots, H_d\}$ is excellent for V, then $V \cap H_1 \cap \cdots \cap H_d = \{0\}$. Thus we have the localized intersection product $(V \bullet H_1 \bullet \cdots \bullet H_d)_0$ in $H_0(\{0\}; \mathbb{Z}) = \mathbb{Z}$, the identification being made by the augmentation, and we may think of it as a positive integer.

Definition 14.8. The *multiplicity* of V at 0 is defined by

$$m(V, 0) = (V \bullet H_1 \bullet \cdots \bullet H_d)_0,$$

where $\{H_1, \dots, H_d\}$ is a collection of d hyperplanes excellent for V.

In the above situation, we set $E = L_{H_1} \oplus \cdots \oplus L_{H_d}$, $s = s_{H_1} \oplus \cdots \oplus s_{H_d}$ and $S = H_1 \cap \cdots \cap H_d$. Then s is a regular section of E with zero set S. We set $E_V = E|_V$ and $s_V = s|_V$. Let $\boldsymbol{Z}$ be the complex space in V defined by s_V. The support of $\boldsymbol{Z}$ is $\{0\}$ and $\boldsymbol{Z}$ defines a class (a number) in $\check{H}_0(\{0\}; \mathbb{Z}) = H_0(\{0\}; \mathbb{Z}) = \mathbb{Z}$, which is also denoted by $\boldsymbol{Z}$.

Proposition 14.18. *We have*

$$m(V,0) = \mathrm{TRes}_{c^d}(s_V, E_V; 0),$$

which is equal to $\boldsymbol{Z}$.

Proof. We have

$$m(V,0) = (V \bullet (H_1 \bullet \cdots \bullet H_d)_S)_0.$$

Noting that the complex space defined by s is reduced, we have, from Proposition 14.2. 2, $(H_1 \bullet \cdots \bullet H_d)_S = S$. The proposition follows from Proposition 14.7. □

From Theorems 13.11 and 13.12, we have:

Corollary 14.2. *In the above situation, suppose* 0 *is an isolated singularity of* V. *Let* ℓ_i *be a defining linear function of* H_i, $i = 1, \ldots, d$. *Then we have*

$$m(V,0) = \mathrm{Res}_0 \begin{bmatrix} d\ell_1 \wedge \cdots \wedge d\ell_d \\ \ell_1, \ldots, \ell_d \end{bmatrix}_V.$$

Moreover, if V *is a CI defined by* $h_1, \ldots, h_k$, $k = n - d$, *it may also be expressed as*

$$m(V,0) = \dim_{\mathbb{C}} \mathscr{O}_n/(h_1, \ldots, h_k, \ell_1, \ldots, \ell_d).$$

Example 14.3. Let V be a hypersurface in U defined by f. Let m be the order of f at 0 (cf. Definition 1.10). By a suitable change of coordinates, we may assume that the order of f in z_n is m (cf. Lemma 1.2). If we let H_i be the hyperplane defined by z_i, $i = 1, \ldots, n-1$, then $\{H_1, \ldots, H_{n-1}\}$ is excellent for V and we see that $(V \bullet H_1 \bullet \cdots \bullet H_{n-1})_0$ is the m-fold point in the z_n-plane (cf. Example 11.7). Thus $m(V,0) = m$.

For example, if V is as in 1, 2 or 3 of Example 14.2, $m(V,0) = 2$.

Here is another expression of the multiplicity at a point.

The projective cone $\boldsymbol{PC}_0(V)$ of V at 0 is the complex space in $\mathbb{P}^{n-1}$ defined by the homogeneous polynomials defining $\boldsymbol{C}_0(V)$. Let $\pi : \tilde{U} \to U$ be the blowing up of U at 0 and D the exceptional divisor, which is biholomorphic with $\mathbb{P}^{n-1}$. Let $\tilde{V}$ denote the proper transform of V (cf. Section 11.6).

Lemma 14.2. *The intersection* $\boldsymbol{E}$ *of* $\tilde{V}$ *and* D *as a complex subspace of* $\tilde{U}$ *is the projective cone* $\boldsymbol{PC}_0(V)$.

Proof. Let $\tilde{U}_\alpha$ be the coordinate neighborhood in $\tilde{U}$ with coordinates $(z_{\alpha+1}, \xi_1^\alpha, \ldots, \xi_{n-1}^\alpha)$ as in the construction of the blowing-up in Section 11.6. Let f be a defining function of V. Let p be the order of f at 0 and write $f(z) = \sum_{\nu \geq p} f_\nu(z)$, where $f_\nu(z)$ is homogeneous of degree ν in $z = (z_1, \ldots, z_n)$. The function $\tilde{f}$ appearing in (11.20) is written

$$\tilde{f}(z_{\alpha+1}, \xi_1^\alpha, \ldots, \xi_{n-1}^\alpha) = \sum_{\nu \geq p} f_\nu(\xi_1^\alpha, \ldots, \xi_\alpha^\alpha, 1, \xi_{\alpha+1}^\alpha, \ldots, \xi_{n-1}^\alpha) \cdot z_{\alpha+1}^{\nu-p}.$$

The intersection of $\tilde{V}$ and D is defined by the ideal generated by the $\tilde{f}$'s and $z_{\alpha+1}$. Thus it is generated by the $f_p(\xi_1^\alpha, \ldots, \xi_\alpha^\alpha, 1, \xi_{\alpha+1}^\alpha, \ldots, \xi_{n-1}^\alpha)$'s and $z_{\alpha+1}$. □

In general, let $\boldsymbol{X}$ be a complex space of dimension d in $\mathbb{P}^n$. If we denote by $L_{\mathcal{H}}$ the hyperplane bundle on $\mathbb{P}^n$, the class $[\boldsymbol{X}] \frown c^1(L_{\mathcal{H}})^d$ is in $H_0(\mathbb{P}^n)$.

Definition 14.9. The *degree* of $\boldsymbol{X}$ is defined by

$$\deg \boldsymbol{X} = \varepsilon_*([\boldsymbol{X}] \frown c^1(L_{\mathcal{H}})^d),$$

where $\varepsilon_* : H_0(\mathbb{P}^n) \to \mathbb{Z}$ denotes the augmentation.

Let $\{\mathcal{H}_1, \ldots, \mathcal{H}_d\}$ be a collection of d hyperplanes in $\mathbb{P}^n$. We say that it is *in general position*, or *general* for short, with respect to $\boldsymbol{X}$ if $Z = X \cap \mathcal{H}_1 \cap \cdots \cap \mathcal{H}_d$ consits of (a finite number of) isolated points. In this case, we have the intersection product $(\boldsymbol{X} \bullet \mathcal{H}_1 \bullet \cdots \bullet \mathcal{H}_d)_Z$ in $H_0(Z)$.

Proposition 14.19. *If $(\mathcal{H}_1, \ldots, \mathcal{H}_d)$ is general with respect to $\boldsymbol{X}$,*

$$\deg \boldsymbol{X} = \varepsilon_*((\boldsymbol{X} \bullet \mathcal{H}_1 \bullet \cdots \bullet \mathcal{H}_d)_Z),$$

where $\varepsilon_* : H_0(Z) \to \mathbb{Z}$ denotes the augmentation.

Proof. By a repeated use of (4.32), we have

$$[\boldsymbol{X}] \frown c^1(L_{\mathcal{H}})^d = [\boldsymbol{X}] \bullet [\mathcal{H}] \bullet \cdots \bullet [\mathcal{H}],$$

which is equal to $[\boldsymbol{X}] \bullet [\mathcal{H}_1] \bullet \cdots \bullet [\mathcal{H}_d]$ in $H_0(\mathbb{P}^n)$. If we let $\iota : Z \hookrightarrow \mathbb{P}^n$ be the inclusion, we have $\iota_*((\boldsymbol{X} \bullet \mathcal{H}_1 \bullet \cdots \bullet \mathcal{H}_d)_Z) = [\boldsymbol{X}] \bullet [\mathcal{H}_1] \bullet \cdots \bullet [\mathcal{H}_d]$ (cf. Proposition 4.12). The proposition follows from the fact that tha augmentations are compatible with ι_*. □

Proposition 14.20. *We have:*

$$m(V, 0) = \deg \boldsymbol{PC}_0(V).$$

Proof. Let $\pi : \tilde{U} \to U$ be the blowing up of U at 0 and D the exceptional divisor, which is biholomorphic with $\mathbb{P}^{n-1}$ (cf. Section 11.6). Let $\{H_1, \ldots, H_d\}$ be a collection of d hyperplanes through 0 excellent for V. Let $\tilde{V}$ denote the proper transform of V and $\tilde{H}_i$ that of H_i for each i. The intersection $\boldsymbol{E}$ of $\tilde{V}$ and D as a complex subspace of $\tilde{U}$ is the projective cone $\boldsymbol{PC}_0(V)$ (cf. Lemma 14.2). By Proposition 14.7, $(\tilde{V} \bullet D)_E = \boldsymbol{E}$. Also the intersection $\mathcal{H}_i$ of $\tilde{H}_i$ and D as a complex subspace of $\tilde{U}$ is a hyperplane in $D = \mathbb{P}^{n-1}$ and $(D \bullet \tilde{H}_i)_{\mathcal{H}_i} = \mathcal{H}_i$.

Noting that $Z = \tilde{V} \cap \tilde{H}_1 \cap \cdots \cap \tilde{H}_{d-1} \cap (\tilde{H}_d \cup D) = E \cap \tilde{H}_1 \cap \cdots \cap \tilde{H}_{d-1}$ consists of a finite number of points, we consider the intersection product $(\tilde{V} \bullet \tilde{H}_1 \bullet \cdots \bullet \tilde{H}_{d-1} \bullet (\tilde{H}_d + D))_Z$ in $H_0(Z)$.

On the one hand, it is equal to the product $(\tilde{V} \bullet D \bullet \tilde{H}_1 \bullet \cdots \bullet \tilde{H}_{d-1})_Z = (\boldsymbol{E} \bullet \tilde{H}_1 \bullet \cdots \bullet \tilde{H}_{d-1})_Z$, which is equal to $(\boldsymbol{E} \bullet_D \mathcal{H}_1 \bullet_D \cdots \bullet_D \mathcal{H}_{d-1})_Z$ by a repeated use of Proposition 14.13 or Proposition 14.15. It is mapped to $\deg \boldsymbol{E}$ by the augmentation $\varepsilon_* : H_0(Z) \to \mathbb{Z}$.

On the other hand, $\tilde{W} = (\tilde{V} \bullet \tilde{H}_1 \bullet \cdots \bullet \tilde{H}_{d-1})_{\tilde{W}}$ is the proper transform of $W = (V \bullet H_1 \bullet \cdots \bullet H_{d-1})_W$. We have, since Z is compact (cf. Remark 14.4. 1),

$$
\begin{aligned}
(\tilde{W} \bullet (\tilde{H}_d + D))_Z &= \underline{\tilde{W}} \frown c^1(L_{\tilde{H}_d + D}|_{\tilde{W}}, \pi^* s|_{\tilde{W}})) \\
&= \underline{\tilde{W}} \frown \varpi^* c^1(L_{H_d}|_W, s|_W),
\end{aligned}
$$

which is sent to $\underline{W} \frown c^1(L_{H_d}|_W, s|_W) = (V \bullet H_1 \bullet \cdots \bullet H_{d-1} \bullet H_d)_0$ by ϖ_*, where ϖ is the restriction of π to $\tilde{W}$. □

Remark 14.8. The independence of $m(V, 0)$ of the choice of a collection of hyperplanes excellent for V in Definition 14.8 can also be seen from the above proposition.

We finish this section by giving some remarks on the multiplicity at a point. Let U and V be as above.

Lemma 14.3. *Let P be a non-singular hypersurface* (*through* 0) *in U. Suppose P intersects transversely with V_{reg} near* 0 *and set $V' = V \cap P$, then*

$$m(V', 0) = m(V, 0).$$

Proof. Since P is non-singular, by a suitable change of coordinates, we may assume that P is a hyperplane. Note that $\dim V' = d - 1$ (cf. Theorem 2.11). If $\{H_2, \ldots, H_d\}$ is a collection of hyperplanes excellent for V', then $\{P, H_2, \ldots, H_d\}$ is a collection of hyperplanes excellent for V. We have

$$m(V', 0) = (V' \bullet H_2 \bullet \cdots \bullet H_d)_0.$$

On the other hand, from the assuptions we have $V' = (V \bullet P)_{V'}$ (cf. Proposition 14.7). □

Let V be a variety of dimension d in a neighborhood U of 0 in $\mathbb{C}^n$ containing 0. Let $\{H_1, \ldots, H_d\}$ be a collection of hyperplanes in U excellent for V. Let $E = L_{H_1} \oplus \cdots \oplus L_{H_d}$ and $s = s_{H_1} \oplus \cdots \oplus s_{H_d}$. For each i, let $\boldsymbol{G}_i$ denote the intersection of V and H_i as a complex space in U. We may think of it as the complex space in V defined by the section $s_{H_i}|_V$ of $L_{H_i}|_V$. If we denote by G_i the support of $\boldsymbol{G}_i$, by Proposition 14.15. 1,

$$m(V, 0) = ((V \bullet H_1)_{G_1} \bullet_V \cdots \bullet_V (V \bullet H_d)_{G_d})_0.$$

Since $\dim G_i = d - 1$, $(V \bullet H_i)_{G_i} = \boldsymbol{G}_i$ so that (cf. Proposition 14.15. 2)

$$m(V, 0) = (\boldsymbol{G}_1 \bullet_V \cdots \bullet_V \boldsymbol{G}_d)_0.$$

Let V' be a subvariety of V containing 0. Let $\{H_2, \ldots, H_d\}$ be a collection of hyperplanes excellent for V' and V and let $\boldsymbol{G}'_i$ be the intersection of V' and H_i as a complex space. Denoting by G'_i the support of $\boldsymbol{G}'_i$, we have

$$m(V', 0) = ((V' \bullet H_2)_{G'_2} \bullet_{V'} \cdots \bullet_{V'} (V' \bullet H_d)_{G'_d})_0.$$

We have $(V' \bullet H_i)_{G'_i} = \boldsymbol{G}'_i = (V' \bullet_V \boldsymbol{G}_i)_{G'_i}$. Thus by Proposition 14.16,

$$m(V', 0) = (V' \bullet_V \boldsymbol{G}_2 \bullet_V \cdots \bullet_V \boldsymbol{G}_d)_0.$$

Lemma 14.3 can also be seen from the above arguments.

14.4 Intersection product in singular surfaces

In this section, we apply the considerations in the previous sections to the case of singular surfaces. A curve or a surface means a variety of pure dimension one or two, respectively.

Let V be a (possibly singular) surface in a complex manifold M. Let D be a Cartier divisor on V and denote by L_D the associated line bundle on V. The bundle L_D admits a natural meromorphic section s_D and its first Chern class $c^1(L_D)$ is localized at the support $|D|$ of D by s_D. The localized class $c^1(L_D, s_D)$ is in $H^2(V, V \smallsetminus |D|)$. We have the associated residue $\mathrm{TRes}_{c^1}(s_D, L_D; |D|)$ as the image of $c^1(L_D, s_D)$ by the Alexander morphism

$$H^2(V, V \smallsetminus |D|) \longrightarrow \breve{H}_2(|D|).$$

By Theorem 14.4, the class of the Weil divisor associated with D is equal to $\mathrm{TRes}_{c^1}(s_D, L_D; |D|)$, which may be identified with $\mathrm{Res}_{c^1}(s_D, L_D; |D|)$.

Let D_1 and D_2 be divisors on V. As we have seen in the previous section, we may define the intersection product $(D_1 \cdot_V D_2)_Z$ in $\check{H}_0(Z)$, $Z = |D_1| \cap |D_2|$, if at least one of them is Cartier. If D_2 is Cartier, the image of $(D_1 \cdot_V D_2)_Z$ by the canonical morphism $\check{H}_0(Z) \to \check{H}_0(|D_1|)$ is given by (cf. (14.25))

$$(D_1 \cdot_V [D_2]_V)_{|D_1|} = D_1 \frown c^1(L_{D_2}|_{|D_1|}). \tag{14.28}$$

It is further sent to $[D_1]_V \cdot_V [D_2]_V = [D_1]_V \frown c^1(L_{D_2})$ by $\check{H}_0(|D_1|) \to \check{H}_0(V)$.

If Z is compact, we have (cf. Proposition 14.17)

$$(D_1 \cdot_V D_2)_Z = \underline{D_1} \frown c^1(L_{D_2}|_{|D_1|}, s_{D_2}|_{|D_1|}) \qquad \text{in } H_0(Z), \tag{14.29}$$

where $\underline{D_1}$ is the class of D_1 in $H_2(|D_1|, |D_1| \setminus Z)$. In this case, we have the intersection number $\sharp(D_1 \cdot_V D_2)$ as defined in (14.24). In particular, if Z consists of a finite number of isolated points, we have (Theorem 14.5)

$$\sum_{p \in Z} (D_1 \cdot_V D_2)_p = \sharp(D_1 \cdot_V D_2).$$

If V has an isolated singularity at p and if D_1 and D_2 are Cartier the intersection number at p is given, for example (cf. (14.26)), by

$$(D_1 \cdot_V D_2)_p = \operatorname{Res}_p \begin{bmatrix} df_1 \wedge df_2 \\ f_1, f_2 \end{bmatrix}_V .$$

If $D_1 = \sum m_i V_i'$ is Weil, we have (cf. (14.27))

$$(D_1 \cdot_V D_2)_p = \sum m_i \operatorname{Res}_p \begin{bmatrix} df_2 \\ f_2 \end{bmatrix}_{V_i'} .$$

If D is a Cartier divisor with compact support in V, then $Z = |D|$ and (cf. (14.28))

$$\sharp(D \cdot_V D) = \varepsilon_*(D \frown c^1(L_D|_{|D|})), \tag{14.30}$$

where $\varepsilon_* : H_0(|D|) \to \mathbb{Z}$ denotes the augmentation. It is referred to as the *self-intersection number* of D.

Effect of blowing-up

Let U be a neighborhood of 0 in $\mathbb{C}^n$ and V a surface in U containing 0. Let $\pi : \tilde{U} \to U$ be the blowing-up of U at 0, $D = \pi^{-1}(0)$ the exceptional divisor, which may be thought of as $\mathbb{P}^{n-1}$, $\tilde{V}$ the proper transform of V. The intersection $\boldsymbol{E}$ of $\tilde{V}$ and D as a complex subspace of $\tilde{U}$ is the projective cone $\boldsymbol{PC}_0(V)$ (cf. Lemma 14.2). By Proposition 14.7, $(\tilde{V} \cdot D)_E = \boldsymbol{E}$. The complex space $\boldsymbol{E}$ is considered as a Cartier divisor on $\tilde{V}$.

Let C be a curve through 0 in V. The proper transform $\tilde{C}$ of C is a curve in $\tilde{V}$ and $Z = \tilde{C} \cap E$ consists of a finite number of points so that we have $(\tilde{C} \cdot_{\tilde{V}} \boldsymbol{E})_Z$ in $H_0(Z)$.

Lemma 14.4. *We have*

$$m(C, 0) = \sharp(\tilde{C} \cdot_{\tilde{V}} \boldsymbol{E}).$$

Proof. Let H be a hyperplane through 0, excellent for C and V, and $\tilde{H}$ the proper transform of H. Noting that $Z = \tilde{C} \cap (\tilde{H} \cup D) = \tilde{C} \cap D = \tilde{C} \cap E$ consists of a finite number of points, we consider the intersection product $(\tilde{C} \cdot (\tilde{H} + D))_Z$ in $H_0(Z)$. On the one hand, by Proposition 14.12, it is equal to $(\tilde{C} \cdot D)_Z = (\tilde{C} \cdot_{\tilde{V}} \boldsymbol{E})_Z$. On the other hand we have, since Z is compact (cf. Remark 14.4. 1),

$$\begin{aligned}(\tilde{C} \cdot (\tilde{H} + D))_Z &= \underline{\tilde{C}} \frown c^1(L_{\tilde{H}+D}|_{\tilde{C}}, \pi^* s|_{\tilde{C}})) \\ &= \underline{\tilde{C}} \frown \varpi^* c^1(L_{H_d}|_C, s|_C),\end{aligned}$$

which is sent to $\underline{C} \frown c^1(L_H|_C, s|_C) = (C \cdot H)_0$ by ϖ_*, where ϖ is the restriction of π to $\tilde{C}$. □

Let $\rho : \tilde{V} \to V$ denote the restriction of π. For a curve C in V through 0, the analytic inverse image $\rho^* C$ may be thought of as a divisor on $\tilde{V}$. In particular, if C is Cartier, so is $\rho^* C$.

Lemma 14.5. *If C is Cartier, the multiplicity $m(C, 0)$ is divisible by $m(V, 0)$ and if we set $m(C, V; 0) = m(C, 0)/m(V, 0)$, we have*

$$\rho^* C = \tilde{C} + m(C, V; 0)\boldsymbol{E}.$$

Proof. Since C is Cartier, we may write $\rho^* C = \tilde{C} + k\boldsymbol{E}$ for some integer k. Let H be a hyperplane through 0, excellent for C and V. Let $\boldsymbol{G}$ be the intersection of V and H as complex subspace of M. We have $(V \cdot H)_G = \boldsymbol{G}$. Also let $\tilde{\boldsymbol{G}}$ be the intersection of $\tilde{V}$ and $\tilde{H}$ as complex subspace of $\tilde{M}$. We have $(\tilde{V} \cdot \tilde{H})_{\tilde{G}} = \tilde{\boldsymbol{G}}$. We set $Z = \tilde{G} \cap |\rho^* C|$, which consists of a finite number of points and compute $(\tilde{\boldsymbol{G}} \cdot_{\tilde{V}} \rho^* C)_Z$ in two ways. On the one hand, since $\tilde{G} \cap \tilde{C} = \emptyset$, we have

$$(\tilde{\boldsymbol{G}} \cdot_{\tilde{V}} \rho^* C)_Z = k(\tilde{\boldsymbol{G}} \cdot_{\tilde{V}} \boldsymbol{E})_Z.$$

On the other hand,

$$(\tilde{\boldsymbol{G}} \cdot_{\tilde{V}} \rho^* C)_Z = \underline{\tilde{\boldsymbol{G}}} \frown \rho^* c^1(L_C, s),$$

which is sent to $\underline{\boldsymbol{G}} \frown c^1(L_C, s) = (\boldsymbol{G} \cdot_V C)_0 = m(C, 0)$ by ρ_*. As in the proof of Proposition 14.20, $m(V, 0) = \sharp(\boldsymbol{E} \cdot \tilde{H}) = \sharp(\boldsymbol{E} \cdot_{\tilde{V}} \tilde{\boldsymbol{G}})$. □

Theorem 14.6. *Let $\rho\colon \tilde{V} \to V$ be the restriction of π as above. Let C_1 and C_2 be curves in V. Suppose $C_1 \cap C_2 = \{0\}$ and C_2 is Cartier. Then*

$$(C_1 \cdot_V C_2)_0 = \sum_{q \in \rho^{-1}(0)} (\tilde{C}_1 \cdot_{\tilde{V}} \tilde{C}_2)_q + m(C_1, 0) \cdot m(C_2, V; 0).$$

Proof. Noting that $Z = \tilde{C}_1 \cap |\rho^* C_2|$ consists of a finite number of points, we consider the product $(\tilde{C}_1 \cdot_{\tilde{V}} \rho^* C_2)_Z$. On the one hand, by Lemmas 14.4 and 14.5, it is equal to

$$(\tilde{C}_1 \cdot_{\tilde{V}} (\tilde{C}_2 + m(C_2, V; p)\boldsymbol{E}))_Z = (\tilde{C}_1 \cdot_{\tilde{V}} \tilde{C}_2)_Z + m(C_2, V; 0)(\tilde{C}_1 \cdot_{\tilde{V}} \boldsymbol{E})_Z,$$

where $(\tilde{C}_1 \cdot_{\tilde{V}} \boldsymbol{E})_Z$ is mapped to $m(C_1, 0)$ by ε_*. On the other hand, it is equal to $\underline{\tilde{C}_1} \frown \rho^* c^1(L_{C_2}|_{C_1}, s_{C_2}|_{C_1})$ (cf. (14.29)), which is mapped to $\underline{C_1} \frown c^1(L_{C_2}|_{C_1}, s_{C_2}|_{C_1}) = (C_1 \cdot_V C_2)_0$ by ρ_*. □

We give the following expression for the multiplicity:

Proposition 14.21.

$$m(V, 0) = -\sharp(\boldsymbol{E} \cdot_{\tilde{V}} \boldsymbol{E}).$$

Proof. We have $\sharp(\boldsymbol{E} \cdot_{\tilde{V}} \boldsymbol{E}) = \varepsilon_*(\boldsymbol{E} \frown c^1(L_{\boldsymbol{E}}|_E))$ (cf. (14.30)). Consider the commutative diagram of inclusions

$$\begin{array}{ccc} E & \xrightarrow{\tilde{\jmath}} & D \\ {\scriptstyle \jmath}\downarrow & & \downarrow{\scriptstyle \imath} \\ \tilde{V} & \xrightarrow{\tilde{\imath}} & \tilde{U}. \end{array}$$

We have $L_{\boldsymbol{E}}|_E = \jmath^*(\tilde{\imath})^* L_D = (\tilde{\jmath})^* \imath^* L_D$. Recall $\imath^* L_D = -L_{\mathcal{H}}$, negative of the hyperplane bundle on $D = \mathbb{P}^{n-1}$. Since we may write $L_{\mathcal{H}} = \imath^* L_{\tilde{H}}$, $L_{\boldsymbol{E}}|_E = -(\tilde{\jmath})^* \imath^* L_{\tilde{H}} = -\jmath^*(\tilde{\imath})^* L_{\tilde{H}} = -\jmath^* L_{\tilde{\boldsymbol{G}}}$, we have $\boldsymbol{E} \frown c^1(L_{\boldsymbol{E}}|_E) = -\boldsymbol{E} \frown c^1(L_{\tilde{\boldsymbol{G}}}|_E)$. □

If C_1 and C_2 are two Cartier curves such that $C_1 \cap C_2 = \{0\}$, then $|\rho^* C_1| \cap |\rho^* C_2| = E$, which is compact. We compute, using Lemmas 14.4 and 14.5 and Proposition 14.21,

$$(C_1 \cdot_V C_2)_0 = \sharp(\rho^* C_1 \cdot_{\tilde{V}} \rho^* C_2). \tag{14.31}$$

Intersection product of Weil curves

Let V be a surface in a complex manifold M. Let C_1 and C_2 be two (distinct) curves in V. If at least one of them is Cartier, We have the local and global intersection numbers of C_1 and C_2. If C_1 and C_2 are only Weil curves, we proceed as follows.

Let p be a point in V. Let $\pi_p : \tilde{M} \to M$ be the blowing-up at p and $\rho_p : \tilde{V} \to V$ the restriction of π_p, where $\tilde{V}$ is the proper transform of V under π_p. Let C be a curve on V through p and $\tilde{C}$ the proper transform of C under π_p.

Definition 14.10. The *total transform* of C under ρ_p is defined by

$$\rho_p^{\star} C = \tilde{C} + \frac{m(C,p)}{m(V,p)} \boldsymbol{E}.$$

If C is Cartier, this coincides with the inverse image $\rho_p^* C$ by Lemma 14.5. In general $m(C,p)/m(V,p)$ is a rational number and $\rho_p^{\star} C$ is different from $\rho_p^* C$.

Let C_1 and C_2 be curves on V through p such that they intersect only at p near p. If either $\tilde{C}_1$ or $\tilde{C}_2$ is Cartier, we may define the intersection number by

$$(C_1 \bullet_V C_2)_p = \sum_{q \in \rho^{-1}(p)} (\tilde{C}_1 \bullet_{\tilde{V}} \tilde{C}_2)_q + \frac{m(C_1,p) \cdot m(C_2,p)}{m(V,p)}, \tag{14.32}$$

which reduces to the formula in Theorem 14.6, if either C_1 or C_2 is Cartier. By Lemma 14.4 and Proposition 14.21, we may write

$$(C_1 \bullet_V C_2)_p = \sharp(\rho_p^{\star} C_1 \bullet_{\tilde{V}} \rho_p^{\star} C_2),$$

which reduces to (14.31), if both C_1 and C_2 are Cartier.

If both $\tilde{C}_1$ and $\tilde{C}_2$ are not Cartier, we define the intersection number by recursion of (14.32).

Note that, if either one of C_1 and C_2 is not Cartier at p, $(C_1 \bullet_V C_2)_p$ is only a rational number, in general, for $m(V,p)$ might not divide $m(C_1,p) \cdot m(C_2,p)$.

In the global situation, if $C_1 \cap C_2$ consists of a finite number of isolated points, we define by recursion

$$\sharp(C_1 \bullet_V C_2) = \sum_{p \in C_1 \cap C_2} (C_1 \bullet_V C_2)_p.$$

More generally, if $Z = C_1 \cap C_2$ is compact, we define

$$\sharp(C_1 \bullet_V C_2) = \sharp(\rho_p^\star C_1 \bullet_{\tilde{V}} \rho_p^\star C_2).$$

In particular, the self-intersection number of a compact curve C in V is given by

$$\sharp(C \bullet_V C) = \sharp(\rho_p^\star C \bullet_{\tilde{V}} \rho_p^\star C).$$

Remark 14.9. 1. In the above, we need not to resolve the singularities of V, we only need to take blowing-ups sufficiently many times so that the curve becomes Cartier.

2. Intersections of Weil divisors are defined by extending the above linearly.

Example 14.4. Let V be defined by $z_1 z_2 - z_3^2 = 0$ in $\mathbb{C}^3 = \{(z_1, z_2, z_3)\}$ (cf. Exercise 11.11). Let C_1 and C_2 be curves in V defined by $z_1 = z_3 = 0$ and $z_2 = z_3 = 0$, respectively. Then C_1 and C_2 are Weil curves (only $C_1 \cup C_2$ is Cartier). Since $m(V, 0) = 2$, $m(C_1, 0) = m(C_2, 0) = 1$, and $\tilde{C}_1$ and $\tilde{C}_2$ are disjoint, we compute

$$(C_1 \bullet_V C_2)_0 = \sharp(\tilde{C}_1 \bullet_{\tilde{V}} \tilde{C}_2) + \frac{m(C_1, 0) \cdot m(C_2, 0)}{m(V, 0)} = 0 + \frac{1 \cdot 1}{2} = \frac{1}{2}.$$

14.5 Excess intersections

Let M be a complex manifold of dimension n and E a holomorphic vector bundle of rank l on M. Also let s be a holomorphic section of E and S the zero set of s. In this situation, we have the residues $\mathrm{TRes}_{c^l}(s, E; S)$ in $\check{H}_{2(n-l)}(S; \mathbb{Z})$ and $\mathrm{Res}_{c^l}(s, E; S)$ in $\check{H}_{2(n-l)}(S; \mathbb{C})$. The complex space $\boldsymbol{S}$ defined by s has S as its support. Recall that $\dim S \geq n - l$, where the equality holds if and only if s is a regular section. In the case $\dim S = n - l$, we may define the homology class of $\boldsymbol{S}$, denoted also by $\boldsymbol{S}$, in $\check{H}_{2(n-l)}(S; \mathbb{Z})$ (cf. Definition 12.6) and we have (cf. Theorem 12.9)

$$\boldsymbol{S} = \mathrm{TRes}_{c^l}(s, E; S),$$

which may also be identified with $\mathrm{Res}_{c^l}(s, E; S)$.

Now we consider the case $\dim S > n - l$. Suppose there exists a subbundle E' of rank $l' = n - \dim S$ of E such that s is a section of E'. Then we may define the class of $\boldsymbol{S}$ in $\check{H}_{2(n-l')}(S; \mathbb{Z})$ and we have $\boldsymbol{S} = \mathrm{TRes}_{c^{l'}}(s, E'; S)$. Let E'' denote the quotient bundle E/E', which is of rank $l'' = l - l'$.

From Proposition 5.10 (see also Corollary 10.2), we have:

Theorem 14.7. *In the above situation, we have:*

$$\mathrm{TRes}_{c^l}(s, E; S) = \boldsymbol{S} \frown i^* c^{l''}(E'') \qquad \text{in } \check{H}_{2(n-l)}(S),$$

where $i : S \hookrightarrow M$ denotes the inclusion.

Let E, s and $\boldsymbol{S}$ be as in the beginning of this section. Let V be a variety of dimension d in M and set $E_V = E|_V$ and $s_V = s|_V$. Let $\boldsymbol{Z}$ be the complex space in V defined by s_V. Its support Z is given by $Z = V \cap S$. Suppose s is a regular section so that $\dim S = n - l$. Then, by Proposition 14.7.1,

$$(V \bullet \boldsymbol{S})_Z = \mathrm{TRes}_{c^l}(s_V, E_V; Z) \qquad \text{in } \check{H}_{2(d-l)}(Z). \tag{14.33}$$

In general, $\dim Z \geq d - l$ by Theorem 2.11. If $\dim Z = d - l$, we may define the homology class of $\boldsymbol{Z}$ (cf. Definition 13.6) and we have $\boldsymbol{Z} = \mathrm{TRes}_{c^l}(s_V, E_V; Z)$ so that (cf. Proposition 14.7.2)

$$(V \bullet \boldsymbol{S})_Z = \boldsymbol{Z}.$$

Now we consider the case $\dim Z > d - l$, which means that s vanishes on V "more than expected" so that V and $\boldsymbol{S}$ intersect in a dimension higher than expected. Suppose there exists a subbundle E'_V of rank $l' = d - \dim Z$ of E_V such that s_V is a section of E'_V. Then we may define the class of $\boldsymbol{Z}$ in $\check{H}_{2(d-l')}(Z; \mathbb{Z})$ and we have $\boldsymbol{Z} = \mathrm{TRes}_{c^{l'}}(s_V, E'_V; Z)$. Let E''_V denote the quotient bundle E_V/E'_V, which is of rank $l'' = l - l'$. From Corollary 14.1, we have

$$c^l(E_V, s_V) = c^{l'}(E'_V, s_V) \smile c^{l''}(E''_V).$$

We also have a statement as in Proposition 4.3 in this situation and we have (cf. the proof of Proposition 5.10):

$$\mathrm{TRes}_{c^l}(s_V, E_V; Z) = \mathrm{TRes}_{c^{l'}}(s_V, E'_V; Z) \frown j^* c^{l''}(E''_V) \qquad \text{in } \check{H}_{2(d-l)}(Z),$$

where $j : Z \hookrightarrow V$ denotes the inclusion. The left-hand side of the above is equal to $(V \bullet \boldsymbol{S})_Z$ by (14.33). Thus we have:

Theorem 14.8 (Excess intersection formula). *In the above situation, we have:*

$$(V \bullet \boldsymbol{S})_Z = \boldsymbol{Z} \frown j^* c^{l''}(E''_V) \qquad \text{in } \check{H}_{2(d-l)}(Z).$$

Notes

The adjunction formula (14.17) was generalized in [Kodaira (1960)] for a possibly singular curve C in a complex surface S as:

$$\chi(\widetilde{C}) = -(K_S + C) \cdot C + \sum_{i=1}^{r} c(C, p_i),$$

where $\widetilde{C}$ is a non-singular model of C and $c(C, p_i)$ is an invariant of C at the singular point p_i, which is related to the Milnor number $\mu(C, p_i)$ by $c(C, p_i) = \mu(C, p_i) + s_i - 1$ with s_i the number of (local) branches of C at p_i. Since $\chi(\widetilde{C}) - \sum_{i=1}^{r}(s_i - 1) = \chi(C)$, the above formula is equivalent to the one in Theorem 14.3.

For details on tangent cones and the multiplicity of a variety at a point, we refer to Chapter 7 of [Whitney (1972)]. The terminology "excellent" appears there. The multiplicity in Definition 14.8 above coincides with the one in there. It also coincides with the one in p.79 of [Fulton (1984)] (cf. Proposition 14.20).

The intersection theory on normal singular surfaces is teated in [Mumford (1961)], see also [Sakai (1984)]. Our approach to the intersection theory on singular surfaces is based on residue theory. It is consistent with the one in the references above and, moreover, allows us to relate the intersection theory to the residue theories of holomorphic foliations and maps on singular varieties. The description follows (and make them more precise) that in [Bracci and Suwa (2004)], for which we also refer to for an application.

The intersection theory in algebraic category is comprehensively presented in [Fulton (1984)]. The content of this chapter corresponds to some parts in there, in fact it is inspired by them. For example, the basic intersection product construction there is done intersecting the cormal cone by the zero section. In our framework, it appears as a residue (see, e.g., Propositions 14.7 and 14.13 in the text). The notion of the order there is defined in terms of the length of a module. It is replaced with some residue in our setting (cf. Definition 14.4). The excess intersection formula is also discussed in generality there. The formula in Theorem 14.8 corresponds to the one in Example 6.1.7 in there.

Chapter 15

Riemann-Roch Theorem

The Riemann-Roch problem originates in finding the dimension of the space of meromorphic functions, on a compact Riemann surface, with prescribed order of zeros and poles. After the work of K. Kodaira for compact complex surfaces, it was formulated and proved for projective algebraic manifolds by F. Hirzebruch as the equality between the alternating sum of the dimensions of cohomology groups with coefficients in the sheaf of holomorphic sections of a holomorphic vector bundle and a certain topological invariant. It was then generalized in two directions; one as a functorial equality by A. Grothendieck and the other as the index theorem of elliptic operators by M.F. Atiyah and I.M. Singer.

In this chapter, we mostly discuss the former, i.e., the Grothendieck-Riemann-Roch theorem (cf. Theorem 15.11). The proof for this is divided into two parts; for embeddings and for projections. We examin the embedding case in detail, applying our localization approach, and prove a prototype of the localized Riemann-Roch theorem for embeddings on the level of Čech-de Rham cocycles (Corollary 15.1). It immediately yields a theorem in the universal situation (Theorem 15.4). We then present three variations of the localized Riemann-Roch theorem for embeddings (Theorems 15.5–15.7). For the projection part, we outline the proof quoting relevant materials. The relation with the Atiyah-Singer index theorem is briefly mentioned in Notes at the end of the chapter.

In the following, we sometimes omit the symbol $\wedge$ for the exterior product of forms and the symbol $\smile$ for the cup product of cohomology classes and simply denote them as products.

15.1 Riemann-Roch problem for curves

Let C be a compact Riemann surface. The original Riemann-Roch problem consists in finding the dimension of the space of meromorphic functions on C with prescribed orders of zeros and poles. To be more precise, let $\mathscr{M}(C)$ be the space of meromorphic functions on C and $D = \sum m_i p_i$ a divisor on C. If we set

$$L(D) = \{ \varphi \in \mathscr{M}(C) \mid (\varphi) + D \geq 0 \},$$

then a meromorphic function φ is in $L(D)$ if and only if, for i with $m_i > 0$, it has a pole of order at most m_i at p_i, and, for i with $m_i < 0$, it has a zero of order at least $-m_i$ at p_i. In terms of bundles and sheaves, we have an isomorphism

$$L(D) \simeq H^0(C; \mathscr{O}(L_D)).$$

The isomorphism is given by assigning to each meromorphic function in $L(D)$ the section which is locally represented by the holomorphic function obtained by multiplying φ by a (local) defining function of D.

We denote by $\mathscr{M}^1(C)$ the space of meromorphic 1-forms on C. For a meromorphic 1-form ω, we may define the associated divisor (ω) as in the case of meromorphic functions. We set

$$I(D) = \{ \omega \in \mathscr{M}^1(C) \mid (\omega) - D \geq 0 \}.$$

Then we have an isomorphism

$$I(D) \simeq H^0(C; \Omega^1(L_D^*)).$$

We denote by $l(D)$ and $i(D)$ the dimensions of $L(D)$ and $I(D)$, respectively. We also let $d = \sum m_i$, the degree of D, and g the genus of C, i.e., $\dim H^0(C; \Omega^1)$ (cf. the paragraph after Corollary 9.6). Then the classical Riemann-Roch theorem says that

$$l(D) = d - g + i(D) + 1. \tag{15.1}$$

We see later that this is a special case of the Hirzebruch-Riemann-Roch theorem (Theorem 15.12).

15.2 Characteristic classes of virtual bundles

K-groups

Let X be a paracompact topological space and let $A^{\mathrm{t}}(X)$ denote the free Abelian group generated by the isomorphism classes of continuous complex vector bundles on X. For simplicity, we denote the isomorphism class of a vector bundle E also by E. The *K-group* of continuous vector bundles on X, denoted by $K^{\mathrm{t}}(X)$, is the Abelian group obtained as the quotient of $A^{\mathrm{t}}(X)$ by the subgroup generated by the elements of the form $E-(E'+E'')$ for which there is an exact sequence

$$0 \longrightarrow E' \longrightarrow E \longrightarrow E'' \longrightarrow 0.$$

In the following, we consider characteristic classes of elements in $K^{\mathrm{t}}(X)$ and, for this purpose, it is not necessary to distinguish the class $[E]$ in $K^{\mathrm{t}}(X)$ of an element E in $A^{\mathrm{t}}(X)$ from E, so that we also denote $[E]$ by E. Thus $E+F = E \oplus F$ in $K^{\mathrm{t}}(X)$, as every short exact sequence of continuous vector bundles on a paracompact space splits (cf. Proposition 3.3). Note that every element in $K^{\mathrm{t}}(X)$ is expressed as $E - F$, which is referred as a *virtual bundle* (cf. Section 8.3). In fact $K^{\mathrm{t}}(X)$ has the structure of a ring with multiplication coming from the tensor products of vector bundles with the identity the class of the product bundle $\mathbb{C} \times X$.

If $f : X' \to X$ is a continuous map, there is a natural ring morphism $f^* : K^{\mathrm{t}}(X) \to K^{\mathrm{t}}(X')$ coming from the pull-back of vector bundles.

If M is a C^∞ manifold we can also construct the K-group $K^{\mathrm{d}}(M)$ of C^∞ vector bundles on M. There is a caninical morphism $K^{\mathrm{d}}(M) \to K^{\mathrm{t}}(M)$, which is in fact an isomorphism (cf. Remark 3.3).

Chern character and Todd class

Let X be a regular cell complex. For a complex vector bundle E of rank l on X, we have the (total) Chern class $c^*(E) = \sum_{i=0}^{l} c^i(E)$ in $H^*(X;\mathbb{Z})$, $c^0(E) = 1$ (cf. Section 5.2). We consider a formal factorization

$$\sum_{i=0}^{l} c^i(E)\, t^i = \prod_{j=1}^{l} (1 + \gamma_j t). \tag{15.2}$$

Thus $c^i(E)$ is the i-th elementary symmetric function in the γ_j's.

Definition 15.1. The *Chern character* $\mathrm{ch}^*(E)$ of E is the class in $H^*(X;\mathbb{Q})$ given by

$$\mathrm{ch}^*(E) = \sum_{j=1}^{l} e^{\gamma_j}.$$

Note that the right-hand side above is a symmetric series in the γ_j's so that it is expressed in terms of Chern classes. More explicitly, if we set

$s^i(E) = \sum_{j=1}^{l} \gamma_j^i$, then

$$\operatorname{ch}^*(E) = l + \sum_{i \geq 1} \frac{s^i(E)}{i!}.$$

The classes $c^i = c^i(E)$ and $s^i = s^i(E)$ are related by Newton's formula:

$$s^i - c^1 s^{i-1} + c^2 s^{i-2} - \cdots + (-1)^i i\, c^i = 0, \qquad i \geq 1. \tag{15.3}$$

While c^* is multiplicative, ch^* is additive, i.e., if

$$0 \longrightarrow E' \longrightarrow E \longrightarrow E'' \longrightarrow 0 \tag{15.4}$$

is an exact sequence of complex vector bundles,

$$c^*(E) = c^*(E') \cdot c^*(E'') \quad \text{and} \quad \operatorname{ch}^*(E) = \operatorname{ch}^*(E') + \operatorname{ch}^*(E'').$$

From the latter, we have a morphism

$$\operatorname{ch}^* : K^{\mathrm{t}}(X) \longrightarrow H^*(X; \mathbb{Q}). \tag{15.5}$$

If we have two complex vector bundles E and F,

$$\operatorname{ch}^*(E \otimes F) = \operatorname{ch}^*(E) \cdot \operatorname{ch}^*(F)$$

so that ch^* in (15.5) is in fact a ring morphism, the product in the right-hand side being the cup product.

In particular, for a virtual bundle of the form $\xi = \sum_{i=0}^{r} (-1)^i E_i$ with E_i complex vector bundles, $i = 0, \dots, r$,

$$c^*(\xi) = \prod_{i=0}^{r} c^*(E_i)^{\epsilon(i)} \quad \text{and} \quad \operatorname{ch}^*(\xi) = \sum_{i=1}^{r} (-1)^i \operatorname{ch}^*(E_i), \tag{15.6}$$

where $\epsilon(i) = (-1)^i$.

Definition 15.2. Using the factorization (15.2), we define the *Todd class* of E in $H^*(X; \mathbb{Q})$ by

$$\operatorname{td}(E) = \prod_{j=1}^{l} \frac{\gamma_j}{1 - e^{-\gamma_j}}.$$

Note that it can be also expressed in terms of Chern classes of E. It is multipicative, i.e., for an exact sequence as (15.4),

$$\operatorname{td}(E) = \operatorname{td}(E') \cdot \operatorname{td}(E'').$$

Note that the series in the right-hand side starts with 1 so that $\operatorname{td}(E)$ is invertible in $H^*(X; \mathbb{Q})$.

For a complex vector bundle E of rank l, we set $\lambda_E = \sum_{i=0}^{l} (-1)^i \bigwedge^i E$ in $K^{\mathrm{t}}(X)$ so that $\operatorname{ch}^*(\lambda_E) = \sum_{i=0}^{l} (-1)^i \operatorname{ch}^*(\bigwedge^i E)$, where we set $\bigwedge^0 E = \mathbb{C} \times X$. With these we have the following fundamental formula:

Theorem 15.1. *For a complex vector bundle E of rank l,*

$$\operatorname{ch}^*(\lambda_{E^*}) = c^l(E) \cdot \operatorname{td}(E)^{-1} \qquad \text{in } H^*(X; \mathbb{Q}).$$

This can be proved using the expression of $\mathrm{ch}^*(\bigwedge^i E^*)$ in terms of the γ_j's. We prove this on the level of differential forms below (cf. Theorem 15.2). We then prove a "localized version" of this, which is one of the essential ingredients in the proof of the localized Riemann-Roch theorem for embeddings (cf. Corollary 15.1).

15.3 Chern-Weil theory for virtual bundles

In this section, we let M denote a C^∞ manifold.

A fundamental relation in linear algebra

Let A be an $l \times l$ complex matrix. We set

$$\lambda^i A = \sum_{j=1}^{i} I \wedge \cdots \wedge \overset{j}{A} \wedge \cdots \wedge I, \qquad \lambda^0 A = I.$$

Note that $(I - e^{-A})/A$ makes sense, even if A is not invertible.

Lemma 15.1. *We have*

$$\sum_{i=0}^{l} (-1)^i \operatorname{tr}(e^{-\lambda^i A}) = \det A \cdot \det\Big(\frac{I - e^{-A}}{A}\Big).$$

Proof. We have

$$\det(I - e^{-A}) = \sum_{i=1}^{l} (-1)^i \operatorname{tr}(\textstyle\bigwedge^i e^{-A})$$

and, if A is in $GL(l, \mathbb{C})$, the identity follows from

$$\textstyle\bigwedge^i e^{-A} = e^{-\lambda^i A}.$$

Since $GL(l, \mathbb{C})$ is dense in $M(l, \mathbb{C})$,we have the lemma. □

Chern character form and Todd form

Let E be a C^∞ complex vector bundle on M. For a connection ∇ for E, let K denote its curvature and set $A = (\sqrt{-1}/2\pi)K$. We represent K locally by a curvature matrix. Recall that the total Chern form is defined by (cf. Section 8.2)

$$c^*(\nabla) = \det(I + A).$$

Definition 15.3. Using the following particular invariant series, the *Chern character form* and the *Todd form* of ∇ are defined by

$$\mathrm{ch}^*(\nabla) = \mathrm{tr}(e^A) \quad \text{and} \quad \mathrm{td}(\nabla) = \det\Big(\frac{A}{I - e^{-A}}\Big).$$

Proposition 15.1. *The images of* $\mathrm{ch}^*(E)$ *and* $\mathrm{td}(E)$ *by the canonical monomorphism* $H^*(M;\mathbb{Q}) \to H^*(M;\mathbb{C}) \simeq H_d^*(M)$ *are represented by* $\mathrm{ch}^*(\nabla)$ *and* $\mathrm{td}(\nabla)$*, respectively.*

The properties of determinant and trace imply that c^* and td are multiplicative, while ch^* is additive on the form level as well, i.e., if $(\nabla', \nabla, \nabla'')$ is a triple of connections compatible with the exact sequence (15.4), we have

$$\begin{aligned} &c^*(\nabla) = c^*(\nabla') \cdot c^*(\nabla''), \qquad \mathrm{td}(\nabla) = \mathrm{td}(\nabla') \cdot \mathrm{td}(\nabla''), \\ &\mathrm{ch}^*(\nabla) = \mathrm{ch}^*(\nabla') + \mathrm{ch}^*(\nabla''). \end{aligned} \tag{15.7}$$

Note that $I - e^{-A}$ is divisible by A and the result is invertible so that

$$\mathrm{td}(\nabla)^{-1} = \det\Big(\frac{I - e^{-A}}{A}\Big)$$

also makes sense.

If we set $s^i(\nabla) = \mathrm{tr}(A^i)$, then it is a closed $2i$-form on M. Denoting by l the rank of E, we have

$$c^*(\nabla) = 1 + \sum_{i=1}^{l} c^i(\nabla) \quad \text{and} \quad \mathrm{ch}^*(\nabla) = l + \sum_{i \geq 1} \frac{s^i(\nabla)}{i!}.$$

The forms $c^i = c^i(\nabla)$ and $s^i = s^i(\nabla)$ are again related by Newton's formula (15.3).

Note that the constant term in $\mathrm{td}(\nabla)$ is 1 and that $\mathrm{td}(\nabla)$ can be expressed as a series (in fact a polynomial) in $c^i(\nabla)$.

We have also the following formula on the form level (cf. Theorem 15.1), which is a direct consequence of Lemma 15.1:

Theorem 15.2. *Let* E *be a* C^∞ *complex vector bundle of rank* l *and* ∇ *a connection for* E*. Then we have*

$$\sum_{i=0}^{l} (-1)^i \,\mathrm{ch}^*(\textstyle\bigwedge^i \nabla^*) = c^l(\nabla) \cdot \mathrm{td}(\nabla)^{-1},$$

where ∇^* *denotes the connection for* E^* *dual to* ∇ *and* $\bigwedge^i \nabla^*$ *the connection for* $\bigwedge^i E^*$ *induced by* ∇^**. Here we set* $\bigwedge^0 E^* = \mathbb{C} \times M$ *and* $\bigwedge^0 \nabla^* = d$*, the exterior differential.*

Characteristic forms of virtual bundles

If we have a C^∞ complex vector bundle E_i on M, for each $i = 0, \dots, r$, we may consider the virtual bundle $\xi = \sum_{i=0}^{r}(-1)^i E_i$ as an element in $K^{\mathrm{d}}(M)$. Letting $\nabla^{(i)}$ be a connection for E_i, we denote by $\nabla^\star$ the family of connections $(\nabla^{(r)}, \dots, \nabla^{(0)})$ and define its total Chern form $c^*(\nabla^\star)$ by

$$c^*(\nabla^\star) = \prod_{i=0}^{r} c^*(\nabla^{(i)})^{\epsilon(i)},$$

where $\epsilon(i) = (-1)^i$. The q-th Chern form $c^q(\nabla^\star)$ is the component of $c^*(\nabla^\star)$ of degree $2q$. More generally, for an invariant series φ, we write $\varphi = P(c^1, c^2, \cdots)$ as before and set $\varphi(\nabla^\star) = P(c^1(\nabla^\star), c^2(\nabla^\star), \dots)$. In particular, we have

$$\mathrm{ch}^*(\nabla^\star) = \sum_{i=0}^{r}(-1)^i \,\mathrm{ch}^*(\nabla^{(i)}).$$

As in the case of vector bundles we have difference forms for families of connections (cf. Proposition 8.3):

Proposition 15.2. *Suppose we have $p+1$ families of connections $\nabla^\star_\nu = (\nabla^{(q)}_\nu, \dots, \nabla^{(0)}_\nu)$, $\nu = 0, \dots, p$. Then we have a form $\varphi(\nabla^\star_0, \dots, \nabla^\star_p)$ alternating in the $p+1$ entries and satisfying*

$$\sum_{\nu=0}^{p}(-1)^\nu \varphi(\nabla^\star_0, \dots, \widehat{\nabla^\star_\nu}, \dots, \nabla^\star_p) + (-1)^p d\varphi(\nabla^\star_0, \dots, \nabla^\star_p) = 0.$$

Proof. This is proved as Proposition 8.3. Thus, consider the product $\mathbb{R}^p \times M$ with projection $\rho : \mathbb{R}^p \times M \to M$. For each $i = 0, \dots, r$, let $\tilde{\nabla}^{(i)}$ be the connection for $\rho^* E_i$ given by

$$\tilde{\nabla}^{(i)} = \Big(1 - \sum_{\nu=1}^{p} t_\nu\Big)\rho^*\nabla^{(i)}_0 + \sum_{\nu=1}^{p} t_\nu \rho^* \nabla^{(i)}_\nu.$$

We set $\tilde{\nabla}^\star = (\tilde{\nabla}^{(r)}, \dots, \tilde{\nabla}^{(0)})$ and define $\varphi(\nabla^\star_0, \dots, \nabla^\star_p) = \rho'_*(\varphi(\tilde{\nabla}^\star))$, where $\rho' : \Delta^p \times M \to M$ is the restriction of ρ. □

For two families of connections $\nabla^\star_0$ and $\nabla^\star_1$, we have

$$\varphi(\nabla^\star_1) - \varphi(\nabla^\star_0) = d\varphi(\nabla^\star_0, \nabla^\star_1).$$

Hence we reprove that the class $[\varphi(\nabla^\star)]$ of the closed form $\varphi(\nabla^\star)$, which is in fact $\varphi(\xi)$, depends only on ξ and not on the choice of $\nabla^\star$. In particular, the total Chern class $c^*(\xi)$ is the class of $c^*(\nabla^\star)$ and is also given by the first identity in (15.6). The q-th Chern class $c^q(\xi)$ is the component of $c^*(\xi)$ in $H^{2q}_d(M)$ and is the class of $c^q(\nabla^\star)$.

Virtual bundle associated with a sequence

Let

$$0 \longrightarrow E_r \xrightarrow{\psi_r} \cdots \longrightarrow E_1 \xrightarrow{\psi_1} E_0 \longrightarrow 0 \tag{15.8}$$

be a sequence of C^∞ complex vector bundles on M and, for each i, let $\nabla^{(i)}$ be a connection for E_i. We say that the family $(\nabla^{(r)}, \ldots, \nabla^{(0)})$ is *compatible* with the sequence if, for each i, the following diagram is commutative:

$$\begin{array}{ccc} A^0(M, E_i) & \xrightarrow{\psi_i} & A^0(M, E_{i+1}) \\ \downarrow{\scriptstyle \nabla^{(i)}} & & \downarrow{\scriptstyle \nabla^{(i+1)}} \\ A^1(M, E_i) & \xrightarrow{1\otimes\psi_i} & A^1(M, E_{i+1}). \end{array}$$

The following two propositions generalize Propositions 8.5 and 8.6:

Proposition 15.3. *Suppose the sequence* (15.8) *is exact. Given a connection* $\nabla^{(0)}$ *for* E_0*, then there exist connections* $\nabla^{(i)}$ *for* E_i*,* $i = 1, \ldots, r$*, such that the family* $(\nabla^{(r)}, \ldots, \nabla^{(0)})$ *is compatible with* (15.8).

Proof. We proceed by induction on r. It holds obviously if $r = 1$. Suppose it holds for $r = k \geq 1$ and prove that it also holds for $r = k + 1$. Identifying E_{k+1} with a subbundle of E_k by ψ_{k+1}, we have the exact sequences:

$$0 \longrightarrow E_{k+1} \longrightarrow E_k \xrightarrow{\varphi} E_k/E_{k+1} \longrightarrow 0 \tag{15.9}$$

and

$$0 \longrightarrow E_k/E_{k+1} \longrightarrow E_{k-1} \longrightarrow \cdots \longrightarrow E_1 \longrightarrow E_0 \longrightarrow 0 \tag{15.10}$$

By induction hypothesis, there exist connections ∇ for E_k/E_{k+1} and $\nabla^{(i)}$ for E_i, $i = 1, \ldots, k-1$, such that $(\nabla, \nabla^{(k-1)}, \ldots, \nabla^{(0)})$ is compatible with (15.10). By Proposition 8.5, there exist connections $\nabla^{(k+1)}$ and $\nabla^{(k)}$ for E_{k+1} and E_k such that $(\nabla^{(k+1)}, \nabla^{(k)}, \nabla)$ is compatible with (15.9). Then $(\nabla^{(k+1)}, \ldots, \nabla^{(0)})$ is compatible with (15.8). □

Proposition 15.4. *Suppose the sequence* (15.8) *is exact. For a family* $(\nabla^{(r)}, \ldots, \nabla^{(0)})$ *of connections compatible with* (15.8),

$$\prod_{i=0}^{r} c^*(\nabla^{(i)})^{\epsilon(i)} = 1,$$

where $\epsilon(i) = (-1)^i$.

Proof. We prove by induction on r. If $r = 1$, the statement is true, as $c^*(\nabla^{(1)}) = c^*(\nabla^{(0)})$. Now suppose the statement is true for $r = k$ and show for the case $r = k+1$. Again, the sequence (15.8) splits into (15.9) and (15.10). Since $\nabla^{(k+1)}$ and $\nabla^{(k)}$ are compatible with ψ_{k+1}, $\nabla^{(k)}$ induces a connection ∇ for E_k/E_{k+1} so that $(\nabla^{(k+1)}, \nabla^{(k)}, \nabla)$ is compatible with (15.9). By Proposition 8.6, we have

$$c^*(\nabla^{(k)}) = c^*(\nabla^{(k+1)}) \cdot c^*(\nabla).$$

On the other hand, as $(\nabla, \nabla^{(k-1)}, \dots, \nabla^{(0)})$ is compatible with (15.10), by the induction hypothesis, we have the proposition. □

From this we have the following:

Proposition 15.5. *Suppose the sequence* (15.8) *is exact. Let* φ *be an invariant polynomial and* $\nabla_k^\star = (\nabla_k^{(r)}, \dots, \nabla_k^{(0)})$, $k = 0, \dots, p$, *families of connections compatible with* (15.8). *Then*

$$\varphi(\hat{\nabla}_0^\star, \dots, \hat{\nabla}_p^\star) = \varphi(\nabla_0^{(0)}, \dots, \nabla_p^{(0)}) \quad \textit{in particular} \quad \varphi(\hat{\xi}) = \varphi(E_0),$$

where $\hat{\nabla}_k^\star$ *denotes the family of connections* $(\nabla_k^{(r)}, \dots, \nabla_k^{(1)})$ *for the virtual bundle* $\hat{\xi} = \sum_{i=1}^r (-1)^{i-1} E_i$. *Similarly for the other "partitions" of the virtual bundle* ξ.

15.4 Local Chern classes and characters

In this section also, we let M denote a C^∞ manifold.

Characteristic cocycles in the Čech-de Rham complex

Using difference forms we may define characteristic classes of virtual bundles in the Čech-de Rham cohomology $H_D^*(\mathcal{U})$ for an arbitrary covering $\mathcal{U}$ of M (cf. Section 8.4). For simplicity we consider coverings $\mathcal{U}$ consisting of two open sets U_0 and U_1. In fact this is the case we consider in the following.

If $\xi = \sum_{i=0}^r (-1)^i E_i$ is a virtual bundle, we take a family of connections $\nabla_\nu^\star = (\nabla_\nu^{(r)}, \dots, \nabla_\nu^{(0)})$ for ξ on each U_ν, $\nu = 0, 1$. For the collection $\nabla_*^\star = (\nabla_0^\star, \nabla_1^\star)$ and a symmetric series φ, we define a cochain $\varphi(\nabla_*^\star)$ in $A^*(\mathcal{U})$ by

$$\varphi(\nabla_*^\star) = (\varphi(\nabla_0^\star), \varphi(\nabla_1^\star), \varphi(\nabla_0^\star, \nabla_1^\star)).$$

It is in fact a cocycle and defines a class $[\varphi(\nabla_*^\star)]$ in $H_D^*(\mathcal{U})$. It does not depend on the choice of the collection of families of connections $\nabla_*^\star$ and corresponds to the class $\varphi(\xi)$ under the isomorphism $H_D^*(\mathcal{U}) \simeq H_d^*(M)$.

Now we prove a formula which describes explicitly the difference between the cocycle for the product of two symmetries series and the product of cocycles for these series. Although it is stated for connections of a single vector bundle, the same formula holds for families of connections of a virtual bundle. Thus let E be a complex vector bundle on M. Note that $\varphi\psi(\nabla) = \varphi(\nabla) \cdot \psi(\nabla)$, for symmetric series φ and ψ and a connection ∇ for E. Let ∇_i be a connection for E on U_i, $i = 0, 1$. We set $\tilde{\nabla} = (1-t)\rho^*\nabla_0 + t\rho^*\nabla_1$ with $\rho : \mathbb{R} \times U_{01} \to U_{01}$, as in the proof of Proposition 15.2.

Lemma 15.2. *For two symmetric series φ and ψ, we have*

$$\varphi\psi(\nabla_0, \nabla_1) = \varphi(\nabla_0) \cdot \psi(\nabla_0, \nabla_1) + \varphi(\nabla_0, \nabla_1) \cdot \psi(\nabla_1) - d\,\tau_{01},$$

where τ_{01} is a form on U_{01} given by

$$\tau_{01} = \rho'_*(\varphi(\rho^*\nabla_0, \tilde{\nabla}) \cdot d\psi(\rho^*\nabla_1, \tilde{\nabla})).$$

Proof. By definition, the left-hand side is equal to $\rho'_*(\varphi(\tilde{\nabla}) \cdot \psi(\tilde{\nabla}))$ and the sum of the first two terms in the right-hand side is equal to

$$\rho'_*(\varphi(\rho^*\nabla_0) \cdot \psi(\tilde{\nabla}) + \varphi(\tilde{\nabla}) \cdot \psi(\rho^*\nabla_1)).$$

We have

$$\begin{aligned} &\varphi(\tilde{\nabla}) \cdot \psi(\tilde{\nabla}) - (\varphi(\rho^*\nabla_0) \cdot \psi(\tilde{\nabla}) + \varphi(\tilde{\nabla}) \cdot \psi(\rho^*\nabla_1)) \\ &\qquad = d\varphi(\rho^*\nabla_0, \tilde{\nabla}) \cdot d\psi(\rho_1^*\nabla, \tilde{\nabla}) - \rho^*(\varphi(\nabla_0) \cdot \psi(\nabla_1)). \end{aligned}$$

If we denote by i the inclusion of the boundary $\{0,1\}$ of $[0,1]$ into $[0,1]$ and by $\partial\rho'$ the restriction of ρ' to $\{0,1\}$, the lemma follows from the identity $\rho'_* \circ \rho^* = 0$, Proposition 6.3 and

$$\begin{aligned} &(\partial\rho')_* \circ i^*(\varphi(\rho^*\nabla_0, \tilde{\nabla}) \cdot d\psi(\rho^*\nabla_1, \tilde{\nabla})) \\ &\qquad = \varphi(\nabla_0, \nabla_1) \cdot d\psi(\nabla_1, \nabla_1) - \varphi(\nabla_0, \nabla_0) \cdot d\psi(\nabla_1, \nabla_0) \\ &\qquad = 0. \end{aligned}$$

□

From Lemma 15.2, we have the following:

Proposition 15.6. *For two symmetric series φ and ψ, we have, in $A^*(\mathcal{U})$,*

$$\varphi\psi(\nabla_*) = \varphi(\nabla_*) \smile \psi(\nabla_*) + D\tau,$$

where $\tau = (0, 0, \tau_{01})$ with τ_{01} a form on U_{01} as given in Lemma 15.2.

Remark 15.1. In the situation of Proposition 15.6, if $\varphi(\nabla_*)$ is in $A^*(\mathcal{U}, U_0)$, i.e., if $\varphi(\nabla_0) = 0$, so is $\varphi\psi(\nabla_*)$, since $\varphi\psi(\nabla_0) = \varphi(\nabla_0) \cdot \psi(\nabla_0)$. The proposition shows that the class $\varphi\psi(E)$ coincides with $\varphi(E) \smile \psi(E)$ in $H_D^*(\mathcal{U}, U_0)$, since τ is also in $A^*(\mathcal{U}, U_0)$.

Localization by exactness

We discuss the localization theory of characteristic classes of virtual bundles by exactness.

Let S be a closed set in M. Letting $U_0 = M \setminus S$ and U_1 a neighborhood of S in M, we consider the covering $\mathcal{U} = \{U_0, U_1\}$ of M. Note that the cup product of a cochain in $A^*(\mathcal{U}, U_0)$ and a cochain in $A^*(\mathcal{U})$ is in $A^*(\mathcal{U}, U_0)$ and it induces a right $H^*(M;\mathbb{C})$-module structure on $H^*(M, M \setminus S;\mathbb{C})$.

Let

$$0 \longrightarrow E_r \longrightarrow \cdots \longrightarrow E_1 \longrightarrow E_0 \longrightarrow 0 \tag{15.11}$$

be a complex of C^∞ complex vector bundles on M which is exact on U_0. Then we will see below that, for each $q > 0$, there is a canonical localization $c_S^q(\xi)$ in $H^{2q}(M, M \setminus S;\mathbb{C})$ of the Chern class $c^q(\xi)$ in $H^{2q}(M;\mathbb{C})$ of the virtual bundle $\xi = \sum_{i=0}^r (-1)^i E_i$. The basic fact is the following "vanishing theorem", which follows from Proposition 15.4:

Lemma 15.3. *If* $\nabla_0^\star$ *is a family of connections on* U_0 *compatible with* (15.11), *then, for each* $q > 0$,

$$c^q(\nabla_0^\star) = 0.$$

In fact, the above holds for the difference form of a finite number of families of connections compatible with (15.11) on U_0. For a symmetric series φ without constant term, we also have a similar vanishing $\varphi(\nabla_0^\star) = 0$.

Let $\nabla_0^\star$ be a family of connections compatible with (15.11) on U_0 and $\nabla_1^\star$ an arbitrary family of connections for $\xi = \sum_{i=0}^r (-1)^i E_i$ on U_1. Then the class $c^q(\xi)$ is represented by the cocycle

$$c^q(\nabla_*^\star) = (c^q(\nabla_0^\star),\, c^q(\nabla_1^\star),\, c^q(\nabla_0^\star, \nabla_1^\star))$$

in $A^{2q}(\mathcal{U})$. By Lemma 15.3, we have $c^q(\nabla_0^\star) = 0$ and thus the cocycle is in $A^{2q}(\mathcal{U}, U_0)$ and it defines a class $c_S^q(\xi)$ in $H^{2q}(M, M \setminus S;\mathbb{C})$. It is sent to $c^q(\xi)$ by the canonical morphism $j^* : H^{2q}(M, M \setminus S;\mathbb{C}) \to H^{2q}(M;\mathbb{C})$. It is not difficult to see that the class $c_S^q(\xi)$ does not depend on the choice of the family of connections $\nabla_0^\star$ compatible with (15.11) or on the choice of the family of connections $\nabla_1^\star$.

If φ is a symmetric series without constant term, we may also define the localized class $\varphi_S(\xi)$ of $\varphi(\xi)$. In particular, noting that the alternating sum of the ranks of E_i is zero, if $M \setminus S \neq \emptyset$, we have the localized Chern character $\mathrm{ch}_S^*(\xi)$ in the relative cohomology $H^*(M, M \setminus S;\mathbb{C})$, which is sent to $\mathrm{ch}^*(\xi)$ by the morphism j^*. It is the class of the cocycle

$$\mathrm{ch}^*(\nabla_*^\star) = (0, \mathrm{ch}^*(\nabla_1^\star),\, \mathrm{ch}^*(\nabla_0^\star, \nabla_1^\star))$$

in $A^*(\mathcal{U}, U_0)$.

Let F be a C^∞ complex vector bundle on M and ∇ a connection for F. Then its Chern character $\mathrm{ch}^*(F)$ is the class of the cocycle

$$\mathrm{ch}^*(\nabla) = (\mathrm{ch}^*(\nabla)|_{U_0}, \mathrm{ch}^*(\nabla)|_{U_1}, 0)$$

in $A^*(\mathcal{U})$. The complex

$$0 \longrightarrow F \otimes E_r \longrightarrow \cdots \longrightarrow F \otimes E_1 \longrightarrow F \otimes E_0 \longrightarrow 0$$

is exact on U_0 and the family $\nabla \otimes \nabla_0^\star = (\nabla \otimes \nabla_0^{(r)}, \dots, \nabla \otimes \nabla_0^{(0)})$ of connections is compatible with the above sequence on U_0. We set $F \otimes \xi = \sum_{i=0}^r (-1)^i F \otimes E_i$ and let $\nabla \otimes \nabla_1^\star$ denote the family $(\nabla \otimes \nabla_1^{(r)}, \dots, \nabla \otimes \nabla_1^{(0)})$. Then $\mathrm{ch}^*(F \otimes \xi)$ is the class of the cocycle

$$\mathrm{ch}^*(\nabla \otimes \nabla_*^\star) = (0, \mathrm{ch}^*(\nabla \otimes \nabla_1^\star), \mathrm{ch}^*(\nabla \otimes \nabla_0^\star, \nabla \otimes \nabla_1^\star)).$$

We have

$$\mathrm{ch}^*(\nabla \otimes \nabla_1^\star) = \mathrm{ch}^*(\nabla) \cdot \mathrm{ch}^*(\nabla_1^\star),$$
$$\mathrm{ch}^*(\nabla \otimes \nabla_0^\star, \nabla \otimes \nabla_1^\star) = \mathrm{ch}^*(\nabla) \cdot \mathrm{ch}^*(\nabla_0^\star, \nabla_1^\star).$$

Hence, recalling the definition of the cup product, we have

$$\mathrm{ch}^*(\nabla \otimes \nabla_*^\star) = \mathrm{ch}^*(\nabla) \smile \mathrm{ch}^*(\nabla_*^\star) \qquad \text{in } A^*(\mathcal{U}, U_0),$$

thus

$$\mathrm{ch}_S^*(F \otimes \xi) = \mathrm{ch}^*(F) \cdot \mathrm{ch}_S^*(\xi) \qquad \text{in } H^*(M, M \smallsetminus S; \mathbb{C}). \tag{15.12}$$

15.5 Universal localized Riemann-Roch theorem

In this section, we let M be a C^∞ manifold and E a C^∞ complex vector bundle of rank l on M.

Koszul complex

Let s be a C^∞ section of E. The *Koszul complex* associated with (E, s) is the sequence

$$0 \longrightarrow \bigwedge^l E^* \xrightarrow{d_s} \cdots \xrightarrow{d_s} \bigwedge^i E^* \xrightarrow{d_s} \bigwedge^{i-1} E^* \xrightarrow{d_s} \cdots \xrightarrow{d_s} \bigwedge^0 E^* \longrightarrow 0, \tag{15.13}$$

where $\bigwedge^0 E^* = \mathbb{C} \times M$ and $d_s = d_s^i : \bigwedge^i E^* \to \bigwedge^{i-1} E^*$ is defined by

$$d_s(\varphi_1 \wedge \cdots \wedge \varphi_i) = \sum_{j=1}^i (-1)^{j-1} \varphi_j(s(x)) \varphi_1 \wedge \cdots \wedge \widehat{\varphi_j} \wedge \cdots \wedge \varphi_i$$

for $\varphi_1, \dots, \varphi_i$ in E_x^*, $x \in M$. It is not difficult to see that it forms a complex of vector bundles, i.e., $d_s^i \circ d_s^{i+1} = 0$. Let S denote the zero set of s.

Proposition 15.7. *The sequence* (15.13) *is exact on* $M \setminus S$.

Proof. Let x be a point in $M \setminus S$ and take a neighborhood U of x in $M \setminus S$ where E is trivial. Since we may take s as a part of a frame of E on U, there exists a frame $(e_1^*, \dots, e_l^*)$ of E^* on U such that $e_1^*(s) = 1$ and $e_j^*(s) = 0$, for $j \geq 2$. Then we see that both $\operatorname{Im} d_s^{i-1}$ and $\operatorname{Ker} d_s^i$ are spanned by the elements $e_{j_1}^* \wedge \cdots \wedge e_{j_r}^*$, $2 \leq j_1 < \cdots < j_r \leq l$. □

From definitions of dual and exterior product of connections (Section 8.3), we have:

Lemma 15.4. *Let* ∇ *be a connection for* E *on* $M \setminus S$. *If* ∇ *is* s*-trivial, then the family* $(\bigwedge^l \nabla^*, \dots, \bigwedge^0 \nabla^*)$ *is compatible with* (15.13) *on* $M \setminus S$.

Exercise 15.1. Show the above lemma.

In the above situation, we consider two localized classes. Thus let $U_0 = M \setminus S$ and U_1 a neighborhood of S and consider the covering $\mathcal{U} = \{U_0, U_1\}$. Let ∇_0 be an s-trivial connection for E on U_0 and ∇_1 an arbitrary connction for E on U_1.

Localized top Chern class of E: Since $c^l(\nabla_0) = 0$, we have the localization $c_S^l(E, s)$ in $H^{2l}(M, M \setminus S; \mathbb{C})$, which is represented by the cocycle

$$c^l(\nabla_*) = (0, c^l(\nabla_1), c^l(\nabla_0, \nabla_1)) \quad \text{in} \quad A^{2l}(\mathcal{U}, U_0).$$

Note that the class $c_S^l(E, s)$ is in the image of $H^{2l}(M, M \setminus S; \mathbb{Z}) \to H^{2l}(M, M \setminus S; \mathbb{C})$, provided that S is a subcomplex of real codimension greater than or equal to 2 with respect to some triangulation of M (cf. Theorem 10.13).

Localized Chern character of λ_{E^*}: We set $\lambda_{E^*} = \sum_{i=0}^l (-1)^i \bigwedge^i E^*$. Since $\operatorname{ch}^*(\bigwedge^\star \nabla_0^*) = 0$ by Lemma 15.4, we have the localization $\operatorname{ch}_S^*(\lambda_{E^*}, s)$ in $H^*(M, M \setminus S; \mathbb{C})$, which is represented by the cocycle

$$\operatorname{ch}^*(\textstyle\bigwedge^\star \nabla_*^*) = (0, \operatorname{ch}^*(\bigwedge^\star \nabla_1^*), \operatorname{ch}^*(\bigwedge^\star \nabla_0^*, \bigwedge^\star \nabla_1^*)) \quad \text{in} \quad A^*(\mathcal{U}, U_0),$$

where $\bigwedge^\star \nabla_\nu^*$ denote the family of connections $(\bigwedge^l \nabla_\nu^*, \dots, \bigwedge^0 \nabla_\nu^*)$ on U_ν, for $\nu = 0, 1$.

Inverse Todd class of E: We also have the inverse Todd class $\mathrm{td}^{-1}(E)$, which is represented by the cocycle

$$\mathrm{td}^{-1}(\nabla_*) = (\mathrm{td}^{-1}(\nabla_0), \mathrm{td}^{-1}(\nabla_1), \mathrm{td}^{-1}(\nabla_0, \nabla_1)) \quad \text{in} \quad A^*(\mathcal{U}).$$

Note that the class $\mathrm{td}^{-1}(E)$ is in the image of the canonical monomorphism $H^*(M;\mathbb{Q}) \to H^*(M;\mathbb{C})$.

In order to apply Proposition 15.6, let $\rho : \mathbb{R} \times U_{01} \to U_{01}$ be the projection and we consider the connection $\tilde{\nabla} = (1-t)\rho^*\nabla_0 + t\rho^*\nabla_1$ for ρ^*E. Let $\bigwedge^* \tilde{\nabla}^*$ denote the family of connections $(\bigwedge^l \tilde{\nabla}^*, \dots, \bigwedge^0 \tilde{\nabla}^*)$. Let $\rho' : [0,1] \times U_{01} \to U_{01}$ be the restriction of ρ. We also have the cup product

$$A^*(\mathcal{U}, U_0) \times A^*(\mathcal{U}) \stackrel{\smile}{\longrightarrow} A^*(\mathcal{U}, U_0).$$

Theorem 15.3. *In the above situation, we have*

$$\mathrm{ch}^*(\textstyle\bigwedge^* \nabla_*^*) = c^l(\nabla_*) \smile \mathrm{td}^{-1}(\nabla_*) + D\tau \qquad \textit{in} \quad A^*(\mathcal{U}, U_0),$$

where $\tau = (0, 0, \tau_{01})$, $\tau_{01} = \rho'_*(c^l(\rho^*\nabla_0, \tilde{\nabla}) \cdot d\, \mathrm{td}^{-1}(\rho^*\nabla_1, \tilde{\nabla}))$.

Proof. We claim that $(c^l \cdot \mathrm{td}^{-1})(\nabla_*)$ is in $A^*(\mathcal{U}, U_0)$ and that

$$\mathrm{ch}^*(\textstyle\bigwedge^* \nabla_*^*) = (c^l \cdot \mathrm{td}^{-1})(\nabla_*) \quad \text{in} \quad A^*(\mathcal{U}, U_0). \tag{15.14}$$

Then the theorem follows from Proposition 15.6 (see also Remark 15.1).

First, we have

$$(c^l \cdot \mathrm{td}^{-1})(\nabla_*) = ((c^l \cdot \mathrm{td}^{-1})(\nabla_0), (c^l \cdot \mathrm{td}^{-1})(\nabla_1), (c^l \cdot \mathrm{td}^{-1})(\nabla_0, \nabla_1)).$$

We have $(c^l \cdot \mathrm{td}^{-1})(\nabla_0) = c^l(\nabla_0) \cdot \mathrm{td}^{-1}(\nabla_0) = 0$. By Theorem 15.2 and definition of difference forms, we have

$$\begin{aligned}
(c^l \cdot \mathrm{td}^{-1})(\nabla_1) &= c^l(\nabla_1) \cdot \mathrm{td}^{-1}(\nabla_1) = \mathrm{ch}^*(\textstyle\bigwedge^* \nabla_1^*), \\
(c^l \cdot \mathrm{td}^{-1})(\nabla_0, \nabla_1) &= \rho'_*(c^l(\tilde{\nabla}) \cdot \mathrm{td}^{-1}(\tilde{\nabla})) \\
&= \rho'_* \mathrm{ch}^*(\textstyle\bigwedge^* \tilde{\nabla}^*) = \mathrm{ch}^*(\bigwedge^* \nabla_0^*, \bigwedge^* \nabla_1^*).
\end{aligned}$$

Hence we have (15.14) and the theorem. □

Corollary 15.1 (Prototype of localized RR for embeddings). *We have*

$$\mathrm{ch}_S^*(\lambda_{E^*}, s) = c_S^l(E, s) \cdot \mathrm{td}(E)^{-1}.$$

This is an essential ingredient for the localized Riemann-Roch theorem for embeddings.

Remark 15.2. 1. The equality in Corollary 15.1 holds *a priori* in $H^*(M, M \setminus S; \mathbb{C})$. However, since the right-hand side is in $H^*(M, M \setminus S; \mathbb{Q})$, so is the left-hand side and the equality is in $H^*(M, M \setminus S; \mathbb{Q})$.

2. If we take the image of the above equality by the canonical morphism $j^* : H^*(M, M \setminus S; \mathbb{Q}) \to H^*(M; \mathbb{Q})$, it becomes the formula of Theorem 15.1.

Universal situation

Let $\pi : E \to M$ be a C^∞ complex vector bundle of rank l, as above. We denote by Σ the image of the zero section. Then we have the Thom class Ψ_E in $H^{2l}(E, E \setminus \Sigma)$ and the Thom isomorphism

$$T : H^*(M) \xrightarrow{\sim} H^{2l+*}(E, E \setminus \Sigma),$$

which may be expressed as $T(\alpha) = \Psi_E \cdot \pi^* \alpha$ (cf. Sections 4.3 and 7.9). Since $\Psi_E = c^l(\pi^* E, s_\Delta)$ (cf. Theorem 10.9), applying Corollary 15.1 to $\pi^* E$ and s_Δ, we have:

Theorem 15.4 (Universal localized RR for embeddings). *In the above situation, we have*

$$\begin{aligned} \mathrm{ch}^*_\Sigma(\lambda_{\pi^* E^*}, s_\Delta) &= \Psi_E \cdot \mathrm{td}(\pi^* E)^{-1} \\ &= T(\mathrm{td}(E)^{-1}) \qquad \text{in } \ H^*(E, E \setminus \Sigma; \mathbb{Q}). \end{aligned}$$

Once we have a section $s : M \to E$, we recover the equality in Corollary 15.1 by pulling back the above by s (cf. Theorem 10.10).

As mentioned before the equality in Corollary 15.1 is an essential step in the proof of the Riemann-Roch theorem for embeddings. The next step is to identify $c^l_S(E, s)$ with the Thom class of a certain vector bundle.

We may interpret Theorem 15.4 as the Riemann-Roch theorem for the embedding $\Sigma \hookrightarrow E$ with $\mathrm{td}(E)$ the Todd class of the normal bundle E of Σ in E.

15.6 Riemann-Roch theorem for embeddings

Let M be a C^∞ manifold and N a C^∞ complex vector bundle of rank k on M. Also let s be a C^∞ section of N and V the zero set of s. Then we have the localized top Chern class $c_V^k(N,s)$ in $H^{2k}(M, M \smallsetminus V)$ and the localized Chern character $\mathrm{ch}_V^*(\lambda_{N^*}, s)$ in $H^*(M, M \smallsetminus V)$, $\lambda_{N^*} = \sum_{i=0}^k (-1)^i \bigwedge^i N^*$.

If we apply Theorem 15.3 in this situation, it already gives a prototype of the Riemann-Roch theorem for embeddings on the level of Čech-de Rham cocycles. On the cohomology level, it reads (cf. Corollary 15.1)

$$\mathrm{ch}_V^*(\lambda_{N^*}, s) = c_V^k(N,s) \cdot \mathrm{td}(N)^{-1} \qquad \text{in } H^*(M, M \smallsetminus V). \tag{15.15}$$

In the following, we examine the above equality more closely in the following three cases:

(a) M is a complex manifold, $\boldsymbol{X} = (X, \mathscr{O}_{\boldsymbol{X}})$ a complex space in M defined by a regular section (cf. Definition 11.32) and $V = X$ the underlying variety.
(b) M is a C^∞ manifold and V a closed submanifold of M such that its normal bundle N_V admits the structure of a complex vector bundle.
(c) M is a complex manifold and V a closed complex submanifold of M.

We will see that, in each of these cases, $c_V^k(N,s)$ may be interpreted as the Thom class of a subspace or a submanifold and that the right-hand side of (15.15) is the image of the inverse Todd class of some bundle by the Thom morphism. We also give an interpretation of the left-hand side in each case. These constitute a Riemann-Roch theorem for embeddings.

I. *Case* (a)

Let M be a complex manifold of dimension n and $\boldsymbol{X} = (X, \mathscr{O}_{\boldsymbol{X}})$ the complex space in M defined by a regular section s of a holomorphic vector bundle N of rank k on M. Thus $\boldsymbol{X}$ is of dimension $d = n-k$. In view of Theorem 12.9, we give the following:

Definition 15.4. The *Thom class* $\Psi_{\boldsymbol{X}}$ of $\boldsymbol{X}$ is defined by

$$\Psi_{\boldsymbol{X}} = c_X^k(N,s) \qquad \text{in } H^{2k}(M, M \smallsetminus X).$$

Let U be a neighborhood of X in M with a deformation retraction $r: U \to X$, for example, taking a triangulation K_0 of M compatible with X, we may take the open star $O_{K'}(X)$ in the second barycentric subdivision K' of K_0 as U. Note that $H^{2k+p}(M, M \smallsetminus X) \simeq H^{2k+p}(U, U \smallsetminus X)$ by excision. We define the *Thom morphism*

$$T_{\boldsymbol{X}} : H^p(X) \longrightarrow H^{2k+p}(M, M \smallsetminus X) \qquad \text{by } \alpha \mapsto \Psi_{\boldsymbol{X}} \smile r^*(\alpha).$$

Then we have the following commutative diagram:

$$\begin{array}{ccc} H^p(X) & \xrightarrow{T_{\boldsymbol{X}}} & H^{2k+p}(M, M \smallsetminus X) \\ {\scriptstyle P_X}\downarrow & \swarrow_{\wr}\ {\scriptstyle A_{M,X}} & \\ \breve{H}_{2d-p}(X), & & \end{array}$$

where P_X is the Poincaré morphism, given by the cap product with X (cf. Section 13.2).

Denoting by $i : X \hookrightarrow M$ the embedding, we define the *Gysin morphism* $i_* : H^p(X) \to H^{2k+p}(M)$ as the composition (cf. Section 13.2)

$$H^p(X) \xrightarrow{T_{\boldsymbol{X}}} H^{2k+p}(M, M \smallsetminus X) \xrightarrow{j^*} H^{2k+p}(M).$$

Remark 15.3. If $\boldsymbol{X}$ is reduced, the above Thom class and Thom isomorphism coincide with those for $V = X$ (cf. Theorem 12.9, Definition 13.1, (13.9) and (13.10)).

If we set $N_X = N|_X$, the bundle r^*N_X admits a C^∞ structure and is isomorphic with $N|_U$ as C^∞ bundle (cf. Remark 3.3). Thus the right-hand side of (15.15) is written

$$T_{\boldsymbol{X}}(\mathrm{td}(N_X)^{-1}).$$

As for the left-hand side, from the Koszul complex for (N, s), we have a sequence

$$0 \longrightarrow \mathscr{O}_M(\textstyle\bigwedge^k N^*) \xrightarrow{d_s} \mathscr{O}_M(\textstyle\bigwedge^{k-1} N^*) \xrightarrow{d_s} \cdots \xrightarrow{d_s} \mathscr{O}_M(N^*) \xrightarrow{d_s} \mathscr{O}_M.$$

The following is proved using the fact that s is a regular section:

Lemma 15.5. *The above sequence is exact.*

Since the image of $d_s : \mathscr{O}_M(N^*) \to \mathscr{O}_M$ is the ideal sheaf $\mathscr{I}$ of $\mathscr{O}_M$ generated by the local components of s and $\mathscr{O}_{\boldsymbol{X}} = i^{-1}(\mathscr{O}_M/\mathscr{I})$, where $i : X \to M$ is the embedding, we have the exact sequence of $\mathscr{O}_M$-modules

$$0 \longrightarrow \mathscr{O}_M(\textstyle\bigwedge^k N^*) \longrightarrow \cdots \longrightarrow \mathscr{O}_M(N^*) \longrightarrow \mathscr{O}_M \longrightarrow i_*\mathscr{O}_{\boldsymbol{X}} \longrightarrow 0. \quad (15.16)$$

In view of the above, we define (cf. (15.21) below)

$$\mathrm{ch}_X^*(i_*\mathscr{O}_{\boldsymbol{X}}) = \mathrm{ch}_X^*(\lambda_{N^*}, s).$$

Then from (15.15), we have:

Theorem 15.5 (RR for embeddings I). *In the case* (a), *we have:*

$$\operatorname{ch}_X^*(i_!\mathcal{O}_{\boldsymbol{X}}) = T_{\boldsymbol{X}}(\operatorname{td}(N_X)^{-1}) \qquad \textit{in } H^*(M, M \smallsetminus X),$$
$$\operatorname{ch}^*(i_!\mathcal{O}_{\boldsymbol{X}}) = i_*(\operatorname{td}(N_X)^{-1}) \qquad \textit{in } H^*(M).$$

Remark 15.4. 1. For a complex vector bundle F on X, the complex

$$0 \longrightarrow r^*F \otimes \bigwedge^k N^* \longrightarrow \cdots \longrightarrow r^*F \otimes N^* \longrightarrow r^*F \longrightarrow 0.$$

is exact on $U \smallsetminus X$. Thus if we set

$$\gamma(F) = r^*F \otimes \lambda_{N^*},$$

then we have the localized Chern character $\operatorname{ch}_X^*(\gamma(F))$ and by (15.12), $\operatorname{ch}_X^*(\gamma(F)) = \operatorname{ch}^*(r^*F) \cdot \operatorname{ch}_X^*(\lambda_{N^*}, s)$. Thus we have:

$$\operatorname{ch}_X^*(\gamma(F)) = T_{\boldsymbol{X}}(\operatorname{ch}^*(F) \cdot \operatorname{td}(N_X)^{-1}) \qquad \text{in } H^*(M, M \smallsetminus X),$$
$$\operatorname{ch}^*(\gamma(F)) = i_*(\operatorname{ch}^*(F) \cdot \operatorname{td}(N_X)^{-1}) \qquad \text{in } H^*(M).$$

2. The above equalities hold *a priori* in the cohomology with $\mathbb{C}$ coefficients, however, since the right-hand sides of the both are in the cohomology with $\mathbb{Q}$ coeffitients, they hold in the cohomology with $\mathbb{Q}$ coeffitients.

II. *Case* (b)

Let M be a C^∞ manifold of dimension m and V a closed C^∞ submanifold of dimension d' of M. We assume that the normal bundle $p : N_{\mathbb{R},V} \to V$ admits the structure of a complex vector bundle of rank k, $2k = m - d'$. Thus we denote it by N_V. Recall that (cf. Section 4.3) we have the Thom isomorphism

$$T_V : H^p(V) \xrightarrow{\sim} H^{2k+p}(M, M \smallsetminus V). \qquad (15.17)$$

The Thom class Ψ_V of V is the image of $[1] \in H^0(V)$ by T_V. Let Z denote the image of the zero section of N_V. By the tubular neighborhood theorem (Theorem 3.10), there exist a neighborhood U of V in M, a neighborhood W of Z in N_V and a C^∞ diffeomorphism $\tau : U \to W$ such that $\tau(V) = Z$ and that $(p \circ \tau)|_V = 1_V$. We have isomorphisms

$$H^*(M, M \smallsetminus V) \simeq H^*(U, U \smallsetminus V) \underset{\tau^*}{\xleftarrow{\sim}} H^*(W, W \smallsetminus Z) \simeq H^*(N_V, N_V \smallsetminus Z).$$

The Thom class Ψ_V of V corresponds to the Thom class Ψ_{N_V} of the bundle N_V under the above isomorphism and the Thom isomorphism (15.17) corresponds to the Thom isomorphism $T_{N_V} : H^p(Z) \xrightarrow{\sim} H^{2k+p}(N_V, N_V \smallsetminus Z)$.

Note that, if we denote by s_Δ the diagonal section of the bundle p^*N_V on N_V, its zero set is Z and we have $\Psi_{N_V} = c_Z^k(p^*N_V, s_\Delta)$ (cf. Theorem 10.9). If we set $N = r^*N_V$, $r = p \circ \tau : U \to V$, it is isomorphic with $\tau^*(p^*N_V|_W)$ (cf. Proposition 3.2. 2) and we have the commutative diagram

$$\begin{array}{ccc} N & \xrightarrow{\sim} & p^*N_V|_W \\ \downarrow & & \downarrow \\ U & \xrightarrow[\tau]{\sim} & W. \end{array} \tag{15.18}$$

If we let s be the section of N corresponding to s_Δ, its zero set is V and we have

$$\Psi_V = c_V^k(N, s).$$

Thus in this cases, (15.15) may be written

$$\mathrm{ch}_V^*(\lambda_{N^*}, s) = T_V(\mathrm{td}(N_V)^{-1}) \qquad \text{in } H^*(M, M \smallsetminus V). \tag{15.19}$$

For a complex vector bundle F on V, if we set $\gamma(F) = r^*F \otimes \lambda_{N^*}$, as in the case (a) (cf. Remark 15.4. 1), we have the localized Chern character $\mathrm{ch}_V^*(\gamma(F))$, which is equal to $\mathrm{ch}^*(r^*F) \cdot \mathrm{ch}_V^*(\lambda_{N^*}, s)$. Combined with (15.19), we have:

Theorem 15.6 (RR for embeddings II). *In the case* (b), *for a complex vector bundle F on V, we have:*

$$\begin{aligned} \mathrm{ch}_V^*(\gamma(F)) &= T_V(\mathrm{ch}^*(F) \cdot \mathrm{td}(N_V)^{-1}) \qquad \textit{in } H^*(M, M \smallsetminus V), \\ \mathrm{ch}^*(\gamma(F)) &= i_*(\mathrm{ch}^*(F) \cdot \mathrm{td}(N_V)^{-1}) \qquad \textit{in } H^*(M). \end{aligned}$$

Remark 15.5. 1. A remark similar to the one in Remark 15.4. 2 applies to this case as well.

2. Taking a connection of F, we may write down the formula on the cocycle level.

Case (c) is a special case of (b) and Theorem 15.6 holds as it is. To say more about the left-hand side, we prepare some materials.

Real analytic functions

Let M be a real analytic manifold. It is known that the sheaf $\mathscr{C}_M^\omega$ of rings of real analytic functions on M is coherent.

If M is a complex manifold, the sheaf $\mathscr{C}_M^\omega$ contains the sheaf $\mathscr{O}_M$ of holomorphic functions as a subsheaf of rings. In the following, we omit the subscript M in the sheaf notation, if there is no fear of confusion. It is known that, if $\mathscr{M}$ is a coherent $\mathscr{O}$-module, then $\mathscr{C}^\omega \otimes_{\mathscr{O}} \mathscr{M}$ is a coherent $\mathscr{C}^\omega$-module and that the assignment $\mathscr{M} \mapsto \mathscr{C}^\omega \otimes_{\mathscr{O}} \mathscr{M}$ is an exact functor from the category of coherent $\mathscr{O}$-modules to that of coherent $\mathscr{C}^\omega$-modules.

Let M be a complex manifold and V a closed complex submanifold of M. We denote by $\mathscr{I}_V$ the ideal sheaf of V in M and by $\mathscr{I}_Z$ the ideal sheaf of Z in N_V, where Z is the image of the zero section of N_V.

We quote the following "complex tubular neighborhood theorem", which is a refinement of Theorem 3.10 in this situation:

Proposition 15.8. *There exist a neighborhood U of V in M, a neighborhood W of Z in N_V and a real analytic isomorphism $\tau : U \to W$ such that $\tau(V) = Z$, $(p \circ \tau)|_V = 1_V$ and that $\tau^{-1}|_{N_{V,z} \cap W}$ is holomorphic for each z in V.*

Moreover, a map τ as above induces an isomorphism

$$(\tau, \tau^*) : (U, \mathscr{C}_U^\omega \otimes_{\mathscr{O}_U} \mathscr{I}_V) \xrightarrow{\sim} (W, \mathscr{C}_W^\omega \otimes_{\mathscr{O}_W} \mathscr{I}_Z).$$

See Definition 11.5 and the subsequent paragraph for what is meant by (τ, τ^*) being an isomorphism.

Characteristic classes of coherent sheaves

Let M be a complex manifold and $\mathscr{M}$ a coherent $\mathscr{O}_M$-module. Suppose there exist, on M, a complex

$$0 \longrightarrow E_r \longrightarrow \cdots \longrightarrow E_1 \longrightarrow E_0 \longrightarrow 0 \tag{15.20}$$

of real analytic vector bundles and an epimorphism $\mathscr{C}^\omega(E_0) \to \mathscr{C}^\omega \otimes_{\mathscr{O}} \mathscr{M}$ such that the sequence

$$0 \longrightarrow \mathscr{C}^\omega(E_r) \longrightarrow \cdots \longrightarrow \mathscr{C}^\omega(E_1) \longrightarrow \mathscr{C}^\omega(E_0) \longrightarrow \mathscr{C}^\omega \otimes_{\mathscr{O}} \mathscr{M} \longrightarrow 0$$

is exact. We reffer to a sequence as above a *locally free resolution* of $\mathscr{M}$ by real analytic vector bundles.

In this situation, we define the Chern character of $\mathscr{M}$ in $H^*(M)$ by

$$\mathrm{ch}^*(\mathscr{M}) = \mathrm{ch}^*(\xi), \qquad \xi = \sum_{i=0}^{r} (-1)^i E_i. \tag{15.21}$$

If S is the support of $\mathscr{M}$, it is an analytic variety in M and the complex (15.20) is exact on $M \smallsetminus S$. Thus we may define the localized class $\mathrm{ch}_S^*(\mathscr{M})$ in $H^*(M, M \smallsetminus S)$.

The definition above does not depend on the choice of the locally free resolution.

Remark 15.6. If $\mathscr{M}$ is a coherent $\mathscr{O}_M$-module, $\mathscr{M}$ admits a locally free resolution by real analytic vector bundles on any compact set in M.

III. *Case* (c)

Let M be a complex manifold and V a closed complex submanifold of codimension k of M. In this case, the normal bundle $p : N_V \to V$ is holomorphic and there exists a tubular neighborhood U with a deformation retraction $r = p \circ \tau : U \to V$ as in Proposition 15.8. Let $i : V \hookrightarrow U \subset M$ and $j : Z \hookrightarrow W \subset N$ denote the embeddings.

As in the case (b), we have the Thom isomorphism (15.17) and the Thom class is given by $\Psi_V = c^k_V(N, s)$. Here N is the bundle on U given by $N = r^* N_V$. Since $r = p \circ \tau$, N is a real analytic vector bundle which is isomorphic with $p^* N_V|_W$, and s is the (real analytic) section corresponding to the diagonal section s_Δ (cf. (15.18)).

Since s_Δ is a regular section of $p^* N_V$, as in the case (a) (let M, N and s be N_V, $p^* N$ and s_Δ, respectively, in (15.16)), we have the exact sequence

$$0 \longrightarrow \mathscr{O}(\textstyle\bigwedge^k p^* N_V^*) \longrightarrow \cdots \longrightarrow \mathscr{O}(p^* N_V^*) \longrightarrow \mathscr{O} \longrightarrow \mathscr{O}/\mathscr{I}_Z \longrightarrow 0,$$

where $\mathscr{O} = \mathscr{O}_W$. By Proposition 15.8, after tensoring the above sequence with $\mathscr{C}^\omega$, we have a resolution of $i_! \mathscr{O}_V$ by the Kozsul complex for (N, s).

Thus (15.15) may be written

$$\mathrm{ch}^*_V(i_! \mathscr{O}_V) = T_V(\mathrm{td}(N_V)^{-1}) \qquad \text{in } H^*(M, M \smallsetminus V).$$

By Proposition 15.8, we may replace λ_{N^*} with $i_! \mathscr{O}_V$ and for a holomorphic vector bundle F on V, we may replace $\gamma(F)$ with $i_! \mathscr{F}$. Thus we have:

Theorem 15.7 (RR for embeddings III). *In the case* (c), *for a holomorphic vector bundle F on V, we have*:

$$\mathrm{ch}^*_V(i_! \mathscr{F}) = T_V(\mathrm{ch}^*(F) \cdot \mathrm{td}(N_V)^{-1}) \qquad \textit{in } H^*(M, M \smallsetminus V),$$
$$\mathrm{ch}^*(i_! \mathscr{F}) = i_*(\mathrm{ch}^*(F) \cdot \mathrm{td}(N_V)^{-1}) \qquad \textit{in } H^*(M).$$

Remark 15.7. 1. A remark similar to the one in Remark 15.4. 2 applies to this case as well.

2. If V is compact, we may take a coherent $\mathscr{O}_V$-module as $\mathscr{F}$ (cf. Remark 15.6).

From the exact sequence

$$0 \longrightarrow TV \longrightarrow TM|_V \longrightarrow N_V \longrightarrow 0,$$

we have

$$\mathrm{td}^{-1}(N_V) = i^* \,\mathrm{td}^{-1}(M) \cdot \mathrm{td}(V)$$

so that we have:

Corollary 15.2. *In the situation of Theorem 15.7, it holds*

$$\mathrm{ch}^*(i_!\mathscr{F}) \cdot \mathrm{td}(M) = i_*(\mathrm{ch}^*(\mathscr{F}) \cdot \mathrm{td}(V)).$$

15.7 Grothendieck-Riemann-Roch Theorem

The original Grothendieck-Riemann-Roch theorem is stated and proved for proper morphisms of non-singular quasi-projective varieties (over an arbitrary algebraically closed field K). Here we make interpretations in our context (in the case $\mathsf{K} = \mathbb{C}$). The key point in this is what is called the "GAGA principle".

GAGA principle

Let M be a quasi-projective manifold. It is equipped with the sheaf $\mathscr{O}^{\mathrm{a}}_M$ of rings of algebraic functions, which is a subsheaf of the sheaf $\mathscr{O}_M$ of holomorphic functions. Recall that (cf. Sections 2.2 and 11.1), if we denote by M_Z and $\mathscr{O}^{\mathrm{r}}$, respectively, the manifold M with Zarisky topology and the sheaf of rings of regular functions on M_Z, then $\mathscr{O}^{\mathrm{a}} = 1^{-1}\mathscr{O}^{\mathrm{r}}$, where $1 : M \to M_Z$ is the identity.

For a coherent $\mathscr{O}^{\mathrm{r}}$-module $\mathscr{F}$, we set $\mathscr{F}^{\mathrm{a}} = 1^{-1}\mathscr{F}$ and

$$\mathscr{F}^{\mathrm{h}} = \mathscr{O} \otimes_{\mathscr{O}^{\mathrm{a}}} \mathscr{F}^{\mathrm{a}}$$

which is a coherent $\mathscr{O}$-module, as taking tensor product is right exact.

Theorem 15.8. *Let M be a projective algebraic manifold.*

1. *For a coherent $\mathscr{O}^{\mathrm{r}}$-module $\mathscr{F}$, there is a canonical isomorphism*

$$H^p(M_Z; \mathscr{F}) \xrightarrow{\sim} H^p(M; \mathscr{F}^{\mathrm{h}}).$$

2. *Let $\mathscr{F}$ and $\mathscr{G}$ be coherent $\mathscr{O}^{\mathrm{r}}$-modules. Then every morphism $\mathscr{F}^{\mathrm{h}} \to \mathscr{G}^{\mathrm{h}}$ comes from a unique morphism $\mathscr{F} \to \mathscr{G}$.*

3. *For every coherent $\mathscr{O}$-module $\mathscr{M}$, there is a coherent $\mathscr{O}^{\mathrm{r}}$-module $\mathscr{F}$ such that $\mathscr{F}^{\mathrm{h}}$ is isomorphic with $\mathscr{M}$. Moreover such a sheaf $\mathscr{F}$ is determined uniquely by $\mathscr{M}$ up to isomorphisms.*

K-group of projective algebraic manifolds

Let M be a projective algebraic manifold of dimension n. We denote by $K^{\mathrm{h}}(M)$ and $K_{\mathrm{h}}(M)$ the K-groups of holomorphic vector bundles and of coherent $\mathscr{O}$-modules on M, respectively. There is a natural morphism $K^{\mathrm{h}}(M) \to K_{\mathrm{h}}(M)$, which is induced by $F \mapsto \mathscr{O}(F)$.

Theorem 15.9. *The above morphism is an isomorphism:*

$$K^{\mathrm{h}}(M) \xrightarrow{\sim} K_{\mathrm{h}}(M).$$

This is proved using the fact that every coherent $\mathscr{O}^{\mathrm{r}}$-module on M_Z admits a locally free resolution by vector bundles and the GAGA principle. Thus we may define the Chern character

$$\mathrm{ch}^* : K_{\mathrm{h}}(M) \simeq K^{\mathrm{h}}(M) \longrightarrow K^{\mathrm{d}}(M) \simeq K^{\mathrm{t}}(M) \longrightarrow H^*(M;\mathbb{Q}).$$

Let $f : M \to M'$ be a holomorphic map of projective algebraic manifolds. We have a morphism $f^* : K^{\mathrm{h}}(M') \to K^{\mathrm{h}}(M)$, which is induced from $E \mapsto f^*E$ for any holomorphic vector bundle E on M'. In view of Theorem 15.9, we also have a morphism $f^* : K_{\mathrm{h}}(M') \to K_{\mathrm{h}}(M)$.

Let $f : M \to M'$ be as above. If $\mathscr{F}$ is a coherent $\mathscr{O}_M$-module, then $R^p f_*\mathscr{F}$ are coherent $\mathscr{O}_{M'}$-modules and $R^p f_*\mathscr{F} = 0$ for $p > \dim M$ (cf. Theorem 11.14 and Remark 11.8 (1)). We have a morphism

$$f_! : K_{\mathrm{h}}(M) \longrightarrow K_{\mathrm{h}}(M'),$$

which is induced from $\mathscr{F} \mapsto \sum_p (-1)^p R^p f_*\mathscr{F}$ (cf. (11.16)). Note that this can also be described using coherent $\mathscr{O}^{\mathrm{r}}$-modules and the GAGA principle.

If we have two such maps $f : M \to M'$ and $g : M' \to M''$, we have

$$(g \circ f)_! = g_! \circ f_!. \tag{15.22}$$

This is proved by applying the spectral sequence $(E_2^{p,q}, H^r)$ with

$$E_2^{p,q} = R^p g_* R^q f_*\mathscr{F} \quad \text{and} \quad H^r = R^r(g \circ f)_*\mathscr{F}$$

(cf. Proposition 11.22).

Suppose we have two maps $f_i : M_i \to M_i'$, $i = 1, 2$. Then there is a morphism

$$K_{\mathrm{h}}(M_1) \otimes K_{\mathrm{h}}(M_2) \longrightarrow K_{\mathrm{h}}(M_1 \times M_2),$$

given by $\xi_1 \otimes \xi_2 \mapsto \xi_1 \times \xi_2 = p_1^*\xi_1 \otimes p_2^*\xi_2$. For the map $(f_1, f_2) : M_1 \times M_2 \to M_1' \times M_2'$, we have

$$(f_1, f_2)_!(\xi_1 \times \xi_2) = (f_1)_!\xi_1 \times (f_2)_!\xi_2. \tag{15.23}$$

For the projective space $\mathbb{P}^N$, we set $\mathscr{O}(n) = \mathscr{O}_{\mathbb{P}^N}({H_N}^{\otimes n})$.
We quote:

Theorem 15.10.
1. *$K_{\mathrm{h}}(\mathbb{P}^N)$ is generated by the classes of $\mathscr{O}(n)$, $0 \leq n \leq N$. We also have*

$$\chi(\mathbb{P}^N; \mathscr{O}(n)) = \binom{N+n}{N}.$$

2. *For any projective algebraic manifold M, the morphism*

$$K_{\mathrm{h}}(\mathbb{P}^N) \otimes K_{\mathrm{h}}(M) \longrightarrow K_{\mathrm{h}}(\mathbb{P}^N \times M)$$

is surjective.

Grothendieck-Riemann-Roch Theorem

Theorem 15.11. *Let $f : M \to M'$ be a holomorphic map of projective algebraic manifolds. Then, for every ξ in $K_{\mathrm{h}}(M)$,*

$$\mathrm{ch}^*(f_!\xi) \cdot \mathrm{td}(M') = f_*(\mathrm{ch}^*(\xi) \cdot \mathrm{td}(M)) \qquad \textit{in } \ H^*(M'; \mathbb{Q}).$$

Proof. The proof goes as follows:

Step I. Suppose we have maps $f : M \to M'$ and $g : M' \to M''$ as above. If the formula holds for f and g, then it also holds for $g \circ f$. Indeed, this follows from $(g \circ f)_* = g_* \circ f_*$ and (15.22).

Step II. Let $f : M \to M'$ be as in the statement of the theorem. Denoting by $i : M \hookrightarrow \mathbb{P}^N$ an embedding, we may express f as the composition

$$M \xrightarrow{\tilde{f}} \mathbb{P}^N \times M' \xrightarrow{p_2} M',$$

where $\tilde{f}(z) = (i(z), f(z))$ and p_2 is the projection onto the second factor. Note that $\tilde{f}$ is a closed embedding.

We have already proved the formula for $\tilde{f}$ (cf. Corollary 15.2), so we are left to prove it for the projection p_2.

Step III. Suppose we have two maps $f_i : M_i \to M_i'$, $i = 1, 2$. If the formula hold for (f_i, ξ_i), then it hold for $(f_1 \times f_2, \xi_1 \times \xi_2)$. This follows from (15.23).

Since we may write $p_2 : \mathbb{P}^N \times M' \to M'$ as $p \times 1_{M'} : \mathbb{P}^N \times M' \to \{pt\} \times M'$, we are left to prove the formula for the case $p : \mathbb{P}^N \to \{pt\}$ and $\xi = \mathscr{O}(n)$, i.e., to prove

$$\mathrm{ch}^*(p_!\mathscr{O}(n)) \cdot \mathrm{td}(\{pt\}) = p_*(\mathrm{ch}^*(\mathscr{O}(n)) \cdot \mathrm{td}(\mathbb{P}^N)) \qquad \text{in } H^*(\{pt\}; \mathbb{Q}).$$

Step IV. Note that $H^*(\{pt\};\mathbb{Q}) = H^0(\{pt\};\mathbb{Q}) \xrightarrow{\sim} \mathbb{Q}$, the isomorphism is being given by the composition $\varepsilon_* P_{\{pt\}}$ of the Poincaré isomorphism for the point and the augmentation $\varepsilon_* : H_0(\{pt\};\mathbb{Q}) \to \mathbb{Q}$. We compute the images of the both sides by $\varepsilon_* P_{\{pt\}}$. In the left-hand side, $\mathrm{td}(\{pt\}) = 1$ and $R^i p_* \mathscr{O}(n)$ is the locally free sheaf on $\{pt\}$ of rank $\dim H^i(\mathbb{P}^N; \mathscr{O}(n))$. Thus $\varepsilon_* P_{\{pt\}}(\mathrm{LHS}) = \chi(\mathbb{P}^N; \mathscr{O}(n))$. On the other hand,

$$\begin{aligned} \varepsilon_* P_{\{pt\}}(\mathrm{RHS}) &= \varepsilon_* p_* P_{\mathbb{P}^N}([\mathrm{ch}^*(\mathscr{O}(n)) \cdot \mathrm{td}(\mathbb{P}^N)]_N) \\ &= \varepsilon_* P_{\mathbb{P}^N}([\mathrm{ch}^*(\mathscr{O}(n)) \cdot \mathrm{td}(\mathbb{P}^N)]_N) \\ &= \langle \mathbb{P}^N, [\mathrm{ch}^*(\mathscr{O}(n)) \cdot \mathrm{td}(\mathbb{P}^N)]_N \rangle, \end{aligned}$$

where the first equality is by definition of p_*, the second by (B.2) and the third by (B.19).

Hence it amounts to prove

$$\langle \mathbb{P}^N, [\mathrm{ch}^*(\mathscr{O}(n)) \cdot \mathrm{td}(\mathbb{P}^N)]_N \rangle = \chi(\mathbb{P}^N; \mathscr{O}(n)).$$

From (9.22), we have $c^*(\mathbb{P}^N) = (1+\gamma)^{N+1}$, $\gamma = c^1(H_N)$. Thus

$$\mathrm{td}(\mathbb{P}^N) = \Big(\frac{\gamma}{1 - e^{-\gamma}}\Big)^{N+1}.$$

On the other hand, $\mathrm{ch}^*(\mathscr{O}(n)) = e^{n\gamma}$ and the equality follows from Theorem 15.10. 1.

□

If we let $M' = \{pt\}$ in Theorem 15.11, by a similar computation as in Step IV above, we have:

Theorem 15.12 (Hirzebruch-Riemann-Roch theorem). *Let M be a projective algebraic manifold of dimension n. For a holomorphic vector bundle E on M, we have*

$$\chi(M; \mathscr{O}(E)) = \langle M, [\mathrm{ch}^*(E) \cdot \mathrm{td}(M)]_n \rangle,$$

where $[\ \ ]_n$ denotes the component in $H^{2n}(M)$.

In particular, let $M = C$ be a compact Riemann surface and $E = L_D$ the line bundle determined by a divisor D on C. Then, in the left-hand side, $\dim H^0(C; \mathscr{O}(L_D)) = l(D)$ and $\dim H^1(C; \mathscr{O}(L_D)) = \dim H^0(C; \Omega^1(L_D{}^*)) = i(D)$, by the Kodaira-Serre duality (cf. (9.13)).

On the other hand, we have $\mathrm{ch}^*(L_D) = 1+c^1(L_D)$ and $\mathrm{td}(C) = 1+\frac{1}{2}c^1(C)$. Thus the right-hand side is given by $\langle C, c_1(L_D) + \frac{1}{2}c_1(C)\rangle$, which is equal to $d - g + 1$ so that we recover the classical formula (15.1).

Notes

As to Kodaira's work on the Riemann-Roch theorem for compact complex surfaces, we refer to [Kodaira (1951)].

We list [Atiyah (1967); Hirzebruch (1966)] as references for Section 15.2. Theorem 15.1 is proved in [Hirzebruch (1966)] and its expression on the form level (Theorem 15.2) in [Harvey and Lawson (1993)]. Theorem 15.3 is a localized version of that in terms of Čech-de Rham cocycles and is an essential ingredient for the localized Riemann-Roch theorem for embeddings in our framework, as noted in the text.

The notion of local Chern character and class appears in [Atiyah and Hirzebruch (1961, 1962)] as the authors introduce "generalized difference bundles". It is then formulated explicitly in [Baum, Fulton and MacPherson (1975); Iversen (1976)]. We deal with the problem, in Section 15.4, using families of connections compatible with sequences of vector bundles and defining the classes in the relative Čech-de Rham cohomology. The classes we define have all the necessary properties and should coincide with the ones in the literatures above. Hence they are in the cohomology $H^*(M, M \setminus S; \mathbb{Q})$ with $\mathbb{Q}$ coefficient. Theorem 15.4 is proved in [Iversen (1976)] by different approaches. Our treatment of local Chern character and class as well as the proof of Theorem 15.3 are taken from [Suwa (2000b)]. Also the Chern-Weil theory for virtual bundles that are necessary for these is taken from [Suwa (1998)].

See (18.D) Corollary in [Matsumura (1980)] as to Lemma 15.5 and (15.16).

For details of the paragraphs "Real analytic functions" and "Characteristic classes of coherent sheaves", we refer to [Atiyah and Hirzebruch (1961, 1962)].

Theorems 15.6 and 15.7 are proved in [Atiyah and Hirzebruch (1962)]. Although it is not stated explicitly, they are proved in localized form. In the algebraic category, the formulas are proved for a locally free $\mathscr{O}_V$-module on an LCI by analyzing the graph construction in [Baum, Fulton and MacPherson (1975)]. These formulas are also proved at the level of differential forms and currents in [Harvey and Lawson (1993)], See also [Bismut (1998)].

The Riemann-Roch theorem for embeddings in localized form is used to compute residues of singular holomorphic distributions or foliations, see [Suwa (2012)].

For the paragraph "GAGA principle", we refer to [Serre (1956)] and for "K-group of projective algebraic manifolds" to [Borel et Serre (1958)], where the Grothendieck-Riemann-Roch theorem is proved.

The Hirzebruch-Riemann-Roch theorem (Theorem 15.12) is due to [Hirzebruch (1966)].

As to the Riemann-Roch theorem for singular spaces, it is proved for quasi-projective varieties in [Baum, Fulton and MacPherson (1975)] and for complex spaces in [Levy (1987)].

Theorem 15.12 may be proved for an arbitrary compact complex manifold as a special case of the Atiyah-Singer index theorem (cf. [Atiyah and Singer (1963)] and subsequent papers). The theorem asserts that, for an elliptic differential operator on a compact manifold, the "analytic index" is equal to the "toplogical index".

For example, let M be an oriented C^∞ manifold of dimension m. Then we have the operator $d : A^r(M) \to A^{r+1}(M)$. If we endow M with a Riemannian metric, we have the adjoint $d^* : A^r(M) \to A^{r-1}(M)$ (cf. Section 9.1). The operator

$$D = d + d^* : \bigoplus_{r:\text{even}} A^r(M) \longrightarrow \bigoplus_{r:\text{odd}} A^r(M)$$

is shown to be elliptic. Suppose that M is compact. Then, on the one hand, the analytic index of D, which is $\dim \operatorname{Ker} D - \dim \operatorname{Coker} D$ by definition, is equal to $\dim \operatorname{Ker} D - \dim \operatorname{Ker} D^*$, where $D^* = d^* + d : \bigoplus_{r:\text{odd}} A^r(M) \to \bigoplus_{r:\text{even}} A^r(M)$. Thus it is equal to

$$\dim \bigoplus_{r:\text{even}} \mathcal{H}^r(M) - \dim \bigoplus_{r:\text{odd}} \mathcal{H}^r(M) = \chi(M),$$

the Euler-Poincaré characteristic of M (cf. Proposition 9.5). On the other hand, if m is even, it is shown that, using (5.6) and Theorem 15.1, the topological index is $\langle M, e(M)\rangle$ with $e(M)$ the Euler class of M. Thus we have the equality (5.18) for the case m is even. Note that, if m is odd, the both indices are zero.

Likewise let M be a complex manifold of dimension n and E a holomorphic vector bundle on M. Then we have $\bar\partial : A^{p,q}(M;E) \to A^{p,q+1}(M;E)$. If we endow M and E with Hermitian metrics, we have the adjoint operator $\bar\partial_h^* : A^{p,q}(M;E) \to A^{p,q-1}(M;E)$ with h the metric on E (cf. (9.11)). The operator

$$D = \bar\partial + \bar\partial_h^* : \bigoplus_{q:\text{even}} A^{0,q}(M;E) \longrightarrow \bigoplus_{q:\text{odd}} A^{0,q}(M;E)$$

is shown to be elliptic. If M is compact, the analytic index is the Euler-Poincaré characteristic $\chi(M;\mathscr{O}(E))$ and the topological index is given by

$\langle M, [\mathrm{ch}^*(E) \cdot \mathrm{td}(M)]_n \rangle$. Thus the equality in Theorem 15.12 holds for an arbitrary compact complex manifold.

We list [Freed (2021)] as a survey article on the Atiyah-Singer index theorem.

Appendix A

Commutative Algebra

In this appendix, we review some materials on commutative algebra necessary for our purpose. In Section A.1, we recall homological algebra and related topics and, in Section A.2, we discuss structures of rings and modules.

In the following, we let R denote a commutative ring with unity 1, which may possibly be equal to 0, and deal with R-modules. In particular, if $\mathsf{R} = \mathbb{Z}$, the ring of integers, they are Abelian groups and if $\mathsf{R} = \mathbb{C}$, they are complex vector spaces. A *morphism* of R-modules is a map $\varphi : \mathsf{M} \to \mathsf{M}'$ of R-modules compatible with the R-module structures. An injective morphism is called a *monomorphism* and a surjective one an *epimorphism.* An *isomorphism* is a bijective morphism φ. In this case, its inverse φ^{-1} is also an isomorphism. If there is an isomorphism between M and M', we write $\mathsf{M} \simeq \mathsf{M}'$. Sometimes a morphism is called a *homomorphism.* In general, for a set X, 1_X denotes the identity map of X.

A.1 Homological algebra

In the following, i, k, p, q and r will denote integers.

Homology

Definition A.5. A *chain complex of* R*-modules* is a collection $\{(K_p, d_p)\}$, where for each p, K_p is an R-module and $d_p : K_p \to K_{p-1}$ is an R- morphism with the property $d_p \circ d_{p+1} = 0$.

Thus a chain complex is expressed by a sequence of R-morphisms

$$\cdots \longrightarrow K_{p+1} \xrightarrow{d_{p+1}} K_p \xrightarrow{d_p} K_{p-1} \longrightarrow \cdots$$

with $\operatorname{Im} d_{p+1} \subset \operatorname{Ker} d_p$. A chain complex as above is written $(K_\bullet, d)$, or simply $K_\bullet$. The morphisms d_p are called *boundary operators*. We set $Z_p(K_\bullet) = \operatorname{Ker} d_p$ and $B_p(K_\bullet) = \operatorname{Im} d_{p+1}$. An element in K_p, $Z_p(K_\bullet)$ or $B_p(K_\bullet)$ is called, respectively, a *p-chain*, a *p-cycle* or a *p-boundary*. The *p-th homology* of $K_\bullet$ is defined by

$$H_p(K_\bullet) = Z_p(K_\bullet)/B_p(K_\bullet),$$

which inherits naturally an R-module structure. For a p-cycle c, we denote by $[c]$ its class in $H_p(K_\bullet)$. If $[c] = [c']$, we write $c \sim c'$ and say that they are *homologous*.

Note that the above notions and namings came originally from topology (cf. Appendix B below).

Chain morphisms: Let $(K_\bullet, d_\bullet)$ and $(L_\bullet, d_\bullet)$ be two chain complexes. A *chain morphism* $\varphi_\bullet : K_\bullet \to L_\bullet$ is a collection $\{\varphi_p\}$ of morphisms $\varphi_p : K_p \to L_p$ such that

$$\varphi_{p-1} \circ d_p = d_p \circ \varphi_p.$$

In this case $\varphi_\bullet$ induces a morphism for each p:

$$\varphi_* : H_p(K_\bullet) \longrightarrow H_p(L_\bullet).$$

The *composition* of two chain morphisms is defined in an obvious manner and its induced morphism on the homology is the composition of the induced morphisms.

A sequence of morphisms

$$J_\bullet \xrightarrow{\psi_\bullet} K_\bullet \xrightarrow{\varphi_\bullet} L_\bullet$$

is said to be *exact* if $J_p \overset{\psi_p}{\to} K_p \overset{\varphi_p}{\to} L_p$ is exact, i.e., $\operatorname{Im} \psi_p = \operatorname{Ker} \varphi_p$, for all p. The complex $(K_\bullet, d_\bullet)$ with $K_p = 0$ and $d_p = 0$ for all p is called the *zero complex* and is denoted by 0.

Proposition A.9. *If the sequence*

$$0 \longrightarrow J_\bullet \xrightarrow{\psi_\bullet} K_\bullet \xrightarrow{\varphi_\bullet} L_\bullet \longrightarrow 0$$

is exact, it induces a long exact sequence

$$\cdots \longrightarrow H_p(J_\bullet) \xrightarrow{\psi_*} H_p(K_\bullet) \xrightarrow{\varphi_*} H_p(L_\bullet) \xrightarrow{d_*} H_{p-1}(J_\bullet) \longrightarrow \cdots .$$

In the above, d_*, which is referred to as the *connecting morphism*, assigns to a class $[c]$ in $H_p(L_\bullet)$ the class $[a]$ in $H_{p-1}(J_\bullet)$ such that $c = \varphi_p(b)$ and $d_p(b) = \psi_{p-1}(a)$ with b in K_p.

Let $(K_\bullet, d)$ be a chain complex. A *subcomplex* of $K_\bullet$ is a collection $\{K'_p\}$ such that K'_p is a sub-R-module of K_p and that $d_p(K'_p) \subset K'_{p-1}$ for all p. In this case, $\{(K'_p, d_p)\}$ is a chain complex. We set $K''_p = K_p/K'_p$. Since d_p induces a morphism $K''_p \to K''_{p-1}$, which we also denote by d_p, $K''_\bullet$ is also a chain complex, called the *quotient complex* of $K_\bullet$ by $K'_\bullet$. In this case we have a short exact sequence of complexes

$$0 \longrightarrow K'_\bullet \longrightarrow K_\bullet \longrightarrow K''_\bullet \longrightarrow 0.$$

Note that, in a sequence as in Proposition A.9, $J_\bullet$ is isomorphic with a subcomplex of $K_\bullet$ and $L_\bullet$ with a quotient complex.

Chain homotopy: Let $(K_\bullet, d)$ and $(L_\bullet, d)$ be two complexes and $\varphi_\bullet$ and $\varphi'_\bullet$ two chain morphisms of $K_\bullet$ to $L_\bullet$. A *chain homotopy* between $\varphi_\bullet$ and $\varphi'_\bullet$ is a collection $\{h_p\}$ of morphisms $h_p : K_p \to L_{p+1}$ such that

$$d_{p+1} \circ h_p + h_{p-1} \circ d_p = \varphi_p - \varphi'_p.$$

If there is a chain homotopy between $\varphi_\bullet$ and $\varphi'_\bullet$, we write $\varphi_\bullet \simeq \varphi'_\bullet$. From the definition we have:

Proposition A.10. *If $\varphi_\bullet \simeq \varphi'_\bullet$, they induce the same morphism on the homology:*

$$\varphi_* = \varphi'_* : H_p(K_\bullet) \longrightarrow H_p(L_\bullet).$$

A chain morphism $\varphi_\bullet : K_\bullet \to L_\bullet$ is said to be a *chain equivalence* if there is a chain morphism $\psi_\bullet : L_\bullet \to K_\bullet$ such that $\psi_\bullet \circ \varphi_\bullet \simeq 1_{K_\bullet}$ and $\varphi_\bullet \circ \psi_\bullet \simeq 1_{L_\bullet}$. As a consequence of Proposition A.10, we have:

Corollary A.3. *If $\varphi_\bullet : K_\bullet \to L_\bullet$ is a chain equivalence, the induced morphism is an isomorphism:*

$$\varphi_* : H_p(K_\bullet) \xrightarrow{\sim} H_p(L_\bullet).$$

Cohomology

As a notion dual to chain complex, we have a *cochain complex of* R*-modules*. It is a collection $\{(K^p, d^p)\}$, where for each p, K^p is an R-module and

$d^p : K^p \to K^{p+1}$ is a morphism with the property $d^p \circ d^{p-1} = 0$. Thus it is expressed by a sequence of morphisms

$$\cdots \longrightarrow K^{p-1} \xrightarrow{d^{p-1}} K^p \xrightarrow{d^p} K^{p+1} \longrightarrow \cdots$$

with $\operatorname{Im} d^{p-1} \subset \operatorname{Ker} d^p$. It is also written $(K^\bullet, d)$, or simply $K^\bullet$. The morphisms d^p are called *coboundary operators.* We set $Z^p(K^\bullet) = \operatorname{Ker} d^p$ and $B^p(K^\bullet) = \operatorname{Im} d^{p-1}$. An element in K^p, $Z^p(K^\bullet)$ or $B^p(K^\bullet)$ is called, respectively, a *p-cochain*, a *p-cocycle* or a *p-coboubdary.* The *p-th cohomology* of $K^\bullet$ is defined by

$$H^p(K^\bullet) = Z^p(K^\bullet)/B^p(K^\bullet).$$

It will be also denoted by $H_d^p(K^\bullet)$, if we wish to make the coboudary operator d explicit. For a p-cocycle u, we denote by $[u]$ its class in $H^p(K^\bullet)$. If $[u] = [u']$, we write $u \sim u'$ and say that they are *cohomologous.*

Morphisms of cochain complexes are defined similarly as for morphisms of chain complexes and we also have:

Proposition A.11. *If the sequence*

$$0 \longrightarrow J^\bullet \xrightarrow{\psi^\bullet} K^\bullet \xrightarrow{\varphi^\bullet} L^\bullet \longrightarrow 0$$

is exact, it induces a long exact sequence

$$\cdots \longrightarrow H^p(J^\bullet) \xrightarrow{\psi^*} H^p(K^\bullet) \xrightarrow{\varphi^*} H^p(L^\bullet) \xrightarrow{d^*} H^{p+1}(J^\bullet) \longrightarrow \cdots.$$

In the above, d^* assigns to a class $[u]$ in $H^p(L^\bullet)$ the class $[w]$ in $H^{p+1}(J^\bullet)$ such that $u = \varphi^p(v)$ and $d^p(v) = \psi^{p+1}(w)$ with v in K^p.

A *subcomplex* of $K^\bullet$ is a collection $K'^\bullet = \{K'^p\}$ such that K'^p is a sub-R-module of K^p and that $d^p(K'^p) \subset K'^{p+1}$. In this case, (K'^p, d^p) is a complex. We set $K''^p = K^p/K'^p$. Since d^p induces a morphism $K''^p \to K''^{p-1}$, which we also denote by d^p, $K''^\bullet$ is also a complex, called the *quotient complex.* In this case we have a short exact sequence of complexes

$$0 \longrightarrow K'^\bullet \longrightarrow K^\bullet \longrightarrow K''^\bullet \longrightarrow 0.$$

The notion of homotopy between two cochain morphisms is also defined as in the case of homology. Thus let $(K^\bullet, d)$ and $(L^\bullet, d)$ be two complexes and $\varphi^\bullet$ and $(\varphi')^\bullet$ two cochain morphisms of $K^\bullet$ to $L^\bullet$. A *cochain homotopy* between $\varphi^\bullet$ and $(\varphi')^\bullet$ is a collection $\{h^p\}$ of morphisms $h^p : K^p \to L^{p-1}$ such that

$$d^{p-1} \circ h^p + h^{p+1} \circ d^p = \varphi^p - (\varphi')^p.$$

If there is a cochain homotopy between $\varphi^\bullet$ and $(\varphi')^\bullet$, we write $\varphi^\bullet \simeq (\varphi')^\bullet$. In this case, they induce the same morphism on the cohomology:

$$\varphi^* = (\varphi')^* : H^p(K^\bullet) \longrightarrow H^p(L^\bullet).$$

A cochain morphism $\varphi^\bullet : K^\bullet \to L^\bullet$ is said to be a *cochain equivalence* if there is a cochain morphism $\psi^\bullet : L^\bullet \to K^\bullet$ such that $\psi^\bullet \circ \varphi^\bullet \simeq 1_{K^\bullet}$ and $\varphi^\bullet \circ \psi^\bullet \simeq 1_{L^\bullet}$. In this case, the induced morphism is an isomorphism:

$$\varphi^* : H^p(K^\bullet) \xrightarrow{\sim} H^p(L^\bullet).$$

Dual complex: If $K_\bullet$ is a chain complex, we may construct an associated cochain complex. Thus we set $K^p = \mathrm{Hom}(K_p, \mathsf{R})$, the R-module of morphisms of K_p to R. We define the *Kronecker product* $\langle\ ,\ \rangle$ as an R-bilinear form

$$K_p \times K^p \longrightarrow \mathsf{R} \qquad \text{given by} \ \ (c, u) \mapsto \langle c, u\rangle = u(c).$$

The coboundary operator $d^p : K^p \to K^{p+1}$ is defined as the transpose of d_p:

$$\langle c', d^p u\rangle = \langle d_p c', u\rangle \qquad \text{for } u \in K^p \text{ and } c' \in K_{p+1}. \tag{A.1}$$

Then we see that $d^p \circ d^{p-1} = 0$ so that $(K^\bullet, d)$ is a cochain complex, called the complex dual to $K_\bullet$. Note that the Kronecker product descends to homology and cohomology, i.e., the R-bilinear form

$$H_p(K_\bullet) \times H^p(K^\bullet) \longrightarrow \mathsf{R} \qquad \text{given by} \ \ ([c], [u]) \mapsto \langle [c], [u]\rangle = \langle c, u\rangle$$

is well-defined.

If $\varphi_\bullet : K_\bullet \to L_\bullet$ is a chain morphism, we have the *transpose cochain morphism* $\varphi^\bullet : L^\bullet \to K^\bullet$, which is defined by $\varphi^p(v) = v \circ \varphi_p$, in other words,

$$\langle c, \varphi^p(v)\rangle = \langle \varphi_p(c), v\rangle \qquad \text{for } \ c \in K_p \ \text{ and } \ v \in L^p.$$

Note that the above relation descends to homology and cohomology.

For two morphisms $\varphi_\bullet$ and $\varphi'_\bullet$, a homotopy between $\varphi_\bullet$ and $\varphi'_\bullet$ induces a homotopy between $\varphi^\bullet$ and $(\varphi')^\bullet$.

Acyclicity: Let $K^\bullet$ be a complexe with $K^p = 0$ for $p < 0$. We say that $K^\bullet$ is *acyclic* if $H^p(K^\bullet) = 0$ for $p \geq 1$. The following is not difficult to see:

Proposition A.12. *If there is a collection $\{h^p\}$ of morphisms $h^p : K^p \to K^{p-1}$ such that*

$$d^{p-1} \circ h^p + h^{p+1} \circ d^p = 1_{K^p} \quad \textit{for} \ \ p \geq 1,$$

then $K^\bullet$ is acyclic.

Spectral sequences

Filtration: Let M be an R-module. A (decreasing) *filtration* on M is a family $(F^p\mathsf{M})_p$ of sub-R-modules of M with $F^p\mathsf{M} \supset F^{p+1}\mathsf{M}$. We set

$$G\mathsf{M} = \bigoplus_p G^p\mathsf{M}, \qquad G^p\mathsf{M} = F^p\mathsf{M}/F^{p+1}\mathsf{M}$$

and call it the graded module associated with $(F^p\mathsf{M})$. For $a \in F^p\mathsf{M}$, we denote by $[a]^p$ its class in $G^p\mathsf{M}$.

Let k_0 be a non-negative integer.

Definition A.6. A *spectral sequence* is a system $(E_k^{p,q}, H^r)$ as follows:

(1) for each p, q and $k \geq k_0$, $E_k^{p,q}$ is an R-module such that $E_{k_0}^{p,q} = 0$ if $p < 0$ or $q < 0$,
(2) there is a morphism $d_k^{p,q} : E_k^{p,q} \to E_k^{p+k,q-k+1}$ satisfying
$$d_k^{p,q} \circ d_k^{p-k,q+k-1} = 0,$$
(3) there is an isomorphism
$$E_{k+1}^{p,q} \simeq \operatorname{Ker} d_k^{p,q} / \operatorname{Im} d_k^{p-k,q+k-1},$$
thus, if $k > \max\{p, q+1\}$, then $E_k^{p,q} \simeq E_{k+1}^{p,q} \simeq \cdots$, which is denoted by $E_\infty^{p,q}$,
(4) for each r, H^r is a filtered R-module such that $H^r = 0$ for $r < 0$ and that the filtration satisfies $F^pH^r = H^r$ for $p \leq 0$ and $F^pH^r = 0$ for $p \geq r+1$, i.e.,
$$H^r = F^0H^r \supset F^1H^r \supset \cdots \supset F^{r+1}H^r = 0,$$
(5) there is an isomorphism $E_\infty^{p,q} \simeq G^pH^{p+q}$.

Note that from (3), if $E_k^{p,q} = 0$, then $E_{k+1}^{p,q} = \cdots = E_\infty^{p,q} = 0$, in particular if $p < 0$ or $q < 0$, $E_k^{p,q} = 0$ for all $k \geq k_0$.

Considering the case $q = 0$, if $k_1 \geq \max\{2, k_0\}$, there is a sequence of epimorphisms

$$E_{k_1}^{p,0} \longrightarrow E_{k_1+1}^{p,0} \longrightarrow \cdots \longrightarrow E_\infty^{p,0} \simeq F^pH^p. \tag{A.2}$$

Also considering the case $p = 0$, if $k_1 \geq \max\{1, k_0\}$, there is a sequence of monomorphisms

$$H^q/F^1H^q \simeq E_\infty^{0,q} \longrightarrow \cdots \longrightarrow E_{k_1+1}^{0,q} \longrightarrow E_{k_1}^{0,q}. \tag{A.3}$$

These are called *edge morphisms*.

Suppose $E_{k_1}^{p,q} = 0$ for $q > 0$. Then it can be easily checked that the morphisms in (A.2) are isomorphisms and that $F^p H^p = H^p$. Likewise if $E_{k_1}^{p,q} = 0$ for $p > 0$, the morphisms in (A.3) are isomorphisms and $F^1 H^q = 0$. Thus we have:

Proposition A.13. 1. *Let $k_1 \geq \max\{2, k_0\}$. If $E_{k_1}^{p,q} = 0$ for $q > 0$, then $E_{k_1}^{r,0} \simeq H^r$ for every r.*

2. *Let $k_1 \geq \max\{1, k_0\}$. If $E_{k_1}^{p,q} = 0$ for $p > 0$, then $E_{k_1}^{0,r} \simeq H^r$ for every r.*

Definition A.7. We say that a spectral sequence $(E_k^{p,q}, H^r)$ *degenerates* at E_{k_1} if $d_k^{p,q} = 0$ for every (p, q) and $k \geq k_1$.

Thus in this case, we have $E_{k_1}^{p,q} \simeq \cdots \simeq E_\infty^{p,q} \simeq G^p H^{p+q}$. For example, in the situations of Proposition A.13, the spectral sequence degenerates at E_{k_1}.

Filtered complexes: Let $(K^\bullet, d)$ be a complex of R-modules with $K^r = 0$ for $r < 0$. A filtration on the complex $K^\bullet$ is a family $(F^p K^\bullet)$ of subcomplexes with $F^p K^\bullet \supset F^{p+1} K^\bullet$. Thus for each r, $(F^p K^r)$ is a filtration on K^r and $d(F^p K^r) \subset F^p K^{r+1}$. We assume that

$$F^p K^r = K^r, \text{ for } p \leq 0 \quad \text{and} \quad F^p K^r = 0, \text{ for } p \geq r + 1. \tag{A.4}$$

We have the quotient complex

$$G^p K^\bullet = F^p K^\bullet / F^{p+1} K^\bullet.$$

For each r, by (A.4), $G^p K^r = 0$ for $p < 0$ or $p > r$.

The filtration $(F^p K^\bullet)$ induces a filtration on the cohomology $H^r(K^\bullet)$:

$$F^p H^r(K^\bullet) = Z^r(K^\bullet) \cap F^p K^r / B^r(K^\bullet) \cap F^p K^r.$$

Proposition A.14. *For a filtered complex $K^\bullet$, there is a spectral sequence $(E_k^{p,q}, H^r)$ with $k_0 = 0$,*

$$E_0^{p,q} = G^p K^{p+q}, \quad E_1^{p,q} \simeq H^{p+q}(G^p K^\bullet) \quad \text{and} \quad H^r = H^r(K^\bullet).$$

Proof. For each triple (p, q, k) of non-negative integers, we set

$$\begin{aligned} Z_k^{p,q} &= \{ a \in F^p K^{p+q} \mid da \in F^{p+k} K^{p+q+1} \}, \\ B_k^{p,q} &= (d(F^{p-k+1} K^{p+q-1}) + F^{p+1} K^{p+q}) \cap Z_k^{p,q} \quad \text{and} \\ E_k^{p,q} &= Z_k^{p,q} / B_k^{p,q}. \end{aligned}$$

We verify the conditions in Definition A.6 successively. In the following, the class of $a \in Z_k^{p,q}$ in $E_k^{p,q}$ will be denoted by $[a]_k$.

(1) We have $Z_0^{p,q} = F^pK^{p+q}$ and $B_0^{p,q} = F^{p+1}K^{p+q}$ so that

$$E_0^{p,q} = G^pK^{p+q}, \qquad [a]_0 = [a]^p, \tag{A.5}$$

and by (A.4), we have the condition (1) in Definition A.6.

(2) Since $dZ_k^{p,q} \subset Z_k^{p+k,q-k+1}$ and $dB_k^{p,q} \subset B_k^{p+k,q-k+1}$, the morphism

$$d_k^{p,q} : E_k^{p,q} \to E_k^{p+k,q-k+1}, \qquad [a]_k \mapsto [da]_k,$$

is well-defined and satisfies $d_k^{p,q} \circ d_k^{p-k,q+k-1} = 0$.

(3) Since $Z_{k+1}^{p,q} \subset Z_k^{p,q}$, we have a morphism $Z_{k+1}^{p,q} \to E_k^{p,q}$, $a \mapsto [a]_k$. As $a \in Z_{k+1}^{p,q}$, we have $d_k^{p,q}[a]_k = 0$ and this is a morphism into $\operatorname{Ker} d_k^{p,q}$. Moreover, if $a \in B_{k+1}^{p,q}$, we see that $[a]_k \in \operatorname{Im} d_k^{p-k,q+k-1}$. Thus we have a well-defined morphism

$$E_{k+1}^{p,q} \longrightarrow \operatorname{Ker} d_k^{p,q} / \operatorname{Im} d_k^{p-k,q+k-1}, \qquad [a]_{k+1} \mapsto [[a]_k], \tag{A.6}$$

which is an isomorphism. Indeed, for the surjectivity, take $[a]_k \in \operatorname{Ker} d_k^{p,q}$, i.e., $a \in Z_k^{p,q}$ and $da \in B_k^{p+k,q-k+1}$. Then we may write

$$da = da_1 + b, \qquad a_1 \in F^{p+1}K^{p+q},\ b \in F^{p+k+1}K^{p+q+1}.$$

We have $a - a_1 \in Z_{k+1}^{p,q}$ and $[a - a_1]_k = [a]_k$ as a_1 is in fact in $B_k^{p,q}$. Thus $[a - a_1]_{k+1} \mapsto [[a]_k]$. For the injectivity, take $[a]_{k+1} \in E_{k+1}^{p,q}$ such that $[a]_k \in \operatorname{Im} d_k^{p-k,q+k-1}$. Then there exists $a' \in Z_k^{p-k,q+k-1}$ such that $a - da' \in B_k^{p,q}$. From this we see that $a \in B_{k+1}^{p,q}$.

(4) We set $H^r = H^r(K^\bullet)$ for the condition (4).

(5) If $k > \max\{p, q+1\}$, then $Z_k^{p,q} = Z^{p+q}(K^\bullet) \cap F^pK^{p+q}$, $B_k^{p,q} = (B^{p+q}(K^\bullet) + F^{p+1}K^{p+q}) \cap Z_k^{p,q}$, $E_k^{p,q} = Z_k^{p,q}/B_k^{p,q}$ and $d_k = 0$. Thus $E_k^{p,q} = E_{k+1}^{p,q} = \cdots = E_\infty^{p,q}$ and it is straightforward to see that there is an isomorphism

$$E_\infty^{p,q} \simeq G^pH^{p+q}(K^\bullet), \qquad [a]_\infty \leftrightarrow [[a]]^p, \tag{A.7}$$

where $[a]$ is the class of $a \in Z_\infty^{p,q} = Z^{p+q}(K^\bullet) \cap F^pK^{p+q}$ in $F^pH^{p+q}(K^\bullet)$. Therefore, $(E_k^{p,q}, H^r)$ as defined above is a spectral sequence.

Finally, we try to find $E_1^{p,q}$. Any element in $\operatorname{Ker} d_0^{p,q}$ may be written $[a]_0$, where $a \in Z_0^{p,q}$ with $[da]_0 = 0$, i.e, $da \in B_0^{p,q+1} = F^{p+1}K^{p+q+1}$. The last condition means $a \in Z_1^{p,q}$ and in the step (3) above, we may set $a_1 = 0$. In the identity (A.5), $d_0^{p,q}[a]_0 = [da]_0 = [da]^p = d[a]^p$, where the last d is the one induced on G^pK^{p+q}. Thus we have

$$E_1^{p,q} \simeq \operatorname{Ker} d_0 / \operatorname{Im} d_0 = H^{p+q}(G^pK^\bullet), \qquad [a]_1 \leftrightarrow [[a]^p].$$

□

Proposition A.15. *Let $(E_k^{p,q}, H^r)$ be the spectral sequence as constructed in Proposition A.14. If there exists k_1 such that $d_{k_1}^{p,q} = 0$ for every (p,q), then it degenerates at k_1, i.e., $d_k^{p,q} = 0$ for every (p,q) and $k \geq k_1$.*

Proof. From (A.6) and the assumption, we have an isomorphism

$$E_{k_1+1}^{p,q} \xrightarrow{\sim} E_{k_1}^{p,q}, \qquad [a]_{k+1} \mapsto [a]_k$$

for every (p,q). By this isomorphism, for $(p+k_1+1, q-k_1)$, $d_{k_1+1}[a]_{k_1+1} = [da]_{k_1+1}$ corresponds to $[da]_{k_1} = d_{k_1}[a]_{k_1}$, which is 0 by the assumption. Thus $d_{k_1+1}^{p,q} = 0$ for every (p,q). Continuing this argument, we have the proposition. □

Double complexes: A *double complex* of R-modules is a collection $K^{\bullet,\bullet} = (K^{p,q}, d', d'')$ such that

(1) for each fixed q, $K^{\bullet,q} = (K^{p,q}, d')$ is a complex,
(2) for each fixed p, $K^{p,\bullet} = (K^{p,q}, d'')$ is a complex and
(3) $d' \circ d'' + d'' \circ d' = 0$.

If we set $K^r = \bigoplus_{p+q=r} K^{p,q}$ and define $d : K^r \to K^{r+1}$ by $d = d' + d''$, we get a complex $(K^\bullet, d)$, called the *associated single complex*. In the following we assume that $K^{p,q} = 0$ if $p < 0$ or $q < 0$. On the complex $K^\bullet$, there are two natural filtrations:

$$'F^p K^\bullet = \bigoplus_{i \geq p} K^{i,\bullet-i} \quad \text{and} \quad ''F^q K^\bullet = \bigoplus_{j \geq q} K^{\bullet-j,j}.$$

For these filtrations we have

$$'G^p K^\bullet \simeq K^{p,\bullet-p} \quad \text{and} \quad ''G^q K^\bullet \simeq K^{\bullet-q,q}.$$

Theorem A.13. *For a double complex $K^{\bullet,\bullet}$, there are two spectral sequences $('E_k^{p,q}, H^r)$ and $(''E_k^{p,q}, H^r)$ with*

$$'E_0^{p,q} \simeq K^{p,q}, \quad 'E_1^{p,q} \simeq H_{d''}^q(K^{p,\bullet}), \quad 'E_2^{p,q} \simeq H_{d'}^p(H_{d''}^q(K^{\bullet,\bullet})),$$
$$''E_0^{p,q} \simeq K^{q,p}, \quad ''E_1^{p,q} \simeq H_{d'}^q(K^{\bullet,p}), \quad ''E_2^{p,q} \simeq H_{d''}^p(H_{d'}^q(K^{\bullet,\bullet})) \quad \text{and}$$
$$H^r = H^r(K^\bullet).$$

Moreover, the both sequences degenerate at E_2 so that

$$H_{d'}^p(H_{d''}^q(K^{\bullet,\bullet})) \simeq {'G^p} H^{p+q}(K^\bullet) \quad \text{and} \quad H_{d''}^p(H_{d'}^q(K^{\bullet,\bullet})) \simeq {''G^p} H^{p+q}(K^\bullet).$$

Explicit correspondences in the above isomorphisms are given in the proof.

Proof. The theorem basically follows from Proposition A.14, except for the last statement. We look at the first sequence, the second one being similar. For $a \in {}'F^pK^{p+q}$, we write $a = \sum_{i\geq p} a^{i,p+q-i}$ with $a^{i,p+q-i} \in K^{i,p+q-i}$.

We have ${}'Z_0^{p,q} = {}'F^pK^{p+q}$, ${}'B_0^{p,q} = {}'F^{p+1}K^{p+q}$ and

$$'E_0^{p,q} = {}'G^pK^{p+q} \simeq K^{p,q}, \qquad [a]_0 \leftrightarrow a^{p,q}.$$

In the last isomorphism, $d_0^{p,q}$ corresponds to d'' on $K^{p,q}$ so that

$$\operatorname{Ker} d_0^{p,q} / \operatorname{Im} d_0^{p,q-1} = H^{p+q}({}'G^pK^\bullet) \simeq H^q_{d''}(K^{p,\bullet}). \tag{A.8}$$

We may write ${}'Z_1^{p,q} = \{\, a \in {}'F^pK^{p+q} \mid d''a^{p,q} = 0 \,\}$ and, by (A.6) and (A.8), we have

$$'E_1^{p,q} \simeq H^q_{d''}(K^{p,\bullet}), \qquad [a]_1 \leftrightarrow [a^{p,q}]. \tag{A.9}$$

We may also write

$$'Z_2^{p,q} = \{\, a \in {}'F^pK^{p+q} \mid d''a^{p,q} = 0 \text{ and } d'a^{p,q} + d''a^{p+1,q-1} = 0 \,\}.$$

Recall that $d_1^{p,q} : {}'E_1^{p,q} \to {}'E_1^{p+1,q}$ is given by $[a]_1 \mapsto [da]_1$ and in the isomorphism ${}'E_1^{p+1,q} \simeq H^q_{d''}(K^{p+1,\bullet})$, $[da]_1 \leftrightarrow [d'a^{p,q} + d''a^{p+1,q-1}] = [d'a^{p,q}] = d'[a^{p,q}]$, where the last d' is the one induced by d' on $H^q_{d''}(K^{p,\bullet})$; $d' : H^q_{d''}(K^{p,\bullet}) \to H^q_{d''}(K^{p+1,\bullet})$. Thus, by (A.6) and (A.9), we have

$$'E_2^{p,q} \simeq H^p_{d'}(H^q_{d''}(K^{\bullet,\bullet})), \qquad [a]_2 \mapsto [[a^{p,q}]]. \tag{A.10}$$

As to the last statement, recall that $d_2^{p,q} : E_2^{p,q} \to E_2^{p+2,q-1}$ is given by $[a]_2 \mapsto [da]_2$. By the isomorphism ${}'E_2^{p+2,q-1} \simeq H^{p+2}_{d'}(H^{q-1}_{d''}(K^{\bullet,\bullet}))$, $[da]_2$ is sent to $[[(da)^{p+2,q-1}]] = [[d'a^{p+1,q-1} + d''a^{p+2,q-2}]] = [[d'a^{p+1,q-1}]] = [d'[a^{p+1,q-1}]] = 0$. Hence $d_2^{p,q} = 0$ for every (p,q) and the sequence degenerates at E_2 by Proposition A.15. Thus ${}'E_\infty^{p,q} \simeq \cdots \simeq {}'E_2^{p,q}$, $[a]_\infty \mapsto [a]_2$, where $a \in {}'Z_\infty^{p,q} = Z^{p+q}(K^\bullet) \cap {}'F^pK^{p+q}$. By (A.7) and (A.10), we have

$$'G^pH^{p+q}(K^\bullet) \simeq H^p_{d'}(H^q_{d''}(K^{\bullet,\bullet})), \qquad [[a]]^p \mapsto [[a^{p,q}]].$$

□

Direct limit

Let X be a set. An *order relation*, or *an ordering*, in X is a relation $\leq$ in X with the following properties:

(1) For every x in X, $x \leq x$.
(2) If $x \leq y$ and $y \leq z$, then $x \leq z$.
(3) If $x \leq y$ and $y \leq x$, then $x = y$.

An *ordered set* is a set together with an order relation. If Y is a subset of an ordered set X, then Y is naturally an ordered set. We say that Y is *cofinal*, if for each x in X, there exists y in Y such that $x \leq y$.

An order relation is *total*, if it also has the following property:

(4) For every x and y in X, either $x \leq y$ or $y \leq x$.

A *directed set* is an ordered set Λ such that for every λ and μ in Λ, there exists ν in Λ with $\lambda \leq \nu$ and $\mu \leq \nu$.

Definition A.8. A *direct system* of R-modules on Λ is a system $(\mathsf{M}_\lambda, f_{\mu\lambda})$, where for each λ in Λ, M_λ is an R-module and for each pair (λ, μ) with $\lambda \leq \mu$, $f_{\mu\lambda} : \mathsf{M}_\lambda \to \mathsf{M}_\mu$ is a morphism such that

(1) $f_{\lambda\lambda} = 1_{\mathsf{M}_\lambda}$,
(2) $f_{\nu\mu} \circ f_{\mu\lambda} = f_{\nu\lambda}$ for $\lambda \leq \mu \leq \nu$.

We introduce a relation $\sim$ in the disjoint union $\bigsqcup_\lambda \mathsf{M}_\lambda$ by saying that $x_\lambda \sim x_\mu$ if there exists ν with $\lambda, \mu \leq \nu$ such that $f_{\nu\lambda}(x_\lambda) = f_{\nu\mu}(x_\mu)$. Then it is an equivalent relation and the *direct limit* of the system is defined by:

$$\mathsf{M} = \varinjlim_{\lambda \in \Lambda} \mathsf{M}_\lambda = \bigsqcup_{\lambda \in \Lambda} \mathsf{M}_\lambda / \sim .$$

For each x_λ, we denote by $[x_\lambda]$ its equivalence class. We may define the addition in M by $[x_\lambda] + [x_\mu] = [f_{\nu\lambda}(x_\lambda) + f_{\nu\mu}(x_\mu)]$, taking ν with $\lambda, \mu \leq \nu$. Multiplication of R on M is naturally defined and M becomes an R-module. For each λ, we have a canonical morphism $f_\lambda : \mathsf{M}_\lambda \to \mathsf{M}$, which assigns $[x_\lambda]$ to x_λ. If $\lambda \leq \mu$, we have $f_\lambda = f_\mu \circ f_{\mu\lambda}$.

If each M_λ is an R-algebra and $f_{\mu\lambda}$ is an R-algebra morphism, the direct limit M also admits the structure of an R-algebra.

If Λ' is a cofinal subset of a directed set Λ, then Λ' is a directed set. Moreover, a direct system $(\mathsf{M}_\lambda, f_{\mu\lambda})$ on Λ defines a direct system on Λ'. There is a canonical isomorphism

$$\mathsf{M}' = \varinjlim_{\lambda \in \Lambda'} \mathsf{M}_\lambda \simeq \mathsf{M} = \varinjlim_{\lambda \in \Lambda} \mathsf{M}_\lambda .$$

Taking direct limits is an exact functor (cf. the item "Abelian categories" below).

Five lemma

The proof of the following is rather straightforward:

Lemma A.6. *Let* R *be a commutative ring with unity and let*

$$\begin{array}{ccccccccc}
\mathsf{M}_1 & \longrightarrow & \mathsf{M}_2 & \longrightarrow & \mathsf{M}_3 & \longrightarrow & \mathsf{M}_4 & \longrightarrow & \mathsf{M}_5 \\
\downarrow{\scriptstyle f_1} & & \downarrow{\scriptstyle f_2} & & \downarrow{\scriptstyle f_3} & & \downarrow{\scriptstyle f_4} & & \downarrow{\scriptstyle f_5} \\
\mathsf{N}_1 & \longrightarrow & \mathsf{N}_2 & \longrightarrow & \mathsf{N}_3 & \longrightarrow & \mathsf{N}_4 & \longrightarrow & \mathsf{N}_5
\end{array}$$

be a commutative diagram of R*-modules with exact rows.*

1. *If* f_1 *is an epimorphism and* f_2 *and* f_4 *are monomorphisms, then* f_3 *is a monomorphism.*
2. *If* f_5 *is a monomorphism and* f_2 *and* f_4 *are epimorphisms, then* f_3 *is an epimorphism.*

Modules

Here we collect some basics on R-modules.

An R-module M is *finitely generated* if there exists a finite number of elements $x_1, \ldots, x_p$ in M that generate M, i.e., every element of M is written $\sum_{i=1}^{p} a_i x_i$, $a_i \in \mathsf{R}$. This is equivalent to saying that there is an epimorphism

$$\mathsf{R}^p \longrightarrow \mathsf{M} \longrightarrow 0,$$

where R^p denotes the direct sum of p copies of R. We think of the zero module as being finitely generated.

An R-module M is *free* if it is isomorphic with a direct sum of copies of R. If M is a finitely generated free module, i.e., $\mathsf{M} \simeq \mathsf{R}^r$ for some r, it is shown that r is uiquely determined by M. It is called the *rank* of M over R. In the case R is a field, it is the dimension of the vector space M over R.

An R-module P is *projective* if for every epimorphism $\varphi : \mathsf{M} \to \mathsf{N}$ and every morphism $f : \mathsf{P} \to \mathsf{N}$ of R-modules, there exists a morphism $g : \mathsf{P} \to \mathsf{M}$ such that $f = \varphi \circ g$:

$$\begin{array}{ccc}
 & & \mathsf{P} \\
 & {\scriptstyle g}\swarrow & \downarrow{\scriptstyle f} \\
\mathsf{M} & \xrightarrow[\varphi]{} & \mathsf{N}.
\end{array}$$

This is equivalent to saying that the functor $\mathrm{Hom}(\mathsf{P}, \cdot)$ is exact (cf. Example A.1.1 below). Free modules are projective. An R-module is projective if and only if it is a direct summand of a free module. For every R-module M, there exist a projective module P and an exact sequence $\mathsf{P} \to \mathsf{M} \to 0$.

An R-module I is *injective* if for every monomorphism $\varphi : \mathsf{N} \to \mathsf{M}$ and every morphism $f : \mathsf{N} \to \mathsf{I}$ of R-modules, there exists a morphism $g : \mathsf{M} \to \mathsf{I}$

such that $f = g \circ \varphi$:

$$\begin{array}{ccc} \mathsf{N} & \overset{\varphi}{\rightarrowtail} & \mathsf{M} \\ \downarrow f & \swarrow g & \\ \mathsf{I} & & \end{array} .$$

This is equivalent to saying that the functor $\mathrm{Hom}(\,\cdot\,, \mathsf{I})$ is exact. It is shown that, for every R-module M, there exist an injective module I and an exact sequence $0 \to \mathsf{M} \to \mathsf{I}$. We quote:

Proposition A.16. *A $\mathbb{Z}$-module, i.e., an Abelian group,* M *is injective if and only if it is divisible in the sense that, for every $x \in \mathsf{M}$ and $n \in \mathbb{Z}$, $n \neq 0$, there exists $x' \in \mathsf{M}$ such that $nx' = x$.*

An R-module F is *flat* if for every monomorphism $\mathsf{M}' \to \mathsf{M}$ of R-modules, $\mathsf{F} \otimes_{\mathsf{R}} \mathsf{M}' \to \mathsf{F} \otimes_{\mathsf{R}} \mathsf{M}$ is a monomorphism. This is equivalent to saying that the functor $\mathsf{F} \otimes_{\mathsf{R}} \cdot$ is exact. Projective modules are flat.

An R-module M is *torsion-free* if $ax = 0$ ($a \in \mathsf{R}$, $x \in \mathsf{M}$) implies that either a is a zero divisor in R or $x = 0$. If R is an integral domain, this is equivalent to saying that $ax = 0$ implies $a = 0$ or $x = 0$. Flat modules are torsion-free. If R is a principal ideal domain (cf. Section A.2), the converse is also true.

Abelian categories

Here we recall some basics on Abelian categories, without going too much in details.

Categories and functors: A *category* is a non-empty collection $\mathcal{C}$ of *objects* such that, for every pair of objects (A, B) in $\mathcal{C}$, associated is a set $\mathrm{Hom}(A, B)$, called the set of *morphisms* of A to B, and, for every triplet (A, B, C), there is a map $\mathrm{Hom}(A, B) \times \mathrm{Hom}(B, C) \to \mathrm{Hom}(A, C)$, called the *composition of morphisms*, with the following properties:

(1) The composition of morphisms is associative.
(2) For every object A, $\mathrm{Hom}(A, A)$ contains an element 1_A, called the *identity morphism*, which is a left and right unity for the composition.

An *isomorphism* is a morphism that admits an inverse. For $u \in \mathrm{Hom}(A, B)$ and $v \in \mathrm{Hom}(B, C)$, we denote the composition by $v \circ u \in \mathrm{Hom}(A, C)$.

Let $\mathcal{C}$ and $\mathcal{C}'$ be categories. A *covariant functor* $F : \mathcal{C} \to \mathcal{C}'$ is a "function" that assigns to each object $A \in \mathcal{C}$ an object $F(A) \in \mathcal{C}'$ and to each morphism $u \in \operatorname{Hom}(A, B)$ a morphism $F(u) \in \operatorname{Hom}(F(A), F(B))$ so that $F(1_A) = 1_{F(A)}$ and $F(v \circ u) = F(v) \circ F(u)$. Likewise a *contravariant functor* $F : \mathcal{C} \to \mathcal{C}'$ is defined by letting $F(u) \in \operatorname{Hom}(F(B), F(A))$ and $F(v \circ u) = F(u) \circ F(v)$. For a sequence $\mathcal{C} \stackrel{F}{\to} \mathcal{C}' \stackrel{G}{\to} \mathcal{C}''$ of functors, the composition $G \circ F$ is defined in an obvious way.

Abelian categories: A category $\mathcal{C}$ is *additive* if, for every pair of objects (A, B), $\operatorname{Hom}(A, B)$ has the structure of an Abelian group and the composition is bilinear. We also assume that, for every pair (A, B), the "direct sum" $A \oplus B$ exists and that there exists an object A ("zero" of $\mathcal{C}$) with $1_A = 0$. For a morphism u in $\mathcal{C}$, the notions of "kernel" $\operatorname{Ker} u$, "cokernel" $\operatorname{Coker} u$, "image" $\operatorname{Im} u$ and "coimage" $\operatorname{Coim} u$ may be defined (they may not exist). There is a canonical morphism $\operatorname{Coim} u \to \operatorname{Im} u$. A functor $F : \mathcal{C} \to \mathcal{C}'$ of an additive category to another is *additive* if $F(u + v) = F(u) + F(v)$ for every $u, v \in \operatorname{Hom}(A, B)$.

An *Abelian category* is an additive category $\mathcal{C}$ with the following properties:

(1) For every morphism u in $\mathcal{C}$, $\operatorname{Ker} u$ and $\operatorname{Coker} u$ exist.
(2) For every morphism u in $\mathcal{C}$, the canonical morphism $\operatorname{Coim} u \to \operatorname{Im} u$ is an isomorphism.

Note that (1) implies the existence of coimage and image for every morphism.

Let $\mathcal{C}$ be an Abelian category. A sequence of morphisms $A \stackrel{u}{\to} B \stackrel{v}{\to} C$ in $\mathcal{C}$ is *exact* if $\operatorname{Ker} v = \operatorname{Im} u$. Let $F : \mathcal{C} \to \mathcal{C}'$ be an additive covariant functor of an Abelian category to another. The functor F is *left exact* if, for every exact sequence of the form $0 \to A' \to A \to A''$, the sequence $0 \to F(A') \to F(A) \to F(A'')$ is exact. It is *right exact* if, for every exact sequence $A' \to A \to A'' \to 0$, the sequence $F(A') \to F(A) \to F(A'') \to 0$ is exact. It is *exact* if it is both left and right exact. This is equivalent to saying that, for every exact sequence $A \to B \to C$, $F(A) \to F(B) \to F(C)$ is exact. Likewise we may define exactness of contravariant functors. For example a contravariant additive functor F is left exact if, for every exact sequence $A' \to A \to A'' \to 0$, the sequence $0 \to F(A'') \to F(A) \to F(A')$ is exact.

Example A.1. 1. Let R be a commutative ring with unity. We have the category $\mathcal{M}od(\mathsf{R})$, whose objects are R-modules and whose morphisms are R-morphisms. For any R-module M, $\operatorname{Hom}_{\mathsf{R}}(\mathsf{M}, \cdot) : \mathcal{M}od(\mathsf{R}) \to \mathcal{M}od(\mathsf{R})$

is a covariant left exact functor and, for any R-module N, $\mathrm{Hom}_{\mathsf{R}}(\cdot\,,\mathsf{N}) : \mathcal{M}od(\mathsf{R}) \to \mathcal{M}od(\mathsf{R})$ is a contravariant left exact functor. Also, $\mathsf{M} \otimes_{\mathsf{R}} \cdot : \mathcal{M}od(\mathsf{R}) \to \mathcal{M}od(\mathsf{R})$ and $\cdot \otimes_{\mathsf{R}} \mathsf{N} : \mathcal{M}od(\mathsf{R}) \to \mathcal{M}od(\mathsf{R})$ are right exact covariant functors.

2. Let X be a topological space. We have the category $\mathcal{S}h(X)$, whose objects are sheaves of Abelian groups on X and whose morphisms are sheaf morphisms. Taking the sections $\Gamma(X;\,\cdot\,) : \mathcal{S}h(X) \to \mathcal{M}od(\mathbb{Z})$ is a left exact covariant functor. See Remark 11.1 for more examples.

Derived functors: Let $\mathcal{C}$ be an Abelian category. If there is an exact sequence of the form $0 \to A' \to A \to A'' \to 0$, we say that A' is a sub-object of A and A'' a quotient of A. In this case we write $A'' = A/A'$. Projective and injective objects in $\mathcal{C}$ are defined as in the case of R-modules. We say that $\mathcal{C}$ has enough projective or injective objects if every object in $\mathcal{C}$ is a quotient of a projective object or a sub-object of an injective object, respectively.

A (cochain) complex in $\mathcal{C}$ is a sequence of morphisms

$$\cdots \longrightarrow K^{p-1} \xrightarrow{d^{p-1}} K^p \xrightarrow{d^p} K^{p+1} \longrightarrow \cdots$$

such that $d^p \circ d^{p-1} = 0$ for all p. In this case $\operatorname{Im} d^{p-1}$ is a sub-object of $\operatorname{Ker} d^p$ and $H^p(K^\bullet) = \operatorname{Ker} d^p / \operatorname{Im} d^{p-1}$ is the p-th cohomology of $K^\bullet$.

Let $F : \mathcal{C} \to \mathcal{C}'$ be a covariant additive functor of an Abelian category to another. Suppose $\mathcal{C}$ has enough injective objects. Then, for each non-negative integer p, we may define the p-th right *derived functor* $R^pF : \mathcal{C} \to \mathcal{C}'$ as follows. For any object $A \in \mathcal{C}$, take an injective resolution, i.e., an exact sequence of the form $0 \to A \to I^0 \to I^1 \to \cdots$, with I^p injective objects, $p \geq 0$, and set $R^pF(A) = H^p(F(I^\bullet))$. It is uniquely determined modulo isomorphisms. Moreover, if F is left exact, $R^0F(A) = F(A)$ and an exact sequence $0 \to A' \to A \to A'' \to 0$ yields an exact sequence

$$\begin{aligned} 0 \to F(A') \to F(A) \to F(A'') \to R^1F(A') \to R^1F(A) \to R^1F(A'') \to \cdots \\ \to R^pF(A') \to R^pF(A) \to R^pF(A'') \to R^{p+1}F(A') \to \cdots. \end{aligned}$$

If F is a contravariant functor, we may define the left derived functor L_pF. That is, for an object A, take an injective resolution as above. Then $F(I^\bullet)$ is a chain complex and we set $L_pF(A) = H_p(F(I^\bullet))$. The left derived functor of a covariant functor or the right derived functor of a contravariant functor is defined by taking projective resolutions. Namely, let $P^\bullet \to A \to 0$ be a projective resolution. If F is covariant, $L_pF(A) = H_p(F(P^\bullet))$ and if F is contravariant, $R^pF(A) = H^p(F(P^\bullet))$.

Example A.2. 1. For a commutative ring R with unity, the category $\mathcal{M}od(\mathsf{R})$ of R-modules has enough projective and injective objects. For an R-module M, the p-th right derived functor of $\mathrm{Hom}_{\mathsf{R}}(\mathsf{M}, \cdot)$ is denoted by $\mathrm{Ext}^p_{\mathsf{R}}(\mathsf{M}, \cdot)$ and, for an R-module N, the p-th right derived functor of $\mathrm{Hom}_{\mathsf{R}}(\cdot, \mathsf{N})$ is denoted by $\mathrm{Ext}^p_{\mathsf{R}}(\cdot, \mathsf{N})$. For any pair (M, N), there is a canonical isomorphism between the $\mathrm{Ext}^p_{\mathsf{R}}(\mathsf{M}, \mathsf{N})$'s defined in two ways.

For example, for a fixed N, $\mathrm{Ext}^p_{\mathsf{R}}(\mathsf{M}, \mathsf{N})$ is defined by taking a projective resolution of M. We have

(1) $\mathrm{Ext}^0_{\mathsf{R}}(\mathsf{M}, \mathsf{N}) = \mathrm{Hom}_{\mathsf{R}}(\mathsf{M}, \mathsf{N})$.
(2) $\mathrm{Ext}^p_{\mathsf{R}}(\mathsf{M}, \mathsf{N}) = 0$ for $p \geq 1$, if M is projective.
(3) A short exact sequence $0 \to \mathsf{M}' \to \mathsf{M} \to \mathsf{M}'' \to 0$ yields a long exact sequence

$$0 \longrightarrow \mathrm{Hom}_{\mathsf{R}}(\mathsf{M}'', \mathsf{N}) \longrightarrow \mathrm{Hom}_{\mathsf{R}}(\mathsf{M}, \mathsf{N}) \longrightarrow \mathrm{Hom}_{\mathsf{R}}(\mathsf{M}', \mathsf{N}) \longrightarrow$$
$$\mathrm{Ext}^1_{\mathsf{R}}(\mathsf{M}'', \mathsf{N}) \longrightarrow \mathrm{Ext}^1_{\mathsf{R}}(\mathsf{M}, \mathsf{N}) \longrightarrow \mathrm{Ext}^1_{\mathsf{R}}(\mathsf{M}', \mathsf{N}) \longrightarrow \cdots .$$

Also, the p-th left derived functors of $\mathsf{M} \otimes_{\mathsf{R}} \cdot$ and $\cdot \otimes_{\mathsf{R}} \mathsf{N}$ are denoted by $\mathrm{Tor}^{\mathsf{R}}_p(\mathsf{M}, \cdot)$ and $\mathrm{Tor}^{\mathsf{R}}_p(\cdot, \mathsf{N})$, respectively. For any pair (M, N), there is a canonical isomorphism between the $\mathrm{Tor}^{\mathsf{R}}_p(\mathsf{M}, \mathsf{N})$'s defined in two ways.

2. For a topological space$X$, the category $\mathcal{S}h(X)$ of sheaves of Abelian groups on X has enough injective objects. The p-th derived functor of $\Gamma(X; \cdot)$ is denoted by $H^p(X; \cdot)$. This is one way of defining the cohomology $H^p(X; \mathscr{S})$ of a sheaf $\mathscr{S}$ on X.

Grothendieck spectral sequence: The theory of spectral sequences may also be developed in Abelian categories. In particular, we quote the following:

Theorem A.14. *Let $\mathcal{C}$, $\mathcal{C}'$ and $\mathcal{C}''$ be Abelian categories and let $F : \mathcal{C} \to \mathcal{C}'$ and $G : \mathcal{C}' \to \mathcal{C}''$ be additive covariant functors. Suppose that $\mathcal{C}$ and $\mathcal{C}'$ have enough injective objects, G is left exact and F transforms injective objects to G-acyclic objects, i.e., if I is injective, $R^qG(F(I)) = 0$ for $q > 0$. Then, for any object A in $\mathcal{C}$, there is a spectral sequence $(E^{p,q}_k, H^r)$ in $\mathcal{C}''$ with*

$$E^{p,q}_2 = R^pG(R^qF(A)) \quad \textit{and} \quad H^r = R^r(G \circ F)(A).$$

For an example, see Proposition 11.22.

A.2 Commutative rings

In the following a ring means a commutative ring with unity 1, which may possibly be equal to 0. On the other hand, for a field we assume that $1 \neq 0$.

Let a and b elements in a ring R. We say that b *divides* a and write $b \mid a$ if $a = bc$ for some c in R.

Integral domains: A *zero divisor* in a ring R is an element a in R such that there is an element $b \neq 0$ in R with $ab = 0$. An *integral domain* is a ring $\mathsf{R} \neq 0$ without non-zero zero divisors, i.e., if $ab = 0$, for a, b in R, then $a = 0$ or $b = 0$. If R is an integral domain, we may form the *fraction field* $F(\mathsf{R})$ of R as follows. We introduce a relation $\sim$ in $\mathsf{R} \times (\mathsf{R} \setminus \{0\})$ by $(a, b) \sim (a', b')$ if and only if $ab' = a'b$. It is an equivalence relation and the equivalence class of (a, b) is denoted by a/b. Then $F(\mathsf{R}) = \mathsf{R} \times (\mathsf{R} \setminus \{0\})/\sim$ becomes a field with the addition $a/b + c/d = (ad + bc)/bd$ and multiplication $a/b \cdot c/d = ac/bd$. Identifying a with $a/1$, we may think of R as a subring of $F(\mathsf{R})$.

Unique factorization domains: We say that an element u in a ring R is a *unit* if there is an element v in R such that $uv = 1$. Let R be an integral domain. An element a in R is *irreducible* if a is not a unit and if the identity $a = bc$ for elements b and c in R implies that either b or c is a unit. Note that 0 is not irreducible. We say that R is a *unique factorization domain*, or a *UFD* for short, if every element a in R which is not 0 or a unit can be expressed as a product of irreducible elements in R and the expression is unique up to the order and multiplications by units. Thus an element a as above may be expressed as

$$a = u\, p_1^{m_1} \cdots p_r^{m_r}, \tag{A.11}$$

where u is a unit, the m_i's are positive integers and the p_i's are irreducible elements such that if $i \neq j$, p_i is not a product of a unit and p_j. An expression as above, which is unique up to the order and multiplications by units, is called the *irreducible decomposition* of a

For example, the ring $\mathbb{Z}$ of integers is a UFD. Also a field is a UFD, since any of its elements is either 0 or a unit. Then ring $\mathcal{O}_n$ of germs of holomorphic functions at 0 in $\mathbb{C}^n$ is a UFD (cf. Theorem 1.11).

Let R be an integral domain. An element a in R is *prime* if a is not 0 or a unit and if $a \mid bc$, for elements b and c in R, then either $a \mid b$ or $a \mid c$. If a is prime, it is irreducible. If R is a UFD, every irreducible element is prime.

Theorem A.15 (Gauss' lemma). *If* R *is a UFD, so is the polynomial ring* $\mathsf{R}[X]$ *in the variable* X.

Let R be a UFD. For non-zero elements a and b there is always the greatest common divisor $\gcd(a,b)$, which is unique up to multiplication by units. We say that a and b are *relatively prime* if $\gcd(a,b)$ is a unit. Note that if either a or b is a unit, then a and b are relatively prime.

Resultant: Let R be an integral domain. For two polynomials $P(X) = \sum_{i=0}^{m} a_i X^i$ and $Q(X) = \sum_{j=0}^{n} b_j X^j$, $a_m b_n \neq 0$, in $\mathsf{R}[X]$, their *resultant* $R(P,Q)$ is an element in R defined by

$$R(P,Q) = \begin{vmatrix} a_m & \cdots & a_0 & & \\ & \ddots & & \ddots & \\ & & a_m & \cdots & a_0 \\ b_n & \cdots & b_0 & & \\ & \ddots & & \ddots & \\ & & b_n & \cdots & b_0 \end{vmatrix}.$$

The *discriminant* $D(P)$ of $P(X)$ is defined by

$$D(P) = (-1)^{\frac{m(m-1)}{2}} \frac{1}{a_m} R(P,P'), \tag{A.12}$$

where $P'(X) = a_1 + \cdots + m a_{m-1} X^{m-1}$.

Theorem A.16. 1. *For* $P(X)$ *and* $Q(X)$ *as above, there exist* $A(X)$ *and* $B(X)$ *in* $\mathsf{R}[X]$ *such that*

$$A(X)P(X) + B(X)Q(X) = R(P,Q).$$

2. *Suppose* R *is a UFD. Then* $P(X)$ *and* $Q(X)$ *are relatively prime if and only if* $R(P,Q) \neq 0$.

Integral extensions: Let R be a subring of a ring S. An element α of S is *algebraic* over R if there is a non-zero polynomial $P(X)$ in $\mathsf{R}[X]$ with $P(\alpha) = 0$. It is *integral* over R if we may take a monic polynomial as $P(X)$ in the above. The ring S is integral over R if every element of S is integral over R. If S is finitely generated as an R-module, S is integral over R.

Let K be a subfield of a field L. In this case an element α in L is algebraic over K if and only if it is integral over K and among the monic polynomials in $\mathsf{K}[X]$ as above, there is a unique polynomial with least degree, which is called the *minimal polynomial* of α.

Let R be an integral domain and $F(\mathsf{R})$ its fraction field. We say that R is *normal* if every element in $F(\mathsf{R})$ which is integral over R is in R. For example, a UFD is normal.

Proposition A.17. *Let R be a normal integral domain and $\mathsf{K} = F(\mathsf{R})$ its fraction field. Let $\mathsf{K} \subset \mathsf{L}$ be a field extension and α an element in L. Suppose α is integral over R, thus in particular α is algebraic over K. In this situation the minimal polynomial of α is in $\mathsf{R}[X]$.*

Principal ideal domains: A *principal ideal domain*, a PID for short, is an integral domain R such that every ideal I is principal, i.e., $I = (a)$ for some a in R. For example, $\mathbb{Z}$ and an arbitrary field are PID. Also the polynomial ring $\mathsf{K}[X]$ of one variable over a field K is a PID. A PID is a UFD.

Proposition A.18. *If M is a free module over a PID, every submodule of M is free.*

Noetherian rings: We say that a ring R is *Noetherian* if every ideal in R is finitely generated, i.e., if I is an ideal in R, there exist a finite number of elements $x_1, \ldots, x_r$ in I such that every element x in I is written $x = \sum_{i=1}^{r} a_i x_i$ with $a_i \in \mathsf{R}$.

Proposition A.19. *For a ring R, the following are equivalent:*

(1) *R is Noetherian.*
(2) *Every ascending sequence*

$$I_1 \subset I_2 \subset \cdots \subset I_n \subset \cdots$$

of ideals in R is stationary, i.e., $I_k = I_{k+1} = \cdots$, for some k.
(3) *Every non-empty set of ideals in R has a maximal element.*

For example, the ring $\mathbb{Z}$ is Noetherian. Also a field is Noetherian, since any of its ideals is either 0 or itself. The ring $\mathscr{O}_n$ is Noetherian (cf. Theorem 1.12).

Theorem A.17 (Hilbert basis theorem). *If R is Noetherian, so is $\mathsf{R}[X]$.*

Maximal ideals: In the following we say that an ideal I in a ring R is *proper* if $I \neq \mathsf{R}$.

We say that an ideal $\mathfrak{m}$ in R is *maximal* if $\mathfrak{m}$ is proper and if there are no ideals I with $\mathfrak{m} \subsetneqq I \subsetneqq \mathsf{R}$. This is equivalent to saying that the quotient $\mathsf{R}/\mathfrak{m}$ is a field. A ring with a unique maximal ideal is called a *local ring.*

For example the ring $\mathscr{O}_n$ is a local ring with the ideal $\mathfrak{m}$ of non-units in $\mathscr{O}_n$ as its unique maximal ideal (cf. Proposition 1.7). There is a canonical isomorphism $\mathscr{O}_n/\mathfrak{m} \simeq \mathbb{C}$.

We quote:

Theorem A.18 (Nakayama's lemma). *Let* $(\mathsf{R}, \mathfrak{m})$ *be a local ring and* M *a finitely generated* R*-module. If* $\mathfrak{m}\mathsf{M} = \mathsf{M}$*, then* $\mathsf{M} = 0$.

There is a canonical isomorphism $\mathsf{M}/\mathfrak{m}\mathsf{M} \simeq \mathsf{R}/\mathfrak{m} \otimes_{\mathsf{R}} \mathsf{M}$ and it is an $\mathsf{R}/\mathfrak{m}$-vector space. As a consequence of the above, we have:

Proposition A.20. *Let* $(\mathsf{R}, \mathfrak{m})$ *and* M *be as above and let* $x_1, \ldots, x_r$ *be elements in* M*. If the classes of* $x_1, \ldots, x_r$ *span the vector space* $\mathsf{M}/\mathfrak{m}\mathsf{M}$*, then* $x_1, \ldots, x_r$ *generate* M *as an* R*-module.*

Prime and primary ideals: For an ideal I in a ring R, its *radical* is defined by

$$\sqrt{I} = \{\, a \in \mathsf{R} \mid a^k \in I \text{ for some positive integer } k \,\},$$

which is an ideal in R containing I.

An ideal $\mathfrak{p}$ in R is said to be *prime* if $\mathsf{R}/\mathfrak{p}$ is an integral domain, i.e., $\mathfrak{p}$ is proper and $ab \in \mathfrak{p}$ implies $a \in \mathfrak{p}$ or $b \in \mathfrak{p}$. If $\mathfrak{p}$ is prime, then $\sqrt{\mathfrak{p}} = \mathfrak{p}$. Every maximal ideal is prime. The ideal $\{0\}$ is prime if and only if R is an integral domain.

If R is an integral domain, for an element $p \neq 0$ in R, p is prime if and only if (p) is prime. Thus, if R is a UFD, p is irreducible if and only if (p) is prime.

We say that an element a in R is *nilpotent* if $a^k = 0$ for some positive integer k.

A proper ideal $\mathfrak{q}$ in R is said to be *primary* if every zero divisor in $\mathsf{R}/\mathfrak{q}$ is nilpotent, i.e., $ab \in \mathfrak{q}$ implies $a \in \mathfrak{q}$ or $b \in \sqrt{\mathfrak{q}}$. This is also equivalent to saying that $ab \in \mathfrak{q}$ implies $a \in \mathfrak{q}$ or $b \in \mathfrak{q}$ or $a, b \in \sqrt{\mathfrak{q}}$.

If $\mathfrak{q}$ is primary, then $\sqrt{\mathfrak{q}}$ is prime.

We quote the following:

Theorem A.19 (Primary decomposition theorem). *If* R *is Noetherian, for every proper ideal* I *in* R*, there exist a finite number of primary ideals* $\mathfrak{q}_1, \ldots, \mathfrak{q}_r$ *such that*

$$I = \mathfrak{q}_1 \cap \cdots \cap \mathfrak{q}_r.$$

Moreover, if I cannot be expressed as the intersection of ideals in a proper subset of $\{\mathfrak{q}_1, \ldots, \mathfrak{q}_r\}$, then the prime ideals $\sqrt{\mathfrak{q}_1}, \ldots, \sqrt{\mathfrak{q}_r}$ are uniquely determined by I.

In the above we call $\sqrt{\mathfrak{q}_1}, \ldots, \sqrt{\mathfrak{q}_r}$ the *associated primes* of I.

Fraction rings: A subset S of a ring R is said to be *multiplicative* if $1 \in S$ and if $s, s' \in S$ implies $ss' \in S$. For such a set S, we construct the *fraction ring*, denoted by $S^{-1}\mathsf{R}$, of R with respect to S as follows. We introduce a relation $\sim$ in $\mathsf{R} \times S$ by $(a, s) \sim (a', s')$ if and only if $s''(as' - a's) = 0$ for some $s'' \in S$. It is an equivalence relation and the equivalence class of (a, s) is denoted by a/s. Then $S^{-1}\mathsf{R} = \mathsf{R} \times S/\sim$ becomes a ring with the usual addition and multiplication of fractions. There is a natural morphism $\mathsf{R} \to S^{-1}\mathsf{R}$ given by $a \mapsto a/1$. It is injective if and only if S does not contain any zero divisor in R. The above morphism gives $S^{-1}\mathsf{R}$ an R-module structure and it is a flat R-module (cf. Section A.1).

If R is an integral domain, $S = \mathsf{R}\setminus\{0\}$ is multiplicative and $S^{-1}\mathsf{R}$ is the fraction field $F(\mathsf{R})$ of R.

In general, for an R-module M and a multiplicative set S in R, we may define $S^{-1}\mathsf{M} = \{x/s \mid x \in \mathsf{M},\ s \in S\}$ the same way as $S^{-1}\mathsf{R}$. It is an $S^{-1}\mathsf{R}$-module and the assignment $x/s \mapsto (1/s) \otimes x$ induces a well-defined $S^{-1}\mathsf{R}$-module isomorphism $S^{-1}\mathsf{M} \xrightarrow{\sim} S^{-1}\mathsf{R} \otimes_{\mathsf{R}} \mathsf{M}$.

Let $\varphi : \mathsf{R} \to \mathsf{R}'$ be a ring morphism. If S is a multiplicative set in R, $S' = \varphi(S)$ is a multiplicative set in R'. The morphism φ gives R' an R-module structure and the assignment $a'/s \mapsto a'/\varphi(s)$ induces a well-defined ring isomorphism $S^{-1}\mathsf{R}' \xrightarrow{\sim} S'^{-1}\mathsf{R}'$.

Proposition A.21. *Let I be an ideal in R. For a multiplicative set S in R, we denote by S' its image in R/I. Then there is a canonical ring isomorphism*

$$S^{-1}\mathsf{R}/I \cdot S^{-1}\mathsf{R} \simeq S'^{-1}(\mathsf{R}/I) \qquad \text{given by} \quad [a/s] \leftrightarrow [a]/[s].$$

Proof. We have

$$S'^{-1}(\mathsf{R}/I) \simeq S^{-1}(\mathsf{R}/I) \simeq S^{-1}\mathsf{R} \otimes_{\mathsf{R}} (\mathsf{R}/I) \simeq S^{-1}\mathsf{R}/I \cdot S^{-1}\mathsf{R},$$

where the last isomorphism is a consequence of the fact that $S^{-1}\mathsf{R}$ is a flat R-module. □

If $\mathfrak{p}$ is a prime ideal in R, then $S = \mathsf{R}\setminus\mathfrak{p}$ is multiplicative. In this case, $S^{-1}\mathsf{R}$ is denoted by $\mathsf{R}_{\mathfrak{p}}$. It is a local ring with $\mathfrak{p}\mathsf{R}_{\mathfrak{p}}$ as its maximal ideal. By Proposition A.21, there is a canonical field isomorphism $\mathsf{R}_{\mathfrak{p}}/\mathfrak{p}\mathsf{R}_{\mathfrak{p}} \simeq F(\mathsf{R}/\mathfrak{p})$.

More generally, let $\mathfrak{p}_1, \ldots, \mathfrak{p}_r$ be prime ideals in R. Then the set $S = \bigcap_{i=1}^r (\mathsf{R} \setminus \mathfrak{p}_i) = \mathsf{R} \setminus \bigcup_{i=1}^r \mathfrak{p}_i$ is multiplicative in R. We set $I = \bigcap_{i=1}^r \mathfrak{p}_i$. In this situation, we have an isomorphism as in Proposition A.21. Note that S' does not contain any zero divisor in R/I in this case.

In the case $\mathsf{R} = \mathscr{O}_n$, R/I and $(S')^{-1}(\mathsf{R}/I)$ are the rings of germs of holomorphic and meromorphic functions, respectively, on the germ of the varaity $V(I) = \bigcup_{i=1}^r V(\mathfrak{p}_i)$.

Height of an ideal: Let R be a ring with $\mathsf{R} \neq 0$. A *prime chain* of length n is a sequence of $n+1$ prime ideals in R:

$$\mathfrak{p}_0 \supsetneqq \mathfrak{p}_1 \supsetneqq \cdots \supsetneqq \mathfrak{p}_n.$$

The *height* of a prime ideal $\mathfrak{p}$, denoted by $\operatorname{ht}\mathfrak{p}$, is the supremum of the lengths of the prime chains with $\mathfrak{p}_0 = \mathfrak{p}$. The height $\operatorname{ht} I$ of a proper ideal I is the minimum of the heights of the prime ideals containing I.

Krull dimension: The *dimension* of a ring $\mathsf{R} \neq 0$, denoted by $\dim \mathsf{R}$, is the supremum of the heights of the prime ideals in R. It is also referred to as the *Krull dimension* of R. From definition, for every proper ideal I,

$$\dim \mathsf{R}/I + \operatorname{ht} I \leq \dim \mathsf{R}.$$

Example A.3. Let $\mathscr{O}_n$ be the ring of germs of holomorphic functions at 0 in $\mathbb{C}^n = \{(z_1, \ldots, z_n)\}$. Then

$$(z_1, \ldots, z_n) \supsetneqq (z_1, \ldots, z_{n-1}) \supsetneqq \cdots \supsetneqq (z_1) \supsetneqq 0$$

is a prime chain of length n. In fact it can be shown that $\dim \mathscr{O}_n = n$.

The *annihilator* of an R-module M is the ideal in R defined by $\operatorname{Ann}(\mathsf{M}) = \{\, a \in \mathsf{R} \mid a\mathsf{M} = 0 \,\}$. If $\mathsf{M} \neq 0$, we define the (Krull) dimension of M by

$$\dim \mathsf{M} = \dim(\mathsf{R}/\operatorname{Ann}(\mathsf{M})).$$

If $\mathsf{M} = 0$, we set $\dim \mathsf{M} = -1$. It is also denoted by $\dim_{\mathsf{R}} \mathsf{M}$.

Regular sequences: Let R be a ring and M an R-module. An element a in R is a *zero divisor* on M if there is an element $x \neq 0$ in M such that $ax = 0$.

A sequence $a_1, \ldots, a_r$ of elements of R is M-*regular* if

(1) a_1 is not a zero divisor on M,
(2) for each $i = 2, \ldots, r$, a_i is not a zero divisor on $\mathsf{M}/(a_1, \ldots, a_{i-1})\mathsf{M}$,
(3) $\mathsf{M} \neq (a_1, \ldots, a_r)\mathsf{M}$.

If the a_i's belong to an ideal I, we say that $a_1, \ldots, a_r$ is an M-regular sequence in I.

For example, if $\mathsf{M} = \mathsf{R} = \mathscr{O}_n$, then $z_1, \ldots, z_n$ is an $\mathscr{O}_n$-regular sequence in the maximal ideal $\mathfrak{m}$. In fact, for any sequence $\nu_1, \ldots, \nu_n$ of positive integers, the sequence $z_1^{\nu_1}, \ldots, z_n^{\nu_n}$ is an $\mathscr{O}_n$-regular sequence in $\mathfrak{m}$.

Depth of an ideal: Let R be a ring and M an R-module. Let I be an ideal in R with $I\mathsf{M} \neq \mathsf{M}$. The *depth* of I on M, denoted by $\operatorname{depth}(I; \mathsf{M})$, is the length of a maximal M-regular sequence in I. The depth of I on R is simply called the depth of I and is denoted by depth I.

Let $(\mathsf{R}, \mathfrak{m})$ and $(\mathsf{S}, \mathfrak{n})$ be Noetherian local rings and $\varphi : \mathsf{R} \to \mathsf{S}$ a local morphism, i.e., a morphism with $\varphi(\mathfrak{m}) \subset \mathfrak{n}$. The morphism φ endows S with an R-module structure. We say that φ is *finite* if S is finitely generated over R. In this case, we have:

$$\dim_{\mathsf{S}} \mathsf{S} = \dim_{\mathsf{R}} \mathsf{S}, \qquad \operatorname{depth}(\mathfrak{n}; \mathsf{S}) = \operatorname{depth}(\mathfrak{m}; \mathsf{S}). \tag{A.13}$$

Cohen-Macaulay modules: Let $(\mathsf{R}, \mathfrak{m})$ be a Noetherian local ring and M a finitely generated R-module. The R-module M is said to be *Cohen-Macaulay*, CM for short, if $\mathsf{M} = 0$, or if $\mathsf{M} \neq 0$ and $\operatorname{depth}(\mathfrak{m}; \mathsf{M}) = \dim \mathsf{M}$. The ring R is CM if it is CM as an R-module.

For example, $\mathscr{O}_n$ is a CM ring.

We have:

Proposition A.22. *Suppose $a_1, \ldots, a_r$ is an M-regular sequence in $\mathfrak{m}$. Then M is CM if and only if $\mathsf{M}/(a_1, \ldots, a_r)\mathsf{M}$ is.*

Proposition A.23. *Let $(\mathsf{R}, \mathfrak{m})$ be a CM local ring. Then:*

1. *For every proper ideal I in R,*

$$\operatorname{ht} I = \operatorname{depth}(I; \mathsf{R}), \qquad \operatorname{ht} I + \dim \mathsf{R}/I = \dim \mathsf{R}.$$

2. *A sequence $a_1 \ldots, a_r$ in $\mathfrak{m}$ is R-regular if and only if $\operatorname{ht}(a_1 \ldots, a_r) = r$. In this case, every associated prime of $(a_1 \ldots, a_r)$ has height r.*

Note that, in general, a sequence being regular depends on the order of the elements, however, in the situation of Proposition A.23, it does not.

From (A.13), we have:

$$\text{If } \varphi : \mathsf{R} \to \mathsf{S} \text{ is finite and if } \mathsf{S} \text{ is a CM ring, then } \mathsf{S} \text{ is a CM } \mathsf{R}\text{-module.} \tag{A.14}$$

Projective dimension: Let R be a ring and M an R-module. The *projective dimension* of M, denoted by $\operatorname{pd}_{\mathsf{R}} \mathsf{M}$, is the minimum of the lengths of projective resolution of M. We quote the Auslander-Buchsbaum formula, which says that if R is a Noetherian local ring, M is finite over R and $\operatorname{pd} \mathsf{M}$ is finite,

$$\operatorname{depth}(\mathfrak{m}; \mathsf{M}) + \operatorname{pd} \mathsf{M} = \operatorname{depth} \mathfrak{m}. \tag{A.15}$$

Determinantal ideals: Let R be a ring and $f : \mathsf{R}^m \to \mathsf{R}^n$ an R-morphism, which may be represented by an $n \times m$ matrix. We assume that $m \geq n$ and denote by $I(f)$ the ideal generated by all the $n \times n$ minors of f. We assume $I(f) \neq \mathsf{R}$. Then we have:

$$\operatorname{ht} I(f) \leq m - n + 1. \tag{A.16}$$

We also have:

$$\text{If } \mathsf{R} \text{ is CM and if } \operatorname{ht} I(f) = m - n + 1, \text{ then } \mathsf{R}/I(f) \text{ is CM.} \tag{A.17}$$

Analytic rings: Let $\mathscr{O}_n = \mathbb{C}\{z_1, \ldots, z_n\}$ denote the ring of convergent power series in n variables. A ring R is an *analytic ring* if $\mathsf{R} \simeq \mathscr{O}_n/I$ for some proper ideal I in $\mathscr{O}_n$ (for some n). The ring $\mathscr{O}_n/I$ is a Noetherian local ring with $\mathfrak{m}/I$ as its unique maximal ideal, where $\mathfrak{m}$ is the maximal ideal in $\mathscr{O}_n$. The ring R inherits the structure of a Noetherian local ring via the above isomorphism. If $\boldsymbol{X} = (X, \mathscr{O}_{\boldsymbol{X}})$ is a complex space, then $\mathscr{O}_{\boldsymbol{X},x}$ is an analytic ring for each point x in X (cf. Section 11.5).

For an ideal I in $\mathscr{O}_n$, we denote by $V(I)$ the germ at 0 (in $\mathbb{C}^n$) of the variety defined by I (cf. Section 2.3). If $\mathsf{R} = \mathscr{O}_n/I$, then $\dim_{\mathsf{R}} \mathsf{R} = \dim V(I)$. We denote by $\dim_{\mathbb{C}}$ the dimension of a complex vector space. By the Nullstellensatz (Theorem 2.8),

$$\dim_{\mathsf{R}} \mathsf{R} = 0 \text{ if and only if } \dim_{\mathbb{C}} \mathsf{R} \text{ is finite.} \tag{A.18}$$

Let $\varphi : \mathsf{R} \to \mathsf{S}$ be a local morphism of analytic rings. The morphism φ induces $\mathbb{C} = \mathsf{R} \otimes_{\mathsf{R}} \mathsf{R}/\mathfrak{m}\mathsf{R} \to \mathsf{S} \otimes_{\mathsf{R}} \mathsf{R}/\mathfrak{m}$. We say that φ is *quasi-finite* if this morphism makes $\mathsf{S} \otimes_{\mathsf{R}} \mathsf{R}/\mathfrak{m}$ a finite-dimensional complex vector space. Clearly a finite morphism is quasi-finite. The coverse is also true:

$$\varphi \text{ is finite if and only if it is quasi-finite.} \tag{A.19}$$

Let $(f, \varphi) : \boldsymbol{X} = (X, \mathscr{O}_{\boldsymbol{X}}) \to T = (T, \mathscr{O}_T)$ be a morphism of complex spaces with T reduced so that we may think of $\mathscr{O}_T$ as the sheaf of holomorphic functions (cf. Section 11.5). For each point x in X, φ induces a local morphism

$$\varphi_x : \mathscr{O}_{T,f(x)} \longrightarrow \mathscr{O}_{\boldsymbol{X},x},$$

which endows $\mathscr{O}_{\boldsymbol{X},x}$ with an $\mathscr{O}_{T,f(x)}$-module structure.

We think of each point t in T as a complex subspace defined by the ideal $\mathscr{I}_t$ of holomorphic functions vanishing at t. The *fiber* $\boldsymbol{X}_t$ of (f, φ) over t is the inverse image of t by (f, φ). Thus it is the complex space with support $X_t = f^{-1}(t)$ and structure sheaf $\mathscr{O}_{\boldsymbol{X}_t} = i_t^{-1}(\mathscr{O}_{\boldsymbol{X}}/(\mathscr{O}_{\boldsymbol{X}} \cdot \mathscr{I}_t))$, where $i_t : X_t \hookrightarrow X$ is the inclusion. For each point x in X_t, we have the exact sequence

$$\mathscr{O}_{X,x} \otimes_{\mathscr{O}_{T,t}} \mathfrak{m}_t \longrightarrow \mathscr{O}_{X,x} \longrightarrow \mathscr{O}_{X,x} \otimes_{\mathscr{O}_{T,t}} \mathscr{O}_{T,t}/\mathfrak{m}_t \longrightarrow 0,$$

where $\mathfrak{m}_t$ is the maximal ideal in $\mathscr{O}_{T,t}$. Thus noting that $(\mathscr{I}_t)_t = \mathfrak{m}_t$, we have

$$\mathscr{O}_{\boldsymbol{X}_t,x} = \mathscr{O}_{\boldsymbol{X},x}/(\mathscr{O}_{\boldsymbol{X},x} \cdot \mathfrak{m}_t) \simeq \mathscr{O}_{X,x} \otimes_{\mathscr{O}_{T,t}} \mathscr{O}_{T,t}/\mathfrak{m}_t \simeq \mathscr{O}_{\boldsymbol{X},x} \otimes_{\mathscr{O}_{\boldsymbol{T},t}} \mathbb{C}.$$

Hence by (A.18) and (A.19), we see that

$$x \text{ is an isolated point in } X_t \text{ if and only if } \varphi_x \text{ is finite.} \tag{A.20}$$

Suppose now that f is a finite map (cf. Definition 2.10) so that every point x in X is isolated in X_t, $t = f(x)$, and $\dim_{\mathbb{C}} \mathscr{O}_{\boldsymbol{X}_t,x}$ is finite. Let t be a point in T. For a point x in X_t, we set $\nu(x) = \dim_{\mathbb{C}} \mathscr{O}_{\boldsymbol{X}_t,x}$ and $\nu(t) = \sum_{x \in X_t} \nu(x)$. We say that (f, φ) is *flat* if $\mathscr{O}_{\boldsymbol{X},x}$ is a flat $\mathscr{O}_{\boldsymbol{T},f(x)}$-module, for every x in X. In this situation, we have:

$$(f, \varphi) \text{ is flat if and only if } \nu(t) \text{ is locally constant.} \tag{A.21}$$

Notes

The contents of Section A.1 may be found in most of the text books dealing with homological algebra. For spectral sequences, in particular the ones in Abelian categories, we refer to [Grothendieck (1957)]. For Abelian categories, we refer to [Grothendieck (1957); Kashiwara and Schapira (1990)].

We list [Bourbaki (1961-65); de Jong and Pfister (2000); Eisenbud (1995); Hotta (1987); Matsumura (1980)] as general references for Section A.2. For (A.16), (A.19) and (A.21), see [Macaulay (1916)], [Narasimhan (1966)] and [Douady (1968)], respectively.

Appendix B

Algebraic Topology

We summarize basic materials in algebraic topology relevant to our purposes. On the way we fix notation and orientation conventions. We adopt the singular homology theory as our reference theory and use cellular decompositions or triangulations of the space for actual computations.

In the following, for a subset A in a topological space X, we denote by $\operatorname{Int} A$ and $\overline{A}$ the interior and the closure of A in X, respectively. A map between topological spaces means a continuous map, unless otherwise stated. We also let R denote a commutative ring with unity. In our case it will be mostly $\mathbb{Z}$, $\mathbb{Q}$, $\mathbb{R}$ or $\mathbb{C}$.

B.1 Singular homology

We refer to Section A.1 for basics on homological algebra.

Affine spaces: An *affine space* is a subset A of $\mathbb{R}^N$ of the form $A = v + \mathsf{V}$ with v a point in $\mathbb{R}^N$ and V a linear subspace of $\mathbb{R}^N$. In this case V is determined uniquely by A and v is determined modulo elements in V. The dimension of A is the dimension of V. For $p+1$ points $v_0, \ldots, v_p$ in A and $p+1$ real numbers $t_0, \ldots, t_p$ with $\sum_{i=0}^{p} t_i = 1$, the point $\sum_{i=0}^{p} t_i v_i$ is in A.

Let $v_0, \ldots, v_p$ be points in $\mathbb{R}^N$. Then the set

$$A(v_0, \ldots, v_p) = \left\{ v \in \mathbb{R}^N \;\middle|\; v = \sum_{i=0}^{p} t_i v_i, \;\; \sum_{i=0}^{p} t_i = 1 \right\}$$

is the minimal affine space containing $v_0, \ldots, v_p$. In fact, letting V be the subspace spanned by $v_1 - v_0, \ldots, v_p - v_0$, we may write $A(v_0, \ldots, v_p) = v_0 + \mathsf{V}$. We say that $v_0, \ldots, v_p$ are *in general position* if $\dim \mathsf{V} = p$.

A map $f : A \to \mathbb{R}^k$ is *affine* if $f((1-t)v + tv') = (1-t)f(v) + tf(v')$, for any points v and v' in A. This is equivalent to saying that the map $\tilde{f} : \mathsf{V} \to \mathbb{R}^k$ defined by $\tilde{f}(v) = f(v_0 + v) - f(v_0)$, for a point v_0 in A, is linear. It is sometimes called linear in the affine sense, or simply linear.

Simplices: Let $v_0, \ldots, v_p$ be points in general position in $\mathbb{R}^N$. The *p-simplex* with vertices $v_0, \ldots, v_p$ is the set

$$\boldsymbol{s} = \left\{ v \in \mathbb{R}^N \;\middle|\; v = \sum_{i=0}^{p} t_i v_i, \;\; \sum_{i=0}^{p} t_i = 1, \;\; t_i \geq 0 \right\}. \tag{B.1}$$

For a point v in $\boldsymbol{s}$, the numbers $(t_0, \ldots, t_p)$ as in (B.1) are uniquely determined by v and are called the *barycentric coordinates* of v. For each i, t_i is the barycentric coordinate of v relative to v_i. The point $\frac{1}{p+1}\sum_{i=0}^{p} v_i$ is called the *barycenter* of $\boldsymbol{s}$ and is denoted by $b_{\boldsymbol{s}}$. Note that in the case $p = 0$, $\boldsymbol{s} = b_{\boldsymbol{s}} = \{v_0\}$.

Singular homology

In $\mathbb{R}^p$, the points $P_0 = (0, \ldots, 0)$, $P_1 = (1, 0, \ldots, 0), \ldots, P_p = (0, \ldots, 0, 1)$ are in general position. The *standard p-simplex* Δ^p is the p-simplex with vertices $P_0, \ldots, P_p$. Let X be a topological space. A *singular p-simplex* of X is a continuous map

$$\sigma : \Delta^p \longrightarrow X.$$

We denote by $S_p(X; \mathsf{R})$ the free R-module generated by all the singular p-simplices of X, i.e., the direct sum $\bigoplus \mathsf{R}\sigma$ with σ all the singular p-simplices of X. If there is no fear of confusion, we omit the coefficient ring R and denote $S_p(X; \mathsf{R})$ by $S_p(X)$. Thus an element c in $S_p(X)$, called a *singular p-chain*, is expressed uniquely as a finite sum $c = \sum a_\sigma \sigma$ with a_σ elements in R and σ singular p-simplices. For $p > 0$ and $i = 0, \ldots, p$, we denote by $\varepsilon_p^i : \Delta^{p-1} \to \Delta^p$ the affine map with

$$\varepsilon_p^i(P_j) = \begin{cases} P_j & \text{for } \; j < i, \\ P_{j+1} & \text{for } \; j \geq i. \end{cases}$$

In short, ε_p^i is the operation that leaves P_i out of Δ^p. For a singular p-simplex σ, $p > 0$, and $i = 0, \ldots, p$, the *i-th face* $d_i\sigma$ is the singular $(p-1)$-simplex defined by $d_i\sigma = \sigma \circ \varepsilon_p^i : \Delta^{p-1} \to X$. We extend d_i as an R-morphism of $S_p(X)$ to $S_{p-1}(X)$ and define the *boundary operator*

$$\partial = \partial_p : S_p(X) \longrightarrow S_{p-1}(X) \qquad \text{by } \; \partial_p = \sum_{i=0}^{p} (-1)^i d_i.$$

We set $S_p(X) = 0$ for $p < 0$ and $\partial_0 = 0$. Then we see that

$$\partial_p \circ \partial_{p+1} = 0$$

and $(S_\bullet(X), \partial)$ is a chain complex (cf. Section A.1). By convention, we set $S_p(\emptyset) = 0$ for all p.

Definition B.9. The *p-th singular homology* $H_p(X; \mathsf{R})$ of X with coefficients in R is the p-th homology of $(S_\bullet(X), \partial)$.

Remark B.8. There is a canonical isomorphism $S_p(X; \mathsf{R}) \simeq \mathsf{R} \otimes_{\mathbb{Z}} S_p(X; \mathbb{Z})$, through which the two R-modules may be identified. However, this relation does not directly descend to the homology level in general (cf. Theorem B.20 below).

We define the *augmentation* ε: $S_0(X) \to \mathsf{R}$ by $\varepsilon(\sum a_\sigma \sigma) = \sum a_\sigma$. Since $\varepsilon \circ \partial : S_1(X) \to \mathsf{R}$ is the zero map, it induces a morphism $\varepsilon_* : H_0(X; \mathsf{R}) \to \mathsf{R}$, which is also called the augmentation. Noting that we may regard a 0-simplex as a point and a 1-simplex as a path in X, we see that, if $X (\neq \emptyset)$ is path connected, ε_* is an isomorphism.

In general, if $X = \bigsqcup_\lambda X_\lambda$ is the decomposition into path components, then $H_p(X; \mathsf{R}) = \bigoplus_\lambda H_p(X_\lambda; \mathsf{R})$.

Example B.4. If $X = \{x\}$, the space consisting of one point x, there is a unique singular p-simplex (the constant map) for each p and we see that

$$H_p(\{x\}; \mathsf{R}) \simeq \begin{cases} \mathsf{R} & p = 0, \\ 0 & p \neq 0. \end{cases}$$

In the following, the coefficients ring R in the homology is also omitted if there is no fear of confusion.

Homotopy: Let $f, f' : X \to Y$ be two maps of topological spaces. We say f is *homotopic* to f' if there is a map $F : X \times I \to Y$ such that $F(x, 0) = f(x)$ and $F(x, 1) = f'(x)$ for all $x \in X$, where I denotes the interval $[0, 1]$. In this case, we write $f \simeq f'$. The map F is called a *homotopy* between f and f'. Setting $f_t = F(\ , t)$, we also say $\{f_t\}_{0 \leq t \leq 1}$ is a homotopy from $f_0 = f$ to $f_1 = f'$.

Let A be a subspace of X. A homotopy F as above is *relative to* A if $F(x, t) = f(x)$ for all $x \in A$ and $t \in I$.

A map $f : X \to Y$ is a *homotopy equivalence* if there is a map $g : Y \to X$ such that $g \circ f \simeq 1_X$ and $f \circ g \simeq 1_Y$. Such a map g is called a *homotopy inverse* of f.

Recall that a map $f : X \to Y$ is said to be *proper* if the inverse image of each compact set in Y by f is compact.

Let $f, f' : X \to Y$ be two proper maps. A *proper homotopy* between f and f' is a map $F : X \times I \to Y$ as above that is proper.

Induced morphism: Let $f : X \to Y$ be a continuous map of topological spaces. It induces a chain morphism $f_\bullet : S_\bullet(X) \to S_\bullet(Y)$ by assigning to $\sigma : \Delta^p \to X$ the composition $f \circ \sigma : \Delta^p \to Y$. It in turn induces a morphism

$$f_* : H_p(X) \longrightarrow H_p(Y).$$

It is compatible with the augmentations, i.e., the following diagram is commutative:

$$\begin{array}{ccc} H_0(X) & \xrightarrow{f_*} & H_0(Y) \\ & \searrow^{\varepsilon_*} & \downarrow{\scriptstyle \varepsilon_*} \\ & & \mathsf{R}. \end{array} \tag{B.2}$$

If two maps $f, f' : X \to Y$ are homotopic, then $f_\bullet, f'_\bullet : S_\bullet(X) \to S_\bullet(Y)$ are chain homotopic so that $f_* = f'_* : H_p(X) \to H_p(Y)$. If $f : X \to Y$ is a homotopy equivalence, $f_\bullet : S_\bullet(X) \to S_\bullet(Y)$ is a chain equivalence so that $f_* : H_p(X) \xrightarrow{\sim} H_p(Y)$ is an isomorphism.

Deformation retract: Let A be a subspace of X and $i : A \hookrightarrow X$ the inclusion. We say that A is a *retract* of X if there is a continuous map $r : X \to A$ such that $r \circ i = 1_A$. Such a map r is called a *retraction.* Also, A is a *deformation retract* of X if there is a retraction r such that $i \circ r \simeq 1_X$. Thus in this case, i is a homotopy equivalence so that $i_* : H_p(A) \xrightarrow{\sim} H_p(X)$ with r_* its inverse. A space X is *contractible* if there is a point x_0 in X such that $\{x_0\}$ is a deformation retract of X.

A *proper deformation retraction* is a deformation retraction that is proper.

Relative homology: Let A be a subspace of X and $i : A \hookrightarrow X$ the inclusion. The complex $(S_\bullet(A), \partial)$ may naturally be thought of as a subcomplex of $(S_\bullet(X), \partial)$ by identifying a singular simplex $\sigma : \Delta \to A$ with $i \circ \sigma : \Delta \to X$. Let $(S_\bullet(X, A), \partial)$ denote the quotient complex, $S_\bullet(X, A) = S_\bullet(X)/S_\bullet(A)$.

Proposition B.24. *For each p, $S_p(X, A)$ is a free* R*-module with basis the classes of singular p-simplices σ of X with $\sigma(\Delta^p) \not\subset A$.*

This can be seen from the decomposition $S_p(X) = S_p(A) \oplus S'_p$, where S'_p is the free R-module generated by singular p-simplices σ of X with $\sigma(\Delta^p) \not\subset A$. Note that $S'_\bullet$ is not naturally a subcomplex of $S_\bullet(X)$ in general.

We denote by $H_p(X, A; \mathsf{R})$ the p-th homology of $(S_\bullet(X, A), \partial)$ and call it the p-th *relative homology of X mod A*, or the p-th *homology of the pair* (X, A). Note that $S_\bullet(X, \emptyset) = S_\bullet(X)$, thus $H_p(X, \emptyset\,; \mathsf{R}) = H_p(X; \mathsf{R})$.

Denoting by $j : (X, \emptyset) \hookrightarrow (X, A)$ the inclusion, we have the exact sequence of complexes:

$$0 \longrightarrow S_\bullet(A) \xrightarrow{i_\bullet} S_\bullet(X) \xrightarrow{j_\bullet} S_\bullet(X, A) \longrightarrow 0. \tag{B.3}$$

From this, we have the long exact sequence (cf. Proposition A.9):

$$\cdots \longrightarrow H_p(A) \xrightarrow{i_*} H_p(X) \xrightarrow{j_*} H_p(X, A) \xrightarrow{\partial_*} H_{p-1}(A) \longrightarrow \cdots . \tag{B.4}$$

We may also describe $H_p(X, A)$ as follows. We set

$$\begin{aligned} Z_p(X, A) &= \{\, c \in S_p(X) \mid \partial c \in S_{p-1}(A)\,\}, \\ B_p(X, A) &= \{\, c \in S_p(X) \mid c \sim c', \quad c' \in S_p(A)\,\}. \end{aligned}$$

Then we have

$$H_p(X, A) \simeq Z_p(X, A)/B_p(X, A).$$

If we have a triple (X, A_1, A_2) of spaces with $A_2 \subset A_1 \subset X$, we have the exact sequence of complexes

$$0 \longrightarrow S_\bullet(A_1, A_2) \xrightarrow{i_\bullet} S_\bullet(X, A_2) \xrightarrow{j_\bullet} S_\bullet(X, A_1) \longrightarrow 0,$$

which yields a long exact sequence:

$$\cdots \to H_p(A_1, A_2) \xrightarrow{i_*} H_p(X, A_2) \xrightarrow{j_*} H_p(X, A_1) \xrightarrow{\partial_*} H_{p-1}(A_1, A_2) \to \cdots . \tag{B.5}$$

If $f : (X, A) \to (Y, B)$ is a continuous map of pairs of topological spaces, i.e., a continuous map $f : X \to Y$ with $f(A) \subset B$, it induces a chain morphism $f_\bullet : S_\bullet(X, A) \to S_\bullet(Y, B)$ by assigning to the class of σ the class of $f \circ \sigma$. Thus it in turn induces a morphism

$$f_* : H_p(X, A) \longrightarrow H_p(Y, B).$$

Singular cohomology

As a notion dual to the singular homology, we have the singular cohomology. Thus, for a topological space X, we set $S^p(X) = \mathrm{Hom}_{\mathsf{R}}(S_p(X); \mathsf{R})$ and define the coboundary operator

$$\delta = \delta^p : S^p(X) \longrightarrow S^{p+1}(X)$$

as the transpose of ∂ (cf. (A.1)):

$$\langle c', \delta u\rangle = \langle \partial c', u\rangle \qquad \text{for } u \in S^p(X) \text{ and } c' \in S_{p+1}(X).$$

Definition B.10. The *p-th singular cohomology* $H^p(X; \mathsf{R})$ is the p-th cohomology of the complex $(S^\bullet(X), \delta)$.

Remark B.9. Let 1 denote the 0-cochain defined so that it assigns 1 to each singular 0-simplex and is extended R-linearly. Then we see that it is a cocycle and we have its class [1] in $H^0(X; \mathsf{R})$.

If $f : X \to Y$ is a continuous map, it induces a cochain morphism $f^\bullet : S^\bullet(Y) \to S^\bullet(X)$ and thus a morphism

$$f^* : H^p(Y) \longrightarrow H^p(X).$$

If two maps $f, f' : X \to Y$ are homotopic, then $f^\bullet, (f')^\bullet : S^\bullet(Y) \to S^\bullet(X)$ are cochain homotopic so that $f^* = (f')^* : H^p(Y) \to H^p(X)$. If $f : X \to Y$ is a homotopy equivalence, $f^\bullet : S^\bullet(Y) \to S^\bullet(X)$ is a cochain equivalence so that $f^* : H^p(Y) \xrightarrow{\sim} H^p(X)$ is an isomorphism.

For a subspace A of X, we let $S^\bullet(X, A)$ denote the cochain complex dual to $S_\bullet(X, A)$. The *p-th relative cohomology* $H^p(X, A; \mathsf{R})$ of X mod A is the p-th cohomology of the complex $(S^\bullet(X, A), \delta)$. From (B.3), for each p, we have the exact sequence (cf. Example A.2. 1)

$$0 \longrightarrow S^p(X, A) \xrightarrow{j^p} S^p(X) \xrightarrow{i^p} S^p(A) \longrightarrow \mathrm{Ext}^1_{\mathsf{R}}(S_p(X, A), \mathsf{R}).$$

Thus we may identify $S^p(X, A)$ with

$$\{\, u \in S^p(X) \mid \langle c, u\rangle = 0 \text{ for } c \in S_p(A) \,\}. \tag{B.6}$$

Moreover $\mathrm{Ext}^1_{\mathsf{R}}(S_p(X, A), \mathsf{R}) = 0$ by Proposition B.24 and i^p is surjective. Thus we have a long exact sequence:

$$\cdots \longrightarrow H^{p-1}(A) \xrightarrow{\delta^*} H^p(X, A) \xrightarrow{j^*} H^p(X) \xrightarrow{i^*} H^p(A) \longrightarrow \cdots. \tag{B.7}$$

The Kronecker product on chains and cochains descends to a product on homology and cohomology, which we still call the Kronecher product and denote by the same symbol:

$$H_p(X, A) \times H^p(X, A) \longrightarrow \mathsf{R}, \qquad \langle [c], [u]\rangle = \langle c, u\rangle.$$

If $f : (X, A) \to (Y, B)$ is a continuous map of pairs of topological spaces, it induces a morphism

$$f^* : H^p(Y, B) \longrightarrow H^p(X, A).$$

In this situation, we have

$$\langle a, f^*(\beta)\rangle = \langle f_*(a), \beta\rangle \qquad \text{for } a \in H_p(X, A), \quad \beta \in H^p(Y, B). \tag{B.8}$$

Excision

A pair (X_1, X_2) of subspaces of X is called an *excisive couple* if the inclusion $S_\bullet(X_1)+S_\bullet(X_2) \hookrightarrow S_\bullet(X_1 \cup X_2)$ is a chain equivalence. It is shown that, if $\operatorname{Int} X_1 \cup \operatorname{Int} X_2 = X$, then (X_1, X_2) is an excisive couple. If (X_1, X_2) is an excisive couple, the morphism $S_\bullet(X_1)/S_\bullet(X_1 \cap X_2) \to S_\bullet(X_1 \cup X_2)/S_\bullet(X_2)$ induced by the inclusion $(X_1, X_1 \cap X_2) \hookrightarrow (X_1 \cup X_2, X_2)$ is a chain equivalence so that there exist canonical isomorphisms

$$\begin{aligned} H_p(X_1, X_1 \cap X_2) &\xrightarrow{\sim} H_p(X_1 \cup X_2, X_2), \\ H^p(X_1 \cup X_2, X_2) &\xrightarrow{\sim} H^p(X_1, X_1 \cap X_2). \end{aligned} \tag{B.9}$$

Let S be a closed set in X and U an open set in X containing S. Then $(U, X \setminus S)$ is an excisive couple and there exist canonical isomorphisms

$$H_p(U, U \setminus S) \xrightarrow{\sim} H_p(X, X \setminus S), \quad H^p(X, X \setminus S) \xrightarrow{\sim} H^p(U, U \setminus S). \tag{B.10}$$

Mayer-Vietoris sequence

Let X_1 and X_2 be subspaces of X. Then there is an exact sequence of chain complexes

$$0 \longrightarrow S_\bullet(X_1 \cap X_2) \xrightarrow{i_\bullet} S_\bullet(X_1) \oplus S_\bullet(X_2) \xrightarrow{j_\bullet} S_\bullet(X_1) + S_\bullet(X_2) \longrightarrow 0,$$

where $i_\bullet(c) = (c, -c)$ and $j_\bullet(c_1, c_2) = c_1 + c_2$. Thus if (X_1, X_2) is an excisive couple, we have the following exact sequence:

$$\begin{aligned} \cdots \longrightarrow H_p(X_1 \cap X_2) \xrightarrow{i_*} H_p(X_1) \oplus H_p(X_2) &\xrightarrow{j_*} \\ H_p(X_1 \cup X_2) &\xrightarrow{\partial_*} H_{p-1}(X_1 \cap X_2) \longrightarrow \cdots, \end{aligned} \tag{B.11}$$

where ∂_* assigns to the class of $c_1 + c_2$ the class of c in $X_1 \cap X_2$ such that $\partial c_1 = c$ and $\partial c_2 = -c$.

Likewise we have the exact sequence in cohomology:

$$\begin{aligned} \cdots \longrightarrow H^p(X_1 \cup X_2) \xrightarrow{j^*} H^p(X_1) \oplus H^p(X_2) &\xrightarrow{i^*} \\ H^p(X_1 \cap X_2) &\xrightarrow{\delta^*} H^{p+1}(X_1 \cup X_2) \longrightarrow \cdots. \end{aligned}$$

In the above, denoting by $i_k : X_1 \cap X_2 \hookrightarrow X_k$ and $j_k : X_k \hookrightarrow X_1 \cup X_2$, $k = 1, 2$, the inclusions, we may write $j^*(\alpha) = (j_1^*(\alpha), j_2^*(\alpha))$ and $i^*(\alpha_1, \alpha_2) = i_1^*(\alpha_1) - i_2^*(\alpha_2)$. For every cocycle $u \in S^p(X_1 \cap X_2)$ there exist cochains $u_k \in S^p(X_k)$, $k = 1, 2$, such that $u = i_1^p(u_1) - i_2^p(u_2)$. The connecting morphism δ^* assigns to the class of u the class of $v \in S^{p+1}(X_1 \cup X_2)$ such that $j_1^{p+1}(v) = \delta u_1$ and $j_2^{p+1}(v) = \delta u_2$.

Example B.5. 1. Consider the unit q-sphere in $\mathbb{R}^{q+1} = \{(x_1, \dots, x_{q+1})\}$:

$$\mathbb{S}^q = \{ x \in \mathbb{R}^{q+1} \mid \|x\|^2 = |x_1|^2 + \cdots + |x_{q+1}|^2 = 1 \}.$$

We may write $\mathbb{S}^q = S^+ \cup S^-$ with $S^+ = \mathbb{S}^q \smallsetminus \{(0, \dots, 0, 1)\}$ and $S^- = \mathbb{S}^q \smallsetminus \{(0, \dots, 0, -1)\}$. Note that both S^+ and S^- are contractible.

In the case $q = 0$, $\mathbb{S}^0 = \{-1, 1\}$ and we have:

$$H_p(\mathbb{S}^0; \mathsf{R}) \simeq \begin{cases} \mathsf{R} \oplus \mathsf{R} & p = 0, \\ 0 & \text{otherwise.} \end{cases}$$

In the case $q \geq 1$, $\mathbb{S}^q \cap \{x_{q+1} = 0\} = \mathbb{S}^{q-1}$ is a deformation retract of $S^+ \cap S^-$. By a repeated use of (B.11), we have:

$$H_p(\mathbb{S}^q; \mathsf{R}) \simeq \begin{cases} \mathsf{R} & p = 0,\ q, \\ 0 & \text{otherwise.} \end{cases}$$

2. Let $\mathbb{B}^q = \{ x \in \mathbb{R}^q \mid \|x\| \leq 1 \}$ be the closed unit q-ball in $\mathbb{R}^q$, $q \geq 1$. Then $\mathbb{S}^{q-1} = \partial \mathbb{B}^q$ is the unit $(q-1)$-sphere. From the contractibility of the ball and (B.4) we have:

$$H_p(\mathbb{B}^q, \mathbb{S}^{q-1}; \mathsf{R}) \simeq \begin{cases} \mathsf{R} & p = q, \\ 0 & \text{otherwise.} \end{cases}$$

Note that the above holds also for $q = 0$, if we set $\mathbb{B}^0 = \{0\}$ and $\mathbb{S}^{-1} = \emptyset$. In this case, $H_0(\mathbb{B}^0, \mathbb{S}^{-1}; \mathsf{R}) = H_0(\{0\}; \mathsf{R})$ has a canonical generator, i.e., the class of the constant map $\Delta^0 \to \{0\}$.

Note also that, if $q > 1$,

$$\partial_* : H_q(\mathbb{B}^q, \mathbb{S}^{q-1}; \mathsf{R}) \xrightarrow{\sim} H_{q-1}(\mathbb{S}^{q-1}; \mathsf{R}) \tag{B.12}$$

is an isomorphism.

Since $\mathbb{R}^q$ and $\mathbb{R}^q \smallsetminus \{0\}$ deformation retract to $\mathbb{B}^q$ and $\mathbb{S}^{q-1}$, respectively, there is a canonical isomorphism $H_p(\mathbb{R}^q, \mathbb{R}^q \smallsetminus \{0\}; \mathsf{R}) \simeq H_p(\mathbb{B}^q, \mathbb{S}^{q-1}; \mathsf{R})$.

For the cohomology we have:

Example B.6. 1. For the unit sphere $\mathbb{S}^q$ in $\mathbb{R}^{q+1}$, $q \geq 1$,

$$H^p(\mathbb{S}^q; \mathsf{R}) \simeq \begin{cases} \mathsf{R} & p = 0,\ q, \\ 0 & \text{otherwise.} \end{cases}$$

2. For the closed unit q-ball $\mathbb{B}^q$ in $\mathbb{R}^q$ and the unit $(q-1)$-sphere $\mathbb{S}^{q-1} = \partial\mathbb{B}^q$, $q \geq 1$,

$$H^p(\mathbb{B}^q, \mathbb{S}^{q-1}; \mathsf{R}) \simeq \begin{cases} \mathsf{R} & p = q, \\ 0 & \text{otherwise.} \end{cases}$$

The above holds also for $q = 0$. In this case, $H^0(\mathbb{B}^0, \mathbb{S}^{-1}; \mathsf{R}) = H^0(\{0\}; \mathsf{R})$ has a canonical generator, i.e., the class of the cocycle that assigns 1 to the constant map $\Delta^0 \to \{0\}$.

Note that, if $q > 1$, $\delta^* : H^{q-1}(\mathbb{S}^{q-1}; \mathsf{R}) \xrightarrow{\sim} H^q(\mathbb{B}^q, \mathbb{S}^{q-1}; \mathsf{R})$ is an isomorphism. There is also a canonical isomorphism $H^p(\mathbb{R}^q, \mathbb{R}^q \setminus \{0\}; \mathsf{R}) \simeq H^p(\mathbb{B}^q, \mathbb{S}^{q-1}; \mathsf{R})$.

Spaces with finite homology type: A pair (X, A) of topological spaces is said to have a *finite homology type* if, for every p, $H_p(X, A; \mathbb{Z})$ is finitely generated. For such a pair, by the structure theorem for finitely generated Abelian groups, we have

$$H_p(X, A; \mathbb{Z}) \simeq \mathbb{Z}^{b_p} \oplus T_p, \qquad T_p = \mathbb{Z}/(t_1) \oplus \cdots \oplus \mathbb{Z}/(t_{r_p}), \tag{B.13}$$

where b_p is a non-negative integer and $t_1, \ldots, t_{r_p}$ positive integers with $t_1 \neq 1$ and $t_1 \mid t_2 \mid \cdots \mid t_{r_p}$. Note that r_p may possibly be 0. The integers b_p and $t_1, \ldots, t_{r_p}$ are uniquely determined by (X, A). We call $\mathbb{Z}^{b_p}$ the *free part* of $H_p(X, A; \mathbb{Z})$ and b_p the p-th *Betti number* of (X, A). We also call T_p the *torsion part* and $t_1, \ldots, t_{r_p}$ the torsion coefficients. The torsion part may also be expressed as

$$T_p = \mathbb{Z}/(p_1^{k_1}) \oplus \cdots \oplus \mathbb{Z}/(p_l^{k_l}),$$

where the p_i's are prime numbers (not necessarily distinct) and the k_i's are positive integers, $i = 1, \ldots, l$. The powers $p_i^{k_i}$ are uniquely determined.

Similar statements hold if we replace $\mathbb{Z}$ with a principal ideal domain.

Universal coefficient theorem

The following is proved using Propositions A.18 and B.24 (cf. Example A.2.1 for Tor and Ext):

Theorem B.20. *Let* R *be a principal ideal domain. Then, for a pair of topological spaces* (X, A) *and an* R*-algebra* S*, there exist canonical exact sequences*

$$0 \to \mathsf{S} \otimes_{\mathsf{R}} H_p(X, A; \mathsf{R}) \to H_p(X, A; \mathsf{S}) \to \mathrm{Tor}_1^{\mathsf{R}}(H_{p-1}(X, A; \mathsf{R}), \mathsf{S}) \to 0,$$

$$0 \to \mathrm{Ext}_{\mathsf{R}}^1(H_{p-1}(X, A; \mathsf{R}), \mathsf{S}) \to H^p(X, A; \mathsf{S}) \to \mathrm{Hom}_{\mathsf{R}}(H_p(X, A; \mathsf{R}), \mathsf{S}) \to 0,$$

both of which split.

Here are some of the consequences of the above:

(1) If S is R-torsion-free, thus R-flat, there exists a canonical isomorphism

$$H_p(X, A; \mathsf{S}) \simeq \mathsf{S} \otimes_{\mathsf{R}} H_p(X, A; \mathsf{R}).$$

(2) If $H_{p-1}(X, A; \mathsf{R})$ is R-free, thus R-projective and R-flat, then there exist canonical isomorphisms

$$H_p(X, A; \mathsf{S}) \simeq \mathsf{S} \otimes_{\mathsf{R}} H_p(X, A; \mathsf{R}), \quad H^p(X, A; \mathsf{S}) \simeq \mathrm{Hom}_{\mathsf{R}}(H_p(X, A; \mathsf{R}), \mathsf{S}).$$

(3) If R is a field, then there exists a canonical isomorphism

$$H^p(X, A; \mathsf{R}) \simeq H_p(X, A; \mathsf{R})^*.$$

Note that the isomorphism is induced by the Kronecker product.

(4) Suppose $\mathsf{R} = \mathbb{Z}$. If S is divisible, thus $\mathbb{Z}$-injective (cf. Proposition A.16), then there exists a canonical isomorphism

$$H^p(X, A; \mathsf{S}) \simeq \mathrm{Hom}_{\mathbb{Z}}(H_p(X, A; \mathbb{Z}), \mathsf{S}).$$

(5) Suppose (X, A) is of finite homology type. Then, for $\mathbb{K} = \mathbb{Q}$, $\mathbb{R}$ or $\mathbb{C}$, we have (cf. (B.13))

$$H_p(X, A; \mathbb{K}) \simeq \mathbb{K}^{b_p}, \qquad H^p(X, A; \mathbb{K}) \simeq \mathbb{K}^{b_p}.$$

We also have

$$H^p(X, A; \mathbb{Z}) \simeq \mathbb{Z}^{b_p} \oplus T_{p-1}.$$

For the last isomorphism, let $\mathsf{R} = \mathsf{S} = \mathbb{Z}$ in the second sequence in Theorem B.20. First note that $\mathrm{Hom}_{\mathbb{Z}}(H_p(X, A; \mathbb{Z}), \mathbb{Z}) \simeq \mathbb{Z}^{b_p}$. To compute $\mathrm{Ext}_{\mathbb{Z}}^1(H_{p-1}(X, A; \mathbb{Z}), \mathbb{Z})$, take a free resolution of $H_{p-1}(X, A; \mathbb{Z})$ of the form

$$0 \longrightarrow \mathbb{Z}^{r_{p-1}} \stackrel{\tau}{\longrightarrow} \mathbb{Z}^{b_{p-1}} \oplus \mathbb{Z}^{r_{p-1}} \longrightarrow H_{p-1}(X, A; \mathbb{Z}) \longrightarrow 0,$$

where τ is a morphism given by $(n_1, \ldots, n_r) \mapsto (0; n_1 t_1, \ldots, n_r t_r)$, $r = r_{p-1}$. Then we see that $\mathrm{Ext}_{\mathbb{Z}}^1(H_{p-1}(X, A; \mathbb{Z}), \mathbb{Z}) \simeq \mathbb{Z}/(t_1) \oplus \cdots \oplus \mathbb{Z}/(t_{r_{p-1}}) = T_{p-1}$.

Products

Cup product: We define the *cup product*

$$S^p(X) \times S^q(X) \xrightarrow{\smile} S^{p+q}(X), \qquad (u,v) \mapsto u \smile v$$

by setting

$$\langle \sigma, u \smile v \rangle = \langle d_{p+1} \cdots d_{p+q}\sigma, u \rangle \langle d_0 \cdots d_{p-1}\sigma, v \rangle \tag{B.14}$$

for a $(p+q)$-simplex σ and extending it R-linearly.

Remark B.10. For a $(p+q)$-simplex σ, $d_{p+1} \cdots d_{p+q}\sigma$ is the "first p-simplex" of σ, i.e., the composition of the affine map $\varepsilon : \Delta^p \to \Delta^{p+q}$ given by $\varepsilon(P_i) = P_i$, $0 \le i \le p$, and σ. Also, $d_0 \cdots d_{p-1}\sigma$ is the "last q-simplex" of σ, i.e., the composition of the affine map $\varepsilon' : \Delta^q \to \Delta^{p+q}$ given by $\varepsilon'(P_i) = P_{p+i}$, $0 \le i \le q$, and σ. If $q = 0$, the former is σ itself and the latter the 0-simplex that assigns to P_0 the point $\sigma(P_p)$ in X.

The cup product is R-bilinear and we have

$$\delta(u \smile v) = \delta u \smile v + (-1)^p u \smile \delta v. \tag{B.15}$$

This shows that the above induces the cup product on the cohomology:

$$H^p(X) \times H^q(X) \xrightarrow{\smile} H^{p+q}(X),$$

which makes $H^*(X) = \bigoplus_{p \ge 0} H^p(X)$ a graded R-algebra with unity. The unity is the class [1] of the cochain that assigns 1 to each 0-simplex. Note that the cup product in $H^*(X)$ is anticommutative, i.e., for $\alpha \in H^p(X)$ and $\beta \in H^q(X)$,

$$\alpha \smile \beta = (-1)^{pq} \beta \smile \alpha.$$

Let A_1 and A_2 be subspaces of X. If the couple (A_1, A_2) is excisive, we see that we also have the cup product

$$H^p(X, A_1) \times H^q(X, A_2) \xrightarrow{\smile} H^{p+q}(X, A_1 \cup A_2).$$

If $f : X \to Y$ is a continuous map, $f^* : H^*(Y) \to H^*(X)$ is a morphism of graded R-algebras with unity.

Cap product: We define the *cap product*

$$S_r(X) \times S^p(X) \xrightarrow{\frown} S_{r-p}(X)$$

by

$$c \frown u = \sum_\sigma a_\sigma \langle d_{p+1} \cdots d_r\sigma, u\rangle\, d_0 \cdots d_{p-1}\sigma \tag{B.16}$$

for an r-chain $c = \sum a_\sigma \sigma$ and a p-cochain u, $r \geq p$ (cf. Remark B.10). Then $c \frown u$ is R-linear in c and u and we have

$$\partial(c \frown u) = (-1)^p(\partial c \frown u - c \frown \delta u). \tag{B.17}$$

We also have

$$c \frown (u \smile v) = (c \frown u) \frown v \tag{B.18}$$

for $c \in S_r(X)$, $u \in S^p(X)$ and $v \in S^q(X)$.

From (B.17) we have the cap product

$$H_r(X) \times H^p(X) \xrightarrow{\frown} H_{r-p}(X).$$

The relation (B.18) holds in homology and cohomology and the cap product makes $H_*(X) = \bigoplus_{r\geq 0} H_r(X)$ a right $H^*(X)$-module.

In the case $r = p$, we have

$$\varepsilon(c \frown u) = \langle c, u\rangle, \tag{B.19}$$

where $\varepsilon : S_0(X) \to \mathsf{R}$ is the augmentation. Thus from (B.18), if $p + q = r$,

$$\langle c, u \smile v\rangle = \langle c \frown u, v\rangle. \tag{B.20}$$

The above identities hold also in homology and cohomology.

Let A be a subspace of X and $i : A \hookrightarrow X$ the inclusion. Then for c in $S_r(A)$ and u in $S^p(X, A)$, $c \frown u = 0$. Thus we have

$$S_r(X, A) \times S^p(X, A) \xrightarrow{\frown} S_{r-p}(X), \tag{B.21}$$

which induces

$$H_r(X, A) \times H^p(X, A) \xrightarrow{\frown} H_{r-p}(X).$$

Also we may define

$$H_r(A) \times H^p(X) \xrightarrow{\frown} H_{r-p}(A) \qquad \text{by } \ a \frown \alpha = a \frown i^*\alpha. \tag{B.22}$$

For a continuous map $f : (X, A) \to (Y, B)$, we have

$$f_*(a \frown f^*(\beta)) = f_*(a) \frown \beta \tag{B.23}$$

for $a \in H_r(X, A)$ and $\beta \in H^p(Y, B)$, which generalizes (B.8).

Remark B.11. The cap product above is what is called the *left cap product.* There is also the *right cap product*

$$S^p(X) \times S_r(X) \xrightarrow{\frown} S_{r-p}(X),$$

which is defined by

$$u \frown c = \sum_{\sigma} a_\sigma \langle d_0 \cdots d_{q-1}\sigma, u\rangle \, d_{q+1} \cdots d_r \sigma$$

for an r-chain $c = \sum a_\sigma \sigma$ and a p-cochain u, $r \geq p$. Then $u \frown c$ is R-linear in u and c and we have

$$\partial(u \frown c) = (-1)^{r-p} \delta u \frown c + u \frown \partial c. \tag{B.24}$$

We also have

$$(u \smile v) \frown c = u \frown (v \frown c) \tag{B.25}$$

for $c \in S_r(X)$, $u \in S^p(X)$ and $v \in S^q(X)$. Moreover we have

$$(u \frown_r c) \frown_l v = u \frown_r (c \frown_l v), \tag{B.26}$$

where $\frown_l$ and $\frown_r$ denote left and right cap products, respectively.

From (B.24) we have the right cap product

$$H^p(X) \times H_r(X) \xrightarrow{\frown} H_{r-p}(X).$$

The relation (B.25) holds in homology and cohomology and the right cap product makes $H_*(X)$ a left $H^*(X)$-module.

Note that, on the homology and cohomology level, the left and right cap products differ by a sign of $(-1)^{p(r-p)}$.

Künneth formula: Let X and Y be topological spaces. Denoting by p_1 and p_2 the projections $X \times Y \to X$ and $X \times Y \to Y$, we define the *cross product*

$$H^p(X) \times H^q(Y) \xrightarrow{\times} H^{p+q}(X \times Y) \qquad \text{by } \alpha \times \beta = p_1^*(\alpha) \smile p_2^*(\beta).$$

Suppose the coefficient ring R is a principal ideal domain. If X is a space of finite homology type and if either $H^*(X)$ or $H^*(Y)$ is R-free, then there is an isomorphism

$$H^*(X \times Y) \simeq H^*(X) \otimes H^*(Y), \tag{B.27}$$

where $\alpha \times \beta$ in $H^*(X \times Y)$ corresponds to $\alpha \otimes \beta$ in $H^*(X) \otimes H^*(Y)$.

B.2 Cell complexes

In this section we review cell complexes and, as a special case, we discuss simplicial complexes. They provide practical ways of representing homology and cohomology of spaces that admit a cellular or simplicial decomposition.

Cell complexes

Let X be a Hausdorff topological space.

Definition B.11. A *cell* in X is a subset $\boldsymbol{e}$ of X that admits a surjective continuous map $\chi : \mathbb{B}^p \to \overline{\boldsymbol{e}}$ such that $\chi(\mathbb{S}^{p-1}) \subset \overline{\boldsymbol{e}} \smallsetminus \boldsymbol{e}$ and that $\chi|_{\mathbb{B}^p \smallsetminus \mathbb{S}^{p-1}}$ is a homeomorphism onto $\boldsymbol{e}$.

A map χ as above is called a *characteristic map* of $\boldsymbol{e}$. The integer p, which is determined uniquely by $\boldsymbol{e}$, is called the dimension of $\boldsymbol{e}$. A cell $\boldsymbol{e}$ of dimension p is called a p-cell and is sometimes denoted by $\boldsymbol{e}^p$. In the above we allow p to be equal to 0 and a 0-cell is a point of X, which has a unique characteristic map.

A subset of the form $\overline{\boldsymbol{e}}$ is referred to as a *closed cell.* In particular, $\mathbb{B}^p$ is a closed cell in $\mathbb{R}^p$ with the identity map as a characteristic map.

Definition B.12. A *cell complex* is a Hausdorff topological space X together with a family $\{\boldsymbol{e}_\lambda\}_{\lambda \in \Lambda}$ of cells in X with the following properties:

(1) $X = \bigsqcup_{\lambda \in \Lambda} \boldsymbol{e}_\lambda$ (disjoint union). We set $X^p = \bigsqcup_{p_\lambda \leq p} \boldsymbol{e}_\lambda$, $p_\lambda = \dim \boldsymbol{e}_\lambda$.
(2) For every $\lambda \in \Lambda$, $\overline{\boldsymbol{e}_\lambda} \smallsetminus \boldsymbol{e}_\lambda \subset X^{p_\lambda - 1}$.

In this case, we say that $\{\boldsymbol{e}_\lambda\}_{\lambda \in \Lambda}$ is a *cellular decomposition* of X. We call X^p the *p-skeleton* of X. A *subcomplex* of a cell complex $(X, \{\boldsymbol{e}_\lambda\}_{\lambda \in \Lambda})$ is a cell complex $(A, \{\boldsymbol{e}_\lambda\}_{\lambda \in \Lambda'})$ such that $A \subset X$ and $\Lambda' \subset \Lambda$. For every p, X^p is a subcomplex. The intersection and union of subcomplexes are subcomplexes.

A cell complex $(X, \{\boldsymbol{e}_\lambda\}_{\lambda \in \Lambda})$ is said to be *finite* if Λ is a finite set. It is *locally finite* if, for every $x \in X$, there exists a finite subcomplex A with $x \in \operatorname{Int} A$. A subcomplex of a locally finite cell complex X is a closed set in X. A locally finite cell complex satisfies the conditions for what is called a "CW-complex" and in the following we only consider locally finite cell complexes.

Homology of cell complexes

First we prove the following:

Lemma B.7. *Let* $\boldsymbol{e}$ *be a* p*-cell in* X. *If* $\chi : (\mathbb{B}^p, \mathbb{S}^{p-1}) \to (\overline{\boldsymbol{e}}, \overline{\boldsymbol{e}} \smallsetminus \boldsymbol{e})$ *is a characteristic map of* $\boldsymbol{e}$, *it induces an isomorphism*

$$\chi_* : H_q(\mathbb{B}^p, \mathbb{S}^{p-1}; \mathsf{R}) \xrightarrow{\sim} H_q(\overline{\boldsymbol{e}}, \overline{\boldsymbol{e}} \smallsetminus \boldsymbol{e}; \mathsf{R}) \qquad \text{for all } q.$$

Proof. Since $\mathbb{B}^p \smallsetminus \{0\}$ deformation retracts to $\mathbb{S}^{p-1}$, there is a canonical isomorphism $H_q(\mathbb{B}^p, \mathbb{S}^{p-1}) \simeq H_q(\mathbb{B}^p, \mathbb{B}^p \smallsetminus \{0\})$, which is canonically isomorphic with $H_q(\mathbb{B}^p \smallsetminus \mathbb{S}^{p-1}, (\mathbb{B}^p \smallsetminus \mathbb{S}^{p-1}) \smallsetminus \{0\})$ by excision. By similar arguments, we see that there is a canonical isomorphism $H_q(\overline{\boldsymbol{e}}, \overline{\boldsymbol{e}} \smallsetminus \boldsymbol{e}) \simeq H_q(\boldsymbol{e}, \boldsymbol{e} \smallsetminus \{x\})$, $x = \chi(0) \in \boldsymbol{e}$. Since $\chi|_{\mathbb{B}^p \smallsetminus \mathbb{S}^{p-1}}$ is a homeomorphism, we have the lemma. □

Thus, for a p-cell $\boldsymbol{e}$ in X (cf. Example B.5. 2, with p and q interchanged),

$$H_q(\overline{\boldsymbol{e}}, \overline{\boldsymbol{e}} \smallsetminus \boldsymbol{e}; \mathsf{R}) \simeq \begin{cases} \mathsf{R} & q = p, \\ 0 & \text{otherwise.} \end{cases} \tag{B.28}$$

Let $(X, \{\boldsymbol{e}_\lambda\}_{\lambda \in \Lambda})$ be a cell complex. For each p-cell $\boldsymbol{e}_\lambda$, the inclusion $i_\lambda : (\overline{\boldsymbol{e}_\lambda}, \overline{\boldsymbol{e}_\lambda} \smallsetminus \boldsymbol{e}_\lambda) \hookrightarrow (X^p, X^{p-1})$ induces a morphism

$$(i_\lambda)_* : H_q(\overline{\boldsymbol{e}_\lambda}, \overline{\boldsymbol{e}_\lambda} \smallsetminus \boldsymbol{e}_\lambda; \mathsf{R}) \longrightarrow H_q(X^p, X^{p-1}; \mathsf{R}).$$

With these, we have:

Theorem B.21. *In the above situation, we have an isomorphism*

$$\bigoplus (i_\lambda)_* : \bigoplus H_q(\overline{\boldsymbol{e}_\lambda}, \overline{\boldsymbol{e}_\lambda} \smallsetminus \boldsymbol{e}_\lambda; \mathsf{R}) \xrightarrow{\sim} H_q(X^p, X^{p-1}; \mathsf{R}),$$

where the direct sum is taken over all the p*-cells* $\boldsymbol{e}_\lambda$ *of* X.

Thus $H_p(X^p, X^{p-1}; \mathsf{R})$ is a free R-module and $H_q(X^p, X^{p-1}; \mathsf{R}) = 0$ for $q \neq p$.

Orientation of cells: Let $\boldsymbol{e}$ be a p-cell in X. Setting $\mathsf{R} = \mathbb{Z}$ in (B.28), we have $H_p(\overline{\boldsymbol{e}}, \overline{\boldsymbol{e}} \smallsetminus \boldsymbol{e}; \mathbb{Z}) \simeq \mathbb{Z}$. An *orientation* of $\boldsymbol{e}$, or of $\overline{\boldsymbol{e}}$, is a generator of $H_p(\overline{\boldsymbol{e}}, \overline{\boldsymbol{e}} \smallsetminus \boldsymbol{e}; \mathbb{Z})$. Thus each cell has two orientations. An *oriented cell* is a cell together with a specified orientation.

Remark B.12. A 0-cell $\boldsymbol{e}^0 = \{x\}$, $x \in X$, has a canonical orientation, i.e., the class of the constant map $\Delta^0 \to \{x\}$ in $H_0(\overline{\boldsymbol{e}}, \overline{\boldsymbol{e}} \smallsetminus \boldsymbol{e}; \mathbb{Z}) = H_0(\{x\}; \mathbb{Z})$.

Homology and cohomology of cell complexes: For a cell complex $(X, \{e_\lambda\})$, we set

$$C_p(X; \mathsf{R}) = H_p(X^p, X^{p-1}; \mathsf{R}).$$

For each p-cell e_λ, we choose and fix an orientation $\mu_\lambda \in H_p(\overline{e_\lambda}, \overline{e_\lambda} \setminus e_\lambda; \mathbb{Z})$ and let $i_\lambda : (\overline{e_\lambda}, \overline{e_\lambda} \setminus e_\lambda) \hookrightarrow (X^p, X^{p-1})$ be the inclusion as before. Then by Theorem B.21, $C_p(X; \mathbb{Z})$ is the free $\mathbb{Z}$-module (Abelian group) generated by the $(i_\lambda)_*(\mu_\lambda)$'s, for all the p-cells e_λ in X. By Theorem B.21 and (B.28), we have $H_{p-1}(X^p, X^{p-1}; \mathbb{Z}) = 0$. Hence by Theorem B.20, we have:

Proposition B.25. *There is a canonical isomorphism*

$$C_p(X; \mathsf{R}) \simeq \mathsf{R} \otimes_{\mathbb{Z}} C_p(X; \mathbb{Z}).$$

Thus intuitively we may say that $C_p(X; \mathsf{R})$ is the free R-module generated by all the oriented closed p-cells in X.

We define the boundary operator (we omit the coefficient R)

$$\partial_p : C_p(X) \longrightarrow C_{p-1}(X)$$

as the connecting morphism

$$H_p(X^p, X^{p-1}) \xrightarrow{\partial_*} H_{p-1}(X^{p-1}, X^{p-2})$$

in the long exact sequence for the triple (X^p, X^{p-1}, X^{p-2}) (cf. (B.5)). Then we have $\partial_{p-1} \circ \partial_p = 0$ and $(C_\bullet(X; \mathsf{R}), \partial_\bullet)$ is a chain complex.

If A is a subcomplex of X, we may think of $C_p(A; \mathsf{R})$ as the free R-module generated by the oriented closed p-cells in A, by Proposition B.25, and identify with a direct summand of $C_p(X; \mathsf{R})$. The boundary operators are compatible with the inclusion $C_p(A; \mathsf{R}) \hookrightarrow C_p(X; \mathsf{R})$ and we have the quotient complex $C_\bullet(X; \mathsf{R})/C_\bullet(A; \mathsf{R})$, which we denote by $C_\bullet(X, A; \mathsf{R})$. Note that we may think of $C_p(X, A; \mathsf{R})$ as the free R-module generated by the closures of oriented p-cells in $X \setminus A$.

Denoting by $C^\bullet(A; \mathsf{R})$, $C^\bullet(X; \mathsf{R})$ and $C^\bullet(X, A; \mathsf{R})$ the cochain complexes dual to $C_\bullet(A; \mathsf{R})$, $C_\bullet(X; \mathsf{R})$ and $C_\bullet(X, A; \mathsf{R})$, respectively, we have the exact sequence, as in the case of singular cochains:

$$0 \longrightarrow C^\bullet(X, A; \mathsf{R}) \longrightarrow C^\bullet(X; \mathsf{R}) \longrightarrow C^\bullet(A; \mathsf{R}) \longrightarrow 0.$$

We have:

Theorem B.22. *For a pair (X, A) of cell complexes, there exist canonical isomorphisms:*

$$H_p(C_\bullet(X, A; \mathsf{R})) \simeq H_p(X, A; \mathsf{R}), \quad H^p(C^\bullet(X, A; \mathsf{R})) \simeq H^p(X, A; \mathsf{R}).$$

Example B.7. The sphere $\mathbb{S}^q$, $q \geq 1$, may be identified with the space obtained from $\mathbb{B}^q$ by shrinking $\mathbb{S}^{q-1} = \partial\mathbb{B}^q$ to a point. Let $\chi : \mathbb{B}^q \to \mathbb{S}^q$ be the projection in this process. Setting $\boldsymbol{e}^0 = \chi(\mathbb{S}^{q-1})$ and $\boldsymbol{e}^q = \mathbb{S}^q \setminus \boldsymbol{e}^0$, we have a cellular decomposition $\{\boldsymbol{e}^0, \boldsymbol{e}^q\}$ of $\mathbb{S}^q$.

Suppose $q \geq 2$. For $\mathbb{B}^q$, we set $\boldsymbol{e}^q = \mathbb{B}^q \setminus \mathbb{S}^{q-1}$ and for $\mathbb{S}^{q-1}$, we take the above decomposition $\{\boldsymbol{e}^0, \boldsymbol{e}^{q-1}\}$. Then $\{\boldsymbol{e}^0, \boldsymbol{e}^{q-1}, \boldsymbol{e}^q\}$ is a cellular decomposition of $\mathbb{B}^q$. Applying Theorem B.22, we come back to the results of Example B.5.

A cell is said to be *regular* if it admits a characteristic map that is a homeomorphism. A cell complex is *regular* if every cell is regular. As particular regular cell complexes, there are "simplicial complexes".

Simplicial complexes

Let $\boldsymbol{s}$ be a p-simplex with vertices $v_0, \ldots, v_p$ in $\mathbb{R}^N$. A *face* of $\boldsymbol{s}$ is a simplex with vertices a subset of $\{v_0, \ldots, v_p\}$. If $\boldsymbol{t}$ is a face of $\boldsymbol{s}$, we write $\boldsymbol{t} \prec \boldsymbol{s}$.

Definition B.13. A *simplicial complex* is a collection K of simplices in $\mathbb{R}^N$ satisfying the following conditions:

(1) Every face of a simplex in K is in K.
(2) If $\boldsymbol{s}$ and $\boldsymbol{t}$ are in K and if $\boldsymbol{s} \cap \boldsymbol{t} \neq \emptyset$, $\boldsymbol{s} \cap \boldsymbol{t}$ is a face of each of $\boldsymbol{s}$ and $\boldsymbol{t}$.
(3) Every point x in $|K| = \bigcup_{\boldsymbol{s} \in K} \boldsymbol{s}$ has a neighborhood intersecting with only a finite number of simplices in K.

The set $|K|$ as above is called the *polyhedron* of K. Let K and K' be simplicial complexes. A *simplicial map* of K to K' is a continuous map $f : |K| \to |K'|$ which carries each simplex of K linearly (in the affine sense) onto a simplex of K'. It is a *simplicial isomorphism* if, in addition, f is a homeomorphism.

A *subcomplex* of K is a subset of K which itself is a simplicial complex. If L is a subcomplex of K, $|L|$ is a closed subset of $|K|$. Let A be a subset of $|K|$. The *star* and the *link* of A in K are defined, respectively, by

$$\mathcal{S}_K(A) = \{\boldsymbol{t} \in K \mid \boldsymbol{t} \prec \boldsymbol{s} \text{ for some } \boldsymbol{s}, A \cap \boldsymbol{s} \neq \emptyset\},$$
$$\mathcal{L}_K(A) = \{\boldsymbol{t} \in \mathcal{S}_K(A) \mid A \cap \boldsymbol{t} = \emptyset\}.$$

They are subcomplexes of K. We set $S_K(A) = |\mathcal{S}_K(A)|$ and $L_K(A) = |\mathcal{L}_K(A)|$ and call them also the star and the link of A. The *open star* of A is $O_K(A) = S_K(A) \setminus L_K(A)$. Note that $|K| \setminus O_K(A)$ is the polyhedron of the subcomplex consisting of simplices that are disjoint from A.

For a simplex $\boldsymbol{s}$, we set $\boldsymbol{e_s} = \boldsymbol{s} \setminus \bigcup_{\boldsymbol{t} \subsetneq \boldsymbol{s}} \boldsymbol{t}$. Thus $\overline{\boldsymbol{e_s}} = \boldsymbol{s}$. If K is a simplicial complex, $(|K|, \{\boldsymbol{e_s}\})$ is a regular cell complex. If L is a subcomplex of K, $|L|$ is a sub-cell complex of $|K|$.

Orientation of simplices: Let $\boldsymbol{s}$ be a p-simplex with vertices $v_0, \dots, v_p$. Assume that $p \geq 1$ for the moment. We introduce a relation $\sim$ in the set of ordered vertices of $\boldsymbol{s}$ by saying that $(v_{i_0}, \dots, v_{i_p}) \sim (v_{j_0}, \dots, v_{j_p})$ if $(i_0, \dots, i_p)$ is an even permutation of $(j_0, \dots, j_p)$. Then it is an equivalence relation and there are exactly two equivalence classes, each of which is called an *orientation* of $\boldsymbol{s}$. An *oriented simplex* is a simplex together with a specified orientation. The specified orientation is called positive and the other one negative. A simplex with the orientation represented by an ordered vertices $(v_0, \dots, v_p)$ is also denoted by $(v_0, \dots, v_p)$ and the simplex with the opposite orientation by $-(v_0, \dots, v_p)$. Then for every permutation ρ of $(0, \dots, p)$,

$$(v_{\rho(0)}, \dots, v_{\rho(p)}) = \operatorname{sgn} \rho \cdot (v_0, \dots, v_p).$$

By convention we think of a 0-simplex $\boldsymbol{s}$ with vertex v as having two orientations and denote $\boldsymbol{s}$ with the "specified orientation" by (v) and $\boldsymbol{s}$ with the "other orientation" by $-(v)$.

Remark B.13. 1. Specifying an orientation of the standard p-simplex Δ^p in $\mathbb{R}^p$, $p \geq 1$, is equivalent to specifying an orientation of $\mathbb{R}^p$ as a vector space (cf. Section 3.7), since every orientation of Δ^p is represented by an ordered vertices of the form $(P_0, P_{i_1}, \dots, P_{i_p})$. The *standard orientation* of Δ^p is the one represented by $(P_0, P_1, \dots, P_p)$. We always endow Δ^p with the standard orientation; $\Delta^p = (P_0, P_1, \dots, P_p)$, which is consistent with Convention 3.1. In the case $p = 0$, we specify an "orientation" of Δ^0 so that we may write $\Delta^0 = (P_0)$.

The boundary $\partial \Delta^p$ is oriented as

$$\partial(P_0, \dots, P_p) = \sum_{i=0}^{p} (-1)^i (P_0, \dots, \widehat{P_i}, \dots, P_p).$$

We may think of Δ^p as a p-dimensional piecewise C^∞ manifold with boundary in $\mathbb{R}^p$ (cf. Definition 5.10). The above orientation of the boundary is consistent with Convention 3.2, modified for piecewise C^∞ manifolds.

2. There is a natural homeomorphism $h : (\Delta^p, \Delta^p \setminus \operatorname{Int} \Delta^p) \xrightarrow{\sim} (\mathbb{B}^p, \mathbb{S}^{p-1})$, $p \geq 0$. Thus $H_p(\Delta^p, \Delta^p \setminus \operatorname{Int} \Delta^p; \mathbb{Z}) \simeq \mathbb{Z}$ (cf. Example B.5. 2). Moreover the homology has a canonical generator, i.e., the class $[1_p]$ of the identity map 1_p of Δ^p. This in turn gives a canonical generator $h_*([1_p])$ of $H_p(\mathbb{B}^p, \mathbb{S}^{p-1}; \mathbb{Z})$.

We always think of $\mathbb{B}^p$ as an oriented closed cell with this orientation. By (B.12), $H_p(\mathbb{S}^p;\mathbb{Z})$, $p \geq 1$, has also a canonical generator.

3. The above definition of orientation of a simplex $\boldsymbol{s}$ is made to be compatible with that of the cell $\boldsymbol{e_s}$ in the following sense. For an oriented p-simplex $\boldsymbol{s} = (v_0, \dots, v_p)$, $p \geq 1$, we define an affine map $\chi : \Delta^p \to \boldsymbol{s}$ by $\chi(P_i) = v_i$. We may think of it as a characteristic map for $\boldsymbol{e_s}$ and, noting that $\overline{\boldsymbol{e_s}} = \boldsymbol{s}$, we endow the cell $\boldsymbol{e_s}$ with the orientation $\chi_*([1_p]) \in H_p(\boldsymbol{s}, \boldsymbol{s} \setminus \boldsymbol{e_s};\mathbb{Z})$ (cf. Lemma B.7).

In the case $\boldsymbol{s}$ is a 0-simplex with vertex v, we specify an orientation of $\boldsymbol{s}$; $\boldsymbol{s} = (v)$. Define $\chi : \Delta^0 \to \boldsymbol{s}$ by $\chi(P_0) = v$, then $\chi_*([1_0]) \in H_0(\boldsymbol{s}, \boldsymbol{s} \setminus \boldsymbol{e_s};\mathbb{Z}) = H_0(\{v\};\mathbb{Z})$ is the canonical orientation of $\boldsymbol{e_s}$ (cf. Remark B.12).

Homology and cohomology of simplicial complexes: Let K be a simplicial complex. By Remark B.13, we may think of the chain group $C_p(|K|;\mathsf{R})$ as the free R-module generated by the oriented p-simplices in K with the boundary operator given by

$$\partial(v_0, \dots, v_p) = \sum_{i=0}^{p} (-1)^i (v_0, \dots, \widehat{v_i}, \dots, v_p). \tag{B.29}$$

If L is a subcomplex of K, we may identify $C_p(|L|;\mathsf{R})$ with a direct summand of $C_p(|K|;\mathsf{R})$ and we have the quotient chain complex $C_\bullet(|K|,|L|;\mathsf{R}) = C_\bullet(|K|;\mathsf{R})/C_\bullet(|L|;\mathsf{R})$. Denoting by $C^\bullet(|K|,|L|;\mathsf{R})$ the cochain complex dual to $C_\bullet(|K|,|L|;\mathsf{R})$, we have canonical isomorphisms (cf. Theorem B.22)

$$\begin{aligned} H_p(C_\bullet(|K|,|L|;\mathsf{R})) &\simeq H_p(|K|,|L|;\mathsf{R}), \\ H^p(C^\bullet(|K|,|L|;\mathsf{R})) &\simeq H^p(|K|,|L|;\mathsf{R}). \end{aligned} \tag{B.30}$$

The above isomorphisms may also be interpreted as follows. Let $\boldsymbol{s}$ be an oriented p-simplex of K. Regarding the map $\chi : \Delta^p \to \boldsymbol{s} \subset |K|$ defined in Remark B.13. 3 as a singular simplex of $|K|$, we have a morphism

$$\eta_p : C_p(|K|,|L|;\mathsf{R}) \longrightarrow S_p(|K|,|L|;\mathsf{R}).$$

We also have its transpose

$$\eta^p : S^p(|K|,|L|;\mathsf{R}) \longrightarrow C^p(|K|,|L|;\mathsf{R}).$$

Then the above morphisms are compatible with the boundary and coboundary operators, respectively, and induce the isomorphisms in (B.30).

Cup and cap products: Let K be a simplicial complex and $\mathcal{V}$ the set of vertices of K.

Definition B.14. An ordering on $\mathcal{V}$ is *simplicial* if its restriction to the set of vertices of each simplex of K is total.

For example, a total ordering on $\mathcal{V}$ is simplicial. The cup and cap products are defined on the level of simplicial chains and cochains. For this we fix a simplicial ordering on $\mathcal{V}$.

For an oriented simplex $\boldsymbol{s}$ of K we denote by $\vartheta(\boldsymbol{s})$ its dual in $C^p(|K|;\mathsf{R})$, i.e., the element defined by

$$\langle \boldsymbol{s}', \vartheta(\boldsymbol{s})\rangle = \begin{cases} 1 & \text{if } \boldsymbol{s}' = \boldsymbol{s}, \\ 0 & \text{otherwise} \end{cases}$$

and extended R-linearly. Then $C^p(|K|;\mathsf{R})$ is the free R-module generated by the $\vartheta(\boldsymbol{s})$'s.

The cup product

$$C^p(|K|;\mathsf{R}) \times C^q(|K|;\mathsf{R}) \xrightarrow{\smile} C^{p+q}(|K|;\mathsf{R})$$

is defined by setting, for oriented p-simplex $\boldsymbol{s}_1$, q-simplex $\boldsymbol{s}_2$ and $(p+q)$-simplex $\boldsymbol{s}$

$$\langle \boldsymbol{s}, \vartheta(\boldsymbol{s}_1) \smile \vartheta(\boldsymbol{s}_2)\rangle = \varepsilon\,\varepsilon_1\,\varepsilon_2 \tag{B.31}$$

if we may write $\boldsymbol{s}_1 = \varepsilon_1(v_0, \ldots, v_p)$, $\boldsymbol{s}_2 = \varepsilon_2(v_p, \ldots, v_{p+q})$ and $\boldsymbol{s} = \varepsilon(v_0, \ldots, v_{p+q})$ with $v_0 < \cdots < v_{p+q}$, setting 0 otherwise, and extending R-linearly.

The cap product

$$C_r(|K|;\mathsf{R}) \times C^p(|K|;\mathsf{R}) \xrightarrow{\frown} C_{r-p}(|K|;\mathsf{R})$$

is defined by

$$\langle c \frown u, v\rangle = \langle c, u \smile v\rangle$$

for an r-chain c, a p-cochain u and an $(r-p)$-cochain v. Thus if $\boldsymbol{s} = \varepsilon(v_0, \ldots, v_r)$ with $v_0 < \cdots < v_r$ and $\boldsymbol{s}_1 = \varepsilon_1(v_0, \ldots, v_p)$,

$$\boldsymbol{s} \frown \vartheta(\boldsymbol{s}_1) = \varepsilon\varepsilon_1(v_p, \ldots, v_r). \tag{B.32}$$

Also, for $\boldsymbol{s}$ and $\boldsymbol{s}_1$ as above,

$$\begin{aligned} \boldsymbol{s} \frown \delta\vartheta(\boldsymbol{s}_1) &= \varepsilon\langle \partial(v_0, \ldots, v_{p+1}), \vartheta(\boldsymbol{s}_1)\rangle(v_{p+1}, \ldots, v_r) \\ &= (-1)^{p+1}\varepsilon\varepsilon_1(v_{p+1}, \ldots, v_r). \end{aligned} \tag{B.33}$$

If L is a subcomplex of K, for c in $C_r(|L|;\mathsf{R})$ and u in $C^p(|K|,|L|;\mathsf{R})$, $c \frown u = 0$. Thus we have

$$C_r(|K|,|L|;\mathsf{R}) \times C^p(|K|,|L|;\mathsf{R}) \xrightarrow{\frown} C_{r-p}(|K|;\mathsf{R}). \tag{B.34}$$

Remark B.14. 1. The above products induce the cup and cap products on the level of homology and cohomology and correspond to the ones on singular homology and cohomology under the isomorphisms of (B.30).

2. The above cap product is the left cap product. The right cap product

$$C^p(|K|;\mathsf{R}) \times C_r(|K|;\mathsf{R}) \xrightarrow{\frown} C_{r-p}(|K|;\mathsf{R})$$

is defined by

$$\langle u \frown c, v\rangle = \langle c, v \smile u\rangle$$

for an r-chain c, a p-cochain u and an $(r-p)$-cochain v. Thus if $\boldsymbol{s} = \varepsilon(v_0,\dots,v_r)$ with $v_0 < \cdots < v_r$ and $\boldsymbol{s}_1 = \varepsilon_1(v_{r-p},\dots,v_r)$,

$$\vartheta(\boldsymbol{s}_1) \frown \boldsymbol{s} = \varepsilon\varepsilon_1(v_0,\dots,v_{r-p}). \tag{B.35}$$

Barycentric subdivision: Let K be a simplicial complex. A *subdivision* of K is a simplicial complex K' such that $|K'| = |K|$ and that each simplex of K' is contained in a simplex of K. A particular type of subdivisions of K is given as follows. For a simplex $\boldsymbol{s}$, we denote by $b_{\boldsymbol{s}}$ its barycenter. The *barycentric subdivision* K' of K is the set of simplices obtained in the following way. Take a sequence of simplices $\boldsymbol{s}_0 \precneqq \cdots \precneqq \boldsymbol{s}_p$ in K. Then the simplices in K' consist of $\{b_{\boldsymbol{s}_0},\dots,b_{\boldsymbol{s}_p}\}$ for all the sequences as above. It is again a simplicial complex.

A subcomplex L of K is *full* if $\boldsymbol{s}$ is a simplex in K whose vertices are all in L, then $\boldsymbol{s}$ is in L. For any subcomplex L of K, L' is full in K', the barycentric subdivision of K.

We give a proof of the following for the sake of completeness:

Proposition B.26. *If L is a full subcomplex of K, $|L|$ is a deformation retract of $S_{K'}(|L|)$.*

Proof. Let x be a point in $O_K(|L|)$. For each vertex v in L, we define a number $x(v)$ as follows. If there is a simplex $\boldsymbol{s}$ in K such that x is in $\boldsymbol{s}$ and v is a vertex of $\boldsymbol{s}$, $x(v)$ is the barycentric coordinate of x relative to v. Note that it is a positive number not depending on such an $\boldsymbol{s}$. If there are no such simplices, we set $x(v) = 0$. Noting that $\sum_{v\in L} x(v)$ is a finite sum, which is non-zero, we set

$$r(x) = \frac{1}{\sum_{v\in L} x(v)} \sum_{v\in L} x(v)v.$$

It is in a face of a simplex of K whose vertices are in L. Since L is full, the point is in $|L|$. If x is in $|L|$, $\sum_{v \in L} x(v) = 1$ and $r(x) = x$. Thus it is a retraction of $O_K(|L|)$ onto $|L|$. Moreover, it is a deformation retraction by the homotopy $r_t(x) = (1-t)x + t\,r(x)$. The restriction of r to $S_{K'}(|L|)$ is a desired deformation retraction. □

Note that the above retraction $r : S_{K'}(|L|) \to |L|$ is proper.

B.3 Homology of locally finite chains

Let X be a locally compact topological space. A *locally finite singular p-chain* of X is a formal sum

$$c = \sum_{\sigma} a_\sigma \sigma,$$

with σ singular p-simplices of X and $a_\sigma \in \mathsf{R}$, such that, for every compact set C in X,

$$\{\sigma \mid a_\sigma \neq 0,\ \sigma(\Delta^p) \cap C \neq \emptyset\}$$

is a finite set. If c is a locally finite singular p-chain as above, its support $|c| = \bigcup_{a_\sigma \neq 0} \sigma(\Delta^p)$ is closed and the boundary ∂c is a well-defined locally finite singular $(p-1)$-chain. The set $\breve{S}_p(X; \mathsf{R})$ of locally finite chains is an R-module and we have a chain complex $(\breve{S}_\bullet(X; \mathsf{R}), \partial_\bullet)$. The homology $\breve{H}_p(X; \mathsf{R})$ of this complex is called the *homology of locally finite singular chains.*

There is a natural inclusion $S_\bullet(X; \mathsf{R}) \hookrightarrow \breve{S}_\bullet(X; \mathsf{R})$, which induces a morphism

$$H_p(X; \mathsf{R}) \longrightarrow \breve{H}_p(X; \mathsf{R}).$$

If X is compact, it is an isomorphism.

If $f : X \to Y$ is a proper map of locally compact spaces, it induces a morphism

$$f_* : \breve{H}_p(X) \longrightarrow \breve{H}_p(Y),$$

which is determined by the proper homotopy class of f.

Remark B.15. The homology of locally finite singular chains is not a homotopy invariant as the following example shows:

$$\breve{H}_p(\mathbb{R}^q; \mathbb{Z}) = \begin{cases} \mathbb{Z} & p = q, \\ 0 & p \neq q. \end{cases}$$

If A is a closed subset of X, then the inclusion $A \hookrightarrow X$ is proper and we may identify $\breve{S}_\bullet(A;\mathsf{R})$ with a subcomplex of $\breve{S}_\bullet(X;\mathsf{R})$. Thus we have the complex $\breve{S}_\bullet(X,A;\mathsf{R}) = \breve{S}_\bullet(X;\mathsf{R})/\breve{S}_\bullet(A;\mathsf{R})$ of relative chains, which defines the relative homology $\breve{H}_p(X,A;\mathsf{R})$ of locally finite singular chains. We have also the exact sequence as (B.4) for this homology theory.

For A as above, we may define the cap product

$$\breve{S}_r(X,A) \times S^p(X,A) \xrightarrow{\frown} \breve{S}_{r-p}(X)$$

as in (B.16), see also (B.21). It induces the cap product

$$\breve{H}_r(X,A) \times H^p(X,A) \xrightarrow{\frown} \breve{H}_{r-p}(X). \tag{B.36}$$

Let $(X, \{e_\lambda\}_{\lambda\in\Lambda})$ be a (locally finite) cell complex. We denote by $\breve{C}_p(X;\mathsf{R})$ the R-module of formal sums of oriented closed p-cells in X with coefficients in R, i.e., the direct product $\prod \mathsf{R}\overline{e_\lambda}$ with $\overline{e_\lambda}$ all the oriented closed p-cells in X. Then we have a chain complex $(\breve{C}_\bullet(X;\mathsf{R}),\partial)$. If A is a sub-cell complex of X, $\breve{C}_\bullet(A;\mathsf{R})$ is a subcomplex of $\breve{C}_\bullet(X;\mathsf{R})$ and we may consider the quotient complex $\breve{C}_\bullet(X,A;\mathsf{R}) = \breve{C}_\bullet(X;\mathsf{R})/\breve{C}_\bullet(A;\mathsf{R})$. Then there is a canonical isomorphism

$$H_p(\breve{C}_\bullet(X,A;\mathsf{R})) \simeq \breve{H}_p(X,A;\mathsf{R}). \tag{B.37}$$

For a simplicial complex K, $\breve{C}_p(|K|;\mathsf{R})$ is the R-module of formal sums of oriented p-simplices in K with coefficients in R. If L is a sub-simplicial complex of K, $\breve{C}_\bullet(|L|;\mathsf{R})$ is a subcomplex of $\breve{C}_\bullet(|K|;\mathsf{R})$ and we have the quotient complex $\breve{C}_\bullet(|K|,|L|;\mathsf{R}) = \breve{C}_\bullet(|K|;\mathsf{R})/\breve{C}_\bullet(|L|;\mathsf{R})$. There is a canonical isomorphism

$$H_p(\breve{C}_\bullet(|K|,|L|;\mathsf{R})) \simeq \breve{H}_p(|K|,|L|;\mathsf{R}). \tag{B.38}$$

The cap product (B.36) may also be described in terms of locally finite chains of K (cf. (B.34)):

$$\breve{C}_r(|K|,|L|;\mathsf{R}) \times C^p(|K|,|L|;\mathsf{R}) \xrightarrow{\frown} \breve{C}_{r-p}(|K|;\mathsf{R}). \tag{B.39}$$

Notes

As a basic reference on general topology, we list [Munkres (1975)]. There are a number of references on algebraic topology. Here we list [Greenberg and Harper (1981); Hattori (1991); Spanier (1966)].

Bibliography

Abate, M., Bracci, F., Suwa, T. and Tovena, F. (2013). *Localization of Atiyah classes*, Rev. Mat. Iberoam. **29**, pp. 547-578.

Abate, M., Bracci, F and Tovena, F. (2008). *Index theorems for holomorphic maps and foliations*, Indiana Univ. Math. J. **57**, pp. 2999-3048.

Aluffi, P. (1999). *Chern classes for singular hypersurfaces*, Trans. Amer. Math. Soc. **351**, pp. 3989-4026.

Angella, D., Suwa, T., Tardini, N. and Tomassini, A. (2020). *Note on Dolbeault cohomology and Hodge structures up to bimeromorphisms*, Complex Manifolds **7**, pp. 194-214.

Atiyah, M.F. (1957). *Complex analytic connections in fibre bundles*, Trans. Amer. Math. Soc. **84**, pp. 181-207.

Atiyah, M.F. (1967). *K theory*, Benjamin, New York.

Atiyah, M.F. and Bott, R. (1967-8). *A Lefschetz fixed point formula for elliptic complexes: I, II*, Ann. Math. **86**, pp. 374-407, **88**, pp. 451-491.

Atiyah, M.F. and Hirzebruch, F. (1961). *Analytic cycles on complex manifolds*, Topology **1**, pp. 25-45.

Atiyah, M.F. and Hirzebruch, F. (1962). *The Riemann-Roch for analytic embeddings*, Topology **1**, pp. 151-166.

Atiyah, M.F. and Singer, I.M. (1963). *The index of elliptic operators on compact manifolds*, Bull. Amer. Math. Soc. **69**, pp. 422-433.

Baum, P. and Bott, R. (1972). *Singularities of holomorphic foliations*, J. Differential Geom. **7**, pp. 279-342.

Baum, P., Fulton, W. and MacPherson, R. (1975). *Riemann-Roch for singular varieties*, Publ. Math. IHES **45**, pp. 101-145.

Bierstone, E. and Milman, P.D. (1988). *Semianalytic and subanalytic sets*, Publ. Math. IHES **67**, pp. 5-42.

Bisi, C., Bracci, F., Izawa, T. and Suwa, T. (2016). *Localized intersection of currents and the Lefschetz coincidence point theorem*, Annali di Mat. Pura ed Applicata **195**, pp. 601-621.

Bismut, J.-M. (1998). *Local index theory and higher analytic tosion*, Documenta Math. Extra Volume ICM 1998, I, pp. 143-162.

Bismut, J.-M., Gillet, H. and Soulé, C. (1988). *Analytic torsion and holomorphic determinant bundles I: Bott-Chern forms and analytic torsion*, Comm. Math. Phys. **115**, pp. 49-78.

Borel, A. et Serre, J.-P. (1958). *Le théorème de Riemann-Roch*, Bull. Soc. Math. France **86**, pp. 97-136.

Bott, R. (1972). *Lectures on characteristic classes and foliations*, Lectures on Algebraic and Differential Topology, Lecture Notes in Math. **279**, Springer, Berlin, pp. 1-94.

Bott, R. and Chern, S.S. (1965). *Hermitian vector bundles and the equidistribution of the zeroes of their holomorphic sections*, Acta Math. **114**, pp. 71-112.

Bott, R. and Tu, L. (1982). *Differential forms in algebraic topology*, Graduate Texts in Math. **82**, Springer, Berlin.

Bourbaki, N. (1961-65). *Éléments de Mathématique, Algèbre Commutative, Chap. 1-7*, Hermann, Paris.

Bracci, F. and Suwa, T. (2004). *Residues for singular pairs and dynamics of biholomorphic maps of singular surfaces*, Intern. J. Math. **15**, pp. 443-466.

Brasselet, J.-P. (1981). *Définition combinatoire des homomorphismes de Poincaré, Alexander et Thom pour une pseudo-variété*, Caractéristique d'Euler-Poincaré, Astérisque **82-83**, Soc. Math. France, pp. 71-91.

Brasselet, J.-P. (1995). *Aspects combinatoires des singularités*, Singularity Theory, World Scientific, Singapore, pp. 74-127.

Brasselet, J.-P. (2022). *Characteristic classes*, Chapter 5 in Handbook of Geometry and Topology of Singularities III, J.L. Cisneros-Molina, D.-T. Lê and J. Seade eds., Springer Berlin.

Brasselet, J.-P., Lehmann, D., Seade, J. and Suwa, T. (2001). *Milnor classes of local complete intersections*, Trans. Amer. Math. Soc. **354**, pp. 1351-1371.

Brasselet, J.-P., Schürmann, J. and Yokura, S. (2010). *Hirzebruch classes and motivic Chern classes for singular spaces*, J. Topology and Analysis **2**, pp. 1-55.

Brasselet, J.-P. et Schwartz, M.-H. (1981). *Sur les classes de Chern d'un ensemble analytique complexe*, Astérisque **82-83**, Société Mathématique de France, pp. 93-147.

Brasselet, J.-P., Seade, J. and Suwa, T. (2009). *Vector fields on singular varieties*, Lecture Notes in Math. **1987**, Springer, Berlin.

Brasselet, J.-P. and Suwa, T. (2021). *Local and global coincidence homology classes*, J. Fixed Point Theory Appl. **23**, 24.

Burgos Gil, J.I., Kramer, J. and Kühn, U. (2005). *Arithmetic characteristic classes of automorphic vector bundles*, Documenta Math. **10**, pp. 619-716.

Burgos Gil, J.I. and Liţkanu, R. (2010). *Singular Bott-Chern classes and the arithmetic Grothendieck-Riemann-Roch theorem for closed immersions*, Documenta Math. **15**, pp. 73-176.

Cartan, H. (1961). *Théorie élémentaire des fonctions analytiques d'une ou plusieurs variables complexes*, Hermann, Paris. English translation : *Elementary Theory of Analytic Functions of One or Several Complex Variables*, Dover Publications, New York, 1995.

Cartan, H. et Serre, J.-P. (1953). *Un théorème de finitude concernant les variétés analytiques compactes*, C. R. Acad. Sci. Paris **237**, pp. 128-130.

Corrêa, M. and Suwa, T. (2020). *Localization of Bott-Chern classes and Hermitian residues*, J. London Math. Soc. **101**, pp. 349-372.

de Jong, T. and Pfister, G. (2000). *Local analytic geometry*, Advanced Lectures in Math., Vieweg, Wiesbaden.

Deligne, P., Griffiths, P., Morgan, J. and Sullivan, D. (1975). *Real homotopy theory of Kähler manifolds*, Invent. math. **29**, pp. 245-274.

Demailly, J.-P. (2012). *Complex analytic and differential geometry*, https://www-fourier.ujf-grenoble.fr/~demailly/manuscripts/agbook.pdf.

Dimca, A. (1992). *Singularities and Topology of Hypersurfaces*, Springer, Berlin.

Douady, A. (1968). *Flatness and privilege*, Enseign. Math. **14**, pp. 47-74.

Ebeling, W. and Gusein-Zade, S. (2001). *On the index of a holomorphic 1-form on an isolated complete intersection singularity*, Doklady Math. **64**, pp. 221-224.

Ehresmann, C. (1950). *Les connexions infinitésimales dans un espace fibré différentiable*, Colloque de Topologie, Bruxelles, pp. 29-55.

Eisenbud, D. (1995). *Commutative Algebra with a View Toward Algebraic Geometry*, Graduate Texts in Math. **150**, Springer, Berlin.

Fischer, G. (1976). *Complex analytic geometry*, Lecture Notes in Math. **538**, Springer, Berlin.

Freed, D.S. (2021). *The Atiyah-Singer index theorem*, Bull. Amer. Math. Soc. **58**, pp. 517-566.

Fritzsche, K. and Grauert, H. (2002). *From holomorphic functions to complex manifolds*, Graduate Texts in Math. **213**, Springer, Berlin.

Frölicher, A. (1955). *Relations between the cohomology groups of Dolbeault and topological invariants*, Proc. Nat. Acad. Sci. **41**, pp. 641-644.

Fujisawa, K. (2018). *Thom form in equivariant Čech-de Rham theory*, Intern. J. Math. **29**, 1850050.

Fulton, W. (1984). *Intersection theory*, Springer, Berlin.

Fulton, W. and Johnson, K. (1980). *Canonical classes on singular varieties*, Manuscripta Math. **32**, pp. 381-389.

Gibson, C.G., Wirthmüller, K., du Plessis, A.A. and Looijenga, E.J.N. (1976). *Topological stability of smooth mappings*, Lecture Notes in Math. **552**, Springer, Berlin.

Gillet, H. and Soulé, C. (1990). *Characteristic classes for algebraic vector bundles with Hermitian metric*, I, Ann. Math. **131**, pp. 163-203.

Godement, R. (1958). *Topologie Algébrique et Théorie des Faisceaux*, Hermann, Paris.

Gómez-Mont, X., Seade, J. and Verjovsky, A. (1991). *The index of a holomorphic flow with an isolated singularity*, Math. Ann. **291**, pp. 737-751.

Goresky, M. and MacPherson, R. (1980). *Intersection homology theory*, Topology **19**, pp. 135-162.

Goresky, M. and MacPherson, R. (1985). *Lefschetz fixed point theorem for intersection homology*, Comment. Math. Helv. **60**, pp. 366-391.

Grauert, H. (1958). *On Levi's problem and the imbedding of real-analytic manifolds*, Ann. Math. **68**, pp. 460-472.

Grauert, H. (1960). *Ein Theorem der analytischen Garbentheorie und die Modulräume komplexer Strukturen*, Publ. math. I.H.É.S. **5**, pp.5-64.

Grauert, H. and Remmert, R. (1984). *Coherent analytic sheaves*, Grundlehren der math. Wissenschaften **265**, Springer, Berlin.

Greenberg, M. and Harper, J. (1981). *Algebraic topology: A first course*, Addison-Wesley.

Greuel, G.-M. (1975). *Der Gauß-Manin Zusammenhang isolierter Singularitäten von vollständigen Durchschnitten.* Math. Ann. **214**, pp. 235-266.

Griffiths, P. and Harris, J. (1978). *Principles of Algebraic Geometry*, John Wiley & Sons, New York.

Grothendieck, A. (1957). *Sur quelques points d'algèbre homologique*, Tôhoku Math. J. **9**, pp. 119-227.

Grothendieck, A. (1960-61). *Techniques de construction en géométrie analytique. II. Généralités sur les espaces annelés et les espaces analytiques*, Séminaire Henri Cartan **13**, n° 1, exp. n° 9, pp. 1-14.

Guillemin, V. and Pollack, A. (1974). *Differential topology*, Prentice-Hall, Englewood Cliffs, New Jersey.

Gunning, R. and Rossi, H. (1965). *Analytic functions of several complex variables*, Prentice-Hall, Englewood Cliffs, New Jersey.

Hamm, H. (1971). *Lokale topologische Eigenschaften komplexer Räume*, Math. Ann. **191**, pp. 235-252.

Hartshorne, R. (1966). *Residues and duality*, Lecture Notes in Math. **20**, Springer, Berlin.

Hartshorne, R. (1977). *Algebraic geometry*, Graduate Texts in Math. **52**, Springer, Berlin.

Harvey, R. and Lawson, H.B. (1993). *A theory of characteristic currents associated with a singular connection*, Astérisque **213**, Soc. Math. France.

Hattori, A. (1991). *Topology*, Iwanami Shoten, Tokyo (in Japanese).

Hauser, H. (2003). *The Hironaka theorem on resolution of singularities* (*Or: A proof we always wanted to understand*), Bull. Amer. Math. Soc. **40**, pp. 323-403.

Hironaka, H. (1964). *Resolution of singularities of an algebraic variety over a field of characteristic zero, I, II*, Ann. of Math. **79**, pp. 109-203, pp. 205-326.

Hirsch, M.W. and Mazur, B (1974). *Smoothings of piecewise linear manifolds*, Ann. of Math. Studies **80**, Princeton Univ. Press, Princeton, New Jersey.

Hirzebruch, F. (1966). *Topological methods in algebraic geometry*, Third ed., Grundlehren der math. Wissenschaften **131**, Springer, Berlin.

Honda, N., Izawa, T. and Suwa, T. (2023). *Sato hyperfunctions via relative Dolbeault cohomology*, J. Math. Soc. Japan **75**, pp. 229-290.

Hotta, R. (1987). *Introduction to algebra - Groups, rings & modules -*, Shokabo, Tokyo (in Japanese).

Husemoller, D. (1994). *Fibre bundles*, Third ed., Graduate Texts in Math. **20**, Springer, Berlin.

Ikeda, T, (2018), *Lectures on enumerative geometry*, Univ. of Tokyo Press, Tokyo (in Japanese).

Iversen, B. (1971). *Critical points of an algebraic function*, Inventiones Math. **12**, pp. 210-224.

Iversen, B. (1976). *Local Chern classes*, Ann. Scient. Éc. Norm. Sup. **9**, pp. 155-169.

Izawa, T. and Suwa, T. (2003). *Multiplicity of functions on singular varieties*, Internat. J. Math. **14**, pp. 541-558.

Kashiwara, M. and Schapira, P. (1990). *Sheaves on manifolds*, Grundlehren der math. Wissenschaften **292**, Springer, Berlin.

Kobayashi, S. and Nomizu, K. (1963). *Foundations of differential geometry, Vol. I*, John Wiley & Sons, New York.

Kodaira, K. (1951). *The theorem of Riemann-Roch on compcat analytic surfaces*, Amer. J. Math. **73**, pp. 813-875.

Kodaira, K. (1954). *On Kähler varieties of restricted type (an intrinsic characterization of algebraic varieties)*, Ann. Math. **60**, pp. 28-48.

Kodaira, K. (1960). *On compact complex analytic surfaces, I*, Ann. Math. **71**, pp. 111-152.

Kodaira, K. (1986). *Complex manifolds and deformation of complex structures*, Grundlehren der math. Wissenschaften **283**, Springer, Berlin.

Lê, D.-T. (1974). *Calculation of Milnor number of isolated singularity of complete intersection*, Funct. Anal. Appl. **8**, pp. 127-131.

Lee, J.M. (2013). *Introduction to smooth manifolds*, Second ed., Graduate Texts in Math. **218**, Springer, Berlin.

Lefschetz, S. (1926). *Intersections and transformations of complexes and manifolds*, Trans. Amer. Math. Soc. **28**, pp. 1-49.

Lefschetz, S. (1937). *On the fixed point formula*, Ann. Math. **38**, pp. 819-822.

Lehmann, D. (1991). *Systèmes d'alvéoles et intégration sur le complexe de Čech-de Rham*, Publications de l'IRMA **23** N° VI, Université de Lille I.

Lehmann, D., Soares, M. and Suwa, T. (1995). *On the index of a holomorphic vector field tangent to a singular variety*, Bol. Soc. Bras. Mat. **26**, pp. 183-199.

Levy, R.N. (1987). *The Riemann-Roch theorem for complex spaces*, Acta Math. **158**, pp. 149-188.

Łojasiewicz, S. (1991). *Introduction to complex analytic geometry*, Birkhäuser.

Looijenga, E.J.N. (1984). *Isolated singular points on complete intersections*, London Math. Soc. Lecture Note Series **77**, Cambridge Univ. Press, Cambridge.

Macaulay, F.S. (1916). *Algebraic theory of modular systems*, Cambridge Tracts **16**, Cambridge Univ. Press, Cambridge.

MacPherson, R. (1974). *Chern classes for singular varieties*, Ann. Math. **100**, pp. 423-432.

Malgrange, B. (1957). *Faisceaux sur des variétés analytiques réelles*, Bull. Soc. Math. France **85**, pp.231-237.

Mather, J. (2012). *Notes on topological stability*, Bull. Amer. Math. Soc. (New Series) **49**, pp. 475-506.

Matsumura, H. (1980). *Commutative algebra*, Second ed., Benjamin/Cummings, Reading, Massachusetts.

Matsushima, Y. (1972). *Differentiable manifolds*, Translated by E.T. Kobayashi, Marcel Dekker, New York.

Milnor, J. (1968). *Singular points of complex hypersurfaces*, Ann. of Math. Studies **61**, Princeton Univ. Press, Princeton, New Jersey.

Milnor, J. and Stasheff, J. (1974). *Characteristic classes*, Ann. of Math. Studies **76**, Princeton Univ. Press, Princeton, New Jersey.

Mumford, D. (1961). *The topology of normal singularities of an algebraic surface and a criterion for simplicity*, Publ. Math. I.H.E.S. **9**, pp. 5-22.

Munkres, J.R. (1966). *Elementary differential topology*, Ann. of Math. Studies **54**, Princeton Univ. Press, Princeton, New Jersey.

Munkres, J.R. (1975). *Topology, a first course*, Prentice-Hall, Englewood Cliffs, New Jersey.

Narasimhan, R. (1966). *Introduction to the theory of analytic spaces*, Lecture Notes in Math. **25**, Springer, Berlin.

Noguchi, H. and Fukuda, T. (1976). *Elementary catastrophes*, Kyoritsuzensho **208**, Kyoritsu Shuppan, Tokyo (in Japanese).

Noguchi, J. (2016). *Analytic function theory of several variables: Elements of Oka's coherence*, Springer, Berlin.

Ohmoto, T. and Shiota, M. (2017). *C^1-triangulations of semialgebraic sets*, J. Topology **10**, pp. 765-775.

Oka, M. (1997). *Non-degenerate complete intersection singularity*, Actualités Mathématiques, Hermann, Paris.

Parusiński, A. (1988). *A generalization of the Milnor number*, Math. Ann. **281**, pp. 247-254.

Parusiński, A. and Pragacz, P. (1995), *A formula for the Euler characteristic of singular hypersurfaces*, J. Algebraic Geom. **4**, pp. 337-351.

Parusiński, A. and Pragacz, P. (2001), *Characteristic classes of hypersurfaces and characteristic cycles*, J. Algebraic Geom. **10**, pp. 63-79.

Rolfsen, D. (2003). *Knots and links*, AMS Chelsea Publishing, Amer. Math. Soc., Providence, Rhode Island.

Rourke, C. and Sanderson, B. (1972). *Introduction to piecewise-linear topology*, Springer, Berlin.

Rudin, W. (1987). *Real and complex analysis*, Third ed., McGraw-Hill, New York.

Sakai, F. (1984). *Weil divisors on normal surfaces*, Duke Math. J. **51**, pp. 877-887.

Schürmann, J.and Yokura, S (2007). *A survey of characteristic classes of singular spaces*, Singularity Theory, Dedicated to Jean-Paul Brasselet on his 60th birthday, World Scientific, Singapore, pp. 865-952.

Schwartz, M.-H. (1965). *Classes caractéristiques définies par une stratification d'une variété analytique complexe*, C.R. Acad. Sci. Paris **260**, pp. 3262-3264, 3535-3537.

Seade, J. (2019). *On Milnor's fibration theorem and its offspring after 50 years*, Bull. Amer. Math. Soc. **56**, pp. 281-348.

Seade, J. and Suwa, T. (1996). *A residue formula for the index of a holomorphic flow*, Math. Ann. **304**, pp. 621-634.

Seade, J. and Suwa, T. (1998). *An adjunction formula for local complete intersections*, Internat. J. Math. **9**, pp. 759-768.

Serre, J.-P. (1955a). *Un théorème de dualité*, Comment. Math. Helv. **29**, pp. 9-26.

Serre, J.-P. (1955b). *Faisceux algébriques cohérents*, Ann. Math. **61**, pp. 197-278.

Serre, J.-P. (1956). *Géométrie algébrique et géométrie analytique*, Ann. Inst. Fourier **6**, pp. 1-42.

Serre, J.-P. (1975). *Algèbre Locale - Multiplicités*, Troisième éd., Lecture Notes in Math. **11**, Springer, Berlin.

Shiga, K. (1976). *Manifold theory* I, Geometry i, Iwanamikôza Kisosûgaku, Iwanami Shoten, Tokyo (in Japanese).

Shiota, M. (1997). *Geometry of subanalytic and semialgebraic sets*, Progress in Math. Birkhäuser, Boston.

Spanier, E.H. (1966). *Algebraic topology*, McGraw-Hill, New York.

Steenrod, N. (1951). *The topology of fiber bundles*, Princeton Univ. Press, Princeton, New Jersey.

Suwa, T. (1996). *Classes de Chern des intersections complètes locales*, C.R. Acad. Sci. Paris **324**, pp. 67-70.

Suwa, T. (1998). *Indices of vector fields and residues of singular holomorphic foliations*, Actualités Mathématiques, Hermann, Paris.

Suwa, T. (2000a). *Dual class of a subvariety*, Tokyo J. Math. **23**, pp. 51-68.

Suwa, T. (2000b). *Characteristic classes of coherent sheaves on singular varieties*, Singularities – Sapporo 1998, 279-297, Adv. Stud. Pure Math., 29, Kinokuniya, Tokyo, 2000.

Suwa, T. (2003a). *Residues of Chern classes*, J. Math. Soc. Japan **55**, pp. 269-287.

Suwa, T. (2003b). *Characteristic classes of singular varieties*, Sugaku Expositions, Amer. Math. Soc. **16**, pp. 153-175.

Suwa, T. (2005). *Residues of Chern classes on singular varieties*, Singularités Franco-Japonaises, Marseille 2002, Séminaires et Congrès **10**, Soc. Math. France, pp. 265-285.

Suwa, T. (2008). *Residue theoretical approach to intersection theory, real and complex singularities*, Contemporary Math., Amer. Math. Soc. **459**, pp. 207-261.

Suwa, T. (2009). *Čech-Dolbeault cohomology and the $\bar{\partial}$-Thom class*, Singularities – Niigata-Toyama 2007, Adv. Studies in Pure Math. **56**, Math. Soc. Japan, pp. 321-340.

Suwa, T. (2012). *Residues of singular holomorphic distributions*, Singularities in Geometry and Topology, Strasbourg 2009, IRMA Lectures in Mathematics and Theoretical Physics **20**, European Math. Soc., pp. 207-247.

Suwa, T. (2014). *GSV-indices as residues*, J. of Singularities **9**, pp. 206-218.

Suwa, T. (2022a). *Relative Dolbeault cohomology*, Riv. Mat. Univ. Parma **13**, pp. 307-352.

Suwa, T. (2022b). *Residues and hyperfunctions*, Chapter 8 in Handbook of Geometry and Topology of Singularities III, J.L. Cisneros-Molina, D.-T. Lê and J. Seade eds., Springer Berlin.

Suwa, T. (2023). *Representation of relative sheaf cohomology*, Forum Math. **35**, pp. 1047-1076.

Tsikh, A.K. (1988). *Weakly holomorphic functions on complete intersections, and thier holomorphic extension*, Math. USSR Sbornik **61**, pp. 421-436.

Vick, J.W. (1994). *Homology theory: An introduction to algebraic topology*, Second ed., Gaduate Texts in Math., Springer, Berlin.

Voisin, C. (2007). *Hodge theory and complex algebraic geometry. I*, Translated from the French by Leila Schneps, Reprint of the 2002 English edition, Cambridge Studies in Advanced Mathematics, **76**, Cambridge University Press, Cambridge.

Weil, A. (1952). *Sur les théorèmes de de Rham*, Comment. Math. Helv. **26**, pp. 119-145.

Wells, R.O. Jr. (2008). *Differential analysis on complex manifolds*, Third ed., Graduate Texts in Math. **65**, Springer, New York.

Whitehead, J.H.C. (1940). *On C^1-complexes*, Ann. Math. **41**, pp. 809-824.

Whitney, H. (1936). *Differentiable manifolds*, Ann. Math. **37**, pp. 645-680.

Whitney, H. (1965). *Tangents to an analytic variety*, Ann. Math. **81**, pp. 496-549.

Whitney, H. (1972). *Complex analytic vrieties*, Addison-Wesley Publishing Co.

Yokura, S. (2020). *Motivic Milnor-Hirzebruch classes revisited*, Singularities – Kagoshima 2017, World Scientific, Singapore, pp. 265-294.

Index

www.ingramcontent.com/pod-product-compliance
Lightning Source LLC
Chambersburg PA
CBHW071248180926
51613CB00002B/171

IN LOVE WITH LIFE

Memoirs of a Lady Doctor

Mrs Grace Moir
6 Wyndham House
Plantation Rd
Ox, 6JJ
OXFORD UK

IN LOVE WITH LIFE

MEMOIRS OF A LADY DOCTOR

Dear Grace and Chussan
with love
Prema
July 90

P.M. Naidu

STERLING PUBLISHERS PRIVATE LIMITED

STERLING PUBLISHERS PRIVATE LIMITED
L-10, Green Park Extension, New Delhi-110016
G-2, Cunningham Apartments, Cunningham Road, Bangalore-560052

Printed in India

Published by S.K. Ghai, Managing Director, Sterling Publishers Pvt. Ltd., L-10, Green Park Extension, New Delhi-110016. Laserset at Vikas Compographics, A-1/256 Safdarjung Enclave, New Delhi-110029. Printed at Gopsons Papers (P) Ltd., Noida, (India).

When I consider how my light is spent
Ere half my days in this dark world and wide
That one talent which is death to hide
Lodged with me useless...

From the sonnet
"On his Blindness" by Milton

FOREWORD

Dr. Mrs. P. M. Naidu is a rare person, at ease in serious work. For about half a century she was leading out children safely into the human world from complicated vaults of maternal grace. She is an enlightened Gynaecologist and had found her company among obstetricians and gynaecologists of various countries which she had visited both to learn and to teach. A good teacher, she was never known to have addressed her students to disengage her mind.

A modern reader, who is being fed on coloured and concocted news and views by dailies and periodicals, finds no thrill in these pages in which she had recorded her experiences as a pilgrim in this world of anxieties and rejoicings and also ignorance and inhibitions. One should attune oneself to the basics of human well-being to enjoy this eminently readable account.

Mrs. Prema Naidu had been so intensely in love with her profession that at one stage her eyesight was badly impaired. But for the human attention of a famous Eye Specialist of London, she would have been stark blind for life. She is spared to remain an intelligent observer of this world, thanks to Mr. Fison, the specialist.

Her experiences serve as treasure items for those who value life and service. She has been leading a simple life, never like some of those snobs in her profession whose intellectual eminence is built on the money they grabbed from the suffering humanity.

About twenty-five years ago I happened to report a Family Planning Seminar in which she had participated. The policy of the establishment was to propagate and popularise "Loop", an alleged improvement on Foam Pills. But Mrs.Naidu was forthright in her disapproval of the use of the Loop, a foreign matter which a human body would endeavour to reject. How could she be so reasonable when the whole administration had fallen for this new birth controller? She was too academic to favour popular crazes. That is precisely one of the reasons for her to rise in our estimate of her professional integrity.

All her education, experience and professional skill has been useful to the needy and the weaker section of our society, namely the women. She is now past eighty years and is able to look into her past with satisfaction and also ruminate for the benefit of her known and unknown friends.

Mrs. P.M. Naidu had not evidently encountered any administrative problems which are so common in these political times. She has been first a teacher and surgeon. She did no influence peddling!

If each person endowed with one or the other marketable skills, follows her example, the world would be a better place for human beings to live.

G. KRISHNA
Editor
National Information Services, Hyderabad

PROLOGUE

In June 1969, I had totally lost my sight. This sudden affliction aroused an irresistible urge in me to write the story of my life. Blindfolded and lying on a stretcher, I was flown to London across the balmy clouds under the blue canopy in my dark wide world. There Dr. Fison, a world-renowned Eye Surgeon, endowed with skill and compassion, operated upon my eyes on the 4th of July. By the grace of God, my sight returned. Many of my good friends in UK hastened my recovery by showering on me their affection and hospitality. My near and dear ones in India gave me strength through their love and prayers from afar. And perhaps my own self-confidence and the joy of living had helped me along the road to recovery.

Although my professional life was intercepted for nearly a year, I was more than thankful to be back at Hyderabad with my sight restored, to serve, once again, young women suffering from a peculiar distress following difficult child-birth.

Writing my story in the English language is not easy. Memories from the past could be pulled out only a handful at a time as from a lucky dip. Recollections from yester-years may move back and forth during this narrative, for which I seek the indulgence of the reader.

An autobiography has to begin at the beginning. I begin to talk about my earliest recollections from the time I wanted to become a doctor. I must have been about five years of age. It is pleasant now to recall my various experiences of growing up with my parents, my education, medical studies, my work and travels which were full of delight and also hardships.

A listener may get tired of hearing me talk about those experiences again and again. Hence, this attempt to record as far as I can vividly remember them.

At this late stage of my life, remembering those past moments, my reading, writing and listening to music is like a second bloom. My inner self earned to do much more. But I am grateful to the creator for every little thing I have been able to do in my life. There might have been commissions and omissions. But I am satisfied that I have done what I had set out to do.

The past, which is indeed the reality, is pushed back into history, the future is obscure and the present moments are fleeting by.

Dost thou love life?
Then do not squander time
For that is the stuff life is made of.

Benjamin Franklin

This narrative was started in 1974 and had gone on till 1988. There are bound to be some repetitions. Describing his recipe "On what is cooking in the White House", a chef is reported to have said that it includes "a dash of reminiscences and a pinch of politics" for the President's lunch. An attempt has been made to separate travel in quest of knowledge from travels for travel's sake, with pleasing events and contacts with people, old and young.

What I planned to do was to reminisce over an intimate past. It is an indulgence, a leisurely recollection of the bounties of my life.

P.M. NAIDU

CONTENTS

ABBREVIATIONS

A.M.C.	..	Andhra Medical College
O.M.C.	..	Osmania Medical College, Hyderabad
S.M.C.	..	Salem Municipal College
W.C.C.	..	Women's Christian College, Madras
H.E.H.	..	His Exalted Highness, Hyderabad
S.M.P.	..	Simpson Memorial Pavillion, Edinburgh
C.M.O.	..	Chief Medical Officer
C.S/M.O.	..	Chief Surgeon and Medical Officer
G.M.H	..	Government Maternity Hospital
P.P.H.	..	Post Partum Haemorrhage
P.N.M	..	Perinatal Mortalities
G.A.	..	General Anaesthesia
L.A.	..	Local Anaesthesia
W & CR Hospital	..	Women & Childrens Hospital, Madras

1

FIRST HOME

We lived in a place called Thottam—a farm house near Salem Town. Our house was near a huge tree with rich foliage which shed pink fluffy flowers all day long, while its green twigs made of flat leaves, folded up at nights, as if they were asleep. This tree is popularly known as the 'Sleepy' tree, but its real name is Rain Tree. Under this tree stood a sprawling thatched house. We lived in that mud-plastered living space that led into a bedroom.

My earliest recollection of myself standing by a low window-sill, picking up mud from it and licking my fingers, is vivid. My mother's soft voice could be heard calling me to stop doing this. I must have been about three years of age at that time.

My grandfather's huge building was on the west of our thatched house. People drew water from a neighbouring open well, with a bucket tied to a rope running over a pulley.

My father, Madras Muthuswami Masilamani, was a lecturer of history at A.M. College, Guntur. I was born there at the Kugler Hospital. Dr. Kugler was a devout missionary. She is said to have knelt down by my mother's table during her travail and uttered a word of prayer for her safe delivery. Later, my father moved to Salem on a similar assignment he had held at Guntur, when I was a mere toddler. While we still lived in our thatched house, he started building a house on its eastern side.

The Thottam, being on the outskirts of the town, was large open area. A cart-road from a nearby village turned into Thottam in front of our house and ran down towards a low-tiled bungalow belonging to a prominent farmer, Mr. K.T. Paul. His mother was known to have been endowed with an exceptionally strong personality. She ran the farm and managed the farm hands with great skill and competence. She had the help of other managerial men who constantly shuffled between her bungalow and the farm lands. The cart-road leading from our house to hers was lined with beautiful trees, one of them, a large tree with lovely strings of flowers turning from a pink shade to dark red at their tips. Each tiny petal was like a cup. That tree in full bloom was the main spot of attraction in the entire landscape.

Another old tree with a gnarled trunk with widespread branches stood near the farm house. That tree showered each morning delicately scented creamy flowers, like little stars, forming a carpet on the ground. These flowers were threaded together into strings to adorn the beautiful black hair of our womenfolk.

The tree provided shelter to all sorts of vehicles. Once I saw the great person, Rajaji (C.R.) and his daughter Namagiri alighting from the side-car of his motorcycle. They had come to meet my farmer uncle. Rajaji later became the first Governor-General of independent India. My uncle also had been invited to the Round Table Conference in London. My father, in a small way, was also an ardent patriot. He dropped his necktie for a closed coat to begin with; and later wore only handspun cloth, khaddar.

Of all the other trees along the cart road, one was tall and yielded wood apples. The kernel of the wood apple, when scooped out and seasoned with brown palm sugar, made a delicious dessert. Another tall tree nearby put out leaf-like seed cases which, when ripe and dry, burst and filled the air with silk cotton. If collected before they lost their contents, the cotton would make soft pillows luxurious.

Other smaller trees with slender branches bore yellow flowers that could be cooked into delicious dishes. The tiny leaves too could be eaten. The long pods appearing at the end of the flowering season made tasty curries.

The fence opposite our houses had creepers entwining them. One of them bore an edible vegetable, which turned scarlet red when ripe. The other creeper had small bean-pods which, when dry, dropped shiny little egg-shaped seeds, with bottoms black and tops red. I have seen the village goldsmiths using them to balance against minute weights of gold.

The big flowering trees and the creepers on the fence were some of the simple charms of our memorable childhood. We also picked greenish gold beetles and kept them in empty match-boxes till they laid little yellow eggs. They were fed on leaves from the nearby fence.

During the harvest season, the farm hands celebrated the festival of Pongal (Sankranthi) with great gusto. They painted the horns of their cattle with gay colours and stamped their skin with rings of paint. The womenfolk boiled the newly harvested rice with palm sugar in mud pots in the open. At this festival, coconuts were broken and distributed to the guests with neatly cut pieces of sugarcane. The young men especially enjoyed the clanging sound made by striking their bronze bowls and passing among the cattle that also shared the joy by nodding their heads and flapping their ears in response to the musical sounds.

Beyond this site was the beginning of the real farm, spreading over several acres. There were green paddy fields edged with tall coconut trees and other palms. The fields were irrigated from large open wells. The farm boys drew water from the wells by an indigenous method. Large containers, made by soldering large metal sheets and opened at both ends were fitted to bottomless leather bags at the narrow ends. These containers collected water when lowered into the well.The containers were fastened to ropes which in turn were attached to two yoked bullocks. The water containers moved in and out of the well as they walked back and forth.

The water was discharged to the fields through a trough. This simple contraption was interesting to an urban onlooker. One could also get to the water below by walking down a flight of rough stone-steps. Young folks often dived and swam in the crystal-clear well-water.

Besides such wells, a stream wound through the farm, carrying along dry leaves and twigs. Large stone slabs across the rivulet served as bridges for men and the cattle. Mango groves abounded along the waterway, particularly near the stone bridges.

The owner of this Thottam (farm) was a widely-travelled and learned person. He was our "farmer uncle" who invited relatives and friends to carnatic music concert and his secretary sang even church music adapted to carnatic ragas. We all sat under the mango trees on rush mats and enjoyed the novel 'Kutchery' with rapt attention. (Even monkeys could be seen perched on the branches with their long tails hanging while clutching their young ones to their bosoms—perhaps also listening to and enjoying the enchanting vocal rendering of music.) He had several children who had developed varied interests in life. One of them was my contemporary; she and I went in quest of different careers.

2

GROWING UP

There can be no proper growing up without abundant love and care from one's parents. My parents were devoted to each other. This enabled us children to have a happy childhood, leading to our fuller development.

I was born at Guntur in 1909, where my father, M.M.M. was a lecturer. I do not remember much of our life at Guntur. After moving to Salem, I can, as a child, remember my father during our short stay at a thatched house; he was an amiable man, kind, deeply concerned and insisted on bringing up me and my brother with affection and firmness.

He maintained an amiable contact with his own sisters and nephews beside members on my mother's side. Young men sought his advice in studies and future careers. He lectured at the Municipal College at Salem, daily cycling up and down.

I was proud of my father's educational attainments. He was already a graduate in European History before he married my mother in 1903. His father was an iron merchant, who though born a Hindu, allowed his son to study at the Madras Christian College. This was followed by his teacher's training at the L.T. College at Saidapet, Madras. That building, a marvel in architecture, and built without concrete or cement, can be recognised on the way to the Madras airport along Mount Road.

Until I was four, we lived in that thatched house using kerosene lights. Once in 1915 my brother and I were taken to Bangalore by our parents for a holiday. We travelled by train. Staying in an airy campus of Theological College, we were amazed to find lights come on by pressing buttons on the wall. Since that place adjoined a British military cantonment, electricity had been bought there early. I remember hearing the metallic rhythmic sound of the running train during the return journey at night which made a deep impression on me as a child.

On our return to Salem, we were very happy to occupy the new house which my father had built at one edge of the Thottam. Centrally placed were three large rooms in tandem and a tile-roofed sloping verandah running all round. The eastern verandah was my father's apartment, his study, a real one with a good writing table with sheets of blotting papers laid out on it as a pad, a tray of pens and clips, and a dainty pair of scales for weighing letters before posting them. Notable was a pigeon-hole on the left side of the table for depositing correspondence. This rested on a shelf for periodicals like *Punch* and others popular English ones. On the right side of the table stood a revolving bookcase with a variety of general books. In my teens, I sat on a canvas chair between the bookcase and a window and read a book called "Daddy Long Legs" found in this case. Later I saw a film based on it.

Opposite the pigeon-hole stood a sturdy wooden almirah. The top shelf contained books on Greek and Roman History. On the shelf below were some books on Indian History. The third shelf was lined with English dictionaries. I still have in my possession one called *Idiomatic Dictionary*, published in 1906, bearing my father's signature and an ornate rubber stamp on it. Incidentally, I had also saved some of his letters written to me while I was at school and University. Among them is an M.O. coupon for Rs. 25, dated November 1930—a few days before he passed away. There is a note on the coupon in his own handwriting saying that he was always thinking of me and praying for me!

In the room adjoining his office, there was a high table with a typewriter on it. When I said I wanted to learn typing, he typed on the machine that he was glad that I wanted to use it. That typewriter had been brought from the USA for him by my uncle, Mr. KTP. But I am sorry that I never got on with any typing at all. It would have helped me all along my career and especially at this juncture.

Coming back to his office I wonder if many professors in the present times would possess such a fine set of books and appropriate furniture. That office led to the front verandah. My father sat on a cane chair in the morning, reading newspapers or correcting University answer papers. There were some chairs for visitors. At the other end of the verandah was his bicycle to be chained to the door nearby after he had returned from his college work.

My father was a very affable man and never minded me or my brother running up to him with our problems at any time of the day.

Father taught us at home till we were 10 or 11 years old. Maybe he felt that he would be able to give us a better foundation than the poorly organised normal elementary schools of the day.

Each day, when my brother and I got back home after our outdoor games, father would spend an hour or two telling us stories from the *Ramayana* and the *Mahabharata*, besides Bible stories. He made us do mental arithmetic based mainly on the local market products. His teaching of history and of local geography impressed us. By pointing to the starry skies, he used to give us an idea about astronomy. He made the constellations come alive by relating each group to a Greek figure. I recall being awakened by him at 4.00 a.m. once, to see the "Southern Cross" hanging low on the horizon. Conversing in English was part of the evening session. It provided us with a good foundation for later education.

Father's deep-seated interest in tribal medicine came in very helpful for villagers coming from far and near with scorpion-stings, snake-bites or worm-infestations.

Being in the teaching profession, he valued leisure at weekends and the long summer vacations. He played chess with us and 'Pachis' with my grandmother. He won a gold medal in tennis doubles at the local Literary Society. There was a violin in the house but I had never heard him play on it.

While I was in the third year of my medical course, my father began having moods of depression. It was probably due to the possibility of the S.M. College appointing a raw post-graduate student, overlooking the seniority of my father who was already the Vice Principal of the College. He tried to overcome this state of mind by studying for law.

He worked hard for the pleadership examination devoting the early hours of the mornings for study, with a hurricane lantern hanging from a wooden pillar. He did write the pleadership examination and passed it with a good rank, in his early fifties. Though immensely pleased with my progress in my medical studies which made his face glow at times with satisfaction, he became increasingly depressed over the staff problems of the College. In December 1930, he passed away quietly in bed. I was distressed that he did not live to witness my efforts at success at the final years of my study. I had to bear the responsibility of the family and care for my mother, brother and sisters.

My mother had only a primary school education even though she was the daughter of a Brahmin Government official. She was 16 and my father

27 when they were married. She was patient, kind and understanding. She gave her children all her best. She never lost her temper. She would put up with my father's occasional bouts of anger, provoked by her brothers' indifference to their studies, since my father was their guardian.

Though the women's liberation was just being talked about in the West, she was already a member of the District Educational Board at Salem and an Honorary Inspectress of Primary Girls Schools. I remember the school teachers submitting reverently their applications to her for sanction of leave in 1924-25. Her selflessness led her once into a state of poor health needing admission in the General Hospital, Madras. She also respected the orthodox feelings of my father's sisters. She used to cook and send the tiffin across to these close relatives. She used to feed us well. Food was not expensive. She would walk miles or go by a bullock cart to the market to buy the best for us.

Our father taught us at home as mentioned before. The village school master, Govindaswami, was in charge of us in the mornings when he helped us to read and write our mother tongue. But he occasionally did find me playing traunt: Once the manager of a circus set an essay contest to be written in one's mother tongue. The village school master helped me with its writing. I got the second prize. My father gave away the trophies. The ring master of the caged wild animals invited 'any brave soul who cared to come into the ring caged'. My mother's brother (he proved a great adventurer later) walked in. When the lion placed his front paws on his right shoulder he sank to his side. He could sit up straight only when the other lion placed her paws on his left shoulder. His mother was naturally much concerned and began shrinking with fear. But he emerged from the ring proclaiming himself as the brave man of the day.

When my father felt that I had reached the right standard of learning at home, he sent me to the SMC where I was tested and admitted into Form I. Two years later, my brother, Thambi, was admitted into Form I. He had a fret work machine at home with which he cut out photo frames and animal figures like dogs, horses or dolls.

While my father's apartments were on the east, my mother's domain was on the west wing with a storeroom, a dining room which had a large three-legged table (the other half of which served as a tailoring table in the next room). The third small room was the wash room. The kitchen was at a lower level with the tile-roof extending over it. One had to descend five steps to reach the floor and the primitive cooking range. The waste water ran out to the fruit trees. This water often became stagnant turning into a bubbling cesspool. My brother and I used to clear this whenever we were free from school work.

A separate room for bathing with a tin roof was where water was heated on firewood and the smoke hurting our eyes while we bathed. But baths we did have! There was a primitive toilet about 30 yards away from the kitchen. A brass pot had to be left outside to indicate that the toilet was occupied. One morning, entering the stone enclosure, I was taken aback to find a large snake stretched along the floor. It slithered away at the sound of my footsteps.

Extending between the kitchen and the toilet enclosure was a stretch of open land on which were planted a few fruit trees and groundnuts. We children helped with the groundnut digging. My brother at 9 years of age was an enthusiastic carpenter and a builder. He made wooden frames and cut out mud bricks (2"x3") and baked them in a make-shift kiln. With these he built a doll's house in the same design of our house, with 3 rooms and a flat roof. This was white-washed before the housewarming party was held.

The cowshed was behind the stone-walled bathroom and it was no doubt a breeding ground for flies and mosquitoes. Once when the cow was unable to give birth to its young one, the vet was called in to help. Though I was not allowed to watch this event then, later in life I became an expert midwife. Then I realised how skilled the vet must have been. In due course, as a Professor at OMC, I was elevated to be a Fellow of the appropriate Royal College in London.

Salem was surrounded by a chain of hills and hillocks. We could see them through all the windows of our house. The one seen in front of our house looked like an elephant in repose. Forest fires were often seen on that hill. The range on the east had a huge manmade caste mark, 'Nammam' called 'Namakkal'. The highest range on the north was the Shevroy hills with a lovely holiday resort, 'Yercaud'. I recall my several trips to Yercaud. The earliest one was with my grandfather's family. While descending the hairpin bends of the road in double bullock-carts, one of them toppled on its side. But no one was seriously hurt. My last trip was in 1974 with my daughter and her two very vivacious children, with her husband expertly driving the car. We stayed in a comfortable tourist home and the young people often went down to the lake on the hill for boating. We also attended a flower show there. My grandson enjoyed watching the 'rain flies with large wings', calling them by various names associated with aeroplanes. Years earlier our stay in a lovely little cottage was cut short by my Scottish friend falling ill with malaria.

With the completion of Form III, I joined a girls' school right in the heart of the town beyond the women's hospital and the bridge on a narrow river. Once this river was in floods, threatening to enter the wards of the hospital. The sturdy Anglo-Indian husband of the woman doctor offered to carry the patients on his back if she took care of their children.

By the time I completed Form III, my brother, Thambi, was admitted into Form I at the SMC. From then on, my co-educational days at the school level were over. I was shifted from SMC complex to a building in Salem town area beyond the narrow river, which was no more than a drain. On one side of the stream was the local hospital for women headed by an Anglo-Indian doctor, who lived in the hospital campus. The D.M.O. lived next to the women's hospital on the same road. Next to these Government quarters was the Imperial Bank of India. After this, the road bifurcated with a pink coloured church in between. The south compound wall of the church faced the newly built general hospital. My brother and I often walked a mile from Thottam to have our sore throats given a glycerine paint!

Next to the bank, came the main campus of the SMC. It consisted of the college and school buildings. My brother and I attented this school up to Form III while my father was the Vice Principal of the college section.

As one continued along the road, another church, yellow in colour, appeared in an extensive open area. At its gate was the building of the YMCA with its reading room, indoor games and sports room. My father was an ardent member of this worldwide institution.

Further along the main road, started a bazaar street (Shivapet). I often walked back from school through this bazaar street, crossed the newly laid railway line and through another lesser bazaar towards Thottam. The other route I took less often was by passing the pink church which led me to the tank bund. The tank was on one side and on the other side were lovely green paddy fields and coconut palms at a lower level by which the railway line was running. On a certain afternoon when I was plodding my way back from school barefoot with my tiffin carrier hanging on to the handle of the umbrella, I heard someone jump off his bicycle behind me. Turning round, I was pleasantly surprised to find that it was none other than my father. After walking a few yards with me, he offered to tell my mother that I would be home shortly and sped away on his bicycle.

Going back to the description of the school north of the stream, it was laid out on an extensive open space where cattle and donkeys grazed, wandered and rested without interference. When I was studying at SMC in 1922-23, the site assumed great excitement and attraction one day when Gandhiji was addressing a mammoth gathering there. He was clad in a loin cloth and sat with his legs folded together on one side. That scene made an indelible impression on me. It was the first and last time I saw Gandhiji in person.

We had very capable male teachers for mathematics and mother tongue. The sprightly mathematics teacher told us of the story of a village Sunday school. One day, the master sporting a square snuff box, asked us what the shape of the globe was. A smart boy jumped up and said, 'Sir, it is square on Sundays and round on other days', since the non-Sunday teacher carried a globular snuff box. A keen sense of observation, indeed!

The girls' school which I joined was, as I have said, in the heart of the town. The school was being upgraded then into a high school. We were only four girls in the Fourth Form—two Brahmin girls, my cousin and I. The two Brahmin girls were very orthodox and would not even drink water in our houses at Thottam. But they would readily accept milk in a silver glass!

In 1923, I went as a residential pupil to a fine school run by the Scottish Mission at Madras. The Principal was tall and stately and greatly contributed to my progress. Finally she retired to Scotland. Years afterwards, I called on her. She was living with her retired brothers and sisters. Those elderly people just adored my 13-year-old-daughter who had accompanied me.

Life at Salem was perennially enchanting to me and even now, the thought of it gives me great comfort. The people and the place itself, I feel, were created for my enjoyment.

Father had engaged a music master to come home and teach me to play harmonium and learn carnatic *kirtans*. I came up to 'Vatapi Ganapathim Bhaje', a composition I found beautiful then as now.

Several decades later when my grandson Ram was playing this *kirtan* on his veena at Madras, the memory of the training I had had so long ago, bound me to him in a strong bond. I was in my seventies and he in his teens, at the time of his training in music.

After a year in the Salem Girls School with its freedom and fun, I had gone to a residential school at Madras. The atmosphere in that school was stimulating and a very enjoyable with several extra-curricular activities like sports and girl-guide programmes. The Principal was generally kind to pupils and especially to me, for she thought that though a newcomer, I was pleasantly active in all the daily activities of the school. She knew that I wanted to be a doctor. One member of the staff, who taught us English was fresh from Oxford. I kept contact with her for over six decades. Of the other two Indian members, one was an active Girl Guide Captain and the other was made my local guardian.

I rang the school bell, took part in serving others food from huge vessels, reported about mosquitoes in bathrooms; dressed in simple clothes, spoke fluently in English and Tamil. Once we had a day's excursion to the bungalow near the Red Hills lake where I offered to help with the cooking which we did in batches. We had games on the roof. I organised the shy girls into assisting the Principal.

Another memorable event was a visit to Ennore by sea where we were divided into groups. We had competitions in discussion, prayers and play. It was there that my firm friendship with the English teacher from Oxford began. I later visited her several times at her home on the beautiful Fowey Harbour in Cornwall. She drove me to beautiful places with lakes and swans. She told me that all swans in the country belonged to the Crown! I had attented her wedding at the Christian College Chapel opposite the Madras High Court. She was 10 years older than I. Her last letter was addressed to my home address at Hyderabad but with a Bombay pincode with Madras underlined! I visited her home in Cornwall once again and celebrated the 50th Anniversary of our friendship with her family at her home on the heavenly looking harbour. Her sons took me and my daughter, studying then at Cambridge, in a lovely yacht to quaint little fishing villages on the Cornish coast. The visit to 'Lands End' via Penzance and St. Ives where the fascinating colours and shadows could be the dream of any artist was memorable.

My schooldays were full of friendly fellowship and were fascinating to me coming as I did from a rural school. We took part in the school duties and in attending chapel services. But it is a pity that I tended to neglect my mother tongue so much that I failed in Tamil though I had passed creditably in other subjects. I was greatly distressed and got over that state only when the University published a timely eligibility list, which secured me admission into Women's Christian College, Madras.

Once I joined the Women's Christian College, life was full of the joy of learning and enlightement. Dr. Miss Mc Dougall, the Principal, was a great scholar in Greek and taught us extracts from the original Greek scriptures.

I enjoyed learning Chemistry, maybe because this disciplines was being taught by a charming little Indian woman with a musical voice, adding to my fascination for the experiments with their fantastic colour changes in test tubes. I scored a distinction in Chemistry in the Intermediate examination. Miss Coon, Professor of Physics, was from the famous Holyoak College in Massachusetts (USA). She was the President of our Star Club. I was already greatly interested in astronomy since my father had acquainted me and my brother with the subject. He used to point at the starry skies at nights to identify for us the various constellations of the Zodiac. I also used to have a book on Constellations, but now it is untraceable. This brings to my memory how in the early sixties, I had pointed out the "Southern Cross" to Prof. J.C. Moir of Oxford during his visit to the Institute of Obstetrics and Gynaecology, Hyderabad. He remained my mentor at UK ever since.

Besides being a member of these various clubs at the Oxford and Cambridge Universities, I was interested in elocution and in the stage. I selected the poem 'Daffodils' by Wordsworth, in which he refers to the daffodils that "flash upon his inward eye, with the bliss of solitude." I did not get the first place, but my competitor did. She had chosen a ponderous and lengthy poem (I cannot recall its name). It was a poem about a priest in Rome who had prayed all his life to get a vision of the Lord God. And when the vision did appear, the priest had to leave the vision in his cell instantly, for the bells began to clang incessantly, as a reminder that he should go and attend to the feeding of the Monks.

On the stage at the Women's Christain College, I not only acted as Bassanio in *Merchant of Venice* but also as Udayanan in "Vasavadatta" in the famous Museum Theatre at Madras in the second year of my medical studies. Sports like American baseball and Physical Training conducted by Mrs. Buck, an energetic physical instructor, attracted me. On one occasion while playing the part of the shooter in a basketball match, I found my opponent was none other than the famous cinema actress, Enakshi Rama Rao, who played opposite Himansu Rai as Gautama Buddha in "Light of Asia". As the lovely young lady kept frequently using her powder compact, I scored beyond expectations, to the joy of my team. Once I was also selected to read the farewell address to the Professor of English, Mrs. Mona Hensman. In her reply, she referred to my 'telling' glances while reading my address. Recently, my grandchildren were amazed to hear that their grandmother had once been a student!

I had a most interesting time at the Women's Christian College with its excellent staff. I must particularly mention Dr. Miss Mason, an American Professor of Biochemistry, who later on, as an eminent Biochemist working in Bombay, remembered me for my part in the college activities. In the late sixties, when I was undergoing a major operation upon my eyes at the Moorfield Hospital, London, Miss Mason gave me a surprise visit. I recall her pertinent remark about tourists who visit only Delhi and Kashmir and miss South India and its colour and culture.

In London on a previous occasion with my 12-year-old daughter, Renuka, we visited Dr. McDougall, the retired Principal of the Women's Christian College, at her English home in Richmond. We enjoyed the visit by walking along River Thames and watching the Weeping Willow dipping into the narrow part of the great river at this site. Our revered hostess was wearing the college emblem on her neck chain depicting an Indian oil lamp and a script which said 'lighted to lighten'. She had believed that, that the Women's Christian College alumni would carry light to those in the dark unlettered parts of the world. A large fine portrait of this great lady hangs in the lobby of the grand science building in the Women's Christian College campus. It was built and inaugurated during her time as the Principal. It was in 1927 and I was fortunate to witness the magnificient function with my friend and cousin, Annama. We parted later when she went to Presidency College to study English, while I went to do medicine.

Annama and I happened to occupy the same room on the third floor of the hostel, overlooking the pond below and the beautiful chapel, beyond which the area had been gifted to the college by an 'Unknown American friend'. We two girls were very much unlike each other physically, temperamentally and emotionally. In our late teens, and dressing alike, we were dubbed as "twins" as we moved about together. She was interested in English poetry and literature and excelled in the 'Poetry Club' discussions at the college. I took science subjects, enjoyed the Star Club meetings on the roof of the college building. She went to Presidency College to do English and won the first prize in English in the

Province while I went into medicine and topped the list in MBBS in due course.

I cannot forget an interesting practice followed in the dining halls of Cambridge and Oxford Colleges. On every Saturday night, a member of the staff joined the students' at one of their tables, while a student was invited to the High Table to dine with the Principal.

Five years after I had left the Women's Christian College I qualified in Medicine in 1932. I was honoured by being invited to the famous High Table!

Thirty-two years later, my granddaughter, Anu was admitted into the Women's Christian College and she also enjoyed life there and read psychology in this great temple of learning.

3

MEDICAL STUDIES

With a well integrated coaching at the Women's Christian College for two years, I had no difficulty in 1927 in getting admission in the Madras Medical College. But the trouble was in finding accommodation in the Girls Hostel, which was just coming up. So, my father and I had to go to the YMCA to seek lodging for me. This was a temporary annexe to the regular hostel. Having arrived at the YMCA straight from the railway station, we had to sit long in a lobby, waiting to meet the Warden. My father was tired and unshaven, sitting with our screw-top water-carrier, we must have made a pathetic picture! Eventually I was shown in as a third member of a hostel room. I was happy enough to accept it. Father had to go to a nearby lodge for rest and food.

He returned next morning refreshed enough to accompany me to the office of the Principal of the Madras Medical College in order to apply for a studentship. Five studentships worth Rs. 50 per month and the cost of books required for the year, were awarded each year to women students, from the so-called Dufferin Fund. The Principal, Col. Hingston, regretted his inability to help me as five girl students had been already selected. On learning this, my father returned to Salem, hoping for some providential source of help for me to carry on. And I started my medical studies in full vigour. The pre-medical course of six months was completed successfully without much difficulty.

At the next stage of studying anatomy and physiology, I found some sympathetic tutors who encouraged me to approach the Professor of Anatomy for the necessary financial assistance to enable me to continue my studies. The Professor, a very senior member of the staff, recommended my case to the Principal, pointing out the wisdom of helping a student, whose proficiency was already established. On this recommendation, I secured a Dufferin Fund award. The Assistant Professor, in his bombastic language, also remarked that students of good calibre should not be allowed to sit "cribbed, cabined and confined" but should be encouraged to continue further studies! Actually, his Chief had examined me on the portions of work completed, and was satisfied that I deserved the studentship award for which he had recommended me to the Principal.

After six months, when a lumpsum of the studentship came into my hands, I sent a good portion of it to my father. He was greatly pleased at my generosity.

Next, on entering the clinical years of learning in the General Hospital wards, I was most thrilled to watch surgical operations on actual patients, having dissected merely the cadavers till then. One day, a senior woman surgeon from Calcutta came into the ante-room of the Operation Theatre. Col. Pandalai, the Chief Surgeon, was pleased to introduce her to us, the students posted under him. She was his "batch mate" when he was working for his surgical fellowship in London. I was inspired by her visit and longed to go abroad for further studies when possible. This was in 1930. But I could not go abroad till 1947.

Surprisingly, the third surgeon, Dr. James, noticing my keenness, asked me one morning to "wash up" and assist him in a major abdominal operation (gastro-jejunostomy). This was followed by my doing an appendisectomy while he assisted me.

The incident gave my father immense joy, for he realised that I had been given an early opportunity which would enable me to get the feel of major surgical procedures.

The teaching rounds even in the Medical wards were not without some fun and frolic. The studies from then on were conducted with a free and easy communication between all medicos with a real feel of co-education. But in the earlier two years, the girls sat in the front row under the very nose of the lecturer, or they only walked in after the lecturer had taken his place on the dais. The girls used to be nervous and considered even the mildest remarks by the boys as ragging.

One morning a very popular additional Professor asked a young medico to place his palm on the praecordial area of a young woman in bed and to report on what he felt. The young man, not knowing what he was expected to find, hesitated to remove his hand and answer. The teacher ejaculated at the dumb chap, "Are you getting a thrill, young man?" The youngster nearly dropped dead, not appreciating that the Professor was referring to a Praecordial thrill on a cardiac patient.

The Professor of Medicine was a tall, charming Englishman. He stammered a bit during his lectures. Whenever possible a junior student (clerk) was tagged on to a senior one and cases were allotted to them, for clinical studies such as history-taking, clinical examination followed by presentation of the case to the group. One morning, the Professor of Medicine, Col. Malcolmson, pointed at me and questioned about the purpose of diagnosis, especially in a case of sprue. Diagnosis is not easy in such cases and the treatment even less satisfactory. I promptly answered that a pragmatic approach would help in the treatment. He seemed satisfied and asked me if I was a senior or a junior student. It encouraged me greatly. The summer of 1930 saw me through my first clinical year. I was pleased with myself and my progress in studies.

But by the end of another six months, a great calamity struck my family. My father suddenly took ill and passed away in December 1930 at our home in Salem. That necessitated our decision to move the family to Madras. We had to sell away the house father had built and we had to shift to another place. After we had shifted, my mother fell ill and had to be hospitalised. When she recovered, my brother fell very ill with pneumonia. In those days, treatment was only symptomatic since no sulpha or penicillin had been discovered. I used to sit for hours together by the side of my brother during that illness.

The staff nurses were all British, as we were under the British Raj in the late twenties. The nurses on night duty were served with a sumptuous supper at midnight. The patients in the neighbouring women's wards could hear the clatter of forks and knives from the supper room. There are no such luxuries in the present day. Our nurses today bring some frugal snacks to satisfy their hunger at midnight.

Happily my brother recovered and came home. He was a great support to me in every way.

When all were well again, we found ourselves living in a house near the sea with a lovely view of ships on the horizon. My mother, brother and two younger sisters made up the family. I had a room for myself since I had to work for the final MBBS examination. But I could never sit at nights to read for long hours. One night, very foolishly I lighted a kerosene table lamp and placed it inside the mosquito net and comfortably lay down. In no time the top of the net caught fire. My mother and brother rushed from the next room and helped to put out the fire.

Our landlord was an indifferent type and would not comply easily with our requests to rectify any defect in the house. For example, the smoke from the kitchen on the ground floor passed through the bathroom and filled it with smoke before escaping to the roof through the chimney. Once when he was inspecting the bathroom to check if our complaint was genuine, he quickly shut him up in the bathroom so that he could appreciate the problem as the smoke entered his eyes. That brought him to his senses, and he took appropriate steps to direct the smoke to the chimney.

When the time came for me to attend the Maternity Hospital for my training in midwifery and gynaecology, we moved again to live near the hospital.

Now I was in the last year of my medical course. It was a most crucial and important year for me. I had come under the influence of the famous teacher and obstetrician Sir A.L Mudaliar. His influence was to last throughout my professional life. The final ordeal of the MBBS examination with all its several sections—written, clinical and oral—was over. I was satisfied with my all-round efforts, not only in the performance of the three major subjects but also in other disciplines.

I topped the list. Sir A.L. Mudaliar was enormously pleased with my performance. And I became eligible for six God Medals—two for the best outgoing student of the year, one each for surgery and medicine and two for being the best student in midwifery. My guru, Sir A.L.M. said that he was the first person to spot my proficiency as a student and claimed that he felt he had grown 6 inches taller after the results of the Final Examination.

Even today I recall with intense nostalgia, my student days under Sir A.L. Mudaliar. Sir A.L. breathed his last on April 15th, 1976, at the age of 86 at Madras.

With his passing away came the end of an era of great progress in obstetric teaching and training and education in general. He was the Vice-Chancellor of the Madras University for 25 years. His imposing statue stands in front of the University Library on the Marina.

To me, he had been an immense source of encouragement in my medical studies and professional life. His fatherly concern during a period of strain in my home life helped me to cross this bridge to veer to a more creative effort. My first encounter as a student with him was in the Cancer

Ward of the Women and Children Hospital, Madras. The patient who was undergoing radium treatment had developed an unpleasant local complication. When his assistant informed him about it during his ward rounds, his eyes scanned the faces of the students around the bed, till they fell, by chance, on mine and he tartly asked me, "What has happened to this patient, Doctor?" (He always addressed even an undergraduate as a "doctor" boosting up the student's moral.) I promptly replied, "Sir, maybe you have given her more radium than you meant to." At this he exclaimed "Ha, Ha! Here is the Delphian Oracle speaking." I was startled for a moment at his retort.

But soon I found myself working harder than ever to prepare for his impromptu questioning. As days spun into weeks of ward posting, I was increasingly alert with appropriate answers, whenever his finger passed along the faces of the batch students around and finally pointed at me saying, "You, come along Professor." That was in 1931-32 when I did not have the foggiest idea that I would be appointed a Professor in Osmania Medical College, Hyderabad, 20 years later.

With such great personal encouragement I came to hero-worship that great teacher. I would flutter when I heard his footsteps as he came the wards wearing his gold-laced turban and time-honoured stiffened silk suit. As he walked on with his staff and students along the colonnades of the hospital, in and out of the labour ward suites, he would refer to this long band of followers swelling from ward to ward as "Robin Hood and his Merry Men". Such a sense of humour directed at his own cost is notable indeed.

In another busy ward, where suspect and overt problem- cases were being observed, there was, on a certain morning, a woman with her foetus in the transverse lie. After Sir A.L. had appraised the case, he stared at me, a mere undergraduate and said, "Tell me, Doctor, what should I do for this lady? Decapitate her or her infant? And I shall carry out your orders". (She had a safe delivery in due course.)

During the last day of our batches posting under Sir A.L. which has remained memorable, Sir A.L., after seeing an antenatal case in bed, pulled me into a discussion saying, "The H.O. says that she is a case of twins but I say that she is one of Hydramnios. What do you say? Let us have a Battle Royal!" I answered calmly "Both of you could be right, Sir." Here I recall vividly his peculiar mannerism of parting his moustache and smoothing it down whenever he was pleased with the student's answer.

He then went on to question me on the recently published work on the anterior and posterior pituitary hormones. When I said I knew a little about it, he remarked, "Your little must be a pretty little."

The final encounter, with Sir A.L. was at the final clinical examination, qualifying for which is also unforgettable. He introduced me to Dr. Lazarus, his co- examinee and a very well-known obstetrician. To my great embarrassment I failed to diagnose a common pelvic mass. Later with an ante-natal case when I was able to clearly explain the condition of asynclitism to Col. Hingston, the Chief Examiner, he called out loudly to Sir A.L. (by his first name) in the same hall at the famous Giffard's School, saying, "This child has her head screwed on in the right place. Would you see that she gets a resident appointment?" Since I had to take two other leftover subjects, in addition to the major ones this time and still obtain the first rank, Sir A.L. is said to have remarked, "Everybody is not a Masilamani!" (thus unforgettably etching my father's name: I had lost my enlightened parent a year earlier). Sir A.L. had recommended *Kelly's Gynaecology* as a reference book when I stood first in the Prize examination. It is still very useful to me.

In June 1932 at the Convocation at the fabulous Senate House on the Marina, when the Chancellor of the Madras University, the Governor, was handed over the pile of medals in their special boxes to be awarded to me, they slipped from his hands. The time spent in collecting them from the floor of the dais was one of great ovation shown by the stamping of shoes by the large crowd of new graduates from the galleries around, while I must have stood bashful and mute on the dais!

As the best outgoing student of the year I was expected to reply to the toast at the College Day function. I became very nervous and took ill. And I could not even attend the Annual celebration of the college at the famous Museum Theatre.

Magnanimously enough Sir A.L. insisted on my going away for a fortnight's holiday at Ooty after graduation. Before a house job, I had that short period of change and got myself refreshed.

Later, during his ward rounds, Sir A.L. met Col. Hingston who recalled my able performance with him and remarked that "this student had swept the deck", which is an expression used by the navy to applaud a good sailor.

The foremost woman gynaecologists of India, Dr. Lazarus of Madras and Dr. Jhirad of Bombay also had conducted the clinical as co-examiners for the year. I met them later for their mature advice which I followed. They advised my attending the hospitals for rectal surgery and bladder conditions when I got to London.

Immediately after the results were declared, the Director of Medical Services came to my house to announce to my mother and brother that I had topped the list in every subject and in two Special Prize Examinations and was to be awarded six gold medals. I happened to be away from the house at that time. And those dear ones who received this news were jubiliant. We could hardly sleep for excitement. We were all filled with joy and gratitude to the invisible spiritual presence for all these blessings.

4
PROFESSIONAL LIFE

From a very young age, I wanted to be a doctor. As I passed from the stage of studying at home before going to school, and from school to the university, my eagerness to enter the medical field grew. Thus Scottish principal of the School at Madras was so happy, she felt as if I had gone straight from her school into the medical profession. Evidently she had followed my career all the way through with real concern and interest. Her great personal interest in me was evident even when I met her after her retirement many years later at Edinburgh.

During the early years of my medical study, the Professor of Anatomy recognised that I not only needed a studentship, but I deserved it on account of my proficiency in my daily studies. He also helped me to secure studentship from the college, which I had not secured at the start of my medical studies.

On my return after holidaying at Ooty, I reported to duty as an HO under Sir A.L. Mudaliar. I had to work in the labour ward watching the action of the uterus in the first stage. It had to be done astutely by periodically palpipating the uterine contractions, their indentibility, duration and frequency. In those days, there were no appliances, like a tocograph or its recording on a kymograph. Since there were no specialised anesthetists in 1930, I learnt to do the 'Spinal' with the patient in the sitting position.

Once I even gave an "Open Ether" for an operation of simple mastectomy performed by Col. Thomas on an American lady. The patient came round safely. It was indeed a thrill for a fresh House Officer! Later, I helped him with an appendisectory on a British woman. He thanked me and handed me an envelope containing a sum of Rs. 15, my first operating fee.

After completing one year as House Officer in 1933, I was to work on a Research Studentship from the Madras University, under Sir A.L.Mudaliar on projects concerning puerperal infections and making rough observations on effects of hormones on paddy or wheat grown in flowerpots and watered with diluted pregnancy urine.

In the meantime, demands of my own family were growing. I answered an advertisement with success. I took up a clinical post in Residency Hospital, Secunderabad in 1934. This was to last for only one year. I was called to Simla for an interview for selection as a Training Reserve in the WMS.

After a short spell at Simla, I spent a year at Agra doing a mix-up of medico-legal along with clinical work. The maternity work was indeed challenging. I encountered severely deformed osteomalacic pelves and invariably also discovered that some Jat women at the Agra O.P. sometimes carried even houseflies inside their garments!

Shortly after this, an invitation from the Residency Hospital at Secunderbad took me back to my own family once again.

After resuming my work at Secunderabad , I managed to help in the delivery of my sister safely. But I blundered later in the case of a VIP by waiting too long to deliver the after coming head. My immediate senior who looked after me during my ante-natal period, helped me during delivery when I was 39 by classical caesarean section.

Later I was able successfully to tyeat a case of neonatal gastro-enteritis with injections of subcutaneous normal saline with glucose. The mother was Scottish. It brought me a salutary gift which enabled me to go abroad for further training.

It that stage of my career, I was more keen on the diagnostic rather than on surgical, aspects in my gynaecological working, on knowing of the patients gynaecological history, before clinical examination. It guided me to spot various types of tubal pregnancies and bring them to the operation table. I earned a commendation from a senior CMO-COP. MacGregor, for carrying out a plastic repair operation on a case of genital prolapse. After that I confidently proceeded to repair my first case of vesicovaginal fistula (V.V.F.) at the same hospital, paying attention to personal post-operative care, especially to keep the catheter clear for bladder drainage.

After treating 18 such cases, I felt I could write my commentary on these in preparation for the MRCOG examination. Setting my mind on that higher examination, I intended to study a corresponding number of obstetric cases. My clinical study consisted of the relation of the foetal head to the pelvis, and the part played by the uterine action besides the patient's own psychological background. I procured books on "How to write Medical Papers".

I was given a chance of doing a Hysterectomy, a major type in gynaecological surgery for a fibroid uterus, with my senior assisting me. But to my great distress, the woman developed "gas gangrene" infection; which did not respond to the appropriate anti-serum, and pencillin was not available in 1934-35.

I thought I should sit first for the MD examination of the Madras University, before going abroad. A real student's life for me began. In 1945, I wrote to Sir A.L. Mudaliar for advice and visited him at the Alma Mater after a lapse of 11 years. Eventually in 1946, eight of us appeared for the MD examination, but "none passed".

Next year, I was well advised to spend two months at the teaching hospital at Madras. So I became a real graduate student again! I took my small Austin-7 car to Madras, and drove it mainly to the University Library on the Marina, whereI spent time in reviewing the original epoch-making contributions of 1933-34 in my field.

One morning at Professor Thomas's theatre annexe I saw what appeared to me I saw a case of ruptured tubal pregnancy, and the patient being sent to the ward for the treatment of an acute Pelvic Inflamatory Disease (PID). In the afternoon I boldly suggested to Col. Thomas that it was most probably a case of ruptured tubal gestation. He came and saw the patient and finding her breathless, pale and in the wrong position, he at once, ordered to prepare the theatre. He asked me to wash up to help him. Of course, the pregant tube had ruptured filling the abdomen with blood. I was happy about my correct diagnosis that saved the patient.

As I had already arranged I did go abroad and completed the RCOG and RCSE examinations within 13 months. It was quite easy for I had worked for 6 years on my own at the Secunderabad hospital with the help of journals and books, consulting them on problem cases.

Soon after my arrival at London in June 1947, I flew to Dublin to attend the bi-centenary celebrations of the Rotunda Hospital. There I saw and heard the Doyens of my field like Munro Kerr of Glasgow, Isidor Rubin from the USA. I also saw slum children with no shoes and in tattered trousers. I was moved by their poverty. No wonder the Irish people evinced so much sympathy for India's freedom struggle.

From Dublin when I got to London, I registered at the Soho hospital for Prof. Browne's group discussions. Professor Browne had a retired from the University of London. I had studied his book *Ante-natal Care* and drew him into discussion on certain moot points in his treatise. He was surprised at my boldness. Through Prof. Browne I visited Prof. Moir at his Nuffield Department, Oxford. Though I could not observe him repair a fistula, he drew diagrams to illustrate his technique which was modifying Marian Sims' original "edgeparing" into a "flat splitting" method.

Then I went to Edinburgh for the Fellowship course. I enjoyed the talks given by the Registrars and the Consultants on the new approaches in general surgery such as preparation of the patient, the surgical technique, complications, if any and their management. But I enjoyed the Ward rounds best of all. I did not witness any surgical demonstrations. It was all talking, discussing and writing on blackboards.

Sir S.Y. Simpson was the first to seek in 1847, relief of pain for women in child birth. Besides a large institution for learning and research at Edinburgh named after Simpson, there stands a large marble monument on the main road. Mr. Liston at Simpson Memorial Pavilion (SMP) was keen on demonstrating his method of assessing the capacity of the sub-public arch. SMP was meant for obstetric and gynaecological cases. Naturally, I was interested to follow the nurses' work as well on their postpartum rounds to note how they attended to the perineal sutures.

Mr. Sterling allowed post-graduates in his operation theatre (OT) to conduct pelvic examinations only after the women had been anaesthetised. Once while diagnosing a certain case I differed from Mr. Sterling. After the operation, he came up and asked me about my work at Hyderabad.

The Fellowship examinations at Edinburgh were very rewarding. I was also happy to gather extra information.

The gynaecological clinicals were held at SMP. They were based on when the ovarian function became evident, i.e., by what age. They showed cases of Ovarian tumours etc., and questioned me on operative gynaecology.

Then I went to Liverpool, where I stayed with the medical students of the Maternity Hospital. I observed several cases of breech deliveries by the Burns-Marshall Technique demonstrated by the Senior Registrar, Mr. Cox. I also read Munro Kerr's "Operative Obstetrics." I also saw the use of nitrous oxide which was first introduced at Liverpool for obstetric analgesia for general use by Prof. Jeffcoate.

Mr. Malpas took me to a London County Council hospital (now abolished under National Health Scheme) to watch him do a caesarean section. Elsewhere, I had watched him operate on a tubo-ovarian mass. One morning at the Liverpool Women's Hospital, I had the opportunity to guide Mr. Marshall to do a flap splitting repair for a VVF while actually I had gone to Liverpool to watch him do a ceasarean section under local anaesthesia.

Later in London I attended a two-week refresher course at Chelsea Hospital for the Membership candidates. After that, I sat for the MRCOG examination. I was confident that had I done well, both at the papers and clinicals held in 1948. Mr. Malpas, who noted my euphoria, advised me not to believe the examiner chaps! I had met Mr. Malpas on other occasions. It was interesting to learn that he succeeded me as a locum consultant at Thurso in Scotland after I had done the assignment myself in 1972.

I sailed for Bombay the very next day after the examinations were over. My result was sent to me on the high seas, by wireless by my friend Dr. Joan.

On my return from UK, I was posted to Victoria Zenana Hospital at Hyderabad as a general surgeon. On a certain day when a patient for exploratory Laparotomy was on the table, I sent for anaesthetic help. A 19-year-old-girl undergoing compounder's training turned up, literally with a 'rag and bottle'. I was aghast. Next morning I brought with me a book called *A Surgeon's Guide to Local Anaesthesia* by Corlette, which I had picked up at Edinburgh. I had anticipated the poor anaesthesia situation at Hyderabad.

I recalled these incidents when I chanced to see in 1948 Corlette's book edited in 1946. I had carefully adopted at the Victoria Zenana Hospital Theatre the steps for premedication. The preparation of the procaine hydrochloride solution was made for each case afresh. Of course, Corlette had described in his book minute details, the injection procedure for various operations. I tried to contact his son in 1968 while attending the V International Congress of Obstetricians and Gynaecologists at Sydney, to tell him of my very gratifying success at Hyderabad with local anaesthesia. But Corlette Junior had already fallen for general anaesthesia and I could not get him to listen to my amazing experience from his father's book, with over 300 cases of gastrectomy. I could do also vaginal plastics and even vaginal hysterectomy with its clear-cut steps. At the Victoria Zenana Hospital, I had performed a partial gastrectomy for a case of cancer of the lesser curvature on a frail woman, a nephrectomy for a case of obstructive hydronephrosis, and a bilateral uretero-colic implantation on an elderly woman suffering from advanced cancer, cervix with involvement of the bladder leading to continuous incontinence of urine.

With an assistant eager to learn the advantages of the "bloodless and shockless" surgery under local anaesthesia, I had done a hemicolectomy successfully. It was presented at a clinical conference at OGH to the pleasant surprise of some of my colleagues in the field. Some, of course, missed the benefits of the surgery done under local anaesthesia and preferred instead, to discuss the pathology of the mass in the caecum for which the operation had been indicated.

During his visit of the Victoria Zenana to the Hospital in 1949, the new Director-General critically evaluated the work being done then. He noticed surgery being performed under local anaesthesia with little or no staff. He appreciated the effort and transferred me as an Obstetrician and Gynaecologist to OGH, and also as a part-time Professor. I became a full-fledged Professor only after four years, as required by the University rules.

Corlette's book is a very valuable surgeon's guide for any enthusiastic young surgeon working especially in nursing homes in the more populated and less affluent areas. In fact, a senior gynaecologist, Dr. Janaki, is working almost single-handed, taking patients for hysterectomies less suitable for general anesthesia. Less enterprising men, though starting with local anasthesia, are soon found to falling for the comfort of someone else administering general anesthesia. But the dictum, if L.A. is safe for the very ill patient, it is certainly safe for the less ill, holds good even today.

There was an unexpected still-birth at OGH with the mother running only a low temperature. I got the Nursing Sister's cooperation for an autopsy on the foetus on a glass top-trolley in the Labour Room annexe with the cooperation of the pathologist (Dr. Aktar Hussain). The microsections showed malarial parasites in the liver and in the placenta. This finding led the Professor of Pathology, Dr. Gopala Rao, to realise it as an uncommon finding and he felt the need to publish this finding widely. The finding of a malarial parasite in the liver of the dead-born foetus and the placenta urged me to start on a study of the causes of foetal deaths. The word "perinatal" referring to such deaths was coined by Pellar in 1948.

We entertained a hard-working house-officer in the unit who was found eligible for a research fellowship from the Indian Council of Medical Research (ICMR) for a project from 1953 to 1958. The study was planned (1) to collect data of previous deliveries and the case history of the woman, (2) circumstances of misdelivery and (3) autopsy findings from the foetus. Thus the work was commenced under the guidance of the Professor of Pathology.

While the study was well under its way, I planned to go abroad, mainly to meet Dr. Edith Potter of Chicago, who had been working on prenatal deaths.

In fact, I was expecting a Colombo Plan award for a short study tour in Canada with the possibility of visiting Dr. Edith Potter at USA. She had already published her material on foetal deaths in the form of books from Chicago.

Another reason was the first World Conference on Medical Education being held in London in 1953. Moreover, my husband was already in UK. An added attraction was the Coronation of Queen Elizabeth II taking place shortly in London.

In the meantime, the large number of maternity cases which I had been handling, provoked me to undertake a study of post-partum blood loss. Liquid ergot was being dispensed orally to the mother at variable periods during the third stage of labour. When a placental stage haemorrhage takes place, there is a greater danger of maternal mortality. Special care is needed to prevent blood loss during post-portum stage after the expulsion of the placenta. The simple harmless Brandt-Andrews method of counter-traction of the uterus upwards through the abdominal wall and cord traction from below was demonstrated to produce hardly any more loss than six ounces of blood as against the test 10-20 ounces considered normal for healthy women of the West. That much loss would be a disaster for our aenemic and ill-nourished women.

Reviewing the literature and detailing the physiology of placental stage caused a widespread interest in the method of checking such blood loss. This led to a statistical study and preparation of histograms. The Indian Congress of Obstetricians and Gynaecologists was due to be held at Bombay in 1955. It was decided to communicate our study conducted with the help of co-authors like Dr. Kameshwari and others at the conference. At the actual presentation of the paper, a lot of interest was evinced in the method of measuring the blood loss by a simple method.

A dozen questions received on slips of paper were answered effectively. This study appeared outstanding as against another from elsewhere dealing with actual cases of great blood loss causing collapse; and heroic attempts at resuscitation with or without success. These two papers created an awareness of the so-called "P.P.H.". The interest shown by the audience was gratifying.

It was here that the renowned Dr. Hilda Lazarus accosted me about our work. From then on we became very friendly. She came to stay with me for her MLC meetings and I stayed with her during the Andhra Medical College, MD examinations. She was devout, hard-working and remarkably energetic. She drove her own Morris car even in her advancing years. The appreciative DMS Major Rao posted her as an Honorary Post-Graduate Professor at Andhra Medical College, Vishakhapatnam. She was the Director of Christian Medical College at Vellore and later Chief of the Womens' Wing of the Army Officers during the war. It was she who fought fiercely to get a couple engaged in medical service, posted in the same place.

Major K.N. Rao, the then DMS, was a great disciple and admirer of Dr. Lazarus. He got her final official posting at AMC Vizag where she had also her own home. Sir A.L. Mudaliar and Dr. H.M. Lazarus deserve to be honoured better and made known to the younger generations.

First World Conference on Medical Education

Since I was already at London in 1953, the State Government deputed me to attend this epoch-making First World Conference on Medical Education. My husband and I attended all the sessions. I benefitted a lot from listening to the views and experiences of the medical teachers from various parts of the world. The President, Sir David Livingston, the Vice Chancellor of the Oxford University, said in his opening remarks that "Education is not pouring in but drawing out" form youthful humanity. I was greatly impressed by those words and had since passed them on to my colleagues at home. Sir A.L. Mudaliar was one of the four Vice-Presidents of the Congress. It was inspiring to meet him between the sessions. Some of the main topics discussed were: choice of a medical student, content of the curriculum, and teaching the teacher.

Since then, several Congresses had followed at different places on changing trends of health problems. But how far have we progressed in these 35 years? We may soar to supersonic heights, but we should not pull away from our roots, from the basic sciences, for they are the backbone of medicine.

In October 1953, I flew to Canada. The Colombo Plan award came right on time. In Toronto, I observed the methods of training under-graduates and holding clinical conferences for graduate students. I spoke about my experience with 100 cases of VVF. After that I was invited to operate on two such cases of fistulae, one of whom had suffered from incontinence for 26 years.

At Montreal, I got interested in Cytology. Above all this, I had the rare opportunity of visiting Professor Hans Selye, a great endrocri-nologist, in his famous Laboratory.

From Canada, I along with Dr.K. from Madras got to the United States by courtesy of the Colombo Plan Officials. We first went to Michigan. Dr. Miller of the Michigan University was also greatly interested in fistulae work. In Chicago, we watched Dr. Allan doing a vaginal myomectomy very deftly through the Pouch of Douglas by an admirable technique. The Boston Lying-in Hospital was a great centre of obstetricians with men like Dr. Delee, of the "accidental Haemorrhage" fame. Dieckmann advocated using pitocin in the prevention of post partum.

My main purpose in visiting the USA, apart from sight-seeing was to meet Dr. Edith Potter at Chicago and haemorrhage and acquaint myself with her work. Meeting her in her famous laboratory was a great privilege. She had been working on Perinatal Mortality and had published books on her findings. She was a slim person totally devoted to research. She welcomed me and Dr. K. and talked about her interest for a while. And then she invited us to her home for supper. She was 'Mrs. Meyer' at home, wife of a sculptor.

She gave us a lovely supper. After supper, Dr. K. and I offered to help and wash up the used dishes. In spite of Mrs. Meyer's 'no' s' we walked into the kitchen and found seven days' junk piled up in the kitchen sink. We settled down to wash up the whole lot. We saw for ourselves how the working women abroad had to fend for themselves in household chores in spite of exemplary academic preoccupations.

Later, when she paid a visit to India, she is said to have asked for me at Madras to talk over the perinatal work in India. I was disappointed to have missed meeting her again. She had made several significant contributions on perinatal mortality. These have led to the awareness of the problem of foetal loss.

We went to Switzerland, Heidelberg (Germany) and Amsterdam to watch gynae operations by Dutch surgeons. From there we got to Jerusalem where we saw thick skin graft being used for the creation of the absent vagina.

Eventually, I returned home to continue the perinatal mortality work with greater zeal with the help of Dr. Gopal Rao and Dr. Zakia, ICMR Fellow. When we completed the work in 1958, the Indian Council of Medical Research published a brochure on this work. This is the first standard work undertaken, still used as a reference across this country and outside.

Transverse Muscular Incision

By 1955, transverse muscle cutting incision for pelvic surgery was adopted in many units. The first woman on whom I used this is still available for a check. She had a caesarean section done in 1955. The abdomen wall showed good integrity of the scar and its advantages could be generally appreciated.

Alridge's Intrafascial Hysterectomy

This gave a deep supple vagina and it excluded injuries to the bladder, rectum and the ureters. I saw Dr. Alridge using his technique in the famous Sims Theatre in the Women's Hospital, New York. The theatre was dedicated to the memory of Sims.

Boston

Dr. Meigs was operating at Boston Mass General Hospital. His technique demanded a precise knowledge of anatomy of pelvic tissues and proved a good exercise for the future operating gynaecologists. I have trained and advocated to several of them to use this incision. Dr. Meigs was very personable and seemed eager to share his knowledge, experience and exquisite technique.

Department Shifted to GMH

In 1957, we had to close admissions to the Obstetric and Gynaecological wards to OGH and Ob/Gyn. I had a typist and a separate office from that of the Superintendent. The typist saw 92 letters signed by me to the Superintendent on the requirements of the teaching facilities.

Suddenly, a certain other equally qualified officer claiming seniority over me, was posted in my place. I had to leave for Niloufer Hospital. I operated and trained many assistants. During my year at Niloufer Hospital, Dr. Sundari was my capable assistant and Dr. Sita was my Registrar.

In July 1958, Mrs. Indira Gandhi, the then Minister for Information and Broadcasting, was to inaugurate the Post-Graduate Institute at Government Maternity Hospital. So I was sent back to Government Maternity Hospital to try to suitably upgrade the hospital for post-graduate teaching and training. Departments of Pathology, Histopathology, Biochemistry, Animal House etc. had to be developed.

While these efforts were in progress, the day-to-day clinical work had to go on. The never-ending stream of vesico-vaginal fistulae cases were waiting for their repair and relief. Usually, these cases were being operated upon in Trendelenburg position working with two assistants standing on either side while the lower extremities of the patient were fixed comfortably on either side and strapped to the crutches. The surgeon sat on a stool with a back rest (where available) during this usually long surgery and high enough to peer into the vaginal passage from the correct level in order to visualise the fistula. After spotting the exact place of the vesico-vaginal fistula, each assistant would have to retract both the lateral vaginal walls during the period when almost always the assistants would not be able to look into the details of the dissection at all. I often quoted to the assistants from Milton's sonnet, "They also serve who only stand and wait".

The surgeon would require certain qualities for successfully carrying out the repair. These are presumed to be "the hands of a woman, the eyes of a hawk, patience of an angel and the determination of a devil."

In 1961, I again went abroad. I was to present a paper called "Cervical Phimosis" at the third International Congress of Obstetricians and Gynaecologists at Vienna. At the Cafeteria in the Holberg Palace where the Congress was being held, Renuka and I met Professor Norman Morris of the London University and his charming wife. Much later, in 1972, I read a paper on "Certain Aspects of Intersexuality" at the Charing Cross Hospital with Prof. Morris in the chair. It was there that I learnt that Queen Elizabeth I (presumably) belonged to the type known as "Testicular Feminisation Syndrome" with buxom busts associated with only a shallow vagina with testicles in unusual positions.

I had gone this time to UK with Renuka to get her into the Newnham College. Afterwards, I often commuted between Oxford and Cambridge and Prof. J.C. Moir frequently. I had the drafts of my two papers on vesico-vaginal fistuale and he had very graciously suggested some changes on these. While I spent time in the Nuffield Department at Oxford, his secretary had typed them for me. The Nuffield Colleagues were surprised at my having two papers in the *British Journal of Obstetrics and Gynaecology* in the same year.

Prof. Moir visited Hyderabad in 1966 for a fortnight's stay as a Government guest. We had a plethora of conferences and lectures besides demonstrations on the repair of fistulae. His book contained clinical photos of our cases of fistulae and a reference to my pull-through operation in an inoperable case of vesico-vaginal fistulae with Dr. Svarna's help. When Prof. Moir learnt about that successful operation, he exclaimed happily, "splendid."

Dr. K. and I flew to Sydney for the fifth International Congress of Obstetricians and Gynaecologists in 1968. My paper on "Vesicular changes in the Villi of 1000 abortuses" was accepted. I received £100 for it in Australian money from the Congress on its acceptance.

After the Congress at Sydney, Dr. K. and I had got to Tokyo with a hop at Manila. At the Tokyo University, we saw a patient in labour being monitored by a sound system as well as a light flicker on a revolving kymograph—a great advance indeed—while the patient was being observed from behind a glass screen. We had the honour of having lunch with the professor of the Tokyo University in the Senate House. Since the main subject of the Congress was trophoblastic diseases the Professor had spoken about his experiences with this condition he had frequently found in the Far East.

While at Sydney during a sight-seeing trip in aerial bus over the pine mountains across a deep gorge, I met Dr. Kathleen Drury from Hongkong. This was followed by her four-day visit to Hyderabad. She talked to our post-graduates a GMH. It was while on that visit to Sydney that my misplaced passport was brought to me miles away by an Australian police motorcycle rider.

After I was operated at Moorfields for retinal detachment (and total blindness) Dr. Katheen Drury took me to her lovely home at Surrey and helped me with my convalescence.

In 1971, I accompanied a diabetic friend as a companion to London for treatment for loss of sight. It was during this period that I contacted a Medical Assistant (from Hyderabad) working at Eastbourne, looking for a locum reliever. This gave me opportunity to do such assignments starting with Eastbourne. I enjoyed doing gynaecological operations and dealing with other obstetric problems. It was here that I had perforated the uterus while doing a termination of pregnancy and had to open the anterior pouch to find the laceration very near the cornu of the uterus. Another time, the dilator had passed through the posterior wall. I opened the Pouch of Douglas to find the site of perforation and repaired it. Dr. Linacre, the senior anaesthetist who took care of both the patients, gave me a book on "England" at parting, pleased, perhaps, with an Indian woman gynaecologist recognising the mishaps and repairing them aptly.

Next I got a consultant locum at Egdeware Hospital in Northern London where I could do external versions with great ease much to the amazement of the black and the white nurses. Once a case of vaginal myomatous polyp of a large firm fibroid was posted for operation. She was an unduly fat woman. On the table one could see a long vertical and three deep and ugly scars running across her abdomen. I washed up and watched the blunt end of the projecting polypus through the vagina. Some onlookers casually wondered who could have operated on her previously.

A smart theatre technician who had watched various hands welding the surgical knife with varying skills promptly quoted the name of "Sainsbury" who are butchers with shops all over UK. These deep transverse abdominal scars on the fat woman certainly could have inspired the theatre technician to utter the near appropriate comparison.

I also met at Egdeware Mr. Picton who had reviewed more recently the management of placental stage by cord traction and counter-traction of the uterus to minimise blood loss during childbirth. I had a chat with him about the benefits of the method, especially for our anaemic women. I attended also a paediatric clinical conference at this teaching hospital.

Then I moved on to North of Scotland, to Dunbar Hospital at Thurso. The nursing staff was extremely cordial. One sister suggested getting my own instrument from London and got the hospital to pay for the postage charges for it. The anesthetists were local practitioners. I carried out the usual gynaecological surgery and inductions of labour with no ado.

I worked at Thurso a whole month attending antenatal clinics at Wick which was the terminus of the railway line to the North. There I did several amniotic punctures for studying the amniotic fluid for various conditions. The Dunbar staff and hospital were extremely friendly and I was looked after very well. On one occasion, Sir Hector MacLennan, former President of the RCOG, visited me and had lunch with me at Dunbar.

I got a short locum at Chasefarm Hospital in London where the Maylard's Incision for hysterectomy and the subcuticular closing with no dressing were greatly appreciated by the theatre staff.

My next locum was at S.E. Kent, covering hospitals at Folkestone, Dover and Wilsburough. I had three calls one night from each of these hospitals by turn, for doing emergency caesarean sections. I got through these calls very satisfactorily. At Dover I was able to convert even a proud Nurse Tutor to my way of teaching.

I had Moir's book on VVF in my hand during a coffee break one morning. My Irish House Officer, who liked my technique and learnt to do a vaginal hysterectomy from me, was also present. Sister Tutor Steele asked me to instruct her pupils on anatomy of the pelvic floor as her regular consultant kept putting off the topic for obvious reasons. I gave her a diagram. Next morning, her drawing on the blackboard was rather distorted, although she had copied from mine. The direction of the vagina was pointing down instead of being parallel to the pelvic brim. I gently corrected the mistake. The girls were then asked to go with me to learn from my clinical rounds.

Sister Miles at Dover assisted me at a caesarean Section. I had almost whistled the foetus out slowly. She shouted, "lovely". This was the usual method which I had taught my post-graduate students at the Institute at Hyderabad.

I left a typed note with each of the case notes after a Maylard's incision, urging the next obstetrician to repeat it without fear of the scar becoming flabby and giving rise to a hernia. Most hospitals where I worked received the instruction well.

During my locum assignment at the Bechanham Maternity Hospital, I was called in to see a woman who had just delivered her infant and collapsed. On reaching the delivery table I found that she had suffered an acute inversion of the puerperal uterus. I got the Bengali Registrar to wash up and place her forearm and fist into the vagina of the partunient and give her a warm saline douche with the nozzle while her fat forearm closed the vaginal orifice tightly. The vaginal distension caused by the saline helped and the patient recovered from the shock.

In between my locum appointments, I was invited by Dr. Anita Hooper and her husband. They made me very comfortable. Anita was a medical graduate from Bombay. Her parents lived in Hyderabad and both were medical practitioners. Anita had married her colleague, an Englishman, and had three fine children. She had a hearty laugh. She enjoyed listening to my experiences at the various hospitals as a locum consultant.

Victoria Zenana Hospital in 1948

I returned to Hyderabad from abroad in 1948, satisfied that I had received the best possible training in obstetrics and gynaecology.

I reported at the Residency Hospital, Secunderabad, to the first Indian Chief Medical Officer, Dr. Khatri in 1948. He not only applauded my successes at Edinburgh and London but said I deserved a promotion in the service. As there was no such provision at the King Edward Memorial Hospital, he offered to get me posted to the Hyderabad service at the Victoria Zenana Hospital in the old city.

At Dr. Khatri's suggestion, I visited the Victoria Zenana Hospital with a friend. I was stunned at the conditions prevailing in that hospital. Each post-partum case was allowed an attendant to sleep under her bed. The new borns were each in a wire cage tacked on to the foot of the mother's bed. I learnt that this protective measure was adopted after a cat had mauled a new-born infant. Families of the delivered would sit on each bed handling the new born as they wished. I decided immediately to work there in order to change these conditions.

When I went there to report on duty and accept responsibility, the beds (over 400) were distributed as maternity, surgical, paediatric and gynaecological, each under the care of a civil surgeon, each a LRCPS. The Superintendent was a Bombay medical graduate with a Diploma in midwifery from London. She apparently did only administration, checked on statistics etc., but did no clinical work. The medical officer in charge of the maternity wards used a finger stall in place of a full-hand glove on her left hand to make internal examination of women in labour.

The Senior Officer mostly sat in her office knitting a woollen garment. The statistics for the year showed 34 Caesarean sections, 348 D & Cs and 400 operations on the vein (IV injections mostly of calcium gluconate).

There was no anaesthetist in that hospital. Luckily I had brought with me from Edinburgh a book called *A Surgeon's Guide to Local Anaesthesia* by Corlette.

The conditions in the operation theatre were appalling. A young girl, a compounder, turned up with her rag and bottle when I had an emergency on the table. The instruments were boiled in a large aluminium vessel placed on a coal oven. As for the surgical instruments, there was no "toothed" forceps at all. I saw one of my new colleagues struggling inside the pelvis with a short toothless forceps, while doing a sub-total hysterectomy. Fortunately, I found three pairs of slender toothed forceps of a special category which could be clamped to hold a layer of tissues delicately. These proved useful in repairing VVF when later I started this type of work. The main redeeming feature was a willing and smiling Theatre Sister who was ready enough to accept the need to improve the conditions.

I must not forget to state that along the series of photographs in the main office of the previous Superintendents who had run this hospital during the British Raj was a picture of Dr. W.M.S. Maybe she was mainly a physician for the theatre was primitive. All the Assistants were of a lower grade qualification with little or no training.

The Theatre Sister and I soon set about ransacking the Hospital Stores for more scientific equipment. Hardly any could be found. There was no IV sets for giving patients any life-saving drips or dispensing blood. Any observer could see the need for acquiring the items useful to modernise the operation theatre as much as possible. This would be within the power of the Superintendent.

As per the advice of Corlette, I had nine grains of procaine hydrochloride in each gelatine capsule prepared at the dispensary and this was dropped into four ounces of N. saline in a heavy glass container (which I still possess after all these 40 years) with my own syringes and needles. I began performing gynaecological operations with the help of the very cooperative Operation Theatre Sister. She and I, with the help of an ayah, performed the first case of abdominal surgery. The patient received glucose from a test tube with a tapering end attached to an 18" rubber tubing flowing into her vein.

When I asked for at least a medical graduate to help me in this work, one young person was sent with a record of 15 hysterectomies to her credit. She walked in with a single layer of cloth mask covering her head up to her chin with an opening for her eyes. Her eyes, I saw, were full of mascara. When I drew her attention to that, tears ran down her cheeks, making a mess of her face with the dissolution of the black stuff.

She was the single gynaecologist using Local Anaesthesia in her small private theatre. She proved alert, too enthusiastic, in fact, and set about to assist me and learn the new techniques of local anaesthesia and its colossal benefits to the poor or the near-poor.

A large number of cases of gynaecological surgery lower class staff of the hospital. It changed the whole atmosphere in the hospital. Major General Bhatia, the Surgeon-General-cum-Medical Secretary, paid a visit to the Victoria Zenana Hospital. During his official rounds, he noticed in the post-operative ward of Surgical Unit (a four-bed ward) a case of nephrectomy, one of partial gastrectomy and one inoperable case of cancer cervix with incontinence besides a patient who has had a total hysterectomy. He was interested to know about the anaesthetic situation and I told him that I had undertaken all those surgical adventures purely under local anaesthesia.

The visit by the Surgeon-General resulted in my transfer to the Osmania General Hospital as a part-time Lecturer and to be in charge of 150 beds of Obstetric and Gynaecology. At the Osmania General Hospital, an additional facility was the availability for consultation of heads of other disciplines.

I continued to use the local anaesthesia technique even on the wives of colleagues when any surgery was indicated. I was happy to have the privilege of teaching the undergraduates. Some of them became Professors in their own right, or medicines, anatomy and made a name for themselves.

To start with, the men colleagues at Osmania General Hospital had their own misgivings about how I would fare in my work and what my general attitude and behaviour would be. Soon they knew that I meant business and that made all the difference to my footing at Osmania General Hospital.

The clinical load of 150 cases was distributed in three large wards with one small room allotted for the staff. The assistants did not necessarily have very high training; but had been qualified in colleges elsewhere. Anyway I did not express my disappointment at their indifference to the importance of basic sciences. I managed to correct them with patience. Often, I had to take mild tranquillizers to keep my cool.

Mrs. Indira Gandhi Inaugurates Post-Graudate Study

In 1957, Victoria Zenana Hospital was named the Government Maternity Hospital and the undergraduates were directed there for their clinical posting. The dynamic DMS of A.P. Medical Services, Major K.N. Rao, also the Head of the Teaching Institutions, with great forethought, had brought me back from Niloufer Hospital to the Government Maternity Hospital which was now to be upgraded into a post-graduate teaching hospital. I began to organise, with the help of the staff, a pathological museum of all the operated specimens, clearly showing the essential pathology for which they were removed. Those specimens were seen and appreciated by eminent visitors from abroad, among them, a famous cancer surgeon of London, Sir Stanford Cade, and a surgeon from Australia. Prof. Sir Stanford jokingly remarked that our pathological museum was better than the one in the Royal College of O/G at London! That was an occasion for extra jubilation.

The Biochemistry Laboratory was picking up more and more work and reporting on the various blood tests. The Biochemist was full of enthusiasm and I visited his department twice or thrice a week to further his enthusiasm. There were charts in the lecture hall on methods of examination, resuscitation of the new born etc., in the Nurses School. An animal-house was set up, as an essential part of hormonal studies, and rats and rabbits were procured and housed comfortably and their feeding was allocated to a reliable technician.

A small Radiology Section was already there. The Radiologist was insistently demanding a larger unit of the X-ray machine for the safety of the foetus during pregnancy and the gonads when the women had to be radiographed. However, with the emergence of ultrasonography, radiography for studying abnormal conditions in obstetrics and gynaecology was discarded.

The Blood Bank had been established just a few years earlier. When everything was well-set, I informed the DMS that we were ready for Mrs. Gandhi's visit to inaugurate the old hospital and the Post-Graduate Institute. I contacted Mrs. Heda, an M.P.'s wife, to ask if she would help us to invite Mrs. Gandhi for breakfast before the inauguration with some of the senior hospital staff and the Matron. When I got the information, the invitations were sent out.

The auspicious day dawned in July 1958. It had rained heavily the previous night, which had left the coloured shamianas in shambles. I apologised to Mr. Gandhi for the havoc caused by the rain. But she said that she welcomed rain which is greatly needed in every part of our country. I received her as she arrived with Mrs. Heda. She was dressed in a pale salmon pink Orissa sari with her head covered. We took her into the breakfast room. We served her with fruit juice and some cereals in case the South Indian *idli* and *vadai* which were to follow were too pungent for her. She graciously ate everything in small quantities.

Then we went to the present Seminar Room. The Chief Minister, Dr. Sanjeeva Reddy was there to grace the occasion. I introduced her to the audience. She spoke about hospitals in general and training of specialists in particular. She also referred to her dear mother Kamala's illness in 1933 at a Swiss hospital and the devotion with which nurses there, looked after patients under their care.

After the inauguration, she was taken to unveil the marble plaque on the wall of the Seminar Hall. Cameras clicked. As she was about to leave, she gave her autograph to those who requested for it.

This was to be followed by our making use of the newly found facilities to teach and train post-graduates for D.G.O. and the M.D. degree to ensure safe deliveries of robust infants.

Cases of Unusual Interest Handled at Hyderabad (1948 - 1967)

On 15th of August, 1948, I was called away from the Independence day parade at Secunderabad to the labour ward only to see a full-time pregnant woman in spurious labour. The foetus was felt under the abdominal wall. Great foetal distress was appreciable. A trolley was pulled up and the patient rushed on to the table. On opening the abdomen, a well-formed foetus was found, among the mother's bowels. The foetus was delivered quickly for resuscitation. Any attempt to separate the placenta from the large vessels seemed unwise. The cord was cut close to the placenta and tied and the abdomen was closed. It was reopened at the woman's insistence, two months later. The placenta came away easily and the original site of pregnancy and its rupture was traced to the left cornu of the uterus. Hysterectomy was done and the specimen was preserved.

An eighteen-year-old girl with was a case of crypomenorrhea was opened through a narrow blind vagina and was drained of stale blood. Later an artificial vagina was created which proved adequate for marital functions. When she got married and became pregnant, she delivered a full-time infant by caesarean section.

A woman of 26 was brought for consultation, with her uterus firm and uniformly enlarged to the size of 26 weeks of gestation. She was complaining of continuous pain. After doing laparotomy she turned out to be a case of adenomyosis. A bold fundectory was done, reuniting the tube-bearing portions of the uterus. She became pregnant soon after and delivered a 34 weeks infant by caeserean section. The next pregnancy (unwarranted) was in one of the artificial cornu. Unbearable distress and poor general health did not help her to survive the third laparotomy.

A charming healthy young mother had undergone a caesarean section and safely left the table. Before I could meet the students waiting in the Seminar room, we got word that she was dead of profuse haemorrage of the nature of clotting defect of blood. This opened our eyes to be prepared with plasma etc. in future.

The general surgeon operating on a man of 28 for inguinal hernia sent an urgent call to the Gynaecologist. He and I discovered the entire organs of a clear-cut hermophrodite in his left scrotum. We removed two gonads, male and female, the uterus, vagina and the vas deference. This was of unusual clinical interest but we left the young man, an eunuch.

Leila Begum, the 35th wife of the erstwhile Nizam of Hyderabad, was examined in an unusually insanitary apartment. She was surrounded by several children and a tall dark steward in charge of supplies to His Highness' household. The Nizam readily agreed to our advice for an emergency operation. It turned out to be a long standing pelvic haematoma from a leaking tubal pregnancy. The operation was successful. But by the end of the first week, she developed symptoms of intestinal obstruction. This was not relieved even by the use of Wangenstein's tube for bowel decompression.

An Officer of the Singareni Coal Mines brought his daughter with a gynaecological problem. She had not menstruated at all though she was 19. She had a laparotory done at Bombay. It was reported that her uterus was enlarged to 14 weeks size and firm. No vaginal examination had been possible because of a very narrow vagina. Under an anaesthetic, it could be made out that it was a case of uterus didelphus with the formation of haematometra in one horn of the uterus. She was operated upon, allowing the escape of the pent up menstrual blood. In due course she became pregnant in the active cornu. A Caesarean section was performed with a happy outcome, the earlier tubectomy on the inert side was confirmed.

A peri-menopausual woman was admitted with overflow incontinence simulating a retroverted gravid uterus. Under general anaesthesia, it was found that a large fibroid (submucous) had got the uterus inverted distending the vagina with utmost stretching of the urethra and disturbing the bladding function. An abdominal hysterectomy solved her problem of incontinence.

Among some of the inoperable cases of VVF one was taken up for a Lovsley Johnson type operation to create a rectal bladder with bilateral implantation of the ureters into the isolated upper end of rectum. The proximal end of the pelvic colon had been pulled through the pelvic floor. The anal sphincter controlled the rectal bladder and the pelvic colon that had also been pulled through the anal sphincter. There was effective control of both the hollow organs by the original and splinter. This woman went back to her village with successful control of micturition and defaecation. Later she happily returned pregnant and had a live infant by caesarean section at near term.

A young girl in her final years of her school had been under our care for metropathic type of menstrual disturbance. One fine day, she was admitted in a state of collapse with signs of large intraperitoneal haemorrhage. On opening the abdomen and passing a hand into the pelvis and pulling up the uterine fundus as one does in a case of ruptured extra uterine gestation, I found both the ovaries bleeding away each from a small follicular cyst. These were speedily resected and the gap in the ovaries sutured while blood was being run into her veins.

This girl, a lecturer now in the Women's College, gives me a shy smile when I happen to meet her anywhere.

Padma, a girl of 26 years, a Bharatanatyam exponent, was brought to me from the Sugar Factory Complex in Nizamabad. She gave a history of severe menorrhagia since two years and the Gynaecologist whom she consulted "prescribed" some hormone tablets but did not even palpate her abdomen on a table. I found a swelling like a 24-week uterus which was rising in the midline from the pelvis. It felt like a fleshy uniformly enlarged uterus. From the history, I suspected a large submucous fibroid. I opened her up and making an incision into the uterus as in a caesarean section, I found the knife going through several fibroids and not through the large submucous single one as suspected. So with two capable assistants, I delved deep, burrowed and extracted fibroids varying in size from a pea to a hen's egg. The usual rubber tourniquet applied at the level of supravaginal cervix going through the broad ligaments was of only a little help.

The cavity had to be opened and in all 99 fibroids were removed filling a kidney tray. It took five of us working for five hours including the anaesthetist. The patient needed five units of blood. One of the assistants was repeatedly suggesting a hysterectomy instead of mangling the muscular organ. When we called it a day, the uterus was of 14 weeks size. The unavoidable drop in blood pressure was scrupulously cared for by the expert anaesthetist. After she was discharged, I put her on 200 mg of depot provera over several cycles to suppress ovulation.

After she left, she gave dance recitals in China and various islands in the Indian Ocean. When I saw her next, she was married to a fine man who was migrating to Canada. She returned five months later with a pregnancy in that much disturbed uterus. At term, a lower segment caesarean section was performed with the birth of a son—a much rewarding experience to all five of us who worked on Padma's uterus.

Dr. Lata having a delayed period complained of pain reference to the "central" point of her perineum. Her husband agreed reluctantly for a laparotomy, not trusting his wife's genuine distress. A ruptured corpus luteum was found to have emptied its contents into the Pouch of Douglas and the sack was tacked on to the second part of her rectum. Later the decision to open her was commended by the dentist husband. She is now free from all discomfort.

My husband and I went for a short week-end break to Medak at the invitation of the Bishop's wife—a good friend and also an ophthalmologist. A young English nurse, on her way from Britain to Burma, was shown to me as suffering from an unusual type of ovarian dysmenorrhoea. An examination showed the only ovary present had grown to the size of a large walnut and cystic.

It was an embarrassing situation. I split the ovary, excised the cyst from the wall, made the balance of what was left into the size of a normal ovary. I could not foretell the result. Later I heard that she went to North Burma working as a nurse, got married and had three children. Removing the only ovary would have left her barren.

Deaths on the Operation Table Recorded during My Service

Sham Bai: Taken up for an operation of breast cancer. Local anaesthesia given by the surgeon. Operation of Radical Mastectomy was done under almost rural conditions. The patient collapsed on the table for want of fluids due to the inadequate set up.

B.N. Age 28 : The patient after an abortion in her rural home, had come with a loop of small bowels showing outside the vagina.

On opening under general anaesthesia, it was found to be a case of pelvic hematoma and not intrauterine pregnancy. As an emergency procedure with the help of a general surgeon the bowel was resected and an hysterectomy done. The patient did not survive the surgery due to her already severely ill condition.

STB : The old case of adenemyosis : Age 26, she had a fundectomy done, followed by a caesarean section for a live mature infant. She had again become pregnant eight weeks later and came with severe colic pain. It was found that the uterus appeared didelphus due to the adhesion of anterior and posterior uterine scars. She died on the operation table during an attempt at hysterectomy.

Yellamma : During her second stage, labour had suddenly stopped. The abdomen was opened and the uterovesical peritoneum was found distended with blood due to incomplete rupture of the full-time uterus. Blood supplied to her was lost due to the clumsy handling of the transfusion set and the abdomen had to be closed. The patient died on the table for want of blood into her circulation though it had been donated for her.

Mallan Bee, 45 : Hyperhydration in a woman undergoing hysterectomy for perimenopausal bleeding. Her Hb was 60 per cent and could not take another transfusion on account of Rh problem. When she showed signs of peripheral failure, the anaesthetist, not exactly the most capable, carried on with several units of IV glucose saline till her lungs were bubbling with rales. Modern method of handling such a complication were not available in 1949. The patient died.

Y.A. aged 50 : After a fairly long interval of giving up local anaesthesia for general use, a demand by post-graduates for demonstration of a hysterectomy under local anaesthesia for endometrical hyperplasia, was undertaken. The pre-medication was left to an over zealous assistant. The woman was deeply doped.

I gave the intercostal block, para-abdominal block between the last rib and the iliac crest and the other necessary local infiltration at the usual sites, such as the round ligaments and infundibulo pelvic ligaments. The hysterectomy was bloodless and shockless as usual. The abdomen was scaphoid and breathing was shallow when I left her in the care of the duty post-graduate. The patient never recovered from the deep stupor. It was very distressing for me, after having performed endless number of hysterectomies under local anaesthesia, to lose a poor thin woman mainly due to an overdose of pre-medication!

Mangamma, Age 35: She was being treated as a case of threatened abortion by a senior gynaecologist outside with hormone injections. Suspecting chorionepithelioma, the abdomen was opened to find tropoblastic perforations all over the fundus and body. A hysterectomy was begun, only to find large wide mouthed vessels from the pelvic floor in place of the uterines, and any clamp application simply made the tissue so very friable that no bleeding mouth of a vessel could be controlled. The patient succumbed to haemorrhage on the table. No methotrixate was available in those days.

5
PERSONAL LIFE

After my final examination I went to Ootacamand where I met Miss Jean Duncan, who had just come from Scotland to teach in a girls' school at Madras. We became great friends.

On my return, I reported to duty as a Resident House Officer at the Madras Hospital. I lived in the alloted quarters in the Giffards school building. Jean often visited me. Jean thoroughly educated me about the areas of travel, music, appreciation of art stage, screen as well as the benefits of reading, of world affairs and in general on emotional and other aspcets of young womanhood. My knowledge was limited to medical disciplines and to hospital work till I met her.

Jean and I learnt of the Tennis Club in the hospital campus. We visited the tennis court in the evenings to watch the young men playing tennis; and desired to learn, if possible to play this elegant game. There I met an energetic co-officer. He served well and was ever alert to receive the ball from the opposite court. He and his lawyer-cousin, Venkat Rao, offered to teach me and my friend how to play tennis. But with my poor forearm muscles I could neither serve, much less return the tennis ball. Jean made some progress.

This tall House Officer, Dr. Naidu and I were once posted on duty together at the heavy admission ward. I found him disappearing for some rest after midnight leaving me to run the admissions. Hence, I asked the RMO to replace him. But later when we became more acquainted with each other, I asked him for being reposted with me. Dr.Naidu and I also met in the hospital premises under various demands of duty.

Not long after I had started work as the house officer under Sir A.L. Mudaliar, he heard that my tennis coach was an all-round sportsman and also sensed the latter's interest not only in house races but also in me!

I was beginning to get into a major distraction for the first time in my life. Following my visits to the tennis club, I felt drawn to my tennis coach. He was a total stranger. He was very different in his habits and family background. Even the financial difficulties he had to face in order to qualify, seemed to evoke my sympathy. His mannerisms and in fact everything he said were fascinating to me. I was feeling elated; Jean, whose friendship was to last our life time, was perhaps responsible for sparking off a dormant instinct in me. Although I could not make much progress in tennis, I seemed to have got deeper into the bonds of friendship. On a certain evening while we were on duty together he had a call to see a very ill patient. Before he left I saw him tear a corner of a case-sheet and write something quickly on it. He left that note on my side of the table and hastened to the ward. In that small note he asked me to marry him, if I cared for him as much as he cared for me. I was stunned.He returned from his call but walked straight ahead. I silently followed him to the open yard near the main gate of the hospital. He turned round suddenly and facing me, said, "We may have to wait for long years, maybe six or seven years." I stood dumbfounded.

We met more often after that. We got to know more of each other. Both of us had big family problems and responsibilities. Sooner or later we needed to part in order to seek work elsewhere. He went to the Andhra Medical College, Waltair (from where he had qualified, to the Physiology Department). And I accepted a Medical Officer's post at the Residency Hospital at Secunderabad. Thus we parted and waited.

World War II broke out, which altered his and my future plans to go abroad for further studies. Instead, we decided to get married.

Major M.G. Naidu, Mrs.Sarojini Naidu's husband, was the P.M.O. of the Nizam's army. I had met him earlier at Madras in the company of Sir A.L.Mudaliar who was attending on Major Naidu's niece. When I told him that I wanted to get married, he jokingly remarked about my being already married to my profession. However I had an occasion to introduce my fiance, Dr.M.V. Naidu to Major M.G.Naidu.

Eventually, through Major M.G.Naidu's gracious help we had a Brahmo Samaj type of ceremony. Major Naidu kindly secured Professor Chandaverker to conduct the wedding and in English. We both went round a lighted temple lamp seven times after taking our vows, while hooking our little fingers. My youngor brother, Thambi gave away the bride which was me. We both recieved the blessings of the elders present on the occasion. The marriage was also registered soon after.

Our precious little daughter was born at the Residency Hospital where I was working, exactly on our first wedding anniversary. We named her Renuka. Her father had enlisted in the medical wing of IAF as a Ft. Lt. when she was a year old. The little one and I missed him much when he left for his training.

During the war years that followed, my husband came home on annual leave or dropped in at Hyderabad on transfer from one IAF station to another. After the war came his sojourn to UK, for further training to be a specialist. On his return in 1957 with his FRCS, he got appointed with the Singareni Collieries, Bellampalli, later Kothagudam, for 10 years. I used to go to him for long weekends when I enjoyed the almost novel experience of being a homely wife who welcomed her husband on his return from work.

Renuka was married in 1961 and returned from Cambridge with a first class Tripos. After Renuka settled down at Madras after her marriage, we either visted her alone or together.

My term with the Osmania Medical College ended in 1967. At the end of my three years of re-employment, I went to Sydney for the World Congress. Then came the desire to continue to train and teach somewhere, like in Bangalore. There I developed detachment of my retinae and was away at London for the treatment of my eyes for a long period.

On return with the second sight, came the eagerness to guard this recovery and also to utilise it for training junior doctors in friendly nursing homes. At this stage, I would like to express that in general for a woman in the higher intellectual pursuits with her partner working elsewhere, it is not easy to have a homely married life, however much both the partners may desire.

Wherever my husband had been, either in training or working, he was always my friend, philosopher and guide. After he passed away in October 1985, I had no desire left to teach, train or operate and I turned to write my autobiography.

My kind mother attended to Renuka when she was born with love and care. She was dressed up into a little pink woolly bundle on her first day. Her father carried her and said that she looked just like him.

Our child was brought up by me with the help of her wise and kind 'second mother' Sundaramma. Renuka's first birthday was an exceptionally pleasant occasion. My artist brother had painted pictures from Pinocchio stories on the sides of her baby bed with the 'Blue Fairy' and the fish 'Cleo' on panels of wood facing each other. Jean too was present adding to the rejoicings. Mrs. Sarojini Naidu had also graced the occasion.

Soon my husband had to leave to report to the IAF (Medical Wing) as World War II was on and in full swing. As he was moving from various IAF stations for his training, viz., Calcutta to Quetta and from there to Cochin, I decided to delve deep into my books once again, and prepare for the M.D. exam in Obstetrics and Gynaecology of Madras University. At this stage, I planned to visit my great Guru Sir A.L. Mudaliar at Madras for his guidance. I had left the Alma Mater 11 years earlier.

Renuka was developing well with charm and grace. She impressed her school (St. Ann's, Secunderabad) with her intelligent approach to school work. We had to shift in 1950 to King Koti Road in Hyderabad, because I had been promoted and posted to Osmania Medical College. And so, she had to seek admission into St. George's Girls' School which was opposite to the house we had rented. The Australian Principal gave her tests and found her actually fit for a higher class than she asked for, as judged by her English Essay on 'Air Travel'.

Time passed quickly and her father had been disbanded along with his batch of officers when the war was over. He worked for a spell at the ENT Department of Osmania General Hospital to refresh his knowledge. He left for UK, in 1952 for further training. When Renuka got into the Junior Cambridge class at St. George's, I planned a trip to UK with her by sea. Documents were easy enough to obtain in those days and I took some leave due to me from Osmania Medical College.

Earlier my husband, working in the IAF was posted to Cochin. Renuka was then 15 months old. We spent a marvellous month with him in the family station on the Willington Island in Cochin. The child was such a delight. She came with us in her mini AIF uniform walking smartly or paddling at the sea shore. She used to watch the large fishing nets with great wonder in her eyes.

After we got back to Secunderabad her nurse/governess, Sundaramma, who was a sensible and a patient person, helped to bring up little Renuka with affectionate care. She was full of concern for her health and welfare, for nearly 20 years, till Renuka left for her studies at the Cambridge University. Now Sundaramma is no more.

Just before I left for UK, Dr. Janaki, the RMO of the Obstetrics and Gynaecology Department organised for me a send-off dinner party at Osmania General Hospital. No substitute was posted in my place when I was away at UK as Janaki was a capable obstertician.

When Renuka and I sailed from Bombay, we had a cabin to ourselves. She walked about merrily along the deck and watched other young children in their swimming pool under the care of their "Babino" (child nurse in Italian). Since it was an Italian ship and doing her maiden voyage everything was spick and span. The food was excellent. We watched from the deck herds of porpoises jumping one after another out of the dark blue ocean. The winged flying fish were seen for several metres in the air which was a marvel.

While Renuka had been a child in the pram and I was at Residency Hospital as an Assistant Medical Officer, my mother, an indomitable and gregarous person, visited the family in the opposite house. Here she found Mr. Ramdas bandaged all over his head after a motor-cycle accident. He had very recently lost his wife. She had left behind four lovely daughters and two able sons. We became great family friends.

My husband, then a Sqn/Ldr was posted to the "sick bay" at Hyderabad. He had many Br.orderlies under him. He had to visit them, if admitted at the B.M.H. Our women folk at home—my brother's wife, my sister and my husband's sister—joined the Women's Voluntary Corps to help the nursing sisters to make bandages etc. They used to wait for their daily transport in their smart blue uniforms. The house we now lived in, proved convenient to entertain even the foreign RAF Officers. They in turn asked me to the Airport Hanger for parties and mess dinners. Visiting the army canteens and stores proved an expensive pastime. My husband and his two brothers were also in the Indian Air Force; one was a pilot and the other an adjutant. All that glamour ended with the war.

I was then working in the Osmania General Hospital, teaching the undergraduates and also carrying on with much clinical work. I met Sir A.L.Mudaliar again during one of his official visits to Hyderabad. I had kept in close touch with Major M.G.Naidu also and visited him at his home. Here I met his illustrious family members. I drove down in my little Austin-7. The traffic in those days was no problem.

It was a privilege to meet Mrs.Sarojini Naidu, "The Nightingale of the East", in her home, the "Golden Threshold". This gorgeous name was carried as the title of her book of poems. Her description of the Hussain

Sagar looking pearly pink at sunrise and golden blue at sunset is still very true with the reflection of the sky.

Her daughter Padmaja was very gentle and charming, affectionate and courteous. But her young sister Leelamani was unpredictable. Once while she was a lecturer at the Women's College at Lahore, her mother was to address the college. Knowing about her illustrious mother's absentmindedness, Leelamani had put out a sari and a handbag to match. But when the mother actually arrived, she had the Bengal printed sari on, but had a 'tea cosy' under her arm for a bag!

As my final homage I attended the solemn occasion when Mrs. Sarojini Naidu's ashes were brought from the Lucknow Government House and immersed in Musi-Esi confluence "Sangam" when numerous garlands floated along the full stream of water on the occasion.

Padmaja later was the Governor of West Bengal, and the President of the Nehru Memorial Fund. She was charmingly courteous when I called on her at her New Delhi residence, where she was in charge of the Fund in the name of the great man. The beautiful house at Hyderabad, "Golden Threshold" had a long carved dining table which used to be laid with fruits, and food received from various nawabs' cuisine. The two bearers attended on one courteously when one called on Major Naidu. I addressed him as 'Father' as his own children did. The glorious history of the "Golden Threshold" would have been lost in the mists of history but for Mrs.Indira Gandhi dedicating the complex to the Hyderabad University for Study and Research.

Religion in My Life

The fundamental urge for man to communicate with something beyond himself is as deep an instinct as the instinct to love, to live, to eat and to survive. Even to the unbeliever, there come moments in life when he wishes he could pray, the heart desires what the lips deny.

But the medical college life demanded all my mornings making no difference between Sunday and Monday. I became deeply involved in medical studies.

Later, I attended on Sunday evenings the discourses by Swami Rangananda at the Ramakrishna Mutt after reading his *Messages of Upanishads*. This has indeed helped me to find great solace knowing the ancient scriptures, music and worship even as music gave me an insight into *Gita* and the *Vedas*.

Musicians of India, Saint Thyagaraja and his contemporaries, raised their praise and prayer to the Omnipotent alike by their music, like fragrant incense. Knowledge and self-sacrifices were nothing unless there is love of one's fellowmen. Neither did Saint Paul say anything different in his letters to the followers of Christ and Christ-like people whatever language they spoke and whatsoever country they belonged to. This makes me recall my grandson, Ram, singing in his melodious voice "Entha Nerchina, Entha Chuchina" by Thyagaraja on his veena the very things that Saint Paul said affirming that without love of our fellowmen, all knowledge is like empty cymbals. My mother had hung a green plaque with silver letters, saying: "If you want to do anything good to your fellowmen, do it now, for, we shall not pass this way again." Neither did I forget that I merely dressed the wound but God healed it.

The *Messages of Upanishads* was difficult to grasp at times and fascinating at others. Listening to Swami Ranganandа at the Ramakrishna Mutt was a most absorbing experience. I often heard him refer to teachings of Jesus in the manner most appropriate. The shehnai music before the discourse and the silence as the Swamiji entered the Hall set one on the correct wave length for his discourse.

Temples are places of worship. No other country can claim more temples than India. People thronging in thousands make the place noisy and slushy as in Puri, Kanchi, Madurai, etc. If all places of worship were to be built retaining the South Indian architecture, the devotees would worship alike in spirit and in truth the great Creator.

The great Indian Epic *Ramayana* comes alive at Salem (my childhood town) where many natural monuments tell the story of Ravana while carrying Sita to Lanka and slashing Jatayu's wing. This can be seen at the Chalk hills now being worked as magnesite mines. A small temple not far from there, stands over the 18-inch imprint of Sri Rama's foot. It was one of my greatest joys to take my daughter, grandchildren, my son-in-law around these mythological places at Salem.

Exposure to Art and Culture

I have been very fortunate to have seen innumerable plays on the stage, from the age of about 10, when I went to the tin-roofed theatre with my parents at Salem to witness impressive Tamil plays in 1920's produced by an eminent lawyer-producer from Madras. I recall seeing 'Nandanar', 'Sarangadhara', 'Nala Damayanthi', 'Dasavatharam' with excellent stage decor and the players in very heavy gold-laced velvet costumes. I used to repeat some of the words of Damayanthi while she carried her dead son, or some of the words and songs from Nandanar, the famous devotee of Siva, who was born a low caste, but was inspired to go to the holy shrine at Chidambaram. Plays from the famous Bengali theatre on the revolving stage at Calcutta were enthralling too. I also saw the Shakespearean plays in UK. Tickets for the Little Theatre at Stratford-on-Avon had to be booked months in advance. Earlier M.R.A.'s 'Forgotten Factor' and 'Jothan Valley' were beautifully staged at Zamrud Mahal here, with a high moral re-armament background.

During the famous opera, I recollect those present at the Vienna World Congress of Obstetricians and Gynaecologists. The galleries, with their grand crown of crystals dazzling were most gorgeous. Singing at the various operas, calling usually for pathos and not often with happy ending, are uppermost in my mind. Another great opera I attended was at Sydney, organised for the delegates of the fifth World Congress of Obstetricians and Gynaecologists. "Fiddler on the Roof" was really an enjoyable opera with a Russian rural setting. Ballet of course, originally a Russian art, is a real thrill to see. The best among them to be watched are at the Covent Garden festivals to which elegant ladies in very elegant style, flock in rain or snow. Of the most famous are the 'Swan Lake', 'Nut Cracke' Suite.

In Edinburgh the Ballet Theatre's entrance faced the gate of the Edinburgh Royal College of Surgeons' Office. One was lucky to get tickets easily for 'Coppelia' and some other dream-like ballets which I saw four decades ago.

Indian dancers like Uday Shanker and Ram Gopal, who took their troupes all over the world were great exponents of Indian Art.

The best concerts of carnatic music are the excellent ones at home in South India through the fine string instruments: veena, violin accompanied with tanpura and mridangam. I have enjoyed these all over the south. The famous three musical savants contemporarily hailing from around Tiruvayur led by Thyagaraja, Dikshitar, Shyama Sastri, composed nearly 2,400 *kirtanas* in eighteenth and early ninteenth centuries. Earlier *vidwans* like Purandaradasa, Swati Tirunal, Annamayya etc. had composed *kirtanas* which rise like incense towards heaven.

In 1948, I climbed many flights of stairs to go to the top stall at the famous Albert and Victoria Hall in London, having paid only a small fare to listen to Western Orchestra 'conducted' by famous Malcolm Sargeant and his successors. I had also attended classical western music concerts at the Music Hall at Vienna. Opening my Exacta Camera, I could get a slide immortalising the interior of the grand Music Hall. Vienna is still the city of anybody's dream where Beethovan, Mozart, Richard Talbot left behind their immortal music. Beethovan's place still stands in good repair at Vienna to remind a music lover that he composed his famous Sonata when he was totally deaf.

A little incident happened when my young daughter and I wanted to visit Vesuvius near Naples which had erupted and led to the last days of Pompeii. But the cunning taximan took us and an Indian co-passenger to a wayside cafe only to make us listen to their guitar music and drink their preferred Chianti, the red wine.

For paintings, the Rex Museum stands topmost at Amsterdam with Rembrandt and Van Gogh standing apart. How could I omit to mention Louvre in Paris with its Mona Lisa and the 'Last Supper'. by Leonardo da Vinci! London is no less glamorous with its Tate gallery and the great National Gallery in Trafalgar Square containing famous paintings not only worth millions of dollars but attracting millions of admirers. A visit to this museum leads you to the British museum which has collections from Egypt to Amarawati and Assyria and from the pyramids objects fashioned in gold from the tombs of Tuten Kathen exhibited at the British museum depicted the wonderful art beyond one's conception.

On the ceilings of Sistine Chapel and Santa Maria Majora are the paintings of Michelangelo in Rome. How can I forget his marble sculpture of 'Pieta' and the innumerable pieces of sculpture in the Uffizi galleries!

Compared to the thrilling experiences in the field music, art and the theatre, the cinema generally pales into nothingness except of course "The King and I" (if one had seen it like I did on the Fifth Ave Stage at New York with Yul Brunner and Deborah Kerr) or "Marie Waleweska", the immortal love story of Napoleon acted by Charles Boyer and Greta Garbo, "The Sound of Music" and "My Fair Lady" (I saw this last one played in the Cambridge Theatre at London). I cannot forget the puppets in the Little Theatre at Munich which pleasing art I am glad is catching up in our schools also today.

As examples of marvels in engineering to mention only two are the mono-rails of Tokyo and the railway bridge on Firth of Forth near Edinburgh in Scotland exercised my mind but not many others.

Did I forget the new class of show called the "Light and Sound"? I saw this most poignantly performed inside the Red Fort with the Moti Masjid background. It brought the Moghul period to life.

A similar show was organised by an enterprising young woman of Hyderabad at the Golkonda Fort, depicting the Qutubshahi regime. It was about a Revenue Officer, Ramdas, being imprisoned for building the Bhadrachalam temple at the expense of the local revenue funds. This performance should be repeated for the benefit of the younger generations to acquaint them with India's history.

I confess that my reading habits developed rather late. This was inevitable since I had to be devoted to books and journals dealing with my profession. Books and monographs have been pouring into the reading markets as science and research made rapid strides in recent three or four decades. Dependance on basic sciences is the backbone of medicine. However I was interested in biographies of great scientists of the world like Schemelweiss, who pointed out that microbes were the cause of illness especially causing disease, deformity and death of women during the child-bearing age. The novel is a foremost example of his struggle to enlighten the resistant medical world on microbes. Decades later, came the true story of Sir Alexander Flemming, into whose culture plate descended a fungus which dissolved colonies of virulent streptococcus. His life story reads like a novel.

Then came the records of artists like Rembrandt or Von Gogh or a sculptor working on marble like Michelangelo depicted poignantly by Irving Wallace in his *The Agony and the Ecstasy* or the story of the scientist turned artist, Leonardo da Vinci who designed both a flying machine and painted the "Mona Lisa" and the "Last Supper". From then on, I became interested in great fiction writers like Dickens, the Bronte Sisters and Jane Austin.

Then I read travelogues, particularly by H.V. Morton and fictions by Dr. J.A. Cronin and Dr. Conan Doyle, Rabindranath Tagore's immortal *Gitanjali, Applecart, Kabuliwala* and others. And I think with pride of the "Nightingale of India" Sarojini Naidu and enjoyed her poems. Leo Tolstoy and Victor Hugo were also among the authors I read.

I read biographies of political figures like Disraeli, Hitler (*Mein Kempf*) Abraham Lincoln and autobiographies of Mahatma Gandhi and Jawaharlal Nehru.I enjoyed the novels of Daphne Du Maurier and the autobiography of Agatha Christie. Both these women commanded my greatest admiration. I have had the good fortune to visit the beautiful house of the former overhanging the harbour on the River Fowey in Cornwall, with the most appropriate setting which could have inspired her to write *Rebecca, Readymoney Cove, The Blithe Spirit* and other novels.

Agatha Christie, who had started with only school poems, short stories and detectives rose to a climax in *Witness for Prosecution* and her own autobiography.

More modern authors have a variety of exciting and daring authentic stories like the Fabulous Moghul (our own late H.E.H., the Nizam), the kings'story of how the Dutchess of Windsor weaned away the British monarch from his throne. Fiction by Morris West and Irving Wallace provide really absorbing reading.

I read books by C.F. Andrews and C.S. Lewis over and over again. And I cannot but mention the books by Eddie Askew the talks of God listening to bird songs, children's laughter and lover's whispers. He worked for leprosy patients in India.

How can I omit to mention the joys of childhood while reading *Alice in Wonderland, Water Babies, Grimms Fairy Tales and Hans Christian Anderson* whose piper's flute attracted all the unwanted rats away from his town into a cave.

Finally, the *Holy Bible* tells the story of the hardships of the Isreali tribes under slavery and how they were released. This can be better seen in the film "Ten Commandments". Stories of their life in the wilderness and their resentment against their only God - Jehovah - finally and how his promise was fulfilled by the birth of Jesus at Bethlehem to a young woman not yet married are of abiding interest.

The collection of 600 and odd coloured slides would show the story of my travels with a friend, daughter or husband made possible, by my carrying an Exacta Camera weighing 3 pounds, on my shoulder like a professional camera person. I started with the camera in order to copy line drawings to use at my teaching adventures to begin with Osmania Medical College way back in 1950's, when there was no photographic department even in a Medical College.

My Pet Dogs

After a lapse of a few years after my travels, a medical representative gave us an Alsatian pup. We called that dog Simba. Renuka doted on him.

Many snaps show them together in various affectionate poses. This Simba (lion) was followed by a female dog at Glendale, who was harmless to visitors and children. She became as fat as she was docile. One day the fat Simba barked at a bullock cart, jumped at the bull and fell down dead probably of a heart attack.

The last dog—my pet Patchy—a country dog, was sold to me for Rs. 3 by my driver. Patchy grew up to be almost human in his behaviour. He would nudge at my feet exactly at 7 p.m. wanting his dinner. When the chapaties were in his bowl, he would look up and wait for the milk from the kitchen. After his meal he would come with me for a walk. He would take a few steps in the garden keeping pace with me but suddenly dash off after frogs in the bushes.

Once Patchy had disappeared for three days leaving us very upset. One morning as I was standing at the Glendale gate, he was seen limping towards me accros the road and put his nose into my sari looking muddy and ill with injuries caused by a rickshaw and lying in a side lane.

But first of all came Titoo, the black spaniel, with her drooping ears and expressive eyes, looking into mine and putting her paw into my hands as if promising to go with me anywhere, if taken. This brings to my mind Rags, a brown mixed breed at Secunderabad, during the IAF days. He was friendly with a gentle red-eyed white rabbit which was Renuka's pet. One fine morning we found the rabbit torn to pieces and the dog staring at me with no sign of recognition. This suggested 'Rabies'.

Coming back to Patchy, who was the most intelligent dog. He suddenly changed his behaviour. He walked round and round the room, and around the house, not heeding to anyone calling his name. Even the vet on a private visit could not unravel his condition. He hardly ate his food which he had enjoyed formerly. The same vet had seen him promptly obey my orders. He was caged in at the dog hospital and jumping at the railings repeatedly, hurt. Since there was no chance of recovery he had to be put to sleep. That was the lost of my pet dogs.

I rettired from Osmania Medical College in March 1967. On an invitation from a Church hospital, I went to Bangalore to start with, to work at this Institution while hoping to get a teaching assignment at St. John's Medical College.

But suddenly one morning some black spots floated in front of my eyes. My good friend Dr. V., formerly of C.M.C., Vellore, advised me to get back to Hyderabad to my friendly Eye Colleagues for my sight check-up. This was in the first week of June 1969. My husband was working with the Collieries hospital as C.S./M.O. at Kothagudam. The Telangana agitation was going on at Hyderabad. On my return to Hyderabad, I had first to go to the Police Control Room to get a permit to get to my own house. Next morning, my senior Eye colleague saw me and thought that I had a haemorrhage into the vitreou and advised iodine injection.

In spite of the sight trouble, I went to Government Maternity Hospital to meet the post-graduates on the morning of 11th June. I felt some brown shadow rising from below into my sight as if my own cheeks were being pushed up.

At this post-graduates session, I was actually joking at my own inability to 'see' which candidate was answering me and which one was not. I got back home, rested and got ready by about 4.00 p.m. to examine an antenatal case. I could only see her face but not her abdomen, certainly not both at the same time. I rang up my helpful brother about this peculiar disturbances in my sight. He, at once, brought his friend Dr. Satyendran, who preferred to bring in a senior ophthalmologist; and fetched Dr. Anna Reddy by 8.00 p.m. Dr. Anna Reddy was able to make out at once with his ophthalomoscope that both my retinae had become detached. He advised that I should lie flat on my back. He showed much concern. It was a case for surgery. Dr. Reddy said he could operate but had no equipment. Sarojini Devi Eye Hospital had the equipment but not much experience in this line.

Next day, the Superintendent of the Osmania General Hospital brought a Physician and the Ophthalmologist. They and the DMS met together at my bedside to discuss the ways and means for my treatment. They felt I should be sent to London. Due to political agitation, not even a typist was available to prepare the Committee's report. My husband was sent for from the Collieries. He told my daughter at Madras on the phone "your mother is unable to see". On hearing that, she was shocked and fainted. She flew to Hyderabad with her husband and her six-year-old daughter, Anu.

Soon Renuka with her husband started corresponding with the specialist at London. She also wrote to my friendly colleague, Prof. Moir of Oxford. They drove about the town, in spite of the risk on the road due to the agitation, to get to the Reserve Bank of India to obtain the required foreign exchange for my treatment abroad. Being endowed with plenty of common sense, Renuka was able to describe in detail to the Office of the Reserve Bank what retinae detachment was and the significance of the detachment I had sustained. She got £ 1000 foreign exchange sanctioned for my trip to London.

On getting the news, my niece, Premu, at far-off Guwahati, immediately, picked up her mother (my sister) from Delhi and arrived at Hyderabad to help me. My other sister also came from Madras. The main question was who was to accompany me to London. Dr. Devyani Dangoria offered to go with me. It was Dr. Satpathi of the Sarojini Devi Hospital who firmly advised that I should go to the Moorfield Eye Hospital in London. The Eye Doctor had witnessed the great retinal Surgeon from London operating at a developing Hospital at Sitapur near Lucknow. So, the decision was taken.

I needed air accommodation which entailed the removal of three seats from the aircraft to make place for my stretcher both from Hyderabad and again from Bombay. It took 20 days in all for these arrangements. The Durgabai Hospital ambulance kindly took me to the airport. At the airport, there were a great many, dear and near showing concern. Naturally, my husband and my brother were most affected. Dr. Devyani had offered to accompany me, if the British High Commission at Madras would permit her to go to UK. This was made possible and Dr. Devyani was to accompany me by the vigorous efforts of my able son-in-law. During the waiting period at home, dozens of friends visited me. But not wanting them to come in to sympathise, I wanted them to be briefed before they came to me, so that I could cheerfully talk to them about their families, etc.

It was Devyani's first flight. After a night at Bombay Airport the next morning, eight sturdy porters came in to carry me to the BOAC aircraft. After we took off, my niece Premu rang up 'all well' to Hyderabad. I heard later that my husband left for Kothagudem and my daughter's family returned to Madras and my niece must have entrained for Delhi. My brother felt most upset as I flew away from Hyderabad blindfolded on a stretcher. But I was very confident that my sight would be restored. All friends and relatives said that they would pray for the early recovery of my sight.

When we landed at Heathrow airport, a kind nursing Sister came to me to check if I was an 'eye' patient. I told her at once to get herself a drink of mango juice from the air hostess, the kind of which I had just then enjoyed. After she too had enjoyed the drink, the English porters took me out from the aircraft very deftly in a hammock to an ambulance stationed on the tarmac.

While I was inside the ambulance lying blindfolded, I heard the friendly voice of Prof. Moir of Oxford University calling my name and saying that "this is not the way to come to England". He handed me a bouquet of flowers from his wife. When I told him that Devyani had come with me, he immediately went to look for her. But she had been whisked away to an isolation hospital for a further check, because of the papers she carried with her. I missed her sorely. The airport ambulance driver remarked that the Professor had come in a tiny car Morris-minor which would not carry even a Pekinese dog. So, Prof. Moir went back to arrange for another ambulance, which came later. Then we set forth to reach the hospital 45 kms away and Prof. Moir followed us in his little car.

I had a drink of water enroute. It was very hot. At the end of the long journey at Moorfields, the porter picked me up like a child in his arms and carried me and placed me on my bed in the allotted ward. London was very warm even on 30th of June, 1969. A nurse with a surly voice turned up. I asked her for a drink of cold water which she handed to me in a feeding cup to hold, in spite of my lying with my eyes bandaged. I was feeling very hot and requested her to remove my socks and sponge my feet, which she did but reluctantly. She grumbled now as she did daily, to attend on me. Prof. Moir was very embarrassed with her behaviour. I learnt later that she was from South Africa on daily wages and not on the rolls of the hospital.

Next morning, a sister asked me to get up to go with her to the Eye Surgeon's office. I fell back when she tried to pull me up I seemed to have lost all my muscle power by having been in bed for 20 days at Hyderabad. Mr. Fison came later with his Registrar to examine my eyes and went away without a word. But he came again later and sat on my bed and enquired more about the loss of my sight and how many specialists saw me before diagnosing the actual condition. He was very kind and compassionate. He left instruction with the nursing staff to make me sit me up in four hours by giving me an additional pillow each hour. Then I was made to walk. He came the next day and examined my eyes, this time with a special ophthalmoscope designed by him. Then he decided to operate upon both my eyes at the same time on 4th July.

Dr. Devyani came to the Eye Hospital on the 3rd day, released from her quarantine. We hugged each other like long-lost friends. By then Prof.

Moir had phoned Mr. Fison about my having been a hard-working woman gynaecologist back home. The sensible aneasthetist put up a label on my bed as "Professor Naidu" in case the Nursing Staff mistook me for a London dock worker's wife. Dr. Devyani was allowed to stay in my room and the nurses looked after her comforts.

Mr. Fison, the very skilled renowned Eye Surgeon, permitted Devyani to witness his operation. The nurse who sponged and prepared me on the morning turned me about from right to left rather vigorously. In order to make her more friendly I told her a story of a scout cub telling his cub mistress that she had a "swinging brick" for a heart. She was amused with my sense of humour and became kinder ever since.

When my trolley was wheeled into the Operation Theatre, my tongue was parched and thorny, due to the injection of atropine. A very kind technician gave me a wet swab to dab on my tongue. Mr. Fison came in and gave me a reassuring tap on my chin to show that he was there in person to operate upon my eyes.

When I woke up the next morning Dr. Devyani said that I had been operated upon the previous day and that I had slept throughout the night. She gave me the details of the operation which had lasted three hours .

On the 7th day Mr. Fison returned from a conference on Ophthalmology at Oxford from where he had also phoned Prof. Moir about my welfare. He opened my bandage and I could see his face and the face of my little travelling time-piece. He was pleased. And I made stready progress. Even the chef who sent me chicken curry and rice used to come up and grin at my enjoying his cooking.

Mr. Fison came daily at the 'crack of dawn' to note my progress and insisted on my eating well. On the 17th day of my operation Mr. Fison asked me about who would take care of me after I left the hospital. He was satisfied when I told him that Dr. Drury, an English friend who had visited me at Hyderabad, would take me to her home at "Surrey". This friend Dr. Drury whom I had first met at Sydney helped to rehabilitate me in several ways, e.g., cooking 'rice' for supper, to make me feel at home. She also encouraged me to wash my small linen and hang them up. Some young Indian friends from Hyderabad staying in London visited me. Two other English friends also visited me. My hostess had given me a free hand to entertain them. It was with her that I saw on the television the "moon landing" and the return of the 'module' to earth.

When I got back to London alighting at Victoria station, a kind couple, formerly of Hyderabad, received me on the platform and took me to their home in North London. My brother had sent through someone a beautifully polished silver tray and a gorgeous Banaras sari which I sent as gifts to Mrs. Fison through Mr. Fison. Young Eye Surgeons of India could do well to emulate Mr. Fison's courteous manners while examining the retina by asking one to look to your right 'please', left 'please' and into your head 'please' etc. His letters, even after I got back to India, were of the same tenor, kindness and concern.

After Moorfield's Hospital

From Moorfields I was picked up by Vatchi, daughter of Dr. Rajoo (who had delivered me of my child) and her husband General Janardhan who had taken me to the BOAC terminal from where Dr. Kathleen Drury drove me to her home at Rheigate (Surrey) for my convalescence. Mr. Fison was relieved to learn that an English friend was taking me home for convalescence. Dr. Drury drove me over 15 miles to her comfortable home with a lawn. She had a hot house where she grew tomatoes. She welcomed me to entertain any friend who came to me.

That was in July 1969. Dr. Mason, formerly Professor of Bio-Chemistry at W.C.C., used to visit me at Moorfields Hospital. She had also very kindly got my Parker pen repaired. She arranged for me to stay with Miss Hester Smith later for my convalescence. The latter was a dear old lady who had worked in a school at Kerala. She cooked for me and even served my breakfast on a trolley in my bedroom. But I became homesick when I got in and the huge wooden door closed in on me, when I found I could not open the door and go out easily.

Miss Hester Smith gave me a lovely bed with linen sheets. She brought in a trolley laid with much kindness and breakfast. After a few days, whatever degree of sight I had regained began to be disturbed. Miss Smith began to see that I had become very agitated. A certain phobia of being confined to the second floor began to grip me. She tried to read to me and tried to teach me to knit. But nothing could calm me down. Then she took me by taxi to see Mr. Fison, who made out that I had developed Iridocyclitis in both my eyes. Mr. Fison was upset himself that my home-going would be delayed and gave me hydrocortisone eyedrops to be used every three hours. That kept me busy and improving. But even then, I would want to open the big street door to go out early at 3.00 a.m. which would be impossible: I became more and more agitated. I ought to have been able to calm myself because of my age and my profession. When we visited Mr. Fison again, he recognised my mental state and advised my going to my good friend, Dr. Joan Ponsford at Hatfield. Miss Smith very kindly took me there.

While worshipping at Church with Miss Smith one Sunday morning, I fainted from my seat on to the floor. Someone noticed me falling and the Minister came to our seat to administer to me. Miss Smith was extremely kind and took me home. I felt happy when she was kind enough to let me meet another younger lady who came to visit me. She later presented me a large print book called *Healing Touch.*

On Mr. Fison's advice Miss Smith took me by taxi to my good friend Dr. Joan Ponsford's house at Hatfield. I began to gradually gain a grip on myself under Joan's very kind attention and hospitality. Her housekeeper, Mrs. Ray also took personal interest and took me out for walks, and marketing. I made friends with the neighbours of that small friendly town with bungalows and gardens. Even in this glow of euphoria, my sight began to fade again. I was developing cataract in one eye after another. Mr. Fison reassured me that this was no unusal sequence after operations on the retinae.

It fell on Joan's shoulder to organise to send me by taxi to Moorfields for the successive cataract extractions. During this period my kind friends, Margaret Williams and Professor Moir paid me visits at Moorfields.

To begin with, I was to stay for a week with Dr. Joan, but I spent in all five months. I had relaxed and had a happy time visiting the Hatfield House where Queen Elizabeth I had been imprisoned there by her father, Henry VIII. Much later, she was declared the Queen of England underneath (the oak tree the stump is still there). That was after Queen Mary of Scots had been beheaded. At the famous Hatfield House, I saw a sword of Tipu Sultan of Mysore on the side of the staircase.

Dr. Devyani now working at Hammersmith Hospital visited me at Hatfield. Prema Rajan would come and take me out for a walk. I wrote long letters home about how my sight was gradually improving. Dr. Joan very kindly took care of my taxi and hospital charges.

Then one day, Sarala Kota of Bangalore phoned me from somewhere in London. She was on her way to India. Joan asked me to invite her for a meal. She enjoyed the chicken roast and apple pudding. Sarala offered to accompany me to Bombay. And I agreed to go with her.

It was during my stay at Hatfield that 'Thambi' wrote about his son passing through on his way to the States. I waited eagerly at the kitchen door for him and my uncle E.K.S. But they never came because of some difficulty over his baggage. I was much disappointed. They phoned me from the airport later and pacified me.

Joan put me in a taxi early in January 1970 and saw me off to the London airport where I met Sarala. We boarded the plane through an aero-bridge. I had no purchases except a car radio which I bought for £ 12 the previous day at Hatfield with Mrs. Ray's help, for my son-in-law.

Return to India

At Bombay it was good to see Mr. A.K., my niece's husband, waiting for me while the customs men ransacked my suitcase in spite of my asserting that I had my eyes operated upon and I had done no purchase apart from the one I had declared already.

Dr. Joan is a warm and sensitive person. Her concern for me knew no bounds. She was unhappy about my having to undergo several eye operations. Our friendship has outlasted these 40 years.

Leaving London

I got by taxi to Heathrow, though in a less robust state than Sarala. I missed the call to the flight departure but got to the aircraft scrambling up the flights of stairs. The BOAC stewardess sat me outside the aero-bridge till I could regain by breath. The flight was pleasant enough till we landed at Bombay.

Then came all the difficulties with our Indian gentlemen in Amercian-chequered suits and stretching their arms over my shoulder, offering their passport for inspection, forgetting all courtesy they have imbibed, if any, in USA. Then the customs men upset the contents of my suitcase in spite of my telling them about the eye operation I had undergone and that I had only a car radio. Sarala went away with her friends, leaving me to fend for myself. Mr. A.K. was outside the barrier to take me to his dear wife for a meal. It was the greatest relief when finally I reached the Madras airport. It was thrilling to land at Madras airport and to find friendly brother Richard walking into the air cabin to help me out on to the tarmac.

Below were standing my precious daughter, her husband and little Anu. They took me to their pleasant little home. Here they nursed me back to my normal life The first operation I was delighted to perform with my newly given sight was to remove a transparent fish bone fròm the cook's thumb.

Gradually, I learned to walk steadily and my kind son-in-law Bonnie drove us to Gandhi Mandapam where I learned to walk along the tiers of wide steps. Then I could read a book by Noroloft which I had left unfinished before my eye operations commenced. I became physically fit and mentally vigourous with their unlimited kindness.

Then I longed to go to Hyderabad. Though Renuka would not hear of it, I phoned my good friend Dr. Kalyani and went in her car to book an air ticket to Hyderabad. I also knew what was happening to my daughter when she began to feel squeamish. Little Ram was taking form inside her. From Begumpet airport Thambi and Saroj took me to their home. Both were kindness personified and cared for all my needs night and day. They let me go to Glendale for a part of the day to begin with. And by and by, they took me back to Glendale to let me stay there.

Nana had come to join me. He retired later in June 1970. Ram was born in November 1970, such a lovely infant he was, and Dr. Jayant Joshi picked him and checked him up and said what a fine fellow he would grow up to be. Those concerned for providing care and comforts for my daughter's delivery gave us the best available at Andhra Mahila Sabha Nursing Home, now know as Srimathi Durgabai Hospital. I felt very grateful to them all.

A few years later in 1971-72, I was doing a locum assignment in UK. While on a visit to Joan at Hatfield, I was alighting from the train on its two-minute stop, on to the platform with only my handbag and my shipping Shamoi-lined cape. In no time I found myself slipping down between the wheel of the train and the platform. I heard a sharp whistle and a porter in blue uniform was beside me. He lifted me up with his two hands holding me from below my seat while I held on to the edge of the platform. I came up unscathed and found my bag and my cape there. It was a double quick work the porter did in stopping the train and saving me. He found me calm and unruffled after the event. I found only a stain on my cardigan, over the right sleeve. He sat with me on a wooden bench and enquired about my next move. I told him that my friend Dr. Joan lived just beyond the long footbridge, very near the Hatfield station. It spanned 12 pairs of railway tracks below going from South to North. We both then got to the Station Master's office and phoned my friend. In no time I saw

Joan running along the footbridge in a red woollen suit. We met and laughed over my escape before getting home to her place. This was a great scrape perhaps meant for something special.

I was enjoying the walks, fresh air and marketing at Hatfields. I had gone through two cataract extractions. Bifocal spectacles had been fitted up according to Mr. Fison's prescriptions.

The Nizam

It was through the generous interest of Major M.G. Naidu that I met the fabulous Mughal, the Nizam. I went with other medical women. We, the four "lady doctors", alighted in front of the main gate of the Nizam's palace, known as Purdan Gate. The road was screened off from the palace by a large jute curtain which could be rolled up or down on a heavy beam.

Major Naidu met us and led us towards the main building. Suddenly, I found my colleagues saluting someone they had sighted ahead. Curious to know what it was all about, I raised my head and saw the smallish 'grand' old gentleman standing on his verandah with a turkish bath towel wrapped round his red fez. He was in his long-sleeved kurta and a short-sleeved yellow overall. I noted that his footware, the chadavs, were folded in under his heels. "So this is the H.E.H.", I said to myself. We were introduced while we all stood on the ground, one step below. He pointedly asked who I was. Major M.G. Naidu in his official headgear and a belt across his shervani (a sign of servitude), said something about my being Dr. A.L. Mudaliar's protegee. He then demanded that we visit someone in the harem and come back to report to him.

It was my first visit to the Nizam. A few years later, on my subsequent visits, he used to offer me a chair and speak to me very cordially in English. When the question of pregnancy for one of his numerous sons was discussed, he snubbed any idea of investigation and slapped his thigh loudly and asked, "Will he produce a child or not?"

I recall attending a banquet in honour of our first President, Sri Rajendra Prasad, at this Palace on the King Koti Road. There I met some of the galaxy of the gorgeously garbed young Begums of the Nizam's harem. The banquet went on well with a clatter of spoons and forks. I was happy to sit next to Laila Begum who called me 'Mummy' since the time of her surgical episode.

In contrast to that banquet, I gathered that the H.E.H. had also entertained VVIPs like the Prince of Wales and General Wavell and others. The banquets were held at the Falaknuma Palace, at the extra long dining table in the fabulously decorated dining room where dinner was served with 13 courses, all in gold plates. It is said that the Nizam would suddenly get up and the guests had to do the same whether they had their fill or not.

When I lived in the rented house, old "Glendale" on K.K.Road, his fleet of 60 cars would speed past daily to call on his mother. From this house which was nearer the road than the present "Gendale", often I could see the small, frail, "richest" man in the world, sitting slouched in the back seat, as usual in his long-sleeved white shirt and shorter-sleeved yellow overall. The red fez on his head would be tossing its fringes about.

I have also seen every morning a large horse-drawn cart with a towering canopy bringing several large brass pots of water from Miralam lake, drawn from a special filter bed (Musa Bom). This lake is lovely to visit in the evening during a colourful sunset with the ragged outline of the hillocks against it. I recall our many picnics with the Ramdas family. The Golconda fort and the Qutub Shahi tombs are visible against the distant skyline. The description of the Golconda Fort with all its legends would be the poorer without the most poignant of all legends when Ramdas the Revenue Officer, was imprisoned in the Fort for using the official funds to build up the Bhadrachalam temple on the banks of River Godavari. He is said to have been at last released by Lord Rama and Lakshman appearing in person with enough gold before Tana Shah Bahadur, the last king of the Qutub Shahi Dynasty.

Sunset at Miralam was greatly admired by Dr. Lazarus (MLC), Auntie Hilda as she preferred to be addressed by us. She belonged to several generations of family healers and this affectionate gynaecologist even gave her healing touch to my daughter's problem with her early pregnancy. The pregnancy had a successful outcome in the form of a lovely child. And this girl, Anu came from Madras to Hyderabad for the celebration of her third birthday on 2nd April. The celebration with a variety of eatables and games concluded with a magic show by my late cousin Mani, the versatile Director of AIR station. His magic word 'Abracadabra' is being remembered by would-be magicians. Mani had a great sense of humour besides being a voracious reader. When I was down, years later, with detachment of my retinae, he found out from his senior colleague in Delhi about the fee the latter had to pay to the London Eye Surgeon who had operated upon one eye of his wife. Mani called me on the phone and said with glee that if a surgeon's charges for one eyes were £.500, naturally it would be double the amount for operating upon my two eyes! But, actually, Mr. Fison, the world-famous Retinal Surgeon, charged me nothing, but gave me all his skill, compassion and encouragement. I can read and write well even now, 20 years after restoration of my sight.

Earlier, when I was actually lecturing to a batch of undergraduates at the Osmania Medical College, a shocking message about the terrible train accident (caused by a flash flood which had washed away the railway bridge) near Jangaon reached me. In this accident, two of Uncle Ramdas's children were involved. The bogie in which they were travelling, turned turtle and floated on its roof till it was stopped by a muddy bank. We were most relieved.

So was Mr. Ramdas and also the young ones who were our weekly playmates at Lalaguda. One of these girls, Jamu, married a fine Muslim young man from Rajahmundry at Boston, USA. Their young son when he visited us, preferred to be known as 'Joe' instead of Salim. Jamu insisted on meeting my brother, Thambi and she did meet him. Thambi, almost my twin with only two years between us, was my playmate and funmate. We had built the doll's house together at Thottam in Salem when young, gone walking to buy mangoes from a nearby market and borrowed small change from the provision storekeeper to buy sweetmeat. We went on long walks with our fat uncle Gopal for picnics. When Thambi grew up into a man, he organised everything in my life which I could not do. He even gave me away as a bride. He had built this large 'Glendale' for us. He loved his sweet, patient wife more than anyone else in the world.

On discharge from the Osmania General Hospital in his last illness and going home in an ambulance, Thambi pulled my hand from the trolley, with his shaky hand, and held mine in such a grip that said all that he wanted to say on his last way home. On arrival in his bedroom, he sat on his familiar chair and wanted to be fed by his wife and no one else. He died in his sleep that night, gone like the 'will of the wisp'. May his soul rest in peace.

On the 16th day of my 80th year, I "washed up" and watched Dr. Devyani operate very deftly on my old case of infertility. Being in the nostalgic O.T. atmosphere, after a pretty long inactive interval, made me feel 50 years young and not any older. Dr. Rajan, the anaesthetist, was watching out to put a stool behind me to sit which was not needed at all.

Dr. Devyani visited me when I was convalescing in UK after the operation of my eyes. Her brown hair and brown coat made every passerby cast eyes on the young Indian in a sari. She visited me at Dr. Joan's home and won her affection. Dr. Joan's hospitality also knew no bounds. I am so happy that Joan later got married to an old family friend from Japan and has been spending six months on each side of the globe every year.

The Nizam's Family

At about the time of Leila Begum's operation in Osmania General Hospital, one of the Nizam's daughters needed caesarean to be performed on her in 1956. After an interval of several years, she became pregnant and no doubt needed a second caesarean section. But the autocratic father sent word that she should have a normal delivery and no cutting up of her abdomen. After all she was much older than before and was not likely to deliver easily. On the day she was to be sectioned, she bravely and steadily walked into the operation theatre, climed up the operation table and lay down ready for the anaesthesia. Afterwards, word came through that her father, the H.E.H. was coming to visit her and her little infant son. She quickly covered her head with a tinsely shawl, reclined on her elbow and tried to chew a pan. The Nizam arrived with due pomp and asked her if she had an easy delivery. She salamed him and said, "Yes Sir" (a blatant white lie). As soon as he left, the nurses helped her to lie down and made her comfortable,

This little son was adopted by the Nizam's eldest unmarried daughter who surrounded the youngest one with all sorts of luxury and toys from all over the world. He grew up to be a charming and courteous young man.

At the age of 18, during one of his scooter rides, he suddenly disappeared. If money and influence could search the world over, it was carried out by all means. But he was never to come back. Rumours are afloat that he is living somewhere under the effect of drugs or he has been murdered.

I would now like to relate what I know about his two sons, the Prince of Berar, Azam Jah and his younger brother Prince Mozam Jah, born to his eldest of the four wives authorised by the Koran. Their mother had a life spent wandering about in a car visiting shops and picking any attractive articles and any child seen by her, for possible adoption.

At the time when Azam Jah and Mozam Jah were ready for marriage, the Caliph of Turkey had been deposed by a political revolution led by

Kamal Ata Turk. The Caliph's young daughter and his brother's daughter were at school in Nice in France. The H.E.H. being a fabulous Mughal, contributed materially to the welfare of the Caliph in his exile. This contact led to his two sons marrying the two very young Turkish princesses. As they knew only French and Turkish, the young princesses had tuition in their adopted country in English and Urdu. They lived in princely palaces built for them by the Nizam. The elder lady, stately and dignified, would even attend the Secunderabad Ladies Clubs in gorgeous saris. Princess Durreshawar was very tall with sharp features. She would patronise the aristocratic Muslim girls' school like the 'Mahboobia' and attend school functions.

At one such function at Mahboobia Girls School, where only girls from aristocratic families were entertained, my sister, aged about 16 or 17, gave a dance performance called the "Dawn". I happened to devise certain movements of her hands and torso to depict the rising of the sun.

The Prince, Azam Jah and his Turkish wife Durreshavar had two sons while the junior Prince and his wife Niloufer had no family. This junior lady was very beautiful and more sociable and showed interest in health welfare activities. Thus, a children's hospital in Hyderabad came to be named after her as "Niloufer Hospital". A large colour photo of her beautiful face hangs there even now.

The sons of Durreshavar are known as Prince Mokarram Jah and Mufakum Jah. They were educated at Cambridge and Oxford respectively. The elder grandson of the Nizam married a turkish co-scholar at Cambridge. They live abroad most of the time.

Mufakum Jah and his Turkish wife, a cousin of Princess Niloufer, lived in a large house at Banjara Hills. The well known senior physician of Hyderabad, Dr. Mannan arranged for me to visit Princess Esne, the wife of the elder one. I found her most charming, speaking chaste English and her residence beautifully decorated. She had a little son and an English nanny when I attended on her a couple of times. She was friendly and communicative and told me about being isolated from the general Muslim families. She said she had not even seen an Indian wedding ceremony nor did any educated Indian lady visit her. This touched me and I suggested to her to invite Mrs. Manorama Bhargava, wife of an R.R.L. Scientist, herself a Home Science graduate, trained in USA. She taught at the Home Science College of Hyderabad. Princess Esme was more than pleased to ask Mrs. Bhargava to her home for mid-morning coffee. This was arranged when Mrs. Bhargava and I arrived. She asked us to a beautiful 'Gazebo' overlooking the blue waters of the lake below. She was a charming hostess and offered us delicious snacks.

My Sister Sushila

My sister, Sushila, 10 years younger to me, is of a very gentle nature and tender. During one of my holidays from medical college I was sewing a dress for her. I had brought the material from Madras. While she was standing by and watching the machine needle during its running stitch, I asked her to bring me a fresh reel of thread. She seemed hesitant, not knowing where to find a new reel. At this, I gently tapped her on her arm when tears streamed down her cheeks. Finding her go emotional, I hugged her promising her never to do it again—never again in our lives. Since that incident, when she was about 12, we have been more than sisters, and more than friends. She has become part of my life. When I came to work at the Secunderabad Hospital in 1934, she came with me to be a constant companion.

Later, when I went to Agra to work, she was with me and turned out to be a favourite with my LSMF girl students. We visited the Taj together. She kept the house for me while living was reasonably cheap and those who served us were moderately honest and trustworthy. When we all got together as a family once again in Secunderabad in 1936, we lived in the Kingsway flat. A friendly colleague brought a fine man working in Delhi in the Railways to marry her. She agreed after much hesitation to part from me. She got married. Parting from her was heart-rending. My mother and I went with her up to Kazipet. When the train left for Delhi, we spread our sheets on the gravel of the platform in 1938 for the night and afterwards returned home feeling lost. She had to follow her star. Later, we got together often. She arrived post-haste when I lost my sight in 1969 and was in bed blindfolded. She gave me her love by offering me abundant kindness and care.

We were together again in Andhra Pradesh in 1980, with our husbands gone, reaching towards each other for need and help. She is in her autumn now and I in my winter.

Then came an invitation for her from her loving daughter Kanaka to come to USA, when she was much distressed with the double loss of her pilot son and her husband who soon followed him after a visit to the Hyderabad airport to see a friend off to Delhi. His watching the aircrafts landing and taking off set his heart recalling his pilot son disappearing over the Bhutan hills while on a sortie to drop food to the advanced army camps on the mountain sides.

In 1974, she suffered the most harrowing sorrow in USA, when Kanaka passed away. I wanted to get to her side and flew to London on 31st August and even booked the Pan Am to reach her. But in the midst of her sobs she was reluctant to let me come to her because of an irksome guest staying with the family. She was apprehensive of his discourtesy to me; I did not get there. I had to come back home.

By fortuitous circumstances, Sushila and I are living within easy distance again. This is obviously God's own plan after several meetings and separations. She is ever so good to me by spending long weekends making al provisions for my care. We both are so grateful for this great privilege of being able to get together again. We live looking forward to our grandchildren's progress and for each other's companionship.

My 75th Birthday

I had to accept with grace, a request from my loyal and affectionate former students led by Dr. Sundari, Dr. Sultana and Dr. Umesh to celebrate my 75th birthday on 31st March 1984. They were my undergraduates, post-graduates and later assistants while I taught at the Osmania Medical College, from 1950 to 1967. They invited me to the dinner and a felicitation at the college with much ardour. They were teachers in their own right and my co-authors for several publications. Hence, I had to comply with the programme they planned so zealously.

Dinner on the 31st at the Banjara Hotel was a sumptuous buffet to which my erstwhile colleagues and present members of the teaching staff of the Osmania Medical College had been invited.

On 31st morning, I visited the Institute of Obstetrics and Gynaecology at the Government Maternity Hospital to meet the nursing and laboratory staff and the Class IV staff who helped to uphold the difficult office of the Superintendent of the Hospital besides offering good service to the patients there. They garlanded me and spoke words of appreciation of my service to the sick.

At 5 p.m. on the same day, there was a fairly large gathering at the Osmania Medical College Hall where under-graduates, post-graduates and professors of all disciplines were present.

What delighted me most was my husband Nana's presence there in his lovely silk 'panchi' and shirt, sitting on a chair nearby, and dear Premu (Renuka could not come). I sat calmly watching his hands resting on his walking stick.

I was thrilled as the young loyal girls in colourful attire surrounded me with thankful expressions and garlanded me for the teaching and discipline they got from me. They ushered me to the dais with other friendly colleagues like Dr. Brahmaji Roa and Dr. Kanakadurga. Dr. Sridevi, the leading educationist and the erstwhile Principal of the Women's College, talked in praise of me at some length.

In turn, I thanked Dr. Sridevi and other guest speakers. This was not all, for Dr. Sundari released a souvenir on the occasion.

6

TRAVELS

I would like to start with a visit to Cape Comorin (Kanyakumari) at the Southern tip of India where I was taken by the Kerala Governor's party, along with my daughter Renuka and niece Premu. Here the three oceans meet. We collected red, white and black sands in small bottles. I had made a few colourful slides of this setting with the Governor's wife and my good friends against the blue sea. This experience was most enchanting. The seashore was almost bare, except for a few palms and the famous Vivekananda Rock standing in close vicinity. Not far from this point was the Kanyakumari temple of the goddess. On our return journey to Trivandrum by car, we crossed several canals edged with coconut palms and saw barges being paddled and tugged along, presenting a typical Kerala scenario. We also enjoyed the train journey back to Madras. Delicious vegetarian meals were served from the station restaurants in shining stainless steel carriers along with rolled up green banana leaves.

Back at Hyderabad, we had two guests—one from Cambridge and another from USA. Renuka took them for lunch to the Taj Mahal Hotel nearby. They were fascinated by the young Brahmin waiters with their sacred threads thrown diagonally over their golden torsos. Later they wrote to us that the vegetarian lunch served in a simple manner was one of the highlights of their world travel!

After the Kerala trip, we visited Kashmir via Delhi and Pathankot in 1959.

Out stay in the houseboat on the river Jhelum was very pleasant. The Kashmiri family, whose small boat was moored to our larger one, looked after us and fed us well. One-year-old infants of the boatmen's family would sit precariously on the edge of their smaller boathouse and the two-year-olds would swim ably. The women of course were of lovely complexion and wore beads on their necks and long headdresses. Often, vendors of flowers and fruits would paddle up to the boat offering their wares at out door. From the boat, we often took the tiny decorated 'shikaras' which were paddled to the shore and then we walked to other places of scenic beauty. We had pony rides to the mountains such as Gulmarg covered with flowers. Playing on the snow at Khilanmarg was exhilirating.

Then we moved on to a tent camp on the banks of River Liddar with its breathtaking beautiful turbulent waters. My camera took perfect pictures of the waters, snow and flowers. The pony ride was exciting, but one tended to hang round the pony boy's neck to start with!

The camp provided us with hot water in canvas buckets and tubs and also good food thrice a day. Srinagar valley had a comfortable climate and grew rice. A woollen cardigan and silk sari were enough to keep us warm. The Kashmiri students (who had come for training to Hyderabad) organised the outings for us. We went for breakfast to their former Professor's house and were served with plain chunks of cooked meat as in most Kashmiri homes.

Our return through the famous Banihal pass was exhilirating and the bus swerved into narrow roads with many a precipitous valley on one side. The Sikh at the wheel drove us down at "breakneck" speed through all these narrow hairpin bends on the mountain roads and brought us in time to board the train leaving for Delhi from Pathankot.

On reaching Delhi next morning, we saw a telegram from by brother announcing that Renuka stood first in B.Sc (Osmania University) and won a gold medal. Two nights and a day on train brought us back to Secunderabad. Her 'Mama' garlanded the bashful medal winner on reaching home.

I went to Calcutta in May 1968 to deliver the Kedarnath Das Memorial Oration on "Certain aspects of intersexuality" to the Obstetrics and Gynaecology Society of Bengal. I received a gold and silver medal embossed with the figure of a mother caressing her infant. Next day, my nephew and I crossed the Ganges at Farakka barrage in a steam boat. We were met on the other bank by Prem and Cana and we boarded the luxurious Railway Saloon to Kathiar, later got to Siliguri en route to Darjeeling. A travel in the little hill train up the mountainside was so very exhilirating. From the mountain-town, the most gorgeous and breath-taking sight were the Kanchanjunga mountain peaks, reflecting gorgeous hues of pink, and the gold sunset. It was most fascinating to see the ladies of the hill town in their long, embroidered dresses. We walked up and down the hillside and watched the life of those people.

Long back on a previous occasion in 1935, going to Simla via-Kalka was interesting for the one travelling alone for an interview. I found several hugely made bearded men crowd into my compartment but they did not alarm me as they kept to themselves. At Simla, the porters carried towering loads of baggage on their backs, strapped around their foreheads.

A conference at Patna was very exciting being so close to the Buddhist University of Nalanda. My Bengali friends, whose guest I was, drove me to visit this ancient place of learning. There were only small walls left standing on the foundations. They were built of well cut bricks. The size of the cells where the monks spent their time in study and prayer was small and square in outline. It made one think of the time spent by the monks in meditation, each isolated from the other. I regretted my not taking advantage of the short flight to Kathmandu though my good friend Dr. K. went with others. I missed meeting my old students working with the Nepal Medical Service.

In July 1948, I returned to India from the UK where I had been for higher studies. My husband eagerly received me at Bombay. I enjoyed immensely the vegetarian meal at the railway restaurant after a year on Western food. We boarded the train to go across the country to get to the Godavari Basin. This journey was tedious, but there was no other way of getting to his home-town where our child was being safely looked after by my mother-in-law. Because of the atrocities during the Razakar movement, the trains were not running through the Nizam State of Hyderabad. After this long journey, the great joy of getting to our seven-year-old daughter awaited us. We crossed the river Godavari in a large country barge, sailing swiftly. The remaining distance to Amalapuram was done by bus. As we got near my husband's mother, we saw little Renuka was dancing like a fairy in her blue skirt. Slowly, she came towards me smiling, as recognition dawned on her!

After a couple of days, my brother surprised us by bringing my anxious mother from Secunderabad. It was a very happy reunion. We returned to Secunderabad by the first train that was run into the Nizam State. We had a very warm reception at the Secunderabad railway station.

Any travelogue would be lacking in its essentials if the visit to the painted caves at Ajanta and the stone carvings at Ellora are not included. These caves were casually discovered years ago by a British soldier. The Ellora rock caves have been cut out and carved from rocks but exposed to the sun. Tourists come from all over the world to see these marvels of Buddhist art.

The colossal statue of Gomateshwara in Karnataka in his magnificent proportions was amazing. This visit was made possible by my daughter's husband driving us by car all the way from Madras. We returned via Bangalore. The enchanting sculpture at the Belagola temple delayed us so much that we had to suffer great embarrassment for our late arrival and spoiling a magnificent dinner party arranged for us by our hostess.

The time for my first departure for UK planned for further studies had arrived. The six weeks after the MD examination at Madras passed very quickly. On the platform of the Secunderabad railway station were gathered the family and a crowd of well-wishers. The two elderly mothers were seen standing with sombre faces, sad at my impending departure to a far-off country. Little Renuka, about 6, was seen in a pink smock-frock and as the train began to move, my dear brother, holding my child's hand, started to run alongside the train on the platform. This sight pulled at my heart-strings and I began to sob as I had never done before. My husband was sitting beside me, looking on helplessly.

We arrived at Bombay and flew to Karachi where we were lodged in an Air-India hotel. I boarded the plane the same night. The aircraft was a "Seaplane". Inside, the comfortable parlour seats were facing each other with only 4 of us in each cabin. The passengers were served each meal in a right royal style. After dinner, the steward converted the two seats into beds and laid feather pillows.

Early next morning, I felt a gentle hand on my shoulder offering me a cup of tea. I was startled thinking that the hand was my mother's. But it was a solicitous steward, offering bed tea in a pink cup and saucer which he left on the ledge by the bed.

Soon we landed on the Gulf waters at Bahrain. We went up the ramp to a small airport, a mere asbestos shed, where we had breakfast. I saw a camel-man leading his animal by the side of the airport. The bell on the camel's neck made a gentle tinkling sound.

On reaching London, I got into a taxi with Miss Singh, who was in the same cabin with me in the plane, speeding along Gloucester Road in South London. We stopped in front of a badly bombed building which she called "Gloucester Hotel". As Miss Singh knew the lady who owned the place, we got a room in the first floor which had the wall plaster coming off in places and the pillars showing cracks. After a cold wash, we got out in search of a room and peered at all the "To Let" notices. We snatched a quick hasty meal somewhere and our search went on till twilight without any luck. We bought a cauliflower and some potatoes and a loaf of sliced bread and went back to out war-torn hotel. Miss Singh cooked our meal on a gas ring in the fire place. Finding no plates to eat our victuals from, we washed two small ash-trays that served as plates for us! That was my first night in the fabulous city of London!

On the third day, my friend Jean of Madras days, suddenly turned up from nowhere. She phoned her old acquaintance Mrs. Ponsford at 37 Woodlane at Highgate and took me there, all the way up the Nothern line. Mrs. Ponsford was known never to say 'No' to anything to anybody. She said a music student was leaving and I was welcome in the YMCA at S.A. She had her own family and also took in students from anywhere in the world. She ran a kindergarten school on the ground floor. The kids howled when their parents left them each morning but later played happily on the swing in the lawn at the backyard. Mrs. P's kitchen was large and welcoming. Two Siamese cats sat in their places with their eyes closed. The students offered her all help such as bringing in the coal in the cold night from outside. She also had four maids, with rotten teeth, speaking cockney English. World War II was just over. Among other scarcities we each had only one egg a week, imported from Denmark. There was a young German refugee working her way in the house and learning English. Soon Dr. Joan arrived from Edinburgh, where she was doing her "Scottish Conjoint" after having been an occupation therapist. Joan and I chummed up very soon and this friendship has lasted these 40 years. We went for long walks among the stately tall trees and ponds called "the woods" behind the house in Woodlane. I learnt from her more of the world and the life in the West and their culture.

I started attending Prof. F.J. Browne's (Retd. from London University) coaching session at Soho Hospital in the afternoons. But I did not enjoy the sessions till Joan told me to have a snack before attending the meetings! That advice made all the difference. Professor Browne and I often went down the escalators together to board the underground train to travel part of the way with each other. We gaily chatted. It was unsual to see other people indulge in any such communication. Other passengers watched with curiosity at the tall guant old Englishman talking so freely with a young Indian woman!

If I bought a copy of the "Daily Mirror" full of horrible incidents of the day, he would snatch it away from my hands saying "Not meant for children!" During our frequent travels together, I learnt of his wife's serious illness and would occasionally send a bouquet of flowers to her through him. He would willingly carry the flowers and bring back a note of 'Thank You' calling himself my 'postman'. In one note the sick lady said that she was surprised at a person like me coming to her country, learning the language and then sit through for the higher examinations. She obviously did not know that we learnt English in British India.

On another day, my senior companion remarked that had he been my husband he would never have let me come alone to London for I used to chat with him on all sorts of humorous topics! Many years later, when Prof. Browne was emigrating to Sydney, he wrote to me, as he passed through Bombay, about his plans to marry Dr. Miss Cuthbert on arrival.

Earlier, through Prof. Browne's kind introduction and Prof. Moir's invitation, I set out to visit Oxford. In my first experience with underground trains in London, I got off at Farrington instead of at Paddington and went round and round the Central Line of the underground train before I finally got to the Radcliff Department at Oxford! My friendship with Prof. and Mrs. Moir became so warm and cordial that both of them visited me at Hyderabad. Visiting the Nagarjunasagar Dam was a fascinating experience for them. The sight of thousands of workmen, carrying building material on their heads or by balancing the baskets on their shoulders, on a rod at its ends made them imagine how the Pyramids in Egypt must have been built several centuries ago.

In the following October (1947) I travelled by train to Endinburgh to attend the fellowship course. Edinburgh is a most beautiful city. On either side of the Prince's street are the marvels of human creation in an unusual natural setting, with a castle on the hill leading down to a valley of flowers and edged along with monuments—examples of human excellence, the marble statue of Sir S.Y. Simposn who found in 1847 the means to relieve pain in childbirth, by his deep-seated sympathy for the mother-to-be, in 1847, and by his untiring effort and energy. The other is the monument of Sir Walter Scott, the greatest novelist of Scotland in its gorgeous floral setting and the pinnacle towering to the skies.

But on arrival I could find accommodation only in a dismal place called the Commonwealth Home with tiny rooms with a small fireplace set on the ground level. One had to sit on the floor and expose one shoulder after another to get warm. The bed was of a black tubular iron, with heavy black rugs, like those in the military barracks. The food was indifferent with kippers or a fried egg for breakfast. My table mate—not a local citizen—turned his plate round and round to separate the yellow of the egg from the white. He was from South Africa! Luckily for me, my friend Joan from London invited me to tea at "Ainslee Place", a posh residential area for professionals.When I got there, her hostess, a staid English lady, set us a typical Scottish tea near the fireplace, warmed with a log fire. Oat cakes is a speciality of Scottish delicacies. We ate them with honey.

This was my day for talking about their Scotland, telling stories from H.V.Morton's travelogues in Scotland. The happy result of that evening was my being invited to live in at lovely place with the hostess and her daughter, who later was to become a great choreographer. Another young English woman student was also there. I recall my life of 8 months at Edinburgh with nostalgia because of my stay here with these fine folk. I had a lovely large room with a white carpet and white blankets overlooking the 700 years old village Dean, with its stone cut rollers and other structures. And the river Dean was flowing far below my window. One could see also the modern part of the village with doors painted red or green and a milk-van running along its road every morning.

I had registered for the Fellowship Course which was being conducted at the Royal Infirmary at one end of the city of Edinburgh. I took the tram, at the Hope Street Post Office, which ran along the Princes' Street to the Royal Infirmary. When the Ticket Collector was approaching for the fare, I looked at my co-passenger to ask about this. She immediately gave him a large copper (6 p) piece from her pocket. This amazed me for I had heard of the thriftiness of the Scots. She smiled when she heard me tell her this.

I passed the statue of Dr. Simpson immortalised in marble. Next came the tall tower of Sir Walter Scott's monument surrounded by colourful flowerbeds. On the other side were all the huge buildings—Binns, the Departmental Stores complex, and the Crawfords-eating place and MacEvens breakfast house with their large signboards. The birthplace of R.L. Stevenson caught my eye. When I went in on another occasion, I saw his nurse's portrait inscribed by him below it as "my first wife and second mother." Next, the tram swerved to the right towards the large teaching hospital. The porter here directed me in inbred Scottish brogue, "The Royal Infirmary Madam? You are right there", rolling his r's.

After a few days, the Castle on the right was floodlit for a special occasion and being set on a hill, it looked like a "castle in the sky." My other visit to a place near Edinburgh was the large railway bridge on the river Forth called the "Firth or Forth". Edinburgh was cold! During my attendances at the old galleries lecture halls I made friends with a young PG from Jaffna (Sri Lanka). We made bus trips to various interesting places around Edinburgh during the week ends, including the old forges, where they fixed the shoes on to the hooves of horses besides enjoying the lovely country side. As Edinburgh was on the East coast of Scotland we visited the seaside and listened to the cackling of the sea gulls. During our exploring trips, we were sumptuously fed by the cottage housewives for a small fare.

The war memorial near the castle had a huge register of many "Macs" who laid down their lives in the battle (Mac is a very common prefix to Scottish names) of Britain.

Walking down the "Royal Mile" in the old town, below the castle hill, we reached Holyrood Place with all its historical stories of intrigues among royal and various princely households.

By and by when Dr. K. and I accepted the invitation from the MacLennens to stay with them at Glasgow just before their visit to India we had an enjoyable time. The most hilarious incident was when the housemaids placed a large oval carved wooden tray full of fruits, nuts and nut crackers just because Dr. K. said that she was a vegetarian! We laughed and laughed, asking if they thought she was a squirrel! Boiled eggs, various sauces, jellies and jams are what we excepted them to serve—Vegetarian restaurants gave us gorgeous seasoned rice with chunks of vegetables. In Canada vegetarian soups, pancakes and sweet or savoury sauces were delicious and sumptuous. German vegetarian soups were thick with vegetables and well seasoned with appetising flavours. At Florence, in Italy, spaghetti with leafy vegetables was tasty enough with tomato sauce, with a bottle of chianti (red). At the "Asia Centre" at Tokyo, the dishes smacked of Indian curries although a real Japanese resturant served dishes of raw fish to contend with. At Bangkok, a week-end market offered frog legs fried in butter and pickled water beetles.

Getting back to our stay at Gllasgow with the Maclenens, we were taken to a lovely piano recital. Later, we visited a famous comedian's home with Elizabethan furniture, and later that night, saw him on the stage with nothing but a bunch of leaves to cover him when he came out with all his comical performances! Driving through the West of Scotland, we passed through Glencoe, an eerie place, described by Walter Scott in *Rob Roy*. But OBAN - one of the places described by my friend, Jean Duncan, was really so very colourful with purple, heather-covered hills, orange sunset and deep blue sea when we got there late one afternoon.

While talking about Scotland, I must mention our stay with our charming hostess and senior colleague at Aberdeen, Dr. Esslemont, the President of the Medical Women's Federation in UK, in her heavily carpeted three-storey house, comfortable and congenial with housemaids to help. Renuka saw a circus at Aberdeen with her father. While we all enjoyed the visit to Braemar where the summer dances (Highland "Jinks") are held by the kilted Scots, the hotel served us with venison for lunch. The Scottish hills around Braemar pesented an impressive and awesome picture of inaccessibility. I am told that the amber-coloured Caren Gorm stones for tie clips and brooches are picked up from scattered rubble on these hills.

The Columbo Plan award was awaiting our return to London. My flight to Montreal in Canada in October 1954 could be arranged forthwith. Renuka was to attend the Crouch End School near 37 Woodlane going there by train. After she had hurriedly dressed and taken her breakfast in her father's room, she dashed down the slope to the Highgate station. She boarded the whistling train just in time. She often paid her train fare of a (6p) copper piece to the T.C. on alighting at Crouch End Station! She enjoyed her purple warm uniform and meeting her new classmates. The school Principal remarked that Renuka brought real knowledge of the East particularly of India to them. When I left for Canada, her life at woodlane was taken care of by Mrs. P. along with her father helping her often even to wash her long hair at the hand wash-basin in his room!

I sent her parcels of Canadian blouses etc. after I got settled in Montreal. The shopping was a great pastime for Dr. K. and me after the former had joined me on a similar fellowship from Madras. We could enjoy the snow after the colourful "Fall" when maple leaves turned yellow and crimson among the yet green ones.

At Montreal, we were accommodated in a hotel on the side of a hill. By the time the night wore on, the place became very noisy with music and dancing making us feel that we were probably in a "Night Club"! We patiently waited for the crack of dawn when I appraised our host at the Royal Victoria Hospital of our situation.

Soon the senior Resident, Dr. Bunker (whom I met again later in the Sydney Congress in 1968), turned up to retrieve us from that awful place. He himself carried our luggage to his car to the road on top. When we expressed surprised, he told us how he drove tourist buses and explained things to the tourists even after he had qualified! What dignity of labour! During our stay in that city, we met a pleasant Russian woman, a technician, who ran the "Cytology Lab". Later we visited the world-famous German Scientist, Dr. Hans Selye of "Alarm Syndrome" fame in his study.

We were also shown the Red Indian camps and hide-outs with wigwams and feathered headgears. Montreal got its name from the huge lighted cross planted on the highest peak, "Mount Royal". Dr. Philpott advised us to cancel our planned trip to Moosjaw and Calgary and pay a visit to Toronto instead. The travel by train on the parlour seats, watching the large barns and extensive farms, was fascinating.

At Toronto, people were friendly, cordial. There was nothing much for sightseeing in Toronto, but teaching methods of UGs and PGs were very enlightening. The Chief was extremely friendly and appreciative of us, the women doctors from India and our efforts to glean more knowledge of Canadian ways of training residents. I asked him if I had bragged about it when I spoke on my 100 cases of VVF repaired at Hyderabad, but he, in his deep gutteral after removing his pipe, said, "You did not brag about it," while driving us somewhere out in his car. From Toronto, a staff member drove us down to Hamilton to my friends from Scotland where we could visit the magnificent Niagara falls with its foam rising miles into the sky. Niagara falls are the largest in the world, vast and awe-inspring.

After watching the Niagara falls to our heart's content, we caught the train to Michigan (USA) via Detroit, by kind permission of the Columbo Plan officials. Detroit was indeed a sea of Ford Company cars laid out in the open in various states of manufacture, for miles on end.

At dusk, the train halted at the small University Station of Michigan. Snowflakes were lightly coming down on us as we tumbled across the two pairs of railway lines to reach the clear railway yard where Dr. Miller was standing—a tall man with his hat in his hand to welcome us. He asked for our baggage 'checks' and brought out four cases, carrying them himself. He placed them in the luggage boot and said, "Come along into my little car," which was several metres long. He drove us into the town saying that he would drive us to his 'little hospital' and took us up by a fast moving elevator to the 28th floor. Dr. K. and I were surprised at the largeness of everything in USA, especially the hospitality.

Next morning saw us in the hospital and some of its research sections. His next-in-command Dr. Behrman showed various machines for placental perfusion and later took us to his OPD where he was doing AID for some infertile women. The lecture halls were vast. Both Dr. K. and I were asked to talk to the students.

Dr. Miller was interested in repairing VVF and creating a new urethra. That is the reason I chose to visit him. He was very simple and friendly and invited us to his home where Mrs. Miller showed us her beautiful kitchen with floral designs painted on the walls. Dr.Miller later showed us the plans for the new University to be built to last for a century. He drove us to the edge of Lake Michigan which was frozen hard. He said sadly that once a husband was pushing his wife in a wheel-chair on the hard ice which at some spot suddenly gave way and the couple went down into their watery grave.

Our next halt was at New York, in late January 1954. This city is called the "Hub" of the world with its skyscrapers and the dazzling Times Square on 5th Avenue etc. New York controlled the world business, it was said. A young friend from Madras received us at the station and drove us to his apartment. Asking us to take our seats, he brought out a couple of tins of vegetables from his cupboard and cooked a delicious South India dinner in a jiffy. It was most delicious to our Madras palates.

Going to the roof of the Empire State Building by express elevators, we could see the Chrysler Tower. Down below at the Empire State, we attended a show called "Step Brothers". The architecture of the lobby was so modelled that vast masses of humanity could leave the hall with no rush or jam. Dr. and Mrs. Alridge invited us to a delicious dinner at their home and took us to see "The king and I' on the stage. Later there was a grand chance to visit the Statue of Liberty and go inside the Riverside Church with its bright streak of light from the spire falling on to the altar. On January 26th, our Republic Day, we went to the India House to meet Americans and Indians. We even saw a Nazi leader face to face.

Boston being the centre of the Medical world, brought us in contact with the famous gynaecological Surgeon, Dr. Joe Meigs, and his exquisite operating techniques. Here we stayed with an American couple, enjoyed their generous hospitality with food and fruit drinks, besides sightseeing. They suddenly heard of a car accident with the loss of their niece and her family. We were aghast at their stoic attitude on receiving the ghastly news. I was cooking a chicken curry on that day for them. They seemed to enjoy this all the same.

On our return journey after the Columbo Plan award (with a parchment to that effect) we left Ottawa airport after the four feet of snow was swept aside by machine sweepers.

Landing safely at London made us feel ever so thankful. The aircraft from Ottawa was unusual for its comforts. We had sleeping berths and curtains drawn for privacy by a black steward. But we were happy to be met by my child and husband at Heathrow, London. Soon the accommodation and food made Dr. K. feel very flat after all the cartons of milk and buttermilk we had enjoyed at America and Canada. She was almost in tears. I cooked an Indian dish for her at 37 Woodlane in my bedroom.

Then we two flew to Paris and stayed at Rue Galilee where we had nothing but a bread roll and coffee for breakfast. We visited the "Arc de Triumph" of Napolean, the Louvre, the famous picture gallery admiring Mona Lisa, the large painting of the "Last Supper" by Leonardo da Vinci. By night, we got to the Opera and climbed up the well to find the actors flying above the stage. After next day's breakfast with the labourers at Eiffel Tower, we flew to Stockholm and stayed with a hospital sister. This city was scrupulously clean and well laid out. The departmental store of 'Nordisca' was beautiful with attractive cut glass crockery but too expensive for our purses. Sister R. and her mother took us to see the Red Brick Palace where Nobel Prizes are awarded. Later, we crossed a strip of sea in a motorboat to visit Skanson, a village depicting the life of Swedes in the bygone days.

Our return flight to London from Stockholm was beset with difficulties. Due to heavy fog over London, our aircraft was diverted to land at Manchester. We were taken to the Manchester Railway Station, but as no reservation was feasible at such an emergency, each one had to scamper for accommodation in the waiting train and we got in somehow. Meanwhile my daughter had left Cambridge by train, hoping to meet us at London. On reaching the 'town' (London), she rang up a friend, Mrs. Morris, telling the purpose of her arrival at London. The child finding the this flight would not land at Heathrow, wisely decided to return to Cambridge. But the friend not hearing from her again before she left was rather apprehensive for her safety in the London fog. Dr. K. and I, at this late hour of our arrival, could not find any accommodation at the YMCA Hostel and so we were walking along Gower Street, London, hoping to find a night's stop. Seeing the two Indian women, looking lost, two of the good cops tried to help us. It was not easy for them either. We walked almost to the end of the street when a bed and breakfast place was found to our great relief.

On an earlier trip, we had disembarked at Geneva and driven to Rome. We then flew to Zurich (Swiss). We lodged in Rome with an Italian family who spoke no English. Suddenly I heard someone speaking on the phone in this familiar language. I waited till he put the receiver down and got to him. He was a German working in the Philippines Airlines. He sounded friendly and got us to go with him to an Espresso coffee place. Dr. K., Renu and I each had a cup of Espresso coffee. Then Norbetto took us to see the beautiful paintings on the ceilings of the Sistine Chapel, Santa Maria Majora and the little chapel one had to climb up to on one's knees. Norberto who was a pilot took us to the C.I.T. (city bus station) to go sightseeing. He took us to the Roman Forum to see the temple of the vestal virgins, the arch of Severus and the platform from where Mark Antony made a speech when Julius Caesar was murdered. We had a picture taken sitting in the temple of the Heavenly Twins and Renuka on a pedestal of a missing vestal virgin.

The bus took us where the catacombs were the burial and hiding place for the early Christians. It was a circular pit on the sides of which were dug out caverns just to fit a coffin. When we came out, an Italian priest spoke to us in Tamil and said he had worked in Madras for a spell. He called us to the top of the queue over American tourists and gave us each a tiny gold medal of the blessed Virgin Mary and the Baby Jesus. Later we visited the Collesium where the Romans watched the gladiators fight other slaves and/or lions.

We had an address for our lodging but pronouncing it in the English way, we got nowhere, till a kind English-speaking person understood us and directed us to this nice place to stay. The receptionist was very helpful and suggested we climb mount 'Richi' and visit the lakes and also find vegetarian eating places. We enjoyed the climb, slipping on the snow with great fun. Renuka actually fell throwing her feet up in the air. I took a snap of her, to her great annoyance.

Another morning found us atop another mountain with patches of snow where girls were sunbathing. From there we could see the lake of William Tell fame.

We left Renuka with a Swiss girlfriend who had attended the First World Congress on Medical Education at London with her father in 1953, and with whom we had visited the Epping forest of Robinhood fame. We crossed the Rhine to go to Heidelberg from Basle. There a memorable incident took place. A young German passport officer, astutely spotted from my passport that it happened to be my birthday that day and greeted me cordially. That gesture made me reflect as to how such tender-hearted young men could have been turned into cruel storm-troopers by Hitler.

Heidelberg, beside the river Necker was so beautiful with simple German folk around. They served delicious thick soups and dishes in their small restaurants.

Our next stop was Amsterdam with its endless canals and windmills. Tulips were aglow all round. The famous Dutch cheeseballs were being paddled in small boats along the canals. The girls were in pretty colourful pinafores and older men in clogs. A visit to the Rex Museum where Rembrandt's and other famous Dutch paintings were exhibited, was breathtaking in colour and expressed in oils. On entering a departmental store, we were amazed to see the blue and white Dutch crockery, famous the world over.

My Jewish friends, Mr. and Mrs. Spira had pressed us to visit them at their "Kibbutz" in Israel. We landed at Tel Aviv and following our friends' directions, got into a charoot (a longish taxi) and got out at a deserted place called "Ma Yans Vi". The taxi dumped us and our luggage and disappeared. We were looking around for help. Suddenly an Anglo-Indian lady walked up to us from nowhere and asked, "Are you looking for the Spiras?" We were surprised to hear this name from a stranger. We followed her to a very tiny cottage from where Mr. Spira came out. Mrs. Spira was in a state of depression and she could not even communicate. Their daughter, Nurse Ruth, popped in and took us to a commune along with her father, where the Jewish girls served us with some snacks of bread, fried fish and fruit. From where we sat we could look down on the shallow Mediterranean shore where fish culture was being carried on. Mrs. Spira made sandwiches with "cheese" and "love", etc., for my trip to Fowey. They were Jewish refugees from Germany and their near and dear ones had been 'gassed' by Hitler. Their children had been sent to English homes to be brought up. We spent a night at Haifa in a doctor's private nursing home and visited Joppa where Dorcas mended clothes for the disciples. I recalled Mary Dorcas, my little nurse at Hyderabad Hospital who helped us at the family deliveries and after-care of the mother and infant. One of those infants flew to London on 25th March to be married to a fine chap from Poland living there.

Then we flew to Munich and Frankfurt as guests of the Hoechst Company and visited the famous factory with its well-known gate, the facsimile of which is embossed on each of the company's tablets. The officers of Hoechst took us to a lovely wayside eating place where they served us with trout caught just then at a nearby stream, freshly fried with a pat of butter and it was delicious. Driving through the countryside gave us a bird's-eye view of rows of colourful cute little houses of their workers.

The Hoechst doctors sent us by car from Munich to Frankfurt. It was a beautiful morning to walk along the Rhine and see the several colourful little boats floating near a bridge. Later, we embarked on a motor vessel to travel along the River Rhine to Bonn. The rive was full with various cargo boats on it and the endless ruins of German castles on either side. We had some delicious snacks on board the ship. We found the ship stopping at several wharfs when passengers disembarked and embarked with so much agility and with no delay. As dusk descended on the Rhine we turned round to watch the passengers. A German gentleman, speaking broken English, got friendly and enquired about arrangements for our stay at Bonn. He offered to take us to a 'pension' known to him. True to his word, he took us to a clean place after we disembarked. Introducing us to the hotelier, he took himself away after nodding us farewell. We had no problem after that.

Next morning, we found ourselves near the famous beautiful Bonn Cathedral. Later in the morning, we took off for London with Renuka in her pink-checked sari. After we flew over the English Channel, we could hear the pattering of raindrops on the wings of the aircraft. We all murmered saying, "So this is England". Nana received us at the Heathrow Airport and took us by bus to London town. We lodged Dr.K. at the London YMCA and we got to 37 Woodlane.

Next day, we visited Prem Moin Shah and Uncle 'Damash' at a lovely lodge. Jean met us there (grilled iron gate).

When Renuka joined the Newnham College in 1961, another phase of life began for her. She had a typicaly small room with a carpet on the floor. The fireplace in the old building was at the ground level. There were cushions strewn around.

A year after I retired from OMC in 1968, my paper was accepted to be presented at the fifth World Congress of Obstetrics & Gynaecology at Sydney. Dr. K. and I went out of India again by air. At the Sydney airport, several of my Australian friends who had worked at Hyderabad received us with great warmth. My friend Dr. K. remarked that it must have seemed like a homecoming for me meeting these people. Unluckily for her, her only packet of lemon pickle being a forbidden import into this country, was taken away. The wife of late Prof. F.J. Browne (Nee Dr. Cuthbert) received us with great cordiality and housed us in a pleasanter place than the YWCA where we had previously arranged to stay.

At the Wentworth Hotel, Sydney where the Congress was being held, we met old friends from all over the world. Prof. & Mrs. Moir had come from Oxford. I also had an opportunity to meet Jerzy Teler from Poland with whom I had exchanged correspondence on case studies of common interest.

We were invited to the homes of former Australian friends. After intensive window-shopping, we flew to Melbourne. Here we had a picnic spot where Kangaroos were seen hopping around among the famous Australian orange-yellow flowering shrubs called wattle. We also saw a swagman with all his household goods on his back moving from one place to another.

From Melbourne we flew along the coral coast and landed at Bangkok where we saw the famous Emerald Buddha in one of the temples. We were taken to see the open market places; they were selling curious foods like frog legs fried in butter and pickled water beetles. A snake farm run by WHO was an unbelievable sight where thousands of cobras could be seen twirling over each other and over other small shrubs, all in a smallish well shaped pond. The men collecting the snake venom were standing among the cobras and feeding them with small chunks of raw flesh by holding their heads. These workers were wearing ordinary leather shoes and moved fearless among the cobras.

Our Indian friends who gave us hospitality had a Thai maid. We were amazed that she was provided a comfortable room with books and magazines. That was unlike what one would see in India.

By night, the floating restaurants in the Hong Kong sea, festooned with multi-coloured lights, were a veritable fairyland, especially looking on to Hong Kong from the high Victoria viewpoint. We got there by trams clanging their way up and down. Dr. K. and I were dumbfounded to see a gorgeous scenario after being used to the plain and simple South India.

Then the Japanese Airlines plane took us to Tokyo, a city of double culture. The women become 'Americans', when they leave their homes for work, but become real Japanese when they step back into their slippers and enter their homes. The cardiologist of St. Luke's hospital, Dr. Hinohara Tokyo, took charge of us. We stayed at the Asia Centre. He phoned us each night to ask about how our day passed. His family had three cars among its four members.

Dr. Hinohara's sons took us to sightseeing and ushered us into the multistoreyed departmental stores to see or buy the best of travel goods, or Seiko watches. We saw how the pearls are cultured by placing a sand bead into an oyster's "inside", before the baskets containing several of them were put back into the sea for culturing the pearls. The sand beads, we were told, came from Mississippi river bed in USA.

We did not miss a visit to the Kabuki Theatre, so different from the conventional ones we knew. The actors' faces were fully painted white with black eyebrows. Some walked among the audience while still playing their parts.

The Buddhist temple had immense boxes receiving offerings, placed in front. The women hurrying to the Shinto temple in the middie of their shopping spree was a common sight. A trip to Kyoto (the old capital) by the fastest train ever, going past the snow-capped Valcano-Fuji Yama and seeing the single tractor working on the farms was amazing. We were invited to join a Japanese 'tea party' in a garden—a special ceremony of serving green tea in little bowls in a very pretty garden setting. The women doing the ceremony were all in kimonos of lovely printed silk.

The most unusual day was when our host took us to a bamboo restaurant which also served raw fish in neatly cut squares with the whorled appearance of the flesh seen clearly. On the last night, we were taken to our host's home for dinner, set for the occasion in the special

'Japanese' mat-covered dining room. We sat on "legless" chairs and ate food made of soya beans, soya sauce, soya milk. The lady of the house bowed and bowed half bending her body as we took leave and thanked her profusely for her hospitality. Since returning to India, I received, on request, two stereo-needles buried in cotton wool and stuck inside a greeting card. How kind indeed of Dr. Hinohara Nagasaki of St. Lukes Hospital. We were very much at home at the Asia Centre and its cafetaria. But it was difficult one night for a taxi-driver to grasp from our way of pronouncing, the name of our destination in Tokyo.

We returned by the Japanese airlines and then changed into the Malaysian airlines. The air hostesses were charmingly dressed in their national costume and offered each passenger their national flower on emplaning. Landing at Singapore filled us with excitement in anticipation of shopping at this 'free port'. A wrist watch which still works, binoculars and opera glasses, besides a child-size doll, walking on a battery-cell was a great joy to my grandchild Anu. She called the doll "Little Anu". A cute railway engine, a tiny model of a real one, with a whistle and a shaking head and turning round by itself when met by an obstruction was a curio even for the adults. Pleased with all our shopping we set out for Madras in a not anyway unusual Air India aircraft. It was thrilling to fly over Madras at night with the familiar landmarks well lighted.

When we finally did alight we were amazed at the mass of ambulances on the tarmac. On reaching the customs department we were surprised at being congratulated by the customs officials instead of being tough as usual while checking our baggage. Then we learnt that the under-carriage of the aircraft did not come down as it should have and hence the presence of the ambulances and fire engines. Both our families were awaiting our exit from the aircraft with great anxiety and were relieved to see us safely back with them. The greatest excitement followed our arrival at home at Madras with all the purchases made at Singapore. But I was able to bring hardly a single item to Hyderabad as it usually happened after each of my seven trips abroad.

I was overjoyed to see my mother at Madras. Joy of seeing my ailing mother at Madras (in October 1968) on my return from the East was immensely gratifying.

After returning from my visit to Sydney in 1968 and Eastern Countries I attempted to work at the CSI hospital at Bangalore. From there

I made a visit to Madras in time to conduct an M.D. examination (clinicals) and I found my mother critically ill on 6th April and my youngest sister who was caring for her was much relieved to see me walk in. I tried to revive my mother but in vain. She breathed her last under my very eyes. I lay in a bed beside her all that night. Next morning I went to the Examination Hall and carried out my part as best as I could. No one could imagine the state of my mind. When my brother could come from Hyderabad we laid her to rest.

Since my mother had expressed her joy at my trying to work in a Church Hospital, I drove again to Bangalore prepared to stay put. But soon, early in June, I returned forthwith to Hyderabad by car intending to see my eye colleague, for detachment of my retinae.

EPILOGUE

I had an urge to write my autobiography ever since I regained my sight, lost in 1969, while I sat alone at Glendale. I was literally being sustained by the affection of my dear children (former students), Dr. Sundari and Dr. Swarna, the alternate beats of the same heart. I recall telling my young medicos that one thinks of old age after seventy. Now in my eighties I should certainly be dubbed as old.

The recent past has, of course, been on borrowed time—given to writing this and if unfinished I trust my precious daughter, Renuka, would complete this narrative as she knows me to be proud and humble, hot tempered and mild.

Today is the 24th March, the day my dear brother (Thambi) had passed away in 1980. Among the endless number of things he had done for me, he had built 'Glendale' to settle my roving spirit. I recall him driving four times one morning in his small car (Austin 7) from Secunderabad to Hyderabad during its construction. And he often had to argue with the moody and not so honest a contractor.

When 'Glendale' was built in 1957, it was admired and compared to an Italian villa with a velvet patch of green lawn in the open court at the back while neel kamal (blue lotus) bloomed in the oval cement pond near the rear wall. Streaming bunches of golden showers (Bogonia Venesta) literally showered on both sides of that wall. Two temple lamps hung on their heavy brass chains from below the window shades.

Many a time my camera had caught smiling faces at a grandchild's birthday party, or when august visitors from abroad had visited us. A rare picture was that of a black butterfly sitting on a crimson pink chrysanthemum which had bloomed to some size.

My selfless mother was occasionally seen in the background at such a party. And my husband would be reluctantly peeping from behind. All this seems years ago though 'Glendale' had been only built just over 30 years ago.

Now I have agreed to sell the foreground, cemented and used now only as a car park, where once a tall bougainvillaea with its white and pink flowers (Mary Palmer) used to shed its blooms from the flat roof above.

Many an illustrious specialist from near or far had been photographed at Glendale's footsteps. There were also younger groups in their colourful summer saris, the wise and loyal maids who cared for me and my child, succession and the all-purpose gardener-peon in residence.

The dear old retired friendly engineer (I called him Nana Abba) made Glendale larger and larger saying 'Allah provides'. A home is a home. Mine was a very small family in a large house with very big rooms. The compound wall had many flowering creepers lined with a rare type of chrysanthemum forming a backdrop for photographing friends, Indian and foreign, young and old.

My wall-shelves inside were packed with books and journals on Obstetrics and Gynaecology. My husband's collection were varied: biography, travel, politics and religion including many editions of the Holy Bible and a copy of the 'Golden Quran', not missing Rajaji's translation of the *Gita* and the *Mahabharata.*

Many a coloured slide would show the dainty lighting on the dining room wall and the Berlin piano. The Jacobean rosewood dinning table set bought from a dear couple who are lost to me now, is still here. My mother's room, the smallest, was fitted with piping on the wall to help her to move about with comfort and safety. Now as I write this, I use my husband's bedroom as my study, writing mostly after midnight to avoid the glare of the brilliant sun from my lensless vision. The screened back verandah with a small makeshift dining table would make do at present.

The two large bedroom suites in the first floor were built especially to house the foreign colleagues and External Examiners who came by the invitation of Osmania University. They accommodated many a fine colleague. Sadly, now most of them have passed away. The erudite and research-minded people like Dr. Malkani of Delhi, Dr. P.K. Devi of Nagpur, Dr.Lazarus of Vizag, Dr. Krishna Menon of Madras, Dr. Kanak Das Gupta of Calcutta, Dr. Thamban of Trivandrum and Dr. Shantha Rao of Bombay are gone. The very tall Scotsman, Sir Hector, using the bath water to pour over himself as we do, running down the steps with the bath water flowing behind him, screaming for help, was also gone.

The spirit of learning and the quest for knowledge lives on and have been passed on to our grandchildren.

It is long past midnight now while I write with a light from a beautiful tablelamp on my left, a gift from a dear friend and on my right stands the 48-year-old child-cot with scenes painted from "Pinocchio" stories by my dear artist brother. It has been used by three other children after mine had grown up with memories of the pleasant Sundaramma with a smile and a giggle each morning to wake up Renuka.

Is it any surprise that I am in the anteroom now waiting for my summons to be received any time, any day, any minute and my saying 'Thank you Lord' for the untiring and long life of service—now content and weary and undishonoured, waiting for the gift of sleep?

POSTSCRIPT

The six grandchildren—two of mine and four of my sister Sushila—have grown up into fine youths. The former two are essentially academic in their outlooks though they have various other interests as well. My sister's granddaughter, charming and comely, attended schools in various parts of North India, passed her M.A. in History and is now happily married to a fine young man from Poland and they live in London. Her younger brother has shaped into a well known singer in Pop music. His name often appears in the Bombay papers.

The other two grandchildren of my sister, though growing up without their mother in USA have more opportunities for studies and better choice of a career. The elder boy is doing journalism and has grown up to be a stalwart, living away from home for the sake of his education. The younger sister, an all-rounder, playing the piano and tennis is now considering the various options of Universities to go to. She has her father's love as all girls deserve to have.

Thambi's two young grandsons Akshay and Pritesh are growing up in the house that he built. They are lean and lanky, clever and full of fun.

My own grandchildren are Anu and Ram. Anu, now an M.A. and M. Phil at 24 in Psychology, loves music and converses beautifully, is clear-headed and confident. The little fellow Ram is a likely Veena Vidwan and excels in his studies as well. He is humble and courteous and easily deserves the name of Shanth Atma Ram. He aims to be an Electronic Engineer.

I had the privilege of bringing all of them into the world.